COMPLIMENTARY

QUANTITY	PRODUCT CODE	DESCRIPTION	TERRITORY	DATE MO	DAY	YR	NUMBER
1	A03876	MCCLINTIC BASIC 2ND ED	1004	1	17	80	U7570
							SUBJ. AREA OR PRICE
							BG10

JOHN WILEY & SONS
CANADA, LIMITED
22 WORCESTER ROAD, REXDALE, ONTARIO M9W 1L1
TEL 675-3580   TELEX 06-989189

BASIC ANATOMY AND PHYSIOLOGY OF THE HUMAN BODY

**John Wiley
& Sons**

New York
Chichester
Brisbane
Toronto

BASIC ANATOMY AND PHYSIOLOGY OF THE HUMAN BODY

J. Robert McClintic, PhD
California State University, Fresno

Second Edition

Copy Editor: Rosemary Wellner
Photo Research: Kathy Bendo

This book designed by Joseph Gillians
Cover illustration by Joseph Gillians
Production supervised by Linda Indig

Library of Congress Cataloguing in Publication Data:
McClintic. J.   Robert, 1928–
    Basic anatomy and physiology of the human body.

    Includes bibliographies and index.
    1. Human physiology.   2. Anatomy, Human.
I.   Title.
QP34.5.M3   1980   612   79-14295
ISBN 0-471-03876-8

Printed in the United States of America
10 9 8 7 6 5 4 3 2 1

This book is dedicated
to my wife
Peggy
whose patience and assistance
aided its completion

and to our daughters
Cathleen, Colleen, Marlene
who make it all worthwhile

Preface

This second edition of *Basic Anatomy and Physiology of the Human Body* retains the correlated presentations of development, anatomy, and physiology of the first edition and incorporates a variety of improvements to make the book even more appropriate for introductory courses in human anatomy, human physiology, or combined courses in these disciplines.

The most striking change is the entirely new art in the chapter on skeletal muscles. Presentation is largely regional in nature, so that the reader may gain an appreciation of the relationships of a given muscle to its neighbors. Where several layers of muscles are present, the art illustrates the muscles as one would actually dissect them. Each functional grouping of muscles is summarized in a nearby table that indicates each muscle's origin, insertion, action, and innervation.

There are three new chapters: one on basic chemical concepts that are utilized throughout the book; one on surface anatomy; and one on basic kinesiology, the study of body movement. The chapter on kinesiology is unique and enables the reader to understand how muscles work together in the body to achieve its often complex movements.

Throughout the book, explanations of anatomical and physiological phenomena have been expanded, and improved correlations have been made between text material, tables, and figures. All material has been updated and expanded, including the readings, and—where appropriate—discussions of neonatal and childhood structure and function are presented. Thus, we can begin to see that the child is not a "small adult."

As in the first edition, "flow-sheet" diagrams and summaries are presented to enable the reader to follow cause and effect. Chapter objectives have been added to this edition and are correlated with the questions at the end of each chapter. The stated objectives are kept to a number that should neither intimidate the reader nor omit any significant concepts.

New and carefully selected disorders of structure and function emphasize what went wrong with normal conditions and how normal function must be restored. Homeostasis thus becomes the unifying theme of the book.

I thank the Wiley staff and Frederick Corey, biology editor, and his predecessor Robert L. Rogers, for their support and encouragement during the preparation of the manuscript, and for their excellent production of the book. To my wife, Peggy, MSN, I am indebted for her typing, proofreading, and constructive criticism. To Professor David J. Mascaro of the Medical College of Georgia (Augusta), who is both artist *and* anatomist, I owe special thanks for the magnificent artwork of the muscle chapter and several other illustrations.

It is my hope that this book will be not only informative, but enjoyable to read, and will impart a concern leading to care of the body from before birth to death.

Responsibility for errors in the text are mine.

J. ROBERT McCLINTIC

Contents

Chapter 9 Articulations

Chapter 10 The Structure and Properties of Muscular Tissue, with Emphasis on Skeletal Muscle

Chapter 11 The Skeletal Muscles

Chapter 16 The Autonomic Nervous System

Chapter 17 Blood Supply of the Central Nervous System; Ventricles and Cerebrospinal Fluid

Chapter 18 Sensation

Chapter 22 The Heart

Chapter 23 The Blood and Lymphatic Vessels; Regulation of Blood Pressure

Chapter 24 The Respiratory System

Chapter 25 The digestive system

Chapter 26 Absorption of Foodstuffs, Metabolism, and Nutrition

Chapter 27 The Urinary System

Chapter 28 The Reproductive Systems

Chapter 29 The Endocrines

Epilogue

BASIC ANATOMY AND PHYSIOLOGY OF THE HUMAN BODY

Chapter 1

An Introduction to the Structure and Functioning of the Human Body

Objectives

After studying this chapter, the reader should be able to:

■ List the disciplines that constitute the science of biology and
   what each contributes to our understanding of anatomy
   and physiology.

■ Outline the levels of organization of the body.

■ List the "criteria of life" common to all living things.

■ Point out and name the major descriptive areas of the body.

■ Define the terms of direction used to locate body structures.

■ Enumerate and describe the planes of section employed in
   the study of portions of the body and its parts.

■ List the body cavities, the organs they contain, and describe
   the projections of organs on the body surfaces.

If one were to consider what the human body is worth, the often quoted figure of about three dollars is no longer valid. Harold J. Morowitz has cal- culated that the raw materials alone contained within the body—that is, its enzymes, hormones, amino acids, inorganic substances, and other materials—would probably carry a price tag of six million dollars at today's prices. The cost of fashioning these materials into the intricate individual functioning cells of the body has been estimated at six thousand trillion dollars, and no dollar value can be assigned to the expense involved in assembling those cells into a functional human being; each human being is priceless. Perhaps such figures will emphasize the wonderful complexity and amazing resiliency of this shell in which we spend our days and point out the necessity for taking proper care of it to ensure its optimum operation during our lives. The body is constructed so as to be virtually indestructible, and research has indicated that human life spans measured in hundreds of years are possibilities in the very near future. Care for our bodies must begin before conception, with proper nutrition of the body of the prospective mother, and at conception of the new individual, a care must begin that will terminate only with the death of that individual. In short, if your body is to serve you well, do not abuse it; it is the only one you will have.

The study of living organisms

The science of biology studies living things, and includes many subdivisions. ANATOMY is the subdivision that deals with the study of the structure of an organism. In the study of the human body, considerable knowledge of structure may be gained by looking with the naked eye; this acquaints us with the *gross anatomy* of the object being viewed. Dissection, which involves cutting and teasing of body parts, aids the viewing of items of interest that lie covered by other structures, and is, in fact, the basis of the word anatomy (G. *ana*, up + *temnein*, to cut). Viewing the smaller units of body organization may require the use of microscopes of various types. *Microscopic anatomy*, including cytology (study of cells), histology (study of tissues and organs), and developmental anatomy or embryology (study of how the body develops and grows), enables us to study the fine structure and origins of the body components.

PHYSIOLOGY studies how the body and its parts work or function. The word physiology means the study of the nature of things (G. *physis*, nature, + *logos*, study) and the discipline draws on many other areas of knowledge to explain body function. For example, physiology draws on *anatomy* for a structural basis of function; on *chemistry* and *physics* to aid in the definition of basic substances the body contains and laws the body operations follow; on *biochemistry* to aid in the understanding of the complex chemical reactions that occur in the body; on *biophysics* to aid in the explanation of electrical and physical phenomena that occur in the body; and on *genetics* and *embryology* to explain the processes involved in determination of body function, growth, and development.

Anatomy and physiology thus combine to give one a broad and exciting view of the makeup and activity of the shell we inhabit during our days on earth (or in space).

Some generalizations about body structure and function

As we proceed with the study of the human body, it is well to keep in mind some basic ideas about the body. These ideas, or generalizations, will give a direction and purpose to our study. Readers are encouraged to add their own generalizations to those given below.

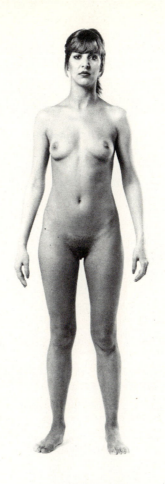

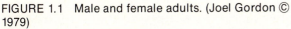

FIGURE 1.1   Male and female adults. (Joel Gordon ©
1979)

1. THE BODY HAS SEVERAL LEVELS OF ORGANIZA-
TION. The basic units of structure and function of
the human body are its CELLS. It has been estimated that there are 1×10^{14} (1 followed by 14
zeros, or 100 trillion) cells in the body. With this
many individual units demanding nutrients and
producing wastes, problems of supply and removal would seem to be too much to overcome,
yet the organization of the body has solved these
problems. Cells that are similar in structure and
function, together with their associated intercellular material (the substance between the cells),
form TISSUES. There are four primary tissue groups
that compose the body.

Epithelial tissues cover and line internal and external body surfaces.
Connective tissues connect and support body parts.
Muscular tissues can shorten or contract to cause
movement.
Nervous tissues conduct nerve impulses through
the body.

Two or more tissues put together in a specific
pattern to carry out a particular job, form an
ORGAN. Several organs working together to carry
out a larger body process form a SYSTEM. Many
systems are combined to form the human body
(Fig. 1.1). Although there are obvious external

TABLE 1.1 Body systems, their organs, and functions

System	Major organs or tissues	Main function(s) of the system
Integumentary	Skin, hair, nails, skin glands	Protection; temperature control
Skeletal	Bones and joints, cartilages	Support; protection; storehouse of minerals; blood cell formation
Muscular	Skeletal muscles	Cause body movement
Circulatory	Blood, heart, blood vessels	Carry nutrients and wastes; circulate the blood; carry the blood
Lymphatic	Lymph, lymph organs (tonsils, nodes, spleen, thymus), lymph vessels	Return tissue fluid to blood vessels; create immunity; form blood cells
Respiratory	Nose, throat, larynx, trachea, lungs	Supply oxygen; eliminate carbon dioxide; regulate acid-base balance
Digestive	Mouth, esophagus, stomach, intestines, liver, pancreas	Digest and absorb nutrients; excrete wastes
Urinary	Kidneys, ureters, bladder, urethra	Excrete wastes and regulate blood composition
Reproductive	Male: testes, ducts, accessory glands	Produce sperm and secretions of the semen
	Female: Ovaries, uterine tubes, uterus, vagina, mammary glands	Produce eggs; nourish offspring
Nervous	Brain, spinal cord, peripheral nerves, organs of special sense	Control many body activities, allow appreciation of changes internally and externally
Endocrine	All glands secreting hormones into the blood stream (e.g.: pituitary, thyroid, parathyroid, adrenals, pancreas, testes, ovaries)	Control metabolism, growth, and development of the body
Reticuloendothelial	The cells of the body (e.g., lymphocytes, plasma cells) that protect from foreign chemicals.	Provide protection against bacteria, viruses, and other disease-causing or potentially harmful agents

differences between the male and female, "inside" we are all nearly the same. The only basic differences between the sexes lie in the organs of the reproductive systems. All other functions are carried out by nearly identical organs with similar functions. Table 1.1 presents an introduction to the systems and organs of the body and their functions.

2. THE FUNCTIONS CARRIED OUT BY THE LOWER LEVELS OF ORGANIZATION ARE ALSO CARRIED OUT BY THE BODY AS A WHOLE. Thus, understanding of the cellular level contributes to knowledge of tissue, organ, and system levels, and ultimately to knowledge of the whole organism. As examples of the dependence of whole body function on cells, we may cite the following comparisons between cellular and organism activity.

Cells are EXCITABLE or capable of responding to changes in the external and internal environments of the body; so the body as a whole responds to change.

Cells INGEST or take in materials; the whole body eats food.

Cells DIGEST foods and metabolize them to release energy for body activities or formation of new materials; the body as a whole carries on digestion and metabolism.

Cells EXCRETE or rid themselves of wastes of their activity; the body rids itself of wastes in the urine and feces, and through the skin.

Cells produce, from body fluids, useful products known as secretions; the process of SECRETION is widely utilized to produce digestive enzymes, hormones, and other materials used in the body.

Most cells REPRODUCE themselves for purposes of repair, growth, and continuance of a given line of cells; the body reproduces itself for continuance of the species.

MOVEMENT of materials occurs within cells, and the whole organism may be caused to move through its environment.

These functions are not merely lists of what the cell or organism does, but are CRITERIA OF LIFE as well. Something that exhibits these activities may properly be considered living.

3. THE BODY, PARTICULARLY ITS FUNCTIONS, IS ORGANIZED TO MAINTAIN HOMEOSTASIS AND ENSURE SURVIVAL OF CELLS. A system of checks and balances operates to maintain body composition and function nearly constant within the very narrow limits necessary for survival of individual cells, and therefore the organism. The term HOMEOSTASIS describes this nearly constant internal state the body normally maintains in such functions as composition of body fluids, body temperature, and levels of acids and bases. Much of physiology deals with discovering and describing the con-

trolling mechanisms that ensure continued body homeostasis. Alterations of homeostasis form the basis for diagnosis of disease and abnormal function, and also the basis for instituting treatment intended to restore normal function.

4. MANY BODY FUNCTIONS ARE DETERMINED BY STRUCTURE, and knowledge of structure may enable prediction of function. A "common sense" relationship often exists between structure and function. For example, muscle cells must be elongate structures or they cannot shorten effectively; bone must be a hard strong tissue to protect and to serve as the support for the body. Many examples of the relationship between structure and function are cited in the chapters that follow.

5. BOTH STRUCTURE AND FUNCTION CHANGE AS THE INDIVIDUAL AGES. A newborn is mostly water and fat, and many body functions are not mature. An adult is different chemically from an infant, containing more solids and relatively less water. Old age is associated with lowered levels of body function and a "return to infant state" in terms of teeth and certain other body structures. Each age group has certain attributes that characterize it, and an infant should not be considered a "small adult" any more than a young or middle-aged adult should be considered to have the same structure and level of function as an octogenarian (someone 80 years of age).

Application of these generalizations to the study of the human body is made in later chapters.

The following sections introduce some basic terminology that will be encountered many times in later chapters. These terms will increase understanding and appreciation of body structure and function.

Gross body areas

An aid to the naming of gross body areas is presented in Figure 1.2. These commonly used terms are often referred to in identifying an area in which organs may be located, or in which there is discomfort or pain in the individual.

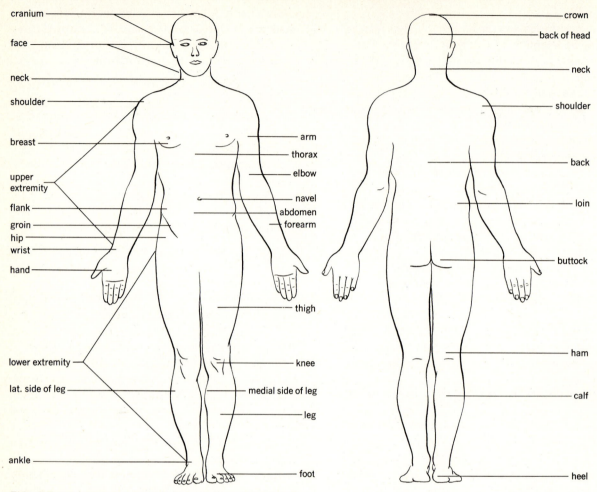

FIGURE 1.2   General descriptive areas of the body.

Terms of direction

The location and relationships of many body parts may be described by words that are used when the body is in ANATOMICAL POSITION (Fig. 1.3). In this position, the body is standing erect, the eyes are level and directed forward, the arms are at the sides with the palms forward, and the feet are parallel, with the heels close together. The terms of direction are:

Anterior or ventral. The front or belly side of the body, or something in front of the original point of reference; for example, the sternum (breastbone) is located on the anterior or ventral part of the thorax.

Posterior or *dorsal.* The back side of the body, or something in back of the original point of reference; for example, the spine is located on the posterior or dorsal part of the body.

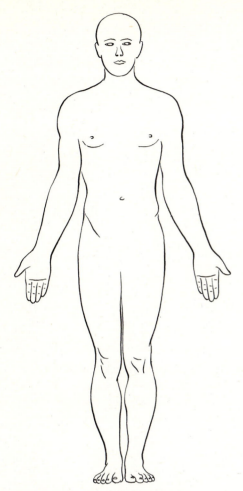 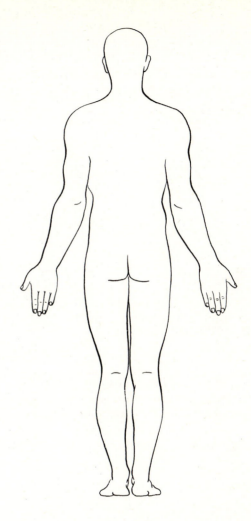

FIGURE 1.3 Anatomical position.

Superior. Above, toward the head, or something higher on the body than the original point of reference; for example, the head is superior to the neck.

Inferior. Below, toward the feet, or something lower on the body than the original point of reference; for example, the feet are inferior to the knees.

Medial. A line running from the center of the forehead to between the feet defines the midline of the body. Medial implies a position closer to the midline; for example, the nose is medial to the eyes.

Lateral. A position away from, or to the side, from the midline; for example, the ear is on the lateral part of the skull.

External. This term is most frequently used to refer to something that is toward the surface of the body or a hollow organ. The term superficial is often used to mean the same thing; for example, the skin covers the external surface of the body.

Internal. This term refers to a position away from the body surface, or toward the interior of a hollow organ. The term deep is used in the same fashion; for example, the internal organs of the abdomen are seen when the abdominal wall is opened.

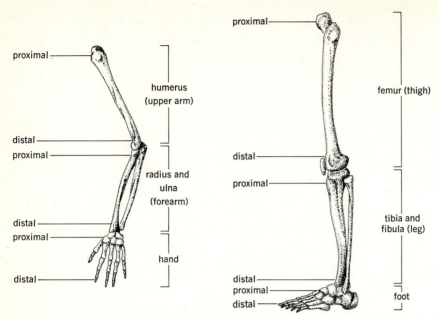

FIGURE 1.4   The use of the terms proximal and distal with reference to the limbs.

Proximal. This term refers to a position closer to the point of attachment of a body part to the midline; for example, the shoulder is proximal to the elbow; the knee is proximal to the ankle.

Distal. The opposite of proximal, distal implies further from the point of attachment or midline; for example, the wrist is distal to the elbow, or, the knee is distal to the hip. The use of the terms proximal and distal is illustrated in Figure 1.4.

Planes of section

Further information about the location of body parts can be provided by the study of sections cut in various directions or planes through the body. These are described by the following terms.

Midsagittal (sagittal) section. The body is divided equally into right and left portions.

Parasagittal plane. Any plane parallel to the midsagittal section, but not in the midline.

Coronal plane. The body is divided into front and back portions.

Transverse (horizontal) plane or cross section. The body is divided into superior (upper) and inferior (lower) portions. The term is also used to describe a section of a body part across its shortest dimension.

These planes are shown in Figure 1.5.

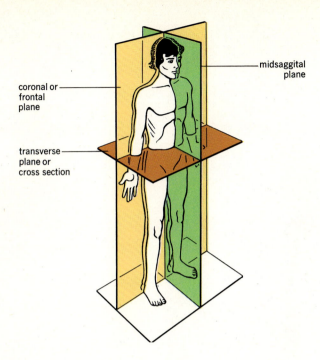

coronal or
frontal
plane

midsaggital
plane

transverse
plane or
cross section

FIGURE 1.5   Planes of section.

Location of organs

Body cavities

Most body organs and systems are found within the body cavities (Fig. 1.6), and location of a given organ does not vary greatly between individuals. A DORSAL CAVITY, located on the posterior or back aspect of the body, contains the brain, spinal cord, and parts of the nerves attaching to these organs. A VENTRAL CAVITY, located on the anterior or belly side of the body is subdivided by the muscular diaphragm into an upper *thoracic* or chest cavity and a lower *abdominopelvic* cavity. The thoracic cavity contains the thymus, esophagus, trachea, bronchi, lungs, heart, and great vessels entering and leaving the heart. The abdominopelvic cavity contains the organs of digestion, excretion, reproduction, several endocrines, and many blood vessels to and

from these organs. These organs together are usually called the VISCERAL ORGANS, or the VISCERA. Some of these body organs are shown projected on the external body surface in Figure 1.7. Many of the organs of the abdominopelvic cavity (e.g., spleen, liver, kidneys, and uterus) may be *palpated* (felt) to determine their size, shape, and location; thus, knowledge of their position is important.

Reference lines

The positions of many body organs may be described by locating them with reference to imaginary lines drawn on the chest and abdomen. The MIDSTERNAL AND MIDCLAVICULAR LINES lie on the

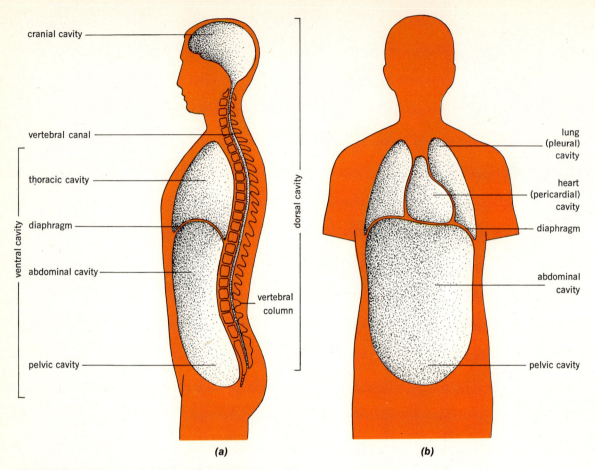

(a)   *(b)*

FIGURE 1.6   The true body cavities. *(a)* Lateral view showing ventral and dorsal cavities. *(b)* Anterior view showing divisions and subdivisions of the ventral cavity.

anterior chest (Fig. 1.8); the ANTERIOR AND POSTERIOR AXILLARY LINES lie on the side of the chest (Fig. 1.9); the VERTEBRAL AND SCAPULAR LINES lie on the back (Fig. 1.10).

On the abdomen, several unnamed lines may be drawn that subdivide the abdomen into QUADRANTS known as the right upper (RUQ), left

upper (LUQ), right lower (RLQ), and left lower (LLQ). Other lines divide the abdomen into NINE SMALLER AREAS. The quadrants and smaller areas are shown on Figure 1.11.

The reader should be able to describe what organs lie in each of these abdominal regions.

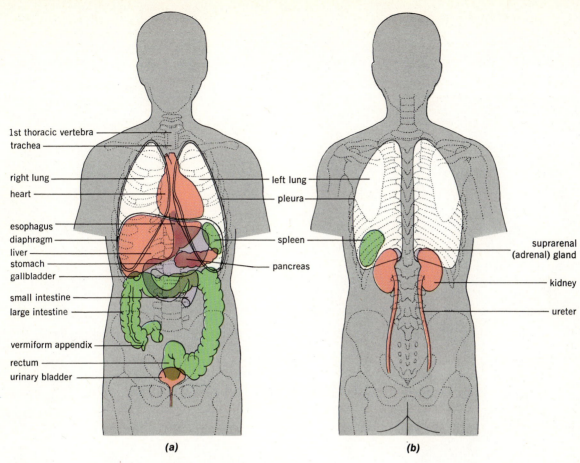

1st thoracic vertebra
trachea

right lung
heart

esophagus
diaphragm
liver
stomach
gallbladder

small intestine
large intestine

vermiform appendix

rectum
urinary bladder

left lung
pleura

spleen

pancreas

suprarenal
(adrenal) gland

kidney

ureter

(a)

(b)

FIGURE 1.7 Surface projections of some visceral organs. (a) Anterior view. (b) Posterior view.

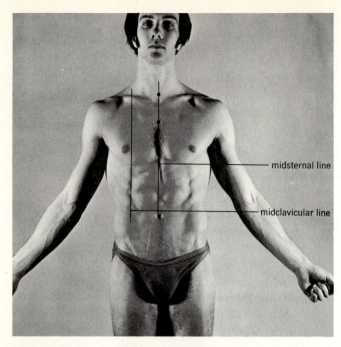

FIGURE 1.8 Anterior thoracic reference lines.

midsternal line

midclavicular line

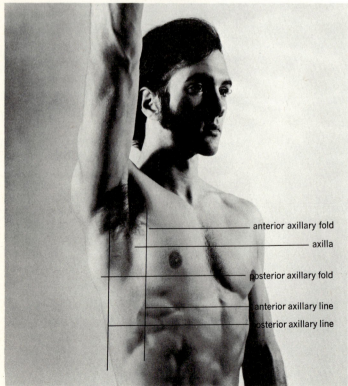

anterior axillary fold

axilla

posterior axillary fold

anterior axillary line

posterior axillary line

FIGURE 1.9 Lateral thoracic reference lines.

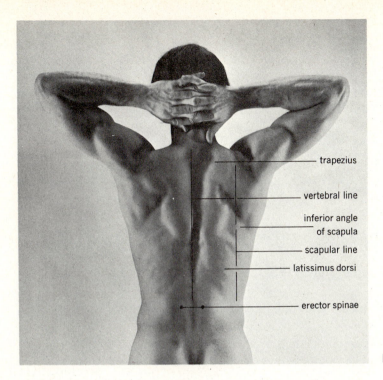

trapezius

vertebral line

inferior angle
of scapula

scapular line

latissimus dorsi

erector spinae

FIGURE 1.10   Posterior thoracic reference lines.

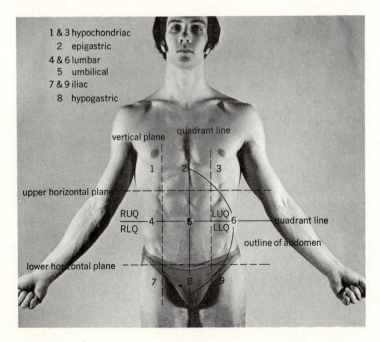

1 & 3 hypochondriac
2   epigastric
4 & 6 lumbar
5   umbilical
7 & 9 iliac
8   hypogastric

quadrant line

vertical plane

upper horizontal plane

RUQ
RLQ

LUQ
LLQ

quadrant line

outline of abdomen

lower horizontal plane

FIGURE 1.11   Abdominal reference lines.

Summary

1. Anatomy.
    a. Studies the structure of living organisms.
    b. Is divided into several areas of study including gross anatomy (what may be seen with the naked eye), microscopic anatomy, and developmental anatomy.

2. Physiology.
    a. Studies the function of living organisms.
    b. Draws on chemistry, physics, anatomy, biochemistry, biophysics, genetics, and embryology to explain body activities.

3. Some generalizations to bear in mind about body structure and function include:
    a. There are several levels of body organization including cells, tissues, organs, systems, and the whole organism.
    b. Functions carried out by lower levels of organization are often repeated at higher levels.
    c. The body is organized to maintain structure and function within narrow limits to ensure cell and organism survival. The maintenance of a nearly constant internal structure and function is called homeostasis.
    d. Structure often determines function.
    e. Structure and function are different according to age.

4. The body may be described by a series of terms that indicate direction and position.
    a. Anterior indicates front; posterior indicates back.
    b. Superior indicates higher; inferior indicates lower.
    c. Medial indicates toward the midline; lateral, away from the midline or to the side.
    d. External indicates toward the surface; internal indicates beneath the surface or inside.
    e. Proximal indicates closest to point of attachment; distal, away from the point of attachment.

5. Planes of section further indicate location of body parts.
    a. A midsagittal plane creates right and left portions.
    b. A coronal plane creates front and back position.
    c. A transverse plane cuts across a structure and creates upper and lower portions.

6. Organs are located within cavities of the body.
   a. Dorsal cavities contain the brain and spinal cord.
   b. Ventral cavities contain the visceral organs.

7. Reference lines provide an additional way of locating organs within the body.

Questions

1. What does the study of each of the following disciplines contribute to our knowledge of anatomy and physiology?
   a. Gross anatomy.
   b. Histology.
   c. Chemistry.
   d. Embryology.

2. What are the various levels of organization in the body, and what is each composed of?

3. What is meant by homeostasis? Name several processes that must be maintained within narrow limits to ensure survival.

4. Using terms of direction, describe the relative positions of the following:
   a. The elbow and the hand.
   b. The "belly button" and the hip.
   c. The thorax (chest) and the head.
   d. The skin and any organ in the abdominal cavity.

5. What is the meaning of each of the following?
   a. Midsagittal plane.
   b. Cross section.
   c. Loin.
   d. Flank.
   e. Cranium.
   f. Leg.

6. What are the body cavities and what does each contain?

7. What organs would be found in:
   a. The RUQ?
   b. The umbilical region?
   c. The right hypochondriac region?

Readings

Devey, Gilbert B., and Peter N. T. Wells. "Ultrasound in Medical Diagnosis." *Sci. Amer. 238*:98, May 1978.

Gordon, Richard. "Image Construction from X-ray." *Sci. Amer. 233*:56, Oct. 1975.

Grobstein, Clifford. "The Recombinant-DNA Debate." *Sci. Amer. 237*:22, July 1977.

Science News. "Artificial Organs and Beyond." *112*:154, Sept. 3, 1977.

Chapter 2
Some Basic Chemistry

Objectives

After studying this chapter, the reader should be able to:

- Define what is meant by the terms atom, element, molecule, and compound, and describe the structure of atoms.

- List the most biologically important elements and, where appropriate, give their chemical symbols, atomic numbers, and atomic weights.

- Define what an electron shell is, and indicate the significance of "outer shell electrons" to reactivity of elements.

- Define an ion, and give examples of ions and their electrical charges.

- Define an isotope, and describe stable and radioactive isotopes. List some of the effects of radioactive isotopes.

- List the emissions that radioactive isotopes release, and the use for these isotopes in studying biological processes. Define what a half-life is and give examples of isotopes with different half-lives.

- Describe the various types of chemical bonds and how they are formed.

- Describe the following and the information they present: molecular formula, structural formula.

- Define what isomers and enantiomers are, and give examples of each and how they are named.

- Distinguish between organic and inorganic compounds, and aliphatic and aromatic hydrocarbons.

- Explain the meanings of saturated and unsaturated with reference to organic compounds.

- Explain the differences between alcohols, ethers, amines, organic acids, ketones, aldehydes, and polymers.

- Describe the several types of chemical reactions, distinguish between exergonic and endergonic reactions, and explain the role of catalysts in influencing the rates of such reactions.

- Explain what is meant by intermediary metabolism and metabolic cycles.

- Distinguish between a solution and a suspension.

- Give the differences between molar, normal, and molal solutions, and list the characteristics of a colloidal suspension.

Atoms, molecules, and compounds

All matter is composed of fundamental building units known as ATOMS. Atoms are the smallest part of a chemical element that can participate in a chemical reaction. An ELEMENT is a substance consisting of only one type of atom. There are about 100 different elements, and each has been assigned a symbol derived from its English or Latin name. These symbols may be used to depict structure of molecules or to show how chemicals react with one another. Table 2.1 lists elements of biological importance. Their names and symbols should be learned.

Atomic structure

Atoms were once thought to be indivisible particles. With the development of increasingly powerful "atom smashers," atoms have been shown to be composed of still smaller particles. About a dozen different particles have been identified as being a part of an atom. Only three are important in our study: the PROTON, NEUTRON, and ELECTRON.

A proton has a mass of *one unit* on an atomic mass scale. It carries a *single positive electric charge*. A neutron has a mass of *one unit* also, but *no electric charge*. These particles are found in an atom in a compact group called the ATOMIC NUCLEUS. An electron has a mass about *1/2000th that of the proton,* and has *one unit of negative electric charge*. According to one view of atomic structure, electrons orbit the nucleus much as the planets orbit the sun. These electrons are arranged around the nucleus in particular energy levels or SHELLS, and each shell can contain only a certain maximum number of electrons. Each shell is designated either by a letter or a number (K or 1, L or 2, M or 3, etc.) from the nucleus outwards. The K shell, nearest the nucleus, can hold a maximum of 2 electrons, the L shell can hold a maximum of 8, and the M shell can hold a maximum of 18 electrons. Table 2.2 shows the numbers of electrons in the various shells of some common elements.

All atoms of a given element have equal numbers of protons and electrons. The element's ATOMIC NUMBER is equal to the number of protons *or* electrons in the atom. The element's ATOMIC

TABLE 2.1   Biologically important elements

Element/radical	Symbol	Atomic number	Atomic weight[a]
Calcium	Ca	20	40.08
Carbon	C	6	12.01
Chlorine	Cl	17	35.45
Cobalt	Co	27	58.93
Copper	Cu	29	63.55
Hydrogen	H	1	1.008
Iodine	I	53	126.90
Iron (ferrous)	Fe	26	55.85
Magnesium	Mg	12	24.31
Manganese	Mn	25	54.94
Nitrogen	N	17	14.01
Oxygen	O	18	15.99
Phosphorus	P	15	30.97
Potassium (kalium)	K	19	39.10
Sodium (natrium)	Na	11	22.99
Sulfur	S	16	32.06
Zinc	Zn	30	65.37

[a] Atomic weights are often "rounded off" to give whole numbers. For example, Ca, 40; Na, 23; K, 39; Fe, 56.

TABLE 2.2   Electrons in the shells of some common elements

Element	K	L	M	N
H	1			
C	2	4		
N	2	5		
O	2	7		
Na	2	8	1	
P	2	8	5	
S	2	8	6	
Cl	2	8	7	
K	2	8	8	1

WEIGHT is the *sum* of the protons and neutrons in the atom.

Isotopes

In some cases, atoms of a certain element may have more neutrons in their nuclei than other atoms of the same element. Most hydrogen atoms have one proton in the nucleus (99 percent have this configuration); some hydrogen atoms (1 percent) have one proton *and* one neutron. Atoms of the same element that differ in neutron number are called ISOTOPES of that element. Many elements of high atomic number and weight are *unstable*. Their various isotopes change or "disintegrate." As they do, they emit small particles, radiation, or both. These elements are said to be RADIOACTIVE. The particles emitted by naturally radioactive elements may be *alpha particles* (a helium nucleus consisting of two protons and two neutrons), or *beta particles* (electrons). High energy radiation called *gamma radiation* (similar to X rays) is also often produced. Biological processes, in general, do not discriminate between isotopes, and thus radioactive isotopes have been widely used as "biological tracers" to follow compounds or elements as they are metabolized by cells. The *energy of emission* is characteristic for each unstable isotope, as is the *rate* at which the isotope "decays" (changes to a more stable form). The term HALF-LIFE ($t_{1/2}$) is used to describe the time it takes for one-half of the atoms in an unstable isotope to decay. Table 2.3 presents some biologically useful isotopes, half-lives for radioactive ones, and some comments regarding the use of these isotopes in living organisms.

TABLE 2.3 Some isotopes useful in biology

Element	Isotope designation		Type of emission	Energy of emission (Mev[c])	Half-life ($t_{1/2}$)	Comments/applications of the isotope
	Stable[a]	Unstable[b]				
Hydrogen	^{2}H (deuterium)		—	—	—	Body water distribution and turnover studies
Carbon	^{13}C		—	—	—	Intermediary metabolism as constituent of organic molecules
Nitrogen	^{15}N		—	—	—	Intermediary metabolism as constituent of proteins, amino acids
Oxygen	^{18}O		—	—	—	Photosynthetic studies, oxidative reactions, water formation
Sulfur	^{34}S		—	—	—	Intermediary metabolism of sulfates, amino acids
Hydrogen		^{3}H (tritium)	β^{-d}	0.016	10.7–12.1 yr	Body water content, photosynthesis
Carbon		^{11}C	β^{+e}	0.97	20 min	CO_2 formation, elimination
Nitrogen		^{13}N	β^+	1.24	10 min	Nitrogen fixation by plants and microorganisms
Sodium		^{22}Na	β^+, γ^f	0.58	2.75 yr	Extracellular fluid volume; distribution in heart disease
		^{24}Na	β^-, γ	1.39	14.8 hr	Electrolyte metabolism, adrenal physiology
Phosphorus		^{32}P	β^-	1.70	14.3 da	Mineral metabolism, bone physiology, intermediary phosphorus metabolism
Chlorine		^{38}Cl	β^-, γ	2.70	37 min	Mineral metabolism
Potassium		^{42}K	β, γ	3.50	12.4 hr	Mineral metabolism
Calcium		^{45}Ca	β^-	0.26	150–180 da	Bone and tooth physiology, mineral metabolism
Iron		^{55}Fe	no particles[g]		4 yr	Blood formation and physiology
Cobalt		^{60}Co	β^-, γ	0.31	5.3 yr	Treatment of tumors
Zinc		^{72}Zn	β^-, γ	0.9 (av)	49 hr	Metabolism of zinc, insulin synthesis

TABLE 2.3   Some isotopes useful in biology (continued)

Iodine	^{131}I	β^-, γ	0.6	8.0 da	Iodine metabolism, treatment of thyroid tumors

[a] Stable isotopes may be detected by means of a mass spectograph that separates isotopes on the basis of mass.

[b] Unstable isotope radiation is detected by Geiger counters, scintillation detectors, and other types of measuring instruments.

[c] Energy of particle emission is determined by the length of its track in a substance and its energy calculated in million electron volts (Mev).

[d] Beta particle.

[e] Positron.

[f] Gamma radiation.

[g] Decays by internal electron rearrangement; no emissions.

Theoretically, half-lives and energy of emission are inversely related. Isotopes with short half-lives are obviously unsuited for long-term metabolic studies, while high energy emission isotopes that deliver a "wallop" for shorter periods of time are more suitable for therapeutic use. Extremely high energy radioactive materials may alter the system they are being used to investigate, and those having very long half-lives may pose problems for future generations because of the problems related to their disposal or storage. Most emissions have definite and usually destructive effects on living tissue: on its DNA, through the production of harmful chemicals, or through cell destruction. Thus, the use of radioactive materials must be tempered by exact knowledge of the effects of the isotope employed.

Chemical reactivity

Atoms have equal numbers of protons and electrons, and thus are *electrically neutral.* Elements whose atoms have eight electrons in their highest occupied shells are unusually stable. Such elements are the noble gas elements (e.g., argon, xenon), and an outer shell containing eight electrons denotes great chemical *unreactivity*.

The number of "unfilled spaces" in the highest occupied shell of a given element — that is, the number actually present as compared to an outer octet — contribute greatly to that element's chemical reactivity. For example, an atom of sodium has a single electron in the M shell, but has eight electrons in the L shell, the next one closer to the nucleus. It can react with other elements to lose the single M shell electron. It then gains a new outer shell, the L shell, and has a stable outer octet configuration. Chlorine has seven electrons in the M shell and can achieve a stable outer octet most directly *not* by losing seven electrons, but by gaining one. If one places sodium metal in contact with chlorine, atoms of sodium lose their outermost electron to chlorine atoms and three things happen: stable configurations (eight electrons in the outer shell) are achieved by both atoms; IONS are formed; and crystals of salt, NaCl or sodium chloride, are formed. Ions are electrically charged particles. If an atom of sodium loses one electron, it will have *one more proton than electrons,* and this new particle will have a single positive charge. We represent it as Na^+. It is not a sodium atom, but a *sodium ion.* Similarly, an atom of chlorine (acquiring an extra electron from sodium) will have *one more electron than protons.* It will now also be an ion, the chloride ion, represented as Cl^-. The electrons that can be easily lost or gained by an atom to give a stable configuration are the VALENCE ELECTRONS of the atom and they determine the manner and degree of chemical reactivity of that atom.

To illustrate:

Sodium (Na) "finds it easier" to lose one electron than to gain seven; its valence is plus 1. The reverse is true of chlorine, hence its valence is minus 1.

Magnesium (Mg; atomic number 12) has an M shell with two electrons in it, but an L shell with eight. To achieve a stable configuration most directly, the atom can lose two electrons. Thus, magnesium has a valence of plus two.

The noble gas elements such as neon or argon already possess full outer octets and react with nothing. Helium, the simplest noble gas, has only two electrons and is also chemically inert. It, however, has a filled K shell (capacity: two electrons). Thus, another condition of stability other than an outer octet is a filled K shell as the outermost level.

Not only does valence suggest degree of reactivity, but it also determines *how many* other atoms can react with a given atom. Magnesium ions, $+2$, can combine with two singly charged negative ions (e.g., Cl^-) to form magnesium chloride ($MgCl_2$). The same principle holds for reactions between other elements. *Their* atoms have the prime objective of acquiring stable outer octets.

Molecules and compounds

The combining of atoms into larger aggregations leads to the formation of MOLECULES. A molecule is an electrically neutral particle made from two or more atoms combined by chemical bonds. Substances consisting of molecules in which the atoms are from *different* elements are called COMPOUNDS.

BONDING. As atoms react with one another to form molecules, they combine in definite and fixed ratios. These *combining ratios* are determined by the capacities of unfilled (valence) electron shells to *lose, acquire,* or *share* electrons, as described below.

There are two basic ways that completed outer octets can be attained: by transferring electrons or by sharing them.

IONIC *(electrostatic)* BONDS are formed when oppositely charged ions *attract* each other. The ions form when electrons are lost by one atom and gained by another. The new ions are attracted to one another because they have opposite electric charges. Sodium chloride (Fig. 2.1) has ionic bonds.

COVALENT BONDS are formed when electrons are *shared* by atoms so as to achieve a statistical completion of their outer octets. The involved electrons are considered to spend part of their time associated with one atom and the rest of the time with the other atom(s). One shared pair of electrons is called a *single covalent bond;* two shared pairs, a *double bond,* and three shared pairs a *triple bond.* Figure 2.2 shows how covalent bonds are attained, using water and carbon dioxide as examples.

Some clusters of several atoms are held together by mutual electron sharing, but the "total package" has a net electrical charge. These clusters are called *polyatomic ions.* Examples are given in Table 2.4.

HYDROGEN BONDS involve what we may simply call a sharing of a hydrogen atom by two other atoms. Hydrogens are commonly shared by a weak form of attraction between oxygen and nitrogen atoms, as, for example, in the hereditary material, deoxyribonucleic acid (DNA) helices. Some examples of hydrogen bonding are:

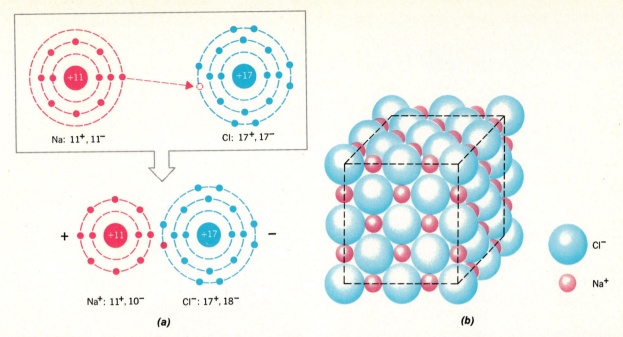

Na: 11^+, 11^- Cl: 17^+, 17^-

$+$ Na^+: 11^+, 10^- Cl^-: 17^+, 18^- $-$

(a)

(b)

Cl^-

Na^+

FIGURE 2.1 *(a)* The formation of an ionic bond between sodium and chloride atoms. *(Upper)* A sodium *atom* has 11 protons and 11 electrons; a chloride atom 17 protons and 17 electrons. *(lower)* Ions are formed by transfer of the single electron from the outer sodium shell to the outer chlorine shell. The positive sodium ion then attracts the negative chloride ion to form a molecule of NaCl. *(b)* The arrangement of sodium and chloride atoms in a cubical crystal of common table salt.

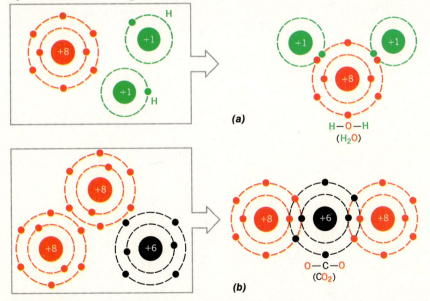

(a)

H—O—H
(H_2O)

(b)

O—C—O
(CO_2)

FIGURE 2.2 The formation of covalent bonds by sharing of electrons in *(a)* the water molecule, *(b)* the carbon dioxide molecule.

TABLE 2.4	Some biologically important polyatomic ions and their valences		
Ion	Symbol	Valence	Atomic weight[a]
Ammonium	NH_4	+1	18
Bicarbonate	HCO_3	−1	61
Hydroxyl	OH	−1	17
Nitrate	NO_3	−1	62
Phosphate	PO_4	−3	95
Sulfate	SO_4	−2	96

[a] Calculated using "rounded off" values for the elements involved.

Because these bonds are relatively weak, highly directional, and can be easily broken, they are found where there is a necessity for easy separation between the bound molecules. Figure 2.3 shows hydrogen bonding between DNA helices in nucleic acids.

Van der Waals forces (Fig. 2.4) also create weak bonding forces holding atoms together. In a given molecule, the shared electrons become unsymmetrical in relationship to the nuclei. One part of a molecule thus tends to have a momentary negative charge, another part tends to have a momentary positive charge. The molecule thus temporarily becomes a *dipole* with negative and positive parts much as a tiny magnet has negative and positive magnetic regions. Opposite ends attract each other, and the forces bind molecules or atoms together. In many substances—

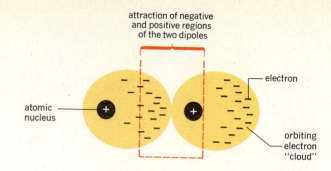

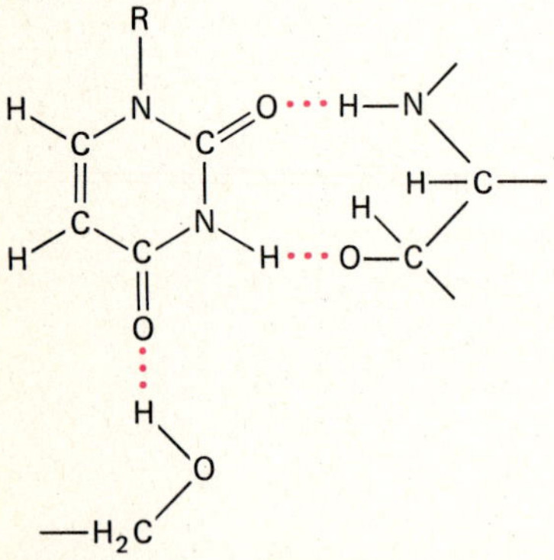

FIGURE 2.4 A diagram to illustrate the formation of dipoles and the resultant van der Waals forces.

water, for example—these dipoles are permanent, since atoms are sometimes attached to one another at specific bond angles.

MOLECULAR STRUCTURE AND FORMULAS. Bonding of atoms together results in the formation of a molecule, and enough identical molecules in one pile gives a visible sample of a compound. The atoms and their numbers that form the compound are listed by chemical symbol in what is called the MOLECULAR FORMULA of the compound. For example, H_2O refers to water, $C_6H_{12}O_6$ to glucose, CO_2 to carbon dioxide. Remember that formation of a molecule requires the formation of an electrically neutral particle. In polyatomic ions, however, the whole unit has a charge, and it reacts as a whole unit with oppositely charged ions. For example, one sulfate ion ($SO_4^=$) reacts with two hydrogen ions (H^+) to give sulfuric acid (H_2SO_4).

STRUCTURAL FORMULAS show not only the atoms forming the molecule, but their relationships to one another. For example, water may be represented as

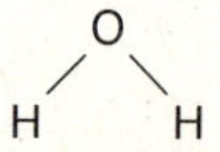

FIGURE 2.3 Hydrogen bonding (dashed lines) between certain components of a nucleic acid. R = an unspecified chemical group. Note sharing of H by two other atoms.

with each line representing a covalent bond. Ethyl alcohol may be shown as

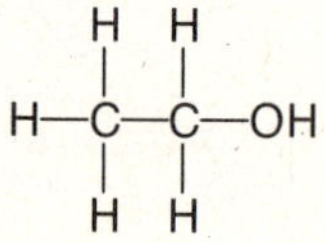

In some cases, special symbols have come into common usage to express structural formulas in highly abbreviated ways. ATP refers to *a*denosine *tri*phosphate, G-6-P designates *g*lucose 6-*p*hosphate. This type of chemical shorthand is used often in later chapters.

Two or more different compounds may contain the same number and kind of atoms—that is, they may have identical molecular formulas and yet have different *arrangement* of atoms in their molecule; they therefore have different chemical properties. For example, the formula $C_6H_{12}O_6$ describes a *series* of 6-carbon sugars called hexoses (6 carbons per molecule). Their structural formulas (Fig. 2.5) indicate that certain sugars have terminal chemical groups that are different from others, and that hydroxyl (—OH) groups are oriented differently. Compounds having the same molecular formula, but different structural formulas are called ISOMERS. If a molecule exists in two nonidentical forms that are *mirror images* of one another, they are called ENANTIOMERS (Fig. 2.6). This kind of isomerism is common among organic compounds and often determines if the molecule can enter a particular chemical reaction. Enantiomers rotate the plane of polarization of a beam of polarized light according to their exact chemical structure, and partly because of this effect the terms D (dextrorotatory) or L (levorotatory) may become part of the name of the substance.

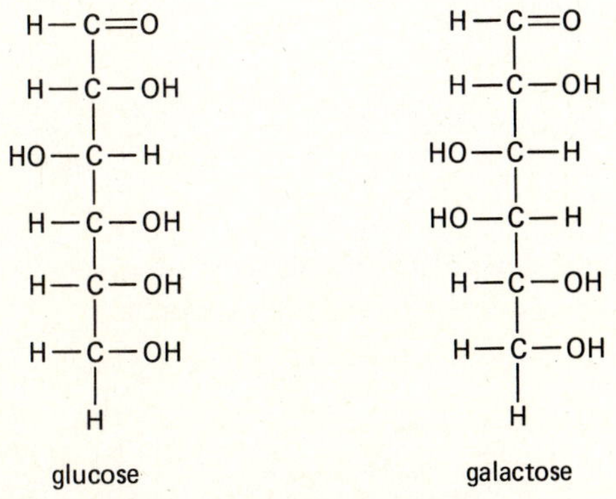

FIGURE 2.5   Structural formulas of two isomers of $C_6H_{12}O_6$ (find the difference).

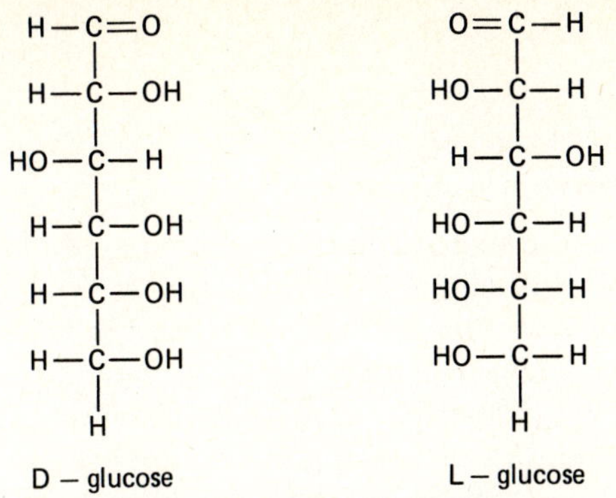

FIGURE 2.6   Enantiomers of glucose.

ORGANIC AND INORGANIC COMPOUNDS. The name *organic chemistry* was originally applied to the study of materials obtained from living or dead plants and animals. Today, many such compounds can be synthesized in the laboratory, and the early definition is no longer valid. The modern definition is that ORGANIC COMPOUNDS are compounds of carbon, excluding cyanides (—C≡N), carbonates ($CO_3^=$), bicarbonates (HCO_3^-), and the oxides of carbon (CO, CO_2). INORGANIC COMPOUNDS comprise all the rest, and are most commonly of mineral origin.

HYDROCARBONS are organic compounds that consist only of C and H. All are combustible and form important energy sources for our economy; examples are propane and gasoline. Hydrocarbons are isolated chiefly from petroleum and coal. *Aliphatic hydrocarbons* have chains or rings of carbons in their molecules. These systems may have saturated (all carbons with hydrogen on them) or unsaturated (double or triple) bonds in them. In general, the more unsaturated bonds in the molecule, the lower its melting point. Unsaturated fats are liquid at room temperature (e.g., a liquid cooking oil). *Aromatic hydrocarbons* have rings in them that are never saturated, and in which the electrons of the ring move in a closed circuit around the ring from bond to bond, giving great stability to the molecule. Molecules of unsaturated compounds have one or more double or triple bonds. For example,

$$H_2C=CH_2$$
ethene

$$H—C\equiv C—H$$
ethyne

If *several* unsaturated carbons are found within the molecule, it is said to be *polyunsaturated*. This term is perhaps familiar to those interested in nutrition who believe consumption of certain fats lessens chances of vascular problems. In cooking, an unsaturated, liquid cooking oil is supposedly preferable to a solid saturated shortening for lowering blood fat levels. ALCOHOLS are aliphatic compounds in which one or more H atoms are replaced by hydroxyl (—OH) groups. ETHERS contain two carbon atoms bonded to the same oxygen $\left(—\overset{|}{\underset{|}{C}}—O—\overset{|}{\underset{|}{C}}—\right)$; they are solvents for fats and oils, and one—ethyl ether—is an important anesthetic. AMINES have one or more hydrogens in ammonia (NH_3) replaced with hydrocarbon groups. The amino acid $R—\overset{H}{\underset{NH_2}{C}}—COOH$ is an important constituent of body proteins. ORGANIC ACIDS contain a carboxyl group $\left(—C\overset{O}{\underset{OH}{}}\right)$, sometimes written as

—COOH) on one or more carbons of the molecule. The carboxyl group commonly loses its hydrogen and becomes an ion, conferring negative electrical charge to the entire molecule

$$—C\overset{O}{\underset{OH}{}} \longrightarrow —C\overset{O}{\underset{O^-}{}} + H^+$$

The O^- group may then combine with other atoms. The carbonyl group $\left(—\overset{|}{\underset{|}{C}}=O\right)$ is found not only in carboxylic acids, but also in KETONES and ALDEHYDES. Aldehydes contain the aldehyde group $\left(—C\overset{O}{\underset{H}{}}\right)$; formaldehyde $\left(H—C\overset{O}{\underset{H}{}}\right)$ is a well-known preservative.

POLYMERS are formed when single units of a substance are bonded together into very large molecules. For example, amino acids form proteins, glucose forms glycogen. In nature, most polymers have structural functions, or may serve as storage forms of the monomers composing them.

Types of chemical reactions

Several types of reactions can occur between elements and compounds to create new products.

In EXCHANGE REACTIONS, one or more atoms, ions, or chemical groups "trade places" with others:

$$Na_2SO_4 + BaCl_2 \longrightarrow BaSO_4 + 2NaCl$$
sodium sulfate · barium chloride · barium sulfate · sodium chloride

In CONDENSATION REACTIONS, two or more atoms, ions, or molecules *combine* to form products:

$$H_2O + CO_2 \longrightarrow H_2CO_3$$
water · carbon dioxide · carbonic acid

or

$$H^+ + Cl^- \longrightarrow HCl$$

hydrogen   chloride   hydrochloric
ion   ion   acid

DECOMPOSITION REACTIONS are basically the reverse of condensation reactions, with a larger molecule being "broken down" into smaller ones:

$$H_2CO_3 \longrightarrow CO_2 + H_2O$$

In REARRANGEMENT REACTIONS, one isomer is converted to another:

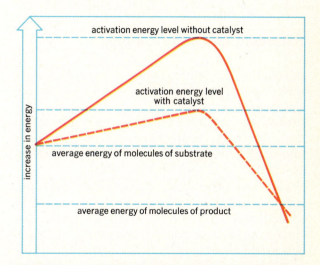

3-phosphoglyceric acid   2-phosphoglyceric acid

Energy exchange in chemical reactions

Every chemical reaction involves energy. Energy is required to break chemical bonds; energy is released when bonds are formed. If energy is released by a chemical change, that reaction is termed EXERGONIC *(energy releasing)*. The reverse situation—that is, the consumption of energy by a chemical change—is termed an ENDERGONIC *(energy requiring)* reaction. Energy input or output may be in the form of heat, light, or other forms of energy; however, energy output is usually measured in heat units (calories).

An important type of reaction occurring in living material is the OXIDATION-REDUCTION reaction. *Oxidation* is basically defined as a loss of electrons, and often occurs when a compound gains oxygen or loses hydrogen. *Reduction* involves gain of electrons, and usually occurs when a molecule loses oxygen or gains hydrogen. In such reactions, one substance is oxidized and the other is reduced, as electron, oxygen, or hydrogen transfer occurs. Oxidative reactions are generally exergonic, while reductions are usually endergonic.

Many chemical reactions involving covalent bonds would require conditions incompatible with life (high temperatures, pressure, etc.), or would proceed too slowly to be of value to the body if it were not for the presence of CATALYSTS. These are substances that, in low concentrations, increase the rates of chemical reactions so that they can proceed under the mild conditions present in the body. A catalyst is not destroyed in the reaction it influences. It works by lowering the

energy of activation for the reaction—that is, by reducing the energy barrier to the reaction (Fig. 2.7). The method by which a catalyst operates may occur as follows: $A + B \longrightarrow AB$

FIGURE 2.7 Graphic representation of an energy-releasing chemical reaction, showing the activation energy levels with and without a catalyst. Only a few of the reacting substrate molecules have a high enough energy to reach the high activating energy level without a catalyst and so the reaction proceeds very slowly. However, by supplying heat or another outside energy source, many more molecules have enough energy to clear the high energy hump, even without a catalyst. If a catalyst is present, its temporary combination with the substrate molecules results in a substance with a lower activation energy level, so the reaction proceeds more rapidly and at a much lower temperature than when no catalyst is present. Enzymes are the principal catalysts in organisms.

(This is the reaction desired, but one with a high requirement for energy input, i.e., the *energy of activation*).

With a catalyst, C,

$$A + C \longrightarrow AC + B \longrightarrow AB + C$$

(The reaction of A and C requires less energy, activates, or raises the energy level of A, and makes it react easily with B).

In the body, ENZYMES act as catalysts.

Reactions in living cells

The preceding sections have described some of the reactions that chemical substances can undergo. In living cells, materials are synthesized for structural and control purposes and metabolized for energy to run body operations and to produce new materials, such as enzymes. These reactions form the basis of BIOCHEMISTRY or INTERMEDIARY METABOLISM. METABOLIC CYCLES are described in later chapters that illustrate the intermediary metabolism of those compounds most essential to maintenance of life.

Solutions and suspensions

A solution is a homogeneous mixture of *solutes* (dissolved substance) in a *solvent* (dissolving medium) where the solute molecules or its parts cannot be distinguished from the solvent by ordinary means. In the body, WATER is the most important solvent; it contains hundreds of solute species.

The CONCENTRATION, or amount of solute per unit of volume of solvent, may be expressed in several ways. A *percentage solution* contains a certain number of grams (percent) of solute per specified weight of solvent. For example, a 5 percent NaCl solution would contain 5 grams of NaCl in 95 grams of solvent, for a total of 100 grams of solution. As one can see, this is a percentage *by weight*. A percentage *by volume* would contain, for example, 70 ml of pure ethyl alcohol and 30 ml of water for a 70 percent alcohol solution.

The MOLAR CONCENTRATION of a solution is defined as the number of moles of solute per liter of solution (or the number of moles dissolved in enough water to make a liter of solution), where 1 mole of any solute is its formula weight expressed in grams (gram molecular weight). A NORMAL SOLUTION contains one gram *equivalent weight* [gram molecular weight divided by the *valence* of the material in question; e.g., Na^+ ($23 \div 1$); Ca^{2+} ($40 \div 2$)] of a substance dissolved in enough water to make one liter of solution. A MOLAL SOLUTION contains a mole of a substance added to 1000 grams of solvent. Thus, the volume of a molal solution would be slightly greater than that of a molar or normal solution; its final volume will be more than 1 liter.

A SUSPENSION—for example, a *colloidal suspension*—contains large molecules such as starch or protein. These molecules do not form a true solution; the solute *can* be distinguished from the solute by turbidity (showing a beam of light), viscosity, and other properties. Particles in a suspension are usually 1 to 100 nanometers (nm)* in size.

A final word

Discussions of electrolytes, acids and bases, and pH are found in appropriate chapters later in the text. Please consult the index if you wish information at this time about these subjects.

*m = meter.
mm = millimeter = 1/1000 m or 10^{-3} m.
μm = micrometer or micron = 1/1,000,000 m or 10^{-6} m.
nm = nanometer or millimicron (mμ) = 1/1,000,000,000 m or
      10^{-9} m.
Å = angstrom = 1/10,000,000,000 m or 10^{-10} m.
pm = picometer = 1/1,000,000,000,000 m or 10^{-12} m.

Summary

1. The three kinds of matter are elements, compounds, and mixtures.

2. Elements may be described by name and chemical symbol, and characterized by atomic number and atomic weight. All the atoms of one element have the same atomic number, which is equal to the number of the atom's protons.

3. An atom is composed of basic particles.
   a. The proton has a mass of 1, and carries a single positive charge.
   b. The neutron has a mass of 1, and no charge.
   c. The electron has a mass of about 1/2000th of a mass unit, and carries a single negative charge.
   d. Protons and neutrons form the atomic nucleus, with electrons orbiting the nucleus.
   e. Electrons are organized in shells or levels around the nucleus with a certain capacity of electrons in each shell.
   f. The element's atomic number is the same as the number of protons or electrons it has, while its atomic weight is the sum of protons and neutrons (the atomic weight is often "rounded off" for convenience in calculations).
   g. Valence electrons determine the types of bonds an atom can form.

4. Isotopes are atoms of identical atomic number but different atomic masses.
   a. Some isotopes are stable, not emitting particles or radiation.
   b. Radioactive isotopes emit particles or radiation (they decay).
   c. The half-life of a radioactive substance is the time required for one-half of the atoms to decay. Half-lives vary widely.
   d. Many radioactive isotopes have biological applications.

5. Molecules are formed by the combining of atoms. They are electrically neutral particles of two or more atoms and are held together by bonds.

6. Various types of bonds are formed between atoms.
   a. Ionic (electrostatic) bonds are forces of attraction between oppositely charged ions, and are formed when electrons are transferred between atoms to create ions.

    b. Covalent bonds are forces of attraction of nuclei toward the shared pairs of electrons between them.

    c. Hydrogen bonds are weak forces of attraction between two polar (charged) molecules in which hydrogen is to some extent "shared" between oxygen or nitrogen atoms.

    d. Van der Waals forces involve weak attraction between dipole molecules.

7. The orientations of atoms in a molecule or compound may be expressed in several ways.

    a. Molecular formulas present the atoms involved and their relative proportions in the molecule.

    b. Structural formulas show general (two-dimensional) spatial arrangement of atoms in the molecule.

8. Isomers are molecules having identical molecular formulas but different structural formulas.

9. Enantiomers are isomers whose molecules are related as mirror images that cannot be superimposed.

10. An organic compound is any compound of carbon except carbonates, bicarbonates, cyanides, and oxides of carbon.

11. Inorganic compounds are all other compounds and are of mineral origin.

12. Hydrocarbons are substances whose molecules contain only carbon and hydrogen.

    a. Aliphatic hydrocarbons are substances whose molecules are straight or branched chains of carbons, or which have rings of carbons.

    b. Aromatic hydrocarbons have ring structures that always contain alternating double and single bonds.

    c. Saturated hydrocarbons have no double or triple bonds in their molecules.

    d. Unsaturated hydrocarbons have double or triple bonds in their molecules.

    e. Polyunsaturated hydrocarbons have multiple double or triple bonds per molecule.

13. Several classes of organic compounds exist.

    a. Alcohols contain hydroxyl groups.

    b. Ethers contain a carbon-oxygen-carbon bond.

    c. Amines have ammonia hydrogens replaced with other groups.

    d. Organic acids contain a carboxyl group(s).

    e. Ketones contain a carbon-double-bonded-oxygen group (carbonyl group).

    f. Aldehydes have a terminal carbon atom to which a double-bonded oxygen and a hydrogen are attached.

    g. Polymers are formed by the joining of many identical molecules.

14. Several types of reactions may occur between molecules.

   a. Exchange reactions involve "trading" of molecular constituents between atoms or ions.

   b. Condensation reactions entail a combination of molecules to give a new product.

   c. Decomposition reactions break down molecules to simple constituents.

   d. Rearrangement reactions change one isomer to another.

15. Chemical reactions are endergonic (energy requiring) or exergonic (energy releasing), in terms of net energy of the reaction. Catalysts accelerate chemical reactions.

16. Intermediary metabolism consists of the chemical reactions occurring in living cells.

17. A solution is a homogeneous mixture of a solute in a solvent.

   a. Percentage solutions contain a given percent (by weight or volume) of one substance in another so as to create a final mixture of 100 gm weight, or 100 ml volume.

   b. Molar solutions contain a certain number of moles in a liter of solution.

   c. Normal solutions contain a gram equivalent weight in a liter of solution.

   d. Molal solutions contain a mole added to 1000 gm of solute.

18. A suspension contains large molecules whose presence may be demonstrated by a variety of simple tests.

Questions

1. Distinguish between molecules and ions; between elements, compounds, and mixtures.

2. What are the most common elements and ions in living material?

3. What are the basic particles composing atoms? How do they differ?

4. What is the relationship between electron shells and chemical reactivity?

5. What is an isotope? A stable isotope? A radioactive isotope?

6. How do radioactive isotopes achieve stability?

7. What is meant by the term half-life?

8. What are some biological applications of radioactive isotopes?

9. How do molecules differ from atoms? Atoms from elements? Molecules from compounds? Elements from compounds?

10. How do ionic, covalent, and hydrogen bonds differ? Describe what each is.

11. If a substance is represented as $C_6H_{12}O_6$, what kind of a formula is this? What are the types of chemical formulas?

12. Distinguish between an isomer and an enantiomer.

13. If someone said, "I have a mixture of an alcohol and a ketone," what would the mixture contain in terms of organic compounds? (Give characteristics of the substances involved.)

14. What happens in a condensation reaction? A rearrangement reaction?

15. How do solutions and suspensions differ?

Readings

Brady, James E., and Gerard E. Humiston. *General Chemical Principles and Structure,* 2nd ed. Wiley. New York, 1978.

Mortimer, Charles E. *Chemistry, A Conceptual Approach.* D. Van Nostrand Co. New York, 1975.

Perl, Martin L., and William T. Kirk. "Heavy Leptons." *Sci. Amer. 238:*50, March 1978.

Peters, Edward I. *Introduction to Chemical Principles.* Saunders. Philadelphia, 1978.

Stryer, Lubert. *Biochemistry.* W.H. Freeman and Co. San Francisco, 1975.

Wilson, Allan C., Steven S. Carlson, and Thomas J. White. "Biochemical Evolution." *Ann. Rev. Biochem.* 46:573, 1977.

Chapter 3

The Chemical and Cellular Levels of Organization and Function

Objectives

After studying this chapter, the reader should be able to:

- List the major elements composing protoplasm and their average percentages of living material.

- List the major groups of substances that form the body and give the general function(s) of each.

- List the main electrolytes of the body, indicating if they are cations or anions, whether they are found as elements or polyatomic ions, and give their functions.

- Define amino acids, peptides, oligopeptides, and poly-peptides, and distinguish between fibrous and globular proteins as to properties and functions.

- List some properties of enzymes and how they are named.

- Give the generalized formula for carbohydrates and distinguish between mono-, di-, and poly- saccharides, giving numbers of "building blocks" in each type and functions of each category.

■ Describe lipid characteristics and list the lipid subdivisions, with examples of each and their functions.

■ List the components of nucleic acids, the two major types, and the functions they serve, and distinguish between nucleotides and nucleosides.

■ Tell what ATP is and its role in the economy of the body.

■ Define a cell and list its three basic subdivisions. Describe the generalized structure of the cell membrane, give its functions, and explain some factors that determine permeability of a membrane.

■ List the common cellular organelles, describe their structure, and give their functions (i.e., endoplasmic reticulum, ribosomes, Golgi complex, lysosomes, mitochondria, central body, microtubules, peroxisomes).

■ Give examples of inclusions and their function(s) in the cell.

■ Describe the structure of the nucleus of a cell and give its functions.

■ Show how form and structure are interrelated in different types of cells.

■ Distinguish between passive and active processes in terms of participation by the cell. Also, describe the various types of each process, with examples of where they occur in the body.

In Chapter 2, we considered inorganic and organic substances as chemical categories. This chapter discusses the more common materials found to compose living units, regardless of their phylogenetic position in the plant or animal kingdoms. A "common denominator" should thus become apparent—the nearly common chemical basis for life. We next consider the organization of these chemicals into functioning units, commonly called cells, from both structural and functional viewpoints. Finally, we examine the methods by which cells acquire the substances they need and those by which they release their products and wastes. These sections should create a picture in the reader's mind of the structure and functioning of a generalized unit, and the individual structural and functional specializations of cells may be more easily understood in later chapters.

The chemical constituents of living material

Four elements compose over 95 percent of the body. These elements and their percentages of body weight are:

Hydrogen (H), 10 percent
Oxygen (O), 65 percent
Carbon (C), 18 percent
Nitrogen (N), 3 percent

Additional substances that may often occur as elements or ions in the body include calcium (Ca), sodium (Na), potassium (K), and chlorine (Cl). Phosphorus (P) and sulfur (S) are most commonly found in polyatomic ions such as phosphate and sulfate.

Hydrogen, oxygen, carbon, nitrogen, phosphorus and sulfur are linked by covalent bonds to form the larger molecules and compounds essential to body structure and function. Water, inorganic substances, proteins (some of which are enzymes), carbohydrates, lipids, trace materials, nucleic acids, and adenosine triphosphate (ATP) are only a few of the substances considered necessary for life. The adult human body contains these major molecular constituents in the following percents.

Material	Percent of body weight[a]	
	Male	Female
Water	62	59
Protein	18	15
Lipid	14	20
Carbohydrates	1	1
Other (nucleic acids, hormones, etc.)	5	5

[a] Sex differences in certain categories of materials may, in part, be accounted for by the presence of greater blood volume in males than females, as well as more muscular tissue and less fat (lipid) under the skin and internally.

Water

Water, a compound formed by the covalent linkage of two hydrogen atoms with an oxygen atom, composes (*on the average*) 55 to 65 percent of the substance of adult cells. It is a good SOLVENT, forming *solutions* with soluble substances, and forming the medium for *suspensions* within the cell. It is NOT TOXIC to the cell when isotonic, and is capable of absorbing a lot of metabolically derived heat; it thus serves as a MEDIUM OF HEAT EXCHANGE AND TRANSFER. When added to water to form a solution, many substances IONIZE, and become more chemically reactive than before. Most water in the body is freely available for movement within and around cells. About 4 percent of the total body water appears to be associated with cell membranes, electrolytes, and other substances as *hydration layers* (layers of water molecules acting as dipoles and "bound" to charged substances by van der Waals or other types of bonds). WATER PARTICIPATES IN CHEMICAL REACTIONS both as a reactant and as a product. In digestion of foods within the organs of the alimentary tract, for example, large molecules are split into smaller units in *hydrolysis reactions* that involve enzymatic addition of water; conversely, when several types of synthesis or polymerization reactions occur in certain body cells, water is a product. Lastly, water, as a constituent of certain types of body fluids called mucus, serous fluids (secreted by linings of closed body cavities), and synovial fluids (in body joints) acts as a LUBRICANT and PROTECTIVE fluid.

Inorganic substances

This group is commonly interpreted to include elements, ions, polyatomic ions (sometimes called *radicals*), and such substances as carbon dioxide. Most, when placed into solution, ionize, and thus the solution conducts electric current. The ionized substances are therefore called ELECTROLYTES. Pos-

TABLE 3.1 The main electrolytes of the body

Substance	Charge	Element (E) or Radical (R)	Anion (A) or Cation (C)	Comments
Sodium	+1	E	C	The major electrolyte of fluids *outside* of cells
Potassium (K)	+1	E	C	The major electrolyte of fluids *inside* of cells
Calcium (Ca)	+2	E	C	Essential for muscle contraction, blood clotting, excitability
Chloride (Cl)	−1	E	A	With Na, a major constituent of fluid *outside* cells
Magnesium (Mg)	+2	E	C	Essential for many enzymatically controlled reactions
Phosphate (HPO_4)	−3	R	A	Many "high energy compounds" contain phosphate
Sulfate (SO_4)	−2	R	A	A product of many chemical reactions involving sulfur-containing amino acids and proteins
Bicarbonate (HCO_3)	−1	R	A	A constituent of fluids *outside* of cells

itively charged materials are attracted to a negatively charged electrical pole (*cathode*) and are called CATIONS; negatively charged materials are attracted to a positively charged electrical pole (*anode*) and are called ANIONS. The most common electrolytes of the body are summarized in Table 3.1.

Among their other functions, electrolytes aid in the establishment of OSMOTIC GRADIENTS (they act as solutes, especially the abundant ions Na^{1+}, K^{1+}, Cl^{1-}), influence enzymes (notably, Mg^{2+}), and aid in the EXCITABILITY and PERMEABILITY of membranes in all body cells (Ca^{2+}, Na^{1+}, Cl^{1-}, K^{1+}).

Organic substances

The reader is reminded that organic compounds contain carbon, hydrogen, and usually oxygen, and form the basis of the structural and functional materials of the body. The ability of carbon to form covalent linkages with a variety of other atoms, or with itself, creates a mind-boggling number of different compounds. This variety, in part, distinguishes humans from other living organisms, and one human from another. Our "chemical fingerprints" are as individual as each human is.

PROTEINS. Proteins are complex molecules containing, on the average, C (50 to 55 percent), H (6 to 8 percent), N (15 to 18 percent), and S (0 to 4 percent). Proteins are composed of smaller building units called AMINO ACIDS. Amino acids are represented by the formula

$$R-\underset{\underset{NH_2}{|}}{\overset{\overset{H}{|}}{C}}-COOH$$

where R may be an aliphatic or aromatic group.

A PEPTIDE is a unit containing fewer amino acids than a protein; an *oligopeptide* contains fewer than 10 amino acids, while a *polypeptide* contains more than 10, but fewer amino acids than the protein it may compose. *Simple proteins* are composed of only amino acids, while *conjugated proteins* contain a protein and a nonprotein substance [e.g., lipoprotein (fat and protein); nucleoprotein (nucleic acid and protein)].

FIBROUS PROTEINS are proteins in elongated or filamentous form. They are relatively insoluble in most dilute solvents and form the basic structure of connective tissue, muscle (contractile proteins), hair, nails, and the outer skin layer (epidermis). GLOBULAR PROTEINS are folded, more soluble, and form important constituents of blood plasma (albumins, globulins), red cells (hemoglobin), and cytochromes (proteins serving in oxidation-reduction reactions in cells). We thus see proteins serving important functions in structure, contraction, protection, and cell metabolism.

ENZYMES, as stated in Chapter 2, are organic catalysts altering the rates of chemical reactions in the body. An enzyme consists of a protein synthesized by a cell that often combines with a nonprotein *coenzyme* to give the active substance. Enzymes are specific in that they usually catalyze a single chemical reaction and work upon specific *substrates* or reactants. They are DESTROYED BY HEAT (heat labile) and may be INHIBITED by other substances that are not their primary substrates. They are often named according to substrate or type of reaction catalyzed, and carry *-ase* as a suffix of their name. Without enzymes, the body would cease to function as a chemical factory. Table 3.2 presents some enzymes and the type of reactions they catalyze.

TABLE 3.2 Partial classification of enzymes			
Enzyme	Substrate (material with which enzyme reacts)	Action	Examples
Hydrolases	None specific. Fats, phosphate-containing compounds, nucleic acids, many others	By adding water (hydrolysis), splits larger molecules into smaller units	Lipases, phosphatases, nucleases
Carbohydrases	Carbohydrates, starches, disaccharides	Fragment carbohydrates by hydrolysis to smaller units	Salivary amylase, pancreatic amylase, maltase, lactase, sucrase
Proteases	Proteins, smaller units containing fewer amino acids	Fragment proteins by hydrolysis to smaller units	Pepsin, trypsin, erepsin
Phosphorylases	Many molecules	Split or activate molecule by addition of phosphate group	Muscle phosphorlyase, glucokinase, fructokinase
Dehydrogenases	Any compound capable of losing 2 H^+	Oxidizes compound by removal of 2 H^+	Succinic dehydrogenase, fumaric dehydrogenase
Oxidases	Any compound that may add oxygen	Oxidizes compound by addition of oxygen	Peroxidase
Aminases	Amine-containing compounds and keto acids	Addition or removal of amine groups	Transaminase, deaminase
Decarboxylases	Acids containing carboxyl group	Remove CO_2 from carboxyl group	Ketoglutaric decarboxylase
Isomerases	Chiefly carbohydrates	Moves a radical to different part of molecule	Isomerase

CARBOHYDRATES. Carbohydrates are, literally, hydrates (H_2O) of carbon (C); they are often abbreviated or represented by the formula CH_2O. The formula suggests that the hydrogen and oxygen in the carbohydrate is always in the ratio of 2:1. As a general rule this is true, but exceptions exist as we shall see. Some of the hydrogens and oxygens are characteristically found as hydroxyl (—OH) groups. These give the substances their "sweet taste." Carbohydrates include sugars and starches, and are conveniently categorized into three groups.

MONOSACCHARIDES are units containing 3 (trioses), 4 (tetroses), 5 (pentoses), 6 (hexoses), or 7 (heptuloses) carbon atoms. As one can see, the suffix -*ose* is applied to many types of carbohydrates. Another name for monosaccharides is *simple sugars;* this suggests that they are the smallest units that retain the characteristics of the carbohydrate, *and* that they can be used as sources for polymerization into larger carbohydrates. For our purposes, we shall concentrate on the 5 and 6 carbon monosaccharides (Fig. 3.1). The most important monosaccharides we shall consider are ribose ($C_5H_{10}O_5$), deoxyribose ($C_5H_{10}O_4$, the best known exception to the CH_2O "rule"), and glucose, fructose, and galactose, isomers with the formula $C_6H_{12}O_6$.

DISACCHARIDES ("double sugars") consist of two simple sugar units joined by a condensation reaction that splits out a molecule of water (Fig. 3.2). Thus, (2) $C_6H_{12}O_6$ become (1) $C_{12}H_{22}O_{11}$ (+ H_2O). Again, there are several important disaccharides:

Maltose (malt sugar) consists of 2 glucose molecules

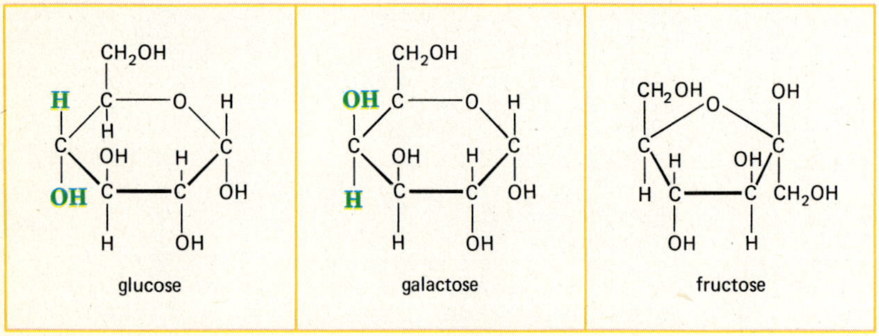

FIGURE 3.1 Monosaccharides (simple sugars). Glucose and galactose differ only in the *position* of the boldfaced chemical groups. Fructose has only five atoms in the ring.

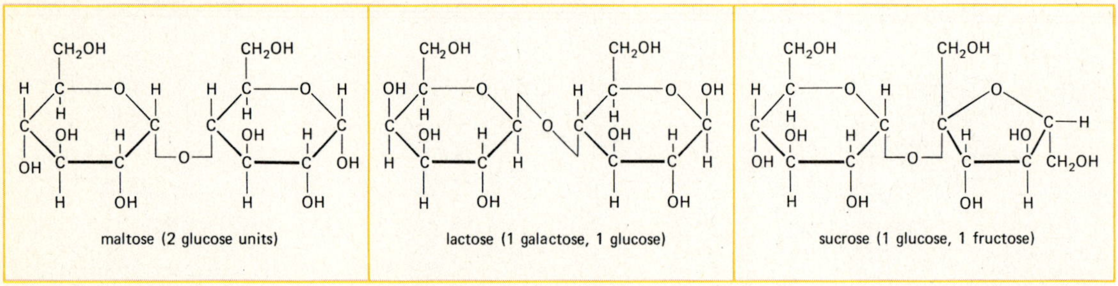

FIGURE 3.2 Some important disaccharides. Maltose is malt sugar. Lactose is milk sugar. Sucrose is cane or beet sugar.

Lactose (milk sugar) consists of a glucose and a
    galactose molecule
Sucrose (cane or beet sugar) consists of a glucose
    and fructose molecule
These disaccharides are also isomers.

POLYSACCHARIDES are polymers formed by the joining of at least three or (usually) many hundreds of
monosaccharides. Glycogen ("animal starch") is the
best known polysaccharide of our bodies (Fig. 3.3).
These carbohydrates may be represented by the formula $(C_6H_{12}O_6)_n$, where *n* is any number above 2. They
usually do not taste sweet and are not readily soluble
in water.

As a group, carbohydrates serve as "quick-
energy" for the body cells. They are easily metabolized, particularly glucose and fructose. They provide energy for the synthesis of ATP, the cells'
chief source of "fuel" for the functioning of cellular machinery. Glycogen forms a "checking account" for glucose; it is a storage form that is
readily available for conversion to glucose. Glycogen is stored chiefly in liver and muscle cells.
Some carbohydrates are joined to noncarbohy-
drate fractions (e.g., glycoproteins) to form structural components for the body, as in cell membranes. Ribose and deoxyribose are constituents
of the nucleic acids that determine our hereditary
potential ("genes").

LIPIDS. Lipids — fatty substances — contain carbon,
hydrogen, and oxygen, but the hydrogen and oxygen are *not* in a ratio of 2:1. Additionally, lipids
are deficient in oxygen compared to the other two
atoms. As a group, they are characterized as being
insoluble in water, but soluble in "fat solvents"
such as ether, chloroform, and alcohol. Three categories of lipids are recognized.

SIMPLE LIPIDS (Fig. 3.4) include fats, waxes, and
oils. Common examples are solid and liquid shorten-
ings, beeswax, and the *triglycerides*. The latter are
compounds of three aliphatic fatty acids attached to a
molecule of glycerol. The triglycerides are the body's

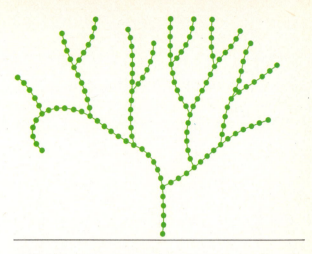

FIGURE 3.3   The glycogen molecule. Each circle represents a glucose molecule.

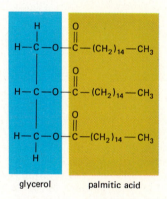

glycerol   palmitic acid

FIGURE 3.4   A triglyceride-tripalmitin- composed of
three palmitic acid molecules attached to glycerol.

storage form of fat in its fat (adipose) cells. Fatty
acids, released from the degradation of triglycerides,
form an important part of the lipids of the bloodstream, and may be a factor in the development of
vascular disease (atherosclerosis). In storage in adipose cells, triglycerides represent a concentrated form
of energy storage; they also insulate the body and
give it shape and form. Unsaturated fatty acids promote the metabolism of saturated fatty acids, and
some of them are called essential fatty acids because
they cannot be made by body cells.

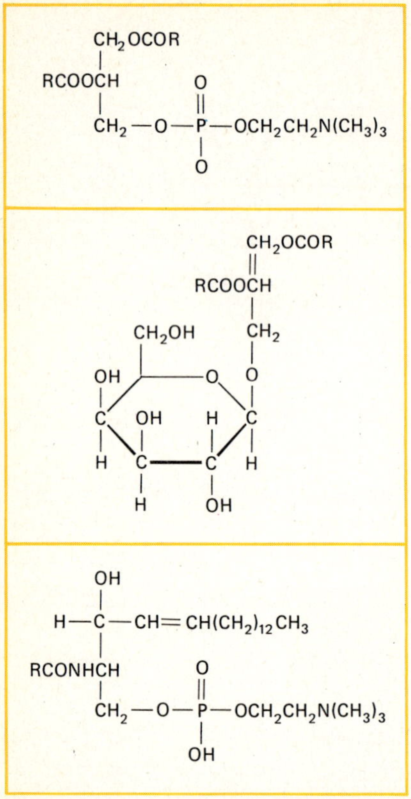

(a)

(b)

(c)

FIGURE 3.5 The basic chemical structures of some representative compound lipids. *(a)* A phospholipid (lecithin, found in membranes). *(b)* A glycolipid (monogalactosyl diacyl glycerol, found in photosynthetic tissue), *(c)* A sphingolipid (sphingomyelin, found in nervous tissue). R = additional chemical groups (not specified).

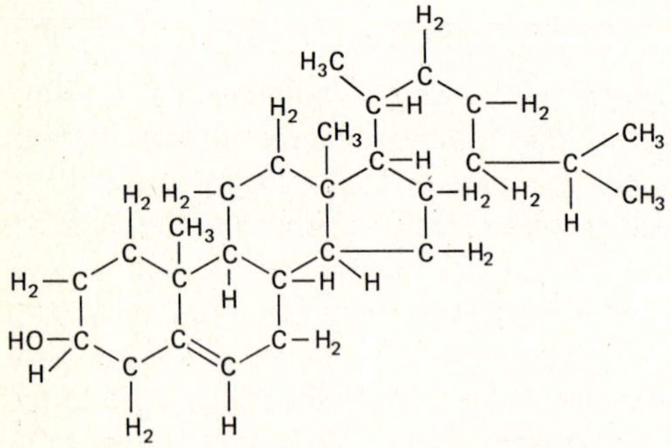

FIGURE 3.6 Cholesterol, a derived lipid.

COMPOUND LIPIDS (Fig. 3.5) consist of a lipid *and* a nonlipid portion, and usually form structural components in the body. For example, phospholipids (lipid and phosphate) are found in cell membranes, glycolipids (lipid and carbohydrate) occur also in cell membrane systems.

DERIVED LIPIDS (Fig. 3.6) include intermediates in lipid metabolism and the *sterols.* The latter are substances derived from cholesterol, and include cholesterol itself and several important steroid hormones (testosterone, estrogens, adrenal cortical hormones). Cholesterol, in combination with proteins (lipoproteins), has been implicated as a factor in atherosclerosis. This condition involves deposition of fats in the walls of the blood vessels, with narrowing of the vessel and decrease of blood flow to vital organs (heart, brain, kidneys).

TRACE MATERIALS. Other organic and inorganic substances are required for normal cellular function, but are NEEDED IN ONLY MINUTE QUANTITIES. Vitamins, certain metals (e.g., Co, Cu, Zn), and several hormones, are included in this category. It is a category whose members might be assigned to other chemical groups, save for their small quantities required by the body. Actions of these substances are considered in greater detail in later chapters.

NUCLEIC ACIDS. NUCLEIC ACIDS (Fig. 3.7) are composed of *nitrogenous bases* (purines and pyrimidines), a *sugar* (ribose or deoxyribose), and *phosphoric acid* (H_3PO_4). Two main categories of nucleic acids are recognized.

DEOXYRIBONUCLEIC ACID (*DNA*) contains the bases adenine (A), guanine (G), cytosine (C), and thymine (T), deoxyribose as the sugar, and is found primarily in the cell nucleus. Here it directs ultimately the production of body proteins. Genes are believed to be sequences of these nucleic acid molecules.

RIBONUCLEIC ACID (*RNA*) contains adenine, guanine, and cytosine, as does DNA, but substitutes uracil (U) for thymine. The sugar in RNA is ribose. RNA is found

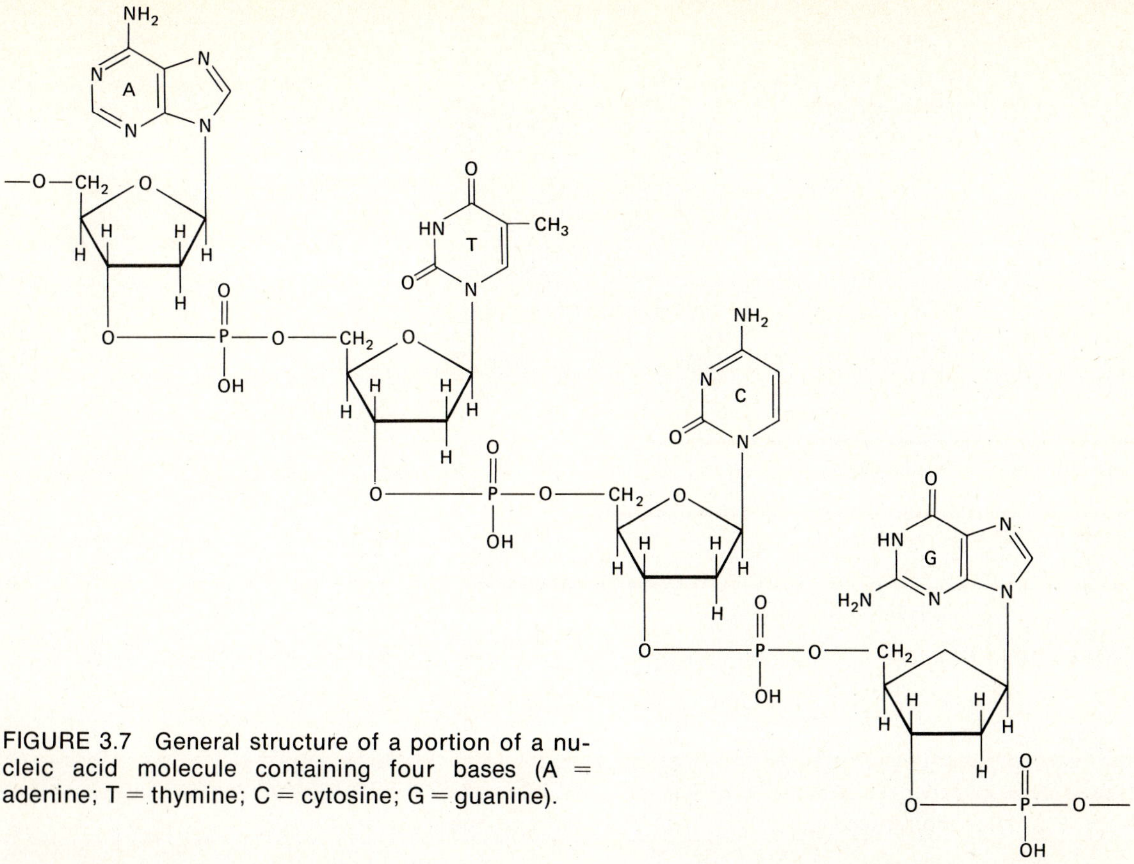

FIGURE 3.7 General structure of a portion of a nucleic acid molecule containing four bases (A = adenine; T = thymine; C = cytosine; G = guanine).

primarily outside the cell nucleus (in the cell cytoplasm), where it carries instructions for protein synthesis and the amino acids required for protein synthesis.

Both types of nucleic acids have, as their basic structural units, molecules called NUCLEOTIDES. These consist of a base, sugar, and phosphoric acid. NUCLEOSIDES lack phosphoric acid, but contain the base and sugar. Polymerization of nucleotides by sugar-phosphate bonds creates the nucleic acid molecules themselves.

ADENOSINE TRIPHOSPHATE (ATP). ATP is an abbreviation for adenosine triphospate (Fig. 3.8). ATP is a chemical that occurs in *all* living material —human, plants, or microorganisms. As seen in Figure 3.8, ATP consists of a nitrogenous base

(adenine), a sugar (ribose), but with three phosphate groups. The last two phosphate groups are attached by "high-energy bonds" ($\sim$) that release four to six times as much energy when broken as other chemical bonds. The energy required to synthesize these high energy bonds is produced primarily by the degradation of carbohydrates and lipids in the body. Thus, it may be said that one reason we eat is to provide energy for ATP synthesis. ATP and its high-energy bonds form the immediate source of energy for fueling body machinery (muscle contraction, activation of molecules for degradation by metabolic cycles, active transport systems, etc.). It is the "chemical currency" for the cell—cash in hand, so to speak.

Table 3.3 summarizes the major chemical constituents of living material.

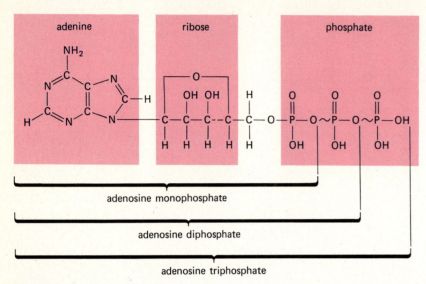

FIGURE 3.8 The nucleotides.

TABLE 3.3 A summary of the chemicals of living material		
Substance	Location in cell	Function
Water	Throughout	Dissolve, suspend, and regulate other materials; regulate temperature
Inorganic salts	Throughout	Establish forces to govern water movement, pH, buffer capacity
Carbohydrates	Inclusions (non-living cell parts)	Preferred fuel for activity
Lipids	Membranes, Golgi apparatus, inclusions	Reserve energy source; give form and shape; protection; insulation
Proteins	Membranes, cytoskeleton, ribosomes, enzymes	Give form, strength, contractility, catalysts, buffering
Nucleic acids		
DNA	Nucleus, in chromosomes and genes	Direct cell activity
RNA	Nucleolus, cytoplasm	Carry instructions; transport amino acids
Trace Materials		
Vitamins	Cytoplasm	Work with enzymes
Hormones	Cytoplasm	Work with enzymes to activate or deactivate enzymes
Metals	Cytoplasm, nucleus	Specific functions in various synthetic schemes, e.g.: synthesis of insulin (zinc), maturation of red cells (cobalt, copper, iron)

Cell structure and function

Knowing the major categories of chemicals that constitute living material, let us next look at the organization of these substances into the structural and functional units of the body: its CELLS.

To better understand the basic organization of cellular parts, we shall describe what is called a *generalized cell* (Fig. 3.9; *see also* Fig. 3.14). Such a cell incorporates the features found in most body cells, those that are essential for cell survival. Later, we discuss some special features particular types of cells possess.

Three basic subdivisions in a cell may be made.

MEMBRANE SYSTEMS are structures that form the external surface of the cell, and which surround the many formed, metabolically active, *organelles* of the cell.

The CYTOPLASM includes all the substances within the external cell surface *except* the nucleus. It is an area of active synthesis and degradation of substances.

The NUCLEUS provides guidance of the cell's overall activity.

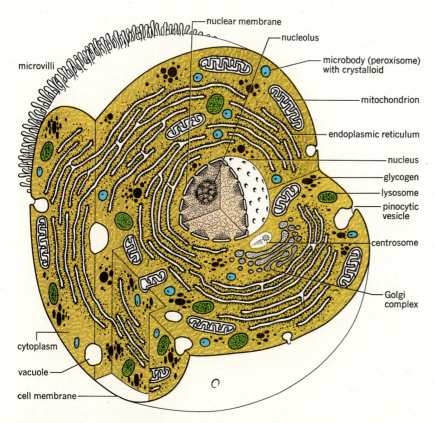

FIGURE 3.9   A generalized cell as it might appear in a three-dimensional electron micrograph.

The membrane systems of cells

The outer limiting membrane of the cell is designated as the PLASMA MEMBRANE. Membranes similar in structure surround the organelles called *mitochondria, endoplasmic reticulum, Golgi apparatus, lysosomes, peroxisomes,* and the cell *vacuoles* and *nucleus.* The structure of all these membranes is *similar* in that they all may be shown to contain both protein and lipid (mainly phospholipid and cholesterol). They *differ* in amounts and type of these components as indicated in Table 3.4. *Arrangement* of the protein and lipid components may also be different in different membranes. The electron microscope shows a trilaminar (three-layered) structure (Fig. 3.10) for the plasma (cell) membrane. Other types of studies suggest that the proteins and lipids of the cell membrane are arranged as shown in Figure 3.11. The lipids are more regularly arranged as a double layer, while the proteins may extend entirely through the lipid layers or only part way. Where there are proteins extending all the way across the plasma membrane, areas of greater ease of passage (permeability) of substances may be present ("pores"). The membrane may also be folded to create *microvilli* on the cell surface, a device to increase absorptive and secretory surface for moving materials into and out of the cell.

Membranes, regardless of where they are found in the cell, CONTROL PASSAGE of substances across themselves. Among the factors involved in determining what will pass (permeability) and its ease and rate of passage are the following.

Membrane thickness. A thicker membrane takes longer to traverse.

Size of substance attempting to cross the membrane. Large molecules (above about 8 Å in diameter and 160 molecular weight) pass with difficulty through a membrane. Water molecules and some noncharged atoms penetrate freely through membranes.

Electric charge. Ions with a charge that is the same as that on the membrane (usually negative at normal body pH) will be repelled. Opposite charge usually means attraction (at least) to the membranes.

Lipid solubility. Since the membrane contains a high concentration of lipids, substances that dissolve easily in lipids (high lipid solubility) pass more rapidly and easily through a membrane.

The presence of ACTIVE TRANSPORT SYSTEMS in the membrane.

As discussed in greater detail later in this chapter, membranes contain devices that can transport molecules through membranes relatively independently of the previously listed factors.

Membranes contain RECEPTOR SITES, particularly numerous on the cell membrane, that can bind substances and aid—perhaps—their passage through the membrane. Many chemicals, among them drugs and hormones, apparently must first be bound to the membrane in order to influence the cell's operations. No binding sites means that

TABLE 3.4 Composition of representative membranes		
Membrane	Protein-lipid ratio (number : 1)	Lipids predominating
Erythrocyte (plasma membrane)	1.5	Phospholipid (55%) Cholesterol (25%)
Mitochondrial (inner membrane)	3.0	Phospholipids (95%) Cholesterol (5%)
Myelin (nerve fibers)	0.25	Phospholipids (32%) Cholesterol (25%) Sphingolipids (31%)

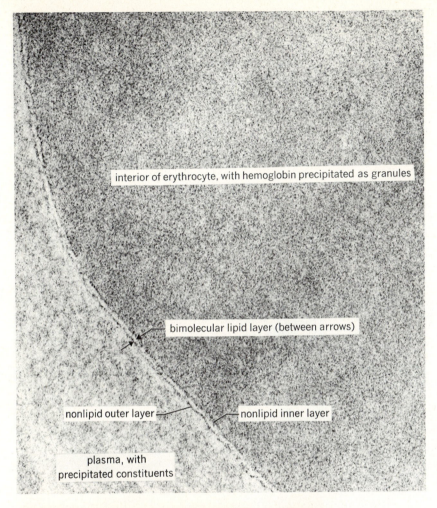

interior of erythrocyte, with hemoglobin precipitated as granules

bimolecular lipid layer (between arrows)

nonlipid outer layer   nonlipid inner layer

plasma, with
precipitated constituents

FIGURE 3.10   The cell membrane of a
rat erythrocyte. Width of lipid bilayer
is 4 nm; width of inner and outer layers
is about 1.2 nm. (Courtesy Dr. Wm. H.
Fletcher, Univ. of Calif., Riverside.)

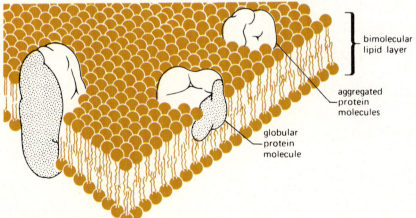

bimolecular
lipid layer

aggregated
protein
molecules

globular
protein
molecule

FIGURE 3.11   The "protein in the sea
of lipids" concept of cell membrane
structure.

that particular chemical will *not* influence the cell machinery.

Lastly, as exemplified by the membranes of the cells lining the small intestine, membranes may CONTAIN ENZYMES. Some of these enzymes actually digest foodstuffs when the foodstuffs contact the membrane, and other enzymes are involved with active transport systems. The value of *microvilli* in creating a greater surface to house such enzymes becomes obvious.

The cytoplasm

The cytoplasm (the area between plasma membrane and nucleus) contains two basic types of formed structures. ORGANELLES are metabolically active, characteristically formed, often self-reproducing units that synthesize and degrade a wide variety of chemical substances. INCLUSIONS are not metabolically active, and include vacuoles for storage of wastes of metabolism or raw materials for cellular activity.

ORGANELLES. The portion of the cytoplasm just beneath the plasma membrane is relatively free of cellular organelles and is known as the *exoplasm.* The *endoplasm* designates the remaining portion of the cytoplasm.

ENDOPLASMIC RETICULUM (*ER*) (Figs. 3.12, 3.15) is a series of fluid-filled, membrane-lined, tubular channels permeating the entire endoplasm and

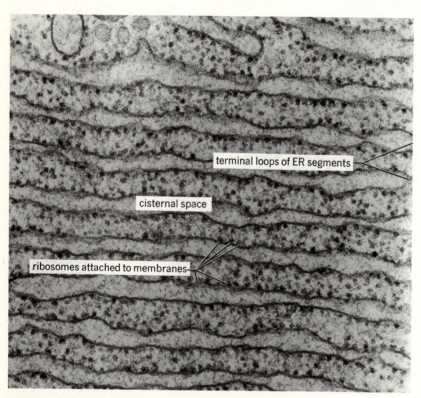

terminal loops of ER segments

cisternal space

ribosomes attached to membranes

FIGURE 3.12 Longitudinal sections of rough endoplasmic reticulum, with ribosomes on the external surface of the membranes. The cisternal space is the interior or cavity of the reticulum tubules. (Courtesy Dr. Wm. H. Fletcher, Univ. of Calif., Riverside.)

reaching both the plasma membrane and the nucleus. Two varieties exist: *rough ER* has ribosomes, described later, attached to the outer surface of the reticulum membranes; *smooth ER* lacks ribosomes. The rough variety, by virtue of its ribosomes, serves as the site of cellular protein synthesis, while the smooth variety functions as a site of synthesis for lipids, steroids, and several types of carbohydrates. An additional function of the ER is to provide a means to "circulate" fluids derived from the cellular environment, along with solutes derived from cell activity, to all parts of the cell.

RIBOSOMES (Fig. 3.13; *see also* Fig. 3.12) are small granules of RNA (two-thirds) and protein (one-third) attached to rough ER or free in the cytoplasm. They are not surrounded by membranes. Protein synthesis occurs on the ribosomes, with the organelle utilizing instructions received from the nucleus about exactly *what* protein is to be synthesized, and "zipping together" the individ-

ual amino acids to form the protein. Groups of ribosomes, known as *polyribosomes,* appear to work together to form the large structural proteins and enzymes that cells use. One theory holds that ribosomes on ER are producing proteins to be "exported" for use elsewhere in the body, and that "free" ribosomes produce proteins for that particular cell's use.

The GOLGI COMPLEX (also: *apparatus* or *body*) (Fig. 3.14) is usually located in one part of the

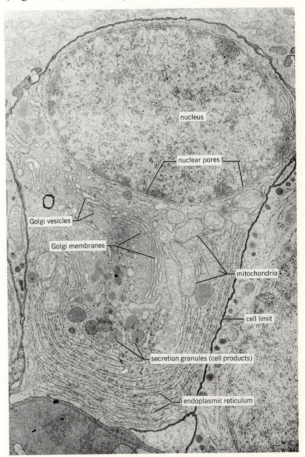

FIGURE 3.14   A prolactin secreting cell from the anterior lobe of a rat pituitary gland to illustrate the Golgi complex and the organization of the cellular organelles. Note that the Golgi complex may be distinguished by its localization within the cell and by the enlarged vesicles at the ends of the membranes. (Courtesy Dr. Wm. H. Fletcher, Univ. of Calif., Riverside.)

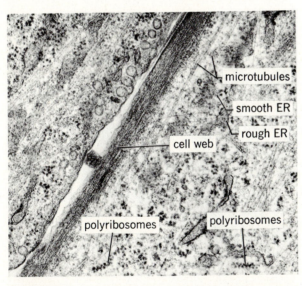

FIGURE 3.13   A cell of a human corpus luteum grown in tissue culture, illustrating polyribosomes and other visible cellular organelles. The cell web is a mass of microfilaments paralleling the cell membrane and perhaps giving strength to the outer portion of the cell. (Courtesy Dr. Wm. H. Fletcher, Univ. of Calif., Riverside.)

cytoplasm close to the nucleus. It consists of flattened agranular "stacks" of membranes (Golgi membranes), with enlarged areas on the ends of the stacks (Golgi vesicles). The Golgi complex synthesizes glycoproteins, glycolipids, protein polysaccharides, and mucus. It also "packages" secretions of the cell for release into the body fluids. This latter activity is associated with surrounding the product with a membrane so that it cannot affect the producing cell and so that it may easily leave the cell by exocytosis (described later).

LYSOSOMES (Fig. 3.15) are membrane-surrounded organelles that contain powerful enzymes capable of digesting large molecules of protein, carbohydrate, and lipid. If a cell is injured, the lysosomes may be released into the cell itself to destroy it. Because of this self-destruct capacity, lysosomes are often referred to as "suicide packets." However, lysosomes more often protect the cell: bacteria enter the organelle and are destroyed by lysosomal enzymes; engulfed cell debris from sites of injury or inflammation are digested by lysosomal enzymes. Phagocytic cells—those that can engulf particulate matter (for example, white blood cells)—scavenge the body continually and digest the engulfed matter using lysosomes.

MITOCHONDRIA (Fig. 3.16; *see also* Fig. 3.14) have a *double* membrane around them. The outer wall is not folded, and contains enzymes that break large molecules (e.g., glucose, 6 carbons) into smaller molecules (such as pyruvic acid, 3 carbons). The inner membrane is folded into shelves or *cristae* that vastly increase the internal surface of the mitochondrion. On the cristae are orderly arrays of enzymes that complete the degradation of molecules like pyruvic acid to CO_2 and H_2O and produce the energy necessary for ATP synthesis. Thus, mitochondria are sometimes called the "powerhouses" of the cell.

The CENTRAL BODY consists of a pair of *centrioles* (Fig. 3.17) and an area of differentiated cytoplasm called the *centrosphere*. Centrioles are involved in cell division, giving rise to tubular structures that aid in separating chromosomes as the cell divides. Certain types of body cells (e.g., mature nerve

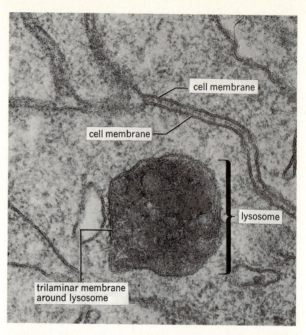

FIGURE 3.15 Cell from a human corpus luteum showing a lysosome. The organelle is distinguished by the dark-staining, uniformly dispersed granular material. (Courtesy Dr. Wm. H. Fletcher, Univ. of Calif., Riverside).

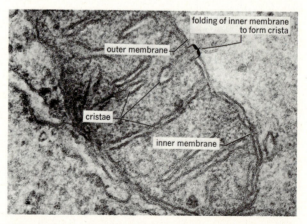

FIGURE 3.16 A mitochondrion from a cell of the human anterior pituitary. Note how the cristae are formed by infolding of the inner mitochondrial membrane. (Courtesy Dr. Wm. H. Fletcher, Univ. of Calif., Riverside).

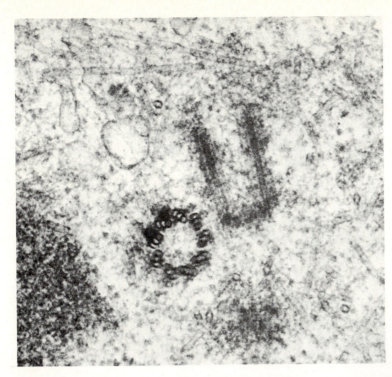

FIGURE 3.17 Centrioles from "3T3 cells" (that resemble fibroblasts). The cross section shows the nine sets of triplet microtubules characteristic of this organelle; the longitudinal section shows the fibrillar nature of the microtubules. (Courtesy G. Albrecht-Buehle, Cold Spring Harbor Laboratory.)

cells) lack or have a nonfunctioning cell center after about five years of age and thus cannot divide to replace damaged or lost cells.

MICROTUBULES (Fig. 3.18; *see also* Fig. 3.13) are tiny (about 250 Å diameter) hollow tubes that are found in all parts of the cytoplasm. They give support (*cytoskeleton*) to the cell and may direct flow of cytoplasmic constituents as they move through the cell. The *spindle,* which attaches to and may be derived from the centrioles and appears at the time of cell division, is composed of microtubules that apparently aid in chromosome movement.

PEROXISOMES (*microbodies*) (Fig. 3.19) are spherical membrane-surrounded organelles shown to contain enzymes (oxidases, catalase). The function of the organelle appears to be concerned with metabolism of hydrogen peroxide (H_2O_2). H_2O_2 is a potent antiseptic agent; it is a normal product of metabolism in white blood cells, where it may aid bacterial destruction. In other cells the organelle may protect the cell from excessive H_2O_2 by breaking it down to water.

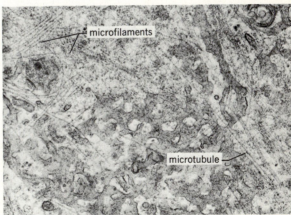

FIGURE 3.18 Cell from a human corpus luteum, illustrating microtubules. Compare size of the tubules to the filaments also visible in the photograph. (Courtesy Dr. Wm. H. Fletcher, Univ. of Calif., Riverside.)

INCLUSIONS. VACUOLES, membrane-surrounded fluid droplets, serve as areas for storage of water-soluble substances. Fat droplets, crystals of glyco-

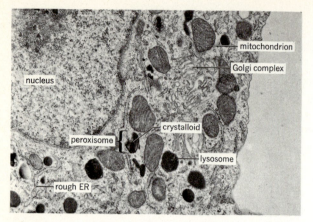

FIGURE 3.19   Cell from the rat anterior pituitary lobe, to illustrate a peroxisome. The organelle typically shows a crystalloid inside. Other organelles are also shown. (Courtesy Dr. Wm. H. Fletcher, Univ. of Calif., Riverside).

gen, and pigment granules are other examples of materials acting as raw materials for cellular activity or products of activity. All these structures are termed inclusions. They are *not* constant features of cells.

The preceding discussion should emphasize the complexity of the cytoplasmic structure and its exquisite organization. Compartmentalization of cellular function is thus achieved to permit various types of cellular reactions from interfering with one another.

The nucleus

The method by which the nucleus governs cellular activity is considered in Chapter 5. Here, we may state that the nucleus (Fig. 3.20), through its DNA,

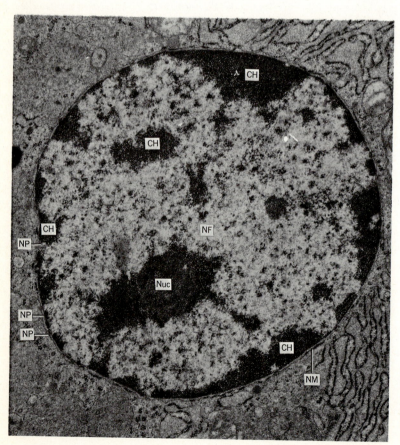

FIGURE 3.20   Nucleus of a tracheal cell. NM, nuclear membrane; Nuc, nucleolus; NP, nuclear pores; CH, chromatin; NF, space containing nuclear fluid × 15,000. (Courtesy Norton B. Gilula, The Rockefeller University.)

in the form of CHROMATIN MATERIAL, provides DIRECTION OF CELLULAR ACTIVITIES such as secretion, protein synthesis, and ATP production. Two MEMBRANES surround the nucleus, one derived from the ER, the other from the nucleus itself. The two membranes are fused at intervals around the nucleus, providing areas of greater permeability called "nuclear pores." Through these "pores" pass large molecules of nucleic acid (RNA) that carry instructions for controlling cell activity. One or more NUCLEOLI (sing.: nucleolus), composed chiefly of RNA, float in the NUCLEAR FLUID.

Specialization of form and function in cells

Although nearly all body cells possess the structures described above, in certain types of cells particular features are present or are more highly developed. Thus, muscle cells contain elongated protein molecules (fibrils) that confer on the cell the ability to CONTRACT or shorten. The cells themselves are elongated, for to shorten effectively they must possess considerable initial length. Secretory cells, such as those that produce mucus in the digestive and respiratory systems, have an extremely well-developed Golgi apparatus. Epithelial cells lining the small and large intestines, and those lining kidney tubules, have especially well-developed microvilli to increase their absorptive surfaces. From centriolelike structures, sperm cells develop long tails composed of microtubules. These *flagella* confer independent movement on the cell, a requisite for moving the sperm to the area where it may fertilize an egg.

In short, function is reflected by structure and vice versa. This point should be kept in mind as the structure and function of the body cells, tissues, and organs are studied in future chapters.

Passage of materials through membranes or cell

Cells and their organelles acquire the substances they need — and pass materials they synthesize — through their membrane and fluid systems. If a substance moves through a membrane without the cell's participation or its expending of energy, a PASSIVE PROCESS has occurred. If the cell *does* have to perform work or expend energy to move a substance across or through a membrane, an ACTIVE PROCESS is said to have occurred.

Passive processes

Passive processes occur because of CONCENTRATION or ENERGY DIFFERENCES on two sides of a membrane or within different parts of a solution.

BULK FLOW. BULK FLOW occurs when an energy or concentration difference causes rapid movement of a considerable volume of solvent through a membrane and solutes are swept through the membrane along with the solvent. The process is relatively nondiscriminatory, with everything moving at once. Bulk flow occurs from capillaries in body tissues and organs, and moves fluid and dissolved materials from the bloodstream to cells for their use.

DIFFUSION. DIFFUSION (Fig. 3.21) is a process that may occur with or without a membrane present in the diffusion system. The basic driving force for diffusion is that molecules are in constant motion, colliding with one another or with a membrane. If there are more solute molecules in an area (more concentrated) or on one side of a membrane as opposed to the other, more collisions occur per unit of time, and the solute will be distributed in the solution until all parts have the same concentration. If a membrane is present, more collisions by solute molecules on the more concentrated side will cause a *net* movement across the membrane to the less concentrated side, again until concen-

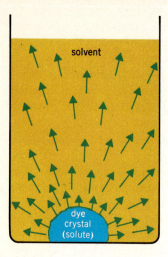

FIGURE 3.21 Diffusion. Molecules of dye move from the area of higher dye concentration to the area of lower dye concentration.

trations are equalized. At that time, movement in both directions will be equal and no further net movement will occur. Diffusion is a process that occurs more rapidly when small molecules are involved because they move more rapidly and are shifted farther by collisions. Gas diffusion (O_2, CO_2) across lung surfaces is a good example of rapid equalization by the process of diffusion.

FACILITATED DIFFUSION. If a substance possesses a concentration difference on two sides of a membrane, and can diffuse through that membrane already, its *rate* of passage may be accelerated by combining with a carrier molecule in the membrane. The carrier actually attaches to the diffusing molecule and speeds (facilitates) its movement through the membrane. Keep in mind that this process *only* speeds movement of something that is already capable of diffusing, and that it occurs only as long as a concentration gradient for the substance is present. Glucose is believed to enter cells by facilitated diffusion.

OSMOSIS. OSMOSIS is a type of diffusion, but the term is used to refer to a net movement of WATER ONLY through a membrane that restricts the pas-

sage of solutes. Movement of water occurs *from* the area of greater water concentration *to* the area of lesser water concentration. The difference in water concentration is established by solute concentration. The greater the solute concentration on one side of the membrane, the less the relative water concentration as solute molecules displace water molecules in a given volume of solution. The *total* solute concentration of a solution is known as its OSMOLARITY. Osmolarity in turn is a function of numbers of particles in the solution and not of their type. For example, a 1 molar (1M) solution of sodium chloride has an osmolarity of 2 osmols, because in solution NaCl dissociates to give 2 particles.

Cells behave like osmotic systems because their membranes permit free movement of water molecules while restricting passage of many solutes (semipermeability). Let us consider what would happen to cells placed in solutions containing various amounts of solute (Fig. 3.22). If a cell is placed in a solution having an osmolarity *equal to* that of the cell (Fig. 3.22*a*), water *and* solute concentrations will be equal inside and outside the cell; water molecules will thus move at equal rates in both directions across the membrane and *no net flow* of water will occur in either direction. The cell will not shrink or swell. In this situation, the cell has been placed in an ISOSMOTIC (*isotonic*) SOLUTION (*iso,* equal). If a cell is put in a solution having less solute than is present in the cell (Fig. 3.22*b*), it will have been placed in a HYPO-OSMOTIC (*hypotonic*) SOLUTION. In this case, water molecules move more rapidly *into* the cell than out, since water concentration is greater outside the cell. The cell will swell and may burst. The reverse situation places a cell in a HYPEROSMOTIC (*hypertonic*) SOLUTION (Fig. 3.22*c*). In this case, water concentration is greater inside the cell than outside, and the cell will suffer a *net loss* of water. It shrinks. Obviously, one important thing the regulatory mechanisms of the body must do is to regulate the concentrations of solutes (particularly the Na and Cl concentrations) in the fluids around body cells so that they do not suffer swelling or

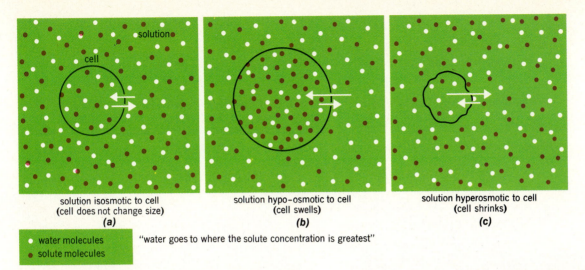

solution isosmotic to cell
(cell does not change size)
(a)

solution hypo–osmotic to cell
(cell swells)
(b)

solution hyperosmotic to cell
(cell shrinks)
(c)

○ water molecules   "water goes to where the solute concentration is greatest"
● solute molecules

FIGURE 3.22 Osmosis. Changes in cell size in *(a)* isosmotic, *(b)* hypo-osmotic, *(c)* hyperosmotic solutions. Open circles represent water molecules, solid circles represent solute molecules. Arrows indicate flow of water, with length denoting greatest direction of flow.

(a) solution isosmotic to cell (cell does not change size)
(b) solution hypo-osmotic to cell (cell swells)
(c) solution hyperosmotic to cell (cell shrinks)
○ water molecules   "Water goes to where the solute
● solute molecules   concentration is greatest"

shrinkage as a result of water shifts. The kidney is an important organ in this regulation.

DIALYSIS. DIALYSIS occurs when there are two or more solutes on one side of a membrane, but the membrane is permeable only to certain ones. If a concentration difference is present for a solute that *can* pass through the membrane, it will diffuse through and thus be separated from the other solutes. The principle of dialysis is utilized in the *artificial kidney* where the patient's blood (500 ml at a time) is passed through a series of dialysing tubes surrounded by a solution that is isosmotic for essential blood constituents, but which contains *zero* concentration of solutes to be removed from the bloodstream. Therefore, a *net outward* diffusion of toxic solutes (urea, for example) occurs from the tubes into the solution. Renewal of the solution keeps the gradient high for the toxic material, and nearly all the toxin can be removed. The process is called *hemodialysis*.

FILTRATION. FILTRATION occurs when there is an *energy difference* on two sides of a membrane. The membrane acts like a sieve, allowing molecules to be forced through the membrane by the energy difference according mainly to their size. Movement of solvents and solutes from capillaries under the energy difference provided by the blood pressure is an example of filtration as it occurs in the human body. This is, as one may recall, an example of bulk flow.

Active processes

Active processes occur where the cell is required to expend energy to cause movement of a substance across a membrane.

ACTIVE TRANSPORT. ACTIVE TRANSPORT — sometimes called *mediated transport* — is distinguished from a passive process in several ways.

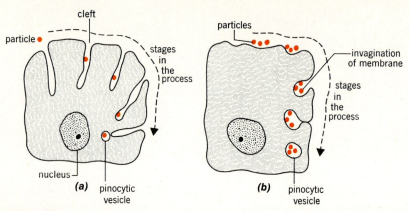

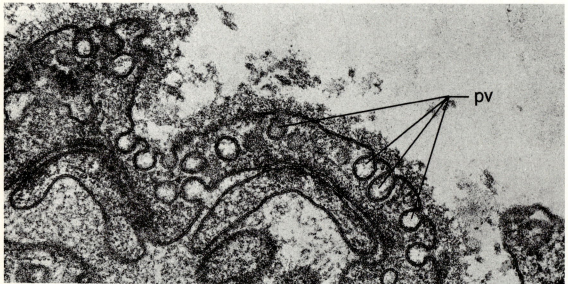

FIGURE 3.23 Pinocytosis. *(a)* The particle enters a cleft and becomes enclosed in a vesicle. *(b)* The particle is adsorbed on the surface of the membrane and is enclosed in a vesicle. *(c)* Electron micrograph of formation of vesicles in skeletal muscle capillary (×22,000); pv = pinocytic vesicle.

Molecules of a SPECIFIC TYPE are transported, apparently by attachment to some sort of "carrier molecule." Glucose, amino acids, and inorganic ions are all transported by *different systems*. Therefore, each substance moves at a rate independent of that of the others.

INHIBITORS that prevent, or interfere with, ATP synthesis by the cell cause the process to cease. This fact defines the role of exergonic reactions of cellular metabolism in the process.

Transport tends to occur at more or less CONSTANT RATES, providing there is enough material present to

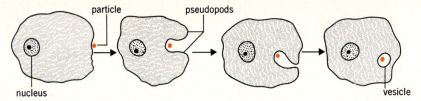

FIGURE 3.24   Phagocytosis. Pseudopods are formed that engulf the particle.

be transported. A rate-limiting reaction in an enzyme controlled system appears to be present.

COMPETITION exists, in that one carrier may carry more than one substance, particularly if they are closely related structurally. If one is carried in preference to another, the transport of one will be reduced (inhibited).

Active transport often moves materials *against* the concentration gradient, *if* energy supply is not diminished. For this reason, active transport systems are often called "pumps."

Thus, we see that the cell is involved in active transport at three points: carrier synthesis, ATP production, and enzyme synthesis. The nature of the carrier molecule has not been determined. Lipid molecules, or "mobile complexes" that coil, uncoil, rotate or "unwind," have been proposed as carriers. Alternatively, "gates" in the membrane that open or close in response to binding of molecules to the cell surface have been suggested.

ENDOCYTOSIS. Cells may acquire macromolecules (viruses, nucleic acids), particulate matter (bacteria), and liquid droplets. The general name for such methods of acquisition is ENDOCYTOSIS. PINOCYTOSIS occurs when the cell membrane sinks as a result of binding or contact of a molecule with the cell surface (Fig. 3.23). An intracellular vesicle (vacuole) is formed that contains the molecule(s). PHAGOCYTOSIS (Fig. 3.24) occurs when a cell engulfs or surrounds a particle and forms a vesicle around it, as when white blood cells take up particles. Destruction of the particle by lysosomal digestion is a common result of phagocytosis.

EXOCYTOSIS. A "reverse pinocytosis" (*emeiocytosis*) occurs if a vesicle formed within the cell moves to the cell membrane, fuses with it, and discharges its contents to the cell exterior. Such activity is called EXOCYTOSIS. It may be utilized to remove excess substances, toxic materials, or products of cell activity from the cell or to transport materials (e.g., fats) through a cell to the exterior.

Active processes imply that the cell can *control* what passes through its walls. Again, the value of such control to maintenance of a cell's internal environment and its functions is obvious.

Summary

1. Four elements compose over 95 percent of the body. They are:

   a. Hydrogen—10 percent

   b. Oxygen—65 percent

   c. Carbon—18 percent

   d. Nitrogen—3 percent

    e. Other elements are found as such or in polyatomic ions to complete the body structure (e.g., calcium, sodium, potassium, chlorine, phosphorus, sulfur, and others, often known as inorganic substances).

2. Compounds are formed by bonding of elements. Some compounds forming living material are:

    a. Water

    b. Proteins

    c. Carbohydrates

    d. Lipids

    e. Trace materials

    f. Nucleic acids

    g. ATP

3. Water

    a. Has the formula H_2O.

    b. Forms an average 55 to 60 percent of living material.

    c. Is a good solvent and suspending medium.

    d. Is generally nontoxic to the cell.

    e. Is a medium of heat exchange and transfer.

    f. Participates in chemical reactions as both a reactant and a product.

    g. Acts as a lubricating and protective fluid.

4. Inorganic substances include elements (Na, K, Ca, Cl, Mg) and polyatomic ions ($PO_4^=$, $SO_4^=$, HCO_3^-) that may exist as ions in a solution. This solution conducts electricity because of the presence of electrolytes (the ions). In general terms, inorganic substances are concerned with osmotic gradients, excitability, and permeability of cells.

5. Proteins are composed of C, H, O, and N, and sometimes S.

    a. The "building units" of proteins are amino acids.

    b. Varying numbers of amino acids form oligopeptides, polypeptides, and proteins.

    c. Simple proteins contain only amino acids; conjugated proteins contain, in addition to a protein, a nonprotein component (e.g., lipid, nucleic acids).

    d. Fibrous proteins are relatively insoluble, are elongated, and form structural and contractile units.

    e. Globular proteins are more soluble, folded, and are utilized in reactions of the cell.

6. Enzymes

    a. Catalyze (change rates) of chemical reactions.

    b. Always have a protein component.

    c. May contain a coenzyme that is not a protein.

    d. Are specific as to the chemical reaction they catalyze.

    e. Are heat labile.

    f. May be inhibited by various chemicals.

    g. Often have the suffix *-ase* on their name, and are named by the reaction or type of reaction they influence.

7. Carbohydrates contain C, H, and O, with H and O often in a 2:1 ratio.

    a. Monosaccharides (simple sugars) contain 3–7 carbon atoms. They form larger units:
        1) Hexoses (glucose, fructose and others) are 6 carbon sugars that are the most common in the body.
        2. Pentoses (ribose, deoxyribose) are 5 carbon sugars found in nucleic acids.

    b. Disaccharides or double sugars (maltose, lactose, sucrose) contain 2 simple sugar units.

    c. Polysaccharides contain 3 to several thousand simple sugar units. Glycogen is a polysaccharide.

    d. Carbohydrates are primarily "quick-energy sources," especially simple sugars.

8. Lipids contain C, H, and a deficiency of O, compared to carbohydrates. They are practically water insoluble, but dissolve in "fat solvents" (ether, chloroform, alcohol).

    a. Simple lipids include triglycerides, waxes, and oils. They act as storage forms in adipose tissue, and are used as energy sources.

    b. Compound lipids (lipid plus nonlipid portions) form structural units in cell membranes.

    c. Derived lipids include cholesterol and its derivatives (hormones, for example).

9. Trace materials are substances required in only minute quantities in the body. Examples include metals, vitamins, and hormones.

10. Nucleic acids contain nitrogenous bases, a pentose, and phosphoric acid.

    a. DNA contains adenine, guanine, cytosine, and thymine as bases, plus deoxyribose and phosphoric acid, and determines our hereditary and biochemical characteristics. It resides primarily in the nucleus.

    b. RNA substitutes uracil for thymine, ribose for deoxyribose, and forms the main nucleic acid of the cytoplasm.

    c. Nucleotides contain all three components (base, sugar, acid); nucleosides lack the acid.

11. ATP is a high-energy compound utilized as the immediate source of energy for cellular activity. Its synthesis occurs using energy from ingested foods.

12. The cell forms the organized structural and functional unit of the body. It has three basic parts:

    a. Membrane systems surround the whole cell and many of its parts.

    b. The cytoplasm, inside the cell membrane, excluding the nucleus.

    c. The nucleus itself.

13. Membranes, regardless of where they occur

    a. Have a protein-lipid composition.

    b. Control passage of materials into and out of the structure surrounded.

    c. Among the factors involved in passage of something through a membrane are:
       1) Thickness of membrane.
       2) Size of substance passing.
       3) Electric charge on substance passing and on membrane.
       4) Lipid solubility of substance passing.
       5) Presence of active transport systems in the membrane.

    d. Membranes also have chemical receptor sites, and often contain enzymes.

14. The cytoplasm contains organelles, and inclusions. Organelles are metabolically active and include:

    a. Endoplasmic reticulum. Rough reticulum has ribosomes on it and synthesizes proteins and circulates materials through the cell. Smooth reticulum lacks ribosomes, synthesizes lipids, steroids, and carbohydrates, as well as circulating materials.

    b. Ribosomes consist of RNA and protein and put amino acids together into proteins.

    c. The Golgi complex synthesizes a variety of materials and packages substances for export from the cell.

    d. Lysosomes contain enzymes and break a variety of large molecules into smaller ones. They are also involved in removing the debris of injury, inflammation, and digest cell engulfed particulate matter.

    e. Mitochondria synthesize ATP.

    f. The central body is involved in cell division and gives rise to microtubules in cilia and flagella.

    g. Microtubules act as supportive structures in cells and form the spindle when a cell divides.

    h. Peroxisomes degrade hydrogen peroxide.

    Inclusions are not metabolically active, and are not found in all cells.

    i. Vacuoles store water soluble materials in the cell.

  j. Fat droplets and crystals are energy sources.

  k. Wastes are inclusions also.

15. The nucleus has a membrane, contains chromatin material, a nucleolus, and nuclear fluid. It directs cellular activity using its DNA.

16. Form and function are interrelated in cells, as exemplified by the long muscle cells that contract, well-developed Golgi apparatus in secretory cells, and flagella on motile cells.

17. Passive processes (cell does no work) and active processes (cell does work) are utilized to move materials through membranes or fluids. Passive processes occur only from a higher to a lower concentration or energy level.

18. Passive processes and their characteristics include:

  a. Bulk flow. This occurs when energy is utilized to force solvent *and* solute at the same time through a membrane (*see also f* below).

  b. Diffusion. Occurs because of a concentration difference and molecular motion. More rapid for small molecules such as gases.

  c. Facilitated diffusion. A substance that can diffuse through a membrane has its passage speeded by attachment to a carrier molecule. The process still "goes" according to a concentration gradient. Example: glucose entering a cell.

  d. Osmosis. Diffusion of solvent through a membrane that restricts solute passage. Occurs by a concentration gradient for the solvent.

   1) Osmolarity refers to solute concentration that in turn determines water concentration and direction of net water movement.

   2) An isosmotic solution is one in which solute and water concentrations equal those of the cell. In such a solution the cell maintains its size.

   3) A hypo-osmotic solution has less solute (and more water) than in the cell; therefore a net flow of water occurs into the cell and it swells.

   4. A hyperosmotic solution has more solute (and less water) than the cell; therefore a net flow of water occurs out of the cell and it shrinks.

  e. Dialysis. Dialysis occurs when some solutes can pass through a membrane and others cannot. A separation of solutes occurs by diffusion. The artificial kidney employs this method to rid the bloodstream of toxic wastes.

  f. Filtration occurs because an energy difference is present on two sides of a membrane. Materials are forced through by pressure in bulk flow, with the membrane acting like a sieve.

19. Active processes and their characteristics include:

  a. Active transport

   1) Requires a carrier molecule, energy source (ATP) and enzymes, supplied by the cell.

   2) Moves molecules of a specific type.

3) Can be inhibited by metabolic poisons or competitively.
4) Moves substances at rather constant rates.
5) Can move materials against gradients.

b. Endocytosis takes in large molecules that cannot be actively transported.

1) Pinocytosis involves membrane invagination and vesicle formation.
2) Phagocytosis involves engulfing of particles by a cell.

c. Exocytosis. A "reverse pinocytosis" in which a vesicle moves to the cell membrane and discharges its contents to the outside.

d. Active processes imply *control* over what enters and leaves the cell.

Questions

1. What are some important elements, polyatomic ions, and compounds that are found in protoplasm? Which are the most abundant?

2. Why is water so important to the body? What are some of its properties and functions?

3. What are electrolytes? What functions do they serve?

4. How can an amino acid be characterized? What do amino acids form, and what functions do these larger units have in the body?

5. What are some of the properties and functions of enzymes?

6. How are carbohydrates characterized? Name some important mono-, di-, and polysaccharides in the body and give their functions.

7. How are lipids characterized? Name several important lipids and give their functions.

8. What are "trace" materials?

9. Of what components are nucleic acids composed? What are their functions?

10. Of what value is ATP to the body? From where is the energy required for its synthesis derived?

11. Describe a generalized cell in general terms.

12. What functions do cellular membrane systems serve?

13. What determines membrane permeability?

14. What is the structure of, and what are the functions of:

a. Endoplasmic reticulum?

b. Ribosomes?

c. Mitochondria?

d. Golgi complex?

e. Lysosomes?

f. The nucleus?

15. What are the differences between active and passive processes without regard to specific types of each?

16. How would each of the following substances most likely cross a membrane or enter a cell? Explain the process.

a. Glucose

b. Water

c. Cholesterol

d. A bacterium

e. A large solid waste

f. An amino acid

Readings

Babior, Bernard M. "Oxygen-Dependent Microbial Killing by Phagocytes." *New Eng. J. Med. 298:*659, March 23, 1978; *298:*721, March 30, 1978.

Capaldi, Roderick A. "A Dynamic Model of Cell Membranes." *Sci. Amer. 230:*26, March 1974.

Clarke, Margaret, and James A. Spudick. "Nonmuscle Contractile Proteins: The Role of Actin and Myosin in Cell Motility and Shape Determination." *Ann. Rev. Biochem. 46:*797, 1977.

Jacobs, John J., and Pedro Cuatrecasas. "Current Concepts: Cell Receptors in Disease." *New Eng. J. Med. 297:*1383, Dec. 22, 1977.

Korenbrot, Juan I. "Ion Transport in Membranes." *Ann. Rev. Physiol. 39:*19, 1977.

Nicholson, Garth L., and George Poste. "Cell-Surface Organization and Modification with Cancer." *New Eng. J. Med. 295:*197, July 22, 1976; *295:*253, July 29, 1976.

Schwartz, R. M., and M. O. Dayhoff. "Origins of Prokaryotes, Eukaryotes, Mitochondria, and Chloroplasts." *Science 199:*395, 27 June 1978.

Science News. "New Theory of Protein Evolution." *111:*228, April 9, 1977.

Science News. "The Bone and Muscle of Cells." *112:*250, Oct. 15, 1977.

Silverstein, Samuel C., Ralph M. Steinman, and Zanvil A. Cohn. "Endocytosis." *Ann. Rev. Biochem. 46:*669, 1977.

Staehelin, L. Andrew, and Barbara E. Hull. "Junctions Between Living Cells." *Sci. Amer. 238:*140, May 1978.

Ulmer, David L. "Current Concepts: Trace Elements." *New Eng. J. Med. 297:*318, Aug. 11, 1977.

Chapter 4
Cell Reproduction and Body Development

Objectives

After studying this chapter, the reader should be able to:

- Describe the stages and end results of the process of mitosis.

- Describe the stages and end results of the process of meiosis.

- Define a neoplasm, and discuss the methods of naming and classifying neoplasms.

- Advance some theories of causation of cancers.

- List some types of treatment employed in combatting cancer.

- Discuss some of the most common types of cancer in terms of cause and symptoms.

- Describe the basic development of the individual in weekly stages, up to the ninth week.

- Describe the events that occur from the ninth week to birth.

Cell reproduction

The body does not spend its life with the same number of or the same cells (nerve cells excepted) with which it was born. Cells wear out and die, are injured, and must often be replaced. Addi- tionally, perpetuation of the species requires the production of specialized haploid (cells containing one-half the chromosome number characteristic for the species) sex cells, whose joining (fertilization) sets the basis for the production of a new organism. Two processes occur in the body to ensure cellular replacement and perpetuation of the species.

Mitosis

MITOSIS (Fig. 4.1) is the type of cell reproduction the body utilizes as the method to replace dead, damaged, or lost cells. In this type of division, the nucleus undergoes a complicated series of changes that *ensure the daughter cells will receive exactly the same complement of DNA as possessed by the diploid parent cell(s).* In this way, both form and function of a particular cell type are maintained. The cytoplasm and organelles undergo an approximately equal division between the daughter cells by simply dividing in half. Mitosis is described

as occurring in STAGES, though the process is continual. Each stage is characterized by particular events.

In *interphase*, the cell appears typical for its type, and is not undergoing division into daughter cells. Epithelial cells look like their particular type, muscle cells resemble other muscle cells of the same variety. This appearance is deceiving, however, for the cells are actively accumulating substances for their own use, they are synthesizing materials, and, most importantly, REPLICATION (duplication) OF DNA is taking place. In short, the cells are carrying out the activities essential for their own life and are preparing for the division to come.

In PROPHASE, the centrioles of the cell center *migrate* to opposite poles (ends) of the cell, and a *spindle* of microtubules forms between the centrioles. The nuclear membrane disappears, and the chromatin

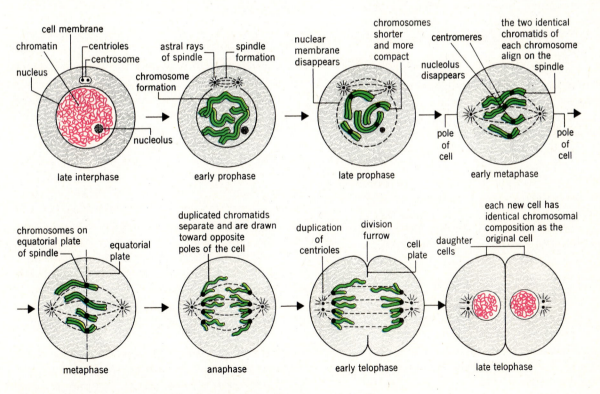

FIGURE 4.1   The major events occurring in mitosis as depicted in a cell having four chromosomes.

material of the nucleus resolves into visible *chromosomes*. At this stage, each chromosome consists of two threads (chromatids) that are copies of one another and are attached to a single portion of the chromosome called a *centromere*. (One centromere is equivalent to one chromosome regardless of how many strands of DNA are attached to it). Thus, we see that the nucleic acid has doubled, though the chromosome number has not.

In METAPHASE, the *chromosomes align on the equatorial plate* of the cell, an imaginary line halfway between the centrioles, perpendicular to the spindle.

In ANAPHASE, the *centromeres divide*, and the duplicated strands of DNA are divided equally between the centromeres. At this point, actual duplication of *chromosomes* has occurred. The chromosomes then separate, moving toward the centrioles. Movement may occur by electrostatic repulsion, by being pulled by the spindle fibers, or because of other factors. (The force causing separation of chromosomes is not known for certain.) The centromeres "lead" the separating chromosomes, and the chromosomes typically present a "V" or "J" shape as their arms lag behind the centromeres.

In TELOPHASE, the *nucleus reorganizes*, including reformation of the nuclear membrane, and chromosomes revert to granular chromatin material. The *centrioles divide*, and the *cytoplasm undergoes division* along the line of the equatorial plate. The cell returns to its original interphase appearance.

Studies of the timing of mitosis in human cells indicates that it has a cyclical duration of about 12 to 24 hours (from interphase to interphase in actively dividing cells such as bone marrow). Actual nuclear division takes only about an hour.

Meiosis

MEIOSIS (Fig. 4.2) occurs in the gonads (ovaries and testes) and produces gametes (ova or eggs, and sperm). Each daughter cell formed by meiosis contains one-half (haploid number) of the chromosome number characteristic for the species (the diploid number). Meiosis thus involves a *reduction division*. If this process did not occur in sex cell formation, the next generation would have twice the normal species number of chromosomes, the generation after that four times the number, and so forth. Such a situation—commonly called *polyploidy*—is not compatible with human survival. In meiosis, *two* divisions occur, one without previous duplication of chromosomes. Also, *exchange of portions of chromosomes* may occur during meiosis. This exchange produces variability in the expression of the characteristics controlled by the genes. In short, the operation of variability may be illustrated by the fact that while we all have the normal number of body parts, we rarely look like one another or like our parents; we resemble them, but are not identical to them (except in the case of identical twins who come from a single egg—even then, the twins are usually mirror images, enantiomorphs if you wish).

Meiosis, like mitosis, is described as occurring in stages.

In MEIOTIC INTERPHASE, the cell is engaged in activities appropriate to its survival and chromatin is duplicated, as in mitosis.

In FIRST MEIOTIC PROPHASE, nuclear changes are the same as in mitosis, and centriole migration and spindle formation occur. A difference occurs in the *pairing of homologous chromosomes* (a pair of chromosomes, consisting of one from each parent, that have the same genes in the same order) and in the CROSSING OVER *(chiasmata)* or SYNAPSIS of portions of the arms of the homologous chromosomes (*see* Fig. 4.2). The crossed arms may actually break off from one chromosome and become attached to the other member of the pair. This step does not occur in mitosis. Thus, a "new chromosome" in terms of variability of gene order has been formed.

In FIRST MEIOTIC METAPHASE, alignment of homologous chromosomes occurs on the cell's equatorial plate.

In FIRST MEIOTIC ANAPHASE, the homologous chromosomes are separated, *without division of centromeres,* and it is at this time that REDUCTIONAL DIVISION occurs, with one member of each pair going to the

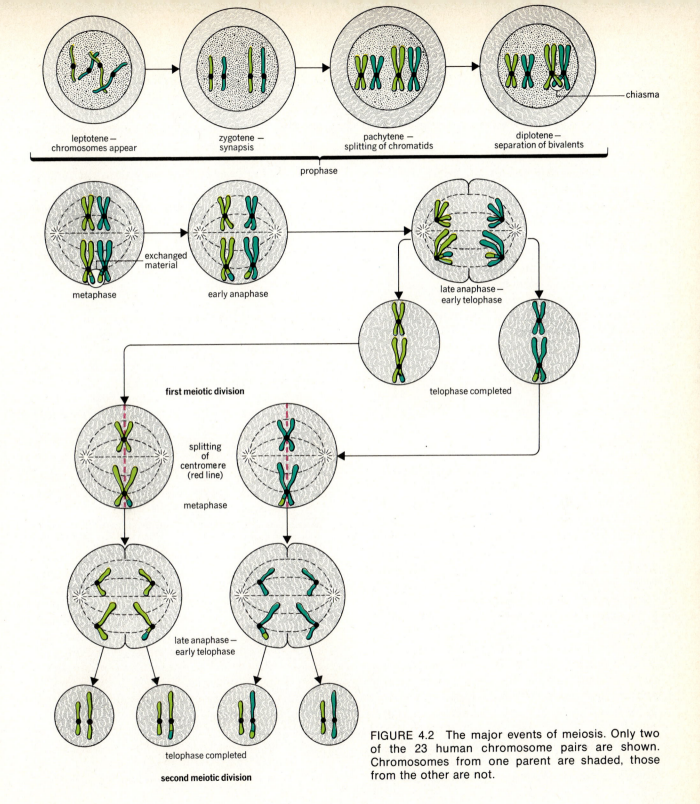

leptotene — chromosomes appear

zygotene — synapsis

pachytene — splitting of chromatids

diplotene — separation of bivalents

chiasma

prophase

metaphase

exchanged material

early anaphase

late anaphase — early telophase

telophase completed

first meiotic division

splitting of centromere (red line)

metaphase

late anaphase — early telophase

telophase completed

second meiotic division

FIGURE 4.2 The major events of meiosis. Only two of the 23 human chromosome pairs are shown. Chromosomes from one parent are shaded, those from the other are not.

opposite poles of the cell. In the human, the diploid number of chromosomes (46) is reduced to the haploid number (23).

In FIRST MEIOTIC TELOPHASE, division of cell cytoplasm and roughly equal division of cytoplasmic mass and organelles occurs.

The SECOND MEIOTIC DIVISION is basically a mitotic division; that is, *centromeres do divide*, and duplicated chromosomes are separated without further change. Two daughter cells are usually formed at each division, giving a total of *four* haploid daughter cells from each parent cell.

Formation of sperm in man has a time cycle of about 74 days. In woman, ova maturation proceeds to first meiotic prophase before birth and then remains "suspended" until puberty and ovulation, about 9 years or so of age. Release of the ova from the ovary may continue until about 50 years, so that the "age" of the ova when they resume development may be 9 to 50 years.

Neoplasms *(cancers)*

Cell division by mitosis is usually balanced to the needs of the body for cell replacement and repair. Uncontrolled cell division may result in the development of NEOPLASMS (literally, new growth) or *cancers*. The term cancer sometimes evoke a mental picture of debility, deformity, and ultimate death. It should be emphasized that this is not necessarily the case. Early detection and adequate treatment are the best insurance for recovery.

Two major types of neoplasms are recognized. BENIGN neoplasms usually grow slowly, are generally limited from surrounding normal tissue by connective tissue capsules (thus making them easier to remove *in toto* by surgical intervention), and are not considered grave threats to the body unless they are in an area where they may grow, undetected, to large sizes. The chief threat from a benign neoplasm is mechanical, for its growth (as in the brain) may compress blood vessels, vital nerve centers, or block a vascular or air passage-

way. MALIGNANT neoplasms grow rapidly, are not usually limited by capsules, and metastasize easily; that is, the neoplasm sheds cells into vascular and/or lymphatic channels. These cells may be trapped in other body areas, and set up secondary neoplasms. Another method by which cancers may be spread is by "seeding" from a surgeon's gloves during operations to remove cancers.

Cancers of all types are the *second leading cause* of death in the United States; the first is death from cardiac and vascular disease. The American Cancer Society suggests seven "danger signals" that *may* signal the presence of a neoplastic process in the body:

Any unusual discharges or bleeding from a body opening (e.g., anus, vagina, mouth, rectum).
A sore that does not heal.
A change in bowel habits (including frequency and consistency) persisting more than three weeks.
Chronic indigestion or difficulty in swallowing.
Any change in the size or color of a wart or mole.
A lump or thickening in the breast or elsewhere.
Persistent hoarseness or coughing.

Again, early detection remains a basic ingredient of successful treatment, and prompt reporting of suspicious symptoms should supplement regular physical examinations for those over 40 years of age.

Prognosis (prediction of the course and outcome of a disease or abnormal process) in an individual with cancer depends on the degree of abnormality between normal and neoplastic cells, and on the length of time the process has operated before discovery. Once discovered, the neoplasm may be GRADED from I to IV according to the degree the neoplastic cells differ in organization and structure from normal cells. A Grade I rating implies minimal differences from normal; a Grade IV rating indicates maximal differentiation. The neoplasm may also be described as to its extent of spread by the term STAGING. Four stages (I to IV) are listed, with Stage I indicating limitation of the neoplasm to the site of origin, and Stage IV indicating widespread metastasis. In neoplasms of the colon and rectum, the lesion may also be TYPED (A,B,C) according to the depth of penetration of the wall of the organ. Type A indicates involvement of mucosal and muscular layers; Type B indicates that the outer wall of the organ has been involved; Type C reflects metastasis to the lymph nodes of the area.

One characteristic—among many others—of a malignant neoplasm is its requirement for nourishment resulting from the high rates of mitosis and requirements for substance. Neoplasms are thus very vascular structures, and they sustain themselves at the expense of normal cells and the host in general. Thus, emaciation, weight loss, anemia, and general fatigue are common symptoms.

SOME THEORIES OF CAUSATION OF CANCERS. A substance capable of inducing the development of a cancer is termed a CARCINOGEN. The most common carcinogens appear to be the substances that enter the body via the *respiratory system* (smog, tobacco smoke, viruses, or fungal spores), *digestive system* (substances in foods and drink), or substances *reaching other organs* after having been absorbed and distributed throughout the body via the bloodstream. Since so many chemicals that are carcinogenic apparently may enter the blood by these routes, the suspicion has arisen that a large part of cancer production may be due to ENVIRONMENTAL CAUSES. Also included as environmental causes of cancer are *radiation*, such as sunlight causing skin cancers, and radiation other than solar (radioactive materials, for example), producing mutations in cells. According to some authorities, there is no "safe" dosage of radiation; these people believe that there is *some* effect of radiation on cells, no matter how small the dose.

Other theories suggest that cancer formation has a GENETIC CAUSE. There is a definite association between Down's Syndrome (mongolism, involving an extra chromosome) and an increased incidence of leukemia; Wilm's tumor is a genetic disorder of the kidney that results in development of a carcinoma from embryonic cells.

Yet another possible cause of cancer is the body's inability to destroy abnormal cells by an immunologic reaction. Cancer cells have antigens on their surfaces that differ from those on normal cells. The normal body response to such abnormal antigens is to produce an antibody that reacts with the abnormal cell and destroys it. The body response to foreign antigens decreases with age, and the response may not be as strong as with other antigens. This may be because the cancer cell antigens may not *differ enough* from normal antigens to call forth a full-fledged response by the immune system.

All in all, most thoughtful individuals agree that about 90 percent of cancer is related to some environmental factor, either natural or artificial, with 10 percent resulting from genetic causes.

AVENUES IN CANCER TREATMENT. CHEMOTHERAPY is one form of treatment employed for cancer. Drugs employed generally interfere with some phase of the cell division process *in all rapidly dividing cells.* Thus, the substance may prove to be toxic to intestinal epithelial cells, spermatogenic cells, skin cells, bone marrow cells, and white blood cells, as well as to the neoplastic cells. Various techniques may be employed to restrict the chemotherapeutic agent to the area where the malignancy resides (such as temporarily occluding blood flow to the area after administration of the substance). Chemotherapeutic agents are divided into several groups according to their probable mode of action.

ALKYLATING AGENTS (nitrogen mustards and related compounds) appear to react with nucleic acids in tumor cells, preventing normal DNA replication.

ANTIMETABOLITES are substances whose chemical structures are similar to but not identical with substances normally used by cells for growth and metabolism. Most interfere with cell utilization of folic acid, a substance required for normal cell growth. Examples include *aminopterin, amethopterin* (methotrexate), and *fluorouracil,* in the treatment of leukemia.

Certain HORMONES may slow tumor growth, possibly by making the tumor environment less favorable for growth. For example, ACTH (adrenocorticotropic hormone) from the pituitary has been shown to slow the progress of leukemia.

RADIOACTIVE MATERIALS and RADIATION kill cells indiscriminately, both tumor cells and normal cells. Use of such measures is thus restricted to those tumors that accumulate a particular isotope (e.g., ^{131}I in the thyroid) or those upon which a beam of radiation can be closely and specifically focused.

SURGERY is a type of treatment that is successful if the entire cancerous mass can be removed, and is often followed by chemotherapy or radiation.

IMMUNE RESPONSE may be stimulated, as by the administration of an agent such as BCG (bacillus Calmette-Guerin). The substance is primarily used as an immunizing agent against tuberculosis, but it also stimulates activity of antibody forming cells. It was postulated that BCG might cause the body to more strongly "attack" the foreign antigens on the surfaces of cancer cells. Results with BCG have not been encouraging.

Survival after five years, following a course of therapy designed to destroy or remove a cancer, is generally taken as a sign of control of the malignancy.

NOTES ON SOME OF THE MORE COMMON CANCERS. LUNG CANCER. No single cause for lung tumors has yet been identified. Tobacco smoking (particularly cigarettes, where the carcinogens are inhaled) slows the ciliary action of epithelial cells lining the tubes of the respiratory system, thus permitting a longer duration for potential carcinogens to act. The activity of pulmonary phagocytic cells is also decreased, with a similar result. Inhalation of asbestos fibers, airborne radiation, and atmospheric pollution have all been suggested as possible additional causes of lung tumors. Cough that is persistent, with copious mucus secretion, a mass on X ray, and bloody sputum are suggestive of a pulmonary lesion. Biopsy (removal of a small portion of the suspected tumor) and microscopic examination may prove the existence of a tumor, as may bronchoscopic examination. Lung cancer is the primary cancerous killer

of males in the United States. Surgery remains the primary treatment and prognosis is generally poor for a patient with this disease, with most surviving a year or less.

BREAST CANCER is the number one cancerous killer of females in the United States. It is a type of cancer in which determination of the distribution of the tumor is a necessary first step. Cancer cells metastasize readily from a mammary cancer, and may be distributed by the extensive lymph drainage of the breast, to the thoracic, axillary (armpit), and neck lymph nodes. Local disease is usually treated by surgery, with radiation to local lymph nodes if they are involved. A current debate in medicine revolves around whether mammograms (X ray of the breasts) should be employed as a routine procedure for detection of tumors. It has been suggested that the X ray itself may trigger tumor development. It seems that the procedure should be avoided in the absence of any other symptoms, unless the female is in an "over 40" category, and only then used on advice of a physician. Another discussion continues concerning the value of conservative versus radical surgical procedures when a tumor *is* discovered. "Lumpectomy" involves removal of the tumor only, while mastectomy involves removal of an entire breast. A radical mastectomy removes not only the breast, but the underlying muscle and all lymph nodes in the area. There is no concrete evidence that one procedure improves chances of survival more than another, *if* widespread metastasis has occurred. Indeed, great psychological damage may occur if a radical procedure is employed when it may not be needed. In women who respond to treatment, prognosis is good if they remain symptom-free for two years or more after treatment.

Cancer of the UTERINE CERVIX is a more common tumor in married or sexually active females. Sexual intercourse may inflict trauma to the cervix and also introduce *Herpes* virus into the damaged cells. The latter is a strain of viruses similar to the organism that causes cold-sores or fever blisters, and may be a factor in causation of cervical cancer. A Pap test (named for George Papanicolaou,

1883–1962), involves taking a smear from the cervix and vagina and examining it for cancerous cells. The test is quick and painless, and may detect cancer very early. Radiotherapy (using radium in "needles") or surgery are the major avenues of treatment in this disease. Over 80 percent survival occurs if the tumor is discovered early.

LEUKEMIA is discussed more fully in the chapter on blood. Here, we may indicate that leukemias may be of bone marrow (myeloid) origin, or from lymphatic tissue (lymphoid). They may be *acute,* a term implying a rapid onset and course, or *chronic,* implying slow development and of longer course. The disease is diagnosed by biopsy of bone marrow or lymph nodes, and may be suspected if there is anemia, high white cell counts (100,000 or more per mm³), and the presence of abnormal white cells in the bloodstream. In subhuman animal forms, a virus has been implicated in development of the disease but evidence is meager for this etiology in humans. There is a greater incidence of leukemia in patients with Down's syndrome ("Mongolism"), a genetic disorder. Exposure to excessive radiation also increases the incidence of leukemia. Treatment generally involves chemotherapy and the use of radiation.

CANCER OF THE DIGESTIVE TRACT is one of the more common malignancies of older people. *Carcinoma of the large intestine* is the fourth most common malignancy of the United States adult population (following skin cancer in both sexes, lung cancer in males, and breast cancer in females). Postulated causes for cancer of the large intestine include *environmental factors* (e.g., low amounts of fiber in the diet that slow passage of potential carcinogens through the tract) and *hereditary factors* (polyploid conditions lead to cancer formation). The presence of a mass on palpation, rectal bleeding, the presence of an antigen called CEA (*carcinoembryonic antigen*) in the bloodstream, and the use of a barium enema (barium is radiopaque and a lesion will be demonstrated by its light color X ray) are methods employed to discover this type of cancer. Proctoscopic examination, in which the lower portions of the large intestine may be visually examined, is a good procedure but it cannot extend beyond about the lower 18 inches or so of intestine. Surgery remains the primary treatment, and recovery is common if early detection has been made.

The reader is referred to the bibliography at the end of this chapter for further discussion of specific cancers.

The basic development of the individual

Human sperm and ova, produced by meiosis, each contain one half the chromosome number characteristic of the species. This halved number of chromosomes is termed the HAPLOID number and is 23 for the human. The haploid number of chromosomes consists of 22 autosomes, and an X or Y (sex) chromosome. The meioses producing sperm create cells with either a 22 + X or 22 + Y content; those producing ova result in a 22 + X content. Restoring normal chromosome number (46) depends on the joining of an egg and sperm to form a ZYGOTE in the process of FERTILIZATION. Fertilization results in a 44 + XY (male) or 44 + XX (female) chromosome constitution in the zygote. Fertilization is the time of conception. Development from conception onward may be described "by the week" in terms of the major processes concerned.

The first week

After fertilization, the zygote undergoes a series of mitotic divisions known as CLEAVAGE. A solid mass of cells, known as a MORULA results. (The morula is about the size of a period in this text.) Little time for growth of cells occurs between cleavages, and the morula is only a little larger

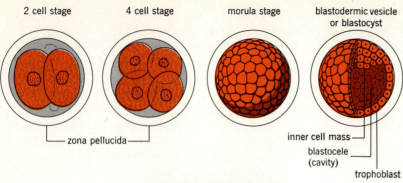

2 cell stage   4 cell stage   morula stage   blastodermic vesicle
                                                    or blastocyst

zona pellucida

inner cell mass
blastocele
(cavity)
trophoblast

FIGURE 4.3.   Embryonic
development during the first
week; cleavage and blastocyst formation.

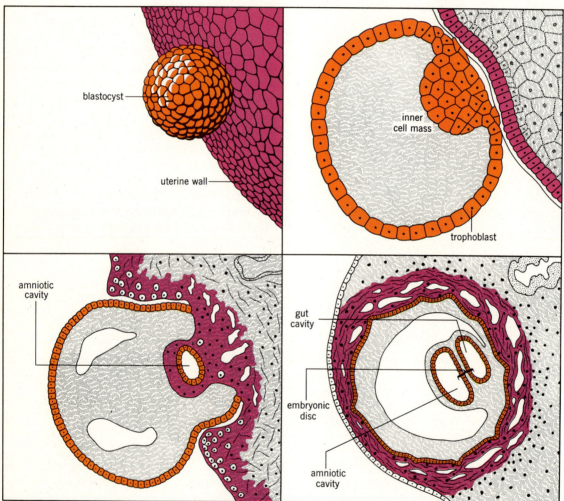

blastocyst

uterine wall

inner
cell mass

trophoblast

amniotic
cavity

gut
cavity

embryonic
disc

amniotic
cavity

FIGURE 4.4 Embryonic development during the
second week; implantation, and the formation of a
two-layered embryo.

than the zygote. Cleavage results in a doubling of cell number for the first 8 to 10 divisions, and thereafter becomes irregular. As cleavage is occurring, the zygote is passing down the uterine (Fallopian) tube from ovary to uterus, a journey that takes about 3 days. Arriving in the uterine cavity, the morula adheres to the lining of the uterus (endometrium) and undergoes a reorganization into a BLASTOCYST. The blastocyst is a hollow structure containing a *cavity,* and an *inner mass* of cells. This reorganization takes 3 to 4 days. The events of the first week are shown in Figure 4.3.

The second week

The blastocyst undergoes further reorganization with the appearance of two cavities in the inner cell mass. The upper one is the AMNIOTIC CAVITY, the lower one, the cavity of the YOLK SAC or gut. A two-layered plate of cells separates the two cavities, and is known as the EMBRYONIC PLATE or DISC. Development to this stage is said to result in the formation of a TWO-LAYERED EMBRYO. The layer of cells in this embryo nearest the amniotic cavity is termed ECTODERM, the other layer, ENDODERM. These layers form two of the three "germ layers" from which all body structures will develop. Finally, the embryo undergoes IMPLANTATION in the wall of the uterus. In this process, the blastocyst "digests" its way into the vascular and glandular endometrium of the uterus to establish a relationship with the mother's blood supply. This relationship assures the nutrients necessary for further growth and development. These processes are shown in Figure 4.4.

TABLE 4.1   Derivatives of germ layers

Ectoderm	Mesoderm	Endoderm
1. Outer layer (epidermis) of skin: Skin glands, Hair and nails, Lens of eye	1. Muscle: Skeletal, Cardiac, Smooth	1. Epithelium of: Pharynx, Auditory (ear) tube, Tonsils, Thyroid, Parathyroid, Thymus, Larynx, Trachea, Lungs, Digestive tube and its glands, Bladder, Vagina and vestibule, Urethra and glands
2. Lining tissue (epithelium) of: Nasal cavities, Sinuses, Mouth: Oral glands, Tooth enamel, Sense organs, Anal canal	2. Supporting (connective) tissue: Cartilage, Bone, Blood	
3. Nervous tissues	3. Bone marrow	2. Pituitary (anterior and middle lobes)
4. Pituitary (posterior lobe)	4. Lymphoid tissue	
5. Adrenal medulla	5. Epithelium of: Blood vessels, Lymphatics, Celomic cavity, Kidney and ureters, Gonads and ducts, Adrenal cortex, Joint cavities	

The third week

The major event occurring at this time is the formation of the third basic germ layer. By mitotic cell division, a layer of cells termed the MESO-DERM is formed between the ectoderm and endoderm of the two-layered embryo. Table 4.1 shows the ultimate derivatives of these three germ layers.

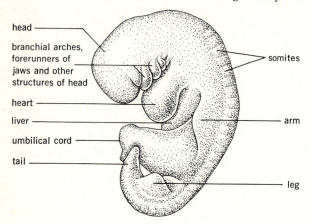

The fourth week

The fourth week is associated with the formation of the NEURAL TUBE, the beginning of the nervous system, and of SOMITES or blocks of mesoderm along the backbone of the embryo. These somites are the forerunners of the bones and muscles of the back. The embryo appears as is shown in Figure 4.5 Notice that development occurs in a head-to-tail (cephalo-caudal) direction, as does maturation of function, once an organ or system is formed.

The fifth through eighth weeks

COMPLETION OF THE EMBRYO occurs during this period, and it assumes a clearly human form.

FIGURE 4.5   The human embryo at four weeks of development.

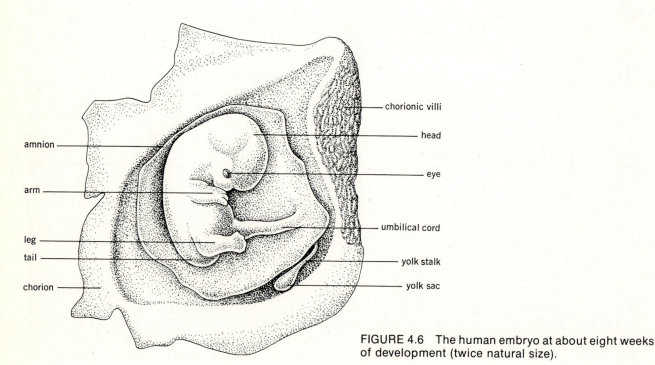

FIGURE 4.6   The human embryo at about eight weeks of development (twice natural size).

The brain is enclosed, the digestive system forms, a heart is formed and circulation of blood in vessels is established. The limbs, eyes, ears, and other features are evident (Fig. 4.6).

The ninth week to birth

This period constitutes the period of the FETUS, a creature that has definite human form and all basic body systems. The fetus is, in some cases, capable of independent survival after about 26 weeks of development, a time that normally permits lung development to be far enough advanced to support life. Table 4.2, Correlated Human Development, indicates that the fetal period is one mainly of growth of the body; by 8 to 12 weeks, all body systems are present and need only to grow and undergo final refinement. The first 12 weeks of development is thus of critical importance for establishing normal organs. Drugs used or diseases (e.g., measles) contracted during this time period may lead to malformations in the develop-

ment of body organs and systems. Congenital (born with) anomalies or malformations are the result of developmental errors. They may be evident at birth, or may cause disorders later in life. Such malformations are the third leading cause of death in the United States between birth and 4 years of age, and the fourth leading cause of death between the ages of 5 to 14 years. Since several organs and systems develop at the same time, a process affecting one system may result in changes in a simultaneously developing system, so that malformations are usually seen in more than one system.

During fetal growth, and after birth, body proportions change (Fig. 4.7). Thus a newborn has a relatively larger head and shorter legs than does the child and adult. Growth in size is the most obvious change (Fig. 4.8) occurring after about 2 months *in utero*.

These sections have given a general view of body formation. The basic development of each system is discussed at the beginning of appropriate chapters that follow.

TABLE 4.2 A reference table of correlated human development (from Arey).

Age in weeks	Size (C R) in Mm	Body form	Mouth	Pharynx and derivatives	Digestive tube and glands	Respiratory system	Coelom and mesenteries
2.5	1.5	Embryonic disc flat. Primitive streak prominent. Neural groove indicated.	—	—	Gut not distinct from yolk sac.	—	Extra-embryonic coelom present. Embryonic coelom about to appear.
3.5	2.5	Neural groove deepens and closes (except ends). Somites 1 – 16± present. Cylindrical body constricting from yolk sac. Branchial arches 1 and 2 indicated.	Mandibular arch prominent. Stomodeum a definite pit. Oral membrane ruptures.	Pharynx broad and flat. Pharyngeal pouches forming. Thyroid indicated.	Fore- and hind-gut present. Yolk sac broadly attached at mid-gut. Liver bud present. Cloaca and cloacal membrane present.	Respiratory primordium appearing as a groove on floor of pharynx.	Embryonic coelom a U-shaped canal, with a large pericardial cavity. Septum transversum indicated. Mesenteries forming. Mesocardium atrophying.
4	5.0	Branchial arches completed. Flexed heart prominent. Yolk stalk slender. All somites present (40). Limb buds indicated. Eye and otocyst present. Body flexed; C-shape.	Maxillary and mandibular processes prominent. Tongue primordia present. Rathke's pouch indicated.	Five pharyngeal pouches present. Pouches 1-4 have closing plates. Primary tympanic cavity indicated. Thyroid a stalked sac.	Esophagus short. Stomach spindle-shaped. Intestine a simple tube. Liver cords, ducts and gall bladder forming. Both pancreatic buds appear. Cloaca at height.	Trachea and paired lung buds become prominent. Laryngeal opening a simple slit.	Coelom still a continuous system of cavities. Dorsal mesentery a complete median curtain. Omental bursa indicated.
5	8.0	Nasal pits present. Tail prominent. Heart, liver and mesonephros protuberant. Umbilical cord organizes.	Jaws outlined. Rathke's pouch a stalked sac.	Phar. pouches gain dors. and vent. diverticula. Thyroid bilobed. Thyro-glossal duct atrophies.	Tail-gut atrophies. Yolk stalk detaches. Intestine elongates into a loop. Caecum indicated.	Bronchial buds presage future lung lobes. Arytenoid swellings and epiglottis indicated.	Pleuro-pericardial and pleuro-peritoneal membranes forming. Ventral mesogastrium draws away from septum.
6	12.0	Upper jaw components prominent but separate. Lower jaw-halves fused. Head becomes dominant in size. Cervical flexure marked. External ear appearing. Limbs recognizable as such.	Lingual primordia fusing. Foramen caecum established. Labio-dental laminae appearing. Parotid and submaxillary buds indicated.	Thymic sacs, ultimobranchial sacs and solid parathyroids are conspicuous and ready to detach. Thyroid becomes solid and converts into plates.	Stomach rotating. Intestinal loop undergoes torsion. Hepatic lobes identifiable. Cloaca subdividing.	Definitive pulmonary lobes indicated. Bronchi sub-branching. Laryngeal cavity temporarily obliterated.	Pleuro-pericardial communications close. Mesentery expands as intestine forms loop.

TABLE 4.2 (continued)

Urogenital system	Vascular system	Skeletal system	Muscular system	Integumentary system	Nervous system	Sense organs	Age in weeks
Allantois present.	Blood islands appear on chorion and yolk sac. Cardiogenic plate reversing.	Head process (or notochordal plate) present.	—	Ectoderm a single layer.	Neural groove indicated.	—	2.5
All pronephric tubules formed. Pronephric duct growing caudad as a blind tube. Cloaca and cloacal membrane present.	Primitive blood cells and vessels present. Embryonic blood vessels a paired symmetrical system. Heart tubes fuse, bend S-shaped and beat begins.	Mesodermal segments appearing (1 — 16 ±). Older somites begin to show sclerotomes. Notochord a cellular rod.	Mesodermal segments appearing (1 — 16 ±). Older somites show myotome plates.	—	Neural groove prominent; rapidly closing. Neural crest a continuous band.	Optic vesicle and auditory placode present. Acoustic ganglia appearing.	3.5
Pronephros degenerated. Pronephric (mesonephric) duct reaches cloaca. Mesonephric tubules differentiating rapidly. Metanephric bud pushes into secretory primordium.	Hemopoiesis on yolk sac. Paired aortae fuse. Aortic arches and cardinal veins completed. Dilated heart shows sinus, atrium, ventricle, and bulbus.	All somites present (40). Sclerotomes massed as primitive vertebrae about notochord.	All somites present (40).	—	Neural tube closed. Three primary vesicles of brain represented. Nerves and ganglia forming. Ependymal, mantle and marginal layers present.	Optic cup and lens pit forming. Auditory pit becomes closed, detached otocyst. Olfactory placodes arise and differentiate nerve cells.	4
Mesonephros reaches its caudal limit. Ureteric and pelvic primordia distinct. Genital ridge bulges.	Primitive vessels extend into head and limbs. Vitelline and umbilical veins transforming. Myocardium condensing. Cardiac septa appearing. Spleen indicated.	Condensations of mesenchyme presage many future bones.	Premuscle masses in head, trunk and limbs.	Epidermis gaining a second layer (periderm).	Five brain vesicles. Cerebral hemispheres bulging. Nerves and ganglia better represented. [Suprarenal cortex accululating.]	Chorioid fissure prominent. Lens vesicle free. Vitreous anlage appearing. Octocyst elongates and buds endolymph duct. Olfactory pits deepen.	5
Cloaca subdividing. Pelvic anlage sprouts pole tubules. Sexless gonad and genital tubercle prominent. Müllerian duct appearing.	Hemopoiesis in liver. Aortic arches transforming. L. umbil. vein and d. venosus become important. Bulbus absorbed into right ventricle. Heart acquires its general definitive form.	First appearance of chondrification centers. Desmocranium.	Myotomes, fused into a continuous column, spread ventrad. Muscle segmentation largely lost.	Milk line present.	Three primary flexures of brain represented. Diencephalon large. Nerve plexuses present. Epiphysis recognizable. Sympathetic ganglia forming segmental masses. Meninges indicated.	Optic cup shows nervous and pigment layers. Lens vesicle thickens. Eyes set at 160° Naso-lacrimal duct. Modeling of ext., mid. and int. ear under way. Vomero-nasal organ.	6

TABLE 4.2 (continued)

Age in weeks	Size (C R) in Mm	Body form	Mouth	Pharynx and derivatives	Digestive tube and glands	Respiratory system	Coelom and mesenteries
7	17.0	Branchial arches lost. Cervical sinus obliterates. Face and neck forming. Digits indicated. Back straightens. Heart and liver determine shape of body ventrally. Tail regressing.	Lingual primordia merge into single tongue. Separate labial and dental laminae distinguishable. Jaws formed and begin to ossify. Palate folds present and separated by tongue.	Thymi elongating and losing lumina. Parathyroids become trabeculate and associate with thyroid. Ultimobranchial bodies fuse with thyroid. Thyroid becoming crescentic.	Stomach attaining final shape and position. Duodenum temporarily occluded. Intestinal loops herniate into cord. Rectum separates from bladder-urethra. Anal membrane ruptures. Dorsal and ventral pancreatic primordia fuse.	Larynx and epiglottis well outlined; orifice T-shaped. Laryngeal and tracheal cartilages foreshadowed. Conchae appearing. Primary choanae rupturing.	Pericardium extended by splitting from body wall. Mesentery expanding rapidly as intestine coils. Ligaments of liver prominent.
8	23.0	Nose flat; eyes far apart. Digits well formed. Growth of gut makes body evenly rotund. Head elevating. Fetal state attained.	Tongue muscles well differentiated. Earliest taste buds indicated. Rathke's pouch detaches from mouth. Sublingual gland appearing.	Auditory tube and tympanic cavity distinguishable. Sites of tonsil and its fossae indicated. Thymic halves unite and become solid. Thyroid follicles forming.	Small intestine coiling within cord. Intestinal villi developing. Liver very large in relative size.	Lung becoming gland-like by branching of bronchioles. Nostrils closed by epithelial plugs.	Pleuro-peritoneal communications close. Pericardium a voluminous sac. Diaphragm completed, including musculature. Diaphragm finishes its descent.
10	40.0	Head erect. Limbs nicely modeled. Nail folds indicated. Umbilical hernia reduced.	Fungiform and vallate papillae differentiating. Lips separate from jaws. Enamel organs and dental papillae forming. Palate folds fusing.	Thymic epithelium transforming into reticulum and thymic corpuscles. Ultimobranchial bodies disappear as such.	Intestines withdraw from cord and assume characteristic positions. Anal canal formed. Pancreatic alveoli present.	Nasal passages partitioned by fusion of septum and palate. Nose cartilaginous. Laryngeal cavity reopened; vocal folds appear.	Processus (saccus) vaginales forming. Intestine and its mesentery withdrawn from cord.
12	56.0	Head still dominant. Nose gains bridge. Sex readily determined by external inspection.	Filiform and foliate papillae elevating. Tooth primordia form prominent cups. Cheeks represented. Palate fusion complete.	Tonsillar crypts begin to invaginate. Thymus forming medulla and becoming increasingly lymphoid. Thyroid attains typical structure.	Muscle layers of gut represented. Pancreatic islands appearing. Bile secreted.	Conchae prominent. Nasal glands forming. Lungs acquire definitive shape.	Omentum an expansive apron partly fused with dorsal body wall. Mesenteries free but exhibit typical relations. Coelomic extension into umbilical cord obliterated.
16	112.0	Face looks 'human.' Hair of head appearing. Muscles become spontaneously active. Body outgrowing head.	Hard and soft palates differentiating. Hypophysis acquiring definitive structure.	Lymphocytes accumulate in tonsils. Pharyngeal tonsil begins development.	Gastric and intestinal glands developing. Duodenum and colon affixing to body wall. Meconium collecting.	Accessory nasal sinuses developing. Tracheal glands appear. Mesoderm still abundant between pulmonary alveoli. Elastic fibers appearing in lungs.	Greater omentum fusing with transverse mesocolon and colon. Mesoduodenum and ascending and descending mesocolon attaching to body wall.

TABLE 4.2   (continued)

Urogenital system	Vascular system	Skeletal system	Muscular system	Integumentary system	Nervous system	Sense organs	Age in weeks
Mesonephros at height of its differentiation. Metanephric collecting tubules begin branching. Earliest metanephric secretory tubules differentiating. Bladder-urethra separates from rectum. Urethral membrane rupturing.	Cardinal veins transforming. Inf. vena cava outlined. Atrium, ventricle and bulbus partitioned. Cardiac valves present. Stem of pulm. vein absorbed into l. atrium. Spleen anlage prominent.	Chondrification more general. Chondrocranium.	Muscles differentiating rapidly throughout body and assuming final shapes and relations.	Mammary thickening lens-shaped.	Cerebral hemispheres becoming large. Corpus striatum and thalamus prominent. Infundibulum and Rathke's pouch in contact. Chorioid plexuses appearing. Suprarenal medulla begins invading cortex.	Chorioid fissure closes, enclosing central artery. Nerve fibers invade optic stalk. Lens loses cavity by elongating lens fibers. Eyelids forming. Fibrous and vascular coats of eye indicated. Olfactory sacs open into mouth cavity.	7
Testis and ovary distinguishable as such. Müllerian ducts, nearing urogenital sinus, are ready to unite as uterovaginal primordium. Genital ligaments indicated.	Main blood vessels assume final plan. Primitive lymph sacs present. Sinus venosus absorbed into right atrium. Atrio-ventricular bundle represented.	First indications of ossification.	Definitive muscles of trunk, limbs and head well represented and fetus capable of some movement.	Mammary primordium a globular thickening.	Cerebral cortex begins to acquire typical cells. Olfactory lobes visible. Dura and pia-arachnoid distinct. Chromaffin bodies appearing.	Eyes converging rapidly. Ext., mid. and int. ear assuming final form. Taste buds indicated. External nares plugged.	8
Kidney able to secrete. Bladder expands as sac. Genital duct opposite sex degenerating. Bulbo-urethral and vestibular glands appearing. Vagina sacs forming.	Thoracic duct and peripheral lymphatics developed. Early lymph glands appearing. Enucleated red cells predominate in blood.	Ossification centers more common. Chondrocranium at its height.	Perineal muscles developing tardily	Epidermis adds intermediate cells. Periderm cells prominent. Nail field indicated. Earliest hair follicles begin developing on face.	Spinal cord attains definitive internal structure.	Iris and ciliary body organizing. Eyelids fused. Lacrimal glands budding. Spiral organ begins differentiating.	10
Uterine horns absorbed. External genitalia attain distinctive features. Meson. and rete tubules complete male duct. Prostate and seminal vesicle appearing. Hollow viscera gaining muscular walls.	Blood formation beginning in bone marrow. Blood vessels acquire accessory coats.	Notochord degenerating rapidly. Ossification spreading. Some bones well outlined.	Smooth muscle layers indicated in hollow viscera.	Epidermis three-layered. Corium and subcutaneous now distinct.	Brain attains its general structural features. Cord shows cervical and lumbar enlargements. Cauda equina and filum terminale appearing. Neurological types begin to differentiate.	Characteristic organization of eye attained. Retina becoming layered. Nasal septum and plate fusions completed.	12
Kidney attains typical shape and plan. Testis in position for later descent into scrotum. Uterus and vagina recognizable as such. Mesonephros involuted.	Blood formation active in spleen. Heart musculature much condensed.	Most bones distinctly indicated throughout body. Joint cavities appear.	Cardiac muscle appearing in earlier weeks, now much condensed. Muscular movements in utero can be detected.	Epidermis begins adding other layers. Body hair starts developing. Sweat glands appear. First sebaceous glands differentiating.	Hemispheres conceal much of brain. Cerebral lobes delimited. Corpora quadrigemina appear. Cerebellum assumes some prominence.	Eye, ear and nose grossly approach typical appearance. General sense organs differentiating.	16

TABLE 4.2 (continued)							
Age in weeks	Size (C R) in Mm	Body form	Mouth	Pharynx and derivatives	Digestive tube and glands	Respiratory system	Coelom and mesenteries
20–40 (5–10 mo.)	160.0–350.0	Lanugo hair appears (5). Vernix caseosa collects (5). Body lean but better proportioned (6). Fetus lean, wrinkled and red; eyelids reopen (7). Testes invading scrotum (8). Fat collecting, wrinkles smoothing; body rounding (8–10).	Enamel and dentine depositing (5). Lingual tonsil forming (5). Permanent tooth primordia indicated (6–8). Milk teeth unerupted at birth.	Tonsil structurally typical (5).	Lymph nodules and muscularis mucosae of gut present (5). Ascending colon becomes recognizable (6). Appendix lags behind caecum in growth (6). Deep esophageal glands indicated (7). Plicae circulares represented (8).	Nose begins ossifying (5). Nostrils reopen (6). Cuboidal pulmonary epithelium disappearing from alveoli (6). Pulmonary branching only two-thirds completed (10). Frontal and sphenoidal sinuses still very incomplete (10).	Mesenterial attachments completed (5). Vaginal sacs passing into scrotum (7–9).

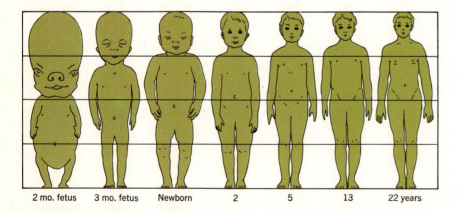

2 mo. fetus   3 mo. fetus   Newborn   2   5   13   22 years

FIGURE 4.7 The changes in the body proportions from before birth to adulthood.

Urogenital system	Vascular system	Skeletal system	Muscular system	Integumentary system	Nervous system	Sense organs	Age in weeks
TABLE 4.2 (continued)							
Female urogenital sinus becoming a shallow vestibule (5). Vagina regains lumen (5). Uterine glands appear (7). Scrotum solid until sacs and testes descend (7–9). Kidney tubules cease forming at birth.	Blood formation increasing in bone marrow and decreasing in liver. (5–10). Spleen acquires typical structure (7). Some fetal blood passages discontinue (10).	Carpal, tarsal and sternal bones ossify late; some after birth. Most epiphyseal centers appear after birth; many during adolescence.	Perineal muscles finish development (6).	Vernix caseosa seen (5). Epidermis cornifies (5). Nail plate begins (5). Hairs emerge (6). Mammary primordia budding (5); buds branch and hollow (8). Nail reaches finger tip (9). Lanugo hair prominent (7); sheds (10).	Commissures completed (5). Myelinization of cord begins (5). Cerebral cortex layered typically (6). Cerebral fissues and convolutions appearing rapidly (7). Myelinization of brain begins (10).	Nose and ear ossify (5). Vascular tunic of lens at height (7). Retinal layers completed and light perceptive (7). Taste sense present (8). Eyelids reopen (7–8). Mastoid cells unformed (10). Ear deaf at birth.	20–40 (5–10 mo.)

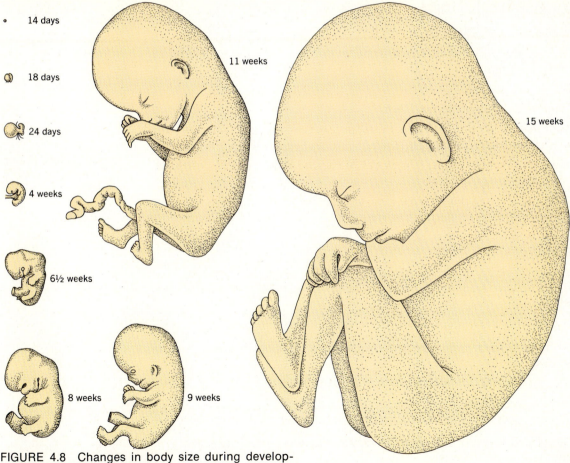

14 days, 18 days, 24 days, 4 weeks, 6½ weeks, 8 weeks, 9 weeks, 11 weeks, 15 weeks

FIGURE 4.8   Changes in body size during development in the uterus. (All figures natural size.)

Summary

1. Division of cells by mitosis and meiosis provides means of replacing dead or lost cells, for purposes of healing, and creates gametes (sex cells) for perpetuation of the species.

2. Mitosis is a process used to create new cells that are genetically identical to the parent cells.

   a. It is used for cell replacement.

   b. The process occurs in stages, with each stage characterized by particular occurrences.

      1) Interphase. DNA duplication.

      2) Prophase. Dissolution of nuclear membrane, centriole migration, spindle formation, chromosome formation.

      3) Metaphase. Alignment of chromosomes on equatorial plate of cell.

      4) Anaphase. Chromosome separation after centromere division.

      5) Telophase. Reconstitution of nucleus; cytoplasm division.

3. Meiosis is a process used to create gametes that are haploid, and therefore genetically different from the parent cells.

   a. It also occurs in stages.

      1) Interphase. Duplication of DNA.

      2) First meiotic prophase. Nuclear dissolution, centriole migration, spindle and chromosome formation. *Pairing of homologous chromosomes.*

      3) First meiotic metaphase. Alignment of chromosome pairs on equatorial plate. Crossing over may occur.

      4) First meiotic anaphase. Separation of chromosomes *without* centromere division. *This is a reduction division.*

      5) First meiotic telophase. Reconstitution of nucleus and cytoplasmic division.

      6) A second series of divisions occurs next. They have the same stages as listed above, and are basically mitotic in nature, with centromere division.

   b. Four cells of haploid constitution (23 for the human) are generally formed.

4. Neoplasms or cancers represent uncontrolled cell division that may be life-threatening.

   a. Benign neoplasms are slow-growing, usually delimited, and are mainly mechanical threats to the body (pressure, blockage).

   b. Malignant neoplasms are fast-growing, metastasize easily, and may kill.

5. There are seven "danger signals" that may indicate the presence of a cancer in the body.

6. Early detection and adequate treatment results in a higher rate of recovery from cancer.

7. Cancers are characterized by degree of difference from normal cells (grade), by extent of spread (stage), and degree of penetration of an organ wall (type). A cancer sustains itself at the expense of the host, commonly causing weight loss, anemia, and fatigue.

8. Hypothesized causes of cancer include environmental carcinogens, radiation, genetic, and immunologic factors.

9. Chemotherapy, radiation, surgery, and stimulation of immune response are types of treatments used on cancers.

10. Several types of cancers are described regarding possible contributing factors in their development, symptoms, and treatments. Included are lung cancer, breast cancer, cervical cancer, leukemia, and cancer of the large intestine.

11. Basic human development may be regarded as occurring "by the week."

    a. The first week is concerned with fertilization, cleavage, and formation of a blastocyst.

    b. In the second week, implantation and formation of a two layered embryo occurs.

    c. Mesoderm formation occurs during the third week, and cells differentiate into lines that will form specific tissues and organs.

    d. The nervous system and somites begin to form by the fourth week.

    e. During the fifth to eighth weeks, body systems are formed and human form is assumed (fetus).

    f. From the ninth week to birth is a time of growth and system refinement. Body proportions and size changes occur.

Questions

1. In what ways are the processes of mitosis and meiosis similar? Different?

2. What are the characteristics of the daughter cells formed by mitosis and meiosis? Where in the body does each process occur?

3. What is a cancer?

4. What are the differences between benign and malignant neoplasms?

5. What are some possible causes of cancer? Does this give you any clues as to why it is taking so long to find a "cure" for cancer?

6. What kinds of treatment are employed in cancer therapy? What dangers to normal cells does each of these treatments pose?

7. What are the major processes that take place in the first four weeks of life *in utero?* Comment briefly about the significance of each process to later development of body systems.

8. How can we account for multiple effects of an agent if it acts on the embryo at a given time of its development?

Readings

Arey, L. B. *Developmental Anatomy,* 7th ed. Saunders. Philadelphia, 1974.

Bryant, Susan V., and Vernon French. "Biological Regeneration and Pattern Formation." *Sci. Amer. 237:66,* July 1977.

Croce, Carlo M., and Hilary Koprowski. "The Genetics of Human Cancer." *Sci. Amer. 238:117,* Feb. 1978.

Jelliffe, Derrick B., and E. F. Patrice Jelliffe. "Current Concepts in Nutrition: Breast is Best; Modern Meanings." *New Eng. J. Med. 297:912,* Oct. 27, 1977.

Kolata, Gina Bari. "Primate Neurobiology: Neurosurgery with Fetuses." *Science 199:960,* 3 Mar. 1978.

Moore, Keith L. *The Developing Human. Clinically Oriented Embryology.* Saunders. Philadelphia. 1973.

Oettgen, Herbert F. "Immunotherapy of Cancer." *New Eng. J. Med. 297:484,* Sept. 1, 1977.

Science News. "Cancer: Clues in the Mind." *113:44,* Jan. 21, 1978.

Science News. "Skin Test Announced for Breast Cancer." *113:180,* March 25, 1978.

Chapter 5

Control of Cellular Activity; Energy Sources for Cellular Activity

Objectives

After studying this chapter, the reader should be able to:

■ List the nitrogenous bases and sugars characteristic of DNA and RNA.

■ Outline what is meant by the genetic code.

■ Define replication, transcription, and translation, and relate these processes to protein synthesis and control of cellular activity.

■ Define what mutations are, in terms of the genetic code, and give examples of the types of errors that can occur.

■ Show how hormones may influence cellular activity.

■ Relate the importance of ATP synthesis to continuance of cellular activity.

■ Describe, in general terms, what each of the following cycles does to the basic foodstuff molecules, and the importance of each to body function:
   a. Glycolysis
   b. The Krebs cycle
   c. Beta-oxidation

d. Biological oxidation

e. The metabolic mill

Growth and development, differentiation of tissues, and metabolism in general, depend on the production of specific chemical compounds to control these processes. The basis of this control appears to be genetic, in that the DNA molecules of the nucleus produce RNA molecules that pass to the cytoplasm and govern the production of proteins. These proteins, as enzymes and hormones, determine the rates and directions of chemical reactions in the body.

Energy for producing such chemicals, and for sustaining cellular activity in general, comes from the metabolism of the three basic foodstuffs (carbohydrates, lipids, proteins), and the synthesis and use of adenosine triphosphate (ATP).

This chapter presents an introduction to control of body operations and the sources of energy utilized to run body operations.

Genetic control by nucleic acids

Both of the nucleic acids (DNA and RNA) contain units known as NUCLEOTIDES (Fig. 5.1). Each nucleotide is composed of a nitrogenous base, a sugar, and phosphoric acid. In DNA, four bases are commonly found: adenine (A); guanine (G); cytosine (C); and thymine (T). The sugar is a five-carbon unit called deoxyribose ($C_5H_{10}O_4$). In RNA, uracil (U) substitutes for thymine, and the sugar ribose ($C_5H_{10}O_5$) is present. The nucleotides are bound together by sugar-phosphate bonds into nucleic acid molecules.

In 1953, Watson and Crick proposed that DNA

consisted of two nucleotide chains, coiled as a double helix, and bound to one another by base to base linkages (Fig. 5.2). Only certain bases are capable of linking to one another (A-T and C-G), in what is called *base pairing*. Thus, specific configurations or sequences of nucleotides are created. A particular sequence of nucleotides may be a gene, the unit controlling the expression of a body characteristic. Watson and Crick further proposed that the four bases of DNA carried information that was coded according to the sequence of bases in the molecule. A sequence of three bases, or a TRIPLET, was suggested to form a code word corresponding to a particular amino acid. The position of that amino acid in a protein molecule is then determined by the position of the triplet in the nucleic acid.

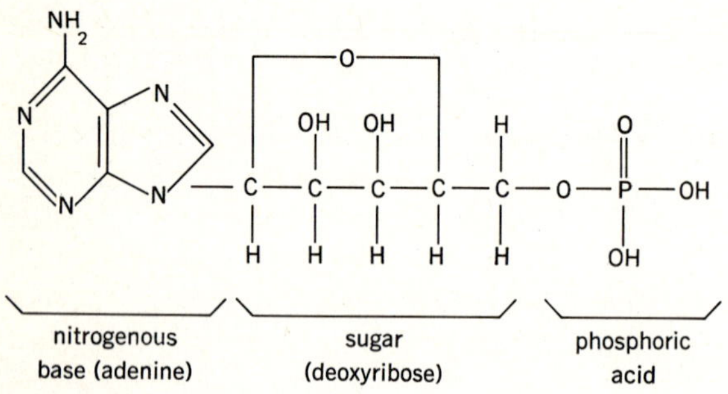

DNA remains within the nucleus, REPLICATING itself, and directs the production of proteins through the synthesis of RNA molecules. In the process known as TRANSCRIPTION, the two joined bases of the double DNA helix are believed to attract a particular nucleotide, free in the nucleus, to a given point along the DNA molecule. Many nucleotides are thus "lined up" in a specific order determined by base pairing between DNA and

FIGURE 5.1   A diagram of the structure of a nucleotide.

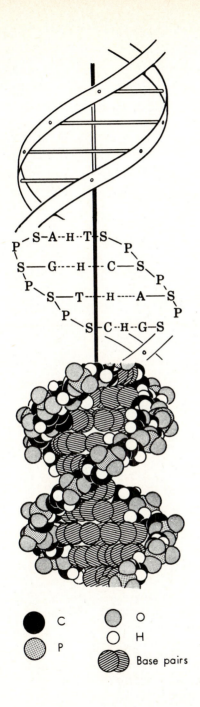

the nucleotide. These nucleotides are then joined to form a molecule of what is termed messenger-RNA (m-RNA). There must be as many m-RNA molecules as there are proteins in the body, since these molecules govern the production of proteins. With three bases available, 64 code words may be created. These 64 code words can form a nearly infinite number of combinations (enough to code for the 100,000± proteins of the body). Messenger-RNA, formed in the nucleus, moves to the cytoplasm through nuclear pores and perhaps the endoplasmic reticulum, and becomes associated with the ribosomes. Another type of RNA, named transfer-RNA (t-RNA), is produced in similar fashion by DNA. T-RNA is a clover-leaf-shaped single helix. It is collected on the nucleolus and passes into the cytoplasm through the nuclear pores. T-RNA attaches to amino acids (the particular one determined by the three bases in the end of the t-RNA molecule) and carries them to the ribosomes for synthesis into proteins. There are as many t-RNA molecules as there are amino acids to be carried, about 20.

Arriving at an m-RNA molecule, the t-RNA molecule, with its attached amino acid, is attracted to a particular point on the m-RNA according to base-paring properties (only A-U and C-G pairings can occur in RNA). Thus the amino acid is "inserted" in proper order for that particular protein. This process is called TRANSLATION (decoding of instructions). Joining of the amino acids forms the protein, which leaves the ribosome, allowing more molecules of protein to be synthesized. T-RNA molecules are also released and pick up more amino acids. These events are presented in Figure 5.3.

MUTATIONS, or changes in the genes and therefore in the proteins they produce, may be interpreted, in the light of this "genetic code," as misspelled words. An abnormal or misplaced triplet will obviously code for an amino acid that does not belong in that region, or for a different amino acid, and an abnormal protein will be produced. Thus, the reaction that the protein is concerned with will not occur, or will occur abnormally. J. C. Kendrew illustrates the manner

FIGURE 5.2 A three-way representation of DNA, showing the double helix, base pairing held by hydrogen bonds, and the arrangement of the atoms in the helices.

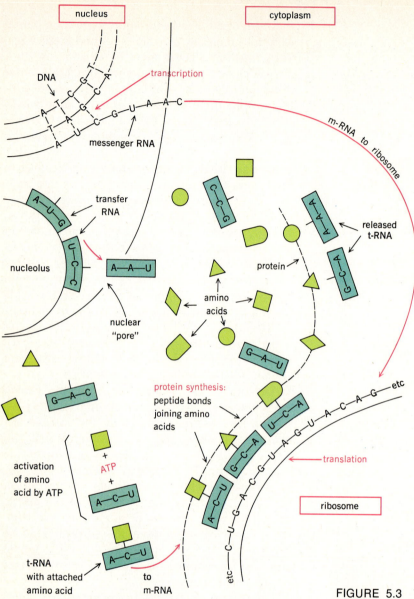

FIGURE 5.3 Transcription, translation, and protein synthesis in a cell.

of production of mutations by the following passages:

A mutation, interpreted in the light of the genetic code, may be likened to a misprint in a line of newspaper type (with a letter representing an amino acid):

"Say it with *glowers*." In this example, there has been a *substitution* of an incorrect for a correct letter.

"The prime minister spent the weekend in the country, shooting *p_easants*." In this case a letter is missing; a *deletion* has occurred.

"The treasury controls the public mon*keys*."

Here, an additional letter is present in what is termed an *insertion*.

"We put our trust in the Un*tied* Nations." Two letters have been reversed in what is termed an *inversion*.

"Each gene consists of *xgmdonrsd ad zbqpt ytrs*." Suddenly, a sequence becomes gibberish, forming *nonsense* words.

It is easy to suspect the havoc that could be wrought in the code by such errors which result in abnormal proteins being produced. Most mutations usually affect only one representative of the pair of genes controlling expression of a given characteristic. The normal gene exerts its effect more strongly (dominance), and the mutation is not expressed. The defect is carried, however, and may express itself at some future time.

Control by hormones

A wide variety of chemical substances are known to exert effects on cellular processes and on the composition of the cellular environment. Among the chemical substances most directly involved in control of cellular activity are the HORMONES, the products of activity of the endocrine glands. Four general theories have been proposed as explanations of how hormones exert their control:

Hormones influence membranes.
Hormones directly influence enzyme activity.
Hormones influence gene activity.
Hormones control the production and/or release or ions or other small molecules that may then exert an influence on cellular activity.

Hormone effects on membranes

Membrane systems surround the cell and many of its organelles. Permeability of these membranes may be altered by the presence of hormones, as described in the following examples.

Insulin, a product of the pancreatic endocrine tissue, is postulated to increase cellular uptake of glucose through increasing plasma membrane permeability to glucose. Parathyroid hormone has been shown to alter mitochondrial membrane permeability to magnesium and potassium, and thus stimulates mitochondrial activity. This may lead to alterations in the rate of oxidative phosphorylation, the activity of the Krebs cycle, and shifts in ion concentrations on either side of the membrane.

Hormone effects on enzymes

Though not definitely proved to have direct effects on enzymes *in vivo* (in the living organism), hormones have been shown to alter the activities of many enzymes *in vitro* (in the test tube). For example, estrogens (female sex hormones) may be shown to influence certain enzymes of the Krebs cycle.

Hormone effects on genes

Many hormones appear to act as gene activators to alter rates of DNA and RNA synthesis or to act as inducers or repressors of gene activity. In the sequence of events occurring in the genetic control of cellular activity, there appear to be four points at which hormones could influence the gene:

By combining with DNA molecules or by controlling gene activators or inactivators within the nucleus.

By combining with the products of the regulator genes and thus acting as inducers or repressors of genes.

By altering the activity of an enzyme required for synthesis of an inducer or repressor.

By altering membrane transport of a particular substance necessary for synthesis of an inducer or repressor.

Examples of hormones acting on or within the nucleus include: the effect of *cortisone* (a hormone of the adrenal cortex) on stimulation of formation of glucose from amino acids (gluconeogenesis) by enhancing the rate of synthesis of enzymes required in the conversion; and the effect of *aldosterone*, an adrenal cortical hormone, on sodium and potassium transport through cell membranes by gene activation and induction of formation of enzymes necessary to the transport process.

While exact mechanisms may not be available to specify the effects of hormones on genes, many (not all) hormones directly or indirectly influence gene activity; resultant changes in enzyme and protein concentrations may be the basis for the physiological effects demonstrated on hormone administration or deficiency.

Hormone effects on small molecules and ions

The effects on a cell, or cell activity, of a hormone may not be through direct action of the hormone

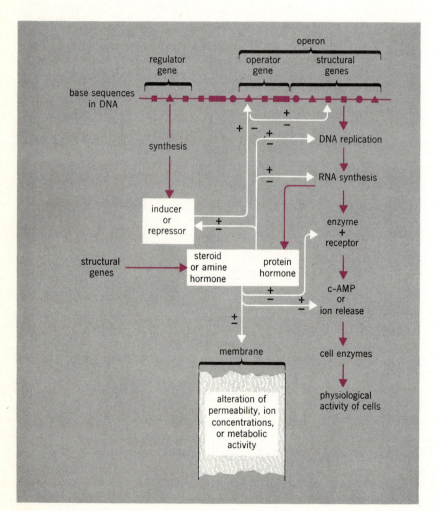

FIGURE 5.4 Genetic and hormonal controls on cellular activity ($+$ = stimulation; $-$ = inhibition).

on a particular gene or enzyme, but on the production or destruction of a molecule that may act as an activator or inhibitor of the gene or enzyme. CYCLIC ADENOSINE MONOPHOSPHATE (*c-AMP*) is produced by the action of several hormones on the enzyme adenylcyclase. This enzyme causes the production of c-AMP from ATP (see Chapter 26). C-AMP results in the activation of enzymes. For example, epinephrine (from the adrenal medulla) results in increased c-AMP production, which activates the enzyme phosphorylase, causing glycogen breakdown to glucose. Similarly, estrogens cause histamine release and influence activity of the uterus.

Ions — particularly calcium and sodium — may be released by hormones from areas of binding or storage. Oxytocin (a posterior pituitary hormone) stimulates calcium ion release, which then causes uterine contraction; aldosterone releases sodium ion that is required for active transport of amino acids into cell nuclei.

The combined action of genetic and chemical factors would thus seem to be necessary for control of cellular reactions. The interrelationships of these various control mechanisms are presented in Figure 5.4.

Energy sources for cellular activity

Energy is necessary to support the many activities the body cells carry out. Energy is supplied by the oxidation of basic foodstuffs such as glucose and fats. Amino acids are usually not utilized to any great extent for energy, since they form the proteins that are the structural and controlling substances of the body. Carbohydrates and fats are degraded in stepwise fashion by metabolic cycles, and the energy released is used to synthesize ATP molecules. ATP (adenosine triphosphate) molecules contain what are called "high energy phosphate bonds." Actually, the last phosphate group in the molecule is transferred to another molecule to raise its energy level and to enable it to continue or start its metabolism. ATP thus acts as a means of transferring energy to a particular physiological process such as muscle contraction, or production of secretions by a cell.

Among the more important metabolic cycles that provide energy for the synthesis of ATP are glycolysis, the Krebs cycle, beta-oxidation, and oxidative phosphorylation.

Glycolysis (Fig. 5.5)

Glycolysis is a scheme by which glucose is broken down in stepwise fashion to two molecules of pyruvic acid, without the presence of oxygen. In the course of the reactions, a total of about 56,000 calories of energy are produced. About 20,000 calories are used to synthesize two new ATP molecules, with the remainder released as heat to maintain body temperature. The cycle also releases four hydrogen atoms. Glycolysis normally catabolizes about three-fourths of the glucose used by the body cells.

The Krebs cycle (Fig. 5.6)

This cycle, after the removal of carbon dioxide from pyruvic acid, converts the resulting acetic acid to carbon dioxide in the presence of oxygen. One new ATP, 2 CO_2, and 8 hydrogens are produced for each molecule of acetic acid that goes through the cycle. A large mass of heat (600,000 + calories/mole or pyruvic acid) is also produced.

Beta oxidation (Fig. 5.7)

Beta oxidation degrades fatty acids to two carbon units of acetic acid that may then enter the Krebs cycle. Hydrogens are also released by this cycle.

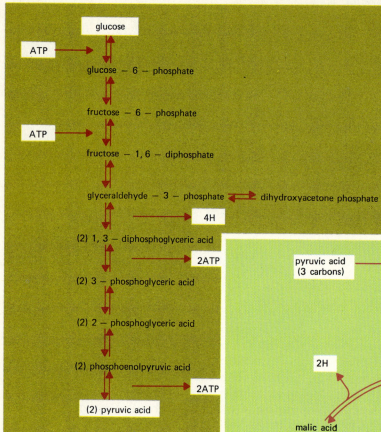

FIGURE 5.5 Glycolysis.

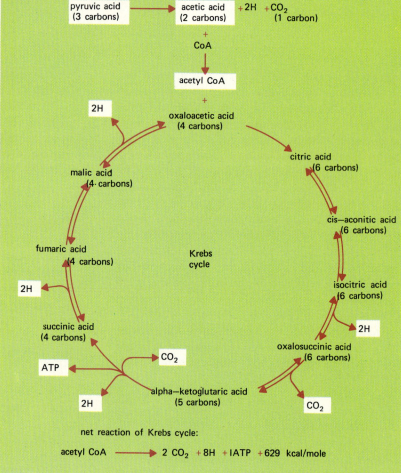

FIGURE 5.6 The conversion of pyruvic acid to acetyl CoA and the combustion of acetyl CoA in the Krebs cycle.

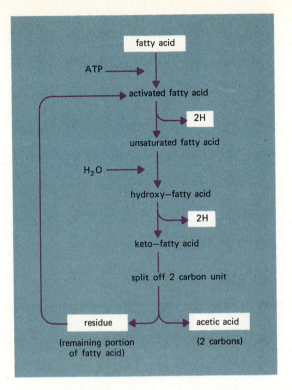

net reaction per split:

fatty acid ⟶ acetic acid + 4H + residue
to other cycles (e.g. Krebs)

FIGURE 5.7 Beta-oxidation, the oxidation of fatty acids.

Oxidative phosphorylation (Fig 5.8)

Hydrogens released from other cycles are carried to this cycle on large molecules known as hydrogen acceptors. Utilizing compounds known as cytochromes, this scheme converts the hydrogens, in the presence of oxygen, to water. It liberates sufficient energy to synthesize either two or three ATP molecules per pair of hydrogens passing through the scheme.

Found within cells are other metabolic schemes for synthesis of glycogen and recovery of glucose from glycogen, and for production of urea, amino acids, and other compounds. Their roles in the body economy are presented in later chapters.

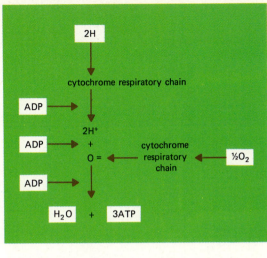

net reaction:

$2H + \frac{1}{2}O_2 + 3ADP \longrightarrow H_2O + 3ATP$

FIGURE 5.8 Oxidative phosphorylation, the biological oxidation of hydrogen.

The metabolic mill

The schemes mentioned above interlock in what is known as the "metabolic mill" (Fig. 5.9). The mill emphasizes that the metabolism of the three basic foodstuffs is interrelated, and that it is possible to make one substance from another, that is,

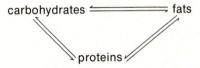

Since ATP may be produced by utilizing the energy from the breakdown of any of the three basic foodstuffs, the cells need never suffer a shortage of ATP.

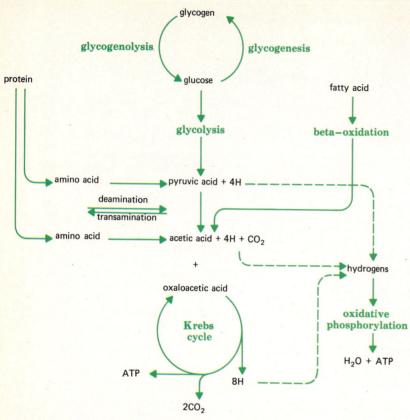

FIGURE 5.9   The "metabolic mill," the interrelationships in the metabolism of protein, fatty acids, and
glucose.

Summary

1. Nucleic acids govern protein synthesis and thus the proteins, as enzymes
and hormones, control the growth, development, and chemical reactions that
occur within the body.

   a. DNA composes the chromosomes and genes and causes the synthesis of
   RNA molecules.

   b. Messenger RNA (m-RNA) is produced by the DNA and goes to the ribosomes to direct protein synthesis.

   c. Transfer-RNA (t-RNA) is produced by the DNA and enters the cytoplasm
   to attach to specific amino acids.

   d. t-RNA carries amino acids to m-RNA on the ribosomes, where they are
   joined to form proteins to govern cell activity.

2. Mutations represent changes in DNA or RNA molecules, which result in synthesis of faulty or abnormal proteins.

3. Hormones exert their control over cellular processes by affecting the cells' membrane systems, enzymes, the genes themselves, or the production of small molecules.

4. Energy sources for cellular activity include ATP, and the carbohydrates, fats, and amino acids that are combusted to give energy to synthesize ATP. Metabolic cycles are utilized to break down the foodstuffs.

   a. Glycolysis takes glucose to pyruvic acid, produces two new ATPs and releases four hydrogens and heat.

   b. The Krebs cycle combusts pyruvic acid to CO_2, and produces eight hydrogens and one ATP for each molecule of pyruvic acid.

   c. Beta-oxidation combusts fatty acids to four hydrogens and a two-carbon unit that may enter the Krebs cycle.

   d. Oxidative phosphorylation converts hydrogens to ATP and water, in the presence of oxygen.

   e. The metabolic mill shows the interrelationships of fat, protein, and carbohydrate metabolism, and the fact that ATP may be synthesized from any of the three basic foodstuffs.

Questions

1. What is the role of DNA in control of cellular activity?

2. What does m-RNA do? What does t-RNA do?

3. What is a mutation? How does it affect cellular activity?

4. What is transcription? Translation?

5. Explain the contribution of the Krebs cycle and oxidative phosphorylation to ATP synthesis.

6. What is the importance of ATP to the cell?

7. In what ways may hormones influence cellular processes?

Readings

Brown, Donald D. "The Isolation of Genes." *Sci. Amer. 229:20,* Aug. 1973.

Clark, Brian F. C., and Kjeld A. Marcker. "How Proteins Start." *Sci. Amer. 218:36,* Jan. 1968.

Cohen, Stanley N. "The Manipulation of Genes." *Sci. Amer. 233*:24, July 1975.

Darnell, J. E. et al. "Biogenesis of m-RNA: Genetic Regulation in Mammalian Cells." *Science 181*:1215, 1973.

Greengard, Paul. "Phosphorylated Proteins as Physiological Effectors." *Science 199*:146, 13 Jan. 1978.

Harper, Harold A. *Review of Physiological Chemistry,* 16th ed. Lange Medical Pubs. Los Altos, Cal., 1977.

Hinkle, Peter C., and Richard E. McCarty. "How Cells Make ATP." *Sci. Amer. 238*:104, March 1978.

Hurwitz, J., and J. J. Furth. "Messenger RNA." *Sci. Amer. 206*:41, Feb. 1962.

Kendrew, John C. *The Thread of Life.* Harvard Univ. Press. Cambridge, Mass., 1968.

Kornberg, A. "The Synthesis of DNA." *Sci. Amer. 219*:64, Oct. 1968.

Marx, Jean L. "Gene Structure: More Surprising Developments." *Science 199*:517, 3 Feb. 1978.

McKusick, Victor A. "The Mapping of Human Chromosomes." *Sci. Amer. 224*:104, April 1971.

Merrifield, R. B. "The Automatic Synthesis of Proteins." *Sci. Amer. 219*:56, March 1968.

Miller, O. L. "The Visualization of Genes in Action." *Sci. Amer. 228*:34, March 1973.

Mills, D. R. et al. "Complete Nucleotide Sequences of a Replicating RNA Molecule." *Science 180*:916, 1973.

Mirsky, Alfred E. "The Discovery of DNA." *Sci. Amer. 218*:78, June 1968.

Ptashne, Mark, and Walter Gilbert. "Genetic Repressors." *Sci. Amer. 222*:36, June 1970.

Tenin, Howard M. "RNA Directed DNA Synthesis." *Sci. Amer. 226*:24, Jan. 1972.

Thompson, J. S., and M. W. Thompson. *Genetics in Medicine.* Saunders. Philadelphia, 1973.

Zubay, G. A Theory on the Mechanism of Messenger-RNA Synthesis. Proceedings, Natl. Acad. Sci. *48*:456, 1962.

Chapter 6

Tissues of the Body; the Skin

Objectives

After studying this chapter, the reader should be able to:

- List the four basic tissue groups composing the body.

- Give the general characteristics of epithelial tissues, those that apply to all members of the group, and describe how epithelia are named in terms of epithelial cell shape and numbers of layers of cells.

- List the types of epithelia found in the body, give their locations, and list some functions of each type.

- Define what is meant by "surface modification" as applied to epithelial cells, list some modifications, and give their functions.

- Define the two types of epithelial membranes and give their location and functions.

- Tell what is meant by simple, compound, exocrine, and endocrine glands, and describe how glands may produce their secretions.

- Give the general characteristics of connective tissues, especially when compared to epithelia, and list, describe, and give functions for the most common connective tissues including: loose connective tissue, adipose tissue, reticular tissue, cartilage, and bone.

- Comment on some factors involved in the formation and maintenance of connective tissues, and describe some disorders of connective tissues.

- List the three types of muscular tissue and where each is found in the body.

- Give the properties of nervous tissue.

- Describe the layers of the skin and their subdivisions, and describe hair and nails regarding structure and function.

- List the functions of the skin and how each is carried out.

- Describe what determines hair and skin color, and the importance of color to the body.

- Describe some disorders associated with the skin, including how burns are designated in extent and depth.

- Point out some considerations in burn treatment.

Chapter 1 has indicated that similar cells form tissues, and that four basic groups of tissues compose the body. The groups are: epithelial, connective (including blood and blood-forming tissues), muscular, and nervous tissues.

Epithelial tissues

Epithelial tissues are the cellular layers that cover and line the outer body surfaces, the interior of hollow organs, and the body cavities. They also act as absorptive and secretory tissues, and generally contain nervous structures, called receptors, for receiving stimuli. Many glands, which secrete a variety of products, are derivatives of epithelial tissues.

Characteristics of epithelia

To form an effective covering, epithelial cells must FIT VERY CLOSELY TOGETHER, or our bodies would be easily invaded from the outside or suffer loss of vital substances from the inside. Epithelia usually attach to underlying connective tissues by a BASEMENT MEMBRANE. Epithelia DO NOT CONTAIN

BLOOD VESSELS, but are nourished by vessels located in the underlying connective tissue. The farther the epithelial cells are from the blood vessels, the less will be their chances of survival. Epithelia may HAVE NERVES that pass between their cells from the underlying connective tissue, and typically show GREAT POWERS OF REGENERATION. For example, it has been estimated that the skin epidermis (outer epithelial layer) is replaced every three weeks, and that intestinal epithelial cells are replaced every two to three days. Epithelia may arise from all three germ layers.

Types of epithelia

If an epithelium contains only one layer of cells, it is called a SIMPLE epithelium. Simple epithelia are well nourished because all cells lie as close as possible to the vessels in the underlying connective tissue. Because of their thinness, simple epithelia are generally the ones most involved in active and passive transfer of substances. If the epithelium contains two or more layers of cells, it is called a STRATIFIED epithelium. The upper layers of cells in a stratified epithelium are far removed from the blood vessels in the underlying connective tissue, and usually show changes characteristic of cell degeneration and death. In many cases, the upper layers of cells in a stratified epithelium are toughened or modified in some way to make the epithelium a good protective layer.

The shape of the cells in an epithelium may be SQUAMOUS or flat, with the cell resembling a piece of tile. Shape may be CUBOIDAL, like a cube of sugar, or it may be COLUMNAR, with the cell shaped like a short pencil. If the epithelium is stratified, it is named according to the shape of the top layer of cells.

There are several types of epithelia that result if we combine the terms describing the layers and shapes given above. The common types are shown in Figure 6.1, and all are described below.

SIMPLE SQUAMOUS EPITHELIUM is one layer of flattened cells. It lines the thoracic and abdominopelvic cavities and their subdivisions (*meso-*

thelium), blood and lymph vessels (*endothelium*), the filtering membrane of the kidney (*glomerulus*), and the air sacs (*alveoli*) of the lungs. It is a tissue adapted, by its thinness, to allow passages of solutes and water across it by physical processes. It is not adapted to resist wear and tear.

SIMPLE CUBOIDAL EPITHELIUM is one layer of cube-shaped cells. It lines many of the kidney tubules, where it actively absorbs and secretes many materials. It covers the surface of the ovary, and lines the follicles of the thyroid gland (an endocrine gland producing hormones).

SIMPLE COLUMNAR EPITHELIUM is one layer of elongated cells. It lines the digestive tube from stomach to anal canal, and forms a secretory and absorptive membrane. It also lines the small tubes of the respiratory tree.

STRATIFIED SQUAMOUS EPITHELIUM consists of several layers of cells, with the top layer flattened. It covers those body surfaces or organs subjected to mechanical wear and tear. Thus the mouth, throat, esophagus, skin, and anal canal have this type of epithelium. The surface cells are easily removed (desquamated) and are rapidly replaced from below, affording protection against wear and tear. The cells may contain a protein, *keratin*; if present, keratin toughens and waterproofs the epithelium, as in the skin. If keratin is present, the epithelium is said to be keratinized or cornified; if absent, the epithelium is said to be noncornified or "soft." This epithelium may become very thick (e.g., calluses) if subjected to great mechanical wear and tear. However, even a newborn has "thick skin" on the palms of its hands and soles of its feet. Such skin is thick because of genetic influences and not "wear and tear."

Stratified cuboidal and columnar epithelia are of rare occurrence (*see* Table 6.1).

It is easy to name the epithelia described above on the basis of number of cell layers and shape of cells. Five other varieties of epithelia are found in the body and they are named by characteristics other than those of cell layers and shape.

PSEUDOSTRATIFIED EPITHELIUM appears stratified, as judged by the many nuclei at different levels. However, all cells touch the basement

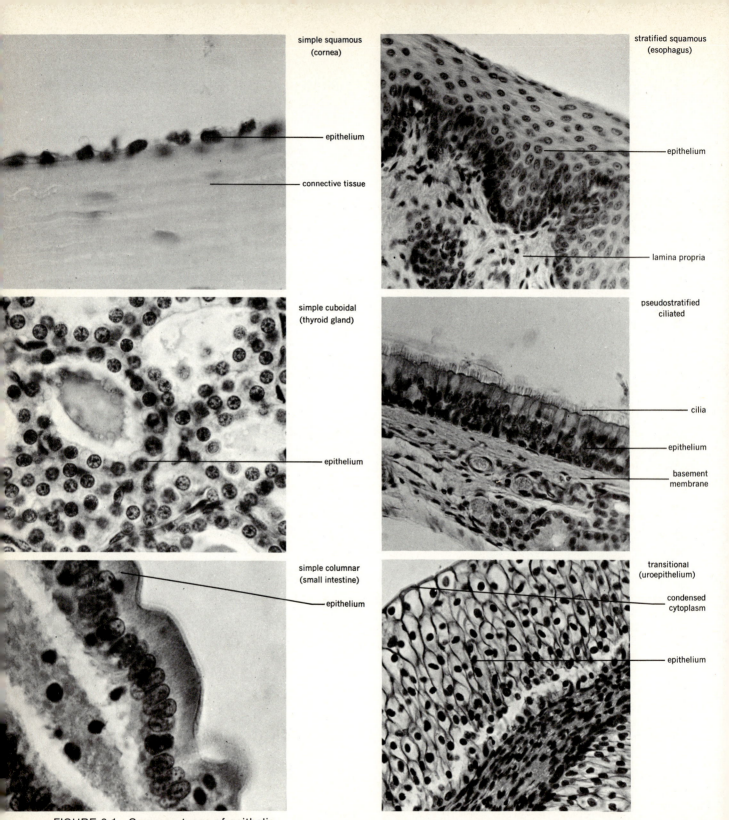

simple squamous
(cornea)

— epithelium

— connective tissue

stratified squamous
(esophagus)

— epithelium

— lamina propria

simple cuboidal
(thyroid gland)

— epithelium

pseudostratified
ciliated

— cilia

— epithelium

— basement
membrane

simple columnar
(small intestine)

— epithelium

transitional
(uroepithelium)

— condensed
cytoplasm

— epithelium

FIGURE 6.1 Common types of epithelia.

TABLE 6.1 Summary of epithelium				
Epithelial type	Characteristics	Examples of locations	Surface modification commonly present	Functions and comments
Simple squamous	One row flat cells, nuclei flattened and parallel to surface	Glomerular capsule, endothelium, mesothelium, Henle's loop of kidney	Microvilli (on mesothelium)	Exchange, because of thinness
Simple cuboidal	One row isodiametric cells, nuclei rounded	Kidney tubules, thyroid, surface of ovary	Microvilli (in kidney)	Secretion and absorption
Simple columnar	One row tall cells nuclei elongated perpendicular to surface	Stomach, small and large intestines, gall bladder, ducts, bronchioles, uterus	Microvilli (in gut) Cilia (in respiratory system)	Secretion and absorption (gut) movement of substances across surface if ciliated
Stratified squamous	Several layers of cells, top layer flattened	Skin (fully cornified), vagina (partially cornified), mouth, esophagus (slight or no cornification)	None	Protection, since cells are easily removed from the surface and replaced rapidly from below
Stratified cuboidal	Several layers of cells, top layer cuboidal	Sweat glands	None	Secretory
Stratified columnar	Several layers of cells, top layer columnar	Larynx, upper pharynx	Cilia	Transition form between stratified squamous and pseudostratified
Pseudostratified	All cells reach basement membrane, not all reach surface. Nuclei at many levels	Nasal cavity, trachea, bronchi, male and female reproductive systems	Cilia	Movement of materials across surface
Transitional	Stratified, top layer not uniform in shape	Kidney pelvis, ureter, bladder	Condensed cytoplasm	Stretches and protects cells from acidic urine
Syncytial	Simple, no membranes between cells	Placental villus	Microvilli	Secretion and protects
Germinal	Several layers showing stages of sperm formation	Tubules of testis	None	Produces sperm
Neuro-epithelium	Nerve cells form part of epithelium	Taste buds, olfactory area, retina, cochlea	None	Sensory, usually as receptors of stimuli

membrane, but not all reach the surface. This epithelium lines most of the respiratory system. It is a type that can change into stratified squamous; this change often occurs in heavy smokers and may give rise to cancers.

TRANSITIONAL EPITHELIUM (*uroepithelium*) is a stratified tissue lining the ureters and urinary bladder. The surface cells are not uniform in shape, and are capable of stretching as the organs expand to pass or contain fluids. The name transitional means "to pass from one state to another"; the cells change from rounded to flattened shape as the epithelium stretches.

SYNCYTIAL EPITHELIUM is a continuous, multinucleated mass with no cell membranes between cells. It covers the villi of the placenta where it is very active in absorption and secretion of materials.

GERMINAL EPITHELIUM lines the tubules of the testis and covers the ovary. In the testes, it is responsible for the production of sperm. In the ovary, it merely forms a covering for the organ.

NEUROEPITHELIUM consists of specialized epithelium in the nose, eye, ear, and tongue. It contains nerve cells specialized for reception of stimuli (environmental changes inside or outside the body).

Surface modifications of epithelia

The term surface modification refers to the presence in the epithelium of something other than the regular epithelial cells, or to a special structure on the surface of the cells. Epithelia may contain one-celled mucus secreting glands known as GOBLET CELLS. Mucus is a thick, moist, and sticky fluid that helps to moisten the epithelium and to protect it from enzyme action (as in the digestive tube), or to aid in cleansing the organ (as in the respiratory system where it "traps" dust and pollen). CILIA, motile (movable) hairlike structures, are found on epithelia where something is to be moved across its surface (e.g., respiratory system, where mucus is moved; uterine tubes,

where ova are moved). MICROVILLI (Fig. 6.2), are tiny fingerlike extensions of the cell surface, and are common on simple cuboidal and simple columnar epithelia (kidney, gut). The extra surface area created ensures rapid absorption and secretion of substances. The epithelium of the urinary system has a thickened layer of cytoplasm (condensed cytoplasm) that protects the cells against the acid urine (*see* Fig. 6.1).

The epithelia are summarized in Table 6.1.

Epithelial membranes

The combination of certain epithelia and their underlying connective tissues form an epithelial membrane. Two types of these membranes are found in the body.

MUCOUS MEMBRANES form linings of body cavities opening on the body surface (mouth, nose, anus), or of organs that form part of the tube opening on a surface (e.g., stomach, intestines, respiratory system). Epithelial cell type is usually pseudostratified, simple columnar, or stratified squamous, and the membrane is always moistened by mucus. The mucus is secreted by goblet cells, or by glands in the connective tissue.

SEROUS MEMBRANES cover the visceral organs and line the true body cavities (those that do not open directly onto a body surface, e.g., thoracic and abdominopelvic cavities). The epithelium is always simple squamous, and is moistened by a watery (serous) secretion formed from the blood vessels in the membrane.

Glands

As mentioned above, secretion is a primary function of epithelia. Glands are specialized cells in the epithelium, groups of epithelial cells that are buried in the connective tissue underlying the epithelium, or groups of cells that originated from an epithelium, but that have lost any obvious or visible relationship to the epithelium. Glands produce substances different from the fluids available to them, and thus utilize active processes to syn-

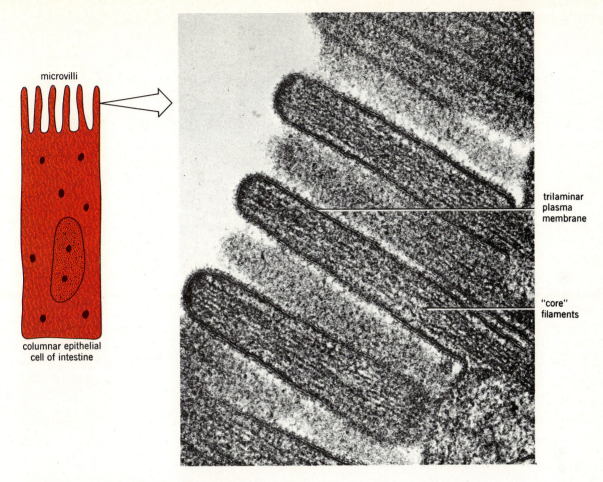

microvilli

columnar epithelial
cell of intestine

trilaminar
plasma
membrane

"core"
filaments

FIGURE 6.2   Microvilli, a surface modification to increase area for absorption and secretion.

thesize and release their products. If the gland
retains a connection, by a duct or ducts, with the
epithelial surface, it is termed an EXOCRINE GLAND.
If it has lost its epithelial connection and secretes
into the bloodstream, it is termed an ENDOCRINE
GLAND. Tables 6.2 and 6.3 summarize facts about
exocrine glands. The endocrine glands are described in Chapter 29.

TABLE 6.2 A summary of exocrine glands

Type of gland*	Diagram^a	Characteristics	Examples
UNICELLULAR			
		One celled, mucus secreting	Goblet cells of resp. and diges. system
MULTICELLULAR			
Simple tubular		One duct—secretory portion straight tube	Crypts of Lieberkuhn of intestines
Simple branched tubular		One duct—secretory portion branched tube	Gastric glands, uterine glands
Simple coiled tubular		One duct—secretory portion coiled	Sweat glands
Simple alveolar		One duct—secretory portion saclike	Sebaceous glands
Simple branched alveolar		One duct—secretory portion branched and saclike	Sebaceous glands
Compound tubular		System of ducts—secretory portion tubular	Testes, liver

TABLE 6.2 (continued)		
Compound alveolar	System of ducts— secretory portion alveolar	Pancreas, salivary glands, mammary
Compound tubulo— alveolar	System of ducts— secretory portion both tubular and alveolar	Salivary glands

^a Simple implies one duct in the gland; compound implies a series of branching ducts in the gland.

TABLE 6.3 Manner of production of secretion by exocrine glands	
Type of gland	Examples
Merocrine—synthesized product independent of basic cell structure	Pancreas, salivary glands
Apocrine—some portion of the cell issued as part of the secretion	Mammary gland
Holocrine—product of gland is a cell, or cell as a whole is shed in the secretion	Testis, sebaceous glands

Connective tissues

Connective tissues are the most abundant and widespread tissues of the body. They connect, support, and offer protection to body organs. They appear to be continuous throughout the body, and if they could be caused to disappear from the body in a wink of the eye, the body would collapse into a pile of many types of cells; all organization would be lost.

Characteristics of connective tissues

Connective tissues consist of LARGE AMOUNTS OF INTERCELLULAR MATERIAL with rather widely scattered cells. The intercellular material is often FIBROUS, and contains collagenous (tough) and elastic (stretchy) fibers. Connective tissues are not generally found on surfaces (except in joints),

are usually vascular, and have a cell that produces and maintains the intercellular material.

Types of connective tissue (Fig. 6.3)

Of the many types of connective tissue present in the body, those of most widespread occurrence, or of greatest importance, are described.

LOOSE (*areolar*) CONNECTIVE TISSUE consists of a three-dimensional network of collagenous (white) fibers and elastic fibers. Collagenous fibers are composed of a protein called collagen, are very strong, and resist stretching; elastic fibers are composed of a protein called elastin, are stretchable, but not very strong. The tissue thus possesses both strength and elasticity "in all directions." FIBROBLASTS produce the fibers of this tissue. Loose connective tissue is found beneath the skin

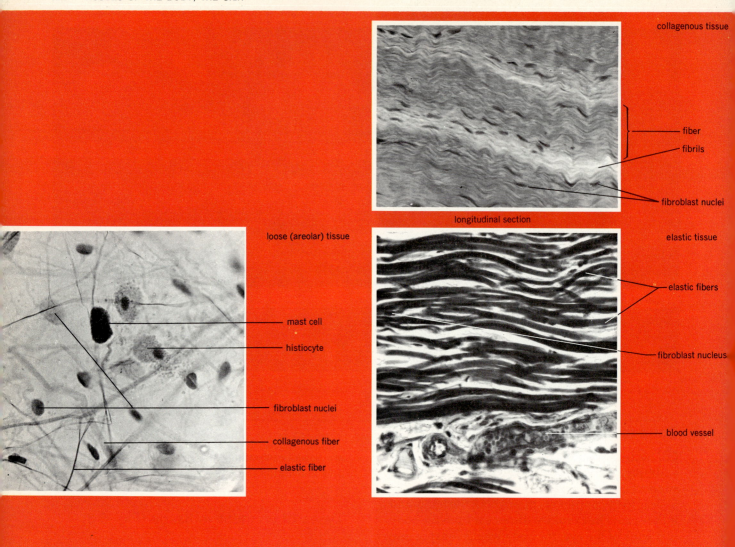

FIGURE 6.3   Common types of connective tissue.

as the superficial fascia or subcutaneous tissue, and occurs around many body organs as a "packing material."

ADIPOSE TISSUE is a connective tissue consisting of large numbers of adipose (fat) cells, which are fibroblasts specialized to store fat. It is found beneath the skin and around many body organs. It serves as an insulating material to help prevent loss of body heat, gives rounded form and shape to the body, stores food energy, and aids in protecting those organs it surrounds (e.g., kidney).

RETICULAR CONNECTIVE TISSUE forms the internal supporting framework of many body organs (e.g., liver, spleen). Its delicate fibers form a three-dimensional support for the cells of these organs.

"FIBROUS CONNECTIVE TISSUE" includes the dense *collagenous tissue* of tendons and ligaments, and the *elastic connective tissue* of the large arteries.

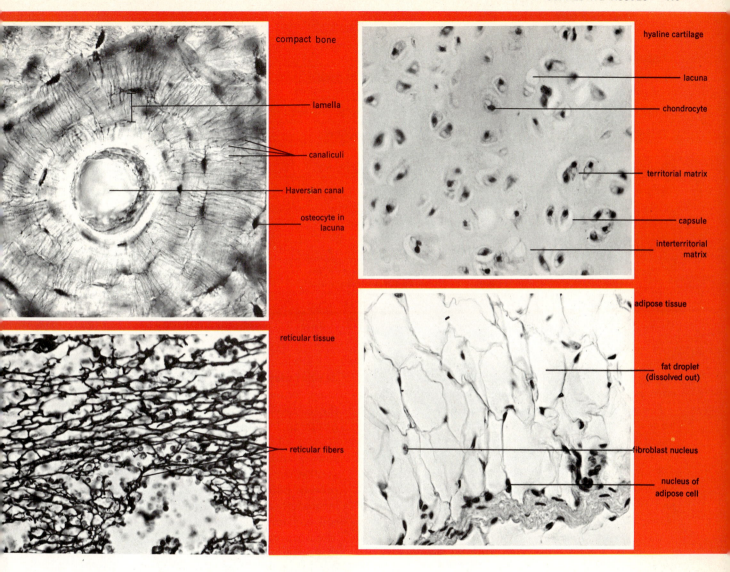

compact bone

lamella

canaliculi

Haversian canal

osteocyte in lacuna

reticular tissue

reticular fibers

hyaline cartilage

lacuna

chondrocyte

territorial matrix

capsule

interterritorial matrix

adipose tissue

fat droplet (dissolved out)

fibroblast nucleus

nucleus of adipose cell

These tissues are composed almost entirely of one type of fiber in a "one-way" pattern of arrangement.

CARTILAGE (*gristle*) is a tough, semisolid tissue that contains CHONDROCYTES that produce the intercellular material. The intercellular material is called the MATRIX, and contains collagenous or elastic fibers. Cartilages are surrounded by a membrane known as the PERICHONDRIUM. The mem-

brane contains cells that can turn into chondrocytes and form new cartilage if the cartilage is broken or damaged. Replacement or healing of cartilage is very slow, probably because of the small number of blood vessels in the tissue. There are three varieties of cartilage.

Hyaline cartilage is a translucent mass forming most of the embryonic skeleton, the cartilages con-

TABLE 6.4 A summary of the main connective tissues (c.t.)

Tissue type	Characteristics	Examples of locations
Loose c.t.	Feltwork of collagenous and elastic fibers	Around organs, subcutaneous tissue
Fibrous c.t. (collagenous)	Wavy bundle of fibers, fibrils present	Ligaments, tendons, sclera of eye
Elastic c.t.	Branching, homogeneous fibers	Aorta, ligamentum nuchae, ligamenta flava
Reticular c.t.	Fine, irregular, branching fibers	Interior of liver, spleen, lymph nodes
Adipose tissue	"signet ring" cells with fat drops inside	Anywhere (almost): subcutaneous tissue of skin
Hyaline cartilage	Clear intercellular material (ICM), cells in lacunae; capsule around lacunae	Embryonic skeleton, costal cartilages, nasal cartilages, articular surfaces
Fibrocartilage	Visible collagenous fibers in ICM; other features same as in hyaline	Intervertebral discs, symphyses
Elastic cartilage	Visible elastic fibers in ICM; other features same as in hyaline	Epiglottis, external ear

necting ribs to sternum (costal cartilages) and the covering of bone ends in freely movable joints.

Fibrous cartilage is very tough because of the presence of many collagenous fibers in the matrix. It is found as the intervertebral discs between the bones of the spine.

Elastic cartilage is a rubbery mass making up the pinna of the ear (the "flap" on the side of the head), and the epiglottis of the larynx. It contains many elastic fibers in the matrix.

BONE is a hard substance forming the major part of the skeleton. Its development, form, structure, and properties are described in Chapter 8. A summary of the main connective tissues appears in Table 6.4.

Formation and physiology of connective tissues

Since the primary component of a connective tissue is the intercellular material, production of that material becomes the primary concern in the development and maintenance of the tissue. In loose connective tissue, and other fibrous tissues, the fibroblast is the cell that produces both the fibers and semiliquid components (ground substance) of the intercellular material. The ribosomes and Golgi apparatus of the fibroblasts synthesize the basic molecules (collagen, elastin, protein polysaccharides) necessary for intercellular materials, and these are then secreted from the cell and formed into fibers or ground substance. The fibroblasts of these tissues are also stimulated to greater activity by foreign substances and by the products of cell injury, so that they become very important in healing of injuries. They also are stimulated by adrenal cortical hormones and play a role in the inflammatory process.

Cartilage relies on its chondrocytes to produce matrix and fibers. Again, the ribosomes and Golgi apparatus appear to be the sites of production of the necessary substances. When damaged, cartilage heals by forming new cartilage at the surface (*appositional growth*) through activity of cells in the perichondrium, or by formation within the mass (*interstitial growth*). Formation of new cartilage is greatly influenced by several vitamins and hormones.

Vitamins A and C are necessary for formation of fibers and ground substance.

Vitamin D is necessary for the absorption of calcium for formation of bone from cartilage.

Growth hormone (from the pituitary gland) and thyroxin (from the thyroid gland) cause stimulation of production of cartilage during body growth.

Clinical considerations

The fibers or matrix or both of connective tissue may be formed abnormally, leading to deformities or abnormalities of the organs in which the connective tissue is found. SCLERODERMA is a condition affecting the skin and blood vessels, in which large amounts of collagenous fibers are formed in the organs. The tissues become hardened and lose elasticity. ACHONDROPLASIA (*chondrodystrophy*) results when cartilage matrix is not formed properly, particularly during skeletal growth. Dwarfing and misshapen bones result. Such disorders appear to have a hereditary basis.

Muscular tissues (Fig. 6.4)

While the structure and functions of muscular tissues are considered in greater detail in subsequent chapters, it may be stated here that the muscular tissues are the CONTRACTILE tissues of the body. By shortening, they cause movement of the body as a whole, or of substances through the body. SKELETAL MUSCLE attaches to, and moves, the bones of the skeleton. CARDIAC MUSCLE is found only in the heart, and causes the circulation of the blood. SMOOTH or VISCERAL MUSCLE is found in internal body organs such as blood vessels, and the stomach and intestines. It is important in control of blood pressure, blood flow, and movement of materials through the alimentary tract.

Nervous tissue (Fig. 6.5)

Nervous tissue is the highly EXCITABLE and CONDUCTILE tissue of the body. It responds to changes (stimuli) in the body's internal and external environments and forms the organs for interpretation and integration of body response to such stimuli.

The skin

The skin and its derivatives, such as hair, nails, and several types of glands, form the integumentary system. The skin is a tough but pliable, durable, and resistant covering for the body. It has, in the adult, a surface area averaging 1.75 square meters (about 3000 square inches), composes about 7 percent of the body weight, and receives about one-third of the blood pumped by the left ventricle of the heart. It is the largest organ of the body!

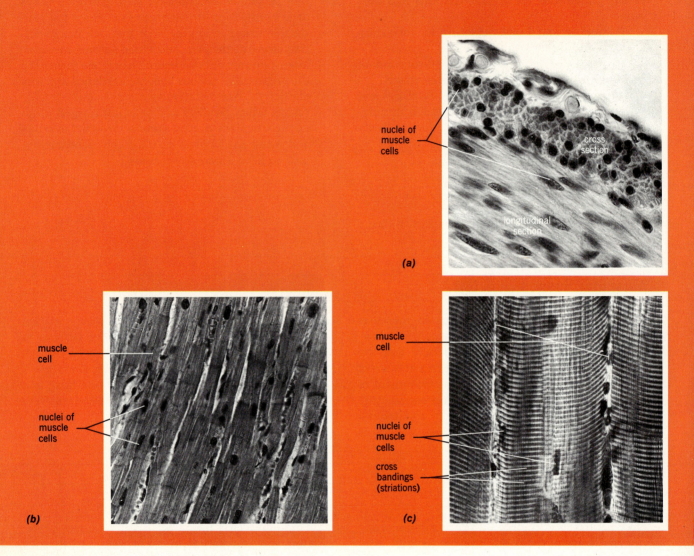

FIGURE 6.4   The three types of muscle. *(a)* Smooth muscle. *(b)* Cardiac muscle. *(c)* Skeletal muscle.

Structure of the skin (Fig. 6.6)

Skin has three major layers: the EPIDERMIS is the outer ectodermally derived epithelial layer; the DERMIS (*corium*) is a layer of connective tissue beneath the epidermis, and contains blood vessels, nerves, and lymphatics; the SUBCUTANEOUS LAYER (*hypodermis*) lies beneath the dermis and contains much adipose tissue. The dermis and subcutaneous layers are of mesodermal origin. The epidermis of the skin contains 30 to 50 layers of cells arranged in several sublayers. From the deepest layer outward, these layers are:

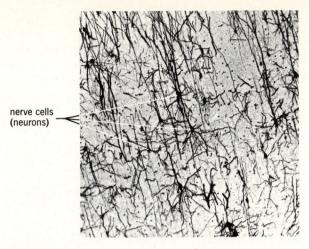

nerve cells
(neurons)

FIGURE 6.5   Nervous tissue, cerebrum.

1. *Statum basale, also known as the stratum cylindricum or stratum germinativum ("basal, cylindrical, or germinal layer").* The cells of this layer form a single layer of columnar-shaped cells that rest on the basement membrane. They undergo continual mitotic division to renew the epidermis. Cells push toward the surface from this layer and are shed (40 to 50 pounds in a lifetime) from the surface.

2. *Stratum spinosum ("prickle cell layer").* The cells of this layer show spinelike projections that attach the cells to one another.

3. *Stratum granulosum ("granular cell layer").* The cells in this layer show changes characteristic of death (lack of nuclei, loss of organelles), and contain dark staining granules. The process of cornification (keratinization) by which the cells become filled with

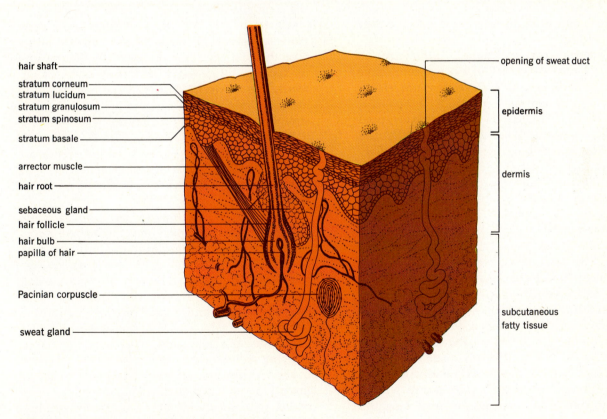

hair shaft
stratum corneum
stratum lucidum
stratum granulosum
stratum spinosum
stratum basale
arrector muscle
hair root
sebaceous gland
hair follicle
hair bulb
papilla of hair
Pacinian corpuscle
sweat gland

opening of sweat duct

epidermis

dermis

subcutaneous
fatty tissue

FIGURE 6.6   Diagram showing the structure of hairy skin.

keratin, is begun in this layer. Death of cells occurs because of the inability of nutrients to diffuse to the cells.

4. *Stratum lucidum ("clear layer")*. The cells of this layer show no nuclei or membranes, and the layer appears as a clear shiny line. The cells contain a chemical known as eleidin, which represents an intermediate stage in formation of keratin.

5. *Stratum corneum ("horny layer")*. This is the surface layer of the epidermis, and consists of many layers of flat, dead, scalelike cells filled with a protein called *keratin*. The keratin waterproofs and toughens the skin, and renders it an effective protective tissue.

The lower surface of the epidermis is folded and interlocks with folds of the underlying dermis. The folds are reflected at the surface as the fingerprints (friction ridges) that are specific for each individual and that aid in picking up or grasping objects.

The DERMIS averages about 3 millimeters in thickness and is composed of a dense connective tissue, containing both collagenous and elastic fibers, many blood vessels, glands, lymphatics, and sensory corpuscles. It is divided into an upper PAPILLARY (folded) LAYER and a lower RETICULAR LAYER. The folded layer contains many capillary networks that nourish the germinal layer and that allow radiation of heat through the surface. This layer also contains touch receptors (Meissner's corpuscles). The reticular layer contains many arteries and veins that may connect directly with one another to form an arteriovenous anastomosis. These anastomoses cause blood to bypass the capillary beds and conserve body heat. Receptors for pressure (Pacinian corpuscles) are found in this layer, as are the sweat and sebaceous glands. Loss of elastic fibers of the dermis occurs with aging, and accounts for the folding (wrinkling) of the skin in older individuals.

The SUBCUTANEOUS LAYER (hypodermis) is composed of loose connective tissue heavily loaded with FAT cells. The layer aids in insulating the body against loss of deep heat, and is loose enough to permit the injection of several milli-

liters of fluid without development of significant pain (subcutaneous injection). The fat in this layer is what gives the general form and shape to the body. It cushions and protects muscles and nerves, and is an area of food storage.

Appendages of the skin

HAIRS (*see* Fig. 6.6) develop as a downgrowth of epidermal cells into the dermis to form a HAIR FOLLICLE that is set at an angle into the skin. Mitosis of the cells in the base of the follicle produces the hair. A HAIR has an expanded *bulb* at its lower end, and the bulb is indented by a *dermal papilla* that supplies blood vessels and nutrients to ensure growth of the hair. The *root* is imbedded in the skin, and the *shaft* is the visible portion of the hair above the skin. *Muscles* (arrector pili) attach to the follicles and pull them into a more vertical position. In furry animals, these muscles erect the hair, thickening the layer, and increasing its capacity to act as an insulating layer. In humans, "goose flesh" occurs as the hair pushes a mound of skin to one side as it is pulled into a straight position. Hair protects the head from sunlight and the eyelashes and eyebrows aid in protecting the eyes from light. When the eyelids are partly closed the eyelashes filter and scatter the light rays entering the eye, reducing their intensity.

NAILS (Fig. 6.7) are appendages useful in grasping and picking up small objects. The nail represents a hardened corneal layer and has a *free edge, body,* and *root*. The nail rests on the *nailbed* (derived from the basal and spiny layers of the epidermis), and has a *cuticle* (eponychium) above the root of the nail, and a *quick* (hyponychium) beneath the free edge. A white half-moon or *lunula,* lies under the root, and represents the active, growing region of the nail.

The GLANDS OF THE SKIN (*see* Fig. 6.6) are of three types: SWEAT (sudoriferous) glands; SEBACEOUS (oil) glands secrete a substance (sebum) that aids in keeping skin and hairs lubricated, pliable, and waterproofed; in the ear canal,

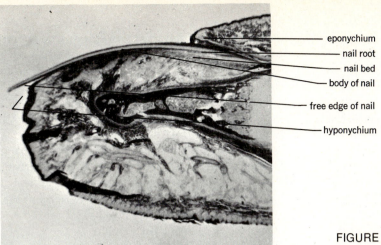

eponychium
nail root
nail bed
body of nail
free edge of nail
hyponychium

FIGURE 6.7 A longitudinal section of the fingertip showing the nail and its associated structures.

CERUMINOUS glands secrete a bitter, brownish, waxlike material that repels insects and lubricates the skin of that area.

SWEAT GLANDS are of two types. *Eccrine glands* are simple coiled tubular glands of widespread occurrence over the body. Their secretion is watery and aids in cooling the body as it evaporates. Eccrine sweat contains inorganic salts. *Apocrine* sweat *glands* are large glands found in the armpits and around the anus. They produce a fluid rich in organic substance, which may be acted on by bacteria to produce odors. Apocrine sweat glands are associated with hairs and begin to function at puberty.

Functions of the skin

The skin is a versatile and active organ. As a covering, it protects the body it encloses, and this forms its primary function. It also acts to aid in regulation of the body temperature, as an absorptive and excretory organ, and in the reception of stimuli that impinge on the body.

PROTECTION. The skin presents several lines of defense against the environment.

A SURFACE FILM composed of water, lipids, amino acids, and polypeptides is derived from the secretions of the sweat and sebaceous glands, and from the breakdown of the cornified surface cells. It has a pH of 4 to 6.8 and forms an effective antiseptic layer, retarding the growth of bacteria and fungi on the skin surface. It also reacts with many potentially toxic materials, preventing their entry through the skin surface. Its water content moistens the skin, and the lipids lubricate and aid in waterproofing the surface. The waterproofing function is not perfect as evidenced by the wrinkles that our skin develops (fingers) on long exposure to water. This is caused by water entering the upper layers of the skin and causing it to swell and fold.

The presence of the horny layer of the skin, with its keratinized cells, acts as a PHYSICAL BARRIER to the penetration of most chemicals, bacteria, parasites, and a good part of the environmental radiation (e.g., sunlight). At the junction between the cornified and noncornified layers of the skin is a region of positive and negatively charged ions that will repel charged substances trying to enter the body.

These devices, which prevent entry of substances into the body from outside, also prevent the loss of essential body constituents from the inside. The body does not "dry out," or lose essential proteins, salts, and other chemicals.

TEMPERATURE REGULATION. Heat production by metabolic processes must constantly be balanced by heat loss to ensure maintenance of temperature homeostasis. Skin contains capillary networks through which blood may be circulated close to the body surface. If the environmental temperature is less than body temperature, RADIATION OF HEAT will occur from the warmer to cooler area. The amount of blood reaching the capillary beds is controlled by the diameter of the arteries leading into the capillaries. The nervous system is responsible for controlling the size of these arteries. Skin also contains the arteriovenous junctions described previously that can bypass many capillary beds. Thus, if the arteries narrow, less blood passes to the upper skin layers, and heat is conserved; opening of these vessels allows greater heat loss.

A second means of temperature regulation is afforded by the SECRETION OF THE ECCRINE SWEAT GLANDS. The watery solution poured on the body surface requires heat to evaporate it. This causes heat loss from the body, and becomes an important method of heat loss when environmental temperature is close to, or above, body temperature.

ABSORPTION. It has been claimed for years that the skin "breathes." Indeed, oxygen, carbon dioxide, and nitrogen pass relatively easily through the skin. Any substance that can dissolve in the surface film of the skin will pass more easily through the skin. A substance that can dissolve the keratin (keratolytic) of the horny layer will also enter the body through the skin. Sex hormones (e.g., estrogens) are fat soluble, and are found in some creams that are applied to the body. Aspirin is a keratolytic, and in salves, may be applied to the skin surface for relief of "aches and pains."

EXCRETION. The excretory function of the skin depends on the skin glands. The eccrine sweat glands continually secrete SWEAT that does not come to our conscious attention (insensible perspiration). This excretion aids in maintaining the body's water balance. The secretion also contains salt (sodium chloride), which if excreted in large amounts (as during heavy work in a hot environment) can lead to muscle cramps. Measurable and significant amounts of METABOLIC WASTES (urea, ammonia) are also eliminated in the sweat. Apocrine sweat glands and sebaceous glands secrete a product rich in organic compounds that may be odoriferous. These compounds are readily attacked and decomposed by the skin bacteria to produce additional compounds that may contribute to "body odor." Suppression or removal of the odors requires removal of the compounds, usually by using soap and water.

RECEPTION OF STIMULI. Sensory receptors for heat, cold, touch, pressure, and pain are found in the skin. Their structure and function are described in Chapter 18. These receptors are essential to the recognition of changes in the environment so that adjustments can be made to maintain homeostasis.

Skin and hair color

Most of us have some color to our skin and hair. The color itself and its intensity are determined by several factors.

PIGMENTS in the skin and hair are produced in skin or hair cells by a variety of biochemical reactions that are genetically determined. *Melanin* is a yellow to black pigment found only in the lower layers of the skin of white people, but in all epidermal layers in the skin of black people. Amount of pigment and, therefore, darkness of skin color is not determined by a single gene, but by as many as eight genes. Thus, the range of skin color is very great. In albinos (L. *albus,* white) a gene has mutated, and melanin cannot be synthesized because of a missing enzyme. Such individuals have white skin (*no* pigment), white hair, and no pigment in the irises of their eyes. Their eyes are very sensitive to light because the lack of iris pigmentation allows much light to enter. *Carotene* is a yellow pigment found in the horny layer of the skin. It is more abundant in the skin of certain

Asian races. Mixtures of the two pigments may give yellows and reds of various intensities.

Pigments in the skin afford protection against solar radiation. Humans are believed to have originated in Africa, where sun and heat is great. Dark skin absorbs much heat, but radiates that heat more efficiently than lighter colored skin, and this gives protection against the effects of heat.

Hair color is determined by the same pigments that are found in the skin, and intensity of color in both skin and hair is correlated. As we age, pigment synthesis in the hair follicles diminishes, and many of the hair cells contain air; this causes graying or whitening of the hair.

The AMOUNT OF BLOOD circulating through the skin and the thickness of the skin determines its "pinkness." Thinner skin allows more color of the blood to show through. The amount of hemoglobin with oxygen on it also determines skin color. Richly oxygenated hemoglobin is bright red; hemoglobin with little oxygen on it is bluish in color. If sufficient "blue hemoglobin" is present in the capillary networks of the skin, a blue color (*cyanosis*) is imparted to the whole skin. *Jaundice* is a yellow color imparted to the skin by excessive red blood cell destruction, or liver malfunction. It results from excessive bilirubin (a pigment) in the bloodstream.

Exposure to ENVIRONMENTAL FACTORS, particularly radiation, can also influence skin color. The ultraviolet radiation of sunlight is damaging to the blood vessels and other tissues of the skin. One response to the radiation is *tanning,* in which the amount of melanin is temporarily increased in the epidermis. The pigment then absorbs more of the radiation. Sunlight also dehydrates the skin and makes it more "leathery," so that a tan may look nice but can be damaging to the health of the skin.

Clinical considerations

The skin is a very sensitive indicator of the presence of abnormal processes at the skin surface, and within the body itself. Rashes or eruptions of the skin may occur as a result of infection or other processes.

LESIONS. Skin eruptions ("breaking out" with a visible change in the skin surface) accompany many diseases. The changes that occur follow a similar series of events regardless of the agent causing them. A given condition may stop in any one of the stages, or progress through them all. The terms below are ones that health personnel use most generally, and are given in order of their appearance. A good example on which to observe the follow-through of each stage is the smallpox vaccination. The various stages through which skin lesions progress are:

Macule A discolored area on the skin, neither raised nor depressed.

Papule A red, elevated area on the skin.

Vesicular Stage

Vesicle A pinhead to split-pea sized elevation filled with fluid.

Pustule A vesicle containing pus.

Bulla A large vesicle or blister.

Crust Dry exudate (weeping of fluid and/or pus) adhering to the skin.

Lichinification Thickening and hardening of the skin.

Scar Replacement of cells by fibrous tissue as healing progresses.

Keloid A large, elevated scar.

COMMON DISORDERS OF THE SKIN. INFECTIONS are perhaps the most common disorders affecting the skin. *Bacteria* (mostly staphylococcus and streptococcus) are the most common infectious agents. Boils are a result of coccal infection. *Viruses* may produce warts or herpes (itching vesicles). *Fungi* produce cracks or fissuring as in athletes foot and scalp infections (ringworm). The fungi thrive in warm, moist environments that makes for an easy transfer from the tennis shoe to the locker room and showers or swim areas. *Insects* such as mites, lice, bees, and spiders may bite or sting, causing eruptions on the skin.

ACNE or "pimples" is a physiological disturbance of puberty and early adulthood and affects more than 80 percent of teenagers. It is caused by the effect of androgenic hormones (male sex hormones) on the sebaceous glands and hair follicles. It occurs in both sexes, because androgens are produced by the adrenal glands, testes, and ovaries. Acne occurs in four grades, depending on severity.

Grade I acne consists of the formation of a fatty-keratin plug, a blackhead, in the opening of a follicle on the skin.

Grade II acne results when the plugged duct ruptures and spills sebum into the surrounding tissues. The sebum irritates the tissue and a pustule forms. Bacteria may also be trapped in the tissue.

Grade III acne occurs when there is tissue destruction by the inflammation, and scarring results.

Grade IV acne results in lesions extending to the shoulders, arms, and trunk, with severe scarring occurring.

Treatment involves washing with a mild soap and warm water as needed, and avoiding foods that may aggravate the condition (e.g., chocolate, nuts, cola drinks). Sun may be helpful, and some of the newer medications are promising. Squeezing or "popping" the pustules should not be attempted since this may increase chances of scarring. Too frequent washing also may traumatize the skin. If the face is disfigured by acne, psychologically sensitive individuals may withdraw from social contact with their fellows, with resultant need for counseling. *Dermabrasion,* in which the skin is "ground down" to the level of the dermal upfolds, may be utilized to remove scars. This method of treatment is done only with the advice of a skin specialist, because not all types of acne scarring will be improved by the procedure.

ECZEMA, SEBORRHEA, and PSORIASIS are other disorders of the skin whose names are commonly read or heard in the mass media.

Eczema refers to a chronic irritation of the skin in which there is redness, eruptions, watery discharge, and formation of crusts and scabs. Its cause is not known, and it probably represents a symptom rather than being a disorder in itself.

Seborrhea is a disorder of the sebaceous glands in which there is excessive production of sebum that may accumulate as crusts and scales on the head, face, or trunk.

Psoriasis results in the formation of silvery scales on scalp, elbows, knees, or the body generally. The cause is unknown, and the disease can occur at any age in both sexes. Treatment is directed toward relief of symptoms, as there appears to be no certain cure.

BURNS. One of the first hazards humans recognized in their efforts to change the quality of their survival was burns. Today nearly one million people are burned severely enough to be hospitalized annually, and about 7000 of these people expire. Burns are one of the major categories included under accidents, the first leading cause of death between 1 to 44 years of age.

Exposure to flame, scalding, hot objects, or some chemicals destroys the skin and its protective functions are lost. Infection and loss of fluids, electrolytes, and blood proteins follow as a result of skin destruction. A burn is described in two ways: first, by the extent of destruction of surface area (the "rule of nines," Fig. 6.8, is a convenient method of estimating the extent of destruction); second, the depth of the burn is described by "degrees."

A FIRST DEGREE BURN involves only the surface layers of the epidermis. The skin is reddened and tender to the touch. A sunburn is a good example of a first degree burn.

A SECOND DEGREE BURN is one in which much of the epidermis is destroyed, but some epidermal remnants are present. It extends into the dermis. There is redness, and blisters are usually present. A burn resulting from briefly touching a hot ob-

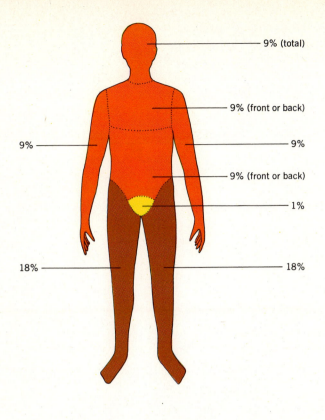

9% (total)

9% (front or back)

9%

9%

9% (front or back)

1%

18%

18%

FIGURE 6.8   The rule of nines, used to calculate the extent of burns.

ject is usually a second degree burn. This type of burn is usually very painful because of irritation of the nerves of the dermis by products of cell destruction.

A THIRD DEGREE BURN involves all skin layers, with no epidermal remnants present in the burned area. Charring of the skin is common, and destruction may involve muscles, tendons, and bones. The skin is insensitive to stimuli because of destruction of nerves in the dermis and hypodermis. On healing, fibrous masses of dead tissue (eschars) may form, which limit movement at joints. Third degree burns may result from prolonged contact with hot objects, from open flame, steam (car radiators), chemicals (strong acids or alkalies), and hot liquids (paraffin, oil).

There are three considerations of primary importance in the treatment of burns.

Prevent, as best as possible, secondary infection in the burned area.

Maintain water and electrolyte homeostasis by administration of appropriate solutions. Urine output is a convenient measure of adequacy of fluid intake.

Encourage adequate nutrition. A burned individual is one who is not interested in eating. Intravenous (IV) feeding may be required. Mobilizing the patient as early as his or her burns permit encourages recovery and feeding.

Summary

1. Cells that are similar form tissues. Four tissue groups compose the body: epithelial, connective, muscular, and nervous tissues.

2. Epithelia have special characteristics.
    a. They cover and line free surfaces of the body.
    b. They are composed of closely packed cells.
    c. They are active as secretory and absorptive tissue.
    d. They rest on a basement membrane and are nourished from blood vessels beneath the epithelium.
    e. They show great powers of regeneration.

3. Epithelia are either one layered (simple) or many layered (stratified), and contain cells that are flat (squamous), cuboidal (cube shaped), or columnar (tall).

   a. Simple squamous epithelium is one layer of flat cells. It is adapted for exchange of materials.

   b. Simple cuboidal epithelium is one layer of cube-shaped cells. It is adapted for absorption and secretion.

   c. Simple columnar epithelium is one layer of tall cells. It is an absorptive and secretory tissue.

   d. Stratified squamous epithelium is several layers of cells, with the top layer flat. It is adapted to resist mechanical wear and tear.

   e. Pseudostratified epithelium lines the respiratory and parts of the reproductive systems. It generally has cilia.

   f. Transitional epithelium lines the urinary system and stretches as the organs expand.

   g. Syncytial epithelium has no membranes between cells; it covers the placenta.

   h. Germinal epithelium lines the testis tubules and covers the ovaries. It produces sperm in the testes.

   i. Neuroepithelium is specialized for reception of stimuli. It is found in the organs of special sense (eye, ear, nose, tongue).

4. Epithelia and their underlying connective tissue form epithelial membranes.

   a. Mucous membranes line cavities that open on a body surface. They are moistened by mucus.

   b. Serous membranes line internal body cavities. They are moistened by a watery secretion.

5. Glands are derivatives of epithelia.

   a. Exocrine glands secrete onto a body surface.

   b. Endocrine glands secrete into the bloodstream.

6. Connective tissues have special characteristics.

   a. They connect, support, and protect body structures.

   b. They have large amounts of intercellular material that is usually fibrous.

   c. They are vascular.

   d. They contain scattered cells.

7. There are several important types of connective tissue.

   a. Loose connective tissue contains strong white fibers and stretchy elastic fibers. The cells called fibroblasts produce fibers. Loose connective tissue occurs under the skin.

   b. Adipose tissue contains many fat cells. It stores energy, insulates, and gives form to the body.

c. Reticular tissue forms the internal framework of body organs.

d. Fibrous connective tissue is dense and forms tendons, ligaments, and is found in blood vessel walls.

e. Cartilage is semisolid and forms part of the skeleton (hyaline), discs between vertebrae (fibrous), and the outer ear (elastic).

f. Bone is hard, forms the skeleton, and protects body organs.

8. The components of the intercellular substance of a connective tissue are produced by the characteristic cells of that tissue. Connective tissues reflect nutritional or hormonal deficiencies very quickly.

9. Connective tissues may form abnormally and result in malformed skeletal structures.

10. Muscular tissue is contractile and occurs in three varieties.

a. Skeletal muscle attaches to and moves the skeleton.

b. Cardiac muscle is found in the heart and circulates the blood.

c. Smooth muscle is found in internal organs and controls blood pressure and movement through hollow organs.

11. Nervous tissue is excitable and conductile and forms sensory, interpretive, and motor pathways.

12. The skin is the major organ of the integumentary system and has three main layers.

a. Five layers occur in the epidermis.

b. The dermis contains fibrous connective tissue, sensory corpuscles, blood vessels, and glands.

c. The subcutaneous tissue contains much fat.

13. Hairs, nails, and glands are appendages (derivatives) of the skin.

a. Hairs lie in a follicle and grow from the follicle.

b. Nails are hardened structures useful in grasping small objects.

c. Sweat glands help cool the body; sebaceous glands lubricate the hair and skin.

14. The functions of the skin include:

a. Protection. A surface film of chemicals acts as an antiseptic and reacts with toxic materials. The horny layer forms a physical barrier to entry of microorganisms and radiation. Loss of body components is also prevented.

b. Temperature regulation. Heat radiates from the blood passing through the skin. Sweat glands produce a fluid that evaporates and cools the skin.

c. Absorption. Gases and lipid soluble materials pass through the skin, as do those substances that dissolve the keratin in the epidermis.

d. Excretion. Water, salts, urea, and ammonia are excreted by the skin glands.

e. Reception of stimuli. Sensory receptors sensitive to heat, cold, touch, pressure, and pain are found in the skin.

15. Skin color depends on:

a. Pigment cells

b. Blood flow

c. Oxygen levels of the blood

d. Bilirubin levels

e. Exposure to sunlight

16. Hair color is determined by pigments or air in the hair cells.

17. The skin is a sensitive indicator of whole body physiology.

a. It may develop lesions; a common series of phases occurs.

b. It may develop disorders including infections, acne, eczema, seborrhea, and psoriasis.

c. It may be burned.

18. Burns are described as to the extent of burning and depth.

a. Extent is referred to by a percent.

b. Depth is referred to as first, second, and third degree. Reddening and tenderness characterizes a first degree burn; destruction, but some epithelial remnants characterizes a second degree burn; no epithelial remnants characterizes a third degree burn.

19. Treatment of burns is centered around prevention of secondary infection, maintenance of fluid and electrolyte balance, and maintenance of adequate nutrition.

Questions

1. Compare and contrast epithelia as to general characteristics, locations, and functions.

2. Describe the relationships between the type of epithelium lining an organ and the function(s) the organ serves. Give examples.

3. What is an epithelial membrane? What function(s) do they serve?

4. How are glands related to epithelia? How are they classified?

5. How does heredity influence connective tissue formation?

6. Compare muscular and nervous tissues as to basic properties.

7. Describe the structure of hairy skin.

8. How is the skin involved in temperature regulation?

9. What changes occur in the skin as it ages?

10. Define: macule, vesicle, scar.

11. What is meant by a "third degree burn over 80 percent of the body"?

Readings

Avioli, Louis V., and Stephen M. Krane. *Metabolic Bone Disorders.* Vols. 1 and 2. Academic Press. New York, 1978.

Bloom, William, and Don W. Fawcett. *A Textbook of Histology,* 10th ed. Saunders. Philadelphia, 1975.

Bourne, Geoffrey H. *The Biochemistry and Physiology of Bone.* Vol. 1. *Structure,* 1972; Vol. 2. *Physiology and Pathology,* 1972; Vol. 3. *Development and Growth,* 1971; Vol. 4. *Calcification and Physiology,* 1976, Academic Press. New York.

Elden, H. R. (ed.). *Biophysical Properties of the Skin.* Wiley. New York, 1971.

Ferriman, D. *Human Hair Growth in Health and Disease.* Thomas Pubs. Springfield, Ill., 1965.

Hardy, J. D., A. P. Gagge, and A. J. Stolwijk. *Physiological and Behavioral Temperature Regulation.* Thomas Pubs. Springfield, Ill., 1970.

Jeghers, H., and L. M. Edelstein. "Pigmentation of the Skin," *Signs and Symptoms,* 12th ed. Edited by C. M. MacBryde, and R. S. Blacklow. Lippincott. Philadelphia, 1970, pp 916–959.

Marback, H. I., and H. S. Gavin. *Skin Bacteria and Their Role in Infection.* McGraw-Hill. New York, 1965.

Marples, M. S. *The Ecology of the Human Skin.* Thomas Pubs. Springfield, Ill., 1965.

Montagna, W. *The Epidermis.* Academic Press. New York, 1964.

Nicoll, P. A., and T. A. Cortese, Jr. "The Physiology of Skin" in: *Annual Review of Physiology,* Vol. 34. Annual Reviews. Palo Alto, Cal., 1972.

Polk, H. C., Jr., and Stone, H. H. (eds.). *Contemporary Burn Management.* Little, Brown. Boston, 1971.

Rook, A. S., and G. S. Walton (eds.). *Comparative Physiology and Pathology of Skin.* Blackwell. Oxford, 1965.

Spearman, R. I. C. *Comparative Biology of Skin.* (Symposia of the Zoologic Society of London. No. 39.) Academic Press. New York, 1977.

Tregear, R. T. *Physical Functions of Skin.* Oxford Univ. Press. New York, 1966.

Chapter 7

The Body as Viewed from the Outside; Surface Anatomy

Objectives

After studying this chapter, the reader should be able to:

- Explain the practical importance of surface anatomy as learning and its value to those who deal with the human body.

- Describe how external appearance of the body changes with age and nutritional state.

- Indicate the significant differences in external appearance of mature male and female bodies.

- Point out the major bony, muscular, vascular, and nervous landmarks on the head, neck, trunk, upper appendages, and lower appendages.

- Outline the triangles of the neck, and the important anatomical features contained within each triangle.

- Comment on the use of superficial blood vessels as pressure points to control bleeding.

A statement of purpose

Having considered the body as an agglomeration of chemicals, cells, and tissues covered by skin, it is now appropriate for us to examine the systems that compose the whole organism.

When viewed from the exterior, the human body presents many LANDMARKS that represent surface features of deeper lying structures. These features may be used in their own right or as *reference points* for locating the deeper lying structures. Medical personnel, in conducting physical examinations, see the body from this viewpoint, and so we begin the systematic consideration of the body by presenting some surface anatomy.

The features described in this chapter may be skeletal, muscular, vascular, or nervous, and it is hoped that we may acquire a better "feel" for body structure by recognizing the body as a living structure—readers are urged to find as many of the landmarks as they can on their own bodies.

Many landmarks are reemphasized in later chapters and the reader should gain a *familiarity* with surface anatomy at this point, without committing to memory each and every feature described.

General surface anatomy

The external surface of the body varies according to age, sex, and state of nutrition. Premature infants (Fig. 7.1.) usually lack the thick subcutaneous layer of adipose tissue found in full-term infants, and the skin appears to hang loosely on the body. Full term infants (Fig. 7.2) have a chubby appearance due to overall deposition of fat, with buccal (cheek), abdominal, and limb deposition particularly obvious. As growth occurs, the distribution of fat becomes more even on the body, and usually is not excessively abundant in any given body area. At the time of puberty and adolescence (Fig. 7.3), sex differences in fat distribution become apparent (*see also* Fig. 1.1). The female accumulates fat pads in the breasts, over the shoulders, buttocks, inner and outer sides of the thighs, lower abdomen, and over the symphysis pubis. In the male, the pads are thinner and more evenly distributed over the body, unless the individual is grossly overweight. In this case, the abdomen usually assumes the greatest role in fat storage. In middle age, there is often a tendency to overweight, while old age (Fig. 7.4) is commonly associated with disappearance of fat and loss of skin elasticity. The skin

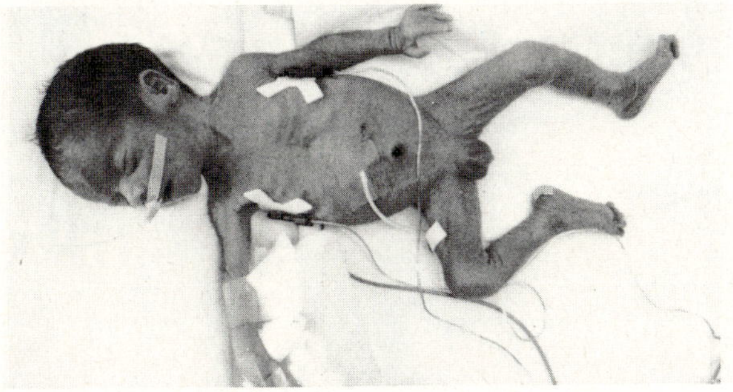

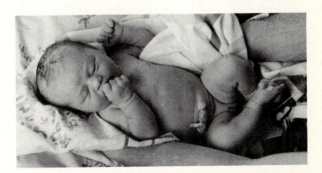

FIGURE 7.1   Premature 28 week fetus. Birth wt. 1.25 kg. Note paucity of skin fat, and "looseness" of the skin. Feeding and monitoring aids are in place.

FIGURE 7.2   A full term infant, 15 minutes after birth. Note the smooth body contours due to well-developed subcutaneous fat layer. (Jim Harrison/Stock, Boston)

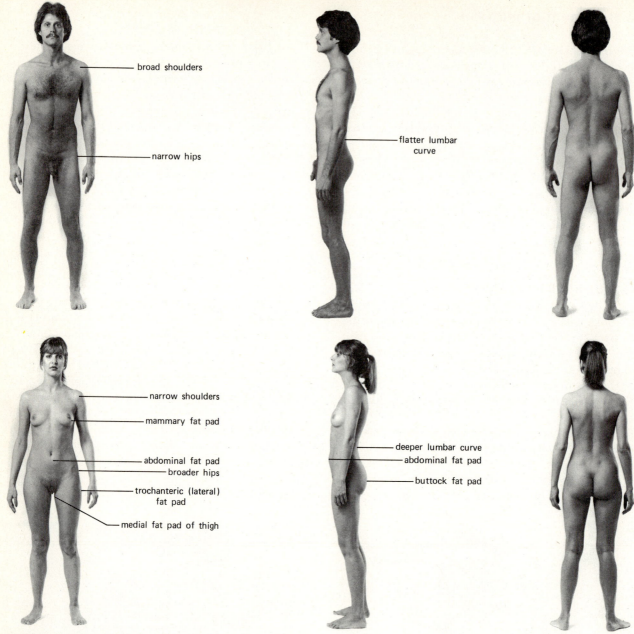

broad shoulders

narrow hips

flatter lumbar
curve

narrow shoulders

mammary fat pad

abdominal fat pad
broader hips

trochanteric (lateral)
fat pad

medial fat pad of thigh

deeper lumbar curve
abdominal fat pad

buttock fat pad

FIGURE 7.3 Adult male and female bodies. (Joel Gordon © 1979)

again tends to hang loosely on the body. The male figure is generally described as being more triangular in shape with broader shoulders and narrower hips. The female body is more diamond shaped, with softer contours and narrower shoulders with broader hips.

FIGURE 7.4 The expressive facial muscles wrinkle the skin at right angles to their lines of pull, while the advancing years, with attendant loss of fat and of the skin's elasticity, accentuate these wrinkles, stamping a characteristic pattern—although it is to be noted that the faces of some old people may remain remarkably smooth. (Monkmeyer Press Photo Service)

Regional surface anatomy

The head

The upper or CRANIAL portion of the head (Fig. 7.5) presents an outline that adheres closely to the structure of the bony parts. There is no muscle over the bones on the upper portion of the skull, and so the skin reflects the bony outline. The FRONTAL EMINENCES lie on the upper part of the forehead. The EXTERNAL AUDITORY MEATUS (mē-ā′tus) marks the opening of the external ear on the lateral sides of the skull. The EXTERNAL OCCIPITAL PROTUBERANCE may be felt on the posterior aspect of the cranium. The MASTOID PROCESS is a prominent elevation below and behind the auricle (ear). On the sides of the cranium, in the area known generally as the temple, the pulsations of the SUPERFICIAL TEMPORAL ARTERY may be felt. The vessel is easily compressed and pressure applied here is often used to help control bleeding on the cranium.

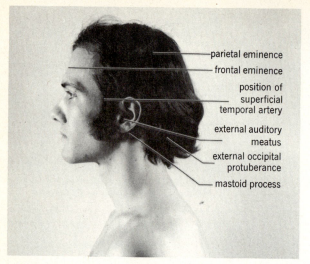

FIGURE 7.5 Some anatomical landmarks on the cranium.

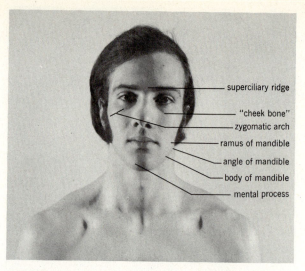

FIGURE 7.6 Some anatomical landmarks on the face.

On the FACE (Fig. 7.6), the bony landmarks are somewhat obscured by muscle and fat. The SUPERCILIARY RIDGES lie above and medial to the orbits housing the eyes. They are generally larger in males than in females, and form one of several criteria used to determine the sex of a skeleton. The "cheekbones" are formed by the zygomatic bones, and extending posteriorly from the cheek is a prominent bony ridge, the ZYGOMATIC ARCH. Rising from the posterior aspect of the arch, and attaching to the lower jaw is the MASSETER muscle. When the jaw is "clenched," the muscle stands out prominently. About one-third of the way from the posterior ANGLE of the mandible toward the chin, pulsations of the FACIAL (*external maxillary*) ARTERY may be felt as it passes over the bone to reach the cheek. This again forms a pressure point for control of bleeding. The INFRAORBITAL NERVES exit below the orbits, supplying the face, anterior teeth, and nose with sensory fibers. Infiltration of this area with anesthetics produces loss of sensation to—perhaps—permit dental repair. The infant skull (Fig. 7.7) presents several landmarks not present in the adult skull. The ANTERIOR (*frontal*) FONTANEL ("soft spot") is represented by a slightly depressed area of skin above the

forehead where several of the cranial bones have not yet knit together. The fontanel is often observed to aid in determination of increased intracranial pressure (bulging skin over the fontanel), or dehydration of the infant (greatly depressed skin over the fontanel). This fontanel normally closes by 18 months of age. The other fontanels shown in Figure 7.7 are normally closed at birth.

The neck.

The bony parts of the neck are well covered by muscle and fascia (connective tissue planes between muscle layers), and are difficult to palpate (feel). In the posterior midline of the neck, at its base, a protuberance may be felt. It is the SPINOUS PROCESS OF THE SEVENTH CERVICAL VERTEBRA (vertebra prominens). Knowing the number of this spine enables one to "count" the spinous processes of the other vertebrae and so determine vertebral levels.

In the anterior midline of the neck, the body of the HYOID BONE may be felt. It lies at the level of the lower border of the mandible. Inferior to

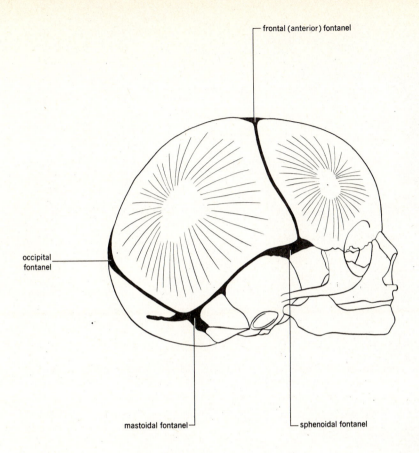

frontal (anterior) fontanel

occipital
fontanel

mastoidal fontanel

sphenoidal fontanel

FIGURE 7.7 Skull of a newborn.

the hyoid bone is the THYROID CARTILAGE OF THE LARYNX ("Adam's apple"). The CRICOID CARTILAGE is a ring-shaped cartilage inferior to the lower border of the thyroid cartilage. Below this, the TRACHEAL RINGS of cartilage may be felt. Obvious muscles of the neck are the paired STERNOCLEI-DOMASTOID MUSCLES, lying on the lateral side of the neck. Turning the head causes the opposite muscle to "stand out." Pulsations of the COMMON CAROTID ARTERY may be appreciated by placing a fingertip just anterior to the sternocleidomastoid muscle about halfway up its length. This is *not* a good pressure point because the vessel cannot be easily compressed against bone.

The muscles of the neck and several bones outline a number of ANATOMICAL TRIANGLES that contain within them important vascular and nervous structures.

The POSTERIOR CERVICAL TRIANGLE (Fig. 7.8) lies between the trapezius muscle posteriorly and the sternocleidomastoid muscle anteriorly and is de-limited by the clavicle at its base. The *omohyoid* muscle passes obliquely through the triangle and divides it into an upper, larger, *occipital triangle,* and a smaller, lower, *omoclavicular triangle.* The occipital triangle passes the 11th cranial nerve (accessory nerve) on its way to innervate the trapezius muscle. The omoclavicular triangle contains the roots of the subclavian artery and vein plus the large nerve trunks of the brachial plexus that supplies the upper appendage. The space behind the clavicle is termed the *supraclavicular notch,* and in this space can be felt the pulsations of the *subclavian artery.* Again, this is a poor pressure point because the vessel cannot be compressed against a bony surface.

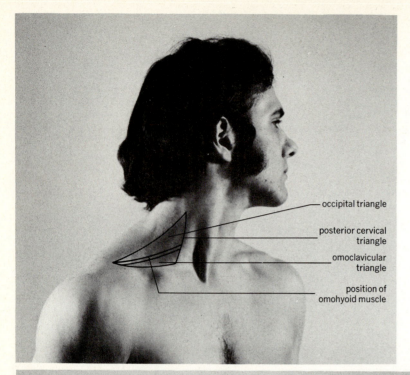

occipital triangle

posterior cervical triangle

omoclavicular triangle

position of omohyoid muscle

FIGURE 7.8   Posterior cervical triangle.

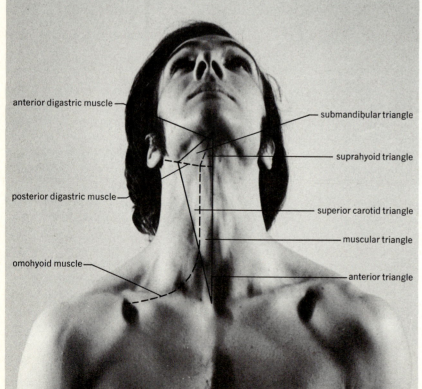

anterior digastric muscle

posterior digastric muscle

omohyoid muscle

submandibular triangle

suprahyoid triangle

superior carotid triangle

muscular triangle

anterior triangle

FIGURE 7.9   Anterior triangle.

The ANTERIOR TRIANGLE (Fig. 7.9) is bounded by the anterior midline of the neck, posteriorly by the sternocleidomastoid muscle, and superiorly by the body of the mandible. It is subdivided into four smaller triangles.

The *inferior carotid (or muscular) triangle* lies between the anterior midline of the neck, and the omohyoid muscle. It contains the thyroid gland, larynx, upper portion of the trachea, the base of the common carotid artery, internal jugular vein, and the first portion of the vagus (10th cranial) nerve.

The *superior carotid triangle* is bounded by the omohyoid, sternocleidomastoid, and posterior digastric muscles. The bifurcation (division) of the common carotid artery lies in this triangle.

The *submandibular (digastric) triangle* is bordered by the anterior digastric and stylohyoid muscles, and by the mandibular body. The facial (7th cranial) nerve, and parotid and submandibular salivary glands lie in this triangle.

The *suprahyoid (submental) triangle* is bounded by the anterior digastric, neck midline, and body of the hyoid bone. Several large lymph nodes and the anterior jugular vein occupy this triangle.

The trunk

The upper portion of the CHEST or THORAX is nearly covered by the large PECTORALIS MAJOR MUSCLE. Laterally, this muscle forms the anterior limit of the axilla or armpit. A line drawn vertically along its border is the ANTERIOR AXILLARY LINE (*see* Fig. 1.9). The POSTERIOR AXILLARY LINE is drawn along the border of the latissimus dorsi, a muscle of the back. The JUGULAR or PRESTERNAL NOTCH, a depression between the inner ends of the clavicles, the STERNAL ANGLE, a ridge about 2 inches below the notch, the XIPHOID PROCESS at the inferior end of the sternum, and the RIBS, mark the major other landmarks of the chest.

The BACK has its midline marked by a furrow, at the bottom of which the SPINOUS PROCESSES of the vertebrae may be felt. The ridges of muscle on either side of the furrow are caused by the EREC-

TOR SPINAE group of muscles that keep the spine straight. The major muscles on the back are the TRAPEZIUS superiorly, and the LATISSIMUS DORSI inferiorly.

The ABDOMEN is delimited primarily by muscle, not bony structures, and the contour established depends largely on the tone of the abdominal muscles and the amount of fat beneath the muscle layers in the abdominal omenta. The upper border of the abdomen is determined by the curve of the ribs and costal cartilages, the lower border by a line running from the iliac crests through the pubis. The anterior wall of the abdomen is formed by one layer of muscle *(rectus abdominus),* while the side has three layers of muscle *(2 obliques* plus the *transverse abdominal).* The floor of the abdominal cavity is a muscular sheet that is penetrated by the openings of the digestive system, urethra, and in the female, the vagina as well.

The upper limb (Fig. 7.10)

The CLAVICLES and SCAPULAE form the bones of the PECTORAL GIRDLE that attach the upper appendages to the thorax. The clavicle is just beneath the skin *(subcutaneous)* throughout most of its length. The scapula is covered with muscle except for the SPINE, that may be felt on the back below the shoulder, and the ACROMION PROCESS that forms the "point" of the shoulder.

The HUMERUS, the bone of the upper arm, is surrounded by muscle except at its proximal and distal ends. The GREATER TUBEROSITY can be felt on the lateral side of the shoulder, and the prominent MEDIAL AND LATERAL EPICONDYLES form protuberances just above the elbow joint. The forearm is composed of a medial ULNA and the laterally placed RADIUS. Posteriorly, the point of the elbow is formed by the OLECRANON PROCESS of the ulna. At the wrist, STYLOID PROCESSES on both radius and ulna form prominent bulges. On the anterior aspect of the wrist, the SCAPHOID and TRAPEZOID bones of the carpals (wrist bones) form the lateral elevations. The PISIFORM (also a carpal) forms a medial elevation, and just distal to it, the

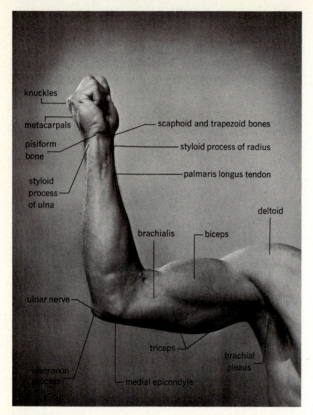

FIGURE 7.10 Some anatomical landmarks of the upper limb.

HAMATE (a carpal) produces a bump easily felt through the muscle. The METACARPALS form the hand, and are best appreciated by palpation on the posterior surface of the appendage. PHALANGES form the digits. There are five digits, four of which are fingers, the fifth being the thumb.

The DELTOID muscle covers the shoulder; the BICEPS is the most obvious muscle on the anterior upper arm. The posterior humerus is covered by the TRICEPS muscle.

Just posterior to the lateral epicondyle, the ULNAR NERVE is subcutaneous. Striking the nerve gives a tingling sensation; the "funny bone" or "crazy bone" is not a bone at all, but the ulnar nerve.

On the anterior forearm, the FLEXOR CARPI RADIALIS TENDON stands out prominently when the wrist is strongly flexed. Just lateral to the tendon, the pulse may be felt on the RADIAL ARTERY. Also on the anterior forearm, the network of superficial vessels formed by the CEPHALIC (laterally) and BASILIC (medially) VEINS are easily recognized. The two vessels are connected across the anterior aspect of the elbow joint by the MEDIAN ANTECUBITAL VEIN, a common site of venipuncture to obtain blood for tests. On the posterior side of the hand, there is also an obvious network of veins. These are called the DORSAL VENOUS NETWORK.

The lower limb (Fig. 7.11)

The ILIAC CRESTS of the os coxae form prominent subcutaneous ridges just below the beltline on the lateral sides of the abdomen. The PUBIC CREST lies in the lower anterior midline of the abdomen. The ISCHIAL TUBEROSITIES lie beneath the buttock muscles, and we sit on them. The GREATER TROCHANTERS of the femurs lie about six inches inferior to the iliac crests on the lateral sides of the upper thigh. "Hip measurements" are commonly made at the level of the trochanters. The shaft of the FEMUR is well covered by muscle, but its distal end presents MEDIAL AND LATERAL EPICONDYLES, easily felt just above the knee. The PATELLA (knee cap) lies anteriorly on the knee joint.

The leg consists of two bones, a medially placed TIBIA that is subcutaneous throughout most of its anterior aspect. The FIBULA is the lateral bone of the leg. Its HEAD may be felt on the lateral aspect of the proximal portion of the leg. At the distal end of the leg, the MEDIAL MALLEOLUS (tibia) and LATERAL MALLEOLUS (fibula) form prominent bumps either side of the ankle, but, though commonly called "ankle bones," do not form parts of the tarsals (ankle bones). The heel is formed by the CALCANEUS bone. Five METATARSALS form the top of the foot, and fourteen PHALANGES compose the toes.

The GLUTEUS MAXIMUS muscle forms the buttocks and below it, on the posterior thigh, are three HAMSTRING MUSCLES. The anterior and lat-

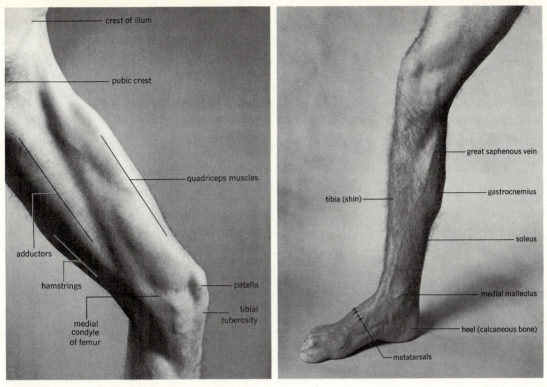

FIGURE 7.11 Some anatomical landmarks on the lower limb.

eral aspects of the thigh are occupied by the QUADRICEPS, a series of four muscles all attaching to the patella. The medial thigh curvature is created by a series of ADDUCTOR MUSCLES.

On the leg, the posterior aspect of the proximal end shows the GASTROCNEMIUS (calf) muscle. It terminates in the heavy TENDON OF ACHILLES (*tendo calcaneus*) attaching to the heel.

Blood vessels of the leg include the SAPHENOUS VEINS, which are superficial vessels especially prominent at the ankle. Arteries may be palpated in the groin (*femoral artery*), at the back of the knee (*popliteal artery*), and on the upper surface of the foot (*dorsis pedis artery*).

Nerves are difficult to palpate on the lower extremity because they are well covered by muscle. Just below the head of the fibula, the COMMON PERONEAL NERVE winds around the lateral aspect of the bone.

Again, it is valuable to be able to recognize these surface features of the body. If someone said "give an injection into the deltoid," or "I just hit my trochanter," one should be able to relate the part mentioned to a particular body area. Learn and find as many of these features as you can. They will help you later in your study of the body.

Summary

1. The study of surface anatomy is used to locate bony, muscular, vascular, and nervous structures in the body and understand their relationships to one another.

2. External body appearance changes with age, mainly as a result of the amount of fat on the body and its deposition.

3. The major landmarks of the head, neck, trunk, and appendages are described.

Questions

1. Describe where one would locate the following arteries:
   a. Dorsis pedis
   b. Superficial temporal
   c. Radial
   d. Common carotid
   e. Subclavian
   Veins:
   a. Saphenous
   b. Internal jugular
   c. Median cubital
   d. Cephalic

2. Forcible clenching of the teeth causes two muscles on the skull to bulge. What are they?

3. Name three bony prominences that may be felt at the elbow region.

4. What are three bony structures that may be palpated at the ankle?

5. Around what bony feature of the femur are hip measurements taken?

6. Name the most obvious muscular structures of:
   a. The upper arm
   b. The calf
   c. The chest
   d. The upper back
   e. The thigh

7. What do fontanels represent, and of what practical importance are they to "the owner" and medical personnel?

8. What is the "crazy bone"?

Readings

Lockhart, R. D. *Living Anatomy*. Faber & Faber Ltd. London, 1970.

Royce, Joseph. *Surface Anatomy*. F. A. Davis. Philadelphia, 1965.

Chapter 8

The Skeleton

Objectives

After studying this chapter, the reader should be able to:

- Describe the structure of bony tissue, in terms of organic and inorganic components.

- Describe the difference in structure, including subunits of structure, between spongy and compact bone.

- Compare and contrast intramembranous and intra-cartilagenous bone formation.

- Discuss what is involved in calcification.

- Explain how bones grow in length and width.

- List the classification of bones by shape.

- Define the terms used to describe surface features of bones.

- Explain how the skeleton is organized and how many bones are found in each portion of the skeleton.

- The remaining major objective is for you to learn the names of the bones of the skeleton and their features. Good luck!

The human skeleton is a living, dynamic structure, which serves as a supporting and protective framework for the body. Muscles attach to the skeleton and cause skeletal movement. Many body organs are suspended from the skeleton. The skeleton provides a storehouse for calcium and phosphate, essential substances for body function. It contains a tissue known as bone marrow, which serves as a reservoir of nutrients and acts as the area of production of several types of blood cells. The bones are thus sites of intense activity, and are not the dry lifeless structures commonly studied in the anatomy laboratory.

The skeleton is composed of osseous tissue (bone), a tissue made hard by the deposition of inorganic substances in it, in the process known as calcification.

The structure of bone

Bone contains about 65 percent inorganic substance, chiefly calcium phosphate $[Ca_3(PO_4)_2]$, and about 35 percent organic substance, consisting of cells and a fibrous protein known as bone collagen. According to the arrangement of these components, three types of bony tissue are described.

CANCELLOUS or SPONGY BONE (Fig. 8.1) consists of interlacing bars and plates of bony tissue with many spaces between. It is found inside many bones, such as the interior of ribs, skull bones, and vertebrae, and the ends of the long bones. Bone cells, or OSTEO-CYTES, lie within cavities or LACUNAE in the bony tissue.

COMPACT BONE is a very dense material with microscopic subunits of structure (Fig. 8.2), known as OSTEONS or Haversian systems. These units are oriented lengthwise in the compact bone. The center of the osteon is formed by the HAVERSIAN CANAL, which carries the blood vessels into the bones. Rings of bony tissue known as LAMELLAE lie around the canal, more or less alternating with rows of LACUNAE containing OSTEOCYTES. Tiny canals or CANALICULI connect lacunae with one another and with the canal.

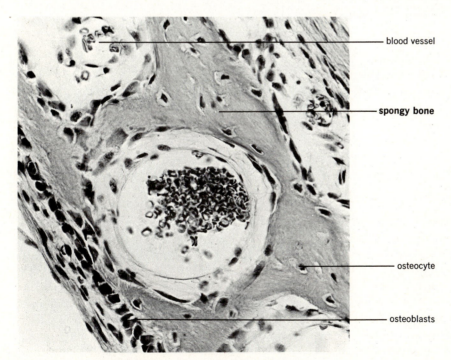

blood vessel

spongy bone

osteocyte

osteoblasts

FIGURE 8.1   Intramembranous bone formation, illustrating spongy bone.

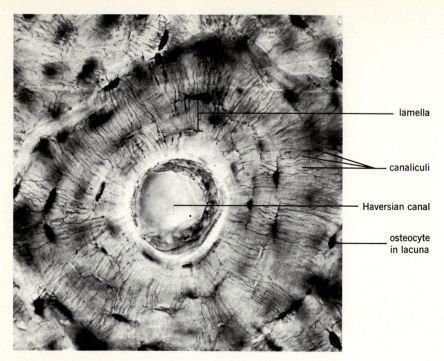

lamella

canaliculi

Haversian canal

osteocyte
in lacuna

FIGURE 8.2   A cross section of ground compact bone, showing the parts of the osteon.

Compact bone forms the shafts of the long bones.

DENSE BONE, without osteons, but containing lacunae and osteocytes, forms the outer covering of the bones.

Formation of bone

Types of formation

Bone is formed two ways in the body.

In INTRAMEMBRANOUS FORMATION (*see* Fig. 8.1), bone is formed "directly" from connective tissue within a fibrous membrane. Mesoderm cells increase in number and size, and the number of blood vessels increases. Some mesodermal cells change into bone-forming cells or OSTEOBLASTS. These osteoblasts form a semisolid mass containing collagenous fibrils known as OSTEOID. The osteoid is then CALCIFIED by deposition of inorganic material to form bone.

In "indirect" bone formation or INTRACARTILAGENOUS (*endochondral*) FORMATION (Fig. 8.3), a MODEL of the bone is first formed in hyaline cartilage. The bones are initially very small in the embryo (*see* Fig. 4.8). Cartilage cells in the center of the model enlarge and become more numerous, squeezing the matrix into thin bars and plates. Outside the model, perichondrial cells are changing into osteoblasts and laying down (intramembranously) a BONE COLLAR around the center of the model to support it. The cartilage matrix is then CALCIFIED. Next, blood vessels

Diagram of the development of a typical long bone as shown in longitudinal sections. *Green*, bone, *blue*, calcified cartilage, *red*, arteries. *a', b', c', d', e'*, cross sections through the centers of *a, b, c, d, e*, respectively. *a*, cartilage model, appearance of the periosteal bone collar; *b*, before the development of calcified cartilage; *c*, or after it, *d; e*, vascular mesenchyme has entered the calcified cartilage matrix and divided it into two zones of ossification, *f; g*, blood vessels and mesenchyme enter upper epiphyseal cartilage; *h*, epiphyseal ossification center develops and grows larger; *i*, ossification center develops in lower epiphyseal cartilage; *j*, the lower and, *k*, the upper epiphyseal cartilages disappear as the bone ceases to grow in length, and the bone marrow cavity is continuous throughout the length of the bone. After the disappearance of the cartilage plates at the zones of ossification, the blood vessels of the diaphysis, metaphysis, and epiphysis intercommunicate. (Bloom and Fawcett, *A Textbook of Histology*, courtesy of W. B. Saunders Co.)

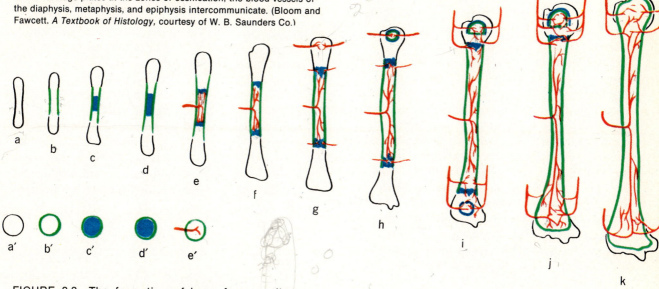

FIGURE 8.3   The formation of bone from cartilage (intracartilagenous bone formation).

penetrate the collar, bringing OSTEOBLASTS into the model. The vessels destroy the cartilage cells; the osteoblasts brought in lay down OSTEOID (see above) on the calcified cartilage plates, and this is CALCIFIED to form bone. Formation of bone from cartilage thus involves two calcifications, with tissue destruction occurring.

In the long bones, formation from cartilage usually begins in the center of the bone, and later in the two ends, so that plates of cartilage (epiphyseal cartilages) remain between the shaft and ends of the bone. As long as these plates are cartilagenous, the bone may grow in length (see below).

Calcification

Deposition of inorganic substance in osteoid to form bone is called calcification. Although it is not definitely known how calcification occurs, it is clear that the calcium and phosphate necessary for the process must first be absorbed from the gut or placenta, until some critical blood value is reached. Calcification requires vitamin D, which accelerates absorption of calcium. Second, the tissue to be calcified must be made calcifiable, that is, capable of having the salts deposited in it. This latter task appears to be carried out by the osteoblasts, which secrete a substance causing binding of phosphate to collagen fibrils. This combination then acts as a center for crystal formation, and deposition of salts may proceed.

Growth of bones

Long bones grow in length by retaining a plate of cartilage between the shaft (diaphysis) and the end (epiphysis) of the bone. Cartilage production must occur more rapidly than bone formation to cause length increase, as new bone is laid down on either side of the epiphyseal cartilage. When the plate itself becomes bony, growth in length ceases. Growth in diameter of bones occurs by intramembranous formation of new bone tissue on the surface of the bone, through the activity of cells located in the PERIOSTEUM, a membrane surrounding all bones. The periosteum forms bone in the intramembranous manner. Growth of bones involves the activity of bone-forming cells, or osteoblasts, and bone remodeling or destroying cells, OSTEOCLASTS. Typically, as bone is added in one area, it is removed in another, so that the thickness of the wall of the bone, or its weight, tends to remain within normal limits, and it usually does not become extremely heavy or thick.

Clinical considerations

Without vitamin D the inorganic constituents are not available for deposition, and bones remain soft or only partially calcified (rickets). Vitamin A deficiency retards maturation and growth of cartilage cells in intracartilagenous formation; vitamin A excess causes accelerated destruction of bone. Vitamin C deficiency results in production of a noncalcifiable matrix.

As bone ages, formation of the organic portion is slowed, and the bones become more brittle. Density decreases, the bones may appear "motheaten," and are easily fractured.

Growth hormone from the pituitary gland is important in controlling production of new cartilage, and in controlling the activity of the periosteum. Parathyroid hormone (PTH) controls the activity of bone-destroying cells (osteoclasts); increased PTH levels are associated with increased bone destruction. Calcitonin, a hormone of the thyroid gland, increases deposition of inorganic salts in bone.

Classification of bones

Bones fall into four general classes according to their shape:

LONG BONES have a greater length than width, have a central SHAFT (diaphysis) composed of compact bone, and proximal and distal ENDS (epiphyses) composed of spongy bone. A central MEDULLARY CAVITY contains bone marrow, and is lined with ENDOSTEUM, a thin membrane. The bones of the fingers and toes (phalanges), upper arm (humerus), lower arm (radius and ulna), thigh (femur), leg (tibia and fibula) are examples of long bones. Figure 8.4 shows a section of a long bone to illustrate its parts.

SHORT BONES are ones in which length and width are not greatly different. The bones of the wrist (carpals) and ankle (tarsals) are short bones.

FLAT BONES consist of inner and outer layers of dense bone (tables) and a central layer of spongy bone (diplöe). The cranial bones and scapulae (shoulder blades) are flat bones.

All other bones are classed as IRREGULAR BONES. They are of complicated shape. The vertebrae and certain skull bones are of this type.

Two other "types of bones" may be mentioned. WORMIAN BONES (sutural bones) are small, irregular pieces of bone found in the course of the major

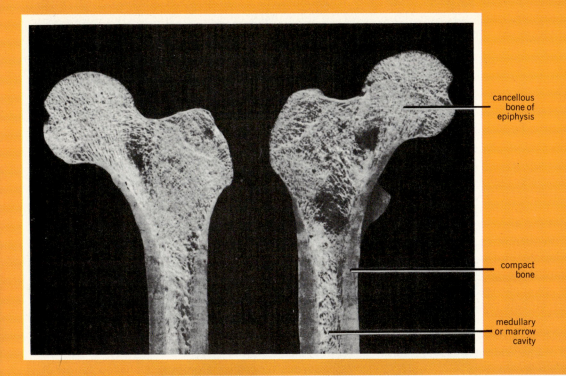

cancellous
bone of
epiphysis

compact
bone

medullary
or marrow
cavity

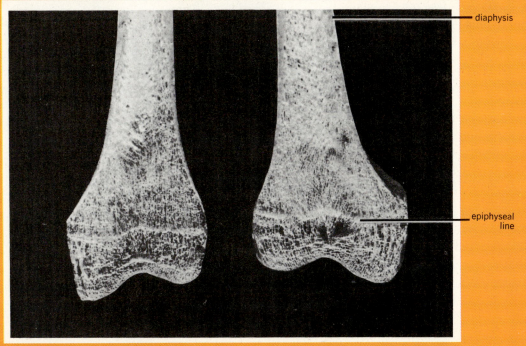

diaphysis

epiphyseal
line

FIGURE 8.4   Photographs of the proximal and distal
ends of a split femur.

sutures of the cranium. They are the result of isolated areas of bone formation separate from those that create the cranial bones, and are of no consequence.

SESAMOID BONES are bones that develop in tendons. The kneecap is a sesamoid bone formed in the patellar tendon. Sesamoid bones often develop where pressure is put on a tendon, as under the "ball" of the foot.

Naming bones and features of bones

Our understanding of bones and their parts can be aided by remembering that most bone names are descriptive of their shape or resemblance to objects around us. For example, the word parietal means, "a wall," and this bone is a major component of the wall of the cranial vault; a bone of the wrist is called the lunate, and it bears a striking resemblance to a (half) moon. Some terms describing surface features are applicable to many bones, and are presented in alphabetical order:

CONDYLE (G. *kondylos*, knuckle). A rounded protuberance at the end of a bone forming an articulation or joint.

CREST (L. *crista*, tuft). A ridge or elongated prominence on a bone.

EPICONDYLE (G. *epi-*, above, + condyle). A projection from a bone, above a condyle.

FACET (Fr. *facette*, small face). A small, smooth, flat or shallow surface on a bone, particularly a vertebra.

FISSURE (L. *fissus*, a cleft). A groove or narrow cleftlike opening.

FORAMEN (L. *foramen*, an opening). A passageway through a bone; a hole.

FOSSA (L. *fossa*, a ditch). A furrow or shallow depression.

FOVEA (L. *fovea*, a pit). A small cuplike depression.

HEAD (A. S. *heafod*, head). The proximal end, or larger extremity of a bone.

LINE (L. *linea*, line). A long, narrow ridgelike structure.

MEATUS (L. *meatus*, passage). A passageway or short canal running within a bone.

NOTCH (A. S. *nocke*, notch). A deep indentation or narrow gap in the edge (or margin) of a bone.

PROCESS (L. *processus*, a going before). A projection or outgrowth from a bone.

SINUS (L. *sinus*, a curve). A cavity within a bone.

SPINE (L. *spina*, spine). A somewhat sharp process from a bone.

SULCUS (L. *sulcus*, a groove). A shallow furrow or groove.

TROCHANTER (G. *trochanter*, a runner). Large processes found only on the proximal end of the femur.

TUBERCLE (L. *tuberculum*, a little swelling). A small rounded eminence on a bone.

TUBEROSITY (L. *tuberositas*, a swelling). An elevated rounded process from a bone. Usually larger than a tubercle (word often used synonomously with tubercle).

One should be aware that it is possible to combine some of these basic terms to give an actual bony part. For example, the condyloid process (of mandible) is a knucklelike process forming a joint.

The organization of the skeleton*

The skeleton (Fig. 8.5) is composed of 206 bones, divided into an AXIAL SKELETON, and an APPENDICULAR SKELETON. The axial skeleton contains 80 bones and consists of the skull, vertebral column, and thorax (ribs and sternum). The appendicular skeleton contains 126 bones, and consists of the limbs and the bones supporting the limbs.

The skull

The skull bones may be divided into two groups: those forming the CRANIUM (the "brain box"), and those forming the FACE. The cranial bones enclose the cranial cavity which is divided into anterior, middle, and posterior portions; those of the face form mainly the anterior part of the skull.

There are 8 CRANIAL BONES:

1 frontal	2 temporals
2 parietals	1 sphenoid
1 occipital	1 ethmoid

The features of each bone appear in Table 8.1.

The FRONTAL BONE (Fig. 8.6) forms the forehead, the floor of the anterior cranial cavity (or roof of the orbit which houses the eye), and part of the roof of the nasal cavities. The bone contains the FRONTAL SINUSES (Fig. 8.7). In the fetus, its two halves are separated by the FRONTAL (*metopic*) SUTURE.

The PARIETAL BONES (Fig. 8.8) form the larger part of the sidewalls and roof of the cranium. They join the frontal bone at the CORONAL SUTURE, and are separated from one another by the SAGITTAL SUTURE.

The OCCIPITAL BONE (Fig. 8.9) forms the posterior and basal portion of the posterior cranial cavity. The LAMBDOIDAL SUTURE separates the

occipital bone from the parietal bones, and commonly contains one or more Wormian bones.

The TEMPORAL BONES (Fig. 8.10) form part of the lateral walls and floor of the cranium. The SQUAMOUS PORTION of the temporal is separated from the parietal bones by the SQUAMOUS SUTURE. The TYMPANIC PORTION lies inferior to the squama, and contains the opening of the ear. The MASTOID PORTION lies posterior to the tympanic portion. The PETROUS PORTION lies mainly within the cranial cavity (Fig. 8.11), where it appears as a ridge of bone housing the delicate structures of the inner ear (cochlea, semicircular canals).

The SPHENOID BONE (Fig. 8.12) is shaped like a butterfly and articulates with all other cranial bones.

The ETHMOID BONE (Fig. 8.13) is an irregularly shaped bone that lies between the orbits and forms a small part of the floor of the anterior cranial cavity.

There are 8 FACIAL BONES:

2 nasals	2 lacrimals
2 zygomatics (malars)	1 vomer
2 maxillae	2 palatines
1 mandible	2 inferior conchae (turbinates)

Table 8.1 presents the features of each bone.

The NASAL bones (*see* Figs. 8.6 and 8.8) are small bones forming the bridge of the nose. What is commonly referred to as the "nose" is mostly cartilage and flesh.

The ZYGOMATIC (*malar*) BONES form the cheek, and part of the ZYGOMATIC ARCH that reaches from cheek to ear. The bone also forms part of the lateral wall and floor of the orbit. Because of its prominence, the zygomatic is very likely to be fractured when blows are delivered to the face (e.g., in automobile crashes).

The MAXILLAE (Fig. 8.14) help form the floor of

(*Text continued on page 150.*)

---

* The figure numbers that appear in the following discussion indicate that picture in which the feature described is most clearly shown. It may appear in other figures as well.

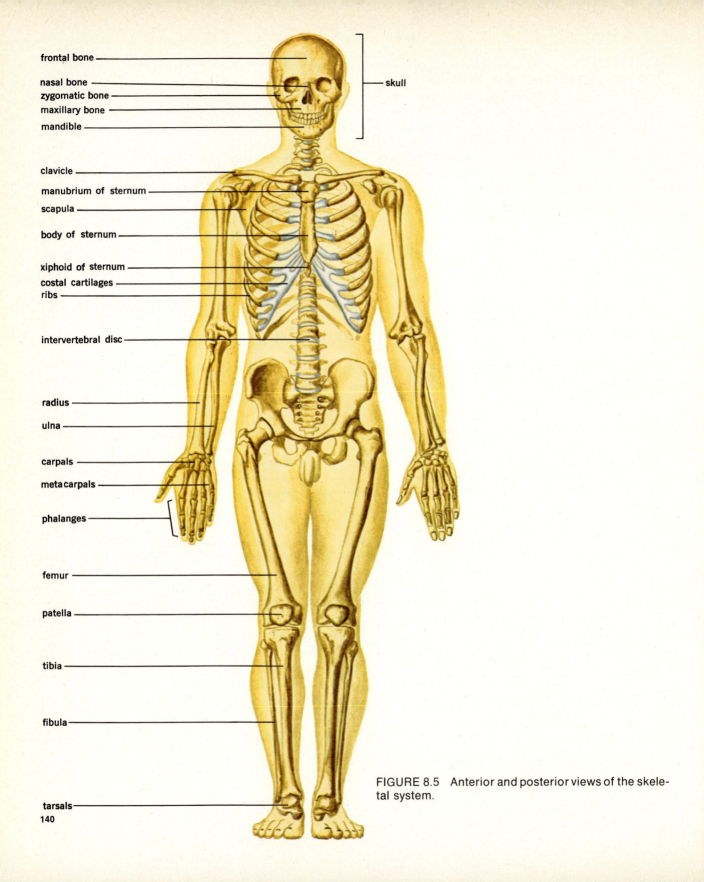

frontal bone

nasal bone
zygomatic bone
maxillary bone
mandible

skull

clavicle
manubrium of sternum
scapula

body of sternum

xiphoid of sternum
costal cartilages
ribs

intervertebral disc

radius

ulna

carpals

metacarpals

phalanges

femur

patella

tibia

fibula

tarsals

140

FIGURE 8.5   Anterior and posterior views of the skeletal system.

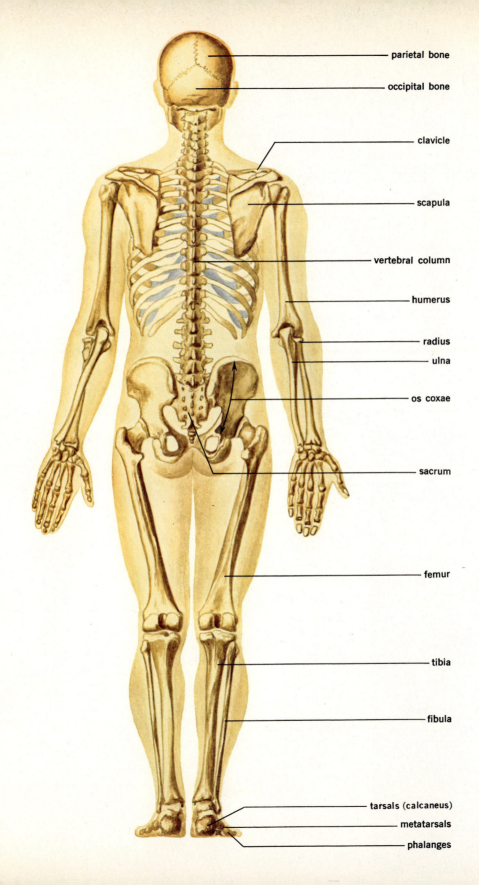

parietal bone

occipital bone

clavicle

scapula

vertebral column

humerus

radius

ulna

os coxae

sacrum

femur

tibia

fibula

tarsals (calcaneus)

metatarsals

phalanges

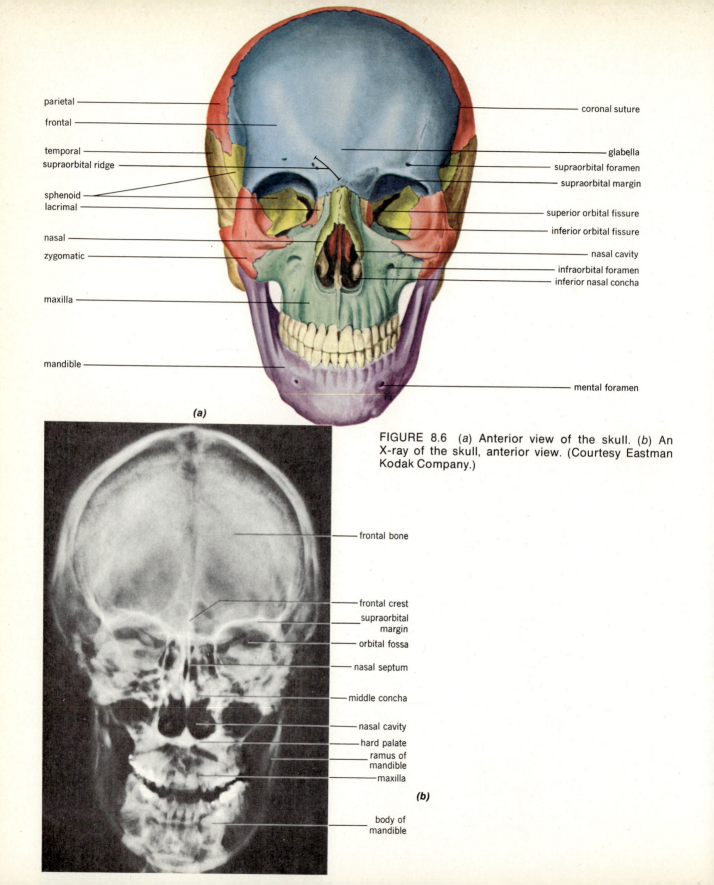

parietal

frontal

temporal

supraorbital ridge

sphenoid

lacrimal

nasal

zygomatic

maxilla

mandible

coronal suture

glabella

supraorbital foramen

supraorbital margin

superior orbital fissure

inferior orbital fissure

nasal cavity

infraorbital foramen

inferior nasal concha

mental foramen

(a)

FIGURE 8.6 (*a*) Anterior view of the skull. (*b*) An X-ray of the skull, anterior view. (Courtesy Eastman Kodak Company.)

frontal bone

frontal crest

supraorbital margin

orbital fossa

nasal septum

middle concha

nasal cavity

hard palate

ramus of mandible

maxilla

body of mandible

(b)

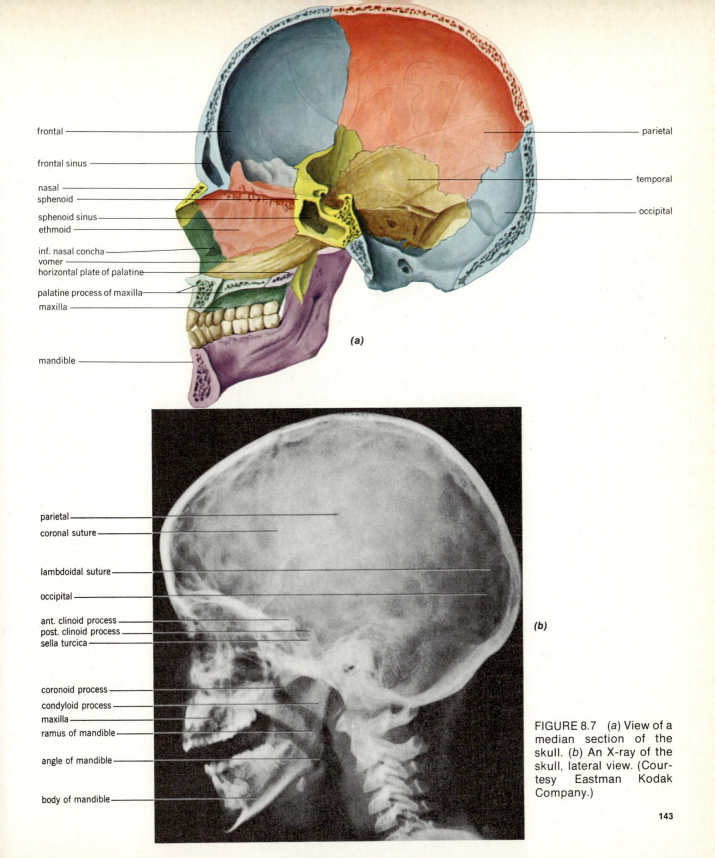

frontal

frontal sinus

nasal
sphenoid

sphenoid sinus
ethmoid

inf. nasal concha
vomer
horizontal plate of palatine

palatine process of maxilla
maxilla

mandible

parietal

temporal

occipital

(a)

parietal
coronal suture

lambdoidal suture

occipital

ant. clinoid process
post. clinoid process
sella turcica

coronoid process
condyloid process
maxilla
ramus of mandible

angle of mandible

body of mandible

(b)

FIGURE 8.7 (a) View of a median section of the skull. (b) An X-ray of the skull, lateral view. (Courtesy Eastman Kodak Company.)

143

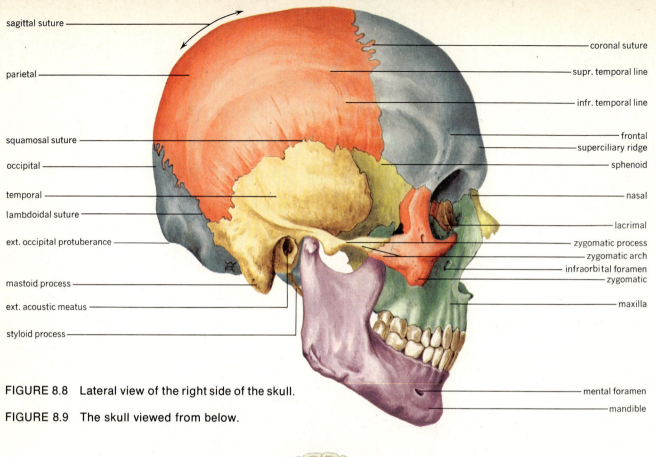

sagittal suture

parietal

squamosal suture

occipital

temporal

lambdoidal suture

ext. occipital protuberance

mastoid process

ext. acoustic meatus

styloid process

coronal suture

supr. temporal line

infr. temporal line

frontal

superciliary ridge

sphenoid

nasal

lacrimal

zygomatic process

zygomatic arch

infraorbital foramen

zygomatic

maxilla

mental foramen

mandible

FIGURE 8.8 Lateral view of the right side of the skull.

FIGURE 8.9 The skull viewed from below.

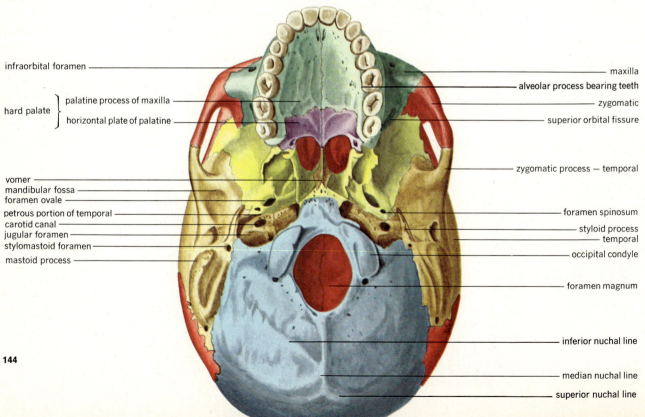

infraorbital foramen

hard palate
{
palatine process of maxilla

horizontal plate of palatine
}

vomer

mandibular fossa

foramen ovale

petrous portion of temporal

carotid canal

jugular foramen

stylomastoid foramen

mastoid process

maxilla

alveolar process bearing teeth

zygomatic

superior orbital fissure

zygomatic process — temporal

foramen spinosum

styloid process

temporal

occipital condyle

foramen magnum

inferior nuchal line

median nuchal line

superior nuchal line

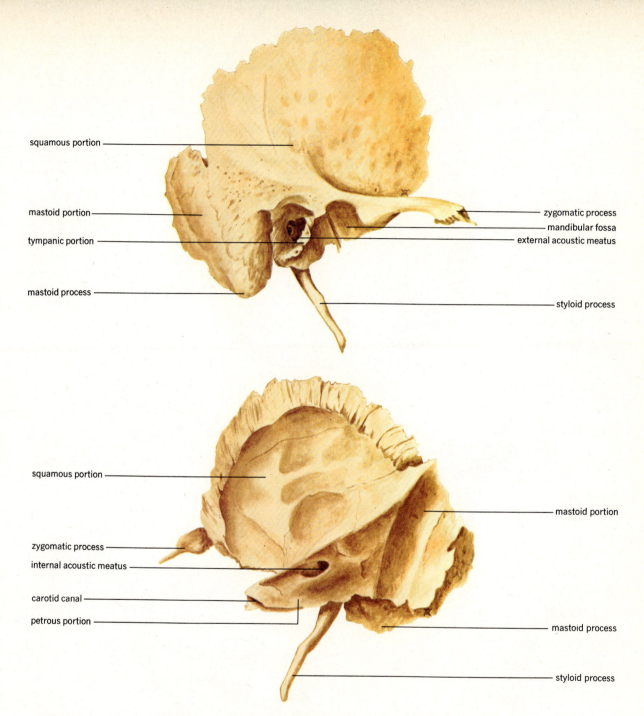

squamous portion

mastoid portion

tympanic portion

mastoid process

zygomatic process

mandibular fossa

external acoustic meatus

styloid process

squamous portion

mastoid portion

zygomatic process

internal acoustic meatus

carotid canal

petrous portion

mastoid process

styloid process

FIGURE 8.10 The right temporal bone. Lateral view (*above*), medial view (*below*).

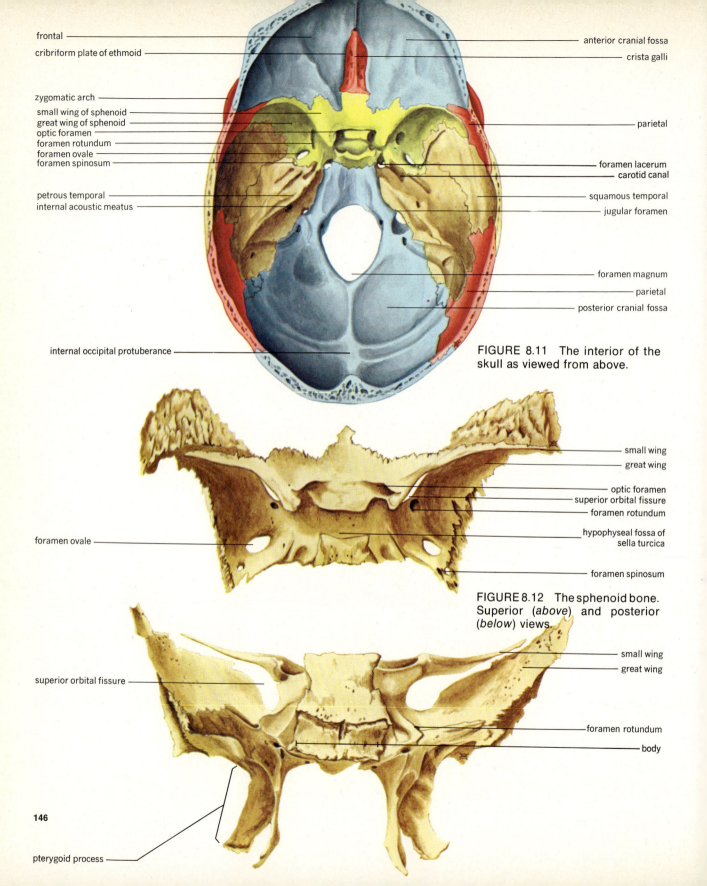

frontal
cribriform plate of ethmoid

anterior cranial fossa
crista galli

zygomatic arch
small wing of sphenoid
great wing of sphenoid
optic foramen
foramen rotundum
foramen ovale
foramen spinosum

parietal

foramen lacerum
carotid canal

petrous temporal
internal acoustic meatus

squamous temporal
jugular foramen

foramen magnum
parietal
posterior cranial fossa

internal occipital protuberance

FIGURE 8.11   The interior of the skull as viewed from above.

small wing
great wing

optic foramen
superior orbital fissure
foramen rotundum

hypophyseal fossa of sella turcica

foramen ovale

foramen spinosum

FIGURE 8.12   The sphenoid bone. Superior (*above*) and posterior (*below*) views.

small wing
great wing

superior orbital fissure

foramen rotundum
body

pterygoid process

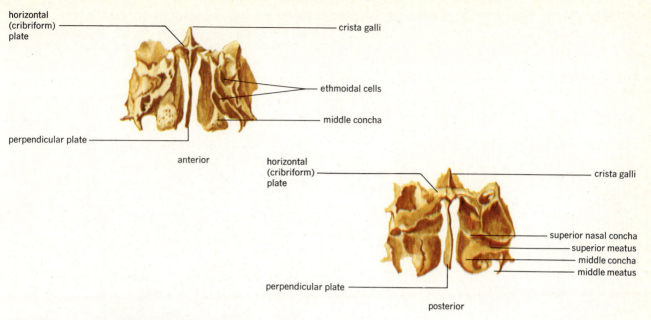

horizontal (cribriform) plate

crista galli

ethmoidal cells

middle concha

perpendicular plate

anterior

horizontal (cribriform) plate

crista galli

superior nasal concha
superior meatus
middle concha
middle meatus

perpendicular plate

posterior

FIGURE 8.13 The ethmoid bone. Anterior and posterior views.

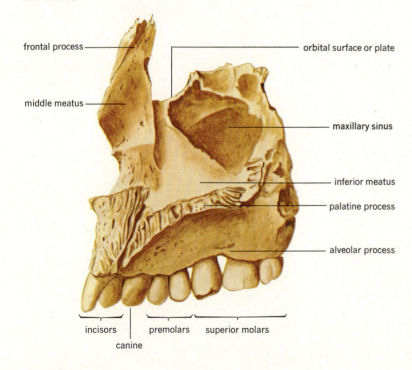

frontal process

orbital surface or plate

middle meatus

maxillary sinus

inferior meatus

palatine process

alveolar process

incisors   premolars   superior molars

canine

FIGURE 8.14 The right maxillary bone viewed from the medial side.

TABLE 8.1   A summary of the skull bones (italics indicate the most important features).

Name of bone and number	Features on the bone	Description/location of bone/feature
Frontal (1)		Forms forehead, roof of orbit and nasal cavities
	Supraorbital margins	Upper edge of orbits
	Superciliary ridges	Ridge above orbits
	Supraorbital foramen/notch	Hole/gap in upper margin
	Frontal sinuses	Cavity inside bone behind ridges
	Glabella	Central flat part of forehead
Parietal (2)		Roof and side walls of cranium
	Temporal lines	Arched lines on side of bone
Occipital (1)		Back and base of skull
	Foramen magnum	Large hole in base of bone
	Occipital condyles	Smooth joint surfaces each side of foramen magnum
	Nuchal lines	Ridges horizontally and vertically on back of bone
	External occipital protuberance	"Bump" on outside of back of bone
	Internal occipital protuberance	"Bump" on inside of bone
Temporals (2)		Side walls and floor of cranium
	Squamous portion (squama)	"Temple" (side) of cranium
	Zygomatic process	Projection from squama, anterior
	Mandibular fossa	At base of zygomatic process; mandible attaches here
	Tympanic portion	Below squama
	External auditory meatus	Opening of ear
	Styloid process	"Spike" extending down from skull
	Mastoid portion	Posterior to ear opening
	Mastoid process	Extends downward behind ear
	Mastoid air cells	Air cavities in mastoid process
	Stylomastoid foramen	Hole between styloid and mastoid processes
	Petrous portion	Ridge inside cranium
	Internal auditory meatus	On anterior medial border of petrous portion
	Jugular foramen	Halfway down joint between petrous temporal and occipital bone
	Carotid canal	Center of base of petrous portion
Sphenoid (1)		"Butterfly shaped" bone in center of skull
	Body	Central part of bone
	Sphenoid sinus	Cavity in body
	Sella turcica	Depression on upper surface of body
	Greater wings	Extend laterally from body
	Lesser wings	Superior to body and greater wings
	Pterygoid processes	Inferiorly from body
	Optic foramen	Hole at base of lesser wing
	Superior orbital fissure	Cleft between greater and lesser wings
	Foramen ovale	Hole at base of greater wing
	Foramen spinosum	Lateral to foramen ovale
	Foramen rotundum	In posterior side of base of greater wing

TABLE 8.1 (Continued)

Name of bone and number	Features on the bone	Description/location of bone/feature
Ethmoid		Lies between orbits
	Horizontal (cribriform) plate	Roof of nasal cavity; full of holes (passes olfactory nerves)
	Crista galli	Crest above horizontal plate
	Perpendicular plate	Separates nasal cavities (upper 2/3)
	Lateral masses (conchae)	Carry two of three turbinates for increasing area of nasal cavities
Nasals (2)		Bridge of the nose
Zygomatic (malar) (2)		Cheekbones
Maxillae (2)		Floor of orbits, upper jaws, part of hard palate
	Orbital plates	Floor of orbits
	Palatine processes	Anterior three fourths of hard palate
	Infraorbital foramen	Hole below orbit
	Maxillary sinus (antrum of Highmore)	Cavity in bone
	Alveolar process	Bears teeth of upper jaw
Mandible (1)		Lower jaw
	Body	Horseshoe shaped part
	Rami (sing.: ramus)	Upwards directed processes
	Angle	Where body and rami meet
	Coronoid processes	Anterior processes of rami
	Condyloid processes	Posterior processes of rami; form joint with temporal
	Mandibular foramen	Hole in medial side of ramus
	Mental foramen	Hole in anterior body
	Mylohyoid line	Ridge from mandibular foramen
	Alveolar margin	Bears teeth of lower jaw
Lacrimal (2)		Fingernail-sized bones in front of medial orbit wall
	Lacrimal canal	Passage through bone
Vomer (1)		Blade-like bone forming lower $\frac{1}{3}$ of nasal septum
Palatine (2)		L-shaped; form posterior $\frac{1}{4}$ of hard palate and sides of nasal cavities
	Horizontal plates	Posterior $\frac{1}{4}$ of hard palate
	Perpendicular plate	Forms part of lateral nasal wall
Inferior conchae (2)		Lower part of lateral walls of nasal cavities
Hyoid (1)		U-shaped bone in upper neck
	Body	Front part of bone
	Greater cornua (horns)	Large, posteriorly projecting processes
	Lesser cornua (horns)	Small, superiorly projecting processes

TABLE 8.1 (Continued)		
Name of bone and number	Features on the bone	Description/location of bone/feature
Ear ossicles (6) [malleus (2), incus (2), stapes (2)]		Found in middle ear cavities
Sutures	Frontal (metopic)	Separates two halves of frontal bone in newborn
	Coronal	Separates frontal and parietal bones
	Sagittal	Separates parietal bones
	Lambdoidal	Separates parietals from occipital bone
	Squamous	Separates parietal and temporal bones
Sinuses or air cells	Frontal sinus	Behind medial parts of superciliary ridges
	Maxillary sinus	Inside maxillary bone
	Sphenoid sinus	In body of sphenoid
	Mastoid air cells	In mastoid process
	Ethmoid air cells	In ethmoid bone
Fontanels (6)		
Frontal (1)		At junction of coronal and sagittal sutures
Occipital (1)		At junction of sagittal and lambdoidal sutures
Sphenoid (2)		At junction of coronal and squamous sutures
Mastoid (2)		At junction of squamous and lambdoidal sutures

the orbit, form the greater part of the hard palate that roofs the oral cavity (floor of nasal cavity), and carry the teeth of the upper jaw.

The MANDIBLE (Fig. 8.15) is the largest of the facial bones, and forms the lower jaw. It also carries teeth.

The LACRIMAL BONES (Fig. 8.16) are the smallest bones of the face, about the size of a fingernail. They lie in the anteromedial wall of the orbit, and contain a lacrimal canal through which the nasolacrimal duct passes. The duct is part of the system for draining tears from eye to nasal cavity.

The VOMER (Fig. 8.17) is shaped like the blade of a plow. It forms about the lower one-third of the bony septum of the nasal cavities.

The PALATINE BONES (Fig. 8.18) are L-shaped bones that form the posterior one-fourth of the hard palate, part of the lateral walls of the nasal cavities, and a small part of the orbit.

The INFERIOR CONCHAE (Fig. 8.19), or turbinates, extend from the lateral walls of the nasal cavities into the lower portion of the nasal cavities. They aid in creating a greater surface area for cleansing inhaled air.

Other bones associated with the skull include the hyoid bone and the ear ossicles. The HYOID BONE (Fig. 8.20) is a U-shaped bone lying in the neck at the level of the mandibular angle. The bone affords attachment for several swallowing muscles. The 6 EAR OSSICLES (3 in each middle ear cavity) (Fig. 8.21) aid in transmission of sound waves from ear drum to cochlea. Their names are MALLEUS (hammer), INCUS (anvil), and STAPES (stirrup). Their functions are described in Chapter 18.

A grand total of 29 bones thus compose or are associated with the skull.

(Text continued on page 154.)

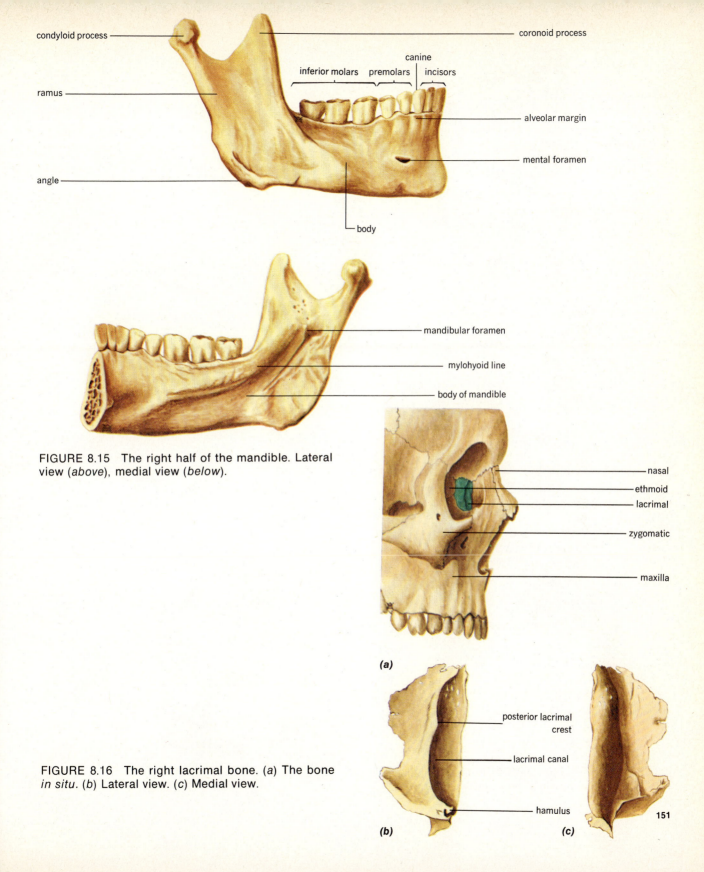

condyloid process

coronoid process

canine

inferior molars   premolars   incisors

ramus

alveolar margin

mental foramen

angle

body

mandibular foramen

mylohyoid line

body of mandible

FIGURE 8.15   The right half of the mandible. Lateral view (*above*), medial view (*below*).

nasal

ethmoid

lacrimal

zygomatic

maxilla

(a)

posterior lacrimal crest

lacrimal canal

hamulus

FIGURE 8.16   The right lacrimal bone. (*a*) The bone *in situ*. (*b*) Lateral view. (*c*) Medial view.

(b)

(c)

151

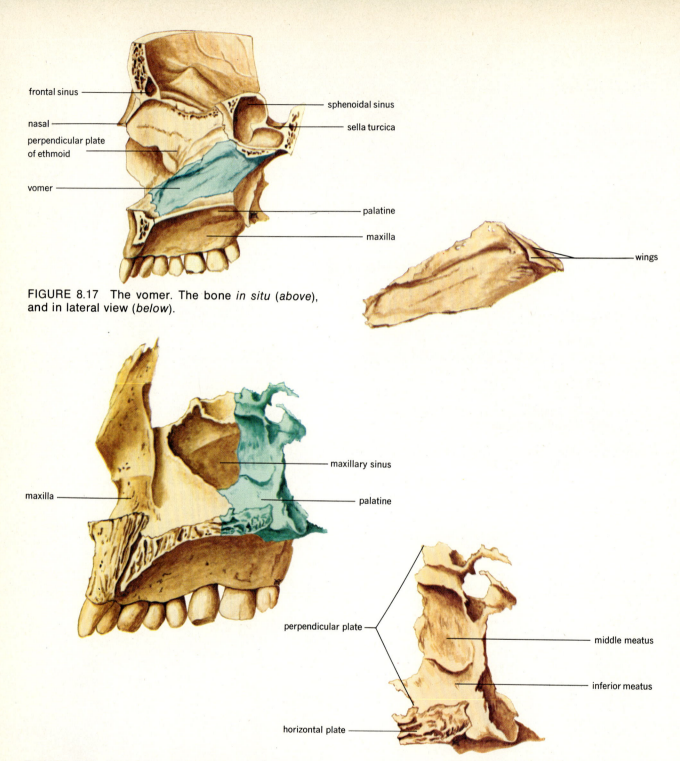

frontal sinus

sphenoidal sinus

nasal

sella turcica

perpendicular plate
of ethmoid

vomer

palatine

maxilla

wings

FIGURE 8.17   The vomer. The bone *in situ* (*above*),
and in lateral view (*below*).

maxillary sinus

maxilla

palatine

perpendicular plate

middle meatus

inferior meatus

horizontal plate

FIGURE 8.18   The right palatine bone. The bone *in
situ* (*above*), and in medial view (*below*).

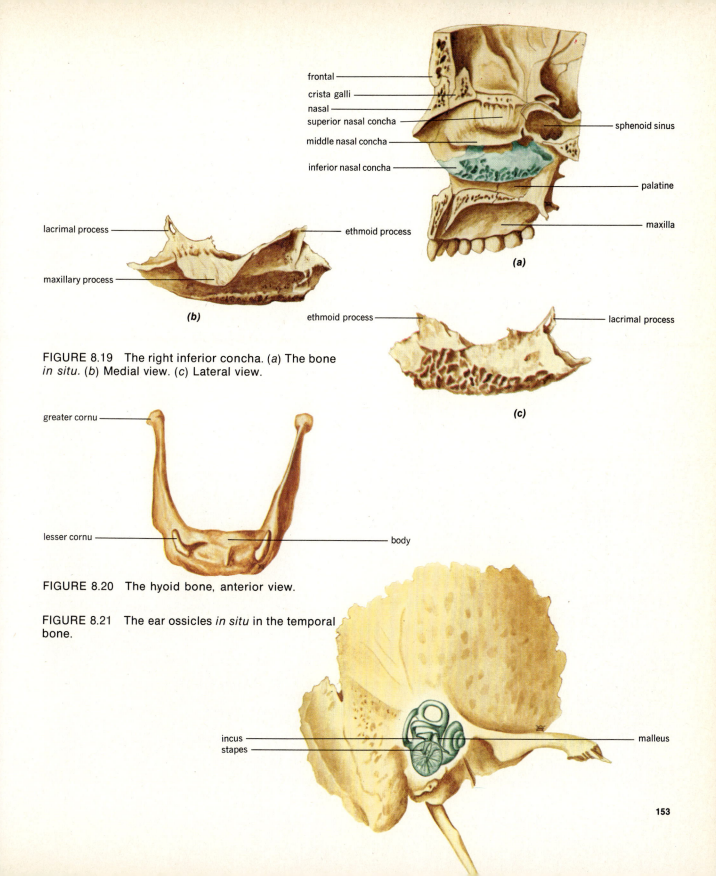

frontal
crista galli
nasal
superior nasal concha
middle nasal concha
inferior nasal concha

sphenoid sinus

palatine

maxilla

(a)

lacrimal process
maxillary process

ethmoid process

(b)

ethmoid process

lacrimal process

(c)

FIGURE 8.19   The right inferior concha. (*a*) The bone *in situ*. (*b*) Medial view. (*c*) Lateral view.

greater cornu

lesser cornu

body

FIGURE 8.20   The hyoid bone, anterior view.

FIGURE 8.21   The ear ossicles *in situ* in the temporal bone.

incus
stapes

malleus

A number of foramina are easily visualized in the basal view of the skull (Fig. 8.9) and in the interior view of the skull (Fig. 8.11). Table 8.2 presents a list of foramina and the structures that pass through them.

FONTANELS. Bone formation is not complete at birth. There are, where the major sutures join one another, large membranous areas known as fontanels. They permit brain growth, since the bones have not yet knit together. The locations and names of the fontanels are shown in Figure 8.22. Some synonomy of terminology exists as is shown below:

Frontal or anterior fontanel or bregma
Occipital or posterior fontanel or lambda
Sphenoid or anterolateral fontanel
Mastoid or posterolateral fontanel

The anterior fontanel measures 4 to 6 centimeters in greatest diameter at birth, and closes between 4 and 26 months of age. Ninety percent close between 7 and 19 months of age. The posterior fontanel measures 1 to 2 centimeters in diameter at birth and usually closes by 2 months of age. The lateral fontanels are usually closed at birth.

The anterior fontanel is commonly observed and palpated (felt) in the infant to give clues as to the state of hydration and intracranial pressure. In a quiet, sitting infant, the fontanel should lie nearly even with the skull bones. An obviously depressed fontanel suggests dehydration, while a bulging fontanel suggests increased intracranial pressure. It should be noted that the fontanel bulges (normally) when the infant cries, coughs, or vomits.

TABLE 8.2 Foramina of the skull of humans		
Foramina of facial bones	Location	Structures passing through
1. Incisive (Stensen; Scarpa, sometimes)	Posterior to incisor teeth in hard palate	Stensen, anterior branches of descending palatine vessels; Scarpa, nasopalatine nerves
2. Greater palatine	Palatine bones at posterior-lateral angle of hard palate	Greater palatine nerve
3. Lesser palatine	Palatine bones at posterior-lateral angle of hard palate	Lesser palatine nerves
4. Supraorbital (or notch)	Frontal bone, above orbit	Supraorbital nerve and vessels
5. Infraorbital	Maxilla, body	Infraorbital nerve and vessels
6. Zygomaticofacial	Zygomatic bone, malar surface	Zygomaticofacial nerve and vessels
7. Zygomaticotemporal	Zygomatic bone, temporal surface	Zygomaticotemporal nerve
8. Zygomaticoorbital	Zygomatic bone, orbital surface	Zygomaticotemporal and zygomaticoorbital nerves
9. Mental	Mandible, lateral anterior surface, inferior to second premolar	Mental nerve and vessels

TABLE 8.2 (Continued)

Foramina of facial bones	Location	Structures passing through
10. Mandibular	Mandible, about center on medial side of ramus	Inferior alveolar vessels and nerves
Foramina of cranial bones	**Location**	**Structures passing through**
1. Olfactory	Ethmoid, cribriform plate	Olfactory nerves (I)
2. Optic	Sphenoid, superior surface	Optic nerves (II); ophthalmic arteries
3. Superior orbital fissure	Sphenoid, between small and great wings	Oculomotor (III); trochlear (IV); trigeminal (V) ophthalmic branch; abducens (VI); orbital branches of middle meningeal artery; superior ophthalmic vein; branch of lacrimal artery
4. Inferior orbital fissure	Sphenoid; zygomatic; maxilla; palatine	Trigeminal (V), maxillary nerve; infraorbital vessels
5. Rotundum	Sphenoid, greater wing	Maxillary nerve (V)
6. Ovale	Sphenoid, greater wing	Mandibular nerve (V); accessory meningeal artery; lesser petrosal nerve
7. Spinosum	Sphenoid, greater wing	Mandibular nerve, recurrent branch (V); middle meningeal vessels
8. Lacerum	Sphenoid, greater wing and body; apex of petrous temporal; base of occipital	Meningeal branch of the ascending pharyngeal artery; internal carotid artery
9. Internal acoustic meatus	Temporal, petrous portion	Facial nerve (VII); acoustic nerve (VIII); internal auditory artery; nervus intermedius
10. Jugular	Temporal, petrous portion; occipital	Glossopharyngeal (IX); vagus (X); accessory (XI); internal jugular vein
11. Hypoglossal canal	Occipital	Hypoglossal nerve (XII); meningeal artery
12. Condyloid canal	Occipital	Vein from transverse sinus
13. Carotid canal	Temporal, petrous portion	Internal carotid artery
14. Stylomastoid	Temporal, between mastoid and styloid processes	Facial nerve (VII)
15. Mastoid	Temporal, mastoid portion	An emissary vein
16. Foramen magnum	Occipital	Medulla oblongata and its meninges; accessory nerves (XI); vertebral arteries

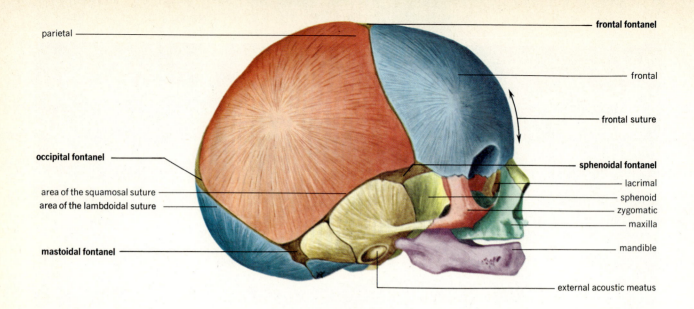

parietal

frontal fontanel

frontal

frontal suture

occipital fontanel

area of the squamosal suture
area of the lambdoidal suture

sphenoidal fontanel

lacrimal
sphenoid
zygomatic
maxilla

mastoidal fontanel

mandible

external acoustic meatus

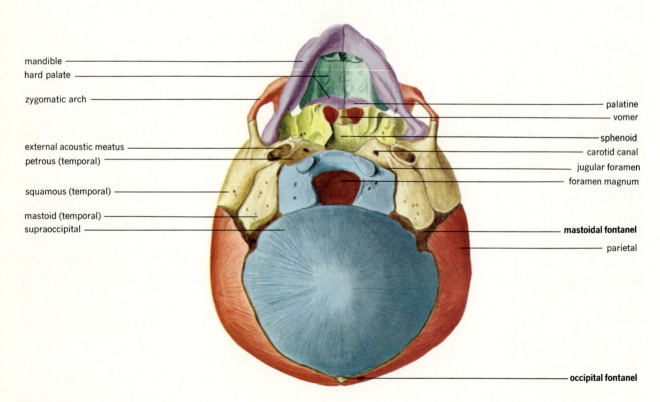

mandible
hard palate

zygomatic arch

external acoustic meatus
petrous (temporal)

squamous (temporal)

mastoid (temporal)
supraoccipital

palatine
vomer

sphenoid
carotid canal
jugular foramen
foramen magnum

mastoidal fontanel
parietal

occipital fontanel

FIGURE 8.22 The infant skull. Lateral view (*above*)
and basal view (*below*).

The vertebral column

The vertebral column (backbone or spine) of the adult consists of 24 separate bones known as VERTEBRAE, plus the SACRUM and COCCYX. The vertebrae are divided into three groups: 7 CERVICAL (neck); 12 THORACIC (chest); and 5 LUMBAR (small of the back). The sacrum is a single bone composed of 5 fused, modified vertebrae, while the coccyx ("tail bone") consists of 3 to 5 fused, small bones. Viewed from the side (Fig. 8.23), the adult column shows a number of CURVATURES. The cervical and lumbar curves are convex anteriorly, while the thoracic and sacral curves are concave anteriorly. The latter two are termed primary curves, the first two secondary curves. At birth there is only one curvature, an anteriorly concave curve extending from head to buttocks. As the child raises its head, and then walks, the secondary curves develop. The curves allow the column to absorb the shocks of locomotion, acting like a "spring." Exaggeration of the thoracic curve is termed kyphosis ("hunchback"); of the lumbar curve, lordosis ("swayback"); a deviation laterally from the normal straight-line of the spine is scoliosis (Fig. 8.24).

Most vertebrae resemble each other in having a BODY, PEDICLES, LAMINAE, TRANSVERSE and SPINOUS PROCESSES, and ARTICULATING PROCESSES (Fig. 8.25). Regionally, there are certain variations (Fig. 8.26).

All cervical vertebrae have, in their transverse processes, a TRANSVERSE FORAMEN that passes an artery to the brain. The first cervical vertebra or ATLAS, lacks a body and is ringlike in shape. It articulates with the skull and permits "yes" movements of the head. The second cervical vertebra, or AXIS (epistropheus), has a DENS (odontoid process) attached to its body. With the atlas, the dens forms a pivot joint permitting the "no" movements of the head.

All thoracic vertebrae have, on body and transverse processes, FACETS for rib attachments.

Lumbar vertebrae are very HEAVY, lack transverse foramina and facets, and have articulating processes that are vertically oriented for strength and weight bearing.

The SACRUM (Fig. 8.27) is designed for strength and support of the body weight above and the lower limbs below. The COCCYX is regarded as the remnant of a tail.

The vertebrae are separated from one another by pads of fibrous cartilage called INTERVERTEBRAL DISCS (see Figs. 8.5 and 8.23b). Each disc has a soft center (nucleus pulposus) and an outer fibrous coat. Rupture of the fibrous coat may allow protrusion of the soft center into the vertebral canal to press on the spinal cord; also, some compression may occur as the vertebra above the ruptured disc "settles" on the one below. Compression of spinal nerves, which exit from between the vertebrae, may occur.

Failure of the laminae of the vertebrae to fuse as the bones develop creates SPINA BIFIDA. The membranes surrounding the cord may protrude through the opening (meningocele), or the membranes and the cord itself may protrude (meningomyelocele).

The thorax

The thorax (Fig. 8.28) is the roughly cone-shaped bony cage formed by the sternum, 12 pairs of ribs, and the costal cartilages. The STERNUM lies anteriorly, and consists of an upper MANUBRIUM, joined to a central BODY at the sternal angle, and a lower XIPHOID (ensiform) PROCESS. The cartilage of the second rib attaches at the angle and enables easy numbering of ribs. The manubrium is subcutaneous (just beneath the skin) on the chest, and affords an easy site for removal of bone marrow samples to aid in diagnosis of some blood disorders.

The 24 RIBS attach posteriorly to the thoracic vertebrae, and curve forward and down. The cartilages of the first 7 pairs attach directly to the sternum and these are known as true ribs. Those of the next 5 pairs attach to the preceding cartilage or not at all, and are called false ribs. The last 2

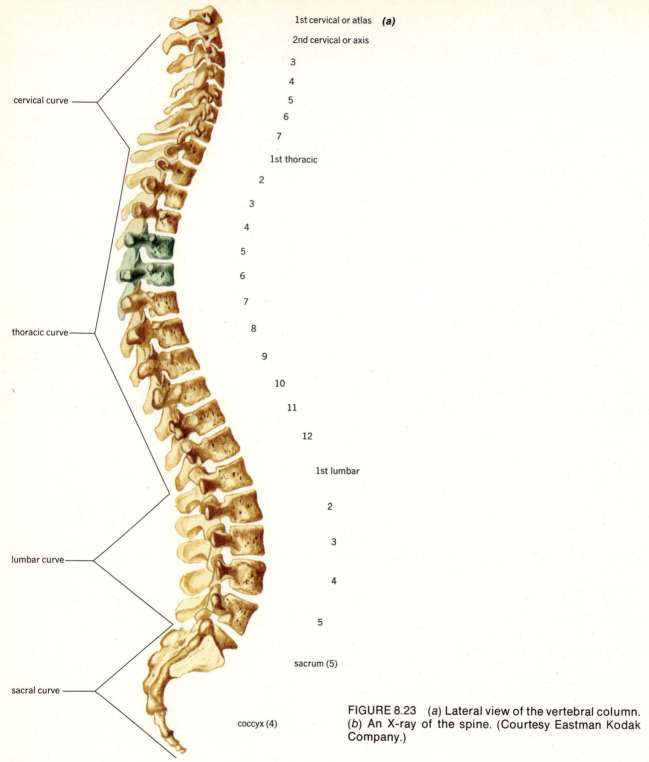

1st cervical or atlas **(a)**

2nd cervical or axis

3

4

5

6

7

1st thoracic

2

3

4

5

6

7

8

9

10

11

12

1st lumbar

2

3

4

5

sacrum (5)

coccyx (4)

cervical curve

thoracic curve

lumbar curve

sacral curve

FIGURE 8.23   (a) Lateral view of the vertebral column. (b) An X-ray of the spine. (Courtesy Eastman Kodak Company.)

(b)

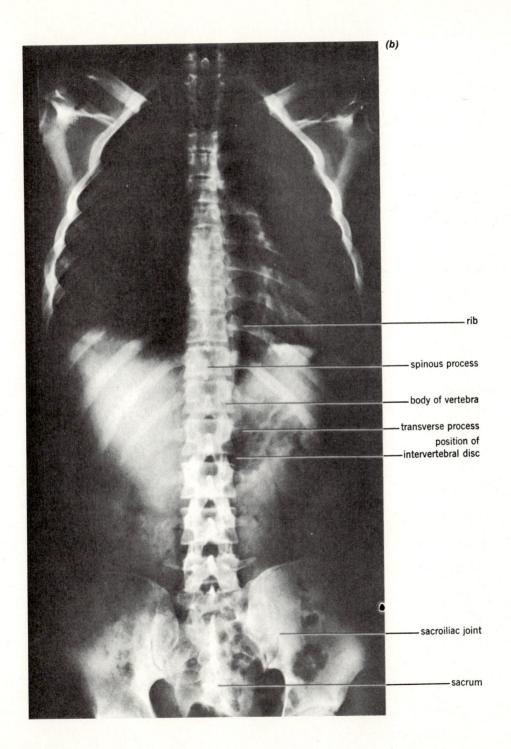

rib

spinous process

body of vertebra

transverse process

position of
intervertebral disc

sacroiliac joint

sacrum

159

pairs of false ribs, which have no anterior attach-ment to the sternum, are *floating ribs*. A typical rib (Fig. 8.29) has a HEAD, NECK, TUBERCLE, SHAFT, ANGLE, and FACET for a costal cartilage.

The appendicular skeleton

THE UPPER APPENDAGE. Each upper appendage contains 32 bones and consists of: the shoulder (pectoral) girdle, composed of scapula and clavi-cle; the humerus (upper arm); radius and ulna (lower or forearm); carpals (wrist); metacarpals (hand); and phalanges (fingers and thumb). The features on the bones of the upper appendages are listed in Table 8.3.

The CLAVICLE (Fig. 8.30) is a doubly curved bone that affords the only direct attachment of the upper limb to the axial skeleton. This attach-ment occurs at the sternoclavicular joint and al-lows great mobility of the upper limb. Bracing the shoulder, the clavicle is often fractured when falling on an extended arm.

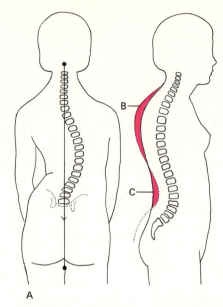

FIGURE 8.24   Abnormal curvatures of the spine. A—Scoliosis. B—Kyphosis. C—Lordosis.

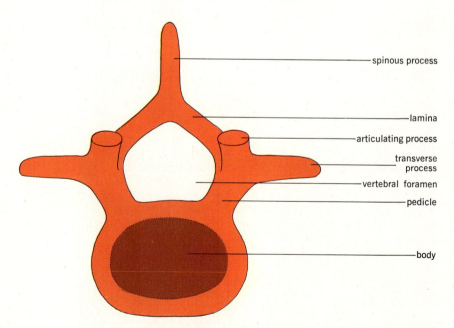

spinous process

lamina

articulating process

transverse process

vertebral foramen

pedicle

body

FIGURE 8.25   The components of a vertebra, superior view.

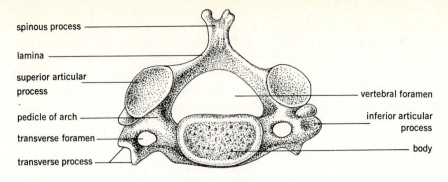

spinous process

lamina

superior articular process

pedicle of arch

transverse foramen

transverse process

vertebral foramen

inferior articular process

body

cervical region

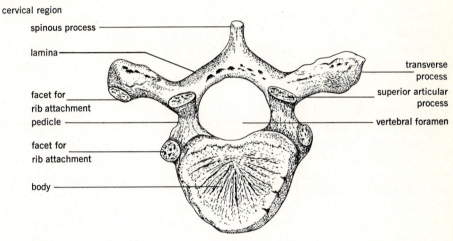

spinous process

lamina

facet for rib attachment

pedicle

facet for rib attachment

body

transverse process

superior articular process

vertebral foramen

thoracic region

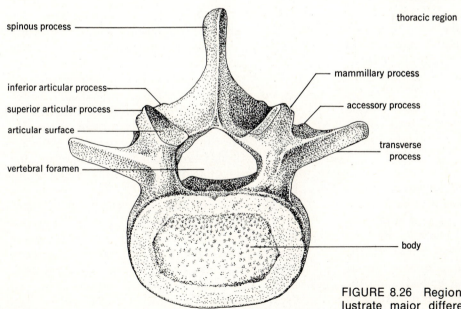

spinous process

inferior articular process

superior articular process

articular surface

vertebral foramen

mammillary process

accessory process

transverse process

body

lumbar region

FIGURE 8.26 Regional variations of vertebrae, to illustrate major differences. (Drawings to the same scale.)

The SCAPULA (Fig. 8.31) floats in muscle, attaching to the clavicle at the acromioclavicular joint, and to the humerus.

The HUMERUS (Fig. 8.32) is the largest bone of the upper limb, and forms the upper arm. It articulates proximally with the scapula, distally with the radius and ulna.

The ULNA (Fig. 8.33) is the *medial* bone of the forearm, and articulates proximally with the trochlea of the humerus, distally with the carpals.

The RADIUS (*see* Fig. 8.33) is a bit shorter than the ulna, and forms the *lateral* bone of the forearm. It articulates proximally with the capitulum of the humerus, distally with the carpals. The most common fracture of the radius is a break in the distal end (Colles' fracture) caused by falling on the hand.

The wrist and hand (Fig. 8.34). Eight (8) CARPAL BONES form each wrist. They are arranged in two rows of four bones each. From the thumb side, the proximal row contains the SCAPHOID or navicular ("boat"), LUNATE (moon), TRIQUETRUM or triquetral (triangular), PISIFORM ("pealike"). The distal row, also from the thumb side contains the TRAPEZIUM (greater multangular, "having many angles"), the TRAPEZOID (lesser multangular), CAPITATE ["having a (rounded) head"], and HAMATE ("hooked" or "hooklike").

Five METACARPALS, numbered 1–5 beginning with the thumb, make up each hand, while 14 PHALANGES compose the 5 digits in each hand. There are 2 phalanges in the thumb, while the remaining digits contain 3 each. Strictly speaking, each hand thus contains 4 fingers and a thumb, not 5 fingers.

THE LOWER APPENDAGE. Each lower appendage contains 30 bones. The os coxae (hip bone) joins the sacrum at the sacroiliac joint, and affords attachment for the lower limb. Each lower limb, in turn consists of a femur (thigh bone), patella (kneecap), tibia and fibula (leg bones), tarsals (ankle bones), metatarsals (foot bones), and phalanges (toe bones). The features on the bones of the lower appendage are shown in Table 8.4.

The OS COXAE or innominate bone (Fig. 8.35) is the bone of the pelvic girdle, and is formed by the fusion of three fetal bones. A superior ILIUM, a posterior ISCHIUM, and an anterior PUBIS form the bone. The suture lines meet in the cuplike ACETABULUM, which articulates with the femur.

The sacrum plus the two os coxae constitute the PELVIS. Males and females show differences in shape and size of the pelvis. The male pelvis is heart-shaped, being narrow and deep. The subpubic angle is usually about 90°. The female pelvis is broader and shallower than that of the male. The subpubic angle is about 120°, creating a larger outlet inferiorly. These differences (Fig. 8.36) appear correlated with the strength usually associated with the male, and the childbearing function usually associated with the female. Also, true and false pelves are shown. The true pelvis is the cavity lying below the iliopectineal (arcuate) line. The true pelvis is marked by the pelvic outlet, which determines the maximum opening through which childbirth may occur.

The FEMUR (Fig. 8.37) is the longest, heaviest, and strongest bone in the body. It transmits, and bears, the shock of locomotion and weight of the body.

The PATELLA (Fig. 8.38), a sesamoid bone, protects the knee joint anteriorly and acts as a pulley to increase the contractile efficiency of certain thigh muscles.

The TIBIA and FIBULA (Fig. 8.39) form the leg, that is, the part of the lower appendage between the knee and ankle. The tibia is the medial bone of the leg, and the fibula is the lateral bone of the leg. A fracture that splits the lateral malleolus from the fibula is common in trauma that violently twists the foot to the outside; it is called Pott's fracture.

The ankle and foot (Fig. 8.40). Seven TARSAL BONES form the ankle. The high points of a medial longitudinal arch, and a transverse arch meet at the navicular bone; this point also represents the point at which the entire body weight is directed. If the arches are maintained, a good degree of "spring" is given; if not, the individual is "flat-footed."

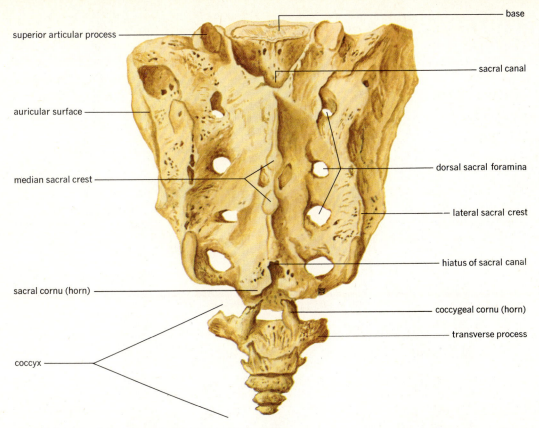

superior articular process

base

sacral canal

auricular surface

median sacral crest

dorsal sacral foramina

lateral sacral crest

hiatus of sacral canal

sacral cornu (horn)

coccygeal cornu (horn)

transverse process

coccyx

FIGURE 8.27   Sacrum and coccyx, posterior view.

Five METATARSALS form the center of the foot; these, like those of the hand, are numbered 1–5 beginning with the big toe. Fourteen PHALANGES form the toes, 2 in the big toe, 3 in each of the 4 smaller toes.

Clinical considerations

FRACTURES. A fracture is a loss of continuity of a bone. It is most frequently due to trauma of some sort, such as falls, accidents, or blows. Pathological causes include tumors or loss of inorganic constituents (osteoporosis) that weaken the bone and render it liable to break under "normal use." According to the manner in which the break occurs, several types of fractures are recognized

(Fig. 8.41). Treatment of the fracture usually involves immobilizing the broken ends after restoring anatomical relationships, to allow healing to occur. Several steps occur as the bone is repaired.

The bone ends are bound together by fibrin from torn blood vessels, lymph, and exudate.

Granulation tissue is formed. This involves invasion of the area by fibroblasts and vascular elements. At the same time, some fibroblasts form osteoblasts. The periosteum is a main source of both cells and vessels.

Callus formation occurs. This involves formation of bony tissue by the endochondral method between the broken ends.

About 4–6 weeks is required for a broken bone to heal completely.

(*Text continued on page 169.*)

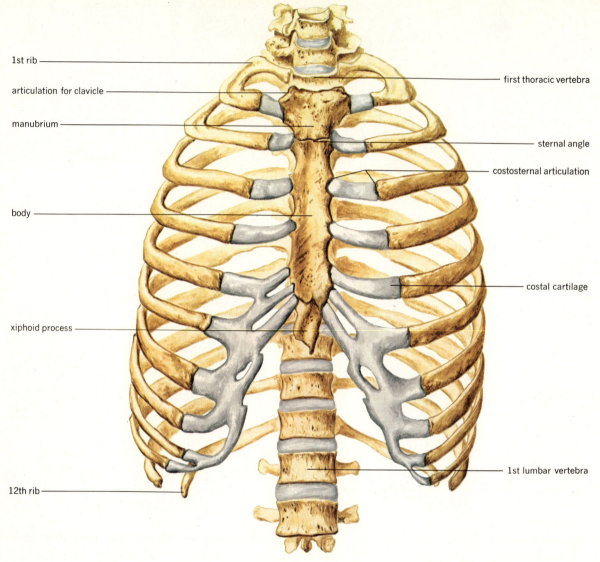

1st rib

articulation for clavicle

manubrium

body

xiphoid process

12th rib

first thoracic vertebra

sternal angle

costosternal articulation

costal cartilage

1st lumbar vertebra

FIGURE 8.28   (a) Thorax, anterior view. (b) An X ray of the thorax. (Courtesy Eastman Kodak Company.)

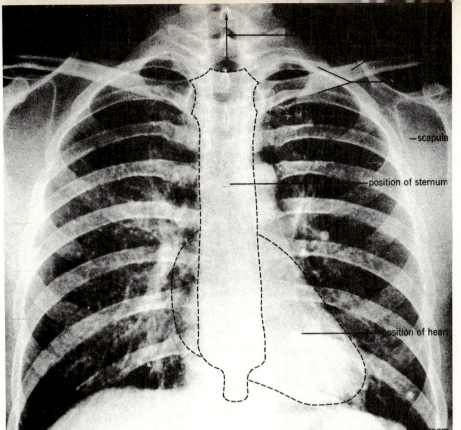

(b)

—scapula

—position of sternum

—position of heart

—diaphragm

TABLE 8.3 A summary of the features of the bones of the upper appendages and girdles (italics indicate the most important features)

Name and number of bones in *each* appendage	Features on the bone	Description/location of bone/feature
Clavicle (1)		Collarbone
Scapula (1)	*Body*	Triangular, flat part of the bone carries most other features
	Superior border	Upper border
	Axillary (lateral) border	Faces the armpit
	Vertebral (medial) border	Faces the backbone
	Spine	Horizontal ridge on posterior surface; carries acromion process
	Supraspinous fossa	Depression above spine
	Infraspinous fossa	Depression below spine
	Subscapular fossa	Depression on anterior surface
	Glenoid fossa	Lateral; joins to humerus
	Acromion process	Projects above glenoid fossa
	Coracoid process	Beaklike; anterior
	Inferior angle	Lower "point"
	Superior angle	Upper and medial "point"
	Lateral angle	"Point" beneath glenoid fossa

TABLE 8.3   (Continued)

Name and number of bones in *each* appendage	Features on the bone	Description/location of bone/feature
Humerus (1)	*Head*	Rounded upper end
	Anatomical neck	Just below head
	Surgical neck	Upper part of shaft
	Greater tuberosity	Lateral to anatomical neck
	Lesser tuberosity	Anterior, below anatomical neck
	Bicipital groove	Lies between tuberosities
	Deltoid tuberosity	Lateral, about $\frac{1}{3}$ the way down bone
	Medial epicondyle	Distal and medial
	Lateral epicondyle	Distal and lateral
	Capitulum⎤are	Round surface on distal end
	Trochlea ⎦condyles	Spool-shaped surface on distal end
	Coronoid fossa	Depression on anterior distal end
	Olecranon fossa	Depression on posterior distal end
	Nutrient foramina	Holes in shaft of bone; allow blood vessels inside
Ulna (1)	*Semilunar notch*	Half-moon-shaped surface, proximal end; joins trochlea
	Olecranon process	"Point" of elbow (posterior)
	Coronoid process	Anterior, above semilunar notch
	Radial notch	Lateral on proximal end; joins radius head
	Styloid process	Distal pointed end
Radius (1)	*Head*	Rounded surface at proximal end; joins capitulum
	Neck	Below head
	Tuberosity	Below neck on anterior surface
	Styloid process	Distal pointed end
Carpals (8)	Navicular ⎤	"Boat"
	Lunate ⎟ Proximal row,	"Moon" (halfmoon)
	Triangular ⎟ lateral to medial	Shaped like a right triangle
	Pisiform ⎦	"Pea"
	Trapezium ⎤	"Little table"
	Trapezoid ⎟ Distal row,	"Table shaped"
	Capitate ⎟ lateral to medial	"Having a head" (rounded)
	Hamate ⎦	"Hooked"
Metacarpals (5)	Form the palm of the hand; numbered 1–5 starting with thumb.	
Phalanges (14)	Form the five digits; 2 in the thumb, 3 in each of the 4 other digits.	

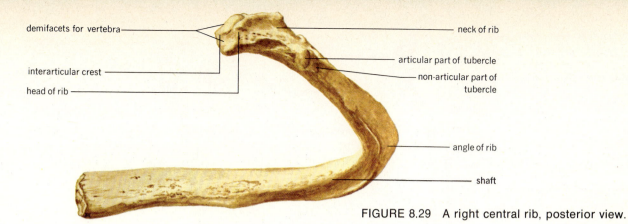

demifacets for vertebra

interarticular crest

head of rib

neck of rib

articular part of tubercle

non-articular part of tubercle

angle of rib

shaft

FIGURE 8.29   A right central rib, posterior view.

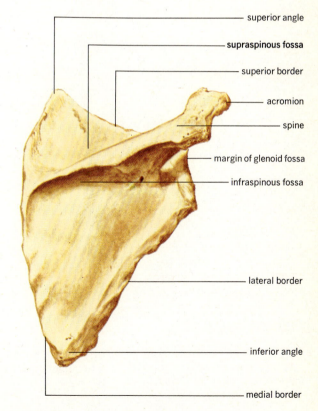

sternal end

acromial end

coracoid tuberosity

FIGURE 8.30   Right clavicle, anterior view.

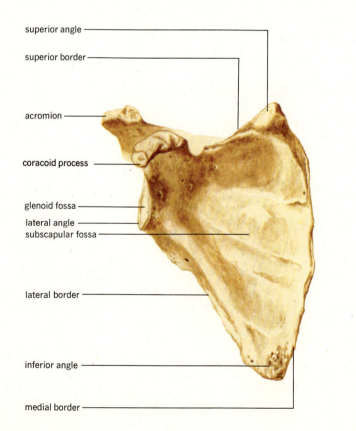

superior angle

superior border

acromion

coracoid process

glenoid fossa

lateral angle

subscapular fossa

lateral border

inferior angle

medial border

superior angle

supraspinous fossa

superior border

acromion

spine

margin of glenoid fossa

infraspinous fossa

lateral border

inferior angle

medial border

FIGURE 8.31   Right scapula. Anterior view *(left)*, posterior view *(right)*.

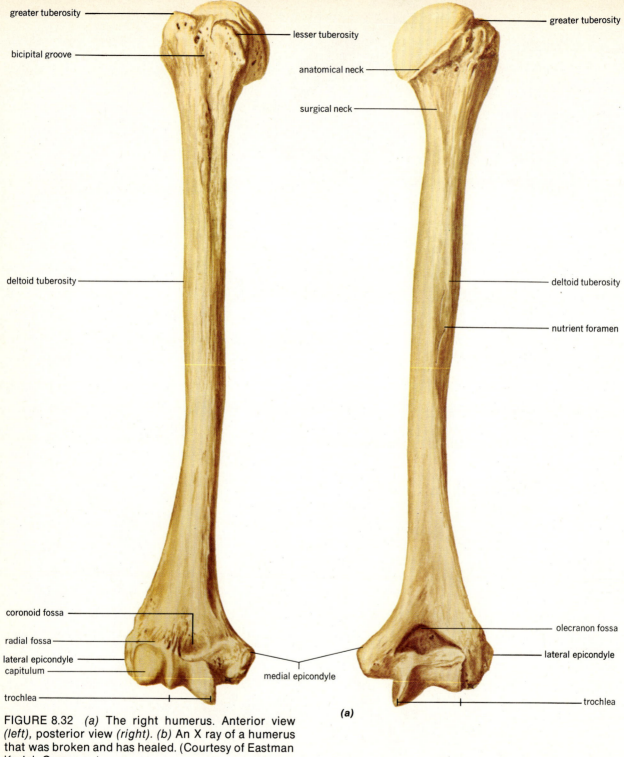

greater tuberosity

lesser tuberosity

bicipital groove

greater tuberosity

anatomical neck

surgical neck

deltoid tuberosity

deltoid tuberosity

nutrient foramen

coronoid fossa

radial fossa

lateral epicondyle

capitulum

trochlea

medial epicondyle

olecranon fossa

lateral epicondyle

trochlea

(a)

FIGURE 8.32 *(a)* The right humerus. Anterior view *(left)*, posterior view *(right)*. *(b)* An X ray of a humerus that was broken and has healed. (Courtesy of Eastman Kodak Company.)

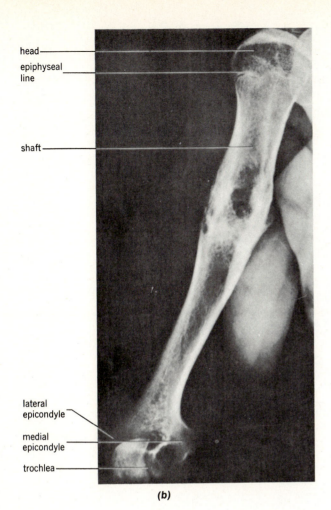

head

epiphyseal
line

shaft

lateral
epicondyle

medial
epicondyle

trochlea

(b)

BONE DISEASES AND DISORDERS. Osteomyelitis is a staphylococcal or streptococcal disease of bone. The bacteria settle in the marrow cavity, causing bone destruction, pus formation, and excruciating pain. Antibiotic therapy usually controls the disease.

Tumors occurring in bone may be benign, malignant, or may arise by metastasis from other tissues and organs. The cause of benign tumors is not known, although some seem to be associated with abnormalities of growth and development of bone. Mild pain or a lump on a bone may signal a benign tumor. Malignant tumors include *osteosarcomas* (bone tumor), *fibrosarcoma* (fibrous tissue tumor), *chondrosarcoma* (tumor of cartilage), and round cell sarcoma, or Ewing's sarcoma. The latter is a tumor of endothelial origin (vascular) in or on a bone. Metastatic tumors form the most common bone tumors. They occur most often in persons over 40, and consist primarily of metastases from breast, prostate, gut, and lung tumors.

No attempt is made to summarize this chapter since it consists mainly of anatomical terms. The glossary and tables will prove helpful if additional aid is required.

(*Text continued on page 181.*)

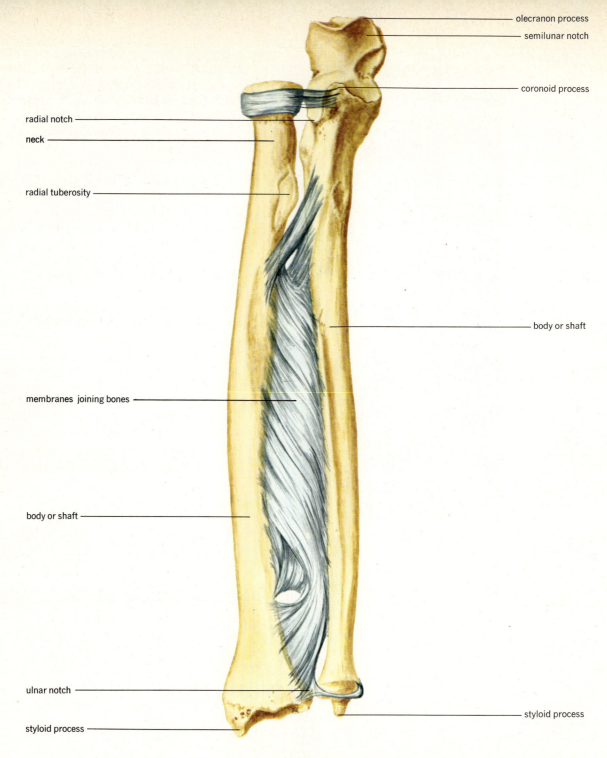

olecranon process
semilunar notch
coronoid process
radial notch
neck
radial tuberosity
body or shaft
membranes joining bones
body or shaft
ulnar notch
styloid process
styloid process

FIGURE 8.33 The right radius and ulna viewed anteriorly. The connective tissue membranes form a syndesmosis between the bones.

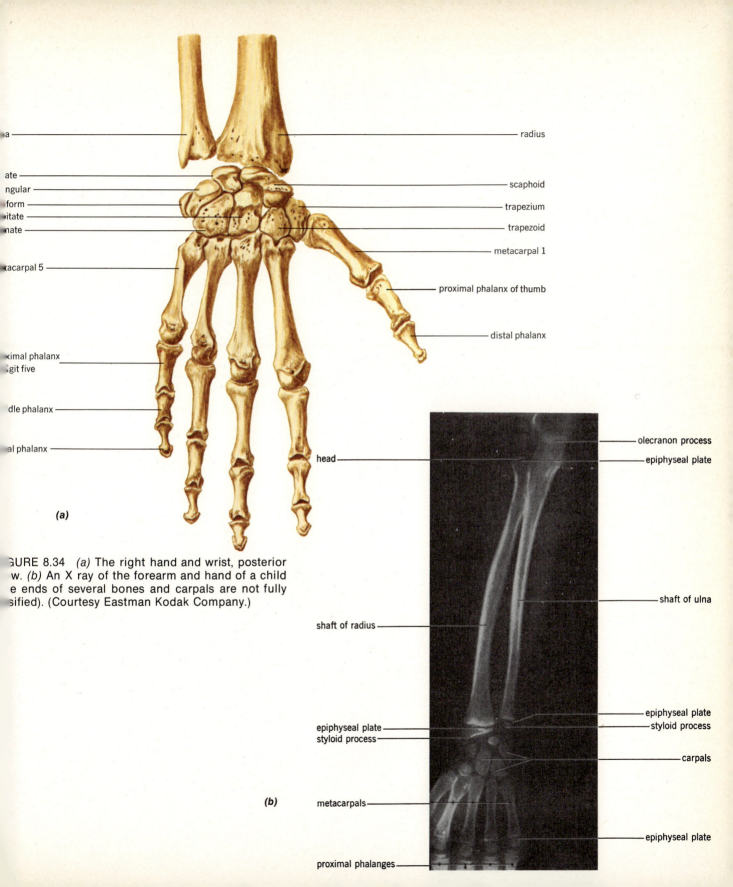

...a ———————————————— radius

...ate ———
...ngular ———
...iform ———
...itate ———
...nate ———

scaphoid

trapezium

trapezoid

...acarpal 5 ———

metacarpal 1

proximal phalanx of thumb

distal phalanx

...ximal phalanx
...git five ———

...dle phalanx ———

...al phalanx ———

(a)

FIGURE 8.34   (a) The right hand and wrist, posterior
...w. (b) An X ray of the forearm and hand of a child
...e ends of several bones and carpals are not fully
...sified). (Courtesy Eastman Kodak Company.)

head ———

olecranon process
epiphyseal plate

shaft of ulna

shaft of radius ———

epiphyseal plate ———
styloid process ———

epiphyseal plate
styloid process

carpals

metacarpals ———

(b)

epiphyseal plate

proximal phalanges ———

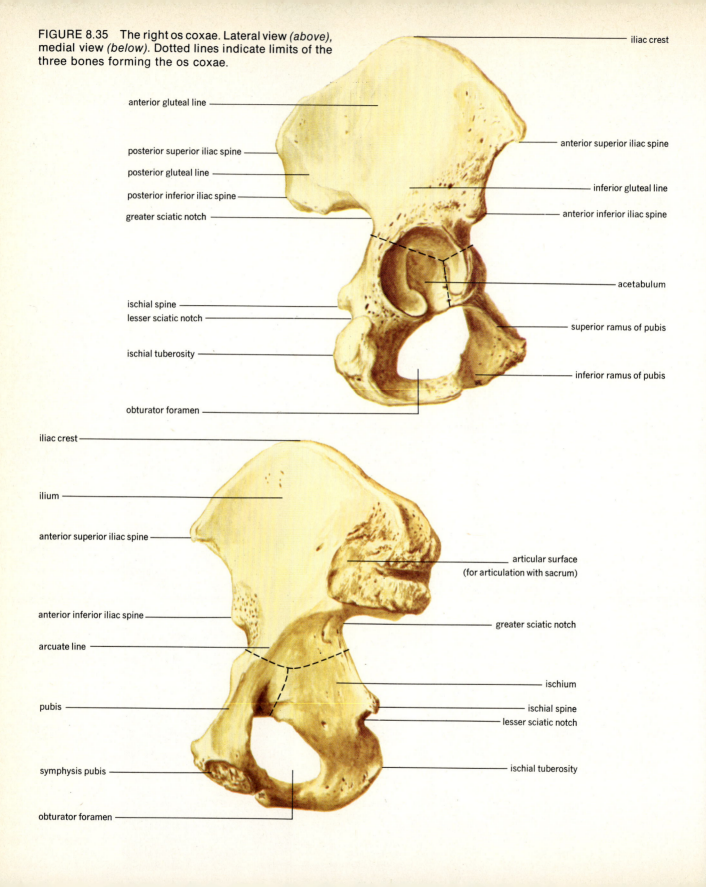

FIGURE 8.35   The right os coxae. Lateral view *(above)*, medial view *(below)*. Dotted lines indicate limits of the three bones forming the os coxae.

iliac crest

anterior gluteal line

posterior superior iliac spine

posterior gluteal line

posterior inferior iliac spine

greater sciatic notch

anterior superior iliac spine

inferior gluteal line

anterior inferior iliac spine

acetabulum

ischial spine

lesser sciatic notch

ischial tuberosity

superior ramus of pubis

inferior ramus of pubis

obturator foramen

iliac crest

ilium

anterior superior iliac spine

articular surface
(for articulation with sacrum)

anterior inferior iliac spine

arcuate line

greater sciatic notch

pubis

ischium

ischial spine

lesser sciatic notch

symphysis pubis

ischial tuberosity

obturator foramen

TABLE 8.4 A summary of the features of the bones of the lower appendage and girdles (italics indicate the most important features)

Name and number of bones in each appendage	Features on the bone	Description/location of bone/feature
Os coxae (1)	Formed from 3 bones	Bone of the pelvic girdle; attaches to sacrum
Ilium		Superior bone of os coxae
	Crest	Upper border of bone
	Anterior superior spine	"Point" of the hip; anterior
	Anterior inferior spine	Below anterior superior spine
	Posterior superior spine	At posterior end of crest
	Posterior inferior spine	Below posterior superior spine
	Gluteal lines	Three curved lines on posterior surface
	Greater sciatic notch	Notch below posterior inferior spine
Ischium		Posterior bone of os coxae
	Tuberosity	"We sit on these"
	Spine	Pointed process above tuberosity
	Lesser sciatic notch	Notch below spine
Pubis		Anterior bone of os coxae
	Symphysis	Joint between the 2 pubic bones
	Rami	Arches above and below symphysis
	Obturator foramen	Large hole enclosed by ischium and pubis
	Acetabulum	The cuplike joint surface on the lateral side of the os coxae (all 3 bones contribute to its formation)
Femur (1)		Bone of the thigh
	Head	Rounded, proximal end
	Neck	Below the head
	Greater trochanter	Projection lateral to neck ("measure hips" here)
	Lesser trochanter	Projection on medial surface below neck
	Body	Length of bone between trochanters and distal end
	Linea aspera	Rough ridge on posterior surface of shaft
	Gluteal tuberosity	Rough line below greater trochanter on posterior aspect of bone
	Medial condyle	Medial articular surface on distal end
	Lateral condyle	Lateral articular surface on distal end
	Intercondylar fossa	Depression between condyles on posterior surface
Patella (1)		Kneecap
Tibia (1)		Medial bone of leg
	Medial condyle	Medial articular surface on proximal end
	Lateral condyle	Lateral articular surface on proximal end
	Intercondylar eminence	Projection between condyles
	Medial malleolus	Large projection on medial aspect of distal end; commonly called "ankle bone"
	Shaft	Length between proximal and distal ends
Fibula (1)		Lateral bone of leg
	Head	Proximal end
	Shaft	Length between proximal and distal ends
	Lateral malleolus	Lateral projection on distal end of bone; commonly called "ankle bone"

TABLE 8.4 (Continued)		
Name and number of bones in each appendage	Features on the bone	Description/location of bone/feature
Tarsals (7)		The ankle bones
	Calcaneus	"Heel bone"
	Talus	Articulates with tibia and fibula
	3 cuneiforms	Arranged across "top of foot"; wedge shaped
	Cuboid	Like a cube) Articulate
	Navicular	Boatlike } with talus
Metatarsals (5)		Numbered 1–5 beginning with great toe
Phalanges (14)		2 in great toe; 3 in each of the other toes

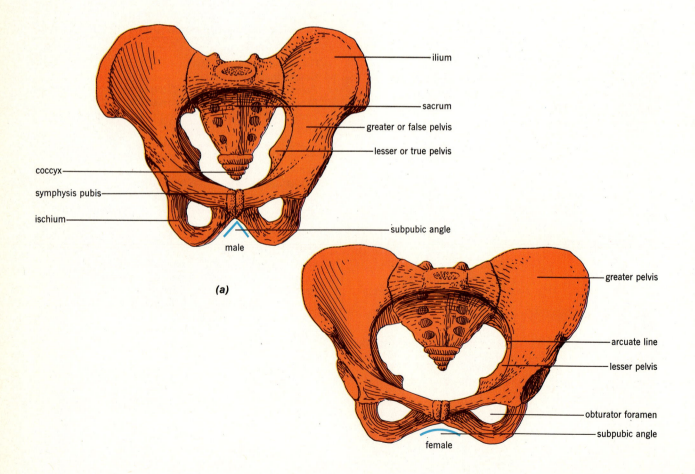

(a)

(b)

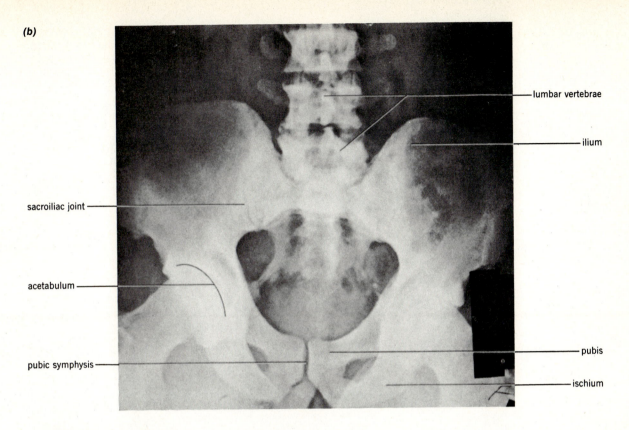

lumbar vertebrae

ilium

sacroiliac joint

acetabulum

pubic symphysis

pubis

ischium

FIGURE 8.36   *(a)* Scheme of male and female pelvis.
(b) An X ray of the pelvic area. (Courtesy Eastman
Kodak Company.)

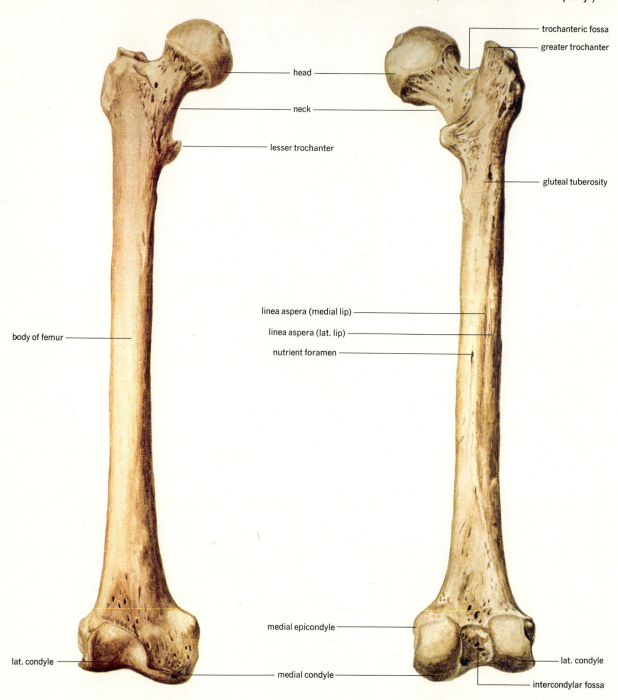

FIGURE 8.37 *(a)* The right femur. Anterior view *(left)*, posterior view *(right)*. *(b)* An X ray of the proximal end of the femur. (Courtesy Eastman Kodak Company.)

head

neck

lesser trochanter

body of femur

lat. condyle

medial condyle

trochanteric fossa

greater trochanter

gluteal tuberosity

linea aspera (medial lip)

linea aspera (lat. lip)

nutrient foramen

medial epicondyle

lat. condyle

intercondylar fossa

(a)

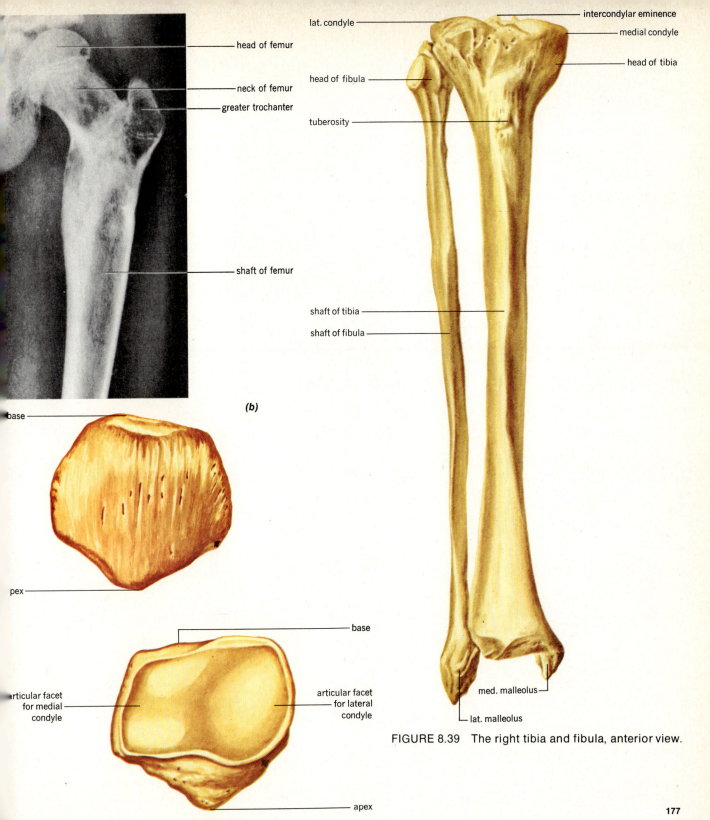

head of femur

neck of femur

greater trochanter

shaft of femur

(b)

base

lat. condyle

head of fibula

tuberosity

intercondylar eminence

medial condyle

head of tibia

shaft of tibia

shaft of fibula

pex

base

articular facet
for medial
condyle

articular facet
for lateral
condyle

apex

med. malleolus

lat. malleolus

FIGURE 8.39   The right tibia and fibula, anterior view.

FIGURE 8.38   The right patella. Anterior view (above),
osterior view (below).

177

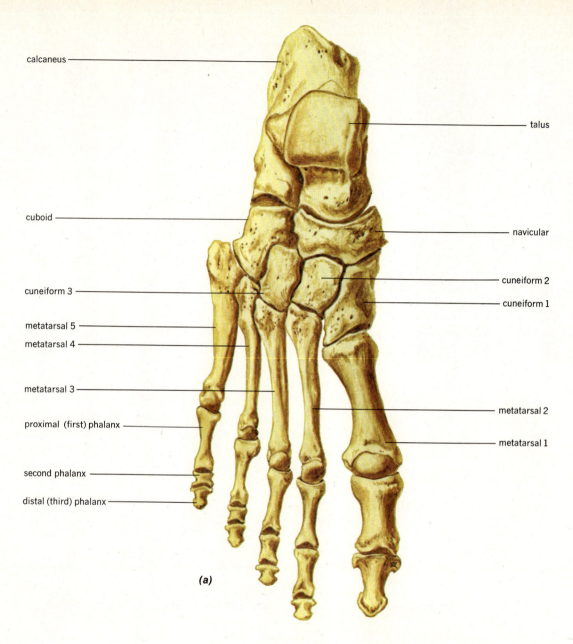

calcaneus

talus

cuboid

navicular

cuneiform 3

cuneiform 2

cuneiform 1

metatarsal 5

metatarsal 4

metatarsal 3

proximal (first) phalanx

metatarsal 2

metatarsal 1

second phalanx

distal (third) phalanx

(a)

FIGURE 8.40 *(a)* The right ankle and foot, superior view. *(b)* X rays of lower leg *(upper),* and foot *(lower).* (Courtesy Eastman Kodak Company.)

(b)

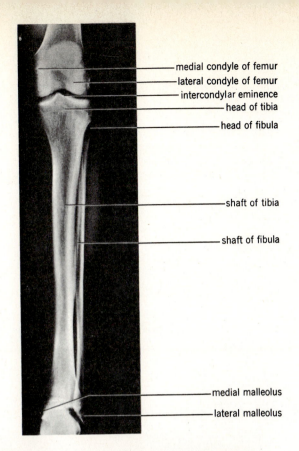

medial condyle of femur
lateral condyle of femur
intercondylar eminence
head of tibia
head of fibula

shaft of tibia

shaft of fibula

medial malleolus
lateral malleolus

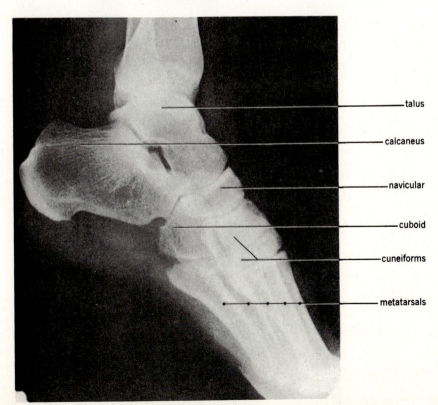

talus

calcaneus

navicular

cuboid

cuneiforms

metatarsals

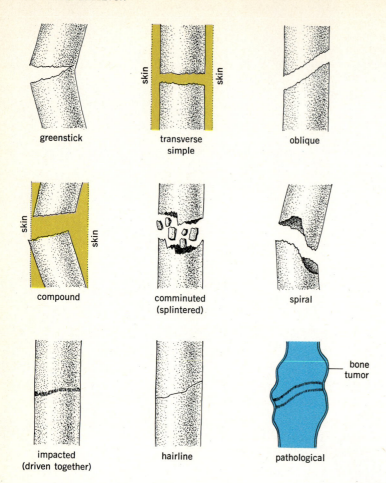

greenstick

transverse
simple

oblique

compound

comminuted
(splintered)

spiral

impacted
(driven together)

hairline

pathological

bone
tumor

FIGURE 8.41   Common types of fractures.

Questions

1. Compare and contrast the two types of bone formation. What is required for calcification to occur?

2. Name and discuss the origin and clinical significance of the fontanels.

3. What bones form the orbit? The nasal cavities?

4. In the disease acromegaly, which occurs in adults, bones grow in width but not in length. Explain how this may occur.

5. What is a "ruptured disc"? What dangers does it present?

6. Compare and contrast spongy and compact bone as to structure, composition, and location within the body.

7. What are the "girdles" of the skeleton? Name the bone(s) composing each.

8. Why is a thumb not a finger, but is a digit?

9. What functions does the skeleton serve?

Readings

Bourne, G. H. (ed.). *The Biochemistry and Physiology of Bone,* Vols. 1, 2, and 3. Academic Press. New York, 1972.

Copenhaver, W. M. et al. *Bailey's Textbook of Histology.* Williams and Wilkins. Baltimore, 1971.

Gray, Henry. *Anatomy of the Human Body,* 29th ed. Edited by C. M. Goss. Lea and Febiger. Philadelphia, 1973.

Hall, M. C. *The Architecture of the Bone.* Thomas Pubs. Springfield, Ill., 1966.

Keim, Hugo A. *The Adolescent Spine.* Academic Press. New York, 1976.

Menczel, J., and A. Harell (eds.). *Calcified Tissue: Structural, Functional and Metabolic Aspects.* Academic Press. New York, 1971.

Milch, H., and R. Milch. *A Textbook of Common Fractures.* Harper. New York, 1959.

Morris' Human Anatomy, 12th ed. Edited by B. J. Anson. McGraw-Hill. New York, 1966.

Nichols, G., and R. H. Wasserman (eds.). *Cellular Mechanisms for Calcium Transfer and Homeostasis.* Academic Press. New York, 1971.

Rodahl, K. et al. *Bone As a Tissue.* McGraw-Hill. New York, 1960.

Zorab, P. A. (ed.). *Scoliosis.* Academic Press. New York, 1973.

Chapter 9

Articulations

Objectives

After studying this chapter, the reader should be able to:

- Define an articulation.

- Give the three major categories of articulations, with types within each category.

- Give examples in the body of each type of articulation.

- Describe the shoulder, sacroiliac, hip, and knee joints concerning finer structure.

- Discuss the composition and functions of synovial fluid.

- Define what the terms dislocation, sprain, and strain mean when applied to joints.

- Define arthritis and its four major categories.

A joint or articulation is formed where a bone joins another bone, or where a cartilage joins a bone. The structure of the joint depends mainly on the function it must serve. Thus, the union may be rigid, or it may permit variable degrees of motion. If permitted, movement may occur in one, two, or three planes of motion, and the joint may be termed uniaxial, biaxial, or triaxial.

Joints depend for their security on closely fitting bony parts, ligaments, or muscles. The closer the fit of the bones, the stronger the joint, but the greater the restriction on number of axes of movement.

Classification of joints

Three categories of joints are recognized.

FIBROUS JOINTS (*immovable joints*) have no joint cavity, and the bones are held together by fibrous membranes. A SYNDESMOSIS (*syn*, with + *des*, bond) occurs when two bones are held firmly together by collagenous connective tissue as between the radius and ulna (*see* Fig. 8.33). This type of joint is often called a *ligamentous union.* SUTURES are found only in the skull, and create rigid, boxlike structures for protection of vital body organs. The bones forming the suture are usually of irregular contour for strength, and have minimal amounts of tissue between the two bones. As these joints age, fibrous tissue may disappear entirely, and a bone-to-bone union called a *synostosis* is formed.

CARTILAGENOUS JOINTS (*amphiarthrosis,* slightly movable joints) have no joint cavity, and hyaline or fibrous cartilage joins the two bones. A SYN-CHONDROSIS involves hyaline cartilage as the joining material. Synchondroses connect the shaft of long bones with their ends (epiphyseal cartilage) in temporary joints, and connect the upper 10 pairs of ribs to the sternum (costal cartilages) in permanent joints. A SYMPHYSIS utilizes fibrous cartilage as the connecting material. Such joints include the symphysis pubis and the intervertebral discs between vertebral bodies.

SYNOVIAL JOINTS (*diarthroses,* freely movable joints) have joint cavities, synovial membranes, synovial fluid, and supporting ligaments (Fig. 9.1). The articulating surfaces are smooth and are covered with articular cartilage for easy movement. Six types of synovial joints are recognized (Fig. 9.2).

A HINGE (*ginglymus*) JOINT is a uniaxial joint and permits only a back-and-forth type of motion. The knee, elbow, finger joints, and ankle are of this type.

A PIVOT (*trochoid*) JOINT permits a turning (rotation) motion, and the joint is uniaxial. Rotation of the head ("no" movements) and turning the forearm occur in pivot joints.

An OVOID (*condyloid*) JOINT is biaxial, permitting side-to-side and back-and-forth motions. The wrist joint is of this type.

A SADDLE JOINT also is biaxial, permitting side-to-side and back-and-forth motions. The carpalmetacarpal joint of the thumb is of this type.

A GLIDING (*arthrodial*) JOINT is biaxial, as in the intercarpal and intertarsal joints, and permits side-to-side and back-and-forth motions.

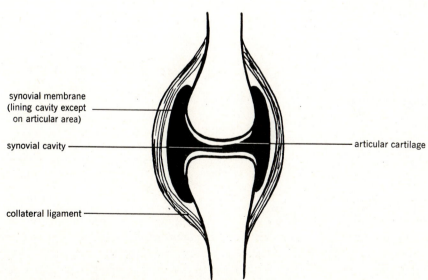

synovial membrane (lining cavity except on articular area)

synovial cavity

collateral ligament

articular cartilage

FIGURE 9.1   The basic structure of a synovial joint.

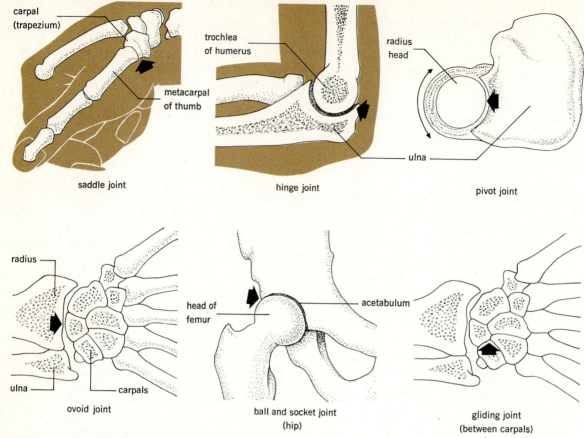

carpal
(trapezium)

metacarpal
of thumb

saddle joint

trochlea
of humerus

ulna

hinge joint

radius
head

ulna

pivot joint

radius

ulna   carpals

ovoid joint

head of
femur

acetabulum

ball and socket joint
(hip)

gliding joint
(between carpals)

FIGURE 9.2   The six basic types of synovial joints, illustrated by joints of the body.

The BALL-AND-SOCKET (*spheroidal*) JOINT is triaxial, permitting side-to-side, back-and-forth and rotational movement. The hip and shoulder joints are of this type.

A summary of these articulations is given in Table 9.1.

Discussion of specific joints

Because of their frequent involvement in sports injuries, or because of congenital malformation, the shoulder, sacroiliac, hip, and knee joints are described in greater detail.

The SCAPULOHUMERAL (*shoulder*) JOINT (Fig. 9.3) is formed by the head of the humerus articulating

with the glenoid fossa of the scapula. The fossa has a low rim of cartilage (the glenoid labrum) to deepen it and is surrounded by several ligaments. The muscles around the shoulder also contribute to its security. A "shoulder separation," or dislocation of this joint, occurs commonly as a result

TABLE 9.1 Summary of articulations				
Major category	Type	Characteristics	Axes of motion permitted	Examples
Fibrous joints		No cavity, immovable	None	Sutures, syndesmoses
	Syndesmosis	Bones held together by connective tissue	None	Tibia-fibula
	Synostosis	Bone-bone junction	None	Arise by aging of fibrous joints, with loss of fibrous tissue
	Suture	Bone-bone joint, minimal c.t. between	None	Skull; commonly becomes synostoses
Cartilagenous joints		No cavity; slightly movable; a bone-cartilage-bone joint	Compression only	Synchondroses, symphyses
	Synchondrosis	Bone-hyaline cartilage-bone joint	None	Shaft and ends of long bones; temporary. Between ribs and sternum (costal cartilages); permanent.
	Symphysis	Bone-fibrocartilage-bone joint	Slight compression	Symphysis pubis, vertebral column
Synovial joints		Cavity, freely movable	One to three	Six types
	Hinge	Spool in halfmoon	Uniaxial; back and forth motion	Knee, elbow, ankle, fingers
	Pivot	Cone in depression	Uniaxial; rotation	Radio-humeral, atlas-axis
	Ovoid	Egg in depression	Biaxial; side-to-side and back-and-forth	Wrist
	Gliding	Nearly flat surfaces opposed	Biaxial; side-to-side and back-and-forth	Intercarpal joints, intertarsal joints
	Ball and socket	Ball in cup	Triaxial; side-to-side, back-and-forth and rotation	Hip, shoulder
	Saddle	"Saddle on horse"	Biaxial: side-to-side and back-and-forth	Carpo-metacarpal joint of thumb

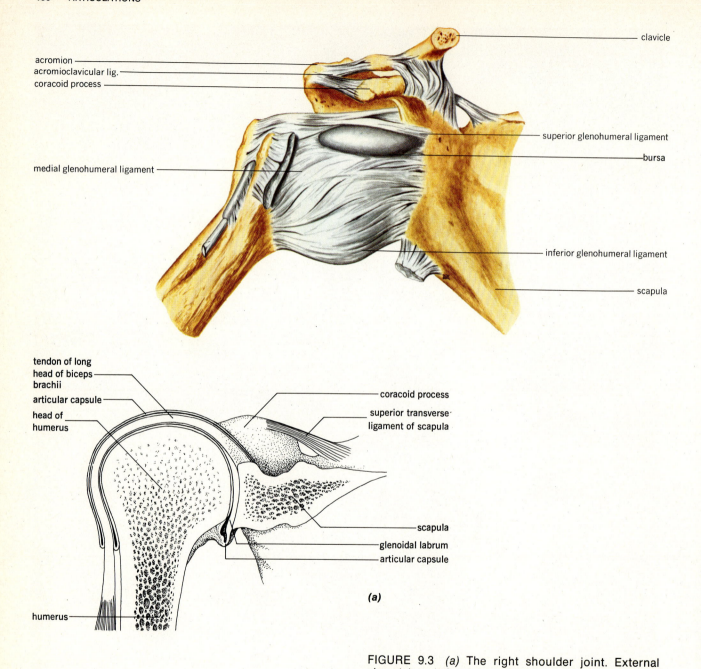

acromion

acromioclavicular lig.

coracoid process

medial glenohumeral ligament

clavicle

superior glenohumeral ligament

bursa

inferior glenohumeral ligament

scapula

tendon of long head of biceps brachii

articular capsule

head of humerus

humerus

coracoid process

superior transverse ligament of scapula

scapula

glenoidal labrum

articular capsule

(a)

FIGURE 9.3 *(a)* The right shoulder joint. External view *(above)*, frontal section *(below)*. *(b)* An X ray of the left shoulder joint. (Courtesy Eastman Kodak Company.)

of falling on the outstretched arm (football and basketball players). In this disorder, the humerus head is driven out of the glenoid fossa, usually in an upward (superior) direction.

(b)

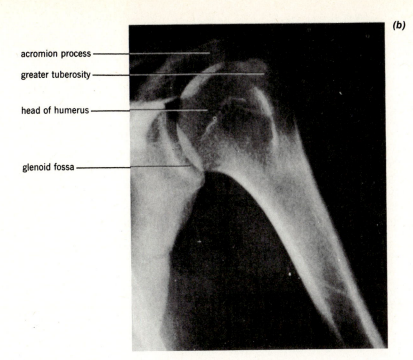

acromion process

greater tuberosity

head of humerus

glenoid fossa

The SACROILIAC JOINT (Fig. 9.4), or union between the sacrum and ilium, is a partially fibrous, partially synovial joint. The joint surfaces are rough and the bones are firmly bound by many ligaments, thus no movement is normally permitted. A "sacroiliac slip" involves stretching of the ligaments holding the joint, and slight movement of the bones may occur. Stretching or tension applied to the sciatic nerve may then result in intense leg pain. "Slips" may occur if heavy loads are carried that put downward pressure on the joint, or in landing on the feet without bending the knees. This tends to push the lower appendages strongly into the hip bones and thus forces upward pressure on the sacroiliac joints.

The HIP JOINT (Fig. 9.5), between the acetabulum and femoral head, is constructed for weight bearing as well as movement. The socket (acetabulum) is deep, and a ligament (ligamentum teres) aids in fixing the head of the femur to the socket. The supporting ligaments are numerous and heavy, and seem to serve to limit motion more than to secure the joint. The joint is rarely dislocated if normal. In congenital malformation of the hip, the acetabular rim is low, and the femoral head slips easily over or out of it. Limited sideward movement of the limb (Fig. 9.6) when hip and knee are flexed (frog position) is usually indicative of congenital malformation. The disorder should be diagnosed in infancy.

The KNEE JOINT (Fig. 9.7) is regarded by many as one of nature's mistakes, because it is basically unstable. Two nearly flat surfaces are held together by a number of ligaments. The CRUCIATE LIGAMENTS tend to limit anterior-posterior movement, while the COLLATERAL LIGAMENTS, with the MENISCI, tend to limit sideward motion. The joint may be disrupted by strong rotation, or by blows to either side, and by blows from the front that tend to overextend it. Such trauma are inherent in the game of football. Crushing of menisci and tearing of ligaments is common, and often requires surgery to repair the damage. Figure 9.8 shows which structures will be disrupted according to the direction of the force imposed on the joint.

(Text continued on page 193.)

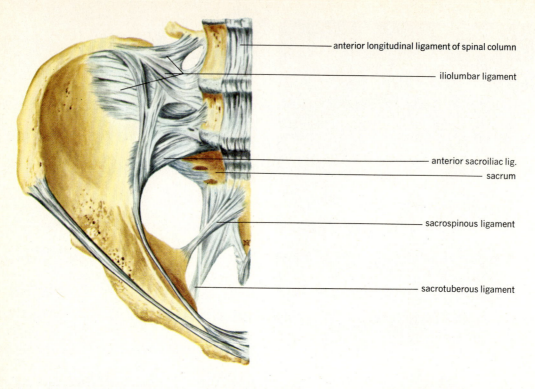

anterior longitudinal ligament of spinal column

iliolumbar ligament

anterior sacroiliac lig.

sacrum

sacrospinous ligament

sacrotuberous ligament

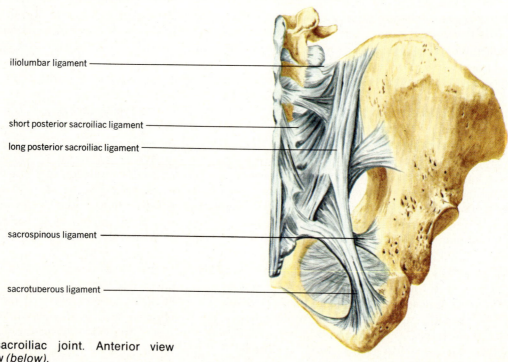

iliolumbar ligament

short posterior sacroiliac ligament

long posterior sacroiliac ligament

sacrospinous ligament

sacrotuberous ligament

FIGURE 9.4 The sacroiliac joint. Anterior view *(above)*, posterior view *(below)*.

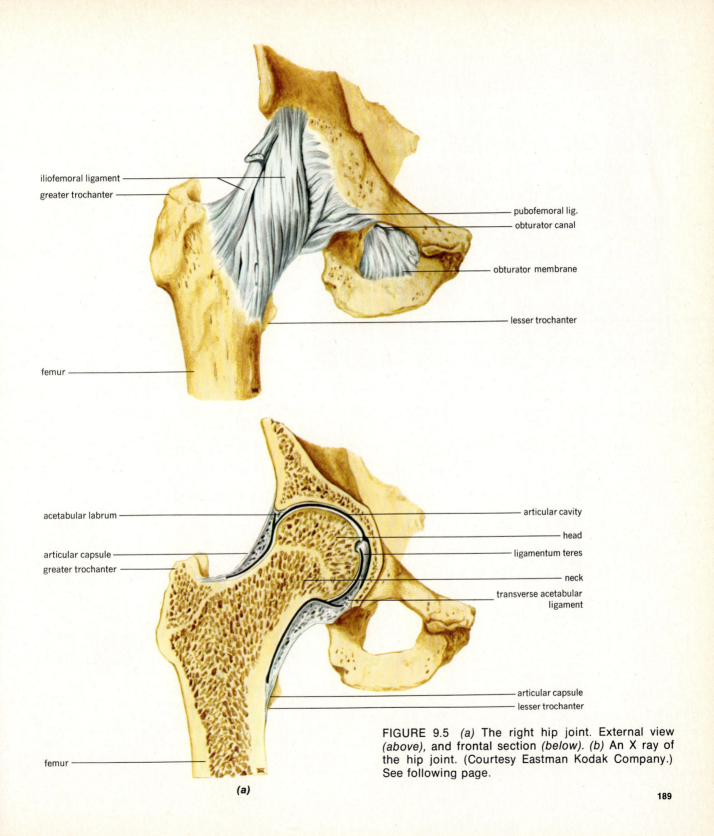

iliofemoral ligament

greater trochanter

pubofemoral lig.

obturator canal

obturator membrane

lesser trochanter

femur

acetabular labrum

articular cavity

head

ligamentum teres

articular capsule

greater trochanter

neck

transverse acetabular
ligament

articular capsule

lesser trochanter

femur

FIGURE 9.5 *(a)* The right hip joint. External view *(above)*, and frontal section *(below)*. *(b)* An X ray of the hip joint. (Courtesy Eastman Kodak Company.) See following page.

(a)

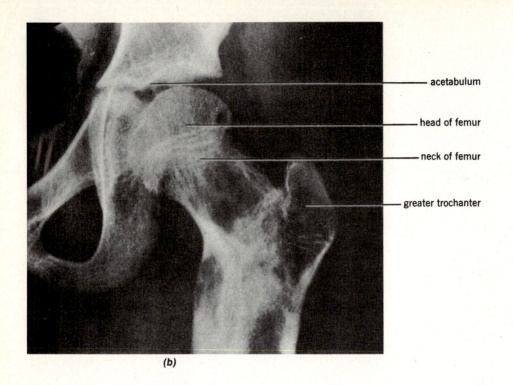

acetabulum

head of femur

neck of femur

greater trochanter

(b)

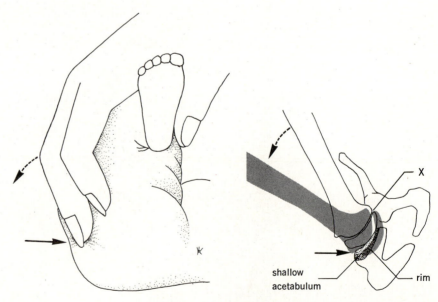

X

shallow
acetabulum

rim

FIGURE 9.6   Testing an infant for congenital dislocation of the hip. Sideward movement is limited when
knee and hip are flexed 90°. A "click" may be felt at
"X" as thigh is abducted.

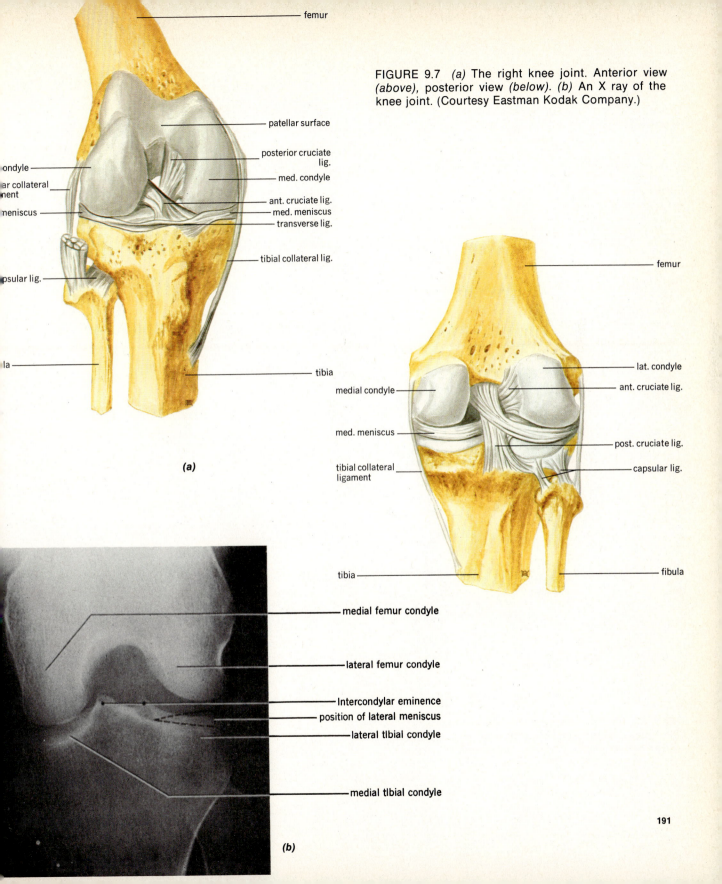

femur

patellar surface

posterior cruciate
lig.

condyle

med. condyle

ar collateral
nent

ant. cruciate lig.

meniscus

med. meniscus

transverse lig.

tibial collateral lig.

psular lig.

la

tibia

(a)

FIGURE 9.7 *(a)* The right knee joint. Anterior view *(above)*, posterior view *(below)*. *(b)* An X ray of the knee joint. (Courtesy Eastman Kodak Company.)

femur

lat. condyle

medial condyle

ant. cruciate lig.

med. meniscus

post. cruciate lig.

tibial collateral
ligament

capsular lig.

tibia

fibula

medial femur condyle

lateral femur condyle

Intercondylar eminence

position of lateral meniscus

lateral tibial condyle

medial tibial condyle

191

(b)

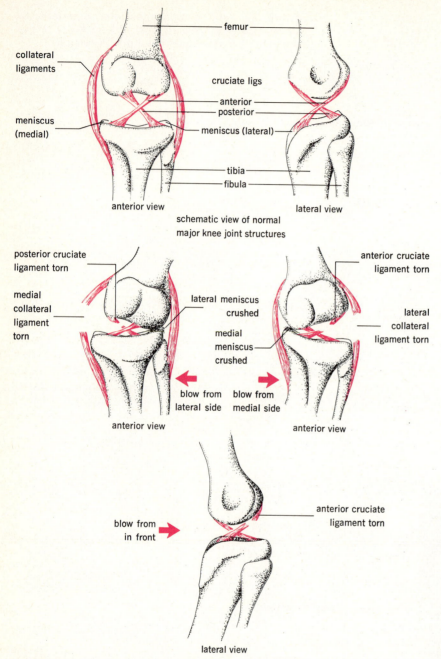

femur

collateral
ligaments

cruciate ligs

anterior
posterior

meniscus
(medial)

meniscus (lateral)

tibia

fibula

anterior view

lateral view

schematic view of normal
major knee joint structures

posterior cruciate
ligament torn

anterior cruciate
ligament torn

medial
collateral
ligament
torn

lateral meniscus
crushed

medial
meniscus
crushed

lateral
collateral
ligament torn

blow from
lateral side

blow from
medial side

anterior view

anterior view

blow from
in front

anterior cruciate
ligament torn

lateral view

FIGURE 9.8 A diagrammatic representation of the
structures of the left knee that will most likely be injured by trauma from various directions.

Synovial fluid

Synovial fluid lubricates synovial joints. It is secreted by the synovial membrane and, in all synovial joints, has a total estimated volume of 100 milliliters. It functions to lubricate the articular surfaces so as to ensure smooth joint action. The fluid contains much mucus and protein-polysaccharides which makes it thicker than other body fluids, and an efficient lubricating material. Excess production of fluid may occur after joint injury, and may limit joint motion (e.g., "water on the knee," inflammation). Joint movement is limited because of the swelling and pressure increase resulting from excessive secretion of synovial fluid.

Clinical considerations

Traumatic damage to a joint

A DISLOCATION is defined as loss of continuity of the joint structures. It may occur as the result of a blow, or because of malformation. A SPRAIN occurs when ligaments and muscles are torn and tendons are stretched as the result of trauma. Pain is common, is often intense, and rapid development of edema (swelling) is usual. Black-blue hemorrhages in the skin (ecchymoses) that change to greenish-brown or yellow with time are also common. Bleeding into the joint produces the common "black-and-blue" mark. As the blood pigments break down, the pigments known as bilirubin (yellow) and biliverdin (green) are produced. Thus the bruise changes color with time. A STRAIN is caused by overstretching or pulling of muscles and tendons, without tearing.

Inflammation of joints

Arthritis is a term used to designate joint inflammation.

RHEUMATOID ARTHRITIS is a disease of connective tissue involving inflammation and destruction of synovial membranes. The joints become enlarged, severely limited in motion, and painful.

OSTEOARTHRITIS is a chronic degenerative disease, especially of weight bearing joints. It is characterized by: destruction of articular cartilage, without great inflammation; overgrowth of bone often with twisting of the affected appendage (e.g., fingers). Pain is minimal, but may increase with hard joint usage.

GOUTY ARTHRITIS is the result of a metabolic error involving the degradation of nitrogenous bases. Uric acid crystals are deposited in certain joints (e.g., ankle, knee, hands), with development of intense pain and swelling.

SEPTIC ARTHRITIS arises as a result of the presence of bacteria or their products in the bloodstream, with involvement of the joints; gonococcal (gonorrhea), nonhemolytic streptococcal (strep), and staphylococcal (staph) infections are most commonly involved as causative agents.

Summary

1. An articulation is a junction between bones, or between a bone and a cartilage.
   a. Joints may permit no movement, slight movement, or free movement.
   b. Joints are secured by bony parts, ligaments, muscles, or combinations of these.

2. Three (3) categories of joints are recognized:
   a. Fibrous joints are immovable and are formed by collagenous tissue or nearly bone-to-bone junctions (sutures of skull). Fibrous joints may become bone-to-bone junctions as they age.
   b. Cartilagenous joints are slightly movable and are joined by hyaline cartilage (synchondrosis) or fibrocartilage (symphysis). Synchondroses may be temporary or permanent.
   c. Synovial joints are freely movable and have a joint cavity, with synovial membranes and fluid, and are of six types:
      (1) Hinge. Allows back-and-forth motion. Fingers, knee, elbow.
      (2) Pivot. Permits turning or rotation. Head ("no movement"), radius on humerus.
      (3) Ovoid. Allows side-to-side and back-and-forth. Wrist.
      (4) Saddle. Allows side-to-side and back-and-forth. Thumb.
      (5) Gliding. Allows side-to-side and back-and-forth. Intercarpal, intertarsal.
      (6) Ball and socket. Permits side-to-side, back-and-forth, and rotation. Shoulder, hip.

3. Synovial fluid is produced by the synovial membrane. It is thick and lubricates synovial joints.

4. Specific joints are briefly discussed: shoulder, sacroiliac, hip, knee.

5. Disorders of joints include:
   a. Dislocations. Loss of continuity of joint structures.
   b. Sprains. Tearing of ligaments and muscles with pain and edema.
   c. Strains. Overstretching of ligaments and muscles without tearing.
   d. Arthritis. Inflammation of a joint. It may be due to reactions secondary to infection, may be a disease of aging, or may be the result of metabolic disorders.

Questions

1. Compare fibrous, cartilagenous, and synovial joints as to structure and freedom of movement.

2. Synovial joints may be described as uniaxial, biaxial, or triaxial. Pick six synovial joints in the body, and classify them according to degrees of movement permitted.

3. Does the skull have only sutures? Explain.

4. Compare dislocations, sprains, and strains.

5. Compare the types of arthritis by cause and symptoms.

Readings

Harris, William H. "Current Concepts: Total Joint Replacement." *New Eng. J. Med. 297*:650, Sept. 22, 1977.

Science News. "A Novel Treatment for Rheumatoid Arthritis." *113*:104, Feb. 18, 1978.

Sokoloff, Leon. *The Joints and Synovial Fluid.* Vols. 1 and 2. Academic Press. New York, 1978.

Sonstegard, David A., Larry S. Mathews, and Herbert Kaufer. "The Surgical Replacement of the Human Knee Joint." *Sci. Amer. 238*:44, Jan. 1978.

Chapter 10

The Structure and Properties of Muscular Tissue, with Emphasis on Skeletal Muscle

Objectives

After studying this chapter, the reader should be able to:

- List the three basic types of muscle found in the body and state their functions.

- Describe the structure of skeletal muscle from the level of the whole muscle to the muscle filaments.

- List the four major proteins composing skeletal muscle and describe their orientation in the muscle filaments.

- Describe the orientation of sarcoplasmic reticulum and T-tubules in the fiber.

- Give the structure of a myoneural junction, how nerve impulses are transduced into chemical stimuli that depolarize muscle membranes, and relate several disorders of muscle to the functioning of the neuromuscular junction.

- Discuss how depolarization of muscle membrane systems causes muscle contraction (excitation-contraction coupling) and depict the events leading to relaxation of a muscle.

- Describe how a muscle maintains its stores of ATP for muscle contraction and the action potential associated with contraction.

- List the types of heat released during muscle activity and relate each to events within the muscle.

- Define subthreshold, threshold, and maximal in terms of stimuli to a muscle and muscle response.

- List some properties of skeletal muscle that enable it to contract most efficiently and smoothly.

- Explain the following as related to skeletal muscle: A twitch, tetanus, treppe, tone, isotonic contraction, isometric contraction.

- Discuss some disorders of the muscular tissue itself.

- Describe the anatomy and types of smooth muscle.

- Tell how smooth muscle is stimulated, the effects of chemical and nervous factors in its operation, and describe some disorders associated with smooth muscle.

It is the function of muscular tissues to CON-TRACT or shorten and so perform work. Contraction then causes MOVEMENT of body parts, or PROPULSION of materials through the body. Muscular activity may also CHANGE THE DIAMETER of organs around which it is found. Thus, an alteration of organ size may occur.

There are *three types* of muscle in the body, each possessing unique properties for its particular tasks.

SKELETAL MUSCLE attaches to and moves the skeleton, allowing the organism to move through space and adjust itself to its external environment. It also forms the type of muscle that causes breathing in mammalian forms. Skeletal muscle is normally stimulated to contract by impulses delivered to it by way of nerves, and is thus called a VOLUNTARY type of muscle. About 40% of the body weight is skeletal muscle.

SMOOTH OR VISCERAL MUSCLE is located around hollow organs of the body, such as the organs of the digestive, urinary, respiratory, and reproductive systems, blood vessels, and the iris of the eye. Contraction of smooth muscle is basically the result of inherent factors, but may be altered by chemical and nervous stimuli.

CARDIAC MUSCLE is located only in the heart, and its contractions propel blood through the body. Like smooth muscle, its contraction is due to inherent factors, although both rate and strength of contraction may be altered by chemical and nervous factors.

Cardiac and smooth muscle together form about 10 percent of the body weight.

The anatomy and physiology of skeletal and smooth muscle are the topics of this chapter. Cardiac muscle is considered in Chapter 22, as part of the study of the heart.

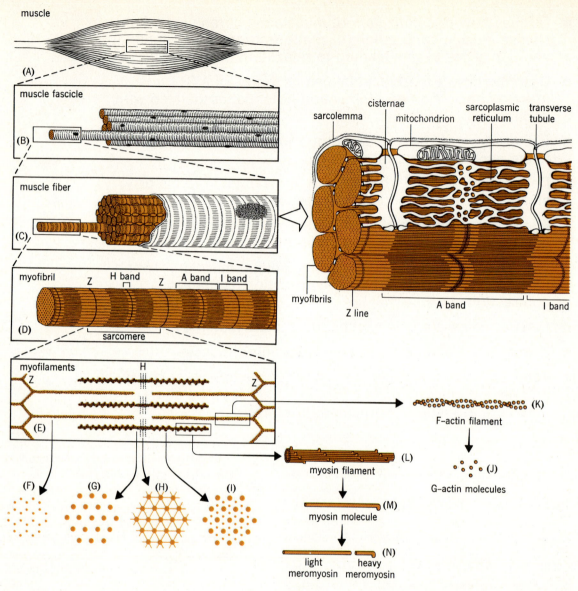

muscle

(A)

muscle fascicle

(B)

muscle fiber

(C)

myofibril   Z   H band   Z   A band   I band

(D)

sarcomere

myofilaments   H

Z   Z

(E)

(F)   (G)   (H)   (I)

sarcolemma   cisternae   sarcoplasmic   transverse
             mitochondrion   reticulum   tubule

myofibrils

Z line   A band   I band

F–actin filament   (K)

G–actin molecules   (J)

myosin filament   (L)

myosin molecule   (M)

light   heavy   (N)
meromyosin   meromyosin

FIGURE 10.1   The structure of skeletal muscle. (F),
(G), (H), and (I) are cross sections of the filaments at
the points indicated.

198

Skeletal muscle

The anatomy of skeletal muscle

Skeletal muscle (Fig. 10.1) is composed of multinucleated, long cylindrical FIBERS or cells. Individual fiber diameters lie between 10 and 100 μm, while length varies between 2 mm and 7.5 cm or more. Small fibers are found in small muscles, such as those turning the eyeball, and larger fibers are located in the muscles of the appendages and trunk. Each fiber has a limiting membrane called the SARCOLEMMA. Lying inside the membrane is the SARCOPLASM, or muscle cytoplasm. Arranged in the sarcoplasm are longitudinal MYOFIBRILS that show alternating light and dark staining STRIATIONS. In a fiber, the light and dark bands on the myofibrils all line up one above the other, giving striations to the entire fiber. The dark bands are called *A lines,* and each is

divided by an *H line.* Light staining lines are designated as *I lines,* and each of these is divided by a *Z line.* The distance between two Z lines is known as a SARCOMERE, a unit that becomes narrower as the muscle shortens.

The myofibrils are in turn composed of FILAMENTS, protein strands involved in muscular contraction. Four major proteins compose these filaments.

MYOSIN (Fig. 10.2) is a large (MW about 500,000) protein possessing enzyme-like properties. The molecules consist of two peptide strands wound around one another in the form of a double helix. At one end of the molecule is an enlarged "head" of protein that can form cross-bridges with one of the other proteins in the muscle. Myosin is restricted to the A line.

ACTIN (Fig. 10.3) is smaller than myosin (MW about

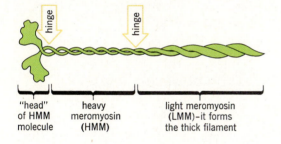

"head" of HMM molecule | heavy meromyosin (HMM) | light meromyosin (LMM)–it forms the thick filament

(a)

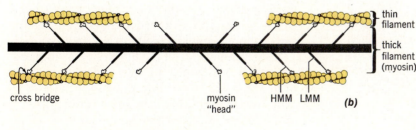

cross bridge | myosin "head" | HMM | LMM | *(b)*

thin filament

thick filament (myosin)

FIGURE 10.2   *(a)* The myosin molecule. "Hinges" are areas of greater flexibility within the molecule where bending may occur. *(b)* The formation of cross bridges between myosin and thin filaments.

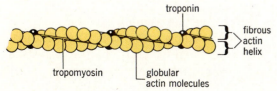

troponin

fibrous actin helix

tropomyosin | globular actin molecules

FIGURE 10.3   The structure of the thin filaments.

60,000) and consists of individual G (globular) actin units polymerized into an F (fibrous) actin strand. F-actin strands are wound into a double helix that attaches to or passes through Z lines. Actin has "reactive sites" on it where it may form bridges with the myosin heads.

TROPOMYOSIN is a filamentous protein with a molecular weight of about 70,000. It is believed to lie twisted around the F-actin helix (*see* Fig. 10.3).

TROPONIN is a globular protein (MW about 50,000) that binds to tropomyosin at certain points along the tropomyosin molecule (*see* Fig. 10.3). Tropin has a strong affinity for binding Ca^{2+}.

Other chemicals in the muscle fiber include creatine phosphate, Ca^{2+}, glucose, K^+, Na^+, and Cl^-.

The sarcoplasm also contains numerous mitochondria, usually alternating in rows with myofibrils.

A SARCOPLASMIC RETICULUM (SR) (*see* Fig. 10.1), corresponding to the ER of other cells, sends longitudinal tubules between myofibrils and bears expanded CISTERNAE at each end. The cisternae are associated with transversely oriented hollow tubules called T-TUBULES. The T-tubules are invaginations of the sarcolemma and form ringlike ANNULI around the myofibrils at junctions of A and I bands. The tubules reach entirely across a fiber and carry tissue fluid along with contained ions and small molecules to the interior of the fiber.

Individual muscle fibers are held to one another by reticular connective tissue and collagenous fibrils forming the ENDOMYSIUM. Several fibers are bound together to form FASCICLES, and each fascicle is surrounded by connective tissue continuous with the endomysium and known as the PERIMYSIUM. Numerous fascicles are bound into the entire MUSCLE, and the latter is surrounded by EPIMYSIUM, also known as the *deep fascia*. Figure 10.4 illustrates how fibers are bound into larger levels of organization.

Excitation of skeletal muscle

THE NEUROMUSCULAR JUNCTION. Skeletal muscle is normally caused to contract by impulses reaching it over nerve fibers. The connection between a nerve fiber and a muscle fiber is made by the NEUROMUSCULAR JUNCTION (Fig. 10.5). The junction has two parts: the *terminal branches of a motor nerve,* and a specialized portion of the sarcolemma known as the MOTOR END PLATE.

The terminal ends of the motor nerves form enlarged bulbs in which are found many mitochondria. These are believed to provide the energy to synthesize a chemical called ACETYLCHOLINE *(Ach)* that is stored in the nerve endings as small membrane-surrounded VESICLES. Each bulb lies in a depression of the sarcolemma called a SYNAPTIC GUTTER that has many FOLDS in it. There is a 200 to 300 Å wide SYNAPTIC CLEFT that separates the end bulb from the gutter. An enzyme called

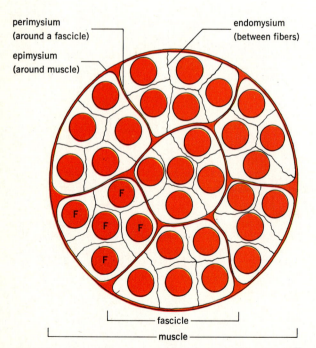

perimysium (around a fascicle)

endomysium (between fibers)

epimysium (around muscle)

fascicle

muscle

FIGURE 10.4 The connective tissue components of a skeletal muscle. F indicates muscle fibers.

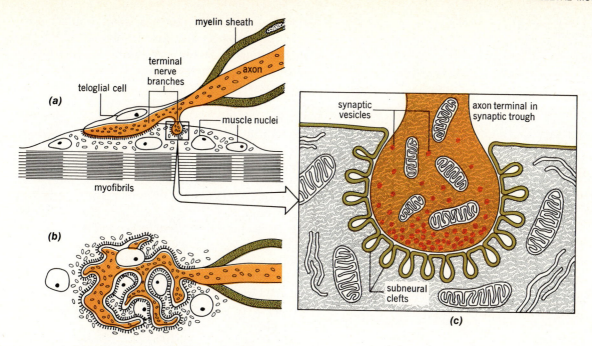

FIGURE 10.5   The neuromuscular junction as seen by light and electron microscopy. *(a)* A section of the junction. *(b)* Surface view of the junction as seen by light microscope. *(c)* Details of the junction as seen by the electron microscope.

CHOLINESTERASE is capable of cleaving acetylcholine and is located in the folds of the synaptic gutter.

A single nerve fiber forms branches that supply up to 1000 muscle fibers. A single nerve fiber and its associated muscle fibers constitute a MOTOR UNIT. Since each muscle contains many (hundreds or thousands) motor units connected to as many nerve fibers, the muscle can alter its strength of contraction, as we shall see later.

The series of events that occur as a nerve impulse travels to a neuromuscular junction may be described as follows.

The impulse causes a membrane permeability change to Ca^{2+} on the knob on the terminal end of a nerve fiber branch.

Ca^{2+}, located in the fluid around the knob (the extracellular fluid or ECF), enters the knob by diffusion.

Calcium either causes rupture of the vesicles containing Ach or brings about actual release of whole vesicles into the synaptic cleft, where they then rupture. The net result of either process is release of Ach into the cleft.

Ach then diffuses to the gutter membrane where it binds to *receptor sites* on the membrane.

This binding causes a change in permeability of the gutter membrane to Na^+, also in the fluids around the junction. Na^+ diffuses into the fiber setting up an electrical current that is ultimately transmitted through sarcolemma, T-tubules, and SR membranes. This wave of depolarization sets the stage for muscle contraction.

Within 2 to 3 milliseconds after contacting the gutter membrane, Ach is cleaved by cholinesterase, and can no longer cause permeability changes in the sarcolemma. Thus, the muscle will relax, unless more impulses arrive over the motor nerve.

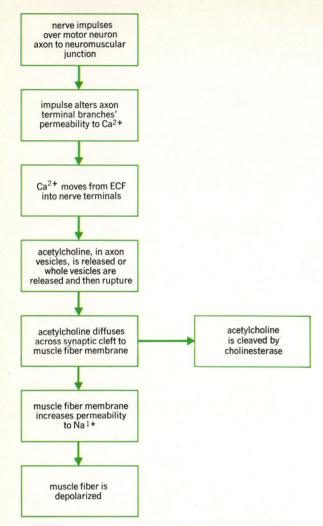

FIGURE 10.6　A summary of the events in neuromuscular junction activity.

These events are summarized in Figure 10.6.

We thus see that the basic function of a neuromuscular junction is to transduce (change) an electrical form of energy (the nerve impulse) to a chemical form, to transmit a disturbance across a gap.

CLINICAL CONSIDERATIONS. MYASTHENIA GRAVIS is a condition characterized by abnormal muscle fatiguability, weakness, and flaccidity (mus-

cles feel like half-filled hot water bottles). The condition has been shown to involve fewer vesicles than normal in motor nerve terminals, fewer than normal Ach receptors on gutter membranes, and the presence of chemicals (antibodies) that "attack" the muscle fibers and junctions themselves to cause destruction. All these features combine to reduce neuromuscular transmission and to cause reduced contraction of the muscle fibers supplied.

PARALYSIS refers to an inability to voluntarily contract a muscle, and may usually be traced to nerve damage.

FATIGUE is a term used in several ways. It may refer to inability of a fiber to contract, and is associated with depletion of vesicles (and therefore Ach) at the neuromuscular junction. It may also be used to refer to the feelings of "tiredness" that occurs after exercise, illness, or psychological upheaval.

Excitation-contraction coupling; the sliding filament model

Excitation-contraction coupling refers to the series of events by which excitation of the muscle fibers' membrane systems leads to contraction.

A series of changes occurs that culminate in contraction. They are described below.

In a nonactive (noncontracting) muscle, there is a measurable electrical difference between the outside surfaces of sarcolemma, T-tubule, and SR membranes, and the fluids they enclose or surround. The membranes are said to be POLARIZED.

Stimulation of the membrane systems, starting with the sarcolemma, causes increased permeability to Na^{1+}. Na^{1+} moves across the membranes, neutralizing or reversing the polarized state. The membrane is said to be DEPOLARIZED.

This depolarized state is transmitted over the T-tubule and SR membranes to deep within the fiber.

Muscle Ca^{2+} "resides" in SR tubules when the muscle is at rest. Depolarization of SR membranes allows Ca^{2+} to flow from the SR to the filaments.

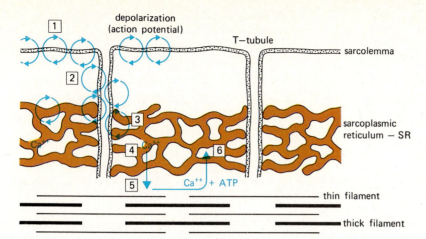

1. depolarization of sarcolemma membrane
2. depolarization of T—tubule membranes
3. depolarization of SR membranes
4. outflow of Ca^{++} from SR
5. activation of myosin; contraction
6. return of Ca^{++} to SR by calcium pump

FIGURE 10.7 Excitation-contraction coupling in skeletal muscle (redrawn from Guyton).

The tropomyosin-troponin complex, described earlier, prevents—in the absence of calcium—the binding of myosin to actin, possibly because the complex actually covers the actin binding sites for myosin. Calcium ions from the SR are bound by troponin, which is hypothesized to undergo a change of shape and pull the entire tropomyosin helix, thereby exposing the actin binding sites.

The enlarged myosin "heads" are attracted to the actin binding sites, and a "cross-bridge" is formed between the two proteins. This binding also causes a change in shape of the myosin heads, so that they tilt and pull on the actin filaments, drawing them closer together.

Once the head has tilted, another reactive site on the myosin head is exposed. ATP binds to that region, according to one theory, or, according to another theory, is already bound to the site.

ATP is now cleaved by the enzyme-like activity of myosin, forming ADP, and releasing energy from the broken high-energy phosphate bond. This energy is used to break the cross-bridge and "cock" the myosin head back to its original position, and it again binds to actin, but this time at a reactive site further along the actin filament than before.

These events are repeated, and the actin filaments are drawn ("slide") toward one another; they may actually overlap as the Z lines are drawn together. Figure 10.7 shows some of the events described above.

As long as calcium ions remain outside the SR, muscle contraction will continue. If depolarization ceases, an active transport mechanism returns Ca^{2+} to the SR. The troponin no longer binds Ca^{2+}, the tropomyosin returns to its original configuration, actin binding sites are covered again, and myosin can no longer bind to it. The muscle relaxes, aided by recoil of the stretched actin filaments.

The entire sequence of events, from nerve stimulation to muscle relaxation, is summarized in Figure 10.8.

Sources of energy for contraction

The role of ATP in muscle contraction has been indicated above. Without ATP, myosin heads cannot be cocked, and thus contraction cannot proceed. It becomes obvious that a continued supply

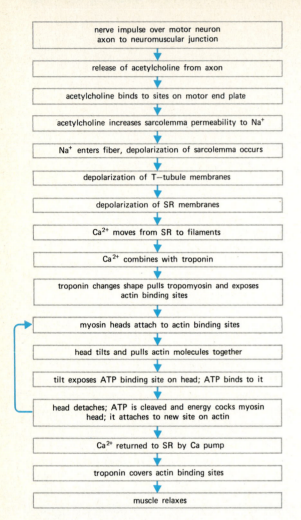

FIGURE 10.8 A summary of the events in muscular activity.

of ATP is essential for contraction. Synthesis of ATP is achieved by degradation of glucose, fatty acids, and creatine phosphate, and utilization of the released energy to combine phosphate and ADP into ATP. The basic cycles involved were presented in Chapter 5 and are reemphasized here.

Anaerobic degradation of glucose (glycolysis):

glucose + 2 ATP $\longrightarrow$ 2 pyruvic acid + 4 ATP + 4H

Aerobic degradation of fatty acids (β-oxidation, Krebs cycle, oxidative phosphorylation):

fatty acid $\longrightarrow$ CO_2 + H_2O + ATP

(The amount of each product depends on the length of the fatty acid chain)

A source of energy unique to muscle is the breakdown of creatine phosphate (CP) by the enzyme creatine phosphokinase (CPK):

creatine phosphate $\xrightarrow{\text{CPK}}$ phosphate + creatine (excreted) + energy

energy + ADP + phosphate $\longrightarrow$ ATP

ATP may be synthesized much more rapidly using creatine phosphate as an energy source than it can through glycolytic pathways.

Other considerations of skeletal muscle contraction

ELECTRICAL EVENTS. An action potential (Fig. 10.9) always accompanies depolarization of muscle membranes. An ELECTROMYOGRAM (*EMG*) is a tracing of these potentials from a contracting muscle. Analysis of a single potential shows that the electrical and mechanical events do not occur simultaneously (Fig. 10.10). A LATENT PERIOD or time lag exists between stimulation of a muscle fiber and its contraction. Spread of the depolarization wave through the fiber, diffusion of Ca^{2+}, and myosin-actin bridge formation take time (about 2 msec), and until these events have been completed, no shortening can occur.

HEAT PRODUCTION. Several "batches of heat" are produced during muscular activity, reflecting the chemical bond-splitting that occurs.

INITIAL HEAT appears with contraction and is believed to be associated with myosin activation and ATP splitting. Oxygen is not required for production of this heat.

RELAXATION HEAT is believed to be associated with the use of ATP to fuel the Ca^{2+} transport mechanism returning Ca^{2+} to the SR.

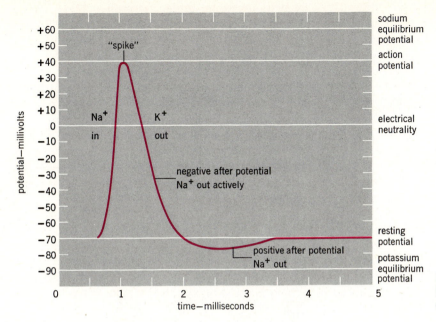

FIGURE 10.9 Changes in membrane potential during development of an action potential.

RECOVERY HEAT is produced for some time after relaxation has taken place and is associated with metabolism of glucose and fatty acids to replenish ATP stores. It does not occur if oxygen is not supplied to the muscle.

The term ACTIVATION HEAT refers to the *total* heat production by a muscle; that is, the *sum* of heats listed above.

PHYSIOLOGICAL PROPERTIES. A muscle will not contract if the stimulus reaching it is too weak to cause depolarization of its membranes. Such a stimulus is called a SUBTHRESHOLD *(subliminal)* STIMULUS. As stimulus strength is raised, muscle fibers with low thresholds may be depolarized, and a barely perceptible reaction will occur. The strength of *a stimulus just sufficient to cause a perceptible response* is called THRESHOLD *(liminal* or *minimal).* Further increase in stimulus strength eventually causes all muscle fibers to contract, and both the stimulus and response is said to be MAXIMAL.

It becomes obvious from the preceding discussion that the STRENGTH of a CONTRACTION CAN

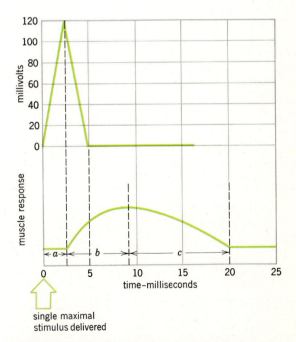

FIGURE 10.10 Electrical and mechanical events correlated for skeletal muscle. Note asynchrony of events. *a*, latent period; *b*, contraction; *c*, relaxation.

BE GRADED according to how many muscle fibers or motor units are activated. Thus, it is possible to match strength to the work required. One learns how to judge weight of objects so that we don't "throw something through the roof" or strain to move it.

A single motor unit, when activated, responds maximally if the stimulus is strong enough to activate it. If not strong enough, no contraction occurs. The *unit* is thus said to follow the ALL-OR-NONE LAW.

Muscles may be caused to contract by direct stimulation, rather than through their nerves. They thus exhibit what is called INDEPENDENT IRRITABILITY. This property is sometimes used to maintain muscle health when muscles have been *denervated,* as by an accident. Electrodes are placed on the skin over the muscle and electric current is applied to cause the muscle to contract. This prevents an *atrophy* (wastage) *from disuse.*

Muscles, if stretched slightly (as they are by their bony attachments in the body), contract with more force and speed than if they are allowed to assume their shortest length or are stretched abnormally. This phenomenon appears to be concerned with providing a maximum number of cross-bridges to be formed during contraction. If shorter or longer than normal, fewer myosin heads are oriented opposite actin binding sites, and fewer bridges are formed, decreasing the strength of the contraction.

Lastly, skeletal muscle has a very short time (*refractory period*) during which it is *not* capable of responding to another stimulus. This permits a muscle to add the results of one contraction to another to cause a sustained contraction called TETANUS. This phenomenon is discussed below.

This section should have indicated to you that skeletal muscle has properties that equip it for the tasks it must perform in the body. For example, it must react quickly, match force to work required, and stay contracted to maintain body posture. You should be able to show how each of these tasks is performed and to appreciate the way in which the muscle properties "fit" various tasks.

Characteristics of the several types of contraction exhibited by skeletal muscle

THE TWITCH. A single maximal stimulus applied to a muscle brings about a single response called a TWITCH (Fig. 10.11). This type of contraction is a laboratory phenomenon, and is used to determine the lengths of the latent period, contraction, and relaxation in a muscle. Study of human muscles discloses that there are two types of muscle in speed of contraction. "Fast muscle," sometimes called "white muscle" is found in the appendages where speed of movement is necessary. "Slow" or "red" muscle is slower and can sustain contractions for a longer period of time without tiring, as in the muscles that maintain posture.

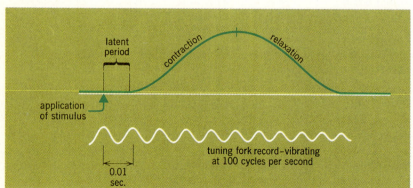

FIGURE 10.11   A single muscle twitch.

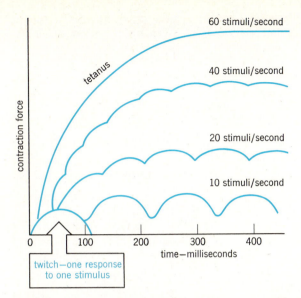

FIGURE 10.12 Wave summation and the genesis of tetanus in a "slow" muscle, as a function of stimulus frequency.

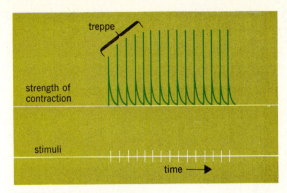

FIGURE 10.13 Treppe.

TETANUS (Fig. 10.12). Repeated stimulation of a muscle by increasing the frequency of stimuli results in a summation (adding together) or fusing of twitches. When a single smooth response occurs to repeated stimuli of a certain frequency (*critical frequency*), TETANIZATION has occurred. This phenomenon allows motions to occur smoothly and not in jerky fashion.

TREPPE (Fig. 10.13). Even with a maximal stimulus, the first few stimuli delivered to a skeletal muscle result in an increased strength of contraction called TREPPE. More calcium flowing into the area of the filaments during the first few stimuli and "thinning out" of muscle sarcoplasm as heat is liberated by the first few contractions have been advanced as theories to explain treppe.

TONE. TONE in a muscle refers to a continuous submaximal contraction that keeps a small degree of tension in a muscle. A denervated muscle has no tone, and thus tone may be the result of nervous stimulation. It apparently keeps the muscle in a state of readiness for contraction.

ISOTONIC AND ISOMETRIC CONTRACTIONS. If, when a muscle contracts and a body part *moves*, an ISOTONIC CONTRACTION has occurred and work has been done (work = force × distance). If tension is developed, but *no* movement occurs, an ISOMETRIC contraction has occurred, and *no work* is performed (distance is zero, therefore work is zero). The term isometric is probably familiar to those who have undertaken training programs to increase muscular strength. In isometric exercises, one group of muscles is pitted against another with no movement allowed. This—by mechanisms largely unknown—causes a more rapid enlargement of fibers than isotonic exercises (e.g., barbell lifting), brings more fibers into play, and causes neuromuscular junctions to increase in size and chemical content faster, bringing stronger stimuli to the muscle.

Table 10.1 summarizes the structure and properties of skeletal muscle.

Clinical considerations

MUSCULAR DYSTROPHY is regarded as an inherited disorder involving enzymes and/or proteins in the muscle. Muscle fibers appear degenerated with obvious differences in size. Degenerated fibers may be replaced with fibrous tissue or fat. Increasing muscular weakness may confine the patient to a wheelchair. Death, if it occurs, is most

TABLE 10.1   Summary of structure and properties of skeletal muscle

Item	Function or structure	Comments
Fiber	The basic unit or "cell" of skeletal muscle	Unit of structure of all named muscles. Size varies according to muscle size.
Sarcolemma	Membrane around fiber	Provides limit to fiber, some control of entry.
Sarcoplasm	The cytoplasm of the fiber	Contains mitochondria, reticulum, myofibrils.
Myofibrils	Longitudinally arranged units of the fiber	Cross banded (striated)
Myofilaments	Protein strands longitudinally arranged inside myofibrils	Contractile units of the muscle. 2 types: thick (contains myosin); thin (contains actin).
ATP, CP, K^+, Ca^{2+}, $PO_4^\equiv$, glucose	Chemicals, within the fiber	Necessary for contraction and nutrition.
Sarcoplasmic reticulum	The "endoplasmic reticulum" of the fiber	Houses Ca^{++} until required for contraction.
T-tubules	Tubules separate from sarcoplasmic reticulum which communicate to fiber exterior	Avenue for passage of substances into fiber.
Endomysium Perimysium Epimysium	Connective tissue binding fibers into a muscle	Endo—binds fibers together. Peri—surrounds fasciculi. Epi—surrounds muscle.
Interdigitating filament model	Theory of arrangement of filaments in myofibrils	Allows explanation of muscle contraction.
Muscle contraction	Shortening of fibers	Ultimate cause of movement.
Creatine phosphate	Compound in muscle releasing energy for ATP synthesis	Immediate source of energy for ATP synthesis.
Glucose, lactic acid, fatty acid.	Combusted to provide energy for ATP synthesis	Provide energy to sustain activity. sustain activity.
Isometric contraction	Tension developed but no shortening	Posture.
Isotonic contraction	Contraction with shortening	Movement.
Twitch	Illustrates phases of muscular activity. 1 response to 1 stimulus.	0.1 second duration (frog muscle). In body, does not normally occur.
Tetanus	Fusion of twitches; sustained maximal contraction	Result of short refractory period allowing twitch fusion.
Tonus	Sustained partial contraction	Depends on intact motor areas in brain, intact upper and lower motor neurons.
Treppe	Increasing strength of contraction with repeated strong stimuli	"Warming up" allows greater contraction.

often due to pulmonary infection as respirations become shallow and weak, or as heart muscle is affected.

In MUSCLE SPASM, a forcible, often painful contraction of a muscle occurs, most commonly in the appendages (e.g., foot). Chemical imbalances are the most common cause, and vigorous massage is often effective in relieving the spasm.

MUSCLE CRAMPS, though similar symptomatically to spasm, have a different explanation. Exercise causes intense afferent (sensory) nerve stimulation in muscles, by chemical or mechanical causes. The spinal cord responds to this input by increasing output to the muscles that causes them to contract more. This leads to more stimulation of the sensory nerves and a "vicious cycle" is established. Relief is often afforded by voluntarily contracting the opposing muscle group while preventing movement of the body part.

"SHIN SPLINTS" is a condition arising in the muscles of the leg after repetitive unusual exertion usually involving running. While there are many theories as to why the condition occurs, it appears to involve ischemia (lowered blood flow) of the involved muscles as a result of swelling when osmotic flow of water occurs into active muscle fibers secondary to metabolite accumulation. The muscles and their blood vessels are compressed because they are contained in closed fascial compartments that permit no expansion. Resting the involved muscles, and a program of graduated exercise to increase blood supply usually relieves or prevents development of the condition.

HERNIA (commonly called "rupture") is a term used to describe the protrusion of abdominal viscera through a defect in the muscle or connective tissue of the abdominal wall, diaphragm, or pelvic floor. Hernias are the result of abdominal weakness, and an intra-abdominal pressure that causes tearing or separation of the components of the abdominal wall. Hernias are more common in obese persons, older individuals, those whose livelihood requires heavy lifting, and in those whose "style of life" results in lack of use and weakening of abdominal muscles.

According to where the hernia occurs, several types are described.

Inguinal hernias account for 80 percent of all hernias. The defect is in the inguinal canal that carries the spermatic cord from the scrotum in the male, and the round ligament of the uterus to the labia majora in the female, Increased intra-abdominal pressure, as in coughing, sneezing, lifting, straining at stool, or in distension caused by fluid or gas, may result in a separation or tearing of the inguinal muscle or abdominal fascia. Abdominal viscera, primarily small intestine, may then protrude into the scrotum (in the male) or labia majora (in the female).

Femoral hernia occurs when there is an enlargement of the femoral ring that normally passes the blood vessels to and from the thigh. A bulging of the skin in the groin is common as intestines push into the area.

Umbilical hernia occurs around the "belly button" (umbilicus) and is more common in infants.

Hiatus hernia is protrusion of the stomach into the chest cavity as a result of a weakness in the opening (hiatus) passing the esophagus through the diaphragm.

Surgical repair is the preferred treatment for those whose activities may lead to difficulties with the hernia. Surgery repairs the defect in the muscle or fascia. After surgery, any type of strenuous activity or lifting should be avoided for several weeks to allow for proper healing and to prevent adhesions from forming.

Smooth muscle

Anatomy

SMOOTH MUSCLE (Fig. 10.14) consists of spindle-shaped fibers 2 to 5 μm in diameter and 50 to 100 μm in length. Each fiber possesses a single nucleus located near its center in the widest portion of the cell. A sarcolemma surrounds the fiber. Myofibrils are few and nonstriated. Smooth muscle is usually said to be involuntary, although its activity may be modified and in some cases controlled by nervous stimulation. The fibers are capable of maintaining a more-or-less constant contraction (tone) regardless of length. Concentrations of actin and myosin are about seven times lower in smooth muscle than in skeletal, as are concentrations of ATP and phosphocreatine. The muscle fibers appear to lack T-tubules and have a poorly developed sarcoplasmic reticulum.

Types of smooth muscle

Two categories of smooth muscle may be distinguished.

UNITARY SMOOTH MUSCLE exhibits spontaneous activity and conducts impulses from one muscle cell to another as though the entire muscle mass was a single cell. This type of muscle is found in the alimentary tract, uterus, ureters, and small blood vessels.

MULTIUNIT SMOOTH MUSCLE does not exhibit spontaneous contraction, usually requires stimulation by nerves, and can grade its strength of contraction. This type of muscle is found in the ciliary muscles and iris of the eye and in larger blood vessels.

Some smooth muscle does not fall clearly into either group. For example, in the bladder, stretch and/or nerve stimulation can cause contraction. Unitary smooth muscle has been studied most extensively and the comments that follow are based primarily on investigation of this type.

Spontaneous activity

Spontaneous activity depends on local depolarization similar to that occurring in cardiac nodal tissue. The membrane potential is about two-thirds that of skeletal muscle and nerve (about 60 mv) and decays so that spontaneous depolarization occurs. This depolarization does not arise in a constant area in the muscle but "migrates" from one place to another. The waves or areas of depolarization are then conducted from cell to cell by way of low resistance "tight junctions" between fused adjacent cell membranes. Contraction of smooth muscle is believed to occur by a process similar to that described for skeletal muscle. Actual contraction time is long, presumably because of the time required for the few myofibrils to develop tension. Strength of smooth muscle contraction depends on the frequency of spontaneous discharge from individual cells and the number of cells contracting together. Smooth muscle contraction, and the degree of tension developed, may be influenced by several factors.

STRETCHING usually serves as an adequate stimulus for depolarization and is followed by contraction.

Chemical factors in smooth muscle contraction

ACETYLCHOLINE stimulates most unitary smooth muscle to increased activity. The mechanism of its action is not clear but may be to increase membrane permeability to Ca^{2+} or Na^{1+}, similar to its effect on skeletal muscle. EPINEPHRINE effects depend on type of muscle and species. Epinephrine usually inhibits contraction of gut musculature and stimulates activity of uterine and vascular muscle. This chemical probably exerts its action by changing membrane permeability to Ca^{2+} or

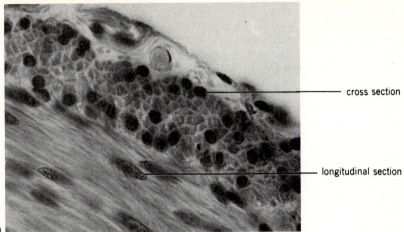

cross section

longitudinal section

(a) View in light microscope X440.

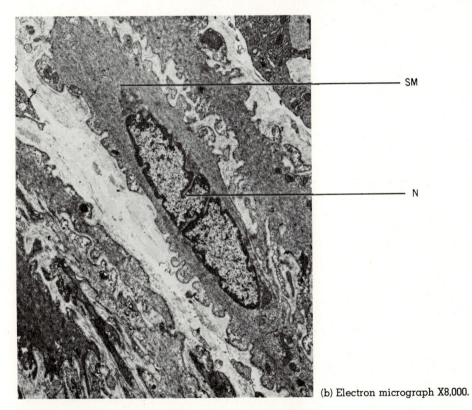

SM

N

(b) Electron micrograph X8,000.

FIGURE 10.14 The morphology of smooth muscle. SM = smooth muscle cell; N = nucleus of smooth muscle cell. (Electron micrograph, courtesy Norton B. Gilula, The Rockefeller University.)

Na^{1+}. NOREPINEPHRINE is similar to epinephrine in action and effect but requires about 100 times the concentration of an epinephrine "dose" to achieve the same degree of effect. ESTROGENIC HORMONES and OXYTOCIN also modify activity of smooth muscle, particularly of the uterus. Estrogen stimulates activity, as does oxytocin. PROGESTERONE tends to quiet or inhibit activity. OTHER CHEMICALS affecting smooth muscle are reserpine, ephedrine, and botulinus toxin. Reserpine, a chemical employed as an antihypertensive agent, causes relaxation of arteriolar smooth muscle. Ephedrine, a substance employed in nasal sprays, constricts vessels in the respiratory system. Botulinus toxin blocks acetylcholine release and leads to muscular paralysis.

Nervous factors in smooth muscle contraction

Autonomic nerves supply most smooth muscle and, by secretion of chemicals at their endings, exert effects similar to those produced by acetylcholine and norepinephrine. Most smooth muscle receives parasympathetic, acetylcholine-secreting nerves and sympathetic, norepinephrine-secreting nerves. No specialized end plate exists to transmit the effects of nerve stimulation to the muscle. Instead, the axons themselves may be seen to contain vesicles that presumably contain the chemicals that are released to influence muscle activity.

We may again note that a double nerve supply enables activity to be regulated to maintain homeostasis of such activities as blood pressure and peristalsis.

Clinical considerations

Nerve plexuses* and fibers are found in the walls of such organs as the small and large intestines. These structures aid in coordinating activity that may originate within the muscle and/or serve to transmit the effects of extrinsic control to the muscle.

CONGENITAL MEGACOLON (Hirschsprung's disease) is enlargement of the colon because of failure of the nervous structures to develop in a section of the organ. The section without nervous structures undergoes a persistent contraction and presents a functional obstruction to passage of materials. The colon proximal to the obstruction enlarges as material accumulates. If the section without nervous structures is short and near the anus, laxatives may provide a means of aiding passage of feces. Most cases, however, require surgical treatment in which the abnormal section is removed and the ends of the colon anastomosed (joined together).

This disorder emphasizes the importance of the intrinsic nervous elements serving smooth muscle in the gut.

* Meissner's plexus (submucosal) and Auerbach's plexus (between smooth muscle layers).

Summary

1. Muscular tissue is contractile, and can shorten and perform work.

2. There are three types of muscular tissue:
    a. Skeletal muscle forms 40 percent of the body weight, attaches to the skeleton and is served by nerves that cause it to contract.
    b. Smooth (visceral) muscle lies around hollow body organs. It may or may not depend on nerves to contract and controls propulsion of materials through organs.

c. Cardiac muscle is found only in the heart, and pumps the blood through the body. Its contraction is controlled by inherent factors.

3. Skeletal muscle cells are called fibers. Each fiber is surrounded by a sarcolemma, and contains sarcoplasm, myofibrils (composed of filaments), mitochondria, T-tubules, and sarcoplasmic reticulum (SR). The fibers show characteristic striations.

4. The myofibrils are composed of four major muscle proteins:

   a. *Myosin.* Composed of light and heavy portions, myosin acts to form bonds to actin during contraction.

   b. *Actin.* A double helix of protein molecules, actin is the molecule drawn together when the muscle shortens.

   c. *Tropomyosin.* This protein lies embedded in the grooves around the actin helix.

   d. *Troponin.* This protein binds Ca^{2+} and is located at repeating distances along the tropomyosin helix.

5. Muscle also contains ATP, Ca^{2+} (in the reticulum), phosphocreatin, glucose, and other substances.

6. The neuromuscular junction serves to transmit nerve impulses to the muscle as follows:

   a. Nerve impulses cause increase of permeability of nerve fibers to Ca^{2+}.

   b. Ca^{2+} causes release of acetylcholine (Ach) stored in nerve endings.

   c. Ach causes depolarization of muscle membrane systems.

   d. An enzyme, cholinesterase, rapidly destroys Ach.

7. Muscle contraction and relaxation involves

   a. Depolarization of sarcolemma, T-tubule membranes, and possibly SR membranes.

   b. Ca^{2+} moves from SR to filaments.

   c. Ca^{2+} binds to troponin, which alters shape and uncovers active sites on actin.

   d. Myosin forms bonds to actin.

   e. ATP attaches to myosin, causing head to oscillate.

   f. Head attaches repeatedly to different actin sites, drawing actin filaments together.

   g. Ca^{2+} is pumped back to SR.

   h. Tropomyosin recovers actin sites.

   i. No more binding of myosin to actin.

   j. Muscle relaxes.

8. ATP forms the immediate source of energy for muscle contraction and return of Ca^{2+} to the SR. Metabolism of creatine phosphate, glucose, and fatty acids furnish energy for ATP synthesis.

9. Electrical events precede a muscle contraction. An action potential forms, recordable as an EMG.

10. Heat production occurs; activation heat refers to all heat produced during one cycle and is subdivided into initial, relaxation, and recovery heats.

11. Some physiological properties of skeletal muscle are
    a. It won't contract until its threshold for depolarization is reached.
    b. Its strength of contraction can be graded according to requirement for tension.
    c. Its motor units follow the all-or-none law.
    d. It contracts more strongly when slightly stretched.
    e. It contracts more rapidly when slightly stretched.
    f. It may be stimulated directly; it shows independent irritability.
    g. It has refractory periods that are short, enabling it to be tetanized.

12. Contraction forms that a skeletal muscle shows include
    a. Isotonic. Shortening occurs; causes movement
    b. Isometric. No shortening occurs; maintains posture.
    c. Twitch. One cycle as a result of one stimulus. Shows phases and timing of activity.
    d. Tetanus. Sustained maximal contraction from fusion or summation of twitches.
    e. Treppe. Increased strength of contraction as muscle "warms up."
    f. Tone. Sustained partial state of contraction maintained without fatigue by alternate contraction of different motor units.

13. Smooth muscle is slow contracting, contains one nucleus per spindle-shaped unstriated cell, and occurs in unitary and multiunit varieties.
    a. Unitary exhibits spontaneous activity and is found primarily in digestive, reproductive, and urinary systems.
    b. Multiunit depends on nerves to contract and is found in the iris of the eye and larger muscular blood vessels.

14. Spontaneous activity originates in a "pacemaker" within the muscle and spreads via tight junctions between cells.

15. Smooth muscle is very sensitive to chemicals
    a. Acetylcholine usually causes unitary muscle to contract.
    b. Epinephrine may cause contraction *or* relaxation depending on location and species.

c. Other chemicals and their effects are presented.

16. Most smooth muscle has a dual nerve supply: one set causes it to contract; the other causes its relaxation. Activity can therefore be closely regulated.

Questions

1. Compare skeletal and smooth muscle as to structure and the properties that fit each of the jobs it does.

2. How does a nerve impulse "get to" a muscle to cause it to react?

3. What would happen to muscle contraction, and why, if Ca^{2+} was not available?

4. What are the roles of the following in muscle contraction/relaxation?
   a. ATP
   b. Myosin
   c. Troponin

5. List the steps involved in muscle contraction.

6. What roles do glucose and phosphocreatine play in muscle metabolism?

7. What effect would you expect a low environmental temperature to have on muscle activity? Explain.

8. Describe one disorder of skeletal muscle and how it interferes with normal function.

9. How do skeletal and smooth muscle react to changes in their chemical environments? Which do you speculate, and why, is more controlled by chemicals?

10. Curare is a chemical that blocks the action of acetylcholine at the neuromuscular junction. Explain, in terms of the junction, why curare is chosen by certain South American Indian tribes as an arrow poison.

Readings

Cohen, Carolyn. "The Protein Switch of Muscle Contraction." *Sci. Amer.* 233:36, Nov. 1975.

Drachman, Daniel B. "Myasthenia Gravis." *New Eng. J. Med.* 298:136, Jan. 19, 1978; 298:186, Jan. 26, 1978.

Felig, Philip, and John Wahren. "Fuel Homeostasis in Exercise." *New Eng. J. Med.* 298:1078, Nov. 20, 1975.

Fozzard, Harry A. "Heart: Excitation-Contraction Coupling." *Ann. Rev. Physiol. 39*:201, 1977.

Homsher, Earl, and Charles J. Kean. "Skeletal Muscle Energetics and Metabolism." *Ann. Rev. Physiol. 40*:93, 1978.

Huddart, Henry, and Stephen Hunt. *Visceral Muscle, Its Structure and Function.* Halstead (Wiley). New York, 1975.

Monster, A. W., H. C. Chan, and D. O'Connor. "Activity Patterns of Human Skeletal Muscles: Relation to Muscle Fiber Type." *Science 200*:314, 21 April 1978.

Science News. "Where the Actin Is." *113*:221, Apr. 8, 1978.

Chapter 11

The Skeletal Muscles

Objectives

The major objectives concerning muscles are for you to learn those muscles—and the facts about them—that your instructor specifies. We suggest you study location and action of those muscles first, then origins, insertions, and innervations. By learning the muscles as organs that operate specific body joints in well-defined patterns, you will increase your appreciation of how movement in general occurs. Also, keep in mind that the action assigned to a particular muscle does not always reflect what that muscle does with other muscles to achieve the often complicated motions of body parts.

Some general statements about the skeletal muscles

Muscles serve as the engines by which movement is achieved. By their ability to contract or shorten, and by their attachments (through the periosteum of a bone) to the skeleton, muscles utilize the bones as levers to achieve motion. Shape, size, power, and speed of the muscle represent a compromise between the location of the muscle on the body and the action it is to cause. Muscles of the appendages are generally *fusiform* or tapered in shape, with a rounded or oval outline in cross section, and tendons at either end for bony attachments. Appendicular muscles usually surround a centrally placed bone. On the thorax and abdomen, the muscles are shaped like a feather *(pennate),* are broad, and present a flattened cross section. This is a form adapted to the large surfaces presented by the rib cage and abdomen and where thick muscles would result in a "lumpy" configuration. Around body orifices, the muscle fibers are arranged so as to close the orifice *(circumpennate)*.

Muscles are named in many ways. For example:

By their shape when compared to geometric figures (e.g., trapezius, rhomboid).
By origin or insertion (e.g., sternocleidomastoid).
By location (e.g., rectus *abdominus*, latissimus *dorsi).*
By action (e.g., *levator* scapulae, *flexor* carpi radialis).

Generally, the name of a muscle may give valuable clues about where it is on the body and what type of movement it causes. Pay attention to what the ancients who named body muscles were trying to tell you.

The term ORIGIN refers to the *less movable bone* to which a muscle attaches; the term INSERTION to the *more movable bone* to which the muscle attaches—in other words, the bone that undergoes the "desired" movement. The term ACTION describes the *type* of movement brought about by the contraction of the muscle. Actions are presented in Table 11.1, and should be thoroughly learned before proceeding with the study of individual muscles. Remember that actions are stated as though the body was in *anatomical position* to start with.

Muscles that cause a given or specified action are termed AGONISTS or PRIME MOVERS. A muscle

TABLE 11.1 Muscle actions	
Action	Definition
Flexion	Decrease of angle between two bones
Extension	Increase of angle between two bones
Abduction	Movement away from the midline (of body or part)
Adduction	Movement toward midline (of body or part)
Elevation	Upward or superior movement
Depression	Downward or inferior movement
Rotation	Turning about the longitudinal axis of the bone
Medial	Toward midline of body ("inward")
Lateral	Away from midline of body ("outward")
Supination	To turn the palm up or anterior
Pronation	To turn the palm down or posterior
Inversion	To face the soles of the feet toward each other
Eversion	To face the soles of the feet away from each other
Dorsiflexion (=flexion)	At the ankle, to move the top of the foot toward the shin
Plantar flexion (=extension)	At the ankle, to move the sole of the foot downward, as in standing on the toes

only pulls, and an opposite action will occur only if there is another muscle, usually on the opposite side of a joint or body part, that can pull the bone in the "other" direction. This muscle must relax to permit the first muscle to cause a given movement and, when it contracts, causes a movement opposite to that of the first muscle. Such a muscle is an ANTAGONIST to the first muscle. Where the opposite action occurs, the previous antagonist becomes the prime mover, and the initial prime mover *now* becomes the antagonist. These facts emphasize that muscles are organized into groups having generally opposite effects on a given body joint and these groups will generally be found on opposite "sides" of a particular bone.

Actions, as given in this text, represent the primary or strongest type of movement a particular muscle causes. It *must* be kept in mind that muscles cooperate to varying degrees to cause particular actions, and that a given muscle may be responsible for a variety of movements depending on the joints involved and the actions of other muscles operating the joint. Thus, where more than one action is given, this reflects the combined efforts of a series of muscles on a body joint.

Let us now proceed to a study of the individual muscles of the body. This book considers the muscles as operating in antagonistic groups where possible and illustrations follow the "classical"

pattern of showing a body area in anatomical position with associated muscles. Where several layers of muscles are present, dissections will usually proceed from superficial to deep, with "new" muscles emphasized in individual drawings as they are discussed. In this way, an appreciation of origin and insertion of the muscle may be acquired more easily. Lastly, each body area is summarized by a table, in which the muscle's name, origin, insertion, action, and innervation appear. Not *every* muscle in the body is considered—the muscles described represent those considered of primary importance in body movement and activity.

Figures 11.1a and 11.1b present a view of the major muscles of the body as they might appear with the skin and fatty layers of the body removed. The superficial muscles are labeled or muscle groups are indicated.

Muscles of the head

Facial muscles (Fig. 11.2)

The facial muscles are the muscles of facial expression. Their contractions move the fleshy parts of the face for articulation of speech, and for the variety of movements that express our emotions. Perhaps the easiest way to learn these muscles is to begin with the circumpennate muscles of the face and then start at the nose and follow the muscles around the corner of the mouth to the chin.

ORBICULARIS OCULI is arranged circularly around the orbit, with fibers passing through the eyelids. The circular *orbital portion* is under strong willful control, and acts to close or wink the eyelids in response to bright light or romantic intent. The *palpebral portion* that transverses the eyelids is under predominately

reflex control that causes a blinking action to spread tears across the eye for lubricating and cleansing actions. This blinking action also moves tears toward the inner margins of the eyelids where they are ultimately drained into the nasal cavities.

ORBICULARIS ORIS encircles the mouth and brings about a closing of the lips and their puckering. Retention of food within the mouth, whistling, and whispering are other activities in which the muscle is involved.

PROCERUS lies between the nasal bone and the skin between the eyebrows. It causes small wrinkles or creases that run across the root of the nose.

COMPRESSOR NARES (nasalis) lies across the fleshy part of the nose and "pinches" the nostrils together.

QUADRATUS LABII SUPERIORIS consists of four small muscles that commence next to the nose and pro-

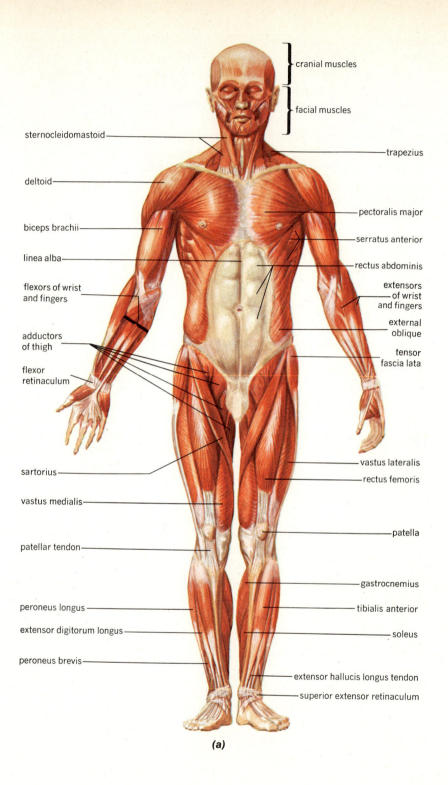

cranial muscles

facial muscles

sternocleidomastoid

trapezius

deltoid

pectoralis major

biceps brachii

serratus anterior

linea alba

rectus abdominis

flexors of wrist and fingers

extensors of wrist and fingers

adductors of thigh

external oblique

flexor retinaculum

tensor fascia lata

vastus lateralis

rectus femoris

sartorius

vastus medialis

patella

patellar tendon

gastrocnemius

peroneus longus

tibialis anterior

extensor digitorum longus

soleus

peroneus brevis

extensor hallucis longus tendon

superior extensor retinaculum

(a)

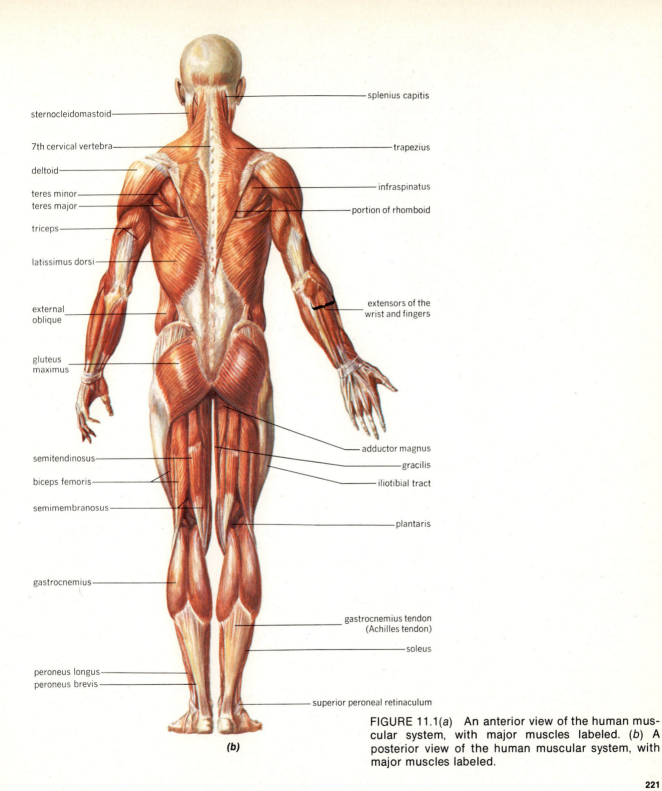

splenius capitis

sternocleidomastoid

7th cervical vertebra

deltoid

teres minor
teres major

triceps

latissimus dorsi

external
oblique

gluteus
maximus

semitendinosus

biceps femoris

semimembranosus

gastrocnemius

peroneus longus
peroneus brevis

trapezius

infraspinatus

portion of rhomboid

extensors of the
wrist and fingers

adductor magnus

gracilis

iliotibial tract

plantaris

gastrocnemius tendon
(Achilles tendon)

soleus

superior peroneal retinaculum

(b)

FIGURE 11.1(a) An anterior view of the human muscular system, with major muscles labeled. (b) A
posterior view of the human muscular system, with
major muscles labeled.

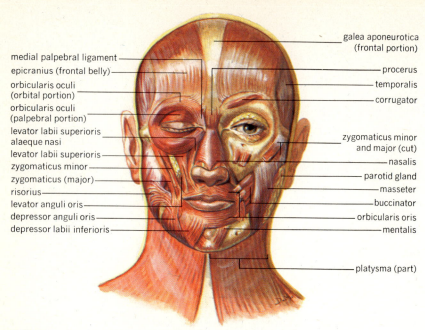

medial palpebral ligament
epicranius (frontal belly)
orbicularis oculi (orbital portion)
orbicularis oculi (palpebral portion)
levator labii superioris alaeque nasi
levator labii superioris
zygomaticus minor
zygomaticus (major)
risorius
levator anguli oris
depressor anguli oris
depressor labii inferioris

galea aponeurotica (frontal portion)
procerus
temporalis
corrugator
zygomaticus minor and major (cut)
nasalis
parotid gland
masseter
buccinator
orbicularis oris
mentalis
platysma (part)

FIGURE 11.2   An anterior view of the muscles of the face and anterior cranium.

ceed to the zygomatic bone. In order from the nose, the muscles are:

levator labii superioris alaeque nasi
levator labii superioris
zygomaticus minor
levator anguli oris

These muscles are the primary elevators of the upper lip when sneering, expressing contempt, or in saying "yeck" or "yuck." Additionally, ZYGOMATICUS MAJOR (zygomaticus), draws the corners of the mouth upwards and backwards in smiling.

RISORIUS lies at the corners of the mouth and draws the corners nearly straight laterally, as in a grimace. When children put their fingers in the corners of their mouths and pull them sideways they are imitating the actions of risorius.

DEPRESSOR ANGULI ORIS or TRIANGULARIS runs at an angle *downwards* from the corners of the mouth. This muscle draws the corners of the mouth down as in expression of sadness ("drooping mouth").

DEPRESSOR LABII INFERIORIS moves the lower lip downwards.

MENTALIS is a small muscle near the mental foramen that protrudes the lower lip, depresses it, and wrinkles the chin as in doubt or contempt.

BUCCINATOR, the "trumpeter's muscle," lies beneath all these muscles and forms the muscle of the cheek. It acts to guide food between the teeth for chewing, to aid circulation of saliva in the mouth, and to assist strongly in blowing, hence the name trumpeter's muscle.

CORRUGATOR *(corrugator supercilii)* produces "frown lines" in the middle forehead.

The facial muscles are presented in Table 11.2.

Cranial muscles (Fig. 11.3)

The cranial muscles lie on the forehead, back of the head, and around the ears. There are *no* muscles on the top and upper sides of the skull; the skin of the scalp lies on the EPICRANIAL APONEUROSIS *(galea aponeurotica)*. The term aponeurosis re-

TABLE 11.2 The facial muscles

Muscle	Origin	Insertion	Action	Innervation
Orbicularis oculi	Medial surface of orbit	Skin of eyelids	Closes or winks eye	7th cranial nerve (facial)
Orbicularis oris	Skin of lips, other facial muscles	Corners of mouth	Closes and puckers lips	7th cranial nerve (facial)
Quadratus labii superioris (a muscle with four parts or heads)	As a whole, from nose to below orbit to zygomatic bone	As a whole, skin of upper lip	As a whole, elevates upper lip	7th cranial nerve (facial)
1. Levator labii superioris alaeque nasi	Nasal process of maxilla	Skin of upper lip	Elevates upper lip, as in sneering	
2. Levator labii superioris	Below orbit			7th cranial nerve (facial)
3. Zygomaticus minor	Anterior aspect of zygomatic bone			
4. Levator anguli oris	Below infraorbital foramen			
Zygomaticus (zygomaticus major)	Lateral aspect of zygomatic bone	Skin of angle (corner) of mouth	Draws mouth up and back, as in smiling	7th cranial nerve (facial)
Risorius	Fascia of masseter muscle	Skin of angle (corner) of mouth	Pulls mouth laterally, as in a grimace	7th cranial nerve (facial)
Triangularis (depressor anguli oris)	Oblique line of mandible	Skin of lower lip at angle (corner) of mouth	Draws corner of mouth downwards, as in sadness	7th cranial nerve (facial)
Depressor labii inferioris	Below mental foramen (mental process)	Skin of lower lip	Depresses lower lip	7th cranial nerve (facial)
Mentalis	Mental symphysis	Skin of chin and lower lip	Depresses lower lip, wrinkles chin	7th cranial nerve (facial)
Buccinator	Alveolar processes of maxillae and mandible	Fibers of orbicularis oris	Compresses cheek, as in blowing (''trumpeter's muscle'')	7th cranial nerve (facial)
Corrugator	Inner end of superciliary ridge	Medial half of eyebrow	Produces vertical lines on forehead, as in frowning	7th cranial nerve (facial)
Procerus	Nasal bones	Skin between eyebrows	Produces transverse wrinkles at root of nose	7th cranial nerve (facial)
Nasalis (compressor naris)	Maxilla, above incisor teeth	Skin along side of nose	Narrows the nostrils	7th cranial nerve (facial)

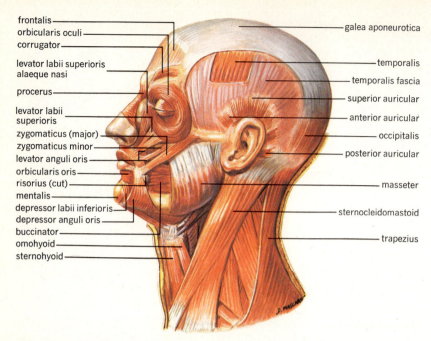

frontalis
orbicularis oculi
corrugator
levator labii superioris
alaeque nasi
procerus
levator labii
superioris
zygomaticus (major)
zygomaticus minor
levator anguli oris
orbicularis oris
risorius (cut)
mentalis
depressor labii inferioris
depressor anguli oris
buccinator
omohyoid
sternohyoid

galea aponeurotica
temporalis
temporalis fascia
superior auricular
anterior auricular
occipitalis
posterior auricular
masseter
sternocleidomastoid
trapezius

FIGURE 11.3 A lateral view of the facial and cranial muscles and several muscles of mastication.

TABLE 11.3 The cranial muscles				
Muscle	Origin	Insertion	Action	Innervation
Epicranius *Occipitofrontal group* Frontalis	Supraorbital margin	Galea aponeurotica	Raises eyebrows (surprise or fright), produces transverse furrows in skin of forehead, and pulls scalp forward	7th cranial nerve (facial)
Occipitalis	Superior nuchal line and mastoid portion of temporal bone	Galea aponeurotica	Pulls scalp backwards	7th cranial nerve (facial)
Temporoparietal group Auriculars Anterior			Draws pinna forwards	
Superior	Fascia of scalp	Pinna of ear	Draws pinna upwards	7th cranial nerve (facial)
Posterior			Draws pinna backwards	

fers to a broad, flat tendon. The galea thus moves as the muscles attached to it contract, and the attached skin is drawn forwards or backwards.

The term EPICRANIUS is applied to the muscles of the scalp as a group. Two subdivisions are recognized in the epicranius: an OCCIPITOFRONTAL GROUP consisting of an anterior *frontalis* on the forehead, and a posterior *occipitalis* on the lateral aspects of the occipital bone; the AURICULAR MUSCLES attach to the pinna *(auricle)* of the ear.

FRONTALIS raises the eyebrows in the expressions of surprise and fright and produces the transverse wrinkles in the forehead associated with thought. Contraction also pulls the scalp forwards.

OCCIPITALIS draws the scalp backwards.

AURICULARIS ANTERIOR moves the pinna forwards, AURICULARIS SUPERIOR elevates the pinna, while AURICULARIS POSTERIOR draws the pinna backwards. Little used in humans, other than to "wiggle the ears," the auriculars probably represent vestiges of muscles used to turn the ears so as to achieve the greatest benefit from sounds.

The cranial muscles are presented in Table 11.3.

Muscles of mastication (Figs. 11.4–11.6)

The muscles of mastication are those that elevate and depress the mandible and move it side-to-side to chew foods. The greatest strength is required to close the jaw; therefore, there are more muscles that elevate the mandible (close the mouth) than depress it (open the mouth).

MASSETER is a thick muscle covering the ramus of the mandible. The muscle is easily felt when the jaw is clenched as it is elevated.

TEMPORALIS occupies the temporal fossa on the sides of the skull. It also elevates the mandible, closing the mouth.

MEDIAL PTERYGOID runs from the medial pterygoid laminae of the sphenoid pterygoid processes to the mandible. It works to elevate the mandible.

LATERAL PTERYGOID has two heads and attaches to the front of the neck of the mandible. These muscles depress and protrude the mandible, and move it in a side-to-side direction during chewing.

Total force exerted by the mandible-elevating muscles has been estimated at a maximum of about 540 pounds when the teeth are clenched.

The muscles of mastication are presented in Table 11.4.

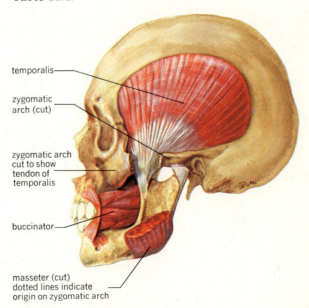

temporalis

zygomatic arch (cut)

zygomatic arch cut to show tendon of temporalis

buccinator

masseter (cut) dotted lines indicate origin on zygomatic arch

FIGURE 11.4   The temporalis and masseter, muscles of mastication (buccinator shown to demonstrate its extent and heads).

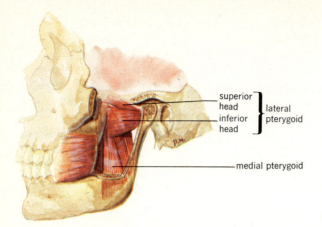

FIGURE 11.5   A lateral view of the skull to show the pterygoid muscles, muscles of mastication (the ramus of the mandible has been partially removed).

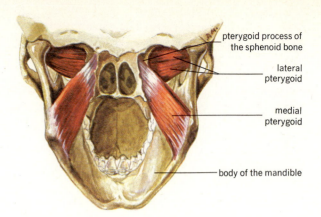

FIGURE 11.6 The pterygoid muscles viewed from below.

TABLE 11.4 The muscles of mastication				
Muscle	Origin	Insertion	Action	Innervation
Temporalis	Temporal fossa	Coronoid process of the mandible	Elevates mandible (and closes mouth)	Mandibular nerve [a branch of 5th cranial nerve (trigeminal)]
Masseter	Zygomatic process of maxilla, and zygomatic arch	Coronoid process, ramus, and angle of mandible	Elevates mandible (and closes mouth)	Mandibular nerve [a branch of 5th cranial nerve (trigeminal)]
Medial pterygoid	Medial pterygoid lamina and pterygoid fossa	Medial aspect of ramus and posterior body of mandible	Elevates mandible (and closes mouth)	Mandibular nerve [a branch of 5th cranial nerve (trigeminal)]
Lateral pterygoid	Lateral pterygoid lamina and greater wing of sphenoid bone	Anterior surface of neck of condyloid process of mandible	Protrudes and opens mouth; moves mandible side to side	Mandibular nerve [a branch of 5th cranial nerve (trigeminal)]

Muscles of the tongue (Fig. 11.7)

The tongue is an important organ: for guiding food between the teeth for chewing, in swallowing, and in the articulation of speech.

The tongue muscles include the *intrinsic muscles* that have their origins and insertions entirely within the tongue and which are responsible for the changes in shape of the tongue involved mainly in speech. These intrinsic muscles form a "three-way" pattern of fibers running longitudinally, transversely, and vertically within the tongue. The *extrinsic muscles* of the tongue have their origins outside the tongue and insertions within the tongue. These muscles are mainly responsible for the "in-and-out" and "side-to-side" movements of the tongue, although they also cause shape changes. The extrinsic muscles carry the suffix *glossus* in their name.

GENIOGLOSSUS raises and protrudes the tongue.

HYOGLOSSUS pulls the tongue upwards and backwards and arches it.

STYLOGLOSSUS pulls the tongue upwards and backwards and raises its sides.

PALATOGLOSSUS raises the base of the tongue, narrowing the opening into the pharynx (throat).

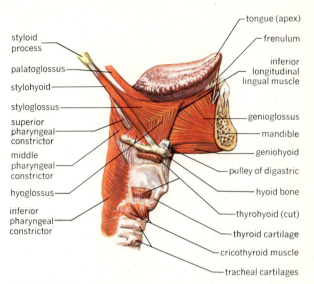

FIGURE 11.7 The extrinsic muscles of the tongue and some associated structures.

The extrinsic muscles of the tongue are presented in Table 11.5.

TABLE 11.5 The extrinsic muscles of the tongue				
Muscle	Origin	Insertion	Action	Innervation
Genioglossus	Genial tubercles on either side of posterior mandibular symphysis	Tongue	Raise tongue and pull it forwards (acting together), protrude tongue to opposite side (acting singly)	12th cranial nerve (hypoglossal)
Hyoglossus	Body and greater horn of hyoid bone	Tongue	Pulls tongue backwards and arches it	12th cranial nerve (hypoglossal)
Styloglossus	Styloid process	Tongue	Pulls tongue upwards and backwards and raises sides of tongue	12th cranial nerve (hypoglossal)
Palatoglossus	Palatine aponeurosis (on posterior aspect of the hard palate)	Tongue	Narrows opening of mouth into pharynx	12th cranial nerve (hypoglossal)

Muscles associated with the hyoid bone and larynx (Figs. 11.8, 11.9)

Muscles associated with the hyoid bone are divided into a *suprahyoid group* that lies mainly above a level with the bone and an *infrahyoid group* that lies below the bone.

Suprahyoid muscles include the following:

MYLOHYOID acts as the floor of the mouth. The muscle extends between the arch of the mandibular body and attaches to the hyoid body, elevating the bone.

DIGASTRICUS has two parts or *bellies.* The anterior runs between the anterior mandible and the hyoid body. This belly draws the hyoid forward. The posterior belly runs from the mastoid process and travels forward to the hyoid body. This belly draws the hyoid upwards and aids in opening the mouth.

STYLOHYOID runs alongside the posterior belly of the digastricus to the hyoid. It elevates and pulls the hyoid backwards.

GENIOHYOID lies above the mylohyoid, running from the hyoid body to the anterior mandible. Its basic action is to elevate the hyoid, but the muscle may also aid in depressing the mandible, opening the mouth.

The suprahyoid muscles are presented in Table 11.6.

The infrahyoid muscles include the following:

STERNOHYOID runs from the manubrium of the sternum to the hyoid bone. The muscle depresses the hyoid bone and steadies it when the suprahyoid muscles are active in swallowing.

STERNOTHYROID runs between the manubrium of the sternum and the thyroid cartilage of the larynx. It depresses the larynx after swallowing and during singing of low notes.

THYROHYOID runs between the thyroid cartilage and

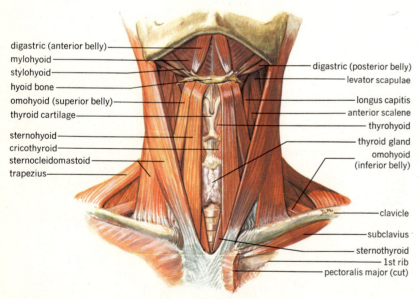

FIGURE 11.8 The superficial muscles concerned with movement of the hyoid bone and larynx, and some adjacent muscles.

the hyoid and aids in elevating the larynx during swallowing. It also depresses the hyoid and steadies it during movements of the tongue.

Oмоhyoid has two bellies or portions, separated by an *intermediate tendon.* As a whole, the muscle depresses the hyoid, and again steadies the bone during swallowing and speech.

The infrahyoid muscles are also presented in Table 11.6.

A diagram presenting a scheme of how the hyoid muscles are named is shown in Figure 11.10.

The supra- and infrahyoid muscles act together, to anchor the hyoid bone to provide a firm platform for the movements of the tongue during mastication, swallowing, and phonation.

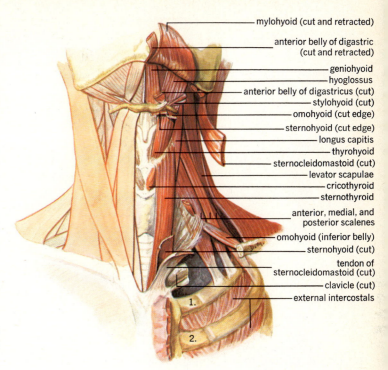

FIGURE 11.9 The deep muscles concerned with moving the hyoid bone and larynx, and some adjacent muscles.

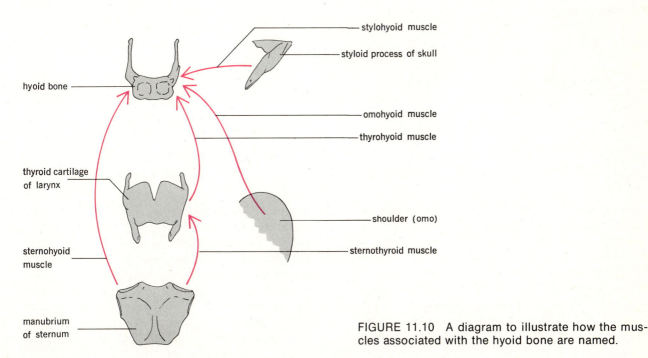

FIGURE 11.10 A diagram to illustrate how the muscles associated with the hyoid bone are named.

TABLE 11.6   The muscles associated with the hyoid bone and larynx

Muscle	Origin	Insertion	Action	Innervation
Suprahyoid group Digastricus Anterior belly	Digastric fossa of mandible (below posterior mandible symphysis)	Body of hyoid bone	Draws hyoid forwards	Mylohyoid branch of mandibular nerve [branch of 5th cranial nerve (trigeminal)]
Posterior belly	Mastoid notch of temporal	Body of hyoid bone	Elevates hyoid and aids mouth opening	Facial nerve (7th cranial nerve)
Stylohyoid	Styloid process of temporal	Greater horn of hyoid bone	Elevates and retracts hyoid	Facial nerve (7th cranial nerve)
Mylohyoid	Mylohyoid lines of mandible	Median raphe in center of muscle	Elevates hyoid bone	Mylohyoid branch of mandibular nerve [branch of 5th cranial nerve (trigeminal)]
Geniohyoid	Genial tubercles on each side of posterior mandibular symphysis	Body of hyoid bone	Elevates hyoid	C 1 and C 2 by way of hypoglossal nerve (12th cranial nerve)
Infrahyoid group Sternohyoid	Manubrium of sternum and medial end of clavicle	Body of hyoid bone	Depresses hyoid	Cervical nerves 1–3
Sternothyroid	Manubrium and first costal cartilage	Thyroid cartilage of larynx	Depress larynx	Cervical nerves 1–3
Thyrohyoid	Thyroid cartilage of larynx	Body and greater horn of hyoid bone	Depresses hyoid	Cervical nerves 1–3
Omohyoid	Superior border of scapula	Body of hyoid bone	Depresses hyoid	Cervical nerves 1–3

Muscles of the neck (Figs. 11.11–11.14; Table 11.7)

The muscles described in this section lie *on* the neck, although not all of them have their major action there. Some reach the face or cranium and thus move the same part of the head as their major action. Others are portions of muscles whose origins lie on the vertebral column below the neck, but send fibers to the neck or head. In general, a muscle with an origin below the neck but an insertion on it bears the term *cervicis* as part of its name. Similarly, the term *capitis* indicates that the muscle inserts on, and primarily moves, the skull. Some of the muscles described in this section are repeated in the next section dealing with the trunk. The reader should consider these two sections together in order to see the interrelationships between the muscles involved. The neck actually is a portion of the spinal column and supports the head, so that movements of one section invariably involve all sections.

PLATYSMA arises from the anterior chest wall and sweeps up each side of the neck, over the mandible, and ends by joining the facial muscle fibers at the corner of the mouth. Contraction of the muscle draws the corners of the mouth downwards as in sadness and when screaming. The muscle is penetrated by any surgical procedure below the skin on the antero-lateral aspect of the neck. The muscle may be absent on one or both sides of the neck.

STERNOCLEIDOMASTOID forms obvious straplike masses on either side of the neck. The muscles incline backwards as they travel from their sternal and clavicular origins to their insertions on the mastoid processes of the skull. It may be recalled that the muscles form the boundaries between the anterior and posterior triangles of the neck. Because of the angle of the muscle pull, it tilts the head to the same side as the muscle contracting, rotates the head to the opposite side (if acting singly), and flexes the neck (if operating together).

LONGUS COLLI (cervicis) lies alongside the upper seven thoracic and the last six cervical vertebral bodies. Fibers pass upwards from the thoracic bodies to insert into the transverse processes and bodies of the cervical vertebra lying above. Some fibers rising from lower cervical bodies reach the atlas. As the muscle contracts, the neck is flexed and rotated.

LONGUS CAPITIS runs between cervical transverse process 3–6, and the occipital bone. It flexes the neck.

RECTUS CAPITIS includes several sections (named by position or size as *anterior, lateralis, posterior minor,* and *posterior major*) that run between the atlas and occipital bone. According to position, the muscles flex the head (anterior), extend the head (posterior), or tilt the head to the side (lateralis).

The SCALENES consist of three muscles on either side of the neck. There is an anterior, middle, and posterior muscle on either side between the first two ribs and the cervical transverse processes. If the ribs are fixed, the muscles bend the neck to the side; if the neck is fixed, the muscles aid in rib elevation, as in heavy breathing.

The SEMISPINALIS CAPITIS and CERVICIS, LONGISSIMUS CAPITUS and CERVICIS, and SPLENIUS are discussed in the next section.

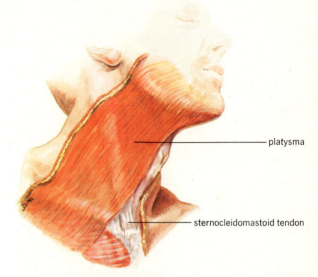

FIGURE 11.11   The platysma muscle.

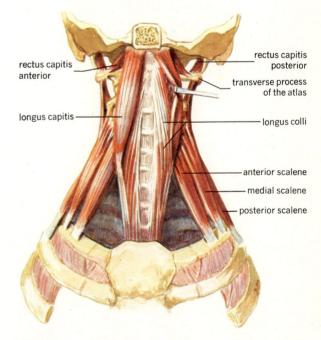

FIGURE 11.12 Some anterior and lateral neck muscles.

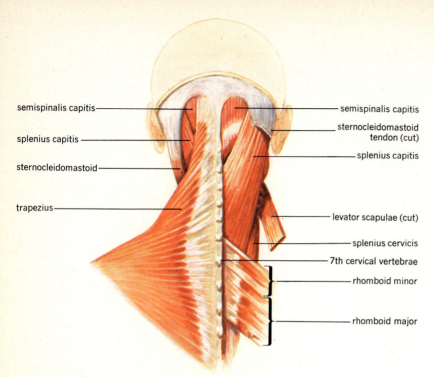

semispinalis capitis

semispinalis capitis

splenius capitis

sternocleidomastoid
tendon (cut)

sternocleidomastoid

splenius capitis

trapezius

levator scapulae (cut)

splenius cervicis

7th cervical vertebrae

rhomboid minor

rhomboid major

(a)

FIGURE 11.13(a)   The superficial muscles of the posterior neck and upper back. (b) The deep muscles of the posterior neck and upper back.

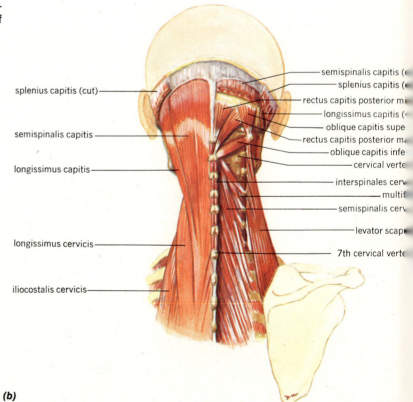

splenius capitis (cut)

semispinalis capitis (c

splenius capitis (c

rectus capitis posterior mi

longissimus capitis (c

oblique capitis supe

rectus capitis posterior ma

oblique capitis infe

cervical verte

semispinalis capitis

interspinales cerv

multif

longissimus capitis

semispinalis cerv

levator scap

longissimus cervicis

7th cervical verte

iliocostalis cervicis

(b)

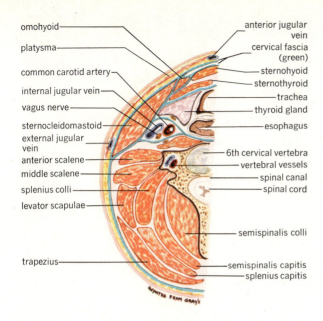

omohyoid
platysma
common carotid artery
internal jugular vein
vagus nerve
sternocleidomastoid
external jugular vein
anterior scalene
middle scalene
splenius colli
levator scapulae
trapezius

anterior jugular vein
cervical fascia (green)
sternohyoid
sternothyroid
trachea
thyroid gland
esophagus
6th cervical vertebra
vertebral vessels
spinal canal
spinal cord
semispinalis colli
semispinalis capitis
splenius capitis

PRINTED FROM GRAYS

FIGURE 11.14 A cross-section of the neck at the level of the sixth cervical vertebra to show arrangement of muscles and fascial planes. (Adapted from Gray.)

TABLE 11.7 The muscles of the neck

Muscle	Origin	Insertion	Action	Innervation
Superficial				
Platysma	Fascia of chest wall below clavicle; acromion process	Muscles of angle of mouth; lower lip; margin of mandible	Pulls corners of mouth down; "screaming"	Facial nerve (7th cranial nerve)
Sternocleido-mastoid	Manubrium of sternum, medial $\frac{1}{3}$ clavicle	Mastoid process of temporal	Flex neck (acting together) Rotate head to opposite side, or tilts to same side (acting singly)	Accessory nerve (11th cranial nerve) and C 2
Deep				
Longus colli (cervicis)	Transverse processes and bodies of vertebrae C 3–T 7	Atlas transverse processes and bodies of cervical vertebrae	Flexes and rotates neck	Cervical spinal nerves (C 2–8)
Longus capitis	Transverse processes C 3–6	Occipital bone	Flexes neck	Cervical spinal nerves (C 1–4)
Rectus capitis	Atlas	Base of occipital bone	Flex, extend, abduct or rotate the head	Cervical spinal nerves (C 1, 2)
Scalenes (anterior, middle, posterior)	Transverse processes, cervical vertebrae	Ribs 1 and 2	Flex neck and bend it to the side; elevate ribs 1 and 2	Cervical spinal nerves (C 3–8)

TABLE 11.7 (continued)				
Transversospinalis	Articular processes of vertebrae C 5–T 12, lumbar vertebrae and sacrum	Spines of vertebrae above and occipital bone	Extend spine and rotate it	Spinal nerves C 2–L 4
Semispinalis (capitus, cervicis, thoracis)	Transverse processes of lower cervical and all thoracic vertebrae	Spinous processes of cervical vertebrae and occipital bone	Extend and rotate spine	Spinal nerves C 1–T 6 (capitus, cervicis) T 7–T 12 (thoracis)
Splenius (capitis, cervicis)	Spinous processes C 8–T 5	Transverse processes C 1–4, superior nuchal line and mastoid process	Extension and rotation of neck and head	Cervical spinal nerves C 2–T 6
Interspinales Intertransversarii	Lie between spinous and transverse processes in virtually only the cervical and lumbar areas.		Steady the column and prevent its "folding"	Spinal nerves C 3–L 4

Muscles of the trunk

The muscles described here are those that flex, extend, bend to the side, and rotate the spinal column (Fig. 11.15). The muscles are shown in Figures 11.16 to 11.20.

TRANSVERSOSPINALIS is the name given to the muscular bundles filling the groove between the transverse and spinous processes of the vertebrae. The deeper fibers are called *multifidus;* the more superficial fibers are called *semispinalis thoracis* or *cervicis* according to whether they arise below or above the sixth thoracic vertebra.

ERECTOR SPINAE or SACROSPINALIS covers the transversospinalis and contains longer muscle bundles running more vertically than the transversospinalis. The most lateral column is the ILIOCOSTALIS, divided into *i. lumborum, i. thoracis,* or *i. cervicis* according to insertion. The largest and intermediate column is

the LONGISSIMUS, divided into *l. thoracis, l. cervicis,* and *l. capitis* according to insertion. The medial column is SPINALIS, divided into *s. thoracis, s. cervicis,* and *s. capitis,* according to insertion. All muscles extend the spine (as in resisting gravity), and bend the spine to the side (especially iliocostalis). Weakness of these muscles is often associated with exaggeration of the normal spinal curvatures and the development of back pain.

SPLENIUS is a broad flat straplike muscle that runs between the cervical and upper thoracic spines and the mastoid process of the skull. The *cervicis* inserts on the cervical vertebrae, the *capitis* on the skull. The muscle extends and rotates the neck and head.

The INTERSPINALES and INTERTRANSVERSARII lie between the spinous and transverse processes (respectively) of the vertebrae. They steady the vertebral column for movement and prevent its buckling.

(*Text continued on page 238.*)

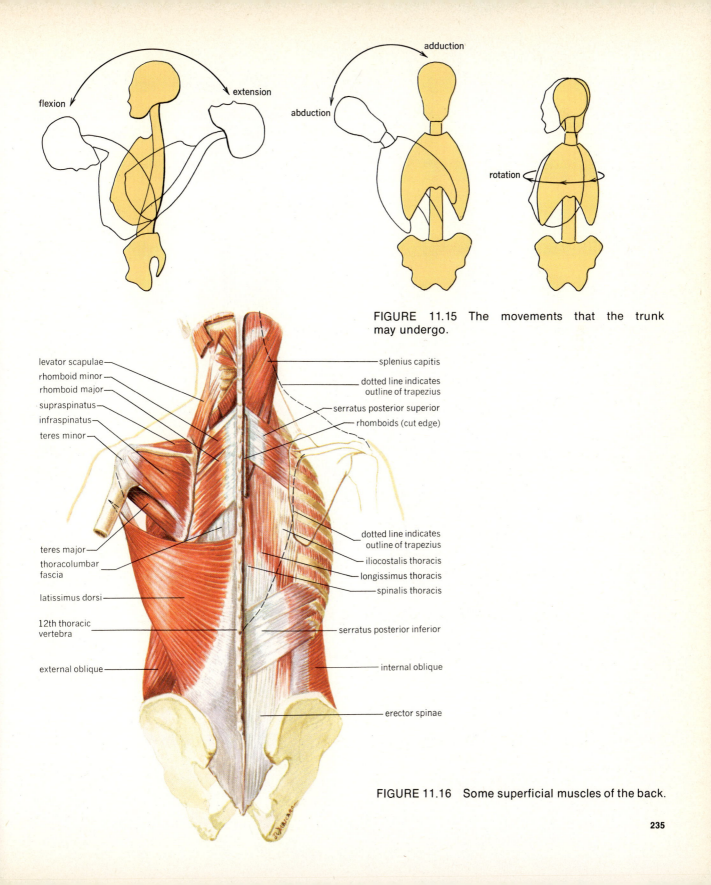

flexion

extension

adduction

abduction

rotation

FIGURE 11.15 The movements that the trunk may undergo.

levator scapulae

rhomboid minor

rhomboid major

supraspinatus

infraspinatus

teres minor

teres major

thoracolumbar fascia

latissimus dorsi

12th thoracic vertebra

external oblique

splenius capitis

dotted line indicates outline of trapezius

serratus posterior superior

rhomboids (cut edge)

dotted line indicates outline of trapezius

iliocostalis thoracis

longissimus thoracis

spinalis thoracis

serratus posterior inferior

internal oblique

erector spinae

FIGURE 11.16 Some superficial muscles of the back.

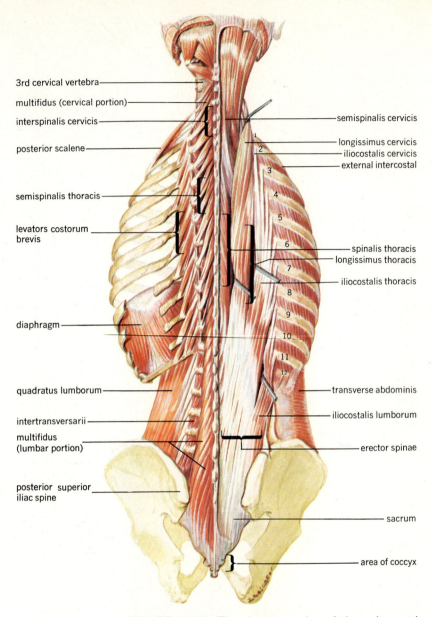

3rd cervical vertebra

multifidus (cervical portion)

interspinalis cervicis

posterior scalene

semispinalis thoracis

levators costorum brevis

diaphragm

quadratus lumborum

intertransversarii

multifidus (lumbar portion)

posterior superior iliac spine

semispinalis cervicis

longissimus cervicis

iliocostalis cervicis

external intercostal

1
2
3
4
5
6
7
8
9
10
11
12

spinalis thoracis

longissimus thoracis

iliocostalis thoracis

transverse abdominis

iliocostalis lumborum

erector spinae

sacrum

area of coccyx

FIGURE 11.17 The deep muscles of the spine and posterior thorax and abdomen.

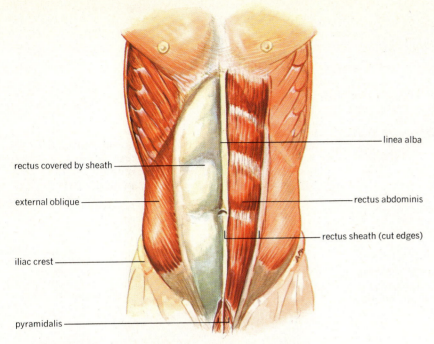

linea alba

rectus covered by sheath

external oblique

rectus abdominis

rectus sheath (cut edges)

iliac crest

pyramidalis

FIGURE 11.18 The superficial muscles of the anterior and lateral abdomen.

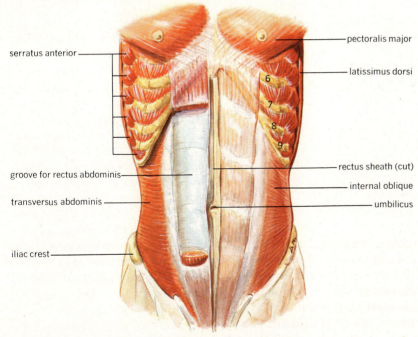

pectoralis major

serratus anterior

latissimus dorsi

6

7

8

9

groove for rectus abdominis

rectus sheath (cut)

internal oblique

transversus abdominis

umbilicus

iliac crest

FIGURE 11.19 The deep muscles of the anterior and lateral abdomen.

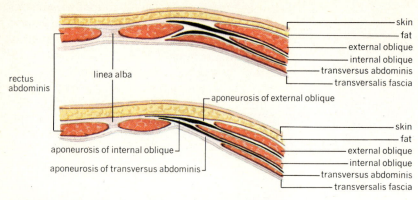

FIGURE 11.20   Cross sections of the anterior abdomen to show the arrangements of muscles and fascial planes. (*Above*) A section above the umbilicus. (*Below*) A section at the level of the iliac crest. Note that the sheath of the rectus abdominis is formed by the aponeuroses of the lateral abdominal muscles, and that—below the iliac crest—the aponeuroses of all three lateral abdominal muscles pass anterior to the rectus abdominus. (Modified from Braune.)

Muscles of the anterior and lateral abdominal wall

These muscles form the soft portion of the anterior and lateral portions of the trunk. The more anterior parts generally flex the spine (as in performing a sit-up), while the more lateral portions are involved in rotating the trunk and compressing the abdomen so as to prevent a "pot" from developing. The manner in which the muscles interdigitate to form the abdominal wall is depicted in Figure 11.20.

RECTUS ABDOMINIS is the only muscle forming the midline anterior portion of the abdominal wall. It arises from the pubic crest and attaches superiorly to the xiphoid process and cartilages of the 5–7 ribs. It is crossed anteriorly by three tendinous insertions, which gives the "segmented" appearance to the anterior abdomen in the living person. Rectus abdominis is the most powerful flexor of the trunk, particularly when the body is supine and one attempts to rise from that position. The small *pyramidalis* muscle, variable in length and often absent, lies in front of the lower part of the rectus abdominis.

The EXTERNAL OBLIQUE is directed downwards and forwards from its origin on the anterior inferior surfaces of the lower eight ribs. Its tendon of insertion forms a sheath for the rectus abdominis and then attaches to the linea alba (the "white line" of the abdominal midline), pubis, and iliac crest. Since the insertion lies medial to the origin of the muscle, contraction of a single muscle rotates the body to the *opposite* side, bringing the *same* shoulder forward as the side of the muscle contracting. Acting together, the muscles of both sides compress the abdomen.

The INTERNAL OBLIQUE is the intermediate layer of the lateral abdominal wall. It takes its origin from the iliac crest and posterior lumbar fascia, and the fibers sweep upwards at about a 90 degree angle to those of the external oblique, to insert into the last three ribs, xiphoid, and linea alba. A moment's reflection should indicate that, if one muscle contracts, the body will be rotated to the *same* side as the muscle contracting, thus bringing the *opposite* shoulder forward. Again, acting together, both muscles compress the abdomen.

The TRANSVERSE ABDOMINAL runs nearly horizontally around the abdomen, between the iliac crest, posterior lumbar fascia and last six ribs, to the pubis, linea alba, and xiphoid. The major function of transverse abdominal is to compress the abdomen.

All these muscles are important in forced expiration, during sneezing or coughing, and in "straining" as during heavy lifting or during defecation.

In the lower part of the abdomen, in the groin area, the tendons of the lateral abdominal muscles sweep around an opening that transmits the spermatic cord in the male, or the round ligament of the uterus in the female. The opening leads to the INGUINAL CANAL. If the tendons surrounding this opening are weakened, the opening may enlarge to permit intestines to pass into the canal, and an *inguinal hernia* has occurred. An *umbilical hernia* occurs in the anterior abdominal wall around the stump of the umbilical cord ("navel").

Muscles of the posterior abdominal wall

QUADRATUS LUMBORUM runs between the iliac crest and lower lumbar vertebrae, to the upper lumbar transverse processes and twelfth rib. Its direction of pull causes the twelfth rib to be drawn closer to the ilium, inclining the trunk to the side of the muscle contracting.

PSOAS MAJOR and ILIACUS run from the lumbar vertebrae and iliac fossa respectively, to the femur. Although their stated action is to mainly flex the thigh on the hip, if the thigh is fixed, the muscles aid in flexing the trunk on the thigh. Because the muscles have a common point of insertion, they are often considered to form a two-headed muscle called *iliopsoas*.

The abdominal muscles are presented in Table 11.8.

TABLE 11.8 Muscles of the trunk and abdomen

Muscle	Origin	Insertion	Action	Innervation
Flexors				
Rectus abdominis	Pubic crest	Xiphoid process, cartilages of 5–7 ribs	Flexes spine (sit-up)	Spinal nerves T 7–T 12
Psoas Iliacus }Iliopsoas	Transverse processes of lumbar vertebrae (psoas) Iliac fossa (iliacus)	Medial part of lesser trochanter	Flex thigh and (if thigh is fixed) flex spine	Spinal nerve L 1 Spinal nerve L 1
Extensors				
Iliocostalis (lumborum, thoracis, cervicis)	Iliac crest and ribs	Ribs and transverse processes of C 4–6	Extend and bend spine laterally	Spinal nerves C 4–L 2
Longissimus (thoracis, cervicis, capitis)	Lumbar vertebrae and sacrum, processes of thoracic and cervical vertebrae	Lower ten ribs, thoracic and cervical transverse processes, mastoid process of temporal	Extend spine, bend it to the sides	C 2–T 6 (capitus, cervicis) T 7–S 3 (thoracis)
Spinalis (thoracis, cervicis, capitis)	Lower thoracic and upper lumbar transverse processes	Spinous processes of thoracic and cervical vertebrae; occipital bone	Extends spine	Spinal nerves C 8–T 12

TABLE 11.8 (continued)

Muscle	Origin	Insertion	Action	Innervation
Abductors/ Adductors Quadratus lumborum	Iliac crest and lower lumbar vertebrae	12th rib and upper lumbar transverse processes	Bends spine to sides	Spinal nerves T 12–L 3
Rotators External oblique	Anterior inferior surfaces of lower 8 ribs	Linea alba, pubis, iliac crest	Singly, rotates spine to bring same shoulder forward; acting together, compresses abdomen	Intercostal nerves (spinal nerves T 2–12)
Internal oblique	Iliac crest, lumbo-dorsal fascia	Lower 3 ribs, linea alba, xiphoid process	Singly, rotates spine to bring opposite shoulder forward; acting together, compresses abdomen	Intercostal nerves (spinal nerves T 2–12)
Transverse abdominal	Iliac crest, lumbar fascia, last 6 ribs	Pubis, linea alba, xiphoid process	Compresses abdomen	Intercostal nerves (spinal nerves T 2–12)

Muscles of respiration

Breathing—particularly the inspiratory phase—involves alterations in the anteroposterior, lateral, and vertical dimensions of the thorax. Elevation of the ribs, because of their shape, causes the increase in anteroposterior and lateral dimensions of the thorax, while contraction and descent of the diaphragm increases the vertical dimension (Fig. 11.21).

The muscles of respiration are shown in Figures 11.22 and 11.23.

The DIAPHRAGM is a dome-shaped muscle with an extensive origin involving the xiphoid, last six ribs, and lumbar vertebrae. Its fibers sweep in a more-or-less radial fashion to be inserted into a *central tendon* in the top of the "dome." Thus, when the fibers contract, the dome is lowered, increasing the vertical dimension of the thorax. Relaxation is associated with return of the dome to its elevated position.

The EXTERNAL INTERCOSTALS form a series of eleven muscles on each side that lie between the outer inferior portions of the rib above and the superior portion of the rib below. Like the links in a chain, contraction of a muscle raises the rib below, with the chain anchored at the first rib. The ribs are elevated and swing upwards and outwards, enlarging the front-to-back and side-to-side dimensions of the thorax.

LEVATOR COSTORUM BREVIS also aids in rib elevation.

The INTERNAL INTERCOSTALS also form a series of eleven muscles on each side, lying internally between the ribs and running from the rib below to the rib above. The "chain" is anchored at the twelfth rib, and thus contraction pulls the ribs downwards, decreasing the thoracic dimensions in the front-to-back and side-to-side directions.

SERRATUS POSTERIOR INFERIOR runs between the lumbar vertebrae and the ribs. Because the ribs incline downwards as they course from spine to sternum, contraction of these muscles also depresses the ribs, aiding expiration. The SUPERIOR elevates the ribs.

The muscles of respiration are presented in Table 11.9.

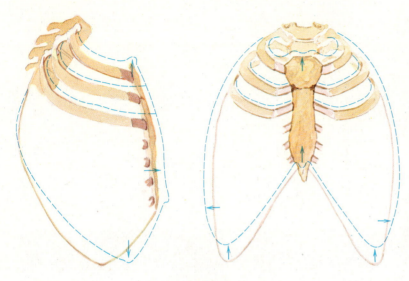

FIGURE 11.21 The movements of the thorax. As the ribs are elevated, the anterior-posterior and transverse dimensions are increased as shown by the dashed lines.

FIGURE 11.22 The diaphragm and some associated muscles and organs. Hiatuses (openings) through the diaphragm are provided for the esophagus, aorta, and inferior vena cava.

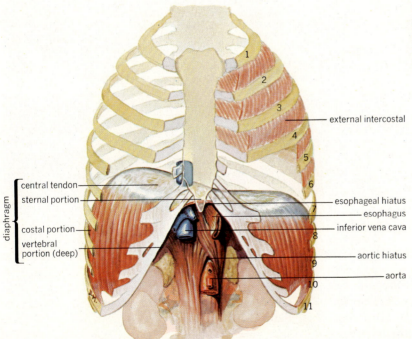

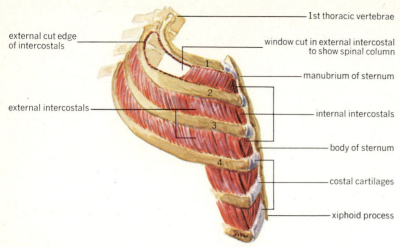

external cut edge of intercostals

external intercostals

1st thoracic vertebrae

window cut in external intercostal to show spinal column

manubrium of sternum

internal intercostals

body of sternum

costal cartilages

xiphoid process

FIGURE 11.23 The intercostal muscles, and associated skeletal structures.

TABLE 11.9 The muscles of respiration				
Muscle	Origin	Insertion	Action	Innervation
Diaphragm	Xiphoid process, last 6 ribs, lumbar vertebrae	Central tendon	Increases vertical dimension of thorax	Phrenic nerve (from cervical plexus, C 4)
External intercostals	Inferior border of upper 11 ribs	Superior border of rib below	Elevate ribs	Intercostal nerves (T 2–12)
Internal intercostals	Inner surface of lower 11 ribs	Superior surface of rib above	Depresses ribs	Intercostal nerves (T 2–12)
Serratus posterior (2 muscles designated s.p. superioris, and s.p. inferioris)	Superioris, between lower ligamentum nuchae and 3rd thoracic spine	Ribs 2–5	Elevates ribs	Spinal nerves C 8–T 6
	Inferioris, spinal processes of T 11 to L 2 vertebrae	Ribs 9–12	Depresses ribs	Spinal nerves T 10–L 4

Muscles of the shoulder (pectoral) girdle

The bones composing the shoulder girdle are the scapulae and clavicles. Of these two bones, only the scapulae are really free to move, with the clavicles acting as the platform from which move-ments of the scapula and humerus may occur. The scapula may undergo the movements shown in Figure 11.24.

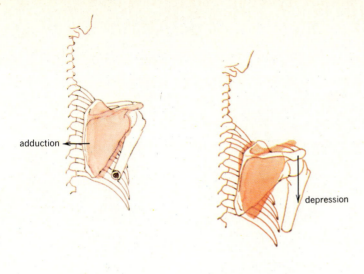

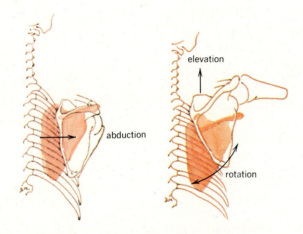

FIGURE 11.24   Movements the scapula may undergo. Colored areas indicate starting positions; uncolored areas and arrows, the movement.

The muscles involved in movements of the shoulder girdle may be conveniently divided into an anterior group, lying on the anterior thorax (or at least taking their origin there), and a posterior group, lying on the upper back or side of the neck (Figs. 11.25, 11.26, 11.27).

The anterior group includes the following:

PECTORALIS MINOR is a small, rather flattened muscle running between the first two ribs and the coracoid process of the scapula. It pulls the scapula for-

wards and assists in elevating the ribs during forced inspiration, if the arm is fixed.

SERRATUS ANTERIOR arises as a series of points (hence, *serrated*) from the lateral surfaces of the upper eight or nine ribs. Its fibers sweep nearly horizontally around the thorax, passing between the ribs and scapula, to insert on the *medial* border of the scapula. Contraction abducts the scapula, as in pushing or punching motions, and aids in keeping the scapula against the thorax. Paralysis of this muscle, or weakness, causes "winging" of the scapula, with

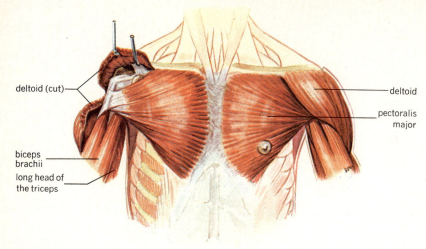

deltoid (cut)

deltoid

pectoralis
major

biceps
brachii

long head of
the triceps

FIGURE 11.25 The superficial muscles
of the upper chest and shoulder, with
part of the muscles of the arm.

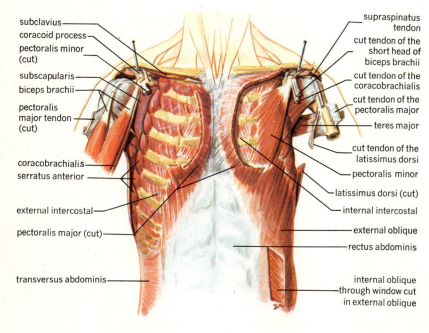

subclavius

coracoid process

pectoralis minor
(cut)

subscapularis

biceps brachii

pectoralis
major tendon
(cut)

coracobrachialis

serratus anterior

external intercostal

pectoralis major (cut)

transversus abdominis

supraspinatus
tendon

cut tendon of the
short head of
biceps brachii

cut tendon of the
coracobrachialis

cut tendon of the
pectoralis major

teres major

cut tendon of the
latissimus dorsi

pectoralis minor

latissimus dorsi (cut)

internal intercostal

external oblique

rectus abdominis

internal oblique
through window cut
in external oblique

FIGURE 11.26 The deep muscles of
the upper chest, and some associated
muscles of the scapula and arm.

the inferior angle and medial border forming a prominent projection on the back.

SUBCLAVIUS lies between the clavicle and first rib.
It steadies the clavicle for movements of scapula and
upper arm.

The posterior group includes the following
muscles:

TRAPEZIUS is the large, superficial, and trapezoid-
shaped muscle of the upper back. It has an extensive
origin from the skull to the twelfth thoracic vertebra,
and its fibers sweep downwards, laterally, and upwards to insert on scapula and clavicle. The superior,
middle, and inferior groups of fibers are sometimes
designated as parts 1, 2, and 3, from top downwards.

Part 1 aids in elevating the scapula, as in "shrugging" the shoulders; part 2 is the strongest and is a powerful adductor of the scapula; part 3 depresses the scapula. In movements of the upper arm that involve its abduction beyond the horizontal position, trapezius aids in scapular rotation.

RHOMBOID is usually considered to be composed of a smaller superior *r. minor,* and a lower larger *r. ma-*

jor. The origin is on the spine, and the fibers incline downwards to the medial scapular border. Thus, contraction moves the scapula upwards *and* inwards.

LEVATOR SCAPULAE runs from the cervical vertebrae almost straight downwards to the superior angle of the scapula. Contraction elevates the scapula.

These muscles are presented in Table 11.10.

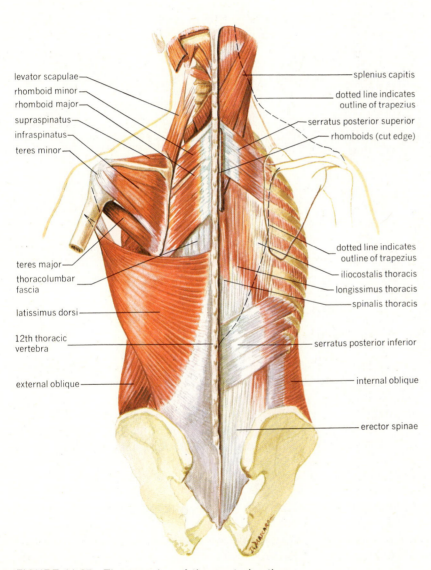

levator scapulae
rhomboid minor
rhomboid major
supraspinatus
infraspinatus
teres minor

teres major
thoracolumbar fascia
latissimus dorsi
12th thoracic vertebra
external oblique

splenius capitis
dotted line indicates outline of trapezius
serratus posterior superior
rhomboids (cut edge)

dotted line indicates outline of trapezius
iliocostalis thoracis
longissimus thoracis
spinalis thoracis

serratus posterior inferior

internal oblique

erector spinae

FIGURE 11.27   The muscles of the posterior thorax.

TABLE 11.10   Muscles of the shoulder (pectoral) girdle

Muscle	Origin	Insertion	Action	Innervation
Anterior				
Pectoralis minor	Outer surface ribs 3–5	Coracoid process	Pulls scapula forwards (if arm is fixed, elevates ribs)	Spinal nerves C 7, 8; T 1
Serratus anterior	Outer surfaces ribs 1–9	Vertebral (medial) border of scapula	Abducts scapula	Spinal nerves C 5–7
Subclavius	Junction of 1st rib and its cartilage	Middle third of clavicle	Depresses scapula	Spinal nerve C 5
Posterior				
Trapezius	Superior nuchal line, ligamentum nuchae, spines of vertebrae C 7–T 12	Lateral third of clavicle, acromion and spine of spacula	"Shrugs" shoulders, adducts and depresses scapula, rotates scapula in elevation of upper arm	11th cranial nerve and spinal nerves C 3, 4
Rhomboid [2 parts: major (upper), minor (lower); separated at T 2 spine]	Lower part of ligamentum nuchae, spines of C 7–T 5	Vertebral (medial) border of scapula between spine and inferior angle	Elevate and adduct scapula	Spinal nerve C 5
Levator scapulae	Transverse processes C 1–4	Between superior angle and spine of scapula on medial border	Elevates scapula	Spinal nerves C 3–5

Muscles moving the shoulder joint (humerus)

The shoulder joint is a triaxial, ball-and-socket joint, and so permits all possible planes of movement (Fig. 11.28). The muscles to be considered in this section are divided into functional groups — that is, flexors, extensors, abductors, adductors, and rotators (Figs. 11.25–11.27, and 11.29–11.31).

PECTORALIS MAJOR is the broad, fan-shaped superficial muscle of the anterior thorax. It has an extensive origin, extending from the clavicle, onto the sternum, and the cartilages of the true ribs. The fibers sweep together into a narrow tendon that forms the anterior axillary fold and inserts into the *lateral* lip of the bicipital groove. Contraction draws the humerus

forwards (flexion), across the chest, and, because of the lateral insertion, medially rotates the humerus. It is an extremely important muscle in throwing, particularly if "spin" is imparted to the object.

CORACOBRACHIALIS runs from coracoid process to midhumerus. A small muscle, it is only a weak flexor of the humerus.

Extensors of the humerus

LATISSIMUS DORSI is the large superficial and fan-shaped muscle of the lower back. Its origin extends from vertebra C6 (thus is overlapped by trapezius 3)

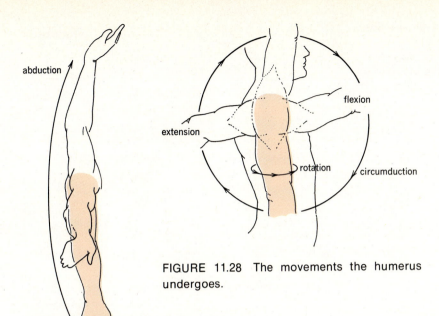

abduction

extension

flexion

rotation

circumduction

adduction

(the upper appendage is viewed from the lateral aspect)

FIGURE 11.28 The movements the humerus undergoes.

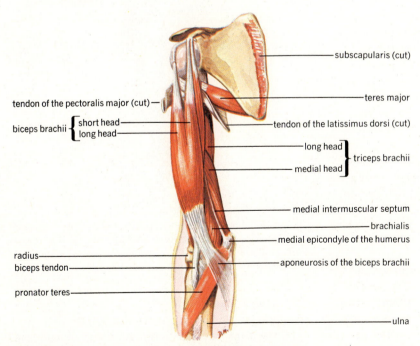

subscapularis (cut)

teres major

tendon of the pectoralis major (cut)

tendon of the latissimus dorsi (cut)

biceps brachii { short head / long head }

long head

medial head
} triceps brachii

medial intermuscular septum

brachialis

medial epicondyle of the humerus

radius

biceps tendon

aponeurosis of the biceps brachii

pronator teres

ulna

FIGURE 11.29 The muscles of the anterior arm, and associated muscles of the scapula and forearm.

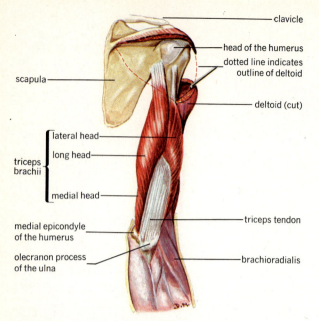

FIGURE 11.30   The muscles of the posterior arm, and associated skeletal structures.

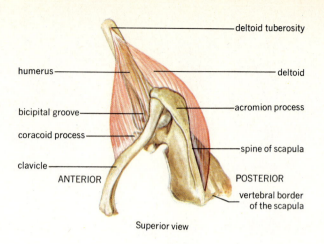

Superior view

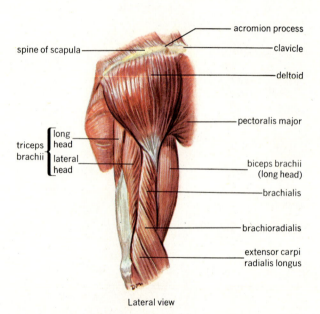

Lateral view

FIGURE 11.31 (*Above*) A superior view of the right shoulder to show the relationships of the deltoid to the scapula, clavicle, and humerus. (*Below*) A lateral view of the right shoulder and arm.

to the sacrum. The fibers form a narrow tendon that passes in front of the humerus to insert into floor of the bicipital groove. Contraction pulls the humerus downwards against resistance, extends it, and—because the tendon passes *in front* of the humerus to its insertion—medially rotates the humerus. It is an important muscle in climbing (as a rope) to pull the humerus downwards, and in the backstroke in swimming.

Teres major runs between the inferior scapular angle, to insert with latissimus on the humerus. The actions of the muscle are the same as the latissimus.

Abductors of the humerus

Deltoid, arising from clavicle and scapula, passes over the top of the shoulder to insert on the deltoid tuberosity of the humerus. The muscle has three sections to it, designated *anterior, middle,* and *posterior.* Acting as a whole, deltoid is the most powerful abductor of the humerus. Contraction of the anterior portion produces flexion, while contraction of the

posterior portion produces extension of the humerus.

Supraspinatus runs from the supraspinous fossa over the top of the shoulder joint to the greater tubercle. It abducts the humerus.

It may be noted that there is or are, no muscle(s) that adduct(s) the humerus. If pectoralis major and latissimus dorsi contract with equal force with the humerus in the horizontal position, the upper arm will be drawn directly inwards. One or the other muscles contracting more strongly will draw the humerus not only inwards, but across the chest or back.

Rotators of the humerus

Pectoralis major, latissimus dorsi, and teres major have already been mentioned as medial rotators of the humerus. Additional rotation is accomplished by the following muscles.

TABLE 11.11   Muscles moving the shoulder joint (humerus)

Muscle	Origin	Insertion	Action	Innervation
Flexors				
Pectoralis major	Medial 1/3 of clavicle, sternum, cartilages of ribs 1–7	Lateral lip of bicipital groove	Flexes, adducts, and medially rotates humerus	Spinal nerves C 5–T 1 (pectoral nerve)
Coracobrachialis	Coracoid process	Midhumerus	Flexes, adducts and medially rotates humerus	Musculocutaneous nerve (C 7)
Extensors				
Latissimus dorsi	Spinous processes T 6–L 5, iliac crest, last 3 ribs	Floor of bicipital groove (intertubercular sulcus) of humerus	Extension, adduction, medial rotation of humerus	Spinal nerves C 6–8 (thoracodorsal nerve, a branch of the posterior cord of the brachial plexus)
Teres major	Inferior angle of scapula	Medial lip of bicipital groove	Extension, adduction, medial rotation of humerus	Spinal nerves C 6–8
Abductors				
Deltoid	Lateral 1/3 of clavicle, acromion, spine of scapula	Deltoid tuberosity	Abducts humerus (acting as a whole)	Axillary nerve
*Supraspinatus	Supraspinous fossa	Greater tubercle of humerus	Abducts humerus	Spinal nerves C 5, 6
Rotators				
*Subscapularis	Subscapular fossa	Lesser tubercle of humerus	Medially rotates humerus	Spinal nerves C 7, 8 (subscapular nerve)
*Infraspinatus	Infraspinous fossa	Greater tubercle of humerus	Laterally rotates humerus	Spinal nerves C 5, 6
Teres minor	Inferior angle of scapula	Greater tubercle of humerus	Laterally rotates humerus	Axillary nerve (C 5)

* "Articular muscles" that aid in stabilizing and holding shoulder joint together.

SUBSCAPULARIS runs from the subscapular fossa across the *front* of the shoulder joint to the humerus. It medially rotates the humerus.

INFRASPINATUS runs from the infraspinous fossa across the *back* of the shoulder joint to the humerus. It laterally rotates the humerus.

TERES MINOR lies inferior to infraspinatus and inserts with infraspinatus on the humerus. It also laterally rotates the humerus.

Supraspinatus, subscapularis, infraspinatus, and teres minor insert so close to the shoulder joint that they have little power to move the humerus. More importantly, they act as *articular muscles* to help hold the shoulder joint together, or to stabilize it.

The muscles operating the shoulder joint are presented in Table 11.11.

Muscles moving the forearm

The muscles moving the forearm take their origins from scapula or humerus, and insert on the radius and/or ulna to cause the forearm to be flexed, extended, or rotated (Fig. 11.32). Remember that in the elbow region there are actually three joints: the ulnohumeral (permitting flexion and extension); the radiohumeral (permitting rotation); the radioulnar (also involved in rotation). The muscles are presented in Figures 11.29–11.31, and 11.35.

Flexors of the forearm

BICEPS BRACHII is a two-headed muscle. The *long head* lies laterally on the anterior humerus and takes its origin from a tubercle above the glenoid fossa (supraglenoid tubercle). The *short head* lies medially, and rises from the coracoid process. Both heads insert together on the radial tuberosity. Contraction flexes the forearm on the arm. If the forearm is pronated, the radial tuberosity is directed posteriorly; thus contraction also supinates the forearm (as when a right-handed person uses a screwdriver).

BRACHIALIS rises from the humerus and inserts on the ulna (a bone that *cannot* rotate). Thus, brachialis only flexes the forearm.

BRACHIORADIALIS runs from humerus to radius and flexes the forearm. It also returns the forearm to anatomical position from either a fully supinated or fully pronated position.

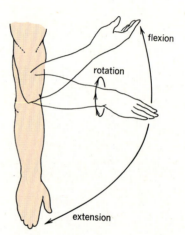

FIGURE 11.32   The movements of the forearm.

Extensors of the forearm

The major extensor of the forearm is the TRICEPS BRACHII. As the name indicates, this a three-headed muscle, with a common insertion on the olecranon process. The *lateral head* rises from the humerus, the centrally placed *long head* from the scapula, and the *medial head* from the humerus. The muscle extends the forearm. ANCONEUS arises from the lateral epicondyle and passes to the olecranon process and aids triceps in extension of the forearm.

TABLE 11.12 Muscles moving the forearm				
Muscle	Origin	Insertion	Action	Innervation
Flexors				
Biceps brachii				
Long head	Supraglenoid tubercle	Radial tuberosity	Flexes and supinates forearm	Musculocutaneous nerve (spinal nerves C 5, 6)
Short head	Coracoid process			
Brachialis	Lower half anterior surface of humerus	Coronoid process of ulna	Flexes forearm	Musculocutaneous nerve
Brachioradialis	Upper 2/3 lateral surface of humerus	Lateral side of radius above styloid process	Flexes, semisupinates, semipronates forearm	Radial nerve (spinal nerves C 5, 6)
Extensors				
Triceps brachii				
Lateral head	Posterior shaft of humerus			
Long head	Infraglenoid tubercle	Olecranon process	Extends forearm	Radial nerve (spinal nerves C 7, 8)
Medial head	Posterior shaft of humerus			
Anconeus	Posterior aspect of lateral epicondyle of humerus	Lateral side of olecranon process	Extends forearm	Radial nerve (spinal nerves C 7, 8)
Rotators				
Supinator	Lateral epicondyle of humerus	Lateral aspect of radius	Supinates forearm	Radial nerve (spinal nerves C 5, 6)
Pronator teres	Medial epicondyle of humerus	Halfway down lateral surface of radius	Pronates forearm	Median nerve (spinal nerves C 6, 7)
Pronator quadratus	Anterior surface of ulna	Anterior surface of radius	Pronates forearm	Spinal nerves C 8, T 1

Rotators of the forearm

As indicated previously, biceps can supinate the forearm, and brachioradialis can both semisupinate and semipronate the forearm.

SUPINATOR runs from the lateral epicondyle of the humerus to half-way down the radius, and—as its name indicates—supinates the forearm.

PRONATOR TERES is a proximal forearm muscle running between the medial distal humerus to the radius. It pronates the forearm.

PRONATOR QUADRATUS is a distal forearm muscle, running between the ulna and radius. It also pronates the forearm.

The muscles operating the forearm are presented in Table 11.12.

Muscles moving the wrist and fingers (Figs. 11.33–11.36, and 11.39–11.41)

Muscles moving the wrist and fingers arise from the distal humerus and/or proximal ulna and radius. The anterior group arises mainly from the medial epicondyle of the humerus, and *flex* something; the posterior group arises mainly from the lateral epicondyle of the humerus, and *extend* something. Abduction and adduction of the wrist are provided by cooperation of the wrist flexors and extensors, and the same movements of the fingers occur by muscles located within the hand. Wrist and finger movements are shown in Figure 11.37.

Flexors of the wrist and fingers

Wrist flexors form a superficial group of three muscles on the anterior aspect of the forearm.

FLEXOR CARPI RADIALIS is the most lateral of the group running between the medial epicondyle of the humerus and the second and third metacarpals. It flexes and slightly abducts the wrist. Its tendon projects prominently from the lateral distal forearm (next to the radial artery where the pulse is taken) when the wrist is forcibly flexed.

PALMARIS LONGUS has a similar origin to the previous muscle, is the central muscle of the forearm, and inserts into a broad flat tendon in the palm of the

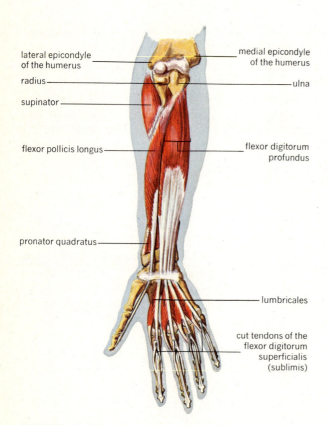

FIGURE 11.33 The deep muscles of the anterior forearm.

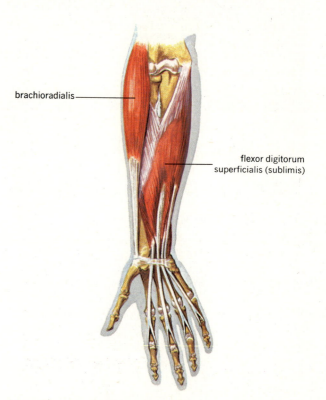

FIGURE 11.34 The intermediate layer of muscles of the anterior forearm.

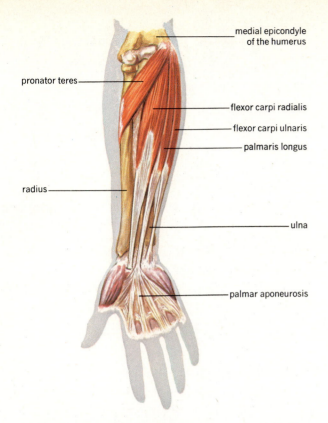

medial epicondyle
of the humerus

pronator teres

flexor carpi radialis

flexor carpi ulnaris

palmaris longus

radius

ulna

palmar aponeurosis

FIGURE 11.35   The superficial layer of muscles of the anterior forearm.

hand called the *palmar aponeurosis*. Wrist flexion is its only action.

FLEXOR CARPI ULNARIS is a two-headed muscle rising from medial epicondyle and ulna. It inserts into the fifth metatarsal and pisiform bone. It flexes and slightly adducts the wrist.

The intermediate layer of the anterior forearm muscles is formed by FLEXOR DIGITORUM SUPERFICIALIS (*sublimis*). The muscle is organized basically as a four-headed muscle, sending tendons to the second phalanx of the fingers. (Remember that there are four *fingers* and a *thumb* on each hand; five digits). Each head operates more-or-less independently of the others, to cause flexion of the fingers at the first interphalangeal joints.

The deep layer of muscle on the anterior forearm is formed by FLEXOR DIGITORUM PROFUNDUS.

The tendons of this muscle pass to the distal phalanx of each finger to flex those joints.

Two muscles are therefore required to achieve complete flexion of the fingers.

Extensors of the wrist and fingers

Six muscles, in a single layer across the posterior forearm, form the extensors of the wrist and fingers.

EXTENSOR CARPI RADIALIS LONGUS is the most lateral of this group, inserts on the second metacarpal, and extends and slightly abducts the wrist.

EXTENSOR CARPI RADIALIS BREVIS inserts on the third metacarpal and has the same action as the previous muscle.

EXTENSOR DIGITORUM COMMUNIS has three heads that send tendons to all the phalanges of the index, middle, and ring fingers, extending them completely.

EXTENSOR INDICIS provides an additional tendon to the phalanges of the index finger for extension.

EXTENSOR DIGITI MINIMI may be a head of the previous muscle, or a separate muscle, and sends its tendons to the phalanges of the little finger to extend it.

EXTENSOR CARPI ULNARIS is the most medial muscle of the posterior group, attaches to the fifth metacarpal, and extends and slightly adducts the wrist.

Abduction and adduction of the wrist

As indicated previously, certain muscles whose primary action is to flex or extend the wrist also slightly abduct or adduct it. More powerful abduction and adduction of the wrist are provided by the cooperation of these muscles, as shown in Figure 11.38. The diagram indicates that abduction is achieved by the flexor carpi radialis and extensor carpi radialis longus, while adduction occurs by the cooperation of flexor carpi ulnaris and extensor carpi ulnaris.

The muscles described above are presented in Table 11.13.

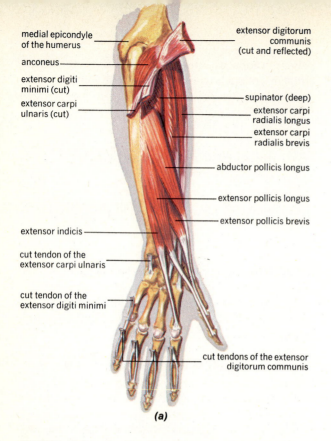

medial epicondyle of the humerus

anconeus

extensor digiti minimi (cut)

extensor carpi ulnaris (cut)

extensor indicis

cut tendon of the extensor carpi ulnaris

cut tendon of the extensor digiti minimi

extensor digitorum communis (cut and reflected)

supinator (deep)

extensor carpi radialis longus

extensor carpi radialis brevis

abductor pollicis longus

extensor pollicis longus

extensor pollicis brevis

cut tendons of the extensor digitorum communis

(a)

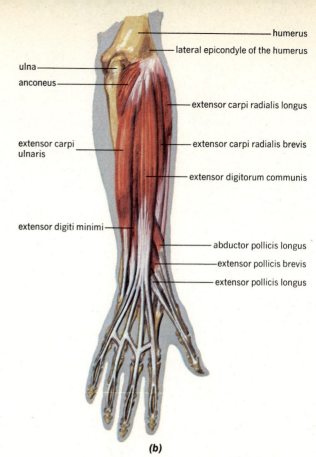

humerus

lateral epicondyle of the humerus

ulna

anconeus

extensor carpi ulnaris

extensor digiti minimi

extensor carpi radialis longus

extensor carpi radialis brevis

extensor digitorum communis

abductor pollicis longus

extensor pollicis brevis

extensor pollicis longus

(b)

FIGURE 11.36 (a) The deep muscles of the posterior forearm. (b) The superficial muscles of the posterior forearm.

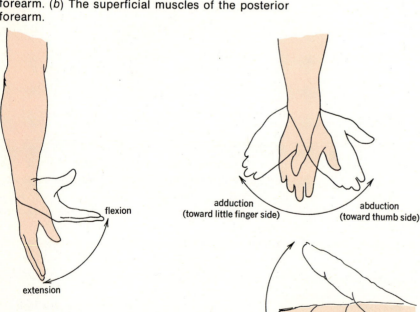

flexion

extension

adduction (toward little finger side)

abduction (toward thumb side)

flexion

extension

FIGURE 11.37 Movements of the wrist and fingers.

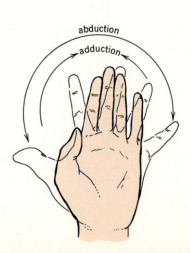

abduction

adduction

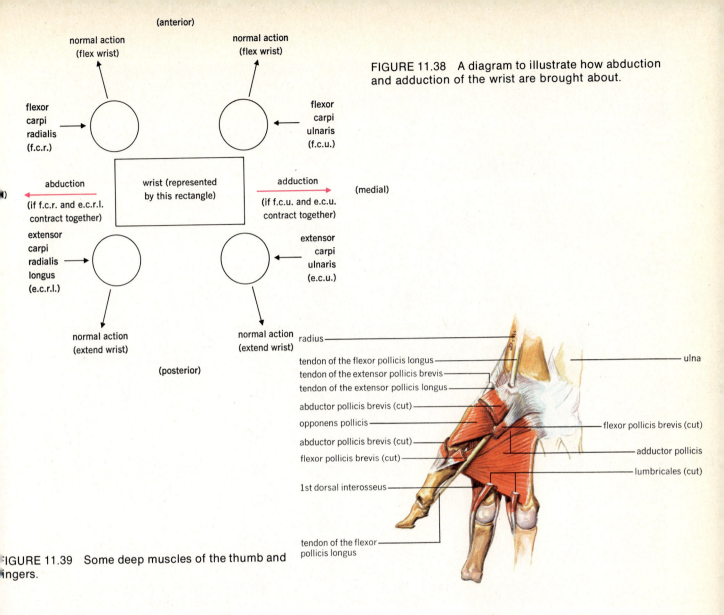

(anterior)

normal action (flex wrist)

normal action (flex wrist)

flexor carpi radialis (f.c.r.)

flexor carpi ulnaris (f.c.u.)

abduction
(if f.c.r. and e.c.r.l. contract together)

wrist (represented by this rectangle)

adduction
(if f.c.u. and e.c.u. contract together)

(medial)

extensor carpi radialis longus (e.c.r.l.)

extensor carpi ulnaris (e.c.u.)

normal action (extend wrist)

normal action (extend wrist)

(posterior)

FIGURE 11.38   A diagram to illustrate how abduction and adduction of the wrist are brought about.

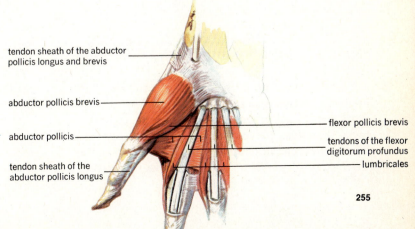

radius

tendon of the flexor pollicis longus

tendon of the extensor pollicis brevis

tendon of the extensor pollicis longus

abductor pollicis brevis (cut)

opponens pollicis

abductor pollicis brevis (cut)

flexor pollicis brevis (cut)

1st dorsal interosseus

tendon of the flexor pollicis longus

ulna

flexor pollicis brevis (cut)

adductor pollicis

lumbricales (cut)

FIGURE 11.39   Some deep muscles of the thumb and fingers.

FIGURE 11.40   Some superficial muscles of the thumb and fingers.

tendon sheath of the abductor pollicis longus and brevis

abductor pollicis brevis

abductor pollicis

tendon sheath of the abductor pollicis longus

flexor pollicis brevis

tendons of the flexor digitorum profundus

lumbricales

255

TABLE 11.13   Muscles moving the wrist and fingers

Muscle	Origin	Insertion	Action	Innervation
Anterior group **Flex wrist**				
Flexor carpi radialis	Medial epicondyle of humerus	Bases of meta-carpals 2, 3	Flexes wrist	Median nerve (spinal nerves C 6, 7)
Palmaris longus	Medial epicondyle of humerus	Palmar aponeurosis	Flexes wrist	Median nerve (spinal nerve C 8)
Flexor carpi ulnaris	Medial epicondyle of humerus and upper 2/3 of posterior ulna	Base of 5th meta-carpal and pisiform bone	Flexes wrist	Ulnar nerve (spinal nerves C 7, 8)
Flex fingers				
Flexor digitorum superficialis (sublimis)	Medial epicondyle of humerus, coronoid process of ulna, shaft of radius	Bases of 2nd phalanges of fingers	Flexes fingers	Median nerve (spinal nerves C 7, 8, T 1)
Flexor digitorum profundus	Proximal 3/4 of ulna and interosseous membrane	Bases of distal (3rd) phalanges of fingers	Flexes fingers	Median (index and middle fingers) and ulnar nerve (ring and little fingers), spinal nerves C 7–T 1
Posterior group **Extend wrist**				
Extensor carpi radialis longus	Lateral epicondyle of humerus	Base of 2nd meta-carpal	Extend wrist	Radial nerve (spinal nerves C 6, 7)
Extensor carpi radialis brevis	Lateral epicondyle of humerus	Base of 3rd meta-carpal	Extend wrist	Radial nerve (spinal nerves C 6, 7)
Extensor carpi ulnaris	Lateral epicondyle of humerus	Base of 5th meta-carpal	Extend wrist	Radial nerve (spinal nerves C 7, 8)
Extend fingers				
Extensor digi-torum communis	Lateral epicondyle of humerus	Phalanges 2 and 3, first 3 fingers	Extend first 3 fingers	Radial nerve (spinal nerves C 7, 8)
Extensor digiti minimi	Lateral epicondyle of humerus	All phalanges of little finger	Extends little finger	Radial nerve (spinal nerves C 7, 8)

Muscles of the thumb (Figs. 11.39, 11.40)

The thumb has its own supply of muscles that enable it to be placed in opposition to and to work with the fingers to perform the complex tasks of which the hand as a whole is capable. There are eight muscles serving the thumb. This discussion concentrates on those most important; these are illustrated in the figures. Table 11.14 presents *all* eight muscles.

A point to be kept in mind when studying the thumb is that it is "set" with its anterior surface at about a 90 degree angle to the same surfaces of the fingers. Thus, flexion of the thumb brings the digit across the palm of the hand, extension moves it laterally away from the fingers, abduction moves it away from the palm of the hand, and adduction brings it toward the fingers.

FLEXOR POLLICIS LONGUS is an anterior forearm muscle flexing the thumb.

EXTENSOR POLLICIS LONGUS is a posterior forearm muscle extending the thumb.

ABDUCTOR POLLICIS BREVIS is a posterior forearm muscle that abducts the thumb.

ADDUCTOR POLLICIS has its belly on the wrist, and adducts the thumb.

TABLE 11.14. Muscles of the thumb				
Muscle	Origin	Insertion	Action	Innervation
Flexor pollicis longus	Anterior midradius and interosseous membrane	Palmar surface of base of distal phalanx of thumb	Flexes thumb and metacarpo-phalangeal joint	Radial nerve (spinal nerves C 8, T 1)
Flexor pollicis brevis	Trapezium	Lateral side of base of 1st phalanx of thumb	Flexes thumb	Median nerve (spinal nerves C 8, T 1)
Abductor pollicis longus	Posterior ulna, interosseous membrane, and middle 1/3 of posterior radius	Lateral side of base of 1st metacarpal	Abducts thumb	Radial nerve (spinal nerves C 7, 8)
Abductor pollicis brevis	Trapezium and scaphoid	Lateral side of base of 1st phalanx of thumb	Abducts thumb	Median nerve (spinal nerves C 8, T 1)
Extensor pollicis brevis	Posterior 1/3 of radius and interosseous membrane	Base of 1st phalanx of thumb	Extends thumb	Radial nerve (spinal nerves C 7, 8)
Extensor pollicis longus	Posterior ulna and interosseous membrane	Base of 2nd phalanx of thumb	Extends thumb	Radial nerve (spinal nerves C 7, 8)
Adductor pollicis	Trapezoid, capitate, 2nd and 3rd metacarpals	Medial side of base of 1st phalanx of thumb	Adducts thumb	Ulnar nerve (spinal nerves C 8, T 1)
Opponens pollicis	Trapezium	Lateral 1/2 of anterior surface of 1st metacarpal	Flexes and adducts thumb (so that thumb may touch any flexed finger tip)	Median nerve (spinal nerves C 8, T 1

Muscles within the hand (Fig. 11.41)

The INTEROSSEI and LUMBRICAL muscles of the hand cooperate to bring about abduction and ad-duction of the fingers at the knuckles, and aid in flexing the fingers.

The INTEROSSEI consist of six muscles, four dorsal (posterior) in placement on the hand, and two palmar in placement. The dorsals fill the spaces between the metacarpals and insert on the first three fingers. The

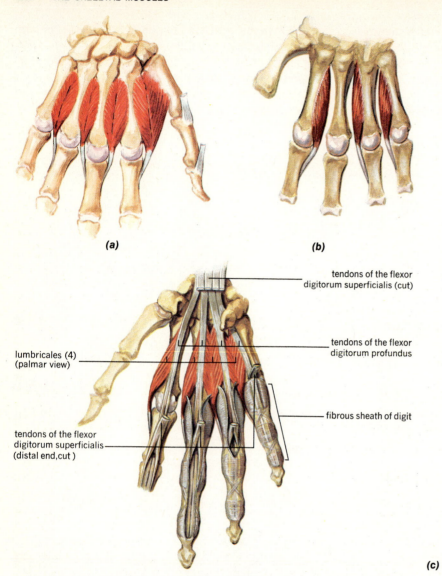

(a)

(b)

lumbricales (4)
(palmar view)

tendons of the flexor
digitorum superficialis
(distal end,cut)

tendons of the flexor
digitorum superficialis (cut)

tendons of the flexor
digitorum profundus

fibrous sheath of digit

(c)

FIGURE 11.41   Muscles of the hand. (*a*) The dorsal interossei. (*b*) The palmar interossei. (*c*) A palmar view of the lumbricales.

palmars arise from the 1st, 2nd, 4th, and 5th metacarpals, and pass to the 1st, 3rd, and 5th digits. The middle finger thus receives two dorsal interossei, ring and index fingers have both a dorsal and palmar muscle, and the thumb and little fingers have only a palmar muscle. Dorsal interossei abduct the digit; the palmars adduct.

The LUMBRICALS consist of four wormlike (hence lumbrical-earthworm) muscles. They are basically palmar in placement. They attach to the dorsal surfaces of the first phalanges of the fingers, and place the hand in a writing position.

These muscles are presented in Table 11.15.

TABLE 11.15. Muscles within the hand				
Muscle	Origin	Insertion	Action	Innervation
Interossei (4 dorsal and 4 palmar muscles; numbered from the thumb side)	*Dorsal*—fill spaces between the metacarpals; from the adjacent surfaces of the bones	Sides of 1st phalanges of first 3 fingers	Abduct fingers	Ulnar nerve (spinal nerves C 8, T 1)
	Palmar—metacarpals 2–5	Sides of 1st phalanges of 1st, 3rd, and 5th fingers	Adduct fingers	Ulnar nerve (spinal nerves C 8, T 1)
Lumbricales (4 muscles)	Tendons of flexor digitorum profundus over palmar surface of metacarpals 2–5	Dorsal surfaces of 1st phalanges of fingers	Flex metacarpophalangeal and extend interphalangeal joints (put hand in "writing" position)	Two medial by ulnar nerve; two lateral by median nerve (spinal nerves C 8, T 1)

Muscles moving the hip and knee joints (Figs. 11.43–11.45)

The hip joint is a triaxial, ball-and-socket joint, permitting all possible movements. It is potentially as mobile as the shoulder joint, but motion is limited by the muscles that attach to the femur and pelvis. The knee is a uniaxial hinge joint permitting flexion and extension. These movements are illustrated in Figure 11.42. The muscles considered in this section may be divided into five functional groups. Some muscles are "two-joint" muscles, operating both the hip and knee.

The first group of muscles lie on the anterior aspect of the posterior pelvic wall and/or anterior thigh. They are flexors of the hip and/or extensors of the knee.

ILIACUS and PSOAS were mentioned in a previous section as aiding trunk flexion. Here, a consideration of their origins and insertion shows that, if the trunk is fixed, the muscles cause the thigh to be flexed on the pelvis and rotated laterally—the latter action occurring because the tendon inclines rearward after passing over the pelvic rim.

The next group of four muscles is called, collectively, the *quadriceps femoris*. The muscles have a common insertion on the patella and, from there, the patellar tendon passes to the tibial tuberosity to allow knee movement.

RECTUS FEMORIS is the most superficial of the group and is placed in the midline of the thigh. It runs from the anterior inferior iliac spine to the patella. A "two-joint" muscle, rectus femoris can both flex the thigh on the trunk *and* extend the knee. A kicking type of movement would illustrate the dual action of the muscle.

VASTUS LATERALIS arises from the lateral lip of the linea aspera on the *posterior* femur and sweeps laterally around to the *front* of the femur to attach to the patella. Since it originates from the femur—not the pelvis—it only extends the knee. The upper and lateral portion of the muscle is often used as an injection site.

VASTUS INTERMEDIUS occupies the anterior aspect of the femur deep to the rectus femoris. It only extends the knee.

VASTUS MEDIALIS rises from the *medial* lip of the linea aspera and sweeps around the inner aspect of the femur to attach to the patella. It also only extends the knee.

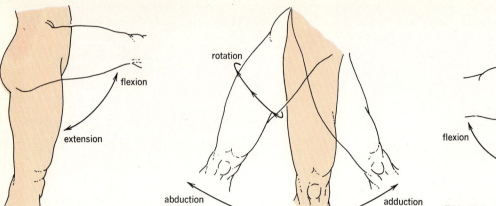

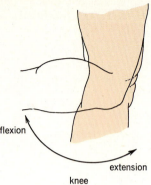

rotation

flexion

extension

abduction

adduction

flexion

extension

knee

FIGURE 11.42   Movements of the hip and knee.

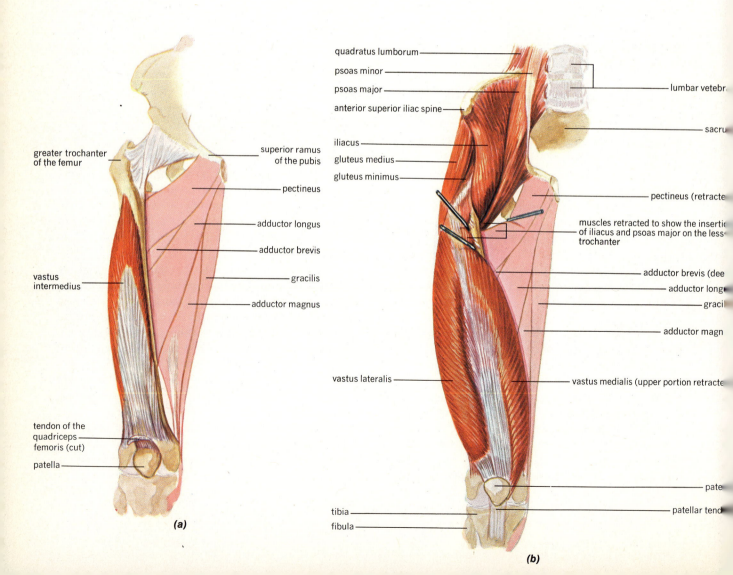

greater trochanter of the femur

superior ramus of the pubis

pectineus

adductor longus

adductor brevis

gracilis

adductor magnus

vastus intermedius

tendon of the quadriceps femoris (cut)

patella

(a)

quadratus lumborum

psoas minor

psoas major

anterior superior iliac spine

iliacus

gluteus medius

gluteus minimus

lumbar vetebra

sacru

pectineus (retracte

muscles retracted to show the insertic of iliacus and psoas major on the less trochanter

adductor brevis (dee

adductor long

graci

adductor magn

vastus lateralis

vastus medialis (upper portion retracte

pate

patellar tend

tibia

fibula

(b)

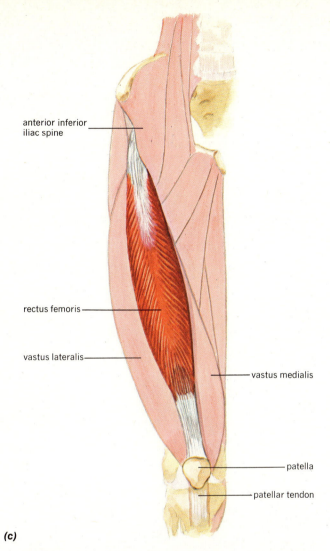

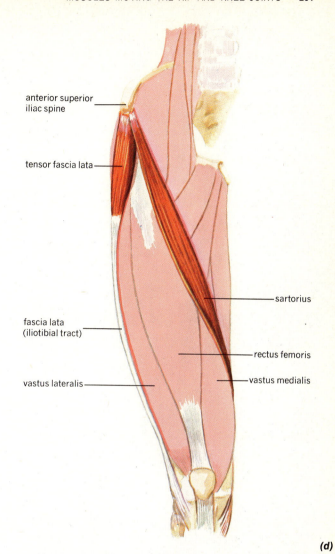

FIGURE 11.43   Muscles of the thigh. (*a*) The vastus intermedius and neighboring muscles. (*b*) Some muscles of the anterior hip and thigh. (*c*) The rectus femoris and neighboring muscles. (*d*) The sartorius, tensor fascia lata, and neighboring muscles.

SARTORIUS is the longest muscle of the body, running from anterior superior iliac spine to the medial tibial head. It crosses from the lateral to medial aspect of the lower appendage, and, when it contracts, flexes both the thigh *and* knee, and rotates the thigh

laterally. The combined movements are well illustrated in crossing one knee over the other or in placing one ankle on top of the opposite knee.

These muscles are presented in Table 11.16A. The medial thigh muscles (Fig. 11.44) are positioned to adduct the thigh as their major action.

ADDUCTOR MAGNUS, as its name suggests, is the largest of the medial group. It is mostly covered by the other adductors; consequently, it is often difficult to appreciate its true size. The ischium serves as

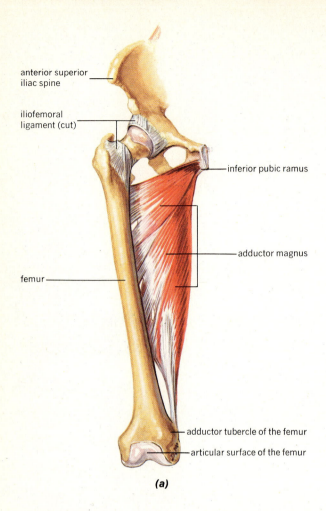

anterior superior iliac spine

iliofemoral ligament (cut)

inferior pubic ramus

adductor magnus

femur

adductor tubercle of the femur

articular surface of the femur

(a)

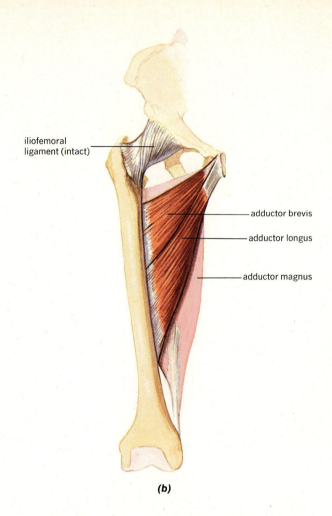

iliofemoral ligament (intact)

adductor brevis

adductor longus

adductor magnus

(b)

its origin, and it inserts on the lower three-fourths of the femur. A powerful adductor of the thigh, the muscle also aids in maintaining the lateral balance of the body when standing.

ADDUCTOR LONGUS lies superior to the magnus, and rises from the pubis. It passes to the linea aspera. Adduction of the thigh is its action.

ADDUCTOR BREVIS lies deep to the longus and runs between the pubis and upper linea aspera. It adducts the thigh.

PECTINEUS is a small muscle running between the pubis and the lesser trochanter. It also adducts the thigh.

GRACILIS lies directly on the medial aspect of the thigh and adducts the thigh.

These muscles are presented in Table 11.16B.

The next group of muscles lies on the posterior aspect of the pelvis and thigh, (Figs. 11.45 *A–D*). Again, there are some "two-joint" muscles in the group, operating both the hip and knee at the same time, extending the thigh, and/or flexing the knee. These muscles are basically antagonists to the first group just described.

GLUTEUS MAXIMUS forms the major muscle of the buttock. It arises primarily from the lateral iliac surface and the fibers pass behind the femur to attach to the gluteal tuberosity of the femur. The muscle is a powerful extensor of the thigh, used in bicycling, stair climbing, and in arising from a squatting or stooping

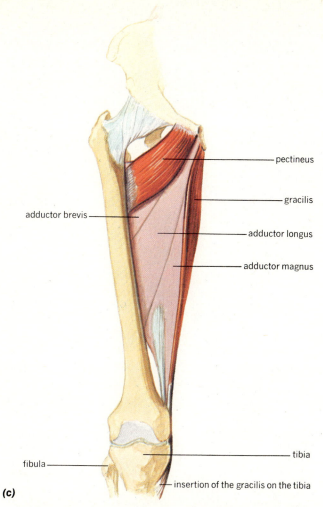

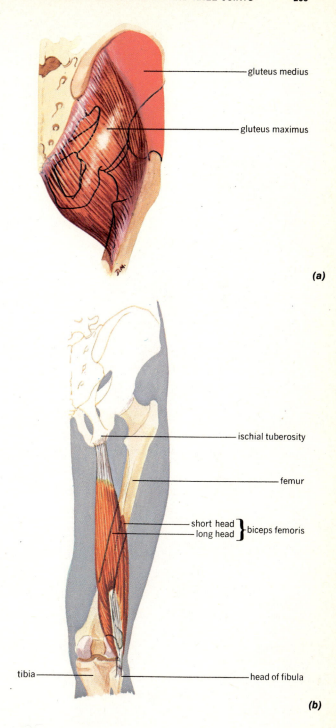

FIGURE 11.44   The medial thigh adductor muscles. (*a*) The adductor magnus. (*b*) The adductors longus and brevis. (*c*) pectineus and gracilis.

FIGURE 11.45   Muscles of the posterior pelvis and posterior thigh regions. (*a*) The gluteus maximus. (*b*) The biceps femoris. (*c*) The biceps femoris and semitendinosus. See following page. (*d*) The semimembranosus, with gracilis as a reference point. (*e*) The gluteus medius and piriformis. (*f*) The gluteus minimus, and the remaining lateral rotators of the thigh.

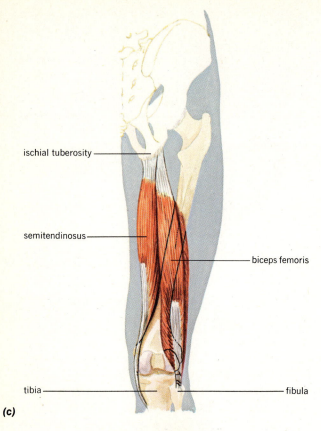

ischial tuberosity —

semitendinosus —

— biceps femoris

tibia —   — fibula

(c)

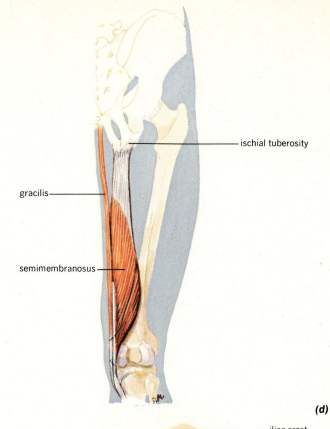

— ischial tuberosity

gracilis —

semimembranosus —

(d)

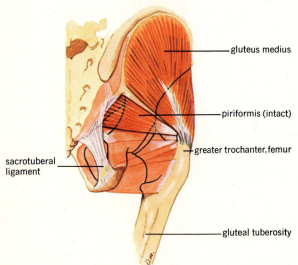

— gluteus medius

— piriformis (intact)

— greater trochanter, femur

sacrotuberal
ligament —

— gluteal tuberosity

(e)

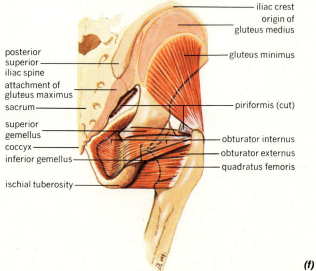

— iliac crest
origin of
gluteus medius

— gluteus minimus

posterior
superior —
iliac spine

attachment of
gluteus maximus

sacrum —   — piriformis (cut)

superior
gemellus —

coccyx —   — obturator internus

inferior gemellus —   — obturator externus
                                                       — quadratus femoris

ischial tuberosity —

(f)

position. It also laterally rotates the femur, because its fibers insert to the lateral side of the posterior midline of the femur. Acting together, the muscles "pinch" the buttocks together.

The next three muscles are collectively called the "hamstrings." From the lateral to medial side of the posterior thigh, the muscles are biceps femoris, semitendinosus, and semimembranosus.

BICEPS FEMORIS has a *long head*, rising from the ischial tuberosity and a *short head* from the femur shaft. Both heads insert on the fibular head. It is a powerful flexor of the knee *and* extends the hip against gravity when the body is standing or running.

SEMITENDINOSUS also rises from the ischial tuberosity. It inserts into the medial tibial head. Action is the same as for biceps femoris.

SEMIMEMBRANOSUS rises from the ischial tuberosity and inserts on the medial tibial head. It has the same action as the other two hamstrings.

When the hamstrings flex the knee, the tendons form the inner and outer margins of a space on the posterior aspect of the knee, known as the *popliteal fossa.*

These muscles are presented in Table 11.16C.

Abductors of the thigh form a group of muscles originating mainly on the posterior surface of the ilium, and running primarily to the greater trochanter.

GLUTEUS MEDIUS is inferior to the maximus, forming the middle layer of the gluteal muscles. It abducts the thigh and medially rotates it.

GLUTEUS MINIMUS is the deepest layer on the posterior ilium. The muscle also abducts and medially rotates the thigh; its tendon passes anterior to the femur to the greater trochanter.

TENSOR FASCIA LATA actually inserts on the iliotibial tract, a band of tissue running along the lateral thigh to the fibular head. As it pulls on the fascia of the tract, its pull is transferred to the fibula, abducting the lower appendage.

These muscles are presented in Table 11.16D.

The last group of muscles run from the sacrum to the greater trochanter. They pass across the *posterior* aspect of the hip joint and laterally rotate the thigh. They also act as articular muscles to stabilize the hip joint. From superior to inferior are the piriformis, gemellus, obturator, and quadratus femoris.

PIRIFORMIS runs from the anterior sacrum to the greater trochanter.

GEMELLUS is divided into *superior* and *inferior* portions. Both act together to laterally rotate the femur.

OBTURATOR has *internal* and *external* portions. Both act to laterally rotate the femur.

QUADRATUS FEMORIS also laterally rotates the femur.

These muscles are presented in Table 11.16E.

TABLE 11.16. Muscles moving the hip and knee joints				
Muscle	Origin	Insertion	Action	Innervation
A. *Anterior group* Flex hip and/or extend knee				
Iliacus ⎫ Iliop-	Iliac fossa	Below lesser trochanter	Flex thigh and rotate hip laterally	Femoral nerve (spinal nerves L 2, 3)
Psoas ⎬ soas	Transverse processes of lumbar vertebrae	Lesser trochanter	Flex thigh and rotate hip laterally	Femoral nerve (spinal nerves L 2, 3)

TABLE 11.16   (continued)

Muscle	Origin	Insertion	Action	Innervation
Rectus femoris	Anterior inferior iliac spine and rim of acetabulum	Patella	Flexes hip and extends knee	Femoral nerve (spinal nerves L 2–4)
Vastus lateralis	Lateral border of linea aspera	Patella	Extends knee	Femoral nerve (spinal nerves L 2–4)
Vastus intermedius	Anterior upper 3/4 of femur	Patella	Extends knee	Femoral nerve (spinal nerves L 2–4)
Vastus medialis	Medial border of linea aspera	Patella	Extends knee	Femoral nerve (spinal nerves L 2–4)
Sartorius	Anterior superior iliac spine	Upper part of medial tibial head	Flexes hip and knee	Femoral nerve (spinal nerves L 2, 3)
B. *Medial group* Adductors				
Adductor magnus	Rami of ischium and pubis	Lower 1/3 of linea aspera	Adducts and aids in flexion and lateral rotation of femur	Sciatic nerve and obturator nerve (spinal nerves L 2–4)
Adductor longus	Pubis, between symphysis and crest	Linea aspera	Adducts and aids in flexion and lateral rotation of femur	Sciatic nerve and obturator nerve (spinal nerves L 2–4)
Adductor brevis	Body and inferior ramus of pubis	Upper 1/2 of linea aspera	Adducts and aids in flexion and lateral rotation of femur	Sciatic nerve and obturator nerve (spinal nerves L 2–4)
Pectineus	Iliopectineal line and pubis	Base of lesser trochanter	Adducts and aids in flexing thigh	Femoral nerve (spinal nerves L 2, 3)
Gracilis	Rami of ischium and pubis	Medial side of tibial head	Adducts, flexes, and medially rotates thigh	Obturator nerve (spinal nerves L 2, 3)
C. *Posterior group* Extend hip and/or flex knee				
Gluteus maximus	Between posterior gluteal line and crest of ilium, sacrum and coccyx	Gluteal tuberosity of femur	Extends thigh	Sciatic nerve (spinal nerves L 5, S 1, 2)
Biceps femoris	Ischial tuberosity (long head) and lateral lip of linea aspera (short head)	Lateral side of fibular head	Extends thigh and flexes knee	Sciatic nerve (spinal nerves L 5, S 1, 2)
Semitendinosus	Ischial tuberosity	Medial surface of upper tibia	Extends thigh and flexes knee	Sciatic nerve (spinal nerves L 5, S 1, 2)
Semimembranosus	Ischial tuberosity	Posterior side of medial tibial condyle	Extends thigh and flexes knee	Sciatic nerve (spinal nerves L 5, S 1, 2)

TABLE 11.16   (continued)

D. *Abductors*

Gluteus medius	Lateral superior surface of ilium	Greater trochanter	Abducts and medially rotates thigh	Sciatic nerve (spinal nerves L 4, 5, S 1, 2)
Gluteus minimus	Lateral inferior surface of ilium	Greater trochanter	Abducts and medially rotates thigh	Sciatic nerve (spinal nerves L 4, 5, S 1, 2)
Tensor fascia lata	Lateral superior surface of ilium above anterior superior spine	Iliotibial tract that inserts into lateral tibial head	Abducts, flexes and medially rotates thigh	Sciatic nerve (spinal nerves L 4, 5, S 1, 2)

E. *Lateral rotators*

Piriformis	Anterior sacrum	Greater trochanter	Lateral rotation of thigh	Sacral spinal nerves

Gemellus
  Superior
  Inferior
	Ischial spine
	Ischial tuberosity
		Greater trochanter	Lateral rotation of thigh	Obturator nerve (spinal nerves L 3, 4)

Obturator
  Internus
 	
  Externus
	Rim of obturator foramen, pubis and ischium
	Rim of obturator foramen
		Greater trochanter	Lateral rotation of thigh	Obturator nerve (spinal nerves L 3, 4)

Quadratus femoris	Ischial tuberosity	Below greater trochanter	Lateral rotation of thigh	Obturator nerve (spinal nerves L 3, 4)

Muscles moving the ankle and foot. (Figs. 11-47–11.50)

These muscles are often termed the *crural* muscles (*crus,* leg). The muscles fall logically into three groups by position: *anterior* (actually anterolateral), *lateral,* and *posterior.* For the most part, the anterior muscles dorsiflex (flex) the ankle, invert the foot, or extend the toes. The lateral group planterflexes (extends) the ankle, and everts the foot, while the posterior group plantar flexes the ankle and flexes the toes. These movements are shown in Figure 11.46.

Anterior crural muscles include the following (Fig. 11.47 *a, b*)

TIBIALIS ANTERIOR lies just lateral to the sharp anterior border of the tibia (the shin). It rises from the lateral tibia and sends its tendon *across* the front of the ankle joint to insert on the *medial* cuneiform and *first* metatarsal. This angle of pull dorsiflexes *and* inverts the foot. The muscle is an important factor in maintaining the transverse arch of the foot, pulling "from above."

EXTENSOR HALLUCIS LONGUS arises from the medial fibular surface. Its tendon crosses to the medial side of the ankle and inserts on the top of the great toe (the *hallux*). Thus, the direction of pull extends the great toe. Continued action also aids in the dorsiflexion of the foot necessary to keep the foot from dragging as it is swung forwards during walking.

EXTENSOR DIGITORUM LONGUS rises from the lateral tibial condyle and the fibula, and four tendons exit

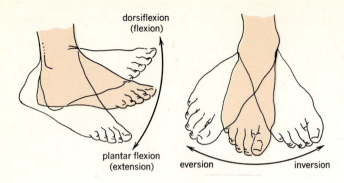

dorsiflexion
(flexion)

plantar flexion
(extension)

eversion   inversion

FIGURE 11.46   The movements of the ankle and foot.

FIGURE 11.47   (a) The superficial anterior crural muscles, and associated musculature. (b) The deep anterior crural muscles, with portions of the superficial muscles removed to show deep muscle bellies.

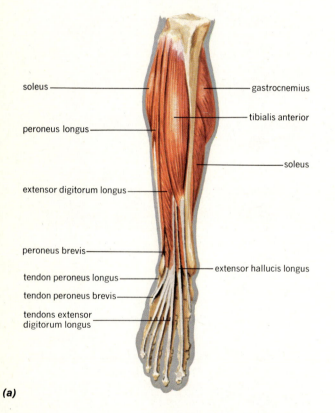

soleus

gastrocnemius

tibialis anterior

peroneus longus

soleus

extensor digitorum longus

peroneus brevis

extensor hallucis longus

tendon peroneus longus

tendon peroneus brevis

tendons extensor
digitorum longus

(a)

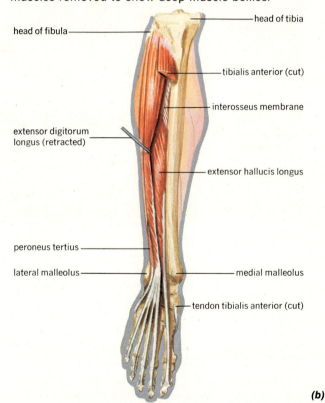

head of fibula

head of tibia

tibialis anterior (cut)

interosseus membrane

extensor digitorum
longus (retracted)

extensor hallucis longus

peroneus tertius

lateral malleolus

medial malleolus

tendon tibialis anterior (cut)

(b)

from it to terminate on the tops of the phalanges of the four outer toes. It may be of interest to note that this arrangement causes the four outer toes to work more or less as a unit. This contrasts with the finger extensors that can be manipulated more independently of one another. The muscle extends the four outer toes and aids in dorsiflexion during walking.

PERONEUS TERTIUS is a muscle of inconstant occurrence. If present, it forms an expansion from the lower portion of the previous muscle and sends its tendon to the top of the fifth metatarsal. It dorsiflexes and slightly everts the foot.

The lateral crural muscles include the following (Fig. 11.48).

The posterior crural muscles include the following (Fig. 11.49*a*–*c*)

GASTROCNEMIUS (the calf muscle) rises by two strong tendons from the posterior aspects of the femur condyles, and inserts, by way of the *tendo calcaneus* (Achilles tendon), into the calcaneus (heel). It is a powerful plantar flexor of the foot, keeping the body balanced against gravity, in pointing the toes, and in giving the "push" necessary for walking, running, jumping, or dancing.

SOLEUS is a thick, flat muscle lying beneath gastrocnemius that rises from the posterior upper tibia and fibula. Its tendon joins that of the gastrocnemius to form the tendo calcaneus. Soleus assists the gastrocnemius in the actions described above.

PLANTARIS is a small and variable muscle beneath the lateral head of the gastrocnemius. It is about four inches long, and has a long slender tendon that joins the tendo calcaneus at the back of the ankle. It gives slight assistance to gastrocnemius.

TIBIALIS POSTERIOR rises from almost the whole posterior surfaces of the tibia and fibula. Its tendon passes beneath the medial malleolus to attach to the navicular and middle cuneiform. It plantar flexes and inverts the foot and gives strong support to the medial longitudinal arch of the foot.

FLEXOR DIGITORUM LONGUS springs from the posterior tibial surface, and its tendon passes beneath the medial malleolus to the sole of the foot. The tendon then divides to give four tendons, each of which ends on the under surfaces of the last phalanges of the four outer toes. Flexion of the toes, as on "pushing off" for walking, and in balancing the body when standing are two of its functions. The muscle also aids in providing a base of support when standing.

FLEXOR HALLUCIS LONGUS rises mainly from the lower posterior fibula, and its tendon passes beneath the talus to insert on the lower surfaces of the phalanges of the great toe. In addition to flexing the great toe, the muscle aids in pushing off, and helps to maintain the longitudinal arch of the foot.

These muscles are presented in Table 11.17.

(*Text continued on page 273.*)

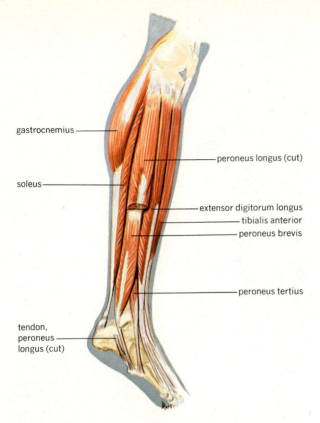

gastrocnemius

peroneus longus (cut)

soleus

extensor digitorum longus
tibialis anterior
peroneus brevis

peroneus tertius

tendon,
peroneus
longus (cut)

FIGURE 11.48   The lateral crural muscles and associated musculature.

PERONEUS LONGUS occupies most of the lateral border of the fibula. Its tendon passes beneath the lateral malleolus, to insert on the first cuneiform and first metatarsal. It may thus be seen that the tendon *crosses* the sole of the foot. Contraction causes plantar flexion, and the medial attachment on the foot creates a direction of pull that everts the foot. Like a bowstring across the sole, the tendon of this muscle assists in maintaining the transverse arch of the foot "from below."

PERONEUS BREVIS rises from below the longus; its tendon accompanies that of the longus beneath the lateral malleolus and inserts on the fifth metatarsal. Again, plantar flexion and eversion are the muscle's actions.

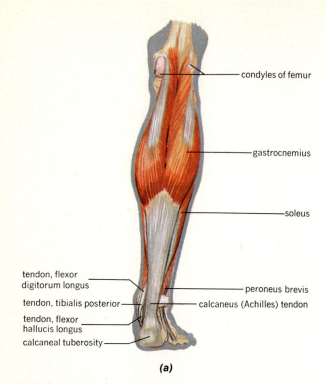

condyles of femur

gastrocnemius

soleus

tendon, flexor
digitorum longus

tendon, tibialis posterior

tendon, flexor
hallucis longus

calcaneal tuberosity

peroneus brevis

calcaneus (Achilles) tendon

(a)

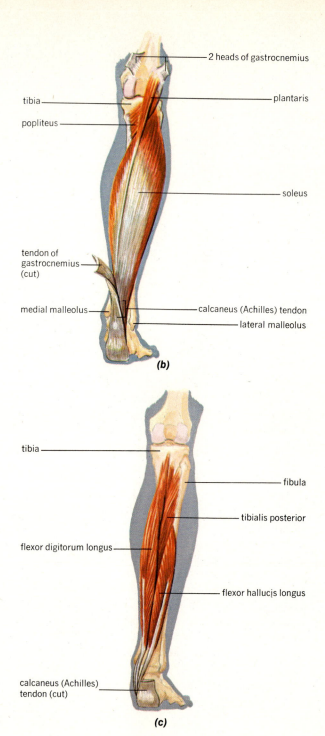

2 heads of gastrocnemius

tibia

popliteus

plantaris

soleus

tendon of
gastrocnemius
(cut)

medial malleolus

calcaneus (Achilles) tendon

lateral malleolus

(b)

tibia

fibula

tibialis posterior

flexor digitorum longus

flexor hallucis longus

calcaneus (Achilles)
tendon (cut)

(c)

FIGURE 11.49 The posterior crural muscles and associated musculature. (*a*) The superficial muscles. (*b*) The intermediate layer of muscles. (*c*) The deep muscles.

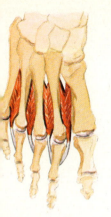

dorsal interossei
(right foot)

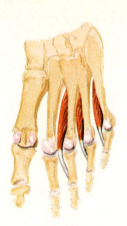

plantar interossei
(right foot)

(a)

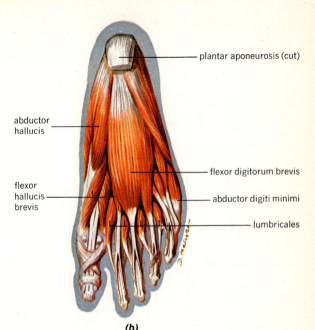

plantar aponeurosis (cut)

abductor
hallucis

flexor digitorum brevis

flexor
hallucis
brevis

abductor digiti minimi

lumbricales

(b)

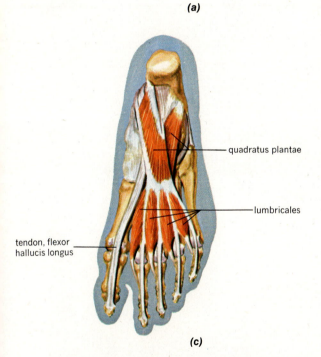

quadratus plantae

lumbricales

tendon, flexor
hallucis longus

(c)

FIGURE 11.50 Muscles within the foot. (*a*) The interossei. (*b*) The intermediate layer of muscles. (*c*) The deep muscles.

TABLE 11.17. Muscles moving the ankle and foot

Muscle	Origin	Insertion	Action	Innervation
Anterior crural Tibialis anterior	Upper 2/3 of lateral tibia	Medial cuneiform and base of 1st metatarsal	Dorsiflexion and inversion of foot	Anterior tibial and common peroneal nerves (spinal nerves L 4, 5)
Extensor hallucis longus	Middle 1/3 of anterior fibula	Top of 2nd phalanx of great toe	Extends great toe	Anterior tibial nerve (spinal nerves L 5, S 1)
Extensor digitorum longus	Lateral condyle of tibia and distal fibula	2nd and 3rd phalanges of 4 outer toes	Extends 4 outer toes	Anterior tibial nerve (spinal nerves L 5, S 1)
Peroneus tertius	Anterior fibula	Dorsal surface of 5th metatarsal	Dorsiflexes and everts foot	Anterior tibial nerve (spinal nerves L 5, S 1)
Lateral crural Peroneus longus	Upper 2/3 of fibula	Medial cuneiform and base of 1st metatarsal	Plantar flexion and eversion of foot	Common peroneal nerve (spinal nerves L 5, S 1)
Peroneus brevis	Lower 2/3 of fibula	Base of 5th metatarsal	Plantar flexion and eversion of foot	Common peroneal nerve (spinal nerves L 5, S 1)
Posterior crural Gastrocnemius	Posterior surfaces of femur condyles	Calcaneus	Plantar flexion	Tibial nerve (spinal nerves L 5, S 1, 2)
Soleus	Upper portions of posterior tibia and fibula	Calcaneus	Plantar flexion	Tibial nerve (spinal nerves L 5, S 1, 2)
Plantaris	Lateral supracondylar ridge	Calcaneus	Plantar flexion	Tibial nerve (spinal nerves L 5, S 1, 2)
Tibialis posterior	Posterior shafts of tibia and fibula	Navicular and medial cuneiform	Plantar flexion and inversion of foot	Posterior tibial nerve (spinal nerves L 5, S 1)
Flexor digitorum longus	Posterior surface of tibial shaft	Distal phalanges of 4 outer toes	Flexes 4 outer toes	Posterior tibial nerve (spinal nerves S 1, 2)
Flexor hallucis longus	Lower 2/3 of fibula	Plantar surfaces of phalanges of great toe	Flexes great toe	Posterior tibial nerve (spinal nerves S 1, 2)

Muscles within the foot (Fig. 11.50*ab*)

There are four DORSAL INTEROSSEI and three PLANTAR INTEROSSEI muscles in the foot. The dorsals lie between the metatarsal bones, with the first and second sending tendons to the second toe, the third and fourth to the third and fourth toes. The plantars rise from the third to fifth metatarsals, and send tendons to the third to fifth toes. The second toe thus receives two dorsal muscles, the third and fourth each a plantar and dorsal, the fifth a plantar only, the great toe none.

The LUMBRICALES of the foot are four small muscles rising from the tendons of flexor digitorum longus. Their tendons pass to the medial sides of the first phalanges of the four outer toes.

Both of these muscle groups flex the metatarsophalangeal joints and extend the interphalangeal joints. These actions aid in creation of a stable platform of support for the erect body.

QUADRATUS PLANTAE is another name for the *flexor digitorum accessorius* that works with flexor digi-torum longus. It rises from the calcaneus, and its tendons insert on the last phalanges of the four outer toes. In addition to flexing the four outer toes, the muscle fixes the toes to give push-off in walking.

ABDUCTOR HALLUCIS runs between the calcaneus and great toe, and ADDUCTOR HALLUCIS from the cuboid and 2–4 metatarsals to the great toe. Their action is inherent in their names.

FLEXOR HALLUCIS BREVIS rises from the cuneiforms and cuboid bone and inserts on the underside of the first phalanx of the great toe. It flexes the great toe.

FLEXOR DIGITORUM BREVIS runs from calcaneus to the four outer toes and aids in flexing them.

ABDUCTOR DIGITI MINIMI abducts the little toe.

These muscles are presented in Table 11.18.

There is no Summary for this chapter and no Questions are provided. Your knowledge of muscles will be based primarily on your ability to correctly name and give origins, insertions, and actions of these muscles.

TABLE 11.18.   Muscles within the foot

Muscle	Origin	Insertion	Action	Innervation
Interossei 4 dorsal	*Dorsals*—between metatarsals arising from facing surfaces of adjacent bones	1st and 2nd to sides of 2nd toe, 3rd and 4th to sides of 3rd and 4th toes	Flex metatarsophalangeal joints and extend interphalangeal joints.	Common peroneal nerve (spinal nerves L 5, S 1)
3 plantar	*Plantar*—medial sides of 3rd to 5th metatarsals	Medial side of corresponding toe	Dorsal abduct and plantar adduct 4 outer toes	
Lumbricales (4 muscles)	From tendons of flexor digitorum longus	Medial sides of 1st phalanges of the 4 outer toes		
Quadratus plantae (flexor digitorum accessorius)	Calcaneus	Distal phalanges of four outer toes	Flex four outer toes	Posterior tibial nerve (spinal nerves S 1, 2)
Abductor hallucis	Calcaneus	Base of first phalanx of great toe	Abducts great toe	

TABLE 11.8 (Continued)				
Muscle	Origin	Insertion	Action	Innervation
Abductor hallucis	Cuboid, metatarsals 2–4	Base of 1st phalanx of great toe	Adducts great toe	Sciatic nerve (spinal nerves L 5, S 1, 2)
Flexor hallucis brevis	Cuneiforms and cuboid	Base of 1st phalanx of great toe	Flexes great toe	
Flexor digitorum brevis	Medial tubercle of calcaneus	Plantar surfaces of middle phalanges of 4 outer toes	Flexes 4 outer toes	Medial plantar nerve (spinal nerves L 5, S 1)
Abductor digiti minimi	Calcaneus	Lateral side, base of 1st phalanx of little toe	Abducts little toe	Lateral plantar nerve (spinal nerves S 1, 2)

Readings

Carter, Barbara J., et al. *Cross-Sectional Anatomy. Computed Tomography and Ultra-sound Correlation.* Appleton-Century-Crofts. New York, 1977.

Clemente, Carmine D. *Anatomy. A Regional Atlas of the Human Body.* Lea & Febiger. Philadelphia, 1975.

Lockhart, R.D., G.F. Hamilton, and F.W. Fyfe. *Anatomy of the Human Body.* J.B. Lippincott. Philadelphia, 1972.

McMinn, R. M. H., and R. T. Hutchings. *Color Atlas of Human Anatomy.* Year Book Medical Publishers. Chicago, 1977.

Chapter 12
Some Basic Kinesiology

Objectives

After studying this chapter, the reader should be able to:

■ Explain how muscles use bones as levers, define the three classes of levers, and calculate the effort required to operate these classes.

■ Tell why muscles use tendons to insert on bones, and what the functions of bursae are.

■ State the basic segments of the body and how they normally relate to one another.

■ Describe what types of motion take place in the sagittal and frontal planes and tell about the vertical and oblique axes of a body part.

■ Explain what the line and center of gravity are.

■ Define base of support and relate line and center of gravity to the base of support and all three to stability and instability of the whole body.

■ Explain how the angle of a muscle's pull is related to movement and joint security and how "pulleys" present on skeletal structures can alter the direction of muscle pull.

■ List some methods by which body stability may be improved.

- State the laws of motion that govern direction and speed of movement of a body part or object.

- Define the term force.

- Define work and how the force a muscle is capable of exerting may be calculated.

- Relate the laws of motion to giving impetus to an object and to receiving a moving object with the least chance of bodily harm.

- Discuss posture from the viewpoints of what it is and how to determine if it needs improvement.

- Derive some "principles" related to posture, its control and correction.

- List some requirements for an exercise program designed to improve physical fitness and devise a conditioning program to meet your needs (if any) for physical fitness.

Having studied the musculoskeletal system and its articulations, we now consider the application of the principles of support and movement to everyday life. The discipline of KINESIOLOGY (*kinesis,* motion + *logos,* study) seeks to analyze and explain how the body moves and adjusts itself to the demands placed on it by daily activity and sports. This chapter indicates the basic mechanical considerations important to prevention of injury and maintenance of balance and equilibrium and delves briefly into the values of exercise to the maintenance of health.

Basic mechanical concepts

The use of bones as levers by muscles

Muscles attach to bones, and their contractions pull on a particular point on the bone, causing it to move. The bones move with reference to a body joint. Remember that the construction of the joint determines what *type* of movement is permitted. The bones and joints therefore form the parts of a system of levers that cause body movement.

Any lever has several basic parts to it (Fig. 12.1).

The FULCRUM (F) is the point about which the level moves, and is provided by a joint in the body.

The POINT OF EFFORT (E) is where the muscle inserts, or where the force of contraction is directed.

The RESISTANCE (R) is the weight that the muscle must overcome to move the lever. Weight is given by the body parts themselves or by objects that are to be manipulated, and is usually considered to be concentrated at a single point on the lever.

The EFFORT ARM (EA) is the *distance* along the length of the bone from the fulcrum to the point of

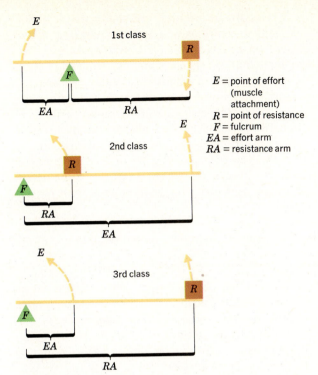

E = point of effort
      (muscle
      attachment)
R = point of resistance
F = fulcrum
EA = effort arm
RA = resistance arm

FIGURE 12.1   The basic parts of a lever, illustrated for the three classes of levers. Note that *EA* and *RA* may overlap, and observe the direction the resistance moves when force is applied.

effort. In general, the shorter the effort arm of the lever, the greater the resistance, and the greater must be the strength of contraction of the muscle to move the lever (*see* Fig. 12.1).

The RESISTANCE ARM *(RA)* is the *distance* along the lever from the fulcrum to the point where the resistance is located. The shorter the resistance arm, the less power required to move the lever.

According to the relative placement of *F, E,* and *R,* three classes of levers are distinguished (*see* Fig. 12.1).

CLASS 1. In this type of lever, *the fulcrum lies between the effort and resistance.* Seesaws, posthole diggers, scissors, and prybars are everyday examples of first class levers. Not too many levers of this class are found in the body because they

would require that a joint be placed *between* the ends of the lever, and most joints occur at the ends of bones. In extension of the forearm by the triceps, the olecranon process extends behind the elbow joint and the forearm forms the lever ahead of the joint. Thus, this is a first class lever.

CLASS 2. In this type of lever, the fulcrum is toward the end of the lever, and the *resistance lies between effort and the fulcrum.* This class requires the least force to move the lever. It is not agreed that there are any second class levers in the body, although some consider that raising oneself on the toes is an example. Here, the effort is applied at the heel, the resistance is the body weight on the ankle joint, and the ball of the foot acts as the fulcrum.

CLASS 3. In this type of lever, the fulcrum is toward the end of the lever, and the *effort is between fulcrum and resistance.* This is the most common type of lever in the body, because there is plenty of bone surface between its ends to attach a muscle. An obvious example of this type of lever is seen in flexion of the forearm by the biceps brachii. The fulcrum is the elbow joint, the resistance is in the hand, and the effort is applied at the radial tuberosity, distal to the elbow joint.

To calculate the force a muscle must provide to overcome a resistance, the formula

$$E \times EA = R \times RA$$

may be applied. What the formula basically demonstrates is that the force is exerted over a distance, so also is resistance; if the forces balance, no movement will occur, but if force is greater than resistance, motion will take place. Let us make some sample calculations, using the formula, for the three examples of muscle action given in the preceding section.

Class 1. $E = ?$
   $EA = 1$ inch (distance from center of elbow joint to tip of olecranon process)
   $R = 10$ pounds

RA = 10 inches (distance from center of elbow joint to center of palm of hand)

$$E \times EA = R \times RA, \text{ or } E = \frac{R \times RA}{EA}$$

thus

$$E = \frac{10 \times 10}{1} = 100 \text{ pounds of effort required to move 10 pounds of resistance.}$$

What would happen to E if the EA were shortened or the RA lengthened?

Class 2. $E = ?$
EA = 10 inches (distance from tip of heel to ball of foot)
R = 10 pounds
RA = 8 inches (distance from center of ankle joint to ball of foot)

again,

$$E = \frac{R \times RA}{EA}$$

thus

$$E = \frac{10 \times 8}{10} = 8 \text{ pounds of effort required to move 10 pounds of resistance.}$$

Class 3. $E = ?$
EA = 2 inches (distance from center of elbow joint to radial tuberosity)
R = 10 pounds
RA = 12 inches (distance from center of elbow joint to palm of hand)

again,

$$E = \frac{R \times RA}{EA}$$

thus

$$E = \frac{10 \times 12}{2} = 60 \text{ pounds of effort required to move 10 pounds of resistance.}$$

Calculate, for the third class of lever, what would happen to E if the EA were shortened to 1 inch.

It becomes obvious that first and third class levers require much more power by the muscles to move them. Although this seems "inefficient,"

it may again be pointed out that body structure basically determines where placement of F, E, and R will occur. An *advantage* to first and third class levers is that the *degree of motion* achieved is great, as seen in the movements of the appendages.

One final point may be made in connection with levers. Muscles have tendons on the ends of their fibers that actually make the connection between muscle fibers (the "belly" of the muscle) and the *periosteum* of the bone. The periosteum in turn inserts its fibers into the bone itself. Tendons provide several advantages over muscle fibers.

The muscular portion may not be long enough to span the distance between origin and insertion.

Tendons are much stronger than muscle tissue.

Tendons can be any length, while muscle bellies are usually restricted in size, conforming to body area.

Tendons often pass over bony prominences that would destroy a muscle fiber as movement occurs.

Tendons are smaller than muscle bellies, requiring less area to make an insertion.

Protection *is* provided for tendons and bellies, where they must pass over bony prominences, by the presence of small fluid-filled sacs called BURSAE, placed between the structure and the bone. BURSITIS is inflammation of a bursa, and the condition usually causes great pain as the joint moves because of the pressure applied to the bursa.

Body segments

The human body may be considered to be composed of many MOVABLE SEGMENTS (Fig. 12.2). The *head* is balanced upon the *neck*, the *trunk* supports both head and neck, the *lower appendages* provide support for the structures above them and allow them to move through space, and the *upper appendages* provide ability to manipulate objects and perform skilled movements. Of course, activities in one segment often require compensating or complimentary actions in others to allow

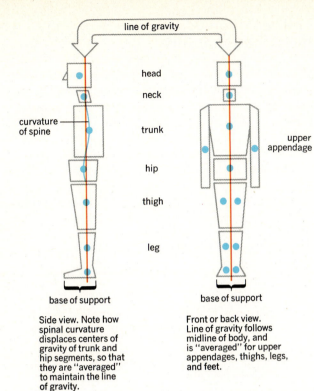

line of gravity

head

neck

curvature
of spine

trunk

upper
appendage

hip

thigh

leg

base of support   base of support

Side view. Note how
spinal curvature
displaces centers of
gravity of trunk and
hip segments, so that
they are "averaged"
to maintain the line
of gravity.

Front or back view.
Line of gravity follows
midline of body, and
is "averaged" for upper
appendages, thighs, legs,
and feet.

FIGURE 12.2   A diagrammatic representation of body
segments, centers of gravity, and line of gravity. What
would happen to stability if the upper appendages
were *unequally* raised to the sides (abducted)?

Side view. Note how spinal curvature displaces centers of gravity of trunk and hip segments, so that they
are "averaged" to maintain the line of gravity.
Front or back view. Line of gravity follows midline
of body and is "averaged" for upper appendages,
thighs, legs, and feet

purposeful actions to occur, and this forms the
basis of skilled motions.

Basic movements of the major body segments

In order to visualize the movements each body
segment can undergo it is easiest to relate them to
a basic starting position. *Anatomical position* provides a reference point to describe these move-

ments. To further aid in understanding these
movements, the reader should review the body
planes, as presented in Chapter 1, and the motions
possible at individual joints of the body as presented in Chapter 9.

MOVEMENTS IN THE SAGITTAL plane include:

FLEXION, in which the angle between a bone and its
reference point is decreased. Forward tipping of the
head, moving the upper arm forward, bending the
spine forward, bending the elbow, moving the thigh
forward, and bending the knee are all examples of
flexion.

EXTENSION moves the body part back to anatomical
position, and is the reverse motion to flexion.

HYPERFLEXION is a term whose use is applied only
to the upper arm. Because of its great mobility, when
the upper arm is flexed beyond the vertical position
it is considered to be hyperflexed. In other joints of the
body, flexion is usually terminated by contact of bony
or muscular structures or is limited by the nature of
the joint(s) involved.

HYPEREXTENSION implies continuation of extension
beyond anatomical position, or beyond a straight line.
For example, if the olecranon process of the elbow is
short, the forearm may be allowed to pass beyond a
straight line before being stopped by contact with the
olecranon fossa.

MOVEMENTS IN A FRONTAL PLANE include:

ABDUCTION, in which a body part moves away from
the centerline of the body or body part in a sideways
motion. Lifting the upper appendages so that they lie
horizontally from the shoulders, lifting the lower appendage to the side, and spreading the fingers are
examples of abduction.

ADDUCTION returns the body part to anatomical
position.

MOVEMENTS AROUND A VERTICAL AXIS are
termed ROTATION, and are subdivided into *lateral*
("outwards") *rotation* and *medial* ("inwards")
rotation.

Movements occurring in other than the planes

described above are termed OBLIQUE MOVEMENTS, and form the basis for manipulation of sports equipment and "nonrobot-like" movement. They comprise combinations of the basic movements and give us our great motor skills that are involved in everyday activity.

The line of gravity and the center of gravity

Movement must always proceed from a balanced starting position. Equilibrium is attained when the LINE OF GRAVITY—an imaginary line drawn connecting the CENTERS OF GRAVITY of the several body segments—falls within the BASE OF SUPPORT provided by the feet (in erect posture). The center of gravity is a point within a body segment where the mass of that segment may be considered to be concentrated. If the line of gravity lies within the base of support, the object is stable, or in equilibrium; if not, it is unstable and will tip over or fall. Figure 12.3 illustrates this principle for the body and some common objects. If the center of gravity of one segment is shifted, that of another must shift in an opposite direction if balance is to be maintained (Fig. 12.4). For example, "slouch-

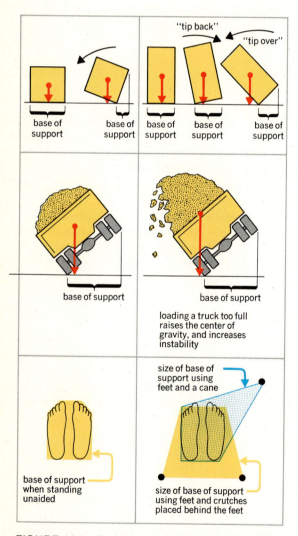

FIGURE 12.3   Stability and instability. As long as the line of gravity (♀) falls within the base of support, there is stability. Base of support may be enlarged by using various devices and stability is improved.

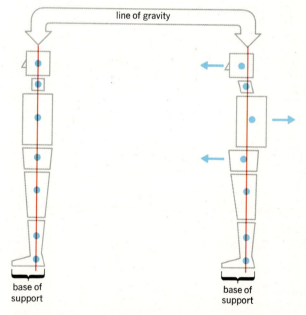

FIGURE 12.4   A diagram to illustrate how shifts in direction of the center of gravity of a body segment must be compensated for by shifts in centers of gravity of other segments in an opposite direction. This keeps the "average" center of gravity over the base of support.

ing," where the head may be thrust forward, is compensated for by moving the trunk back and perhaps also moving the hips forward.

The height from the feet, of the total body center of gravity, for females is about 55 percent of their standing height, and about 56 percent for males. It appears to be more variable in females, perhaps because of different degrees of muscular development in their thoracic area.

The direction of muscle pull

Maintenance of proper alignment of body segments depends on muscles pulling on bones and maintaining joints in proper positions, and on the motions the joints permit. The influences arriving at a muscle from, ultimately, the brain will determine its strength of contraction and how it affects alignment. Thus, a given muscle may have a primary action on a joint, but may do other jobs as the joint moves (Fig. 12.5). For example, the clavicular portion of the pectoralis major flexes the humerus until the bone assumes a horizontal position, then it actually helps abduct the humerus. This fact emphasizes that muscles cooperate with one another to achieve the great range of motion the body has.

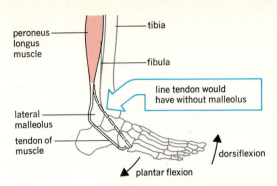

FIGURE 12.6 Tendons of muscles may use bony parts as "pulleys" to change their direction of pull. With the malleolus, the muscle plantar flexes the foot; without it, the muscle would dorsiflex the foot.

Direction of muscle pull may also be altered by using bony parts as pulleys (Fig. 12.6). The tendon of the peroneus longus passes beneath the lateral malleolus, converting its pull to one that plantar flexes the foot, rather than dorsiflexion. The superior oblique of the eyeball uses a projection on the wall of the orbit that turns the eyeball downwards rather than upwards.

Stability and motion

A moving human body is constantly faced by

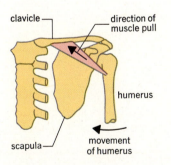

with the limb at the side, the muscle draws the arm forward and across the chest

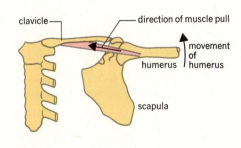

with the limb abducted above a horizontal position, the muscle further abducts the arm

FIGURE 12.5 How a muscle's direction of pull can change, and the bone movement, according to limb position. The example used is the clavicular portion of the pectoralis major (after Wells).

changes in lines and centers of gravity and by changes in the size of the base of support. By using your arms to enfold an object as you carry it you are substantially raising the center of gravity of the body and making it more unstable. Carrying suitcases or similar objects with the arms hanging down lowers the center of gravity and makes the body more stable. Assuming a squatting or crouching position also increases body stability by lowering the center of gravity. If it is not feasible to lower the center of gravity, the base of support may be widened by using a cane, spreading the feet more widely, or assuming a "three or four point" stance involving the upper appendage(s).

Stability in motion also involves the feet — they must have sufficient friction on whatever they stand on to provide a surface for pushing off or stopping. Thus, when walking on icy surfaces the stride is shortened and the arms are brought into play as balancing structures, just as when a tightrope walker leans to one side and lowers the tip of a balance pole on the opposite side to preserve balance.

Stability in the face of external forces impinging on the body is maintained by compensatory reactions of the whole body. Thus, we lean "into a wind," to the inside of a curve, or forwards on acceleration of a vehicle that does not possess a seat against which the body may press. The adjustments required to maintain balance in the face of external forces are handled by structures in the inner ear and within the muscles themselves. Impulses are sent from these areas to the brain for analysis and coordination and from there to the muscles for contraction.

Principles of motion

Movement of body parts are subject to the physical laws that govern movement of all objects.

The LAW OF INERTIA states that an object at rest tends to remain at rest unless acted on by an external force. The muscles provide the external force required to cause movement and to impart motion to objects such as balls and other sports equipment. Additionally, once an object is caused to move, it tends to keep moving in a straight line at uniform speed unless acted on by an external force. An arm will move in a direction determined by the contraction of a given muscle, and will only change direction if other muscles are brought into play. Other forces that may act on a moving object to change its course include friction, gravity, air resistance, or water resistance.

The LAW OF ACCELERATION suggests that a change in velocity (speed) of an object is directly proportional to the force causing that change, and inversely proportional to the mass of the object. Thus, it takes more "brakes" to slow an automobile from 40 miles per hour to 20 miles per hour than from 40 to 30, and an iron ball requires more force to give it the same speed as a wooden ball. One of the functions of muscles that act at angles to — or are antagonistic to — other muscles is to control velocity of moving body parts and change their direction of movement.

The LAW OF REACTION states that for every action there is an equal and opposite reaction. It is this law that causes recoil in the firing of weapons; this law enables us to walk, by pushing the feet against the ground and thus making the body move in the opposite direction. In order to "allow" the law to work, sufficient friction must exist to enable a push to be generated or the body may slip, as when trying to walk on ice.

Force and work

A force may be characterized by its strength, the direction in which it acts, and its point of application to an object.

In movement, force applied by a muscle is determined by the number and size of the fibers within that muscle. The direction of application of the force is determined by the origin and insertion of the muscle and by the motion permitted by the joint involved in the movement; the point of application is the tendinous attachment to the bone.

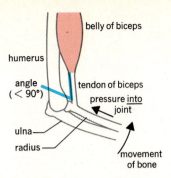

FIGURE 12.7   The biceps pulling on the radius at an angle of less than 90 degrees stabilizes the joint while causing flexion of the elbow (after Wells).

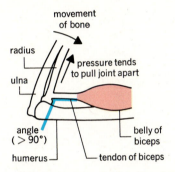

FIGURE 12.8   The biceps pulling on the radius at an angle greater than 90° tends to pull joint apart while causing flexion of the elbow (after Wells).

As a movement occurs, the angle between the bones involved changes and thus the muscle's angle of pull also changes. Only when the muscle pull is at a right angle to the lever (bone) being moved is the entire muscle force being directed to moving the lever. When not at a right angle, the force has two components: a rotatory component that is directed perpendicular to the lever, and which moves it, and a nonrotatory component that is directed parallel to the lever. In the body, most pull is directed to a lever at an angle that is less than 90 degrees and tends to remain less than 90 degrees throughout the motion (Fig. 12.7). Therefore, the nonrotatory component pulls the lever toward the joint and helps stabilize it. If the angle of pull becomes greater than a right angle (Fig. 12.8), the nonrotatory component becomes a

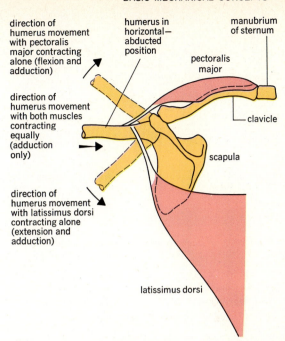

FIGURE 12.9   A diagram to illustrate how muscles having different major actions may cooperate to cause yet another action. The shoulder is viewed as from above, with the latissimus viewed as from behind.

potential dislocating force tending to pry the joint apart. Thus, any motion involves both movement and (usually) stabilization.

If two or more forces act on an object, the speed and direction of movement will be determined by the strength of the forces and their angles of application. Consider, for example, that the deltoid has abducted the upper appendage to a position horizontal to the shoulder. How can the appendage be forcibly adducted back to the body? There is no single muscle that can handle the job. Two muscles acting together—the pectoralis major and latissimus dorsi—can adduct the arm (Fig. 12.9). With the arm in fully abducted position and the pectoralis only contracted, the arm would move downwards, but *across the chest* (flexion). If latissimus only contracted, the arm would move downwards and *across the back* (extension). If however, both muscles contract with equal force, the arm will be adducted "straight in." Obviously, if the forces are not applied equally,

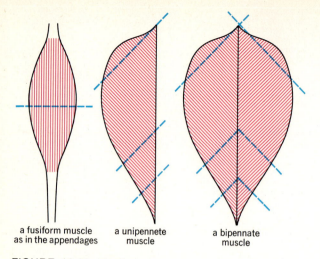

a fusiform muscle   a unipennete   a bipennate
as in the appendages   muscle   muscle

FIGURE 12.10   A diagram to illustrate how physiological cross section of a muscle may be determined. Note that each fiber is "cut" — but only *once* — by each line. See text for explanation (after Wells).

the arm will move in a direction governed by the application of the greater force. Thus, we see:

The resultant direction of two forces is not halfway between them unless the two forces are of equal magnitude.

The angle of application of the two forces will determine the resultant direction.

Only if two forces are applied in the same direction will the resultant magnitude be the sum of the two forces. It will always be something *less* than the sum of the two forces if not applied in the same direction.

Action of two forces on an object acts to steady and direct the movement as well as simply to cause movement.

Work is the product of force and distance, that is, $W = F \times D$. To make a calculation of work a given muscle can do, it is necessary to know to what degree the muscle is capable of shortening, its original length, and its physiologic cross section (PCS). To calculate PCS, one must know the thickness of the muscle and its resting length. Lines are drawn so as to cut all fibers in the muscle and their lengths determined. (Fig. 12.10).

Assuming a muscle of 7 inches resting length and $\frac{3}{4}$ inch average thickness, and that it takes three lines of 3, 4, and 5 inches to cut all fibers in the muscle, the PCS would be calculated as

$$PCS = \frac{3}{4} \text{ inch} \times (3 + 4 + 5 \text{ inches}) = 9 \text{ in.}^2$$

Assuming an average force (determined experimentally) of 90 pounds of force per in.^{2}, and that the muscle can shorten to $\frac{1}{2}$ its original length, the work in *foot pounds* may be calculated by the equation

$$W = \frac{90 \, (\text{PCS in in.}^2) \, (\frac{1}{2} \text{ original fiber length in in.})}{12 \, (\text{to convert inch pounds to foot pounds})}$$

for the muscle above

$$W = \frac{90 \times 9 \text{ in.}^2 \times \frac{1}{2} \, (7)}{12} = 236.25 \text{ ft lbs}$$

Giving impetus and receiving moving objects

Everyday life provides countless examples of the necessity for giving impetus to objects. We walk, pick things up, throw, kick, open windows, close doors, and so forth. Conversely, catching a ball, receiving a block as in football, or falling are examples wherein an object must be brought to a safe, stopped position.

An object will move only if the force brought to bear on it exceeds its inertia and all other forces acting on it such as friction, gravity, or air or water resistance. The greater the force, the farther and faster the object will move. If force is to be increased on an object, it becomes necessary to do one or more of the following:

Increase the strength of muscles by increasing their PCS since length cannot be changed.

Utilize a lever (e.g., prybar, bat, golf club, etc.) to increase the force at its point of application.

Deliver the force at an angle calculated to move the object in the desired direction.

To combine these factors into a maximum performance may require skill and training, especially in activities involving implements. One also learns to utilize the center of gravity of the body, as in pushing or lifting large objects, by placing the body close to the object and not using relatively small or weak segments (e.g., arms alone).

Receiving moving objects, or slowing the entire body's falling, has as its major objective the reduction of the momentum of the object over a period of time, and not instantaneously, or a change in direction of the momentum. This avoids breakage of bones or bruising of flesh. One can also pad the part of the body most likely to receive moving objects to reduce the shock—

hence the baseball glove and padding on football players.

If a moving object is caught with "give," as when moving the hands a short distance in line with the caught object, the shock of contact is lessened. Falling with a rolling motion changes the direction of the force and also results in less initial shock. Lastly, distributing the force of contact over the largest possible area diminishes the shock felt at any given point on the body. Falling on the shoulder rather than on the hand is an example of this type of action.

Postural concepts and principles

When one speaks of posture, what usually comes to mind is the picture the body presents when standing erect in a balanced and more or less "untense" position. Actually, there is no "normal" or "average" posture to which any given individual's standing position may be adequately compared. Each person is an individual, not only genetically but in terms of what is for them a comfortable standing posture. There are also many varieties of human physique, muscular strength, and bone structure, so that any attempts to determine whether posture is what it should be compares an individual human to some mythical norm.

It must also be realized that, to physical educators, activity postures that protect and place the body in the most advantageous position to compete in sports activities should be of greater concern than static poses.

A standing posture that maintains body segments in a comfortable relationship one to another with minimal strain on ligamentous and muscular structures and that permits—or at least does not hinder—visceral activity should become the norm for each individual.

The key to "good posture" in many cases becomes the gradual changing of an habitual body "set" that may reflect not only neuroskeletomuscular factors but psychological factors as well.

A key principle to remember in judging the quality of posture is that there be no *exaggerated* position in any body segment, such as hyperextended knees, tipped pelvis, exaggerated thoracic curve, or a forward thrusted head. Exaggerated positions put undue strain on ligamentous structures that may stretch. Once stretched, most ligaments never regain their original length and this may lead to further exaggeration of a defect.

If there is a desire to improve a given posture, the evaluating individual *must* recognize the genetic influence evidenced by bone structure and muscle development, the psychological influence related to shyness, early development of secondary characteristics that the individual may want to minimize or actually hide, and the person's desire to change the posture. It must also be realized that change is slow, and a desire to maintain a program of corrective exercises must be continued for many weeks or months. Lastly, one looks for *improvement*, not necessarily *cure*, to the point that fitting some mold must be achieved.

The thoughts contained in this section may be summarized by the following statements.

Body segments should be aligned with respect to one another so that strain is minimized.

Good standing posture represents extension of weight bearing joints in a comfortable and not exaggerated position.

Good posture should be one in which energy cost is minimal in order to maintain good segment alignment.

Good posture should permit mechanically efficient joint action with proper angles of pull for muscles.

Good posture is not detrimental to organic function.

There is a definite relationship between posture and personality in terms of eating habits, self-appreciation, and desire to correct or improve posture.

Each person must be treated individually in terms of improving posture, and not forced to meet some hypothetical goal.

Before programs for improvement are recommended, attention should be paid to genetic influences that concern skeletal and muscular development.

Posture should be judged by how well it enables the individual to meet the demands of his/her particular life style throughout a lifetime.

Some comments on conditioning

This section indicates some principles that should be followed in exercise programs designed to improve the total functioning of the human body. This should become the ultimate goal of any exerciser, unless the person is specifically trying to develop a particular body area.

Any program should pay attention to all aspects of fitness: endurance, strength, flexibility, co-ordination, and efficient body mechanics applied to all activities. It follows that improvement in posture, self-image, and total function will occur. CALISTHENICS form an appropriate title for any program of exercises designed to meet all these objectives at once. If any questions arise concerning basic state of health before embarking on an exercise program, see your physician.

This author has been impressed by the type of program for physical fitness that appears in *The Royal Canadian Air Force Exercise Plans for Physical Fitness* (*see* Readings). This particular plan takes into account several important considerations:

It is a *graded* series of exercises that one may enter at a level at which competence is proven.

It specifies *different levels* of activity according to age and sex.

It *develops*, by repetition, *a given level of competence* before allowing advancement to new or more severe exercises.

It is designed to *increase organic function* (circulation and respiration primarily) as well as flexibility and strength.

It requires *at most*, 15 minutes a day, while providing as much beneficial exercise as one might get by jogging, tennis, or other sports.

It may be carried on in the home without investment in expensive equipment.

Initially, the exercise program should provide exercises that stretch muscles. Back bends, toe touching, and lateral movements with the arms stretched over the head are examples of such activity. Sometimes, passive stretching is employed, such as hanging by the hands from a bar, or placing the hands behind the neck and leaning into the elbow placed against a door jam. It must be remembered that an exercise designed to stretch a particular muscle should *not* require that muscle be contracted as part of the exercise — this may defeat the object of the activity.

Next, strengthening a part may be undertaken by exercises that utilize specific muscle groups. For example, sit-ups, pull-ups, push-ups, knee bends (not *deep* knee bends, they are very hard on the knees), and leg lifts (prone and on the back) exercise specific groups of muscles.

Cardiovascular and respiratory function are improved by exercises that put moderate strain on the involved organs. Running in place, jogging (if your feet and ankles are in good shape), swimming, or vigorous walking are good for improving these functions. It is sometimes suggested that a "sprint" of short duration be included in a running type of exercise. There are several studies indicating that periodic stress is particularly beneficial to the heart.

Again, if any benefit is to be achieved from *any* exercise program, it must be continued. One can probably expect breathlessness and a few sore muscles for the first few days of a newly begun program. Do not be discouraged; as fitness improves and one gets second wind, the program will become enjoyable and not just work.

Summary

1. Muscles utilize the bones as levers to cause movement.
   a. A lever has several parts to it.
      1) A fulcrum, usually provided by a joint, about which the lever moves.
      2) A point of effort at which the muscle force is applied.
      3) A resistance to be moved.
      4) An effort arm, a distance over which the effort is applied.
      5) A resistance arm, a distance over which the resistance is offered.
   b. There are three classes of levers.
      1) A first class has fulcrum placed between effort and resistance.
      2) A second class has resistance between effort and fulcrum.
      3) A third class has effort between fulcrum and resistance.

2. Tendons make attachments to bones, rather than fleshy portions of muscles where space, strength, and protection against wear are considerations. Bursae cushion tendons or muscle from wear by bony parts.

3. The human body consists of many movable segments that work together for posture and movement.

4. Basic movements of the body segments occur in planes of motion and utilize anatomical position as the starting point.
   a. Movements in a sagittal plane include flexion and extension.
   b. Movements in a frontal plane include abduction and adduction.
   c. Movements around a vertical axis are rotational.
   d. All other movements are called oblique movements, and usually are combinations of the basic movements.

5. The line of gravity connects centers of gravity of the body segments, and if it falls within the base of support, the body is stable or balanced.

6. To cause movement, muscles pull on bones.

   a. One muscle may cause one or more movements depending on its angle of pull.

   b. Bony projections may be used as pulleys to alter the angle of muscle pull and therefore the action the muscle has.

7. Stability is assured by changes in the base of support, such as widening it, or adding to it as by use of a cane.

8. Several laws or principles of motion affect the movement of body parts.

   a. Law of inertia. An object at rest tends to remain at rest unless acted on by an external force, and tends to move in a straight line at a constant speed.

   b. Law of acceleration. Change in speed of an object is directly proportional to force causing change, and inversely proportional to its mass.

   c. Law of reaction. For every action there is an equal and opposite reaction.

9. Force required to cause movement is provided by the muscles.

   a. Magnitude of force depends on size of muscle fibers.

   b. Direction of force application depends on insertion of the muscle and its angle of pull.

   c. Force applied by muscles usually has a rotatory component that moves the bone and a nonrotatory component that usually stabilizes the joint.

   d. When two muscles act on the same joint, the resultant movement will depend on the motion the joint permits, the angle of muscle pull, and the magnitude of each force acting on the bone.

10. Work is the product of force and distance, and may be calculated for a muscle if its cross section can be determined, and assuming an average force.

11. Giving impetus to objects requires that their inertia be overcome. Force imparted may often be increased by the use of an implement or piece of sports equipment.

12. Receiving moving objects or falling requires an effort to slow down velocity not too suddenly, to change the direction of the object, to pad the receiving part or area fallen upon, and to spread the shock over as large an area as possible.

13. Posture implies a standing position that allows good visceral function, no strain on ligaments and muscles, and minimal metabolic activity to maintain alignment of body segments. It should be an individual thing.

14. Conditioning implies a program of activity that improves endurance, strength, flexibility, and organic function simultaneously. Several exercises designed as examples for each of these goals are described.

Questions

1. If the body segment composed of the head and neck were thrust forward, what might happen to the body segments below and why?

2. Why might oblique movements be considered the basis of skilled movements by the body?

3. How is stability related to the line of gravity falling within the base of support?

4. Why would you shorten your steps when walking on a slippery surface?

5. What laws or principles of motion are involved in striking a thrown ball with a bat?

6. How can the force applied to a given movement be increased?

7. What is "physiological cross section" as applied to muscular force? What is its importance to work that a muscle does?

8. How can the shock of falling or catching something be lessened?

9. In your view, what constitutes good posture?

10. Why should one strive to improve all aspects of fitness instead of, for example, just strength?

11. Describe a series of exercises designed to improve overall physical fitness.

Readings

Cooper, John M., and Ruth B. Glassow. *Kinesiology,* 4th ed. Mosby. St. Louis, 1976.

Hinson, Marilyn. *Kinesiology.* Wm. C. Brown. Boston, 1977.

Kelley, David L. *Kinesiology: Fundamentals of Motion Description.* Prentice-Hall. Englewood Cliffs, N.J., 1971.

Rasch, Philip J., and Roger K. Burke. *Kinesiology & Applied Anatomy: The Science of Human Movement.* Philadelphia, Lea & Febiger. Philadelphia, 1974.

Royal Canadian Air Force. *Exercise Plans for Physical Fitness.* Revised Edition. Simon & Schuster. New York, 1962.

Schutz, Noel W., Jr. *Kinesiology: The Articulation of Movement.* Humanities Press. Atlantic Highlands, N.J., 1976.

Wells, Katherine E., and Kathryn Luttgens. *Kinesiology: Scientific Basis of Human Motion,* 6th ed. Saunders. Philadelphia, 1976.

Chapter 13

The Basic Organization and Properties of the Nervous System

Objectives

After studying this chapter, the reader should be able to:

- Define excitation and conduction.

- Explain some of the mechanisms by which cells are rendered excitable and polarized.

- Explain how a membrane becomes depolarized, what ion(s) move(s), and how an impulse is created.

- Explain what repolarization is, and how it occurs.

- Diagram and explain an action potential.

- Explain how an impulse is conducted.

- Diagram a "typical" neuron.

- List and explain neuronal properties in terms of de- and repolarization.

- Give functions and properties of glial cells, as well as list important types.

- Describe the structure of a synapse.

- Describe how a synapse transmits impulses.

- List and explain synaptic properties in terms of a chemical mode of transmission of impulses.

- List the parts of a reflex arc.

- Explain the characteristics of reflex activity, and the value of such activity to the maintenance of homeostasis.

- Comment briefly on degeneration and regeneration in the central and peripheral nervous systems.

Development of the nervous system

The nervous system is the first system to develop from the ectodermal germ layer (Fig. 13.1). At about 18 days of embryonic development, a thickened layer of ectoderm called the NEURAL PLATE develops. A NEURAL GROOVE appears in the plate and its lateral margins form two NEURAL FOLDS. The folds enlarge, grow, and eventually meet in the midline to form a hollow NEURAL TUBE. Masses of ectoderm separate from the upper and lateral portions of the neural tube. These masses are called NEURAL CREST material, and this material will ultimately form portions of the sensory division of the nervous system, parts of the autonomic system, the adrenal medulla, and melanocytes (pigment cells). The walls of the neural tube differentiate into three layers: an *outer* MARGINAL LAYER is the forerunner of the white matter (fibers) of the nervous system; an INTERMEDIATE LAYER *(mantle layer)* develops cells that will form the gray matter (nerve cells and glia) of the nervous system. An inner layer of cells called the VENTRICULAR *(ependymal)* LAYER forms an epithelium lining the neural tube and its outgrowths.

As development proceeds, the neural tube closes in a head-to-tail direction, and three enlargements develop in its anterior end (*see* Fig. 13.1). These three enlargements, or *primary en-*

TABLE 13.1 Derivatives of the primary neural tube enlargements		
Primary enlargement	Subdivision(s)	Structure(s) arising from subdivision
Forebrain (Prosencephalon)	Telencephalon	Cerebrum Basal ganglia
	Diencephalon	Thalamus Hypothalamus
Midbrain (Mesencephalon)	None	Midbrain
Hindbrain (Rhombencephalon)	Metencephalon	Pons Cerebellum
	Myelencephalon	Medulla

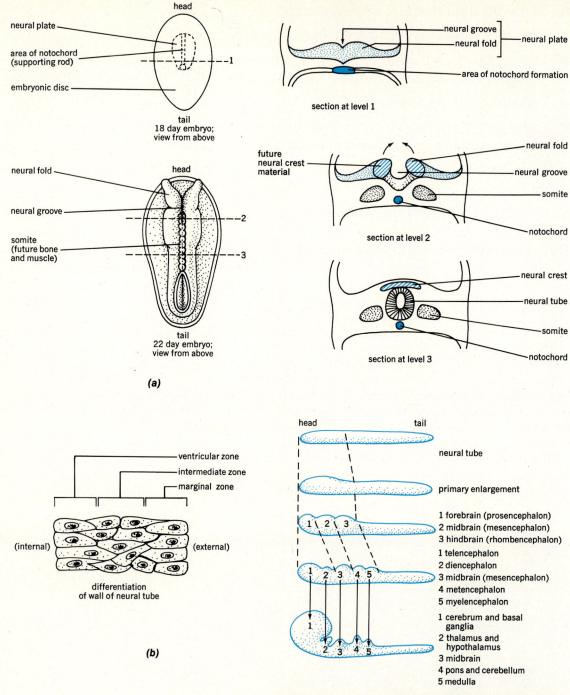

FIGURE 13.1 *(a)* The early development of the nervous system. *(b)* The development of the spinal portion of the peripheral nervous system (w = white matter).

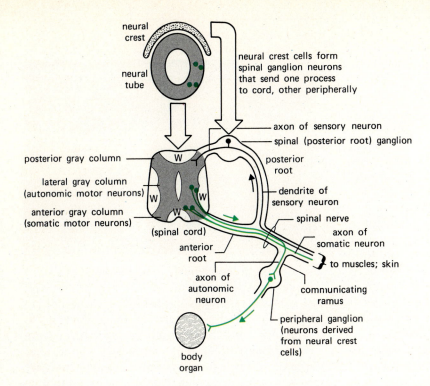

neural crest

neural crest cells form spinal ganglion neurons that send one process to cord, other peripherally

neural tube

axon of sensory neuron

spinal (posterior root) ganglion

posterior gray column

posterior root

lateral gray column (autonomic motor neurons)

dendrite of sensory neuron

anterior gray column (somatic motor neurons)

spinal nerve

axon of somatic neuron

(spinal cord)

to muscles; skin

anterior root

communicating ramus

axon of autonomic neuron

peripheral ganglion (neurons derived from neural crest cells)

body organ

largements, are subdivided as shown in Figure 13.1 and form the portions of the brain listed in Table 13.1. Thus, the individual parts of the brain are formed by 5 to 6 weeks of embryonic life. One reason, perhaps, for the early development of the nervous system is its complexity. An "early start" permits more time for the various parts of the system to attain an organizational level sufficient to sustain the infant after birth.

Organization of the nervous system

As may be appreciated from the previous section, the human nervous system has several parts, distinguished by location and function.

The CENTRAL NERVOUS SYSTEM (*CNS*) consists of the *brain* and *spinal cord* that lie enclosed by the bony structures of the skull and vertebral column. The PERIPHERAL NERVOUS SYSTEM includes all nervous tissues other than brain and spinal cord. Its divisions include: the *somatic system* supplying skin and skeletal muscles; the *autonomic system* supplying smooth and cardiac muscle and glands in the body. The autonomic system is in turn divided into a *parasympathetic* (craniosacral) *division,* which functions to conserve body resources and maintain normal levels of function, and a *sympathetic* (thoracolumbar) *division* that prepares the body to resist stress (the "fight-or-flight" reaction). Most body organs receive nerve fibers from both divisions of the autonomic nervous system, and so their activities may be finely controlled.

Membrane potentials

It may be shown that electrical potentials exist across cell membranes in practically all body cells. The magnitude of that potential is different in different cells. Nerve and muscle cells have potentials that are among the highest in the body. If a cell can establish an electrical potential across its plasma membrane, it becomes EXCITABLE, or capable of responding to changes (stimuli) in its environment. The response includes an alteration in the electrical potential of the membrane, and this response may be CONDUCTED along the cell or to its interior.

The ability to achieve the excitable state, and to conduct a disturbance in the potential, depends primarily on the presence of ions in the extracellular fluids (ECF) outside of the cell and in the intracellular fluid (ICF) within the cell. Of all the ions available in these fluids, those shown in Table 13.2 are the most plentiful and the ones most directly involved in the creation of a membrane potential and the excitable state.

It may also be recalled that the ultimate distribution of such ions inside and outside the cell depends on a variety of passive processes (e.g., diffusion, dialysis), on the permeability of the

TABLE 13.2	Major electrolytes in the body fluids	
Electrolytes	Concentration in extracellular fluid meq/L[a]	Concentration in intracellular fluid meq/L
Sodium	145	12
Potassium	4	155
Chloride	120	4

[a] One equivalent weight of a substance is its atomic weight divided by its valence. It represents the number of grams of an element that will combine with 8 grams of oxygen or 1.008 grams of hydrogen. A milliequivalent is 1/1000 of an equivalent weight.

membrane to substances, and on the presence of active transport systems in the cell membrane. In general, resting membranes, or those not responding to stimuli, are more permeable to potassium ions than to sodium ions. Also, there are nondiffusible ions—chiefly proteins and organic phosphate and sulfate ions—that are present within the cell. Such nondiffusible ions also contribute to production of the membrane potential.

With these facts in mind, let us next examine *how* the excitable state is achieved.

Excitability

How membrane potentials are established

If a *difference* in electrical potential is to be established across a membrane, and the items available are charged particles, it seems reasonable to assume that an *unequal distribution* of charges must be created on the two sides of the membrane. If charges were the same on both sides, no difference would result. There would appear to be two basic ways by which such an unequal distribution of ions may be achieved: differential permeability of membranes to different ions and active transport of ions.

Differential permeability of membranes to ions can create small inequalities in ion distribution because not all membranes are *equally* permeable to all ions. For example, sodium has a more difficult time diffusing through a resting membrane than potassium. Thus, less sodium will enter a cell, while more potassium will cross the membrane—providing that there is a diffusion gradient for the ion in question.

In living cells, the presence of large nondiffusible ions (usually more concentrated *within* the cell) can cause retention in the cell of oppositely charged ions. At normal cellular pH, proteins are usually negatively charged, and can, like tiny magnets, attract positively charged ions, such as potassium. Thus, a mechanism exists for "holding" potassium within the cell.

Through the operation of such mechanisms, an unequal distribution of ions on either side of the membrane may be achieved so as to create a potential difference of about 30 millivolts (mv) across the membrane. Actual potentials on highly excitable cells such as nerve and muscle cells are 70 to 90 mv. Thus, something else must be operating to increase the potential.

Active transport systems would appear to account for the remainder of the membrane potential. A coupled Na-K pump such as the one illustrated in Figure 13.2 can achieve the degree of ionic separation required for the measured potential. What the pump does is to remove from the cell most of the Na^+ that has diffused into the cell and to "recapture" K^+ that has diffused out of the cell.

The resting potential

By such passive and active mechanisms as those just described, a potential called the RESTING POTENTIAL is created. The ion most responsible for the resting potential is believed to be K^+, for the following reasons:

By active transport, sodium is being kept *outside* the cell and potassium *inside* the cell.

A resting cell membrane is 50 to 100 times *more* permeable to K^+ than to Na^+, and so a greater diffusion of K^+ *from* the cell occurs; a surplus of positive charges (Na^+ *and* K^+) accumulates outside the cell.

A membrane with this type of ion distribution is said to possess a "potassium potential," and is POLARIZED (has a certain potential across it).

These relationships are presented in Figure 13.3. Conventionally, if the outside of a cell is electrically positive to the inside, a *negative sign* is placed in front of the measured potential; for example, −70 mv or −90 mv.

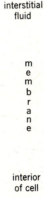

FIGURE 13.2   A coupled sodium-potassium "pump." Sodium combines with a specific carrier *(y)* and is transported to the interstitial fluid. Carrier *y* is enzymatically transformed (E_1) into carrier *x*, which brings potassium in. Again, at the inside of the membrane, enzymatic transformation (E_2) changes *x* to *y* and the process is repeated.

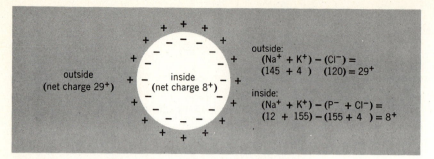

FIGURE 13.3 The separation of ions and charges resulting during attainment of the excitable state in a cell.

The formation of transmissible impulses

If the permeability of the membrane to sodium ion should suddenly increase, sodium would diffuse *into* the cell, following its large diffusion gradient. When a cell is stimulated, this is exactly what happens; there is a 500 fold increase in Na^+ permeability, and also a 30 to 40 fold increase in K^+ permeability. The actual mechanism by which such permeability changes occur is unknown. Suggestions include interruption of the Na-K pump, or the "opening of pores" for sodium in the membrane. The nature of the stimulus causing the permeability changes may be thermal, mechanical, electrical or chemical; all lead to an increased permeability to the ions.

If the fiber is polarized, as shown below, the stimulation results in a greater *inflow* of sodium ion than potassium ion *outflow*, at the point of stimulation. The resting potential may be reversed as positive charges flow into the cell. The cell is now DEPOLARIZED, and since the main ion moving is Na^+, the membrane is now a "sodium membrane" and we have a "sodium potential" in the stimulated or depolarized area. These changes are also shown below.

A depolarized area now lies adjacent to a still polarized area on the cell. Utilizing the ion-laden ECF and ICF, a flow of electrical current occurs between the polarized and depolarized areas, as though two tiny batteries had been connected to one another:

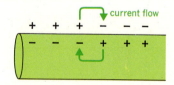

Indeed, the term "battery effect" is often used to describe this current flow. In any event, the current flow represents a measurable phenomenon that, in nerve cells, constitutes a NERVE IMPULSE. (In other cells, it may be referred to as a "depolarization wave.")

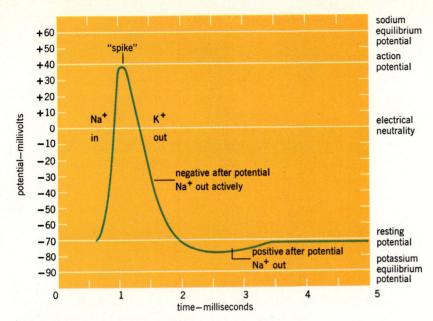

FIGURE 13.4   Changes in membrane potential during development of an action potential.

The action potential

As membrane permeability changes after stimulation, and as ions move, the associated electrical changes may be recorded as an ACTION POTENTIAL (Fig. 13.4). The potential has several parts.

The SPIKE POTENTIAL is the initial large change. It usually lasts (in large nerve cells) about 0.4 milliseconds (msec) and is associated with the large inflow of sodium as permeability to that ion changes.

(The degree by which the spike reverses beyond electrical neutrality is termed the REVERSAL POTENTIAL.)

A NEGATIVE AFTER-POTENTIAL reflects the return of membrane permeability to the original resting values, the active movement of Na^+ out of, and K^+ into the cell, and buildup of positive charge outside the cell.

A POSITIVE AFTER-POTENTIAL occurs as the resting potential becomes a bit higher than it was originally, as though the pump had overshot its mark. This after-potential may last 50 msec to several seconds as the pump equilibrates.

Conductility

If the current flow or impulse described earlier can be transmitted or caused to move along the cell membrane systems, CONDUCTILITY results. In nerve cells, this phenomenon is utilized to "carry messages" over nerve fibers, to and from the CNS, so that muscles may be caused to contract, glands to secrete, or sensations to be appreciated.

The steps in conduction of an impulse are believed to occur as follows.

The "battery effect" creates an electrical field that acts as a stimulus to depolarize an adjacent region of the membrane.

Depolarization of that next region allows ion flow to occur again, with another battery effect being set up a short distance away from the original point of disturbance. The new electrical field depolarizes yet another section of the membrane, and the disturbance "advances" along the membrane.

Meanwhile, as the disturbance moves away from depolarized areas, REPOLARIZATION, or a return to the original resting state, takes place.

These changes occur, in nerve cells, in about 1 msec, so that the cell rapidly returns to its original state, ready to conduct another impulse. The transmitted impulse *does not decrease in strength* as it passes along a normal nerve cell, since it depends upon the activity of living material; its conduction is said to be *decrementless* (without loss).

The cells of the nervous system

Development

ECTODERM provides the germ layer from which most of the nervous system develops. Within the cells of this layer, two types of cells will differentiate and eventually give rise to most of the cells composing the system: NEUROBLASTS will form NEURONS, the highly *excitable* and *conductile* cells of the system; SPONGIOBLASTS will give rise to most of the GLIAL CELLS of the system, cells classically regarded as connective, supportive, and nutritive cells of the system.

The structure of neurons

There are many different shapes and sizes of neurons in the nervous system (Fig. 13.5), but there are enough common features between all neurons to permit a description of a multipolar motor neuron and a sensory neuron of the peripheral nervous system to serve as the basis for all neuron anatomy.

The neuron CELL BODY is surrounded by a typical plasma membrane. The membrane encloses the NEUROPLASM, or neuron cytoplasm. The cytoplasm contains the usual cell organelles (mitochondria, ER, etc.) but after about five years of age the cell center either disappears or becomes nonfunctional. Thus, after this time, neurons are not replaced mitotically if lost. In the neuroplasm are NISSL BODIES (masses of granular ER) that are involved in protein synthesis, and hollow microtubules known as NEUROFIBRILS. The latter may act as routes of distribution of materials through the cell or perform supporting roles. A large NUCLEUS, surrounded by a thin NUCLEAR MEMBRANE, contains the KARYOPLASM or nuclear protoplasm. The nucleus contains one or more NUCLEOLI, composed mostly of RNA, and granular CHROMATIN MATERIAL, composed of DNA.

The cell body gives rise to one or more extensions or PROCESSES. AXONS are usually long (to several feet in length), generally single, sparsely branched, uniform-diameter processes conducting nerve impulses away from (*efferent*) the cell body. DENDRITES are shorter, usually multiple, irregular, and highly branched processes that conduct impulses toward (*afferent*) the cell body. A motor neuron shows both types of processes; a sensory neuron, such as that depicted in Figure 13.5, has processes that are structurally axons, but one will *function* as a dendrite, the other as an axon in terms of direction of impulse conduction.

Both types of processes may have SHEATHS or coverings on them, if built like axons. The MYELIN

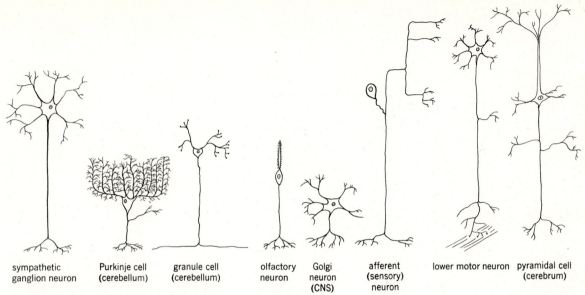

FIGURE 13.5   Some different forms of neurons from the human nervous system (not to the same scale).

SHEATH (Fig. 13.6) is a segmented fatty covering acting to insulate the fiber against current loss and is responsible for saltatory conduction (described later) in nerve fibers. The sheath permits a more rapid conduction of impulses. The NEURILEMMA *(Schwann sheath)* is a thin membrane covering forming part of the myelin lamellae and acting as a guiding tube through which regenerating fibers may grow if they have been injured. (If cell bodies are lost, there is no regeneration; processes can regenerate). SCHWANN CELLS, whose membranes form the neurilemma and myelin lamellae, are also responsible for *production* of myelin in the peripheral nervous system. They are believed to "wrap around" the process (Fig. 13.7). A generalized scheme of what sheaths are found where is presented in Figure 13.8.

Speed of conduction of impulses not only depends on whether or not a myelin sheath is pres-

ent but also on overall fiber diameter, as shown in Table 13.3.

In SALTATORY CONDUCTION, depolarization occurs as described in a previous section. However, the only places that current flow or the "battery effect" can occur is at the NODES *(of Ranvier)* or thin points in the myelin sheath. Because the nodes lie 1 to 3 mm apart along the sheath, the impulse "jumps" from node to node, passing more rapidly along the fiber.

Sheaths are lost before the fiber forms its end branches *(telodendria* or *terminal arborizations)*. Special types of structures may be found on these terminal branches, such as synaptic knobs (described later) or part of a neuromuscular junction (Chapter 10).

Nerve fibers (axons and dendrites) are microscopic in size. They are bound together into larger units, called NERVES, that may become visible to

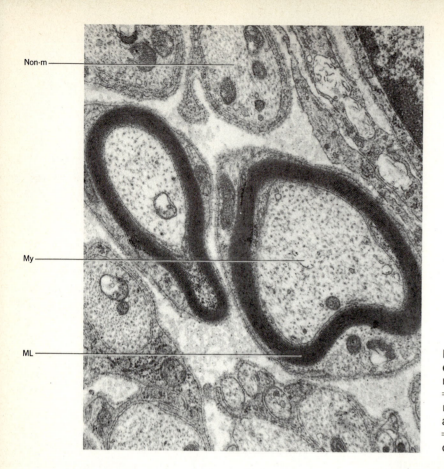

Non-m ——

My ——

ML ——

(a)

FIGURE 13.6 The myelin sheath as seen under th[e]
electron microscope. (a) Myelinated and nonmye[li-]
nated axons from the trachea of lab mouse. Non-[m]
= nonmyelinated axon. My = myelinated axon; ML [=]
myelin lamellae × 24,000. (b) Myelin surrounding [an]
axon in the trachea of the lab mouse. Ax = axon; M[L]
= myelin lamellae × 142,000. (Photographs courtes[y]
of Norton B. Gilula, The Rockefeller University.)

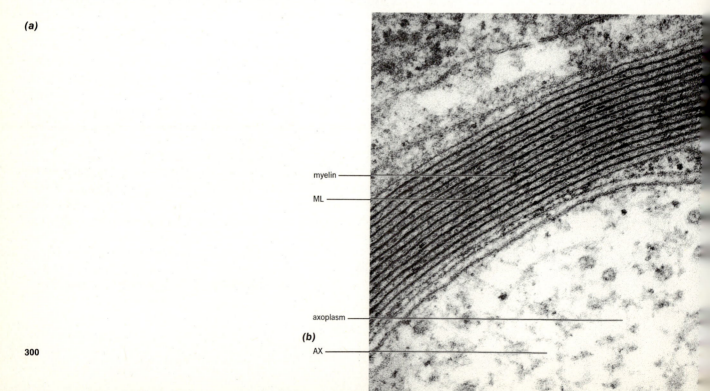

myelin ——

ML ——

axoplasm ——

(b)

AX ——

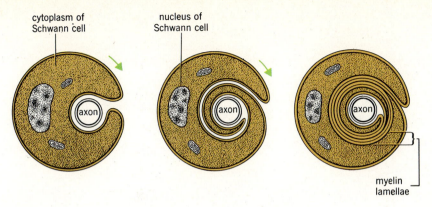

FIGURE 13.7 The formation of myelin lamellae around an axon. The Schwann cell is believed to "wrap around" the axon.

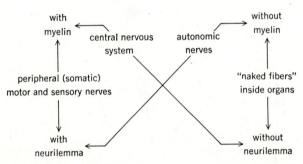

FIGURE 13.8 Combinations of nerve fiber sheaths in the nervous system.

the naked eye. Figure 13.9 shows the telephone-cablelike organization of a nerve. Note the similarity to the organization of muscle fibers into a muscle; the connective tissue component however, receives different suffix names.

Properties of neurons

In addition to showing excitability and conductility, as described earlier, neurons show additional properties, described on page 302.

TABLE 13.3 Characteristics and types of nerve fibers according to diameter				
Type	Diameter (μm)	Myelinated	Conduction velocity (m/sec)	Occurrence
A fibers	1–20	Yes	5–100	Motor and sensory nerves
B fibers	1–3	Some	3–14	Autonomic system
C fibers	<1	No	<3	Skin, viscera

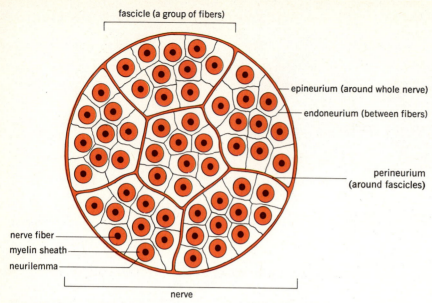

fascicle (a group of fibers)

epineurium (around whole nerve)

endoneurium (between fibers)

perineurium
(around fascicles)

nerve fiber

myelin sheath

neurilemma

nerve

FIGURE 13.9   Diagram showing the connective tissue components of a large peripheral nerve (cross section).

Neurons follow the all-or-none law. For a given strength of stimulus, the fiber either depolarizes completely *(all)* or does not *(none)*. Subthreshold stimuli— those that do not cause depolarization—do, however, alter the state of the membrane, and if such stimuli are applied rapidly enough to a fiber, their individual effects may be added together *(summation)* to cause depolarization. This situation is analogous to driving a nail with several blows instead of driving it home with a single massive stroke; the end result will be the same.

Neurons have a very short refractory period or time that they are in a depolarized state and cannot form or conduct another impulse. Three msec or less is re-

quired to repolarize the fiber after depolarization, and thus a given fiber may conduct many impulses one after the other.

Neurons exhibit a threshold (or rheobase) for stimulation. What this implies is that it requires a certain strength and duration (chronaxie) of a stimulus to cause depolarization. Strength and duration of effective stimuli are generally inversely related; a stronger stimulus is effective over a shorter duration and vice versa. Also, thresholds on the same nerve may be different at different times, depending on the environment of the cell, and they are different in different nerve fibers. A control over response is thus afforded.

Neurons show accommodation. If a stimulus of a given strength is applied continuously to a fiber, its threshold rises. This is accommodation. A fiber thus may "ignore" a stimulus if its characteristics do not change, once a response has been made. It is this property that enables us to wear clothes, rings, or watches on our bodies without being constantly aware of them, and to no longer appreciate an odor in a room after having smelled it at first.

Neurons are very sensitive to environmental factors. Central neurons appear to use only glucose as a fuel source; peripheral neurons use glucose and can also utilize fats. Amino acids are required to synthesize neuronal proteins, and oxygen requirements are about 30 times that of an equivalent mass of muscle. Acidity (pH) of the body fluids and temperature also affect neuron activity; lowered body temperature, as in hypothermia before surgery, will decrease general metabolism and lower the oxygen requirement, and may permit a longer time to be taken for a surgical procedure.

The structure and functions of glia

Glial cells outnumber neurons about five fold in the nervous system. They have processes, but not axons or dendrites, and they do not conduct impulses as neurons do. They do possess the ability to divide throughout life.

Figure 13.10 shows the morphology of several of the glial cells. ASTROCYTES are cells that form supporting networks for neurons and fibers in the central nervous system and transfer nutrients from blood vessels to neurons. OLIGODENDROCYTES are also found in the CNS and synthesize myelin in the CNS (no neurilemmas are found in the CNS). EPENDYMA line the cavities (ventricles and central canal) of the central nervous system, and their cilia may aid transport of molecules from one part of the brain to another. MICROGLIA are also found in the CNS, are of *mesodermal* origin, and are ameboid and phagocytic cells that cleanse the CNS of dead cells and products of disease. SATELLITE CELLS form coverings for neurons in peripheral ganglia, and SCHWANN CELLS, it may be remembered, form the neurilemma and myelin sheaths in the periphery.

Glia have high levels of RNA and protein in them. One theory suggests that memory storage occurs with RNA synthesis and the formation of specific "memory proteins." At the least, glia have properties other than simply acting as connective tissue for the nervous system.

FIGURE 13.10 Some types of glia in the central nervous system. *(a)* Ependyma and neuroglia in the region of the central canal of a child's spinal cord: *A*, ependymal cells; *B* and *D*, fibrous astrocytes; *C*, protoplasmic astrocytes. Golgi method. *(b)* Interstitial cells of the central nervous system; *A*, protoplasmic astrocyte; *B*, fibrous astrocyte; *C*, microglia; *D*, oligodendroglia.

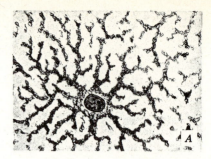

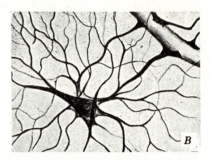

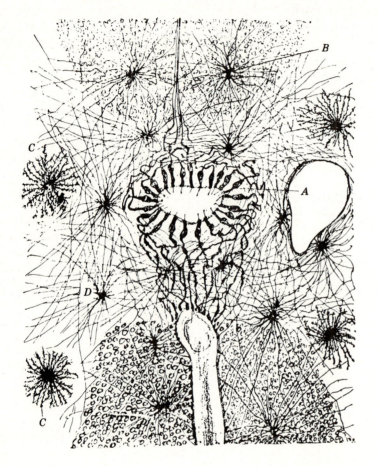

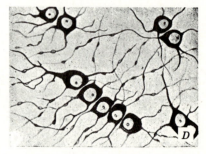

(a)

(b)

The synapse

A synapse is defined as an area of functional but not anatomical continuity between the axonal terminations of one neuron (*presynaptic neuron*) and the dendrites, cell body, or axon of another neuron (*postsynaptic neuron*). It is an area where much *control* over impulses is exerted in terms of blockage, passage, or alteration in nature of the impulse. The synapse also governs direction of passage of impulses across itself.

Structure

The terminal ends of axons may be enlarged to form end feet, end knobs, or other specialized structures (Fig. 13.11). Within these structures are SYNAPTIC VESICLES containing chemicals that may be different in different synapses. A nerve impulse arriving at the end structures increases membrane permeability to Ca^{2+}, which causes re-

lease of whole vesicles or of their contained chemical. The chemical diffuses across a synaptic cleft to cause depolarization of the postsynaptic neuron. A method for destroying the chemical is provided so that it does not exert a continual effect. If all this sounds familiar, recall that the neuromuscular junction, described in Chapter 10, operates in essentially the same manner as does a synapse.

A variety of chemicals acting as synaptic transmitters and methods of destroying them are presented in Table 13.4. A chemical method of enabling an impulse to cross a synapse endows that synapse with properties different from those of the neurons composing it.

Physiological properties

ONE WAY CONDUCTION. Because only the axon terminals of the presynaptic neuron contain the synaptic vesicles, an impulse must travel a presynaptic neuron–synapse–postsynaptic neuron

FIGURE 13.11 The anatomy of the synapse, as visualized under the electron microscope. Notice absence of vesicles in dendrite.

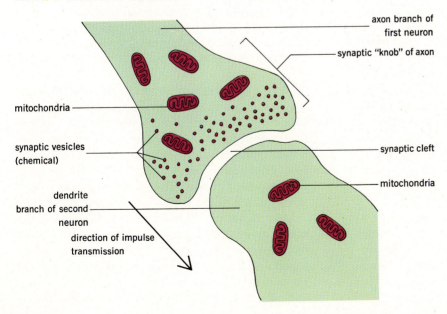

TABLE 13.4 Synaptic transmitters			
Substance or agent	Enzyme or process destroying the agent	Where found, highest concentration	Comments
Acetylcholine	Cholinesterase	Neuromuscular junctions; peripheral synapses between neurons; brain stem, thalamus, cortex, retina	In CNS, not distributed evenly. Perhaps serves specific functions in each area
Norepinephrine	Monoamine oxidase (MAO) or catechol-o-methyl transferase (COMT)	Postganglionic sympathetic neurons; midbrain, pons, medulla	Low concentration in cerebrum
Serotonin	MAO	Hypothalamus, basal ganglia, spinal cord	May be involved in expression of behavior
Histamine	Imidazole-N-methyl transferase, histaminase	Hypothalamus	In mast cells (tissue basophils)
Amino acids, for example GABA (gamma aminobutyric acid)	Oxidation or conversion	Cortex, visual area	GABA is inhibitory to most synapses

course. An impulse can be conducted the "wrong way" up a postsynaptic neuron, but is blocked at the synapse. Prevention of "short circuits" in neuronal pathways is the obvious advantage of 1-way conduction.

FACILITATION. Synaptic transmitters may lower synaptic threshold (*hypopolarization*) by producing an *excitatory postsynaptic potential* (EPSP) that may allow another impulse traveling the same pathway to more easily induce a change in the postsynaptic neuron (*facilitation*). Repeated use of such a pathway may render it one of least resistance, something of value in the learning or conditioning process.

INHIBITION. If the transmitter chemical makes it more difficult to induce a change in the postsynaptic neuron, it is said to produce an *inhibitory postsynaptic potential* (IPSP) by raising the threshold of depolarization (*hyperpolarization*). Thus, trivial or nonessential information may be "filtered out" of the system.

SYNAPTIC DELAY. The events leading to synaptic transmission (release and diffusion of the chemical transmitter, depolarization) all require time in which to occur. It thus takes *more time* for an impulse to traverse a pathway of a given length, if there is a synapse in the path, than if there was no synapse. The extra time represents SYNAPTIC DELAY, and amounts to 0.5 to 1.0 msec per synapse.

SUMMATION. Both TEMPORAL and SPATIAL SUMMATION occur at synapses. A subthreshold stimulus releases a small amount of chemical from the presynaptic axons. If another such stimulus releases more chemical into the synapse before the first chemical is destroyed, the additive effect may produce enough chemical to depolarize the postsynaptic neuron. This is temporal summation. Spatial summation occurs if several neurons terminating on a single neuron each simultaneously release small amounts of chemical, which, added together, amount to enough to depolarize the postsynaptic neuron.

Again it may be noted that the synapse EXERTS

CONTROL over what happens to the impulses reaching it, leading to much control over body activity in general.

Enlargement of the axonal end feet in the presynaptic neuron, as that synapse is used repeatedly, has been observed. Such enlarged feet contain more chemical than smaller ones and presumably release more as impulses reach the synapse. This may contribute to the establishment of the habitual or preferred pathways essential to learning.

DRUGS have great effects on synapses. Hypnotics, analgesics (pain killers), and anesthetics such as bromides, aspirin, morphine, and opium, create IPSP's and decrease synaptic activity.

Curare blocks neuromuscular transmission by preventing acetylcholine from reacting with postsynaptic structures. Physostigmine and strychnine inactivate cholinesterase and permit continued action of the transmitter agent.

CHANGES IN pH alter synaptic activity. Increase of pH increases ease of transmission, while decrease of pH depresses it.

HYPOXIA (lowered O_2 levels) produce cessation of synaptic activity.

The synapse thus seems to be quite responsive to many factors in its environment, a fact that may become extremely important during disease states that may alter the synaptic environment. Such disorders are discussed in later chapters.

The reflex arc: properties of reflexes

The reflex arc (Fig. 13.12) is the simplest functional unit of the nervous system capable of detecting change and causing a response to that change. Many body activities are controlled reflexly. This term implies an *automatic* adjustment to maintain homeostasis without conscious effort. A reflex arc always has five basic parts to it:

A *receptor* to detect change.

An *afferent neuron* to conduct the impulse, resulting from stimulation of the receptor, to the central nervous system.

A *center* or *synapse* where a junction is made between neurons. An internuncial neuron (a neuron interposed between two other neurons) is commonly found at this point, which is often within the central nervous system.

An *efferent neuron* to conduct impulses for appropriate response to an organ.

An *effector* or *organ* that does something to maintain homeostasis.

Functionally, reflexes may be classed as exteroceptive, if the receptor is at or near a body surface;

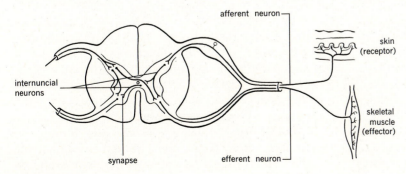

FIGURE 13.12   A reflex arc.

interoceptive, if the receptor is within a visceral organ; proprioceptive, if the receptor is in a muscle or tendon.

Activity that results from passage of impulses over a reflex arc is called the REFLEX ACT or the REFLEX. Reflex acts, regardless of what the stimulus and response are, have several characteristics in common.

Activity of the effector depends on stimulation of a receptor. In short, the arc serving the act must be complete: if any part of the arc is not working, the effector will not respond.

Reflex activity is *involuntary*, but can often be stopped by an act of will; if we expect a certain stimulus we need not respond to it. Reflex activity is *stereotyped* (repetitive); stimulation of a given receptor always causes the same effector response.

Thus, if the patellar tendon is tapped, the knee *always* extends. It is thus *predictable*.

Reflex activity is *specific* in terms of the stimulus and the response. A particular stimulus causes a particular response; the response is usually *purposeful*.

The value of these reflex properties to the body is obvious. A particular stimulus causes a predictable and appropriate response, which will usually result in or aid in maintenance of homeostasis. If responses were different to the same stimulus, the body would have a difficult time maintaining its posture, balance, and overall functions.

Examples of reflexes, especially as they are involved in homeostatic control are presented in subsequent chapters.

Clinical considerations

Degeneration and regeneration in the nervous system

Injury to neuron cell bodies is usually followed by death of the entire neuron. Damage to an axon results in changes in the cell body (the axon reaction) and in the axon distal to the point of injury (Wallerian degeneration).

In the AXON REACTION, the cell body swells, and the Nissl substance fragments into fine granules. In WALLERIAN DEGENERATION, the distal segment of the axon swells, the myelin sheath degenerates, the axon itself degenerates, and only the outer layer of the neurilemma or glial cells remains. The products of degeneration are removed by glial cells.

In peripheral axons, ameboid, bulbous ends develop on the injured axon, and these grow through the neurilemmal tubes at about 2 millimeters per day. Reestablishment of connections, if it occurs, is more widespread than originally, and control is less precise than before.

In the central nervous system, regeneration is more limited than peripherally. Since central axons lack neurilemmal sheaths, reestablishment of normal connections is haphazard and is complicated by the fact that millions of axons are present in the cord and brain. Thus, chances of making proper connections are low.

Disorders

MULTIPLE SCLEROSIS is a disease of unknown origin, and has no cure. It is characterized by degeneration of myelin sheaths in the brain and spinal cord. Changes in function that commonly occur include: lack of ability to concentrate, speech disturbances, visual changes, and increased vigor of knee and ankle jerk reflexes.

Before the availability of vaccines for POLIO-MYELITIS, the viruses often attacked and destroyed the cell bodies of the somatic motor neurons to skeletal muscle. Paralysis followed.

Summary

1. Cells, by achieving an unequal separation of ions, may create an excitable state and an electrical potential across their membranes. The ions responsible for these potentials are sodium and potassium.

    a. Diffusion, of differential nature for different ions, plus a semipermeable membrane, creates a part of the membrane potential.

    b. Active transport, of sodium and potassium primarily, creates a larger part of the membrane potential.

2. Polarization results when ions are separated.

    a. A resting potential of -70 to -90 mv is created by the processes indicated in 1a, 1b. It is due primarily to "leakage" of K^+ from inside to outside of the cell.

3. Impulses are formed when membrane permeability to Na^+ is greatly increased.

    a. A stimulus causes increased ($500 \times$) permeability to sodium ion, sodium enters the fiber, the membrane potential is reversed and the fiber is depolarized.

    b. Current flows between the polarized and depolarized regions and this is the impulse.

    c. Repolarization occurs on return of Na^+ permeability to the original state and pumping out of Na^+. Potassium conductance increases as Na^+ conductance decreases, and this and active K^+ transport return K^+ to the fiber, and slow its outward movement.

4. An action potential is produced as ions move through a depolarized membrane.

    a. The spike potential correlates with sodium entry.

    b. The reversal potential correlates with the reversal of membrane potential as large quantities of Na^+ enter the fiber.

    c. After potentials correlate with active movement of Na^+ and K^+, and return of membrane permeability to normal.

5. Conductility occurs as the current flow depolarizes the next section of the fiber and moves, by repeating the process, down the fiber.

6. Neurons, the highly excitable and conductile cells of the nervous system, arise from neuroblasts of the ectodermal germ layer. Glia arise from spongioblasts of ectoderm.

7. The structure of a neuron is described. The involvement of a fatty myelin sheath in speeding impulse transmission (saltatory conduction) is discussed.

8. Physiological properties of neurons are presented.

    a. Response of a neuron is either maximal or none (all-or-none response).

   b. Refractory periods exist during which the neuron cannot be stimulated or is more difficult to stimulate.

   c. Threshold. The minimum strength of stimulus or the minimum voltage required to depolarize a neuron.

   d. Summation. Subthreshold stimuli may "pool their influence" to cause depolarization.

   e. A stimulus must last a certain time (duration) to cause depolarization. Weaker stimuli must last longer and vice versa.

   f. Continued stimulation causes a rise of neuron threshold called accommodation.

9. Several types of glial cells are described.

   a. Glia play no part in transmission of impulses.

   b. They may be involved in neuron nutrition, form supporting networks for neurons, synthesize myelin, and are phagocytic. They may be involved in formation of memory traces.

10. A synapse is a functional junction between neurons. Its properties are different from the neurons composing it.

   a. Synaptic transmission is by means of chemical substances, which are then destroyed to prevent continued action.

   b. One-way conduction is assured by having the transmitter substance on only one side of the synapse.

   c. Facilitation means an easier passage for an impulse resulting from production of an EPSP.

   d. Inhibition means more difficult passage for an impulse due to production of an IPSP.

   e. Synaptic delay means a slower passage of an impulse across a synapse than down a nerve fiber because of time required for chemical release and diffusion.

   f. Temporal and spatial summation occur as small amounts of chemical are added together.

   g. Drugs may affect synapses to increase or decrease impulse transmission. pH and oxygen levels also influence synaptic activity.

   h. Synapses control their activity to a large degree.

11. The reflex arc is the simplest unit of the nervous system that can automatically maintain homeostasis.

   a. Its parts include: a receptor to respond to change; an afferent neuron to conduct impulses to the central nervous system; a synapse, with the possibility of an internuncial neuron, in the central nervous system; an efferent neuron to conduct impulses away from the central nervous system; an effector to respond.

    b. Reflex acts are involuntary, stereotyped, predictable, and usually purposeful. They automatically control homeostasis.

12. If neuron cell bodies are destroyed, the entire neuron dies. A fiber may regenerate.

    a. Changes called the axon reaction and Wallerian degeneration occur on injury to a fiber.

    b. Regenerating peripheral fibers develop ameboid tips that can reestablish a more generalized connection than before.

    c. Central regeneration is limited and rarely are connections reestablished.

Questions

1. How do excitation and conduction differ in the mechanisms by which they occur?

2. How are ions separated so as to create the excitable state?

3. Explain why a resting potential is a "potassium potential."

4. Explain why an action potential is a "sodium potential."

5. How does saltatory conduction differ from conduction in a nonmyelinated fiber? How are they similar?

6. Compare and contrast neuronal and synaptic properties. Account for them in terms of the methods of neuronal and synaptic transmission.

7. Of what value are reflex arcs to the control of homeostasis?

8. What are the parts of a reflex arc, and what does each contribute to the reflex?

9. From your own knowledge or experience, give some examples of reflex activity that contribute to your homeostasis.

Readings

Axelrod, Julius. "Neurotransmitters." *Sci. Amer. 230:*58, June 1974.

McEven, Bruce S. "Interactions Between Hormones and Nervous Tissue." *Sci. Amer. 235:*24, Aug. 1976.

Nathanson, James A., and Paul Greengard. "Second Messengers in the Brain." *Sci. Amer. 237:*108, Aug. 1977.

Science News. "Multiple Sclerosis: A Closer Look." *113:*183, March 25, 1978.

Sonyen, George C. "Electrophysiology of Neuroglia." *Ann. Rev. Physiol.* Vol. 37. 1975.

Chapter 14

The Spinal Cord and Spinal Nerves

Objectives

After studying this chapter, the reader should be able to:

- Describe the location and size of the adult spinal cord and the locations and functions of the meninges around the spinal cord.

- Describe the gross anatomy of the spinal cord.

- Differentiate between the terms sulcus and fissure as applied to the spinal cord, and locate the various sulci and the fissure.

- Describe the internal structure of the spinal cord.

- Define what a spinal tract is, and list the major tracts of the cord with origin, termination. and function

- Show how impulses arrive at, and leave, the cord.

- Define an internuncial neuron and describe some of its functions in the cord.

- Discuss briefly the role of the cord in reflex activity, explaining the value of: myotatic reflexes, flexion and extension, and righting reflexes.

- Describe the structure and function of a muscle spindle.

- Tell about the spinal nerves as to number, point of origin, connections with the cord, and functional components.

- List the major plexuses formed by the spinal nerves, the nerves forming each, and body area served.

- Discuss some of the effects of injury to the spinal cord.

- Define what spinal shock is, and how function returns after shock.

- Explain what a spinal puncture is, where it is done, and what it is used for.

- Relate the location of various large nerves in the body to the use of intramuscular injections.

The spinal cord serves as the first area of the nervous system where impulses from peripheral sense organs located below the neck are organized and routed, where reflex activity is integrated, and where motor impulses are passed to peripheral effectors. It also serves as a transmitting structure carrying impulses to the brain for further processing and impulses from the brain to muscle and glands. The spinal nerves form a network for carrying nerve impulses to and from peripheral organs.

The location and structure of the cord (Fig. 14.1)

The spinal cord is the portion of the central nervous system enclosed by the neural arches (laminae and pedicles) of the vertebral column. It lies in the VERTEBRAL CANAL formed by these arches. In the adult, the cord has a length of about 45 cm (18 in.) in length, and a diameter of about 2 cm (4/5 in.). Obviously, the cord is smaller in diameter than the canal containing it; the "extra space" is occupied by the MENINGES, three membranes that surround not only the spinal cord, but the brain as well, and by connective tissue lying externally to the meninges on the cord. An outer *dura mater* is a tough, protective membrane composed of collagenous connective tissue. It is separated from the next membrane by a narrow *subdural space*. A middle *arachnoid* is a delicate membrane enclosing the *subarachnoid space*, a larger space filled with *cerebrospinal fluid.* An inner *pia mater* lies on the organs of the central nervous system and carries blood vessels into the cord and brain. Loose connective tissue lies between the dura mater and the walls of the vertebral canal.

In the adult, the substance of the cord proper ends at the level of the second lumbar vertebra (L2). It is this fact that enables a spinal tap or puncture, discussed later in this chapter, to be made.

Grossly, the cord may be seen to be of unequal diameter throughout its length; there is a CERVICAL ENLARGEMENT that marks the entry of the nerves from the upper limb into the cord, and a LUMBAR ENLARGEMENT marking the entry of the nerves from the lower limb. (The portions of the cord—i.e., cervical, thoracic, lumbar, and sacral—

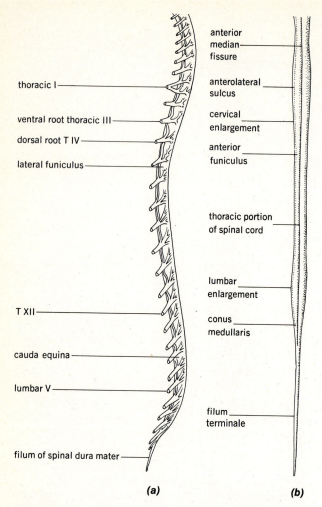

anterior median fissure

thoracic I

ventral root thoracic III

dorsal root T IV

lateral funiculus

anterolateral sulcus

cervical enlargement

anterior funiculus

thoracic portion of spinal cord

lumbar enlargement

T XII

conus medullaris

cauda equina

lumbar V

filum terminale

filum of spinal dura mater

(a)   *(b)*

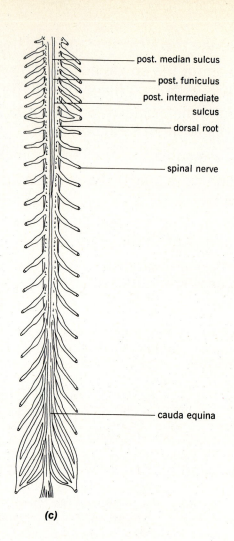

post. median sulcus

post. funiculus

post. intermediate sulcus

dorsal root

spinal nerve

cauda equina

(c)

FIGURE 14.1 Three views of the gross anatomy of the spinal cord. *(a)* Lateral view. *(b)* Anterior view. *(c)* Posterior view.

are designated primarily by the point of exit of the spinal nerves between the vertebrae of a particular area, even though the cord itself does not extend through the entire vertebral canal.) The tapering end of the cord is the CONUS MEDULLARIS, from which the FILUM TERMINALE, a fibrous derivative of the pia mater, extends into the sacral canal to anchor the cord. The mass of nerves that course through the vertebral canal *below* the conus to eventually exit through lumbar and sacral inter-

vertebral foramina form the CAUDA EQUINA, so named because of its resemblance to the hair in a "horse's tail."

The cord is grooved throughout its length by shallow indentations called SULCI (sing., *sulcus*). Each is named according to its *position* on the cord; we may distinguish a POSTERIOR (*dorsal*) MEDIAN SULCUS, two POSTERIOR INTERMEDIATE SULCI, two POSTERIOR LATERAL SULCI, and two ANTERIOR LATERAL SULCI. A deeper indentation,

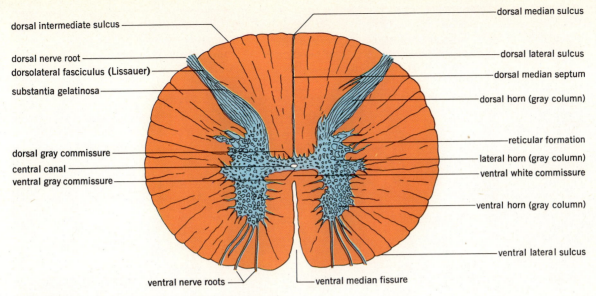

FIGURE 14.2   A cross section of the spinal cord.

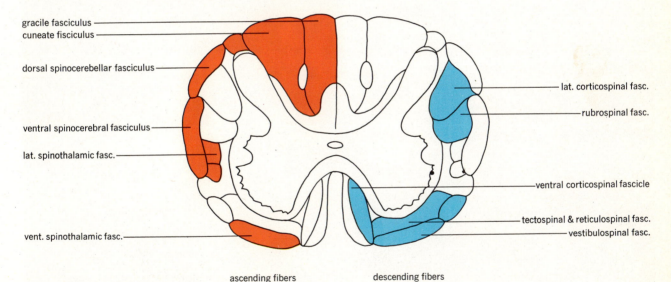

FIGURE 14.3   The major spinal tracts or fasciculi. (Red) sensory tracts; (blue) motor or descending tracts.

the ANTERIOR MEDIAN FISSURE, lies on the anterior midline of the cord. The posterior- and ventral-lateral sulci mark the entry and exit of the spinal nerve roots on the cord.

A cross section of the spinal cord (Fig. 14.2)

illustrates several of the features just described. It also indicates that *within* the cord there is a definite organization of parts. An inner H-shaped mass of GRAY MATTER consists mostly of neuron cell bodies and synapses. It is subdivided into

TABLE 14.1   The major spinal tracts

	Name of tract	Area of origin, or region from which tract receives fibers	Area of termination	Crossed or uncrossed	Function or impulse(s) carried
D E S C E N D I N G	Corticospinal Lateral Ventral	Motor areas of cerebral cortex	Synapses in cord	Crossed in medulla, 80% of fibers Uncrossed, 20% of fibers	Carry voluntary impulses to skeletal muscles
	Rubrospinal	Red nucleus of basal ganglia in cerebrum	Synapses in cord	Crossed in brain stem	Involuntary impulses to skeletal muscles concerned with tone, posture.
	Reticulospinal	Reticular formation of brain stem	Synapses in cord	Crossed in brain stem	Increases skeletal muscle tone and motor neuron activity
	Spinothalamic Lateral	Skin	Thalamus, relays to cerebral cortex	Crossed in cord	Pain and temperature
	Ventral	Skin	Thalamus, relays to cerebral cortex	Crossed in cord	Crude touch
A S C E N D I N G	Thalamus, Spinocerebellar Dorsal Ventral	Muscles and tendons	Cerebellum Cerebellum	Uncrossed Uncrossed	Unconscious muscle sense for controlling muscle tone, posture.
	Gracile Cuneate	Skin and muscles	Medulla, relays to cerebral cortex	Uncrossed Uncrossed	Touch, pressure, two-point discrimination, conscious muscle sense concerned with appreciation of body position

posterior, lateral, and anterior columns (or horns), connected across the midline by the *gray commissures* (the "bar" of the H). The CENTRAL CANAL, a remnant of the hollow of the neural tube, separates a *posterior gray commissure* from an *anterior gray commissure*. Surrounding the gray matter is the WHITE MATTER, composed mainly of myelinated nerve fibers. Again, several areas or *funiculi* are named according to their position as the *posterior, lateral*, and *anterior funiculi* of the white matter.

The white matter is further divided into *functional areas* known as TRACTS (Fig. 14.3), although no anatomical evidence exists for these tracts. Experimentation has shown that damage in particular areas of the white matter is associated with loss of particular types of sensation or motor ability. The names of the major tracts, their areas of origin and termination, and their functions are presented in Table 14.1. Descending tracts are motor in function, ascending ones are sensory.

Functions of the cord

Input to the cord

Incoming impulses arrive at the cord over the POSTERIOR ROOTS (Fig. 14.4). The neuron cell bodies of these roots are located in the SPINAL GANGLIA that lie next to the cord. The root

branches to give a number of *rootlets* that enter the cord along the posterior lateral sulcus. Synapses may be formed directly with neurons that leave the cord through the anterior rootlets and roots (*see* Fig. 14.4), or, more commonly, one or more INTERNUNCIAL *(association)* NEURONS are interposed between the incoming and outgoing nerves.

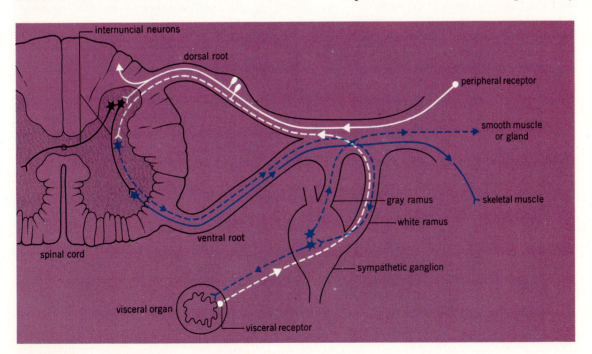

FIGURE 14.4   The connections of sensory and motor neurons to the spinal cord.

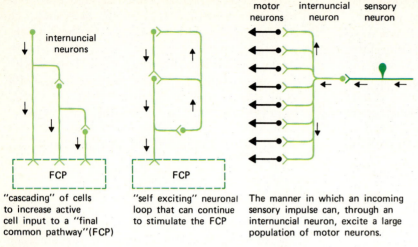

motor internuncial sensory
neurons neuron neuron

"cascading" of cells to increase active cell input to a "final common pathway"(FCP)

"self exciting" neuronal loop that can continue to stimulate the FCP

The manner in which an incoming sensory impulse can, through an internuncial neuron, excite a large population of motor neurons.

FIGURE 14.5 Mechanisms of action of internuncial neurons.

Internuncial neurons

Internuncial neurons form the bulk of the neurons of the cord, outnumbering other neurons by about 30 to 1. By their physical presence, they permit incoming impulses to travel pathways other than one directly from incoming to outgoing neurons; for example, from one side of the cord to the other, to different levels of the cord, or to the brain itself. They also serve as amplifiers of the intensity of incoming impulses, by *cascading*, utilizing *neuronal loops*, or by *distributing impulses* to a wider area to increase total neuron discharge. These devices are shown in Figure 14.5.

The internuncial neurons may also serve as "gates" or "valves," passing or blocking incoming impulses. Thus, a degree of *control* over what next happens to an incoming impulse is afforded.

Output from the cord

Impulses from the brain, peripheral receptors, or internuncial neurons ultimately converge on the motor neurons of the cord, located in the lateral and anterior gray columns. These neurons thus serve as the FINAL COMMON PATHWAY (FCP) for distributing impulses to the muscular tissues and glands of the body by way of the anterior roots (*see* Fig. 14.4).

The cord and reflex activity

In infrahuman animal forms, the spinal cord may be an important area for origination of impulses that lead to complex motor activity. In the human, the cord serves mainly as a reflex center and as a transmitter of impulses to and from the brain. Several types of reflex activities are *controlled* by the cord.

MYOTATIC REFLEXES. The term MYOTATIC REFLEX (*stretch reflex*) refers to the reflex contraction of a skeletal muscle brought about by applying a sudden stretching force to the muscle or its tendon. We are all familiar with the "knee jerk" brought about by striking the patellar tendon with a reflex hammer and the resulting contraction of the quadriceps that extends the knee. Such reflexes would seem to be primarily concerned with maintaining the body erect against the force of gravity

TABLE 14.2 Some reflexes used to assess cord function

Name of reflex	Level of cord involved	How elicited	Interpretation
Jaw jerk	C3	Tap chin with percussion hammer; jaw closes slightly.	Marked reaction indicates upper neuron lesion.
Biceps jerk	C5,6	Tap biceps tendon with percussion hammer; biceps contracts.	If reflex is absent or greatly exaggerated, cord damage, or damage to sensory or motor nerves is probable.
Triceps jerk	C7,8	Tap triceps tendon with percussion hammer; triceps contracts.	
Abdominal reflexes	T9-L2	Draw key or similar object across abdomen at different levels; muscles contract.	
Knee jerk	L2,3	Tap patellar tendon with percussion hammer; quadriceps contract to extend knee.	
Ankle jerk	L5	Tap Achilles tendon; gastrocnemius contracts to plantar flex foot.	
Plantar flexion	S1	Firmly stroke sole of foot from heel to big toe; toes should flex.	If toes fan and big toe extends (Babinski sign), upper neuron lesion is present.

and in rhythmical movements such as walking and chewing. Such reflexes are usually two neuron reflexes (*no* internuncial neuron), and one served by particular levels of the spinal cord. They are often used by the neurologist to assess cord function. Table 14.2 presents several myotatic reflexes and how they are used to assess cord function.

The primary receptor for a myotatic reflex is the muscle spindle (Fig. 14.6), located within the belly of a muscle. *Intrafusal fibers,* located within the spindle, are wrapped by spiral endings of sensory neurons known as *1a afferent neurons.* The intra-fusal fibers receive *gamma efferent* neuron fibers that control the degree of their contraction. The gamma fibers, by controlling intrafusal contraction, determine how much force will be required to elongate the intrafusal fibers; in general, the greater the intrafusal fiber contraction, the less tension required to create a nerve impulse in the sensory nerves. The whole system controls muscle in terms of the loads it must bear to maintain normal orientation of body parts.

FLEXION AND EXTENSION REFLEXES. These types of reflexes originate from receptors in the skin and deeper tissues (excluding muscle) that respond to touch, pressure, heat, cold, and pain.

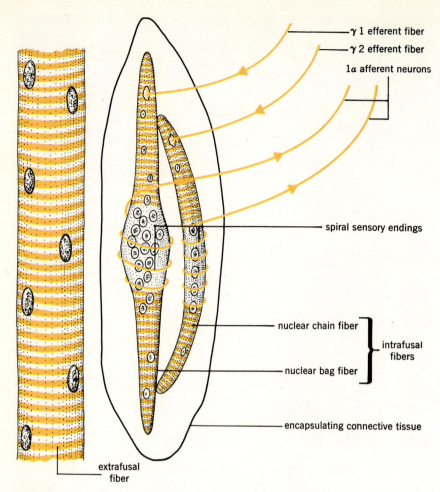

FIGURE 14.6   The structure and innervation of the muscle spindle.

They are primarily protective in that they cause withdrawal of a body part from a potentially dangerous stimulus. They are more complex than the myotatic reflexes because they usually involve internuncial neurons and more than one cord segment.

FLEXOR REFLEXES result when the skin is stimulated (as by touching a hot object), and the cord distributes impulses to all the flexor muscles of the affected limb. The limb is withdrawn from the stimulus.

As a stimulated limb is withdrawn from a stimulus, the limb(s) on the *opposite* side of the body may need to be extended to maintain body balance (as in stepping on a tack). This type of reflex is called a CROSSED EXTENSION REFLEX. Such a response requires an internuncial neuron or neurons crossing to the opposite side of the cord (Fig. 14.7). This system is also important in rhythmical movements such as walking, in which one limb is moving forward and the other backward, because of contractions of opposite groups of muscles.

RIGHTING REFLEXES form a type of activity par-

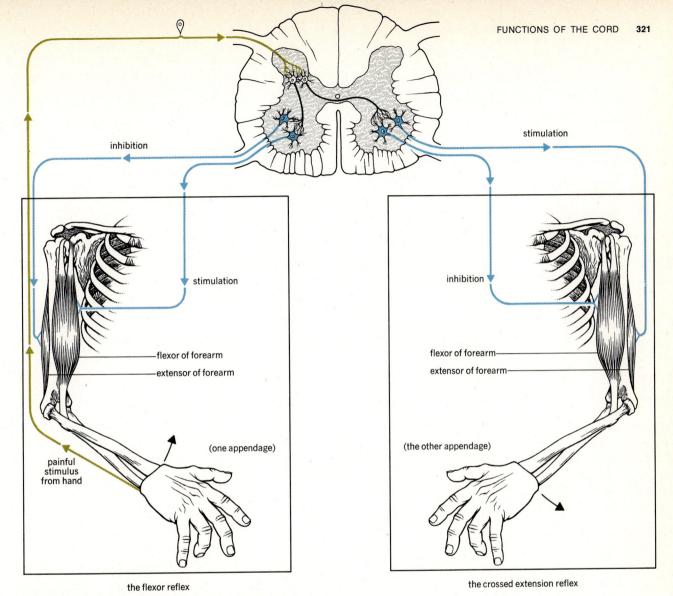

inhibition

stimulation

stimulation

inhibition

flexor of forearm

extensor of forearm

flexor of forearm

extensor of forearm

(one appendage)

(the other appendage)

painful
stimulus
from hand

the flexor reflex

the crossed extension reflex

FIGURE 14.7   The interrelationships of the flexor reflex, the crossed extension reflex, and the phenomenon of reciprocal inhibition (the opposite effect on the same muscles in opposite appendages).

tially served by the cord. Sensory information received from the skin and inner ear structures when the body is out of balance or resting on its side usually result in a return to normal posture. Such movements may still occur with an intact cord, but require the cerebellum to *coordinate* the movements.

The cord may thus be seen to be an important center for controlling those types of muscular activity necessary for proper posture and response to dangerous stimuli.

The spinal nerves

Number and location

The cord gives rise to 31 pairs of spinal nerves, divided into *8 cervical* (C), *12 thoracic* (T), *5 lumbar* (L), *5 sacral* (S), and *1 coccygeal.* They are distributed to the skin and muscles of particular body regions (Fig. 14.8) with overlap of about 30 percent by neighboring nerves. Thus, if a given nerve is cut or damaged, complete function in a body area will not usually be lost.

The spinal nerves exit from the vertebral canal through the *intervertebral foramina* between vertebral pedicles. In the upper portions of the cord, the foramina of exit are nearly opposite the cord segment of nerve origin, and the nerves exit immediately to the side. In the lower segments, since the cord is shorter than the spinal column, nerves travel downward within the canal until they find their foramina of exit. Large numbers of these nerves from the *cauda equina* (horse's tail) of the cord (*see* Fig. 14.1).

Connections with the spinal cord (*see* Fig. 14.4)

Each spinal nerve has a *dorsal root (sensory)*, which conveys impulses to the cord. Cell bodies of these neurons are located in a *dorsal root ganglion* of each nerve. A *ventral root (motor)*

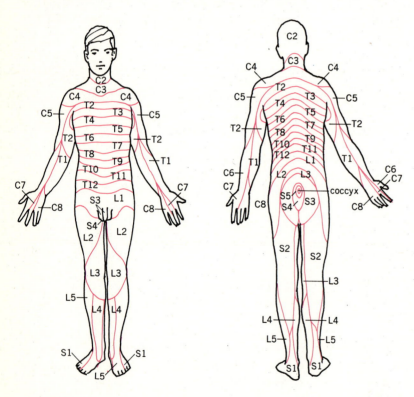

FIGURE 14.8 The segmental distribution of spinal nerves to the body.

conveys impulses from cell bodies in the cord to muscles and glands. *Communicating rami* allow connections for autonomic fibers. Peripherally, *somatic* nerves form anterior and posterior branches to skin and skeletal muscles. The ventral somatic roots to the skeletal muscles constitute the *lower* motor neurons of the motor system.

Components of the spinal nerves

Four functionally distinct types of fibers compose the spinal nerves. They are named SOMATIC AFFERENTS, SOMATIC EFFERENTS, VISCERAL AFFERENTS, and VISCERAL EFFERENTS.

SOMATIC AFFERENTS. Somatic afferent fibers are those carrying sensory impulses from receptors other than those serving sight, hearing, smell, and taste. The *spinal* components of the somatic efferent system serve all but the superior cranial and facial areas of the body in a segmental or regional fashion. The somatic afferents' receptors are located primarily in the skin and are activated by alterations in the external environment, such as temperature changes, pressure on the skin, or damage to the body surface. These receptors are sometimes called *exteroceptive receptors,* and their sensations of touch, pressure, warmth, cold, and pain may be designated *exteroreceptive sensations.* Also conveyed by spinal nerves as part of the somatic afferent component are sensory fibers arising from receptors in tendons, joints, and *skeletal* muscles. These receptors are designated *proprioceptive receptors,* and their sensations *proprioceptive sensations.* Information from these receptors informs us of body position, and direction, rate, and force of movements.

SOMATIC EFFERENTS. Somatic efferents are the axons of the neurons having their cell bodies in the ventral gray columns of the spinal cord. They are the lower motor neurons of the system supplying skeletal muscles with their activating fibers. They terminate on skeletal muscles by way of neuromuscular junctions. It may be recalled that these neurons respond to both "voluntary" and "involuntary" impulses reaching them from several areas of the brain, and that their loss results in a flaccid paralysis.

VISCERAL AFFERENTS. Visceral afferent fibers are derived from receptors located in the walls of abdominal and thoracic blood vessels and organs. Their cell bodies are also in the dorsal root ganglion. These fibers convey information regarding the pressure within the organs, tension in their walls, chemical composition of fluids circulating within the walls, and volume of fluid in a hollow organ. Such receptors may be called *enteroceptive receptors,* and they are activated by changes within the body. They convey *enteroceptive impulses* that are extremely important in maintaining body homeostasis of blood pressure, blood gas composition, and blood flow. These fibers belong to the autonomic nervous system to be described later in this chapter.

VISCERAL EFFERENTS. These fibers also belong to the autonomic nervous system, and are derived from neurons whose cell bodies are located in the lateral gray columns of the spinal cord. The fibers terminate on smooth muscle of the body and on certain body glands, such as sweat glands.

Segmental distribution of spinal nerves

Each pair of spinal nerves supplies a particular region of the body. The development of the muscles and skin and their innervation by the spinal nerves creates a SEGMENTAL DISTRIBUTION of spinal nerves that is retained as development proceeds (Fig. 14.8). A DERMATOME is defined as a skin area supplied by a single spinal nerve's dorsal root. The word implies that the muscles are innervated in a similar segmental pattern. Dermatomes overlap one another by about 30 percent, so that if a given spinal nerve is damaged, total loss in its dermatome will not occur. Knowing the pattern of spinal nerve distribution makes it

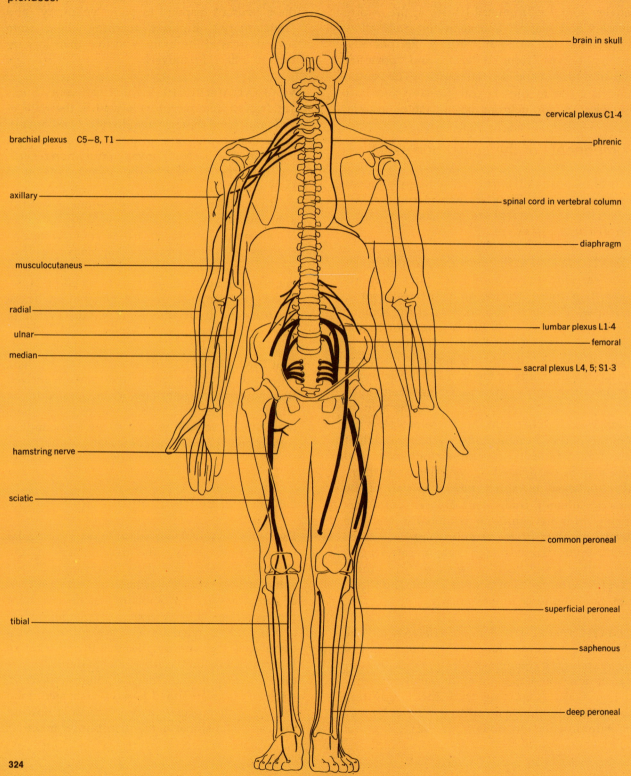

FIGURE 14.9 The organization of spinal nerves into plexuses.

brain in skull

cervical plexus C1-4

brachial plexus   C5—8, T1

phrenic

axillary

spinal cord in vertebral column

diaphragm

musculocutaneus

radial

lumbar plexus L1-4

ulnar

femoral

median

sacral plexus L4, 5; S1-3

hamstring nerve

sciatic

common peroneal

superficial peroneal

tibial

saphenous

deep peroneal

easier to understand why there is specific area loss of sensation and/or movement if spinal nerves or the cord itself are damaged.

Organization of spinal nerves into plexuses

In several regions of the body, the spinal nerves are "collected" into groups that form a plexus. From these plexuses, the nerves are distributed to muscles and skin (Fig. 14.9).

THE CERVICAL PLEXUS (Fig. 14.10). Cervical nerves 1–4 combine to form the cervical plexus that

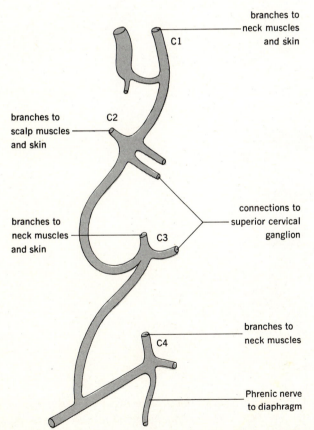

FIGURE 14.10   The cervical plexus. C, cervical spinal nerves.

lies in the upper part of the neck under the sternocleidomastoid muscle. Nerves pass from the plexus to muscles and skin of the scalp and neck, and to the diaphragm by way of the phrenic nerves. A broken neck may thus result in respiratory paralysis through injury to the phrenic nerves.

THE BRACHIAL PLEXUS (Fig. 14.11). The anterior roots of spinal nerves C5-T1 combine to form the brachial plexus that lies in the anterior base of the neck just superior to the clavicle. It sends nerves to the skin and muscles of the upper appendage. Five major nerves exit from the plexus.

The *axillary nerve* supplies the shoulder joint and deltoid muscle.
The *musculocutaneous nerve* supplies the structures of the anterior upper arm.
The *median nerve* supplies the anterior forearm.
The *ulnar nerve* serves the same areas as the median nerve.
The *radial nerve* supplies the posterior upper arm, forearm, and hand.

The nerves are superficial at the axilla. Injury here may result in paralysis. The "crazy bone" is the ulnar nerve as it passes over the medial epicondyle of the humerus.

THE INTERCOSTAL NERVES. Thoracic nerves 2–12 form the intercostal nerves which supply the intercostal and abdominal muscles. A named plexus is not formed.

THE LUMBAR PLEXUS (Fig. 14.12). Lumbar nerves 1–4 combine to form the lumbar plexus that lies within the posterior part of the psoas muscle. Three main nerves exit from the plexus.

The *femoral nerve* supplies the skin and muscles of the anterior thigh.
The *lateral cutaneous nerve* supplies the skin and muscles of the upper lateral thigh.
The *obturator nerve* supplies the skin and muscles of the medial thigh.

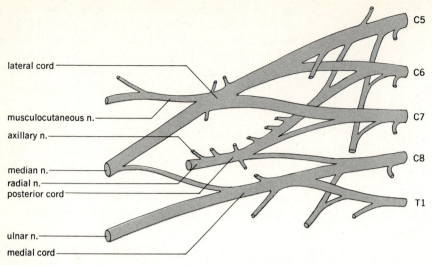

FIGURE 14.11   The brachial plexus. C, cervical; T, thoracic spinal nerves.

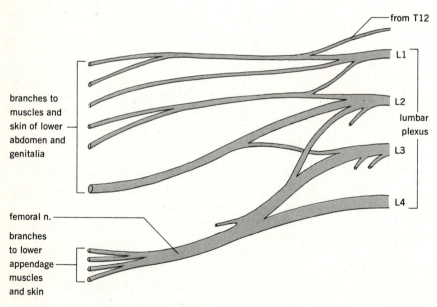

FIGURE 14.12   The lumbar plexus. T, thoracic; L, lumbar spinal nerves.

THE SACRAL PLEXUS (Fig. 14.13). Lumbar nerves 4 and 5, and sacral nerves 1–3 combine to form the sacral plexus which lies in the middle portion of the false pelvis. Some anatomists consider the lumbar and sacral plexuses to be one large plexus, which they call the *lumbosacral plexus*. There is some variation as to which nerves contribute to the two plexuses, particularly the lower lumbar

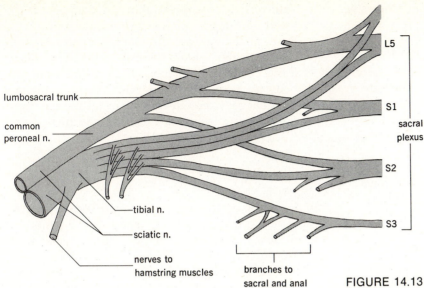

lumbosacral trunk

common
peroneal n.

tibial n.

sciatic n.

nerves to
hamstring muscles

branches to
sacral and anal
muscles

L5

S1

sacral
plexus

S2

S3

FIGURE 14.13 The sacral plexus. L, lumbar; S, sacral spinal nerves.

nerves. The major nerve of the sacral plexus is the *sciatic nerve*, which forms several important branches.

The *nerve to the hamstring muscles* ("hamstring nerve") arises from the sciatic nerve in the thigh and supplies the skin and muscles of the posterior thigh.

The *tibial nerve* arises as one branch from the sciatic just above the knee and supplies the skin and muscles of the knee joint and posterior leg.

The *common peroneal nerve* is the other branch of the sciatic just above the knee. It continues down the leg, giving branches that supply the skin and muscles of the knee, anterior and lateral leg, and foot.

Clinical considerations

Cord and nerve injury

Dislocation of vertebrae and stab or bullet wounds may result in damage to the cord with consequent loss of function. Specific loss depends on what particular cord area and tract(s) are involved. If the cord is severed (complete section) all sensory and motor function will be lost below the cut. Paralysis is of the *spastic* type, characteristic of an upper motor neuron lesion.

Damage to the lower motor neurons results in a *flaccid* paralysis, restricted to the muscles served by the damaged nerve. Regeneration of upper motor neurons occurs, but the chances that normal connections will be reestablished are mimimal; thus, restoration of function is poor. Peripheral nerves may regenerate, but the new connections established are more widely distributed and less precise than originally, so that function may return but will be less exact than before.

HEMISECTION of the cord (cutting halfway through the cord) produces the *Brown-Sequard syndrome*. This is characterized by:

Loss of voluntary muscle movement, touch, and pressure on the same side of the cut;

Loss of pain, heat, and cold on the opposite side of the cut.

Remembering the spinal tracts and the areas supplied by the spinal nerves, it is easy to determine the spinal level of damage and the side of damage.

Spinal shock

If the cord is completely cut or is severely traumatized (injured), certain losses or alterations of function will result.

Permanent loss of voluntary muscular activity, and sensation will occur in the body segments below the level of the cut.

The development of *spinal shock* will occur. It is a *temporary synaptic depression* that results in loss of reflex activity below the level of the cut. The length of time the state lasts depends on the position of the animal in the phylogenetic scale, and extends from minutes for an animal like the frog to months in the human. As recovery from shock proceeds, there is a typical pattern of reappearance of reflex activity.

Knee jerk and other simple reflexes reappear.

The flexion and crossed extension reflexes reappear.

Visceral reflexes, including bladder evacuating reflexes, and sexual reflexes reappear (erection may occur but rarely ejaculation).

The human is subject to a postural hypotension (low blood pressure) that results as body position changes from lying, to sitting or standing postures. The individual whose cord has been damaged no longer has connections with the vasoconstrictor centers that are located in the lower part of the brain stem (medulla oblongata). Postural changes are not met by vasoconstriction (narrowing of blood vessels) to maintain blood pressure. Since there is no sensation from the body parts below the level of section, the individual may also develop skin ulcerations from pressure and ischemia and not be aware of them.

Spinal puncture (tap)

Since the cord stops at about the second lumbar vertebra, it is possible to insert a needle into the subarachnoid space and withdraw spinal fluid. The puncture is usually done at the level of the fourth lumbar vertebra. The fluid may be withdrawn for analysis, or may be replaced with an equivalent volume of anesthetic to create *spinal anesthesia*. Puncture must be done carefully to avoid damaging nerves of the cauda equina.

Intramuscular injections

These injections involve giving medications by insertion of a needle into a muscle. Large nerves may lie in the area into which the injection is to be made, and their path and size must be recalled to avoid injuring them. Branches of the axillary nerve lie beneath the deltoid muscle, and injections should be made in the upper half of this muscle; the sciatic nerve lies beneath the buttock and lateral thigh muscles and, injections should be made in the superior lateral quadrant of the muscle; the femoral nerve lies under the anteromedial thigh muscles. Paralysis or sensory loss may result if the nerve is damaged.

Summary

1. The spinal cord serves as an area for organization of sensory input, reflex control, transmission of impulses to and from the brain, and for sending impulses to muscles and glands.

2. The spinal cord
   a. Is located in the vertebral canal.
   b. Has, in the adult, a length of 45 cm, a diameter of 2 cm.
   c. Is surrounded by meninges: dura mater, protective; arachnoid, enclosing a fluid; pia mater, a vascular membrane on the organ.
   d. Has two enlargements, seven sulci, and one fissure.
   e. Shows gray and white matter in cross section; each has several columns and funiculi respectively.
   f. Contains functional areas known as tracts (see Table 14.1)

3. Impulses reach the spinal cord over peripheral nerves carrying sensory information; the posterior roots and posterior lateral sulci mark entry of such nerves.

4. Internuncial neurons are found within the cord; they provide amplification of input, inhibit passage of impulses, and allow a wider response to stimuli.

5. Output from the cord is over anterior roots to muscle and glandular tissue.

6. The cord is a primary integrator and controller of reflex activity.
   a. Myotatic reflexes are two neuron reflexes, have a muscle spindle as the primary receptor, and cause contraction of muscles mainly to support the erect body.
   b. Flexion and extension reflexes originate in skin and body organs, and are mainly protective in nature.
   c. Righting reflexes are partly controlled by the cord, but require the cerebellum for coordination.

7. The cord gives rise to 31 pairs of spinal nerves that connect to the cord through the posterior and anterior roots.
   a. They are segmentally distributed to the body.
   b. They form four major plexuses, that give rise to important nerves. They are:
      1) Cervical plexus: nerves to neck and shoulder muscles; phrenic nerve to diaphragm.
      2) Brachial plexus: axillary, musculocutaneous, median, radial and ulnar nerves to the upper appendage.
      3) Lumbar plexus: femoral, lateral cutaneous, and obturator nerves to thigh.
      4) Sacral plexus: sciatic, with branches to hamstrings; tibial and common peroneal nerves to thigh and leg.

8. The spinal nerves contain four functional components.
   a. Somatic afferents: sensory for senses other than sight, hearing, smell, and taste.
   b. Somatic efferents: to skeletal muscle.

  c. Visceral afferents: from blood vessels and body organs.

  d. Visceral efferents: to smooth and cardiac muscles and glands.

9. Cord injury produces motor and sensory losses in specific patterns according to tract involved and extent of injury.

10. Spinal shock is characterized by loss of sensory and motor function, then return of cord-controlled activity in a specific order.

11. Spinal puncture is used for diagnostic and/or anesthetic purposes.

12. Intramuscular injections should be made only by one who knows the location of the nerves in the area.

Questions

1. Describe a cross section of the cord, including anatomical features and the locations of the major spinal tracts.

2. If an individual suffers loss of pain, heat, and cold on his right side, muscle paralysis on his left side, and these changes are restricted to his anterior forearm, predict at what level and which side his spinal cord has been injured.

3. Describe some reflexes controlled by the cord.

4. What are the names of the functional components of the spinal nerves, and what does each supply?

5. What are/is:
   a. Dorsal roots.
   b. Plexuses.
   c. Conus medullaris.
   d. Anterior median fissure.
   e. Arachnoid.

6. Describe how it is possible to deliver a spinal anesthetic.

7. What cautions must be observed in giving an intramuscular injection. Relate this to the nerves exiting from the spinal nerve plexuses.

Readings

Anson, Barry J. *Morris' Human Anatomy*. McGraw-Hill. New York, 1966.

Austin, G. *The Spinal Cord*. Thomas Pubs. Springfield, Ill. 1971.

Baker, A. B. *An Outline of clinical Neurology*. Wm. C. Brown Book Co. Dubuque, Iowa, 1965.

Barr, Murray L. *The Human Nervous System; An Anatomical Viewpoint*. Harper & Row. New York, 1972.

Chapter 15

The Brain and Cranial Nerves

Objectives

After studying this chapter, the reader should be able to:

- Define what the term brain includes and list its three major parts.

- Name, briefly describe the structure, and give the functions of the medulla, pons, midbrain, and diencephalon.

- Distinguish between vital and nonvital centers of the medulla and explain why they are so named.

- Point out some effects of lesions of the medulla.

- Describe the importance of the midbrain colliculi in reflex movements associated with vision and hearing.

- Comment briefly on the organization of the thalamic nuclei, their functions, and the importance of the thalamus to sensory functions.

- Explain why the hypothalamus is an important center for homeostasis of body functions.

- Describe the structure of the cerebellum and its functions in motor control.

- Point out some signs of disordered cerebellar function.

- List the lobes and fissures of the cerebrum and, for the functional areas of the cerebral lobes, assign numbers and give the major functions centered in each region.

- Describe the types of fibers composing the cerebral medullary body and their functions.

- Name the major basal ganglia and give their functions.

- Tell what the limbic system does for the body.

- Explain the difference between pyramidal and extrapyramidal systems.

- Name the 12 pairs of cranial nerves, give their composition and functions, and show how to test for cranial nerve function.

The term "brain" is generally used to refer to that part of the central nervous system enclosed within the cranial cavity of the skull. It weighs about $1\frac{1}{3}$ kg, or about 3 pounds. It is divided into three major portions, each with several subdivisions. At this time, the reader may wish to review the basic development of the nervous system as presented in Chapter 13, with particular attention to the primary subdivisions of the primitive brain and their derivatives.

The BRAIN STEM is that part of the brain continuous with the spinal cord through the foramen magnum. From inferior to superior, the subdivisions of the stem include the *medulla oblongata*, the *pons*, the *midbrain*. The DIENCEPHALON includes the *thalamus* and *hypothalamus*. The stem is the smallest of the three major brain divisions in terms of bulk and supports the other portions. It also acts as the point of origin or termination for many cranial nerves.

The CEREBELLUM lies on the posterior aspect of the lower brain stem. It is a portion of the brain concerned with coordination of muscular activity, and is second in mass to the brain stem.

The CEREBRUM lies atop the brain stem, forms by far the largest portion of the brain, and is composed of two hemispheres. It is this portion of the brain that sets us off from infrahuman forms, with the abilities to think and manipulate our environment.

The various subdivisions of the brain, with additional features to be discussed later, are shown in Figure 15.1.

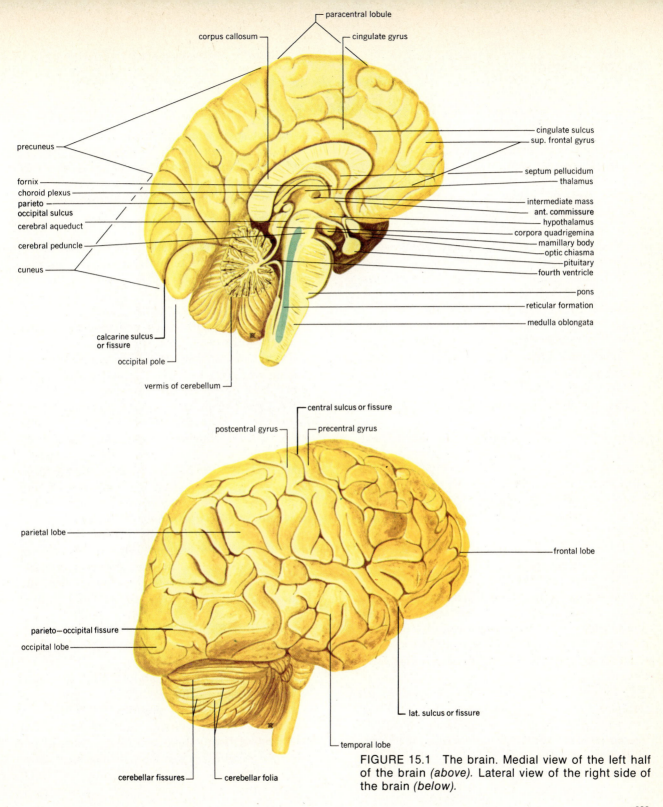

paracentral lobule

corpus callosum

cingulate gyrus

cingulate sulcus
sup. frontal gyrus

precuneus

septum pellucidum
thalamus

fornix
choroid plexus
parieto
occipital sulcus
cerebral aqueduct

intermediate mass
ant. commissure
hypothalamus
corpora quadrigemina
mamillary body
optic chiasma
pituitary
fourth ventricle

cerebral peduncle

cuneus

pons
reticular formation
medulla oblongata

calcarine sulcus
or fissure

occipital pole

vermis of cerebellum

central sulcus or fissure

postcentral gyrus

precentral gyrus

parietal lobe

frontal lobe

parieto—occipital fissure

occipital lobe

lat. sulcus or fissure

cerebellar fissures

cerebellar folia

temporal lobe

FIGURE 15.1 The brain. Medial view of the left half
of the brain *(above)*. Lateral view of the right side of
the brain *(below)*.

The brain stem

The medulla oblongata

STRUCTURE. The medulla oblongata, sometimes referred to simply as the medulla, forms the inferior 3 cm or so of the brainstem, and is continuous with the spinal cord. Spinal tracts that ascend from the cord, or descend from higher parts of the brain, undergo an extensive rearrangement in the medulla. There is no specific organization of gray and white matter as was seen in the cord, but are "all mixed up." The decussation (crossing) of the lateral corticospinal tract occurs in the medulla, and this is responsible for the fact that one side of the brain controls muscles on the other side of the body. Ascending tracts may also undergo crossing in the medulla; this occurs with the tracts carrying the fibers of the posterior funiculus. Sprinkled among the fibers of these tracts are groups of neuron cell bodies that form the VITAL CENTERS and NONVITAL CENTERS of the medulla. Cranial nerves 9–12 attach to the medulla.

FUNCTIONS. The vital centers of the medulla include groups of neurons that control processes essential for survival of the organism. Among these are:

The CARDIAC CENTERS. There are two types of cardiac centers: *cardioaccelerator centers* and *cardio-inhibitory centers*. They receive input from the body's blood vessels, heart, lungs, cerebrum, and other areas. They provide for reflex control of heart rate according to body activity levels, blood pressure, and levels of carbon dioxide and oxygen in the bloodstream.

The RESPIRATORY CENTERS. Two of several paired respiratory centers lie in the medulla. The *inspiratory centers* provide stimuli to the diaphragm and external intercostal muscles that ultimately cause the drawing of air into the lungs (inspiration). *Expiratory centers* provide part of the influence necessary to interrupt inspiratory activity and permit air to be expired or driven out of the lungs.

VASOMOTOR CENTERS include *vasoconstrictor* and *vasodilator centers* that, respectively, cause a narrowing and opening of muscular blood vessels in the body. Important control of the blood pressure is thus provided.

Nonvital centers located in the medulla include those for integration of such activities as *sneezing, coughing,* and *vomiting.*

CLINICAL CONSIDERATIONS. Damage to the medulla may occur with blows delivered to the back of the head or neck. If severe enough, damage to the vital centers may occur, with development of respiratory paralysis, fall of blood pressure and shock, and death. Before the advent of vaccines against poliovirus, one area of infection by the virus was the medulla and *bulbar polio* was said to have occurred. Respiratory paralysis, through involvement of the inspiratory centers, often resulted in the victim's confinement to an "iron lung" a device that alternately compressed and decompressed the chest to cause air movement into and out of the lungs. If the sensory and motor tracts of the medulla are involved, there will obviously be sensory loss and deficits in muscle movement, with the severity and exact nature depending on extent and location of the trauma.

The pons

STRUCTURE. The pons is about 2.5 cm in length and is superior to the medulla. It forms a conspicuous bulge on the anterior brain stem. A BASAL or ANTERIOR PORTION receives fibers from one cerebral hemisphere and sends fibers to the opposite cerebellar hemisphere via the CEREBELLAR PEDUNCLES, stout masses of nerve fibers.

The posterior portion or TEGMENTUM of the pons contains ascending and descending fiber tracts, and the neuron groups forming the nuclei of cranial nerves 5–8. A respiratory center called the PNEUMOTAXIC CENTER is also located in the posterior pons.

FUNCTIONS. The basal portion of the pons acts as a synaptic or relay station for motor fibers conveying impulses from the cerebrum to the cerebellum. The cranial nerve nuclei provide for sensory and motor innervation of the skin and certain muscles of the head, and for hearing, taste, and eye movement. The pneumotaxic center is part of the system regulating the rhythm and depth of breathing.

CLINICAL CONSIDERATIONS. Lesions within the basal portion of the pons will lead to motor deficits—that is, loss of coordination of voluntary muscle movement—and posture becomes difficult to maintain. Interruption of fiber tracts of the tegmentum will cause motor deficits if the descending fibers are interrupted, and sensory loss if the ascending fibers are damaged. Specific loss will depend on extent and location of the lesion. Involvement of the pontine pneumotaxic center will lead to the development of *apneustic breathing;* this is characterized by sustained inspirations, then a sudden expiration, with another inspiration that is held for a period of time. The loss of the rhythm-setting of breathing by this center in part explains this type of breathing.

The midbrain

STRUCTURE. The midbrain is a wedge-shaped portion of the brainstem between the pons and thalamus. Its anterior portion consists chiefly of two large bundles of nerve fibers called the CEREBRAL PEDUNCLES. These peduncles carry the bulk of the efferent motor fibers from the cerebrum to the cerebellum and spinal cord. The posterior portion of the midbrain contains the paired SUPERIOR and INFERIOR COLLICULI (collectively called the *corpora quadrigemina*), and the nuclei of cranial nerves 3 and 4.

FUNCTION. The midbrain, by way of the peduncles, serves as a motor relay station for the fibers from cerebrum to cerebellum and spinal cord. The colliculi integrate a variety of visual (superior colliculi) and auditory (inferior colliculi) motor reflexes, including those concerned with avoiding objects that are seen and turning the head to achieve the greatest benefit from sights and sounds. The midbrain is also involved, to some extent, in righting reflexes that require visual stimuli.

CLINICAL CONSIDERATIONS. Lesions in the anterior midbrain produce disturbances in voluntary movement by way of peduncle involvement. Posterior midbrain involvement is evidenced by inability to respond normally to auditory and visual stimuli and by inability to properly maintain normal orientation of the body in space.

The diencephalon: thalamus and hypothalamus

The thalamus

STRUCTURE. Each thalamus measures about 3 cm from front to back and 1.5 cm in height and width, and together they comprise about four-fifths of the diencephalon. The two thalami are connected across the midline by the *intermediate mass.* De-velopmentally, three portions differentiate in the thalamus primordia to give an upper *epithalamus,* a central *dorsal thalamus,* and a lower *ventral thalamus.* Several groups of nuclei (neuron cell bodies) are found in each area, with the dorsal and ventral thalami (the parts most often referred to when simply using the word *thalamus*) containing

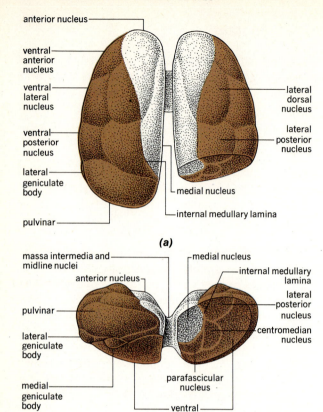

(a)

(b)

FIGURE 15.2 The major thalamic nuclei. *(a)* View from the dorsolateral direction. *(b)* Posterior view.

the greatest number (Fig. 15.2). Fibers entering the thalamus are primarily sensory in nature. Outgoing fibers pass primarily to the sensory and motor regions of the cerebral cortex.

FUNCTIONS. The epithalamus includes the PINEAL BODY *(epiphysis)* and the paired HABENULAR NUCLEI anterior to the pineal body. The epithalamus receives fibers from the olfactory regions of the cerebrum and the limbic system (the latter concerned with emotions and their expressions). The epithalamus sends fibers to the midbrain and the reticular formation so that olfactory stimuli may influence emotional behavior

such as that involved in food seeking or mating. The pineal body is about 7 mm long by 5 mm in diameter, and may be concerned with time of onset of puberty and establishment of circadian rhythms.

The dorsal and ventral thalami act primarily as receiving stations for fibers carrying visual and auditory stimuli and those concerned with pain, heat, cold, touch, and pressure.

It would be of no use to list all the nuclei of the thalamus and their functions. Several nuclei should, however, be emphasized to illustrate and amplify the functions mentioned previously.

The MEDIAL GENICULATE BODY receives fibers from the cochlea and relays impulses for hearing to the cerebral auditory cortex.

The LATERAL GENICULATE BODY receives fibers from the retina and relays impulses for vision to the cerebral visual cortex.

The RETICULAR NUCLEUS forms part of a system designed to maintain the waking or conscious state of the organism.

The VENTRAL POSTERIOR NUCLEUS receives the spinothalamic tracts carrying pain, heat, and cold sensations, and fibers conveying touch and pressure from the medulla.

The VENTRAL LATERAL and VENTRAL ANTERIOR NUCLEI receive fibers from the cerebellum and basal ganglia and send fibers to the cerebrum. These impulses influence *motor activity.*

The PULVINAR, LATERAL POSTERIOR, MEDIAL, and ANTERIOR NUCLEI are concerned with reception and analysis of sensory information that leads to expression of emotions, intellectual function, memory storage, and "moods" and "feelings." Retention of sensory input as memory traces would appear to be essential in order to be able to compare stimuli to see if they are new (experience?).

In summary, the thalamus operates primarily in the reception and sorting out of *sensory* information at primary *and* intellectual levels.

CLINICAL CONSIDERATIONS. Damage to the *sensory* portions of the thalamus produces the THALAMIC SYNDROME. Lesions of this part are most commonly vascular in origin (a vessel blocked or ruptured), and the result is little or no appreciation of sensations until a high and critical point is reached. Then the sensation "bursts," as through a wall, and may be intolerable when this occurs (especially the sense of pain). Selective destruction of the ventral posterior nucleus may ameliorate the condition. In older individuals, vascular lesions of the motor portions of the thalamus may cause the development of tremor and muscular rigidity (e.g., Parkinson's disease). The administration of L-DOPA (*l*evorotatory *d*ihydroxy*p*henyl*a*lanine) may alleviate such symptoms.

The hypothalamus

STRUCTURE. The hypothalamus weighs about 4 grams and contains many nuclei, of which the most important ones are shown in Figure 15.3. It receives fibers from the thalamus, cerebral cortex, brainstem, and limbic system, and sends fibers or chemicals to the pituitary gland and the nuclei in the brainstem controlling visceral func-

tion (e.g., the vital centers of the medulla and cranial nerve nuclei).

FUNCTIONS. The cells of the hypothalamic nuclei not only have nerve connections with many body areas but respond to the properties of the blood reaching them. Among the blood properties to which the area responds are pH, osmotic pressure, and glucose levels. Output from the hypothalamus is designed to control many body processes concerned with MAINTENANCE OF HOMEOSTASIS.

Functions of the hypothalamus include:

Temperature regulation. Human body temperature is normally maintained within a degree or two of 98.6°F or 37°C. A proper balance between heat production and loss is maintained by *heat loss and heat gain centers* in the hypothalamus. The heat loss center causes dilation of skin blood vessels, sweating, and decreased muscle tone, all of which increase heat loss or reduce heat production. The heat gain center causes skin vessel constriction, shivering, and cessation of marked sweating, all of which conserve body heat or increase heat production.

Regulation of water balance. Hypothalamic neurons continually monitor blood osmotic pressures and adjust the tonicity of the body fluids by *ADH* (anti-

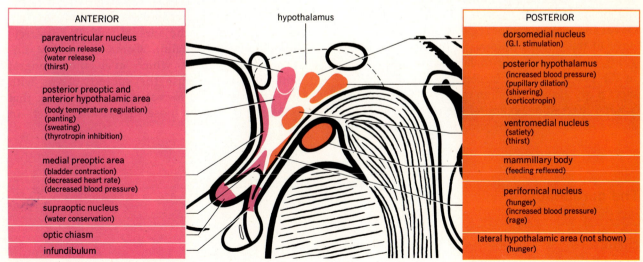

ANTERIOR

paraventricular nucleus
(oxytocin release)
(water release)
(thirst)

posterior preoptic and anterior hypothalamic area
(body temperature regulation)
(panting)
(sweating)
(thyrotropin inhibition)

medial preoptic area
(bladder contraction)
(decreased heart rate)
(decreased blood pressure)

supraoptic nucleus
(water conservation)

optic chiasm

infundibulum

hypothalamus

POSTERIOR

dorsomedial nucleus
(G.I. stimulation)

posterior hypothalamus
(increased blood pressure)
(pupillary dilation)
(shivering)
(corticotropin)

ventromedial nucleus
(satiety)
(thirst)

mammillary body
(feeding reflexed)

perifornical nucleus
(hunger)
(increased blood pressure)
(rage)

lateral hypothalamic area (not shown)
(hunger)

FIGURE 15.3   The major nuclei of the hypothalamus.

diuretic hormone) production. The hormone permits greater water reabsorption from the kidney tubules.

Control of pituitary function. At last count, nine chemicals influencing pituitary function have been isolated from the hypothalamus. Properly termed *pituitary regulating factors*, these chemicals are produced in the hypothalamus as a result of blood-borne and nervous stimuli, are passed by blood vessels to the pituitary, and there stimulate or inhibit pituitary production and release of hormones.

Control of food intake. Initiation and cessation of feeding is controlled by hypothalamic *feeding and satiety centers.*

Regulation of gastric secretion. The amount of gastric juice produced by the stomach is increased by hypothalamic stimulation. Thus, emotions may trigger release of gastric juice when there is no food in the stomach and may lead to ulcer development.

Emotional expression. The hypothalamus is one part of a system necessary for expression of reactions of rage and anger. It also appears necessary for sexual behavior.

CLINICAL CONSIDERATIONS. Depending on where damage occurs in the hypothalamus, and its extent, a wide variety of symptoms may develop. Those associated with disturbances of water balance, temperature regulation, and emotions are the most common. *Diabetes insipidus* is the production of a large volume of very dilute urine as a result of diminished ADH secretion; inability to maintain near constant body temperature indicates hypothalamic damage.

The reticular formation

Located in the brainstem (*see* green area in Fig. 15.1), extending from the medulla to the lower border of the thalamus, is a column of gray matter distinct from that described in previous sections. This column is called the RETICULAR FORMATION and in it are found the *reticular nuclei,* designated specifically by their position and the portion of the brain in which they lie. The formation receives fibers from motor regions of the brain and most of the sensory systems of the body. Its outgoing fibers pass to the thalamus and from there to the cerebral cortex in general. Stimulation of most parts of the formation results in an immediate and marked activation of the cortex, a state of alertness and attention, and, if the organism is sleeping, immediate awakening. The upper portion of the formation *plus* its pathways to thalamus and cerebral cortex have been designated the RETICULAR ACTIVATING SYSTEM (RAS). Maintenance of the waking state is achieved by a positive feedback system in which RAS stimulation causes the cortex to raise the activity level of the RAS. This stimulates muscle tone, visceral activity, and causes the adrenal medulla to secrete epinephrine (adrenalin). Coma, a state of unconsciousness from which even the strongest stimuli cannot arouse the subject, occurs with brainstem reticular formation damage. Drugs that make us sleepy, or those that stimulate, act—in part—on the RAS.

The cerebellum

Structure

The cerebellum (*see* Fig. 15.1) lies on the posterior aspect of the pons and medulla. Grossly, it may be seen to be composed of a centrally placed VERMIS and two CEREBELLAR HEMISPHERES. It is divided into two major LOBES, *anterior* and *posterior,* and several smaller LOBULES, by fissures and sulci (Fig. 15.4). The surfaces of the organ are folded to create numerous small *folia,* each of which contains an outer CORTEX of gray matter and an inner MEDULLARY BODY. The term *arbor vitae* refers to the tree-like arrangement of the cerebellar white matter (*see* Fig. 15.1).

Microscopically, a section of a folium shows the same structure anywhere in the organ (Fig. 15.5). The characteristic cell of the cerebellum is the flask-shaped PURKINJE CELL, whose dendrites form the bulk of the substance of the outer MOLECULAR LAYER of the cerebellar cortex. An inner GRANULAR LAYER is composed of small *granule cells.* Some 30 million FUNCTIONAL UNITS (*see* Fig. 15.5) are formed by the cells of the cortex. Input to these functional units occurs over what are called *climbing* and *mossy fibers* arriving at the unit from the cerebral cortex, basal ganglia, and sensory pathways from inner ear and skeletal muscles. Deep within the medullary body are the CENTRAL NUCLEI of the cerebellum (Fig. 15.6). Axons of the Purkinje cells terminate in these nuclei, and the nuclei in turn send fibers to the brainstem. From the brainstem, fibers are sent to the cerebral cortex, thalamus, basal ganglia, and—over spinal tracts—to skeletal muscles.

Functions

In general terms, the cerebellum operates entirely at the subconscious level, continually monitoring and adjusting motor activities originating in brain or peripheral receptors. It adjusts the rate and speed of movements, the direction of a movement, the force required for the movement, and stops the motion at the desired point.

FUNCTIONS IN VOLUNTARY MOVEMENT. The impulses that initiate voluntary movements probably originate in the basal ganglia or cerebral cortex, and constitute the *intent* of the movement.

Signals from these areas pass also to the cerebellum, which compares the actual movement, using information provided via receptors in the muscles themselves, to intent. The cerebellum corrects the rate, force, and direction of the movement as required. Such a system provides several functions:

Error control. By comparing intent and performance, the cerebellum ensures that the movement is performed accurately in regard to force and direction.

Damping functions. Most body movements are pendular in nature, and oscillations would occur if a movement was not stopped at its intended termination. The cerebellum cancels out the oscillatory nature of a movement and halts it at its intended point.

Prediction. The cerebellum predicts *when* a motion should be slowed so as to have it stop at its intended point, and ensures that proper progression of muscular contraction occurs so that the motion is both directed and purposeful.

FUNCTIONS IN INVOLUNTARY MOVEMENT. The "involuntary functions" of the cerebellum are concerned with maintenance of muscle tone, posture, equilibrium, and orientation of the body in the face of forces tending to upset our relationship to the external world. The "services" provided here are basically the same as those provided for voluntary activity: error control, damping, and prediction. This time, the response is a reflex that maintains our balance and equilibrium. We thus "lean into a turn," forwards on acceleration or backwards on deceleration. The input leading to these responses is provided mainly by the structures of the inner ear.

Clinical considerations

Because the structure of the cerebellum is essentially the same in all parts, damage to the organ is reflected by *extent,* instead of *location,* of the damage. In short, all parts are involved in the previously described functions, and severity of symptoms depends on how much tissue is de-

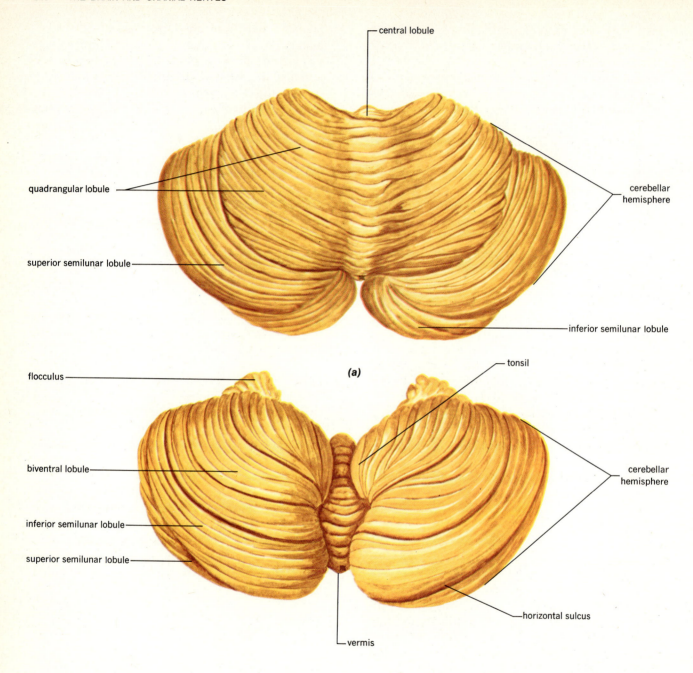

FIGURE 15.4   The cerebellum. (a) Posterior surface. (b) Anterior surface.

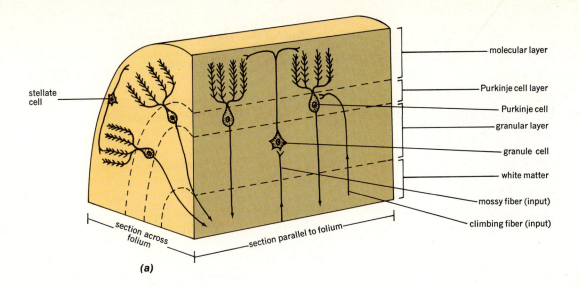

stellate cell

molecular layer

Purkinje cell layer

Purkinje cell

granular layer

granule cell

white matter

mossy fiber (input)

climbing fiber (input)

section across folium

section parallel to folium

(a)

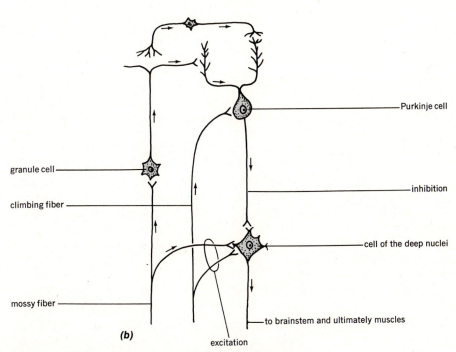

Purkinje cell

granule cell

climbing fiber

inhibition

mossy fiber

cell of the deep nuclei

to brainstem and ultimately muscles

(b)

excitation

FIGURE 15.5 (a) The basic histology of the cerebellum. (b) The functional unit of the cerebellum. [(a) redrawn from Barr.]

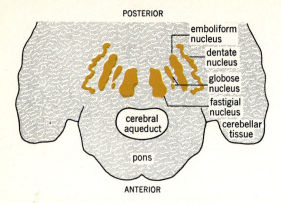

FIGURE 15.6   The deep nuclei of the cerebellum as seen in a cross section at the level of the pons.

stroyed. The body *is* represented in a particular fashion in the cerebellum, with upper parts anteri-

orly, and lower parts posteriorly, so that different regions of the body may be involved with damage.

As expected, signs of cerebellar dysfunction appear in coordination and accuracy of movements, and in muscle tone and equilibrium. Some signs of cerebellar disorder are presented below.

Asthenia. This term refers to lack of muscular strength.

Hypotonia. The muscles are flabby, feeling like half-filled hot water bottles.

Dysmetria. A movement "overshoots" its intended end point.

Ataxia. Muscle incoordination, with tremors, improper progression of components of a motion, and inability to perform rapid movements (e.g., tapping a finger) smoothly.

The cerebrum

The cerebrum is the largest portion of the brain, accounting for about 80 percent of the total brain weight. It is convoluted as viewed from the outside, with upfolds known as GYRI (sing., gyrus), and shallow indentations known as SULCI (sing., sulcus). At intervals, deep indentations, called FISSURES, separate the cerebral hemispheres into LOBES. These features are shown in Figure 15.7. A layer of gray matter, varying in thickness between 2.5 and 4 mm, forms the outer CEREBRAL CORTEX, and encloses a MEDULLARY BODY of white matter. Buried deep within the medullary body lie the BASAL GANGLIA of the cerebrum.

The cerebral cortex

STRUCTURE. Many years ago, Korbinian Brodmann attempted to correlate cellular differences in the cerebral cortex with localization of function. His numbered maps (Fig. 15.8) of areas of the cerebral cortex are used today when referring to specific cortical areas, even though

there may not be a clear-cut structural-functional relationship that can be demonstrated. The areas, as described by Brodmann, *do* imply that a given area is involved with a particular aspect of function by virtue of the connections they make.

Within the cortex, neurons appear to be arranged in vertical columns (Fig. 15.9), with spread to adjacent columns provided by cells oriented between the columns at the surface. The columns analyze and store information and provide us with our various intellectual capacities.

FUNCTIONS. Combining the lobes of the cerebrum with the Brodmann numbered areas, and using the results of stimulation experiments and injury, the following regions of the cortex may be described according to function.

The frontal lobes. Anterior to the central fissure is area 4, designated as the PRIMARY MOTOR AREA. This area controls the contraction of individual muscles or groups of muscles in the body. The body is represented upside-down in area 4, with

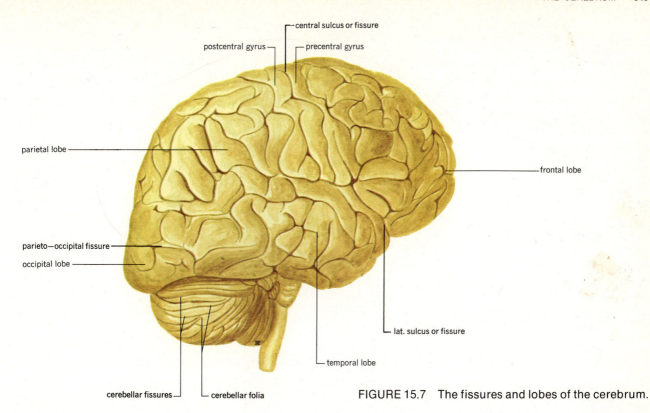

central sulcus or fissure

postcentral gyrus — — precentral gyrus

parietal lobe

frontal lobe

parieto—occipital fissure

occipital lobe

lat. sulcus or fissure

cerebellar fissures — — cerebellar folia

temporal lobe

FIGURE 15.7   The fissures and lobes of the cerebrum.

those areas requiring more complex movements receiving a larger representation (Fig. 15.10). A SUPPLEMENTARY MOTOR AREA lies on the medial surface of each hemisphere (areas 24, 31). Both areas control muscles on the opposite side of the body, as the bulk of their outgoing fibers decussate (cross) in the medulla oblongata.

Area 6 is designated as the PREMOTOR AREA, and may be concerned with *learned* motor activity, because lesions here interfere with performance rather than overt movement.

Area 8, if stimulated, causes scanning types of *eye* movements, and is called the FRONTAL EYE FIELD.

Areas 9–12, lying in the anterior part of the frontal lobes, are designated as the PREFRONTAL CORTEX. In humans, these areas appear to be concerned with intellectual function, emotional display, and social and moral values. Lesions here alter the personality of the individual.

The parietal lobes. Behind the central fissure lie areas 3, 1, and 2, designated as the GENERAL SENSORY AREA. These areas serve as the termination for impulses of pain, heat, cold, touch, and pressure, where the nature of the stimulus is determined and its origin localized. The body is represented upside-down in this area, as in area 4 (Fig. 15.11). Again, those body areas having a greater density of sensory receptors are given a greater area of representation. PARIETAL ASSOCIATION AREAS (5, 7a, b), lying posterior to the primary sensory areas, provide interpretation of textures, shapes, and degrees of sensation (as in degrees of heat or cold).

The occipital lobes. The VISUAL AREA (17) occupies the greater part of the occipital lobe. Fibers from the retina terminate here. VISUAL ASSOCIATION AREAS (18, 19) are concerned with visual experiences, and integrate eye movements designed to fix the eyes on objects viewed.

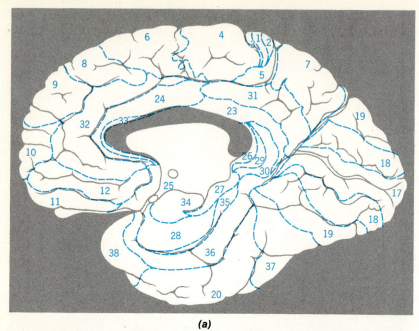

(a)

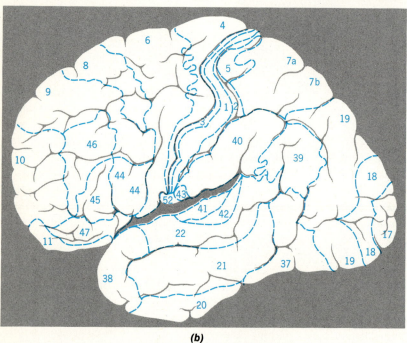

(b)

FIGURE 15.8 The cytoarchitectural areas according to Brodmann. Within certain numbered regions are specific functional areas. *(a)* Medial aspect of cerebrum. *(b)* Lateral aspect of cerebrum.

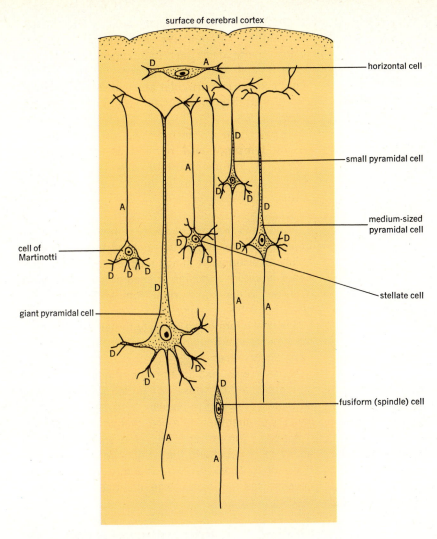

surface of cerebral cortex

horizontal cell

small pyramidal cell

medium-sized pyramidal cell

cell of Martinotti

stellate cell

giant pyramidal cell

fusiform (spindle) cell

FIGURE 15.9   The cellular arrangement of the cerebral cortex. A = axon; D = dendrite.

The temporal lobes. Areas 41 and 42 are designated as the AUDITORY AREAS and AUDITORY ASSOCIATION AREAS. The auditory areas receive fibers from the organ of hearing, the *cochlea*. The association areas are important components of the systems for language function (to a large degree, we imitate what we hear to acquire language functions), and they also store memories of both auditory and visual stimuli.

Aiding in language functions are the BROCA SPEECH AREAS, 44 and 45. The areas include portions of the primary motor areas, to enable articulation of speech, and parts of the auditory areas. Ensurance of the proper use of the organs of speech is thus provided.

CLINICAL CONSIDERATIONS. Lesions or damage to any motor area will interfere with the ability

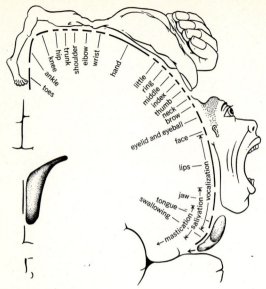

FIGURE 15.10 Cortical location of motor functions; motor homunculus.

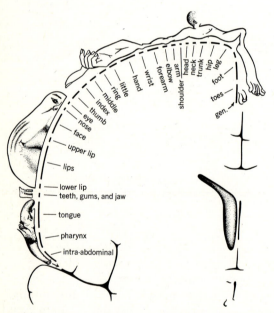

FIGURE 15.11 Cortical location of sensory functions; sensory homunculus.

to perform movements. A primary motor area lesion leads to *paralysis,* or the inability to volun-

tarily move muscles. Sensory area lesions destroy the ability to perceive sensations. Prefrontal lesions were described in connection with that area. Visual and auditory area lesions destroy ability to see in particular areas of the visual field, and the ability to perceive sounds of particular pitches.

Of particular interest are the effects of damage to the temporal and parietal association areas. Temporal lesions produce loss of memory of things learned, and occur in one or more of three forms.

AGNOSIA is the loss of ability to recognize familiar objects. Failure may manifest itself through lack of ability to distinguish things touched *(astereognosis),* heard *(auditory agnosia),* or seen *(visual agnosia),* or in failure to distinguish right from left *(autotopognosis).*

APRAXIA refers to an inability to perform voluntary movements, particularly of speech, when no true paralysis is present.

APHASIA is an inability to use or comprehend ideas communicated by spoken or written symbols. It may be motor, resulting in inability to *express* oneself, or sensory, in that there is a *deficit in understanding.*

ALEXIA, a form of aphasia, refers specifically to inability to comprehend the written word *(word blindness)* or to speak words correctly. It is usually discovered first in school-age children, who exhibit reading defects.

Parietal lobe lesions produce agnosia if the lesion is on the "dominant side" of the cerebrum (usually the left side in right-handed persons and vice versa). If on the "nondominant side," a lesion produces AMORPHOSYNTHESIS, characterized by defective perception of one side of the body.

The medullary body and basal ganglia

STRUCTURE. Each cerebral hemisphere includes a large mass of white matter consisting of three types of fibers.

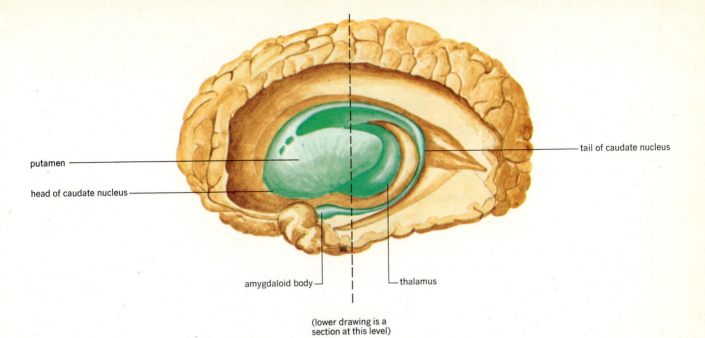

putamen

head of caudate nucleus

tail of caudate nucleus

amygdaloid body

thalamus

(lower drawing is a
section at this level)

FIGURE 15.12 The basal ganglia. Several major ganglia projected on the cerebral hemisphere *(above).* Frontal section of the cerebrum *(below).*

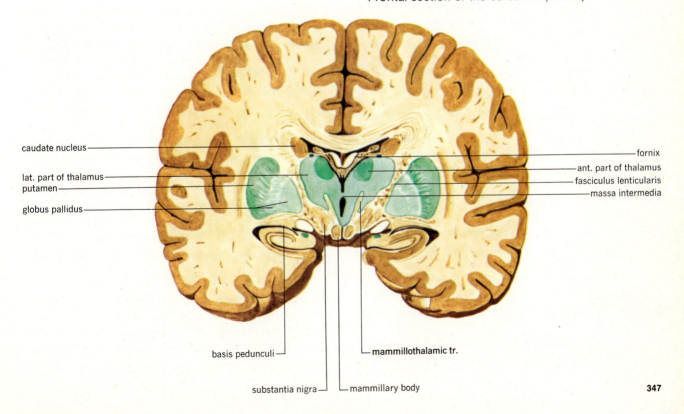

caudate nucleus

lat. part of thalamus

putamen

globus pallidus

fornix

ant. part of thalamus

fasciculus lenticularis

massa intermedia

basis pedunculi

mammillothalamic tr.

substantia nigra

mammillary body

COMMISSURAL FIBERS connect the two cerebral hemispheres through the *corpus callosum* (*see* Fig. 15.1). These fibers allow one side of the cerebrum to communicate with the other.

PROJECTION FIBERS are fibers that come *to* the cortex from noncerebral regions or which pass *from* the cortex to go to other brain or body areas.

ASSOCIATION FIBERS stay within a given hemisphere, connecting one cortical area with another. Many of these fibers form bundles, to which specific names have been assigned.

Buried deep within the white matter of each cerebral hemisphere are several masses of gray matter collectively known as the BASAL GANGLIA (Fig. 15.12). The largest of these is called the CORPUS STRIATUM, which consists of the *caudate nucleus, putamen,* and *globus pallidus.* Lying beneath the thalamus are a group of SUBTHALAMIC NUCLEI, including the *substantia nigra* and *red nucleus.* The *amygdaloid nucleus* lies in the temporal lobe at the tip of the tail of the caudate nucleus and, while it is considered a basal ganglion, it is involved in the limbic system.

FUNCTIONS. The basal ganglia represent, in such animals as fish, amphibians, and reptiles, *the* highest motor center of their brain. In birds and mammals, cerebral development has resulted in much motor function moving out of the ganglia into the cerebral cortex. The question then is: what contributions to motor function remain in the human basal ganglia? Perhaps the most information concerning function comes from degenerative disorders that involve the ganglia; in such disorders, involuntary movements figure prominently.

Stimulation of the caudate nucleus produces inhibition of movement and muscle tone. Lesions of caudate, putamen, globus pallidus, or substantia nigra produce involuntary and uncontrollable tremor, suggesting that normal movement requires an inhibitory function of the ganglia.

CLINICAL CONSIDERATIONS. Lesions of the various types of medullary fibers cause deficiencies characteristic of the fiber tracts interrupted.

Interruption of the commissural fibers results in the creation of essentially "two brains" within the same skull. Each hemisphere can still accumulate and analyze information, but information received by one half of the cerebrum cannot be transferred directly to the other half. Nevertheless, through "lower" centers in the brain, such as the brainstem, *compensation* for such losses can occur to a great degree.

Interruption of incoming projection fibers results primarily in loss of ability to perceive sensations, while interruption of outgoing motor projection fibers produces signs of an *upper motor neuron* lesion: paralysis of voluntary movement with muscles in a contracted state (*spastic paralysis*); also, there is exaggeration of muscular reflexes.

Interruption of association fibers results in inability to integrate activity on one side of the cerebrum. For example, motor responses to sensory input (sensorimotor integration) become defective.

Choreiform movements are rapid, jerky, and purposeless movements most pronounced in the trunk and appendates. *Sydenham's chorea* (acquired) and *Huntington's chorea* (congenital) are specific examples of choreiform disorders. Lesions in the corpus striatum are most commonly associated with such disorders.

Athetoid movements are slow sinous movements most common in the outer (distal) parts of the appendages. Globus pallidus lesions may cause such movements.

The substantia nigra is the most commonly involved nucleus in PARKINSON'S DISEASE (paralysis agitans, shaking palsy). This disease makes its appearance most often in persons between 50 and 65 years of age. There is a fixation of facial expression, tremor of the limbs at rest, muscular rigidity, slow movements, and postural abnormalities. The use of L-DOPA (*levo*rotatory *d*ihydroxy*p*henyl*a*lanine), an amino acid, may reduce the severity of the disease's symptoms by increasing the synthesis of certain synaptic transmitters in the central nervous system.

The limbic system

The LIMBIC SYSTEM is composed of *hypothalamus, thalamus,* the *amygdaloid nuclei,* and several fiber tracts connecting these areas (Fig. 15.13). The entire system is concerned with emotions and motivation. Expression of rage and sexual behavior are among the functions these structures serve. "Pleasure centers" that create positive drives toward stimuli are also found in these areas.

Functionally, two more groupings of efferent fibers from the brain may be made.

The PYRAMIDAL SYSTEM refers to motor fibers originating from the cerebral cortex and passing through the pyramids. About 40 percent of the fibers in this system come from the primary somatic motor area (area 4); the remainder come from other cortical regions.

The EXTRAPYRAMIDAL SYSTEM consists of motor fibers from basal ganglia, brain stem, cerebellum, and other noncortical areas. Although the pyramidal system has traditionally been considered the "voluntary motor system" of the body, this is not strictly true, since only about 40 percent of its fibers come from area 4. The extrapyramidal system is, however, "involuntary" in the sense that it operates without acts of will, and controls tone and reflex movement.

Damage to limbic structures produces exaggerated responses of rage and hypersexuality.

Damage to pyramidal or extrapyramidal systems or both produces motor losses characterized by spastic paralysis. If the damage occurs above the level of the brain stem, the opposite side of the body is usually affected; below the brain stem, the same side is usually affected. This occurs because crossing of motor fibers (e.g., lateral corticospinal tract, rubrospinal tract) occurs in the brain stem.

Electrical activity in the cerebrum

The brain exhibits spontaneous rhythmical electrical activity that may be recorded as an ELECTROENCEPHALOGRAM or EEG. Figure 15.14 shows a normal EEG with several types of wave patterns.

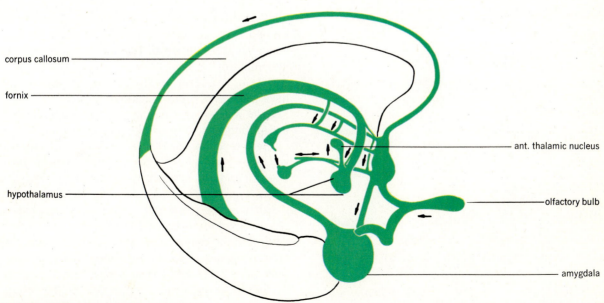

FIGURE 15.13   The limbic system.

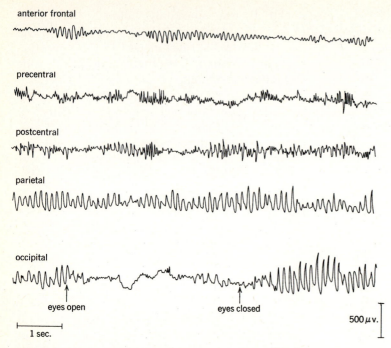

anterior frontal

precentral

postcentral

parietal

occipital

eyes open   eyes closed

1 sec.   500 μv.

FIGURE 15.14   Normal EEG patterns from different regions of the cortex. Alpha waves predominate in parietal and occipital areas, beta waves in precentral area. Alpha waves are blocked when eyes are opened.

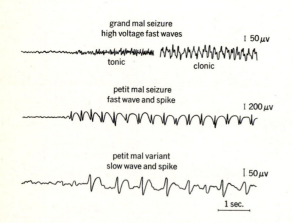

grand mal seizure
high voltage fast waves   I 50 μv
tonic   clonic

petit mal seizure
fast wave and spike   I 200 μv

petit mal variant
slow wave and spike   I 50 μv
1 sec.

FIGURE 15.15   Some abnormal EEGs. Note large spikes and fast rhythm.

Alpha waves have a frequency of 10 to 12 per second, and voltages of about 50 microvolts (1 microvolt = 1/1,000,000th volt). They are obtained from an inattentive or "at rest" brain.

Beta waves occur 15 to 60 times per second and have voltages of 5 to 10 microvolts.

Delta waves occur during sleep and occur 1 to 5 per second with voltages of 20 to 200 microvolts.

Theta waves occur 5 to 8 times per second with voltages of about 10 microvolts. They are common in children and during times of emotional stress.

CONVULSIVE DISORDERS or seizures are characterized by alterations in the EEG (Fig. 15.15) and alterations in the state of consciousness, muscular activity, and sensory phenomena. The subject may show involuntary contractions of the muscles of the trunk and appendages, which constitute the most obvious signs of a seizure.

Fever, infections, lack of oxygen to the brain, tumors, allergic reactions, trauma to the brain, and drugs have all been implicated as causes of convulsive seizures. In other cases, there is no obvious cause of the seizure.

If there is a recurring pattern to the seizures, the subject is usually said to have epilepsy. Epilepsy affects about 0.5 percent of the population and is divided into two general types:

Symptomatic (Jacksonian) *epilepsy* may be demonstrated to have a cause, such as a brain lesion.

Idiopathic epilepsy occurs without any demonstrable cause.

According to the type of symptoms the subject shows, there are several kinds of epilepsy described.

Grand mal epilepsy results in loss of consciousness, falling, and spasm of the muscles, often preceeded by an hallucination of a disagreeable odor.

Petit mal epilepsy clouds the consciousness and is not associated with loss of consciousness or muscular spasms. It produces a "blank" for 1 to 30 seconds in the subject's behavior.

Psychomotor epilepsy lasts 1 to 2 minutes and is characterized by loss of contact with the environment, staggering, muttering, and mental confusion for several minutes after the attack is over.

Infantile spasm occurs in the first three years of life, and may be replaced by other types of seizures later in life. It is characterized by flexion of trunk and arms, and extension of the legs.

If a lesion such as a tumor or scar may be shown to cause the seizure, surgery may be indicated. Management of seizures is often achieved by the administration of anticonvulsant drugs alone or in combinations (e.g., phenobarbital, diphenylhydantoin, bromides).

The cranial nerves

The brain gives rise to 12 pairs of cranial nerves that supply motor and sensory fibers to structures in the head, neck, and shoulder regions. These fibers form, with the spinal nerves, the peripheral nervous system.

The relationships of the nerves to the brain are shown in Figure 15.16. A general description of each nerve follows, and a chart of the nerves is shown in Table 15.1.

I. *Olfactory nerve.* The nerve of smell, the olfactory nerve passes from nasal cavities to the olfactory bulb of the cerebrum.

II. *Optic nerve.* The nerve of sight, the optic nerve conveys impulses from retina to brain stem. These impulses are then relayed to the occipital lobe over the optic radiation.

III. *Oculomotor nerve.* Motor fibers in this nerve control four of the six eye muscles that turn the eye ball, cause pupillary size changes, and aid in focusing. A sensory component relays impulses from the muscles, iris, and ciliary body.

IV. *Trochlear nerve.* One of the six eye muscles is supplied by motor and sensory fibers of this nerve.

V. *Trigeminal nerve* (Fig. 15.17). The largest cranial nerve, it supplies sensory fibers to the anterior cranium and face, and motor fibers to the chewing muscles. Ophthalmic, maxillary, and mandibular branches form the nerve, and supply sensory fibers to all structures within their area of distribution.

VI. *Abducent (abducens) nerve.* This nerve supplies motor and sensory fibers to one extrinsic eye muscle.

VII. *Facial nerve* (Fig. 15.18). The facial nerve supplies motor fibers to the muscles of the face, and the salivary glands. Sensory fibers for taste are derived from the anterior two-thirds of the tongue.

VIII. *Vestibulocochlear* (statoacoustic, acoustic, or

Nerve	Composition M = motor; S = sensory	Origin	Connection with brain or peripheral distribution	Function
I. Olfactory	S	Nasal olfactory area	Olfactory bulb	Smell
II. Optic	S	Ganglionic layer of retina	Optic tract	Sight
III. Oculomotor	MS	M—Midbrain	4 of 6 extrinsic eye muscles (superior rectus, medial and inferior rectus, inferior oblique)	Eye movement
		S—Ciliary body of eye	Nucleus of nerve in midbrain	Focusing, pupil changes, muscle sense
IV. Trochlear	MS	M—Midbrain	1 extrinsic eye muscle (superior oblique)	Eye movement
		S—Eye muscle	Nucleus of nerve in midbrain	Muscle sense
V. Trigeminal	MS	M—Pons	Muscles of mastication	Chewing
		S—Scalp and face	Nucleus of nerve in pons	Sensation from head
VI. Abducent	MS	M—Nucleus of nerve in pons	1 extrinsic eye muscle (lateral rectus)	Eye movement
		S—1 extrinsic eye muscle	Nucleus of nerve in pons	Muscle sense
VII. Facial	MS	M—Nucleus of nerve in lower pons	Muscles of facial expression	Facial expression
		S—Tongue (ant. 2/3)	Nucleus of nerve in lower pons	Taste
VIII. Vestibulocochlear (Statoacoustic, acoustic, auditory)	S	Internal ear: balance organs, cochlea	Vestibular nucleus, cochlear nucleus	Posture, hearing
IX. Glossopharyngeal	MS	M—Nucleus of nerve in lower pons	Muscles of pharynx	Swallowing
		S—Tongue (post. 1/3), pharynx	Nucleus of nerve in lower pons	Taste, general sensation
X. Vagus	MS	M—Nucleus of nerve in medulla	Viscera	Visceral muscle movement
		S—Viscera	Nucleus of nerve in medulla	Visceral sensation
XI. Accessory	M	Nucleus of nerve in medulla	Muscles of throat, larynx, soft palate, sternocleidomastoid, trapezius	Swallowing, head movement
XII. Hypoglossal	M	Nucleus of nerve in medulla	Muscles of tongue and infrahyoid area	Speech, swallowing

TABLE 15.1   Summary of the cranial nerves

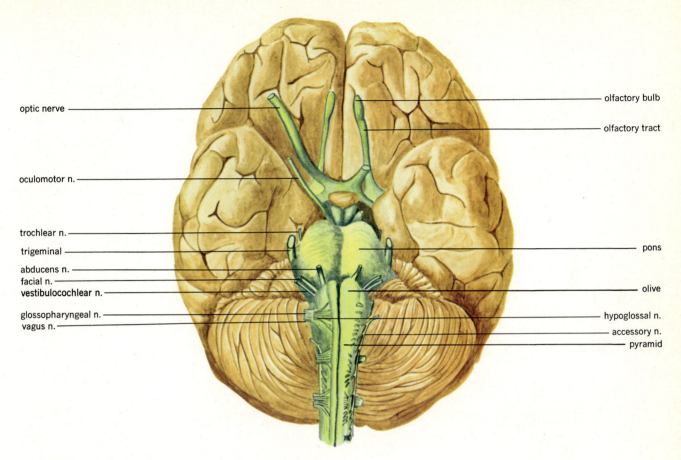

optic nerve

oculomotor n.

trochlear n.
trigeminal
abducens n.
facial n.
vestibulocochlear n.

glossopharyngeal n.
vagus n.

olfactory bulb

olfactory tract

pons

olive

hypoglossal n.
accessory n.
pyramid

FIGURE 15.16   Basal view of the brain showing cranial nerves.

auditory) *nerve.* This nerve carries sensory fibers from the cochlea and organs of equilibrium of the inner ear (semicircular canals, utriculus and sacculus), to the temporal lobe.

IX. *Glossopharyngeal nerve.* This nerve supplies sensory fibers to the posterior one-third of the tongue for taste, and motor fibers to the throat muscles.

X. *Vagus nerve* (Fig. 15.19). The vagus nerve is a very important component of the autonomic system. It supplies motor and sensory fibers to nearly all thoracic and abdominal viscera.

XI. *Accessory (spinal accessory) nerve.* This nerve supplies motor fibers to muscles of the throat, larynx, soft palate, and to trapezius and sternocleidomastoid muscles.

XII. *Hypoglossal nerve.* This nerve supplies some of the tongue muscles and the infrahyoid muscles.

Clinical considerations

Functions of the cranial nerves are easily assessed by simple procedures, and these form an important part of a neurological examination. The various procedures are presented in Table 15.2.

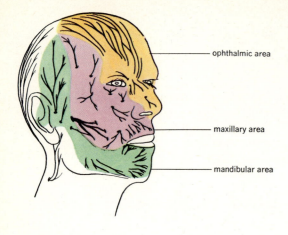

ophthalmic area

maxillary area

mandibular area

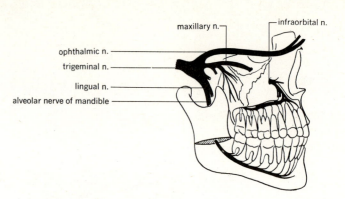

maxillary n.

infraorbital n.

ophthalmic n.

trigeminal n.

lingual n.

alveolar nerve of mandible

FIGURE 15.17   The distribution of the Vth cranial nerve (trigeminal).

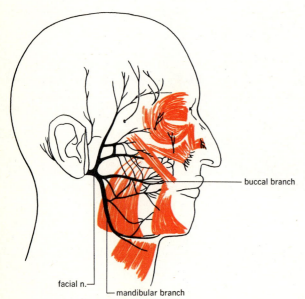

buccal branch

facial n.

mandibular branch

FIGURE 15.18   The distribution of the VIIth cranial nerve (facial).

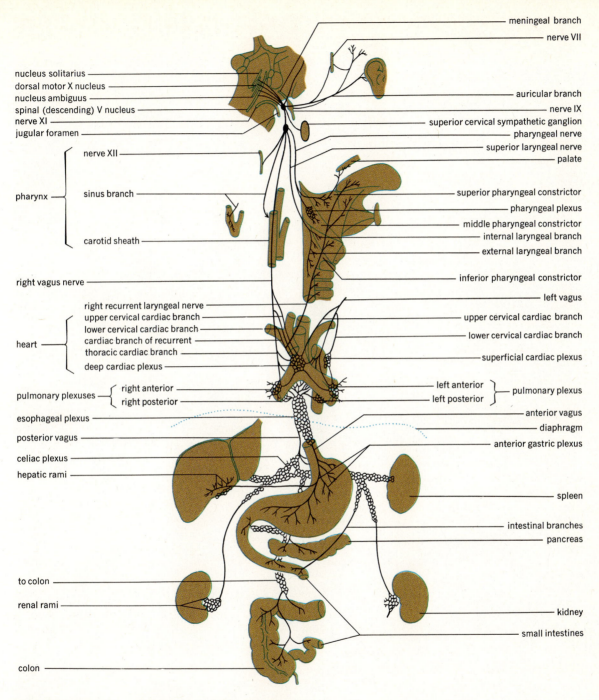

FIGURE 15.19 The distribution of the Xth cranial nerve (vagus).

TABLE 15.2   Methods of assessing cranial nerve function

Nerve being tested	Method	Comments
Olfactory	Ask patient to differentiate odors of coffee, bananas, tea, oil of cloves, etc., in each nostril.	Test may be voided if patient has a cold or other nasal irritation.
Optic	Ask patient to read newspaper with each eye; examine fundus with ophthalmoscope.	Perform test with lenses on, if patient wears them.
Oculomotor	Shine light in each eye separately and observe pupillary changes; look for movement by moving finger up, down, right, and left; have subject's eyes follow finger movements.	Look for defects in both eyes at same time. Light in one eye causes other one to constrict to lesser degree.
Trochlear		Requires special devices. If eye movement is normal, nerves are probably all right.
Trigeminal	Motor portion: have patient clench teeth and feel firmness of masseter; have patient open jaw against pressure.	Subject should exhibit firmness and strength in both tests.
	Sensory portion: test sensations over entire face with cotton (light touch) or pin (pain).	Deficits in particular areas indicate problems in one of the three branches.
Abducent	Perform test for lateral movement of eyes.	Must be very careful to observe any deficit in lateral movement.
Facial	Have subject wrinkle forehead, scowl, puff out his cheeks, whistle, smile. Note nasolabial fold (groove from nose to corners of mouth).	Nerve supplies all facial muscles. Fold is maintained by facial muscles and may disappear in nerve damage.
Vestibulocochlear (statoacoustic, acoustic, auditory)	Cochlear portion: test auditory acuity with ticking watch; repeat a whispered sentence; use tuning fork. Compare ears. Examine with otoscope.	Ears may not be equal in acuity; note.
	Vestibular portion: not routinely tested.	If subject appears to walk normally and keep his balance, nerve is usually all right.
Glossopharyngeal	Note disturbances in swallowing, talking, movements of palate.	
Accessory	Test trapezius by having subject raise shoulder against resistance; have subject try to turn head against resistance.	Strength is index here.
Hypoglossal	Have subject stick out tongue. Push tongue against tongue blade.	Tongue should protrude straight; deviation indicates same side nerve damage. Pushing indicates strength.

Summary

1. The brain includes all nerve structures in the cranium. It is divided into:
   a. Brain stem: medulla, pons midbrain.
   b. Cerebellum.
   c. Diencephalon: thalamus, hypothalamus.
   d. Cerebrum.

2. The medulla is the lowest part of the brain stem.
   a. It contains many important reflex centers for respiration, heart rate, and blood vessel size.
   b. Motor tracts from the cerebrum cross here.
   c. Damage to the medulla may interfere with breathing.

3. The pons lies above the medulla.
   a. It contains one of several respiratory centers (pneumotaxic).
   b. It connects to the cerebellum by way of the cerebellar peduncles.
   c. Damage here produces motor and respiratory deficits.

4. The midbrain lies above the pons.
   a. It shows the cerebral peduncles, receiving fibers from the cerebrum.
   b. It contains the corpora quadrigemina that connect eye and ear with motor fibers for righting and adjustments to visual and auditory stimuli.
   c. Damage here produces deficits in maintenance of balance and equilibrium.

5. The diencephalon includes the thalamus and hypothalamus.
   a. The thalamus has a number of nuclei that are sensory and motor in function. Sensory nuclei provide a means of organizing sensory impulses and crude awareness of sensation. Motor nuclei are concerned with motor activity of emotions. Damage results in the thalamic syndrome.
   b. The hypothalamus contains many nuclei. These nuclei control body temperature, water balance, the pituitary gland, food intake, gastric secretion, and emotional response. Damage produces deficits in the above functions.

6. The reticular formation is concerned with arousal and alerting of the organism. It is part of the reticular activating system.

7. The cerebellum lies posterior to the medulla and pons.
   a. It has a vermis and two hemispheres, an outer gray cortex and inner white matter.
   b. It coordinates and controls muscular movement and tone.

8. The cerebrum is the largest part of the brain.

   a. Each hemisphere contains several lobes, separated by fissures. The hemispheres are folded.

   b. The basal ganglia are nuclei deep within the cerebral hemisphere.

   c. The limbic system is composed of cerebral and lower level structures.

9. The cerebrum contains motor areas, sensory areas, visual and auditory areas, and association areas concerned with sensory interpretation and behavior. Damage interferes with movement, sensory reception, and understanding.

10. The basal ganglia control muscle tone, inhibit movement, and control muscle tremor. Damage produces tremor and loss of tone.

11. The limbic system contains centers for expression of emotion, and pleasure centers. Damage produces exaggerated display of emotions and sexual behavior.

12. The pyramidal system includes nerve fibers causing and modifying voluntary movement; the extrapyramidal system controls muscle reflexes and tone. Damage to either system produces spastic paralysis.

13. Electrical activity in the brain shows four basic wave patterns.

    a. Alpha waves: frequency 10–12/sec; voltage, 50μv

    b. Beta waves: frequency 15–60/sec; voltage 5–10μv

    c. Delta waves: frequency 1–5/sec; 20–200μv

    d. Theta waves: frequency 5–8/sec; 10μv

14. Altered EEG is associated with development of epilepsy.

15. Twelve pairs of cranial nerves arise from the brain. They supply head and neck with motor and sensory fibers.

    a. For functions, see Table 15.1.

    b. For testing cranial nerves, see Table 15.2.

Questions

1. Give the functions centered in

   a. The medulla.

   b. Midbrain.

   c. Cerebellum.

   d. Thalamus.

2. What might be expected to be the primary signs of damage to the following areas?

   a. Hypothalamus.

   b. Primary cerebral motor area.

   c. Temporal association area.

3. Compare the contributions of the cerebellum, basal ganglia and cerebrum to movement.

4. What cranial nerves would be involved if the following symptoms were presented?

   a. Inability to smile.

   b. Deviation of the tongue to one side when protruded.

   c. Can't turn eyes upward.

   d. Can't turn eyes laterally.

   e. Can't maintain balance.

   f. No sensation on lower jaw.

   g. Can't swallow properly.

Readings

Dement, William C. *Some Must Watch While Some Must Sleep.* Freeman. San Francisco, 1974.

Dinarello, Charles A., and Sheldon M. Wolff. "Pathogenesis of Fever in Man." *New Eng. J. Med. 298:*607, March 16, 1978.

Galaburda, Albert M., Marjorie LeMay, Thomas L. Kemper, and Norman Geschwind. "Right-Left Asymmetries in the Brain." *Science 199:*852, 24 Feb. 1978.

Hart, Leslie A. *How the Brain Works: A New Understanding of Human Learning, Motivation, and Thinking.* Basic Books. New York, 1975.

Llinás, Rodolfo R. "The Cortex of the Cerebellum." *Sci. Amer. 233:*56, Jan. 1975.

Pappenheimer, John R. "The Sleep Factor." *Sci. Amer. 235:*24, Aug. 1976.

Science News. "Enkephalins: Pleasures and Seizures." *113:*280, April 29, 1978.

Science News. "Psychosurgery at the Crossroads." *111:*314, May 14, 1977.

Science News. "Schizophrenia and Brain Imbalance." *113:*276, April 29, 1978.

Chapter 16

The Autonomic Nervous System

Objectives

After studying this chapter, the reader should be able to:

- List the types of fibers that compose the autonomic nervous system.

- Show how autonomic nerves connect to the spinal cord, specifically, the functions of the gray and white rami.

- Differentiate between pre- and postganglionic neurons of the autonomic system as to length and mode of transmission of impulses at synapses and junctions with organs.

- List the several groups of autonomic ganglia and locate them.

- Enumerate the subdivisions of the autonomic system and the outflows composing each division.

- Give the effects of each division on body organs.

- Account for the differential effects of epinephrine and norepinephrine on body organs.

- Relate surgical procedures or drug administration to clinical disorders of the autonomic nervous system.

Chapter 14 discussed the spinal nerves and Chapter 15 examined the cranial nerves. The autonomic nervous system involves spinal *and* cranial nerves, and the appropriate parts of these two chapters should be reviewed to achieve the most benefit from the discussion that follows.

The autonomic nervous system has been classically defined as that part of the peripheral nervous system supplying *motor* fibers to smooth and cardiac muscles, and glands, and operating "automatically" at the reflex and subconscious levels. Such a definition is too limited, for the system contains not only motor but also sensory fibers that pass from various body organs to the central nervous system *(CNS)*. These sensory fibers are essential to the operation of a multitude of autonomic reflex arcs. Today we recognize that there are two types of fibers composing the nerves of the autonomic system.

VISCERAL AFFERENTS carry sensory impulses from organs to the CNS.

VISCERAL EFFERENTS carry motor impulses from the CNS to body organs.

Connections of the system with the central nervous system (Fig. 16.1)

Afferent pathways

Nerve impulses of sensory nature originate in a variety of receptors in body viscera, including those sensitive to tension or pressure, pain, heat and cold, and to levels of blood gases. The visceral afferent fibers conveying such impulses usually join a *spinal nerve* by way of the WHITE RAMUS, and proceed toward the central nervous system over the posterior root of the spinal nerve. A single neuron spans the distance from organ to CNS, and has its cell body in the POSTERIOR ROOT *(spinal)* GANGLION. Synapses between these afferent neurons and internuncial neurons occur in the dorsal gray columns of the spinal cord. If afferent impulses are brought to the CNS over cranial nerves, the sensory components of those nerves form the incoming pathway and cell bodies and synapses are found or are made in the nuclei of the appropriate cranial nerves in the brainstem.

Efferent pathways

Visceral efferent fibers have their cell bodies in the lateral gray columns of the spinal cord, or in the motor nuclei of the cranial nerves. Those that leave the spinal cord pass through the ANTERIOR SPINAL NERVE ROOTS, in company with somatic efferents, and leave the spinal nerve by way of the white rami. They may undergo synapses in AUTONOMIC GANGLIA (see below), close to the cord, or pass through such ganglia to undergo synapses in ganglia closer to or within the organ innervated. Those fibers that *do* synapse in ganglia close to the spinal cord have the second neuron in the chain rejoin the spinal nerve by way of the GRAY RAMUS. Such fibers pass usually to blood vessels of skeletal muscle and glands and blood vessels of the skin. In most cases, *two* neurons are thus seen to extend from CNS to the organ innervated. The one passing from CNS to synapse in a ganglion—wherever that may be—is called the PREGANGLIONIC NEURON; the one passing from ganglion to organ is called the POSTGANGLIONIC NEURON. Transmission from pre- to postganglionic neurons is chemical, utilizing acetylcholine. The endings of postganglionic neurons secrete either acetylcholine or norepinephrine at their endings on an organ or tissue, and different chemicals account for the different effects exerted on the organs by the two divisions of the autonomic system.

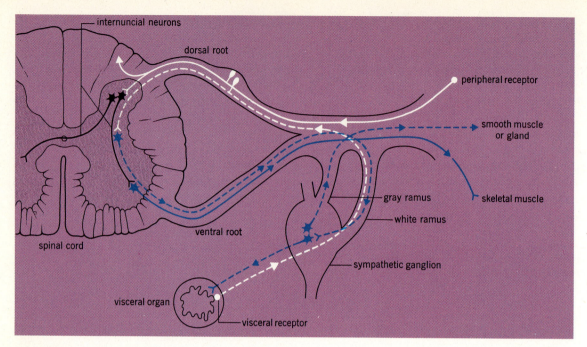

FIGURE 16.1   The connections of the autonomic nervous system to the cord (shown by dashed pathways).

Autonomic ganglia and plexuses

The term GANGLION is usually used to refer to masses of nervous tissue that contain nerve cell bodies and/or synapses, and that are located in the periphery—that is, outside the CNS. The term PLEXUS usually refers to a network of nerve fibers, blood vessels, or lymphatic vessels. (Recall, for example, the cervical, lumbar, and sacral plexuses described in Chapter 14). In some cases, the terms ganglia and plexus are used interchangeably.

Synapses between pre- and postganglionic neurons occur in one of three areas in the periphery.

The LATERAL or VERTEBRAL GANGLIA, also called the *sympathetic* ganglia (*see* Fig. 16.1), form paired chains of 22 ganglia that lie alongside the vertebral bodies in the thoracic and abdominal cavities. They receive preganglionic fibers from the thoracic and lumbar spinal nerves.

The COLLATERAL or PREVERTEBRAL GANGLIA form several groupings of nerves and synapses in association with visceral organs or large blood vessels. The *cardiac*, *celiac* (solar), and *mesenteric plexuses* are three important groupings located near the heart, spleen, and urinary bladder respectively. This is an example where the terms ganglion and plexus are used synonomously. These plexuses or ganglia receive preganglionic fibers from the thoracic, lumbar, and sacral portions of the CNS.

TERMINAL GANGLIA or plexuses lie close to or within the organ supplied, as, for example, the plexuses (Meissner's and Auerbach's) in the wall of the small intestine. These ganglia receive preganglionic fibers from the cranial and sacral portions of the CNS.

Divisions of the system

Structure

Preganglionic autonomic fibers form three out-going groups, or *outflows*. The CRANIAL OUTFLOW is composed of the motor fibers of cranial nerves 3, 7, 9, and 10. The THORACOLUMBAR OUTFLOW consists of the visceral efferents of all thoracic and lumbar spinal nerves. The SACRAL OUTFLOW consists of visceral efferents derived from the second to fourth sacral spinal nerves. Cranial and sacral preganglionic neurons tend to be longer than those of the thoracic and lumbar neurons, and generally synapse in collateral and terminal ganglia. Thus, postganglionic neurons are generally shorter than the preganglionic neurons. Thoracic and lumbar preganglionic neurons often synapse in the vertebral ganglia, and are shorter than the postganglionic neurons.

The fibers are grouped into a *craniosacral* or PARASYMPATHETIC DIVISION, and a *thoracolumbar* or SYMPATHETIC DIVISION, with the outflows involved stated by the italicized names. The two divisions provide a double innervation for most (but not all) body organs (Fig. 16.2).

Functions

The *parasympathetic division,* formed from cranial and sacral outflows, supplies fibers to all autonomic effector organs *except* the adrenal medulla, sweat glands, smooth muscle of the spleen, and blood vessels of skin and skeletal muscles. Postganglionic fibers secrete acetylcholine at their endings on the effector, and are known as *cholinergic* fibers. This division of the system tends to protect and conserve body resources and to preserve normal resting body functions.

The *sympathetic division,* formed from thoracic and lumbar outflows, innervates the same organs as does the parasympathetic division, and is the *only* supply to adrenal medulla, sweat glands, spleen, and skin and skeletal muscle blood vessels. Postganglionic fibers of this division secrete norepinephrine (noradrenalin) at their endings and are called *adrenergic* fibers. We thus see that in the organs listed previously, control of their activity depends on changes in level of activity of only the sympathetic division. Activity in the sympathetic division tends to increase the rate of utilization of body resources and accelerate activity.

Operation of both divisions is TONIC or continuous, so that many body functions, even at rest, reflect a balance between the activity of both divisions. Table 16.1 summarizes the effects of activity of the two divisions on specific body activities or organs.

Norepinephrine is a chemical that is very similar to epinephrine, the hormone secreted by the medulla of the adrenal gland. The two substances have similar, but not identical, effects on the body. To account for the differences in effects of the two chemicals, it has been postulated that, to affect a cell, the chemical must be first bound to a membrane receptor on the cell surface. It is suggested that there are two types of receptors present: *alpha receptors* are affected by both epinephrine and norepinephrine and *beta receptors* are affected only by epinephrine. Alpha receptors promote vasoconstriction (narrowing of blood vessel diameter) pupillary dilation (enlargement), and relaxation of intestinal smooth muscle. Beta receptors promote vasodilation, increase of heart rate and strength of contraction, dilation of heart and skeletal muscle blood vessels, and relaxation of uterine muscle. Thus, a given organ may respond differently to the two chemicals.

Drugs exist that may compete for the alpha and beta receptor sites on effector organs, and may thus block the expression of the effect of norepinephrine or epinephrine. *Phenoxybenzamine* blocks alpha receptors and the effects of both chemicals. It is useful in controlling high blood pressure

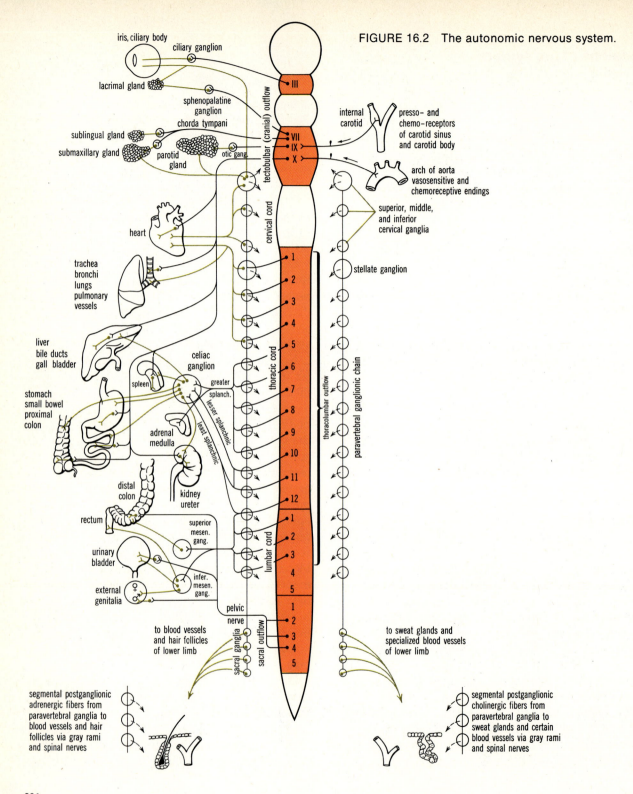

iris, ciliary body
ciliary ganglion
lacrimal gland
sphenopalatine ganglion
chorda tympani
sublingual gland
submaxillary gland
parotid gland
otic gang.

tectobulbar (cranial) outflow

III
VII
IX
X

internal carotid
presso- and chemo-receptors of carotid sinus and carotid body

arch of aorta vasosensitive and chemoreceptive endings

cervical cord

heart

trachea
bronchi
lungs
pulmonary vessels

superior, middle, and inferior cervical ganglia

stellate ganglion

liver
bile ducts
gall bladder

celiac ganglion

spleen
greater splanch.

stomach
small bowel
proximal colon

lesser splanchnic

adrenal medulla

least splanchnic

distal colon

kidney
ureter

rectum

superior mesen. gang.

urinary bladder

external genitalia

infer. mesen. gang.

pelvic nerve

sacral ganglia
sacral outflow

thoracic cord
lumbar cord

thoracolumbar outflow
paravertebral ganglionic chain

1
2
3
4
5
6
7
8
9
10
11
12
1
2
3
4
5
1
2
3
4
5

FIGURE 16.2   The autonomic nervous system.

to blood vessels and hair follicles of lower limb

to sweat glands and specialized blood vessels of lower limb

segmental postganglionic adrenergic fibers from paravertebral ganglia to blood vessels and hair follicles via gray rami and spinal nerves

segmental postganglionic cholinergic fibers from paravertebral ganglia to sweat glands and certain blood vessels via gray rami and spinal nerves

(hypertension) that results from hypersensitivity to both chemicals. *Propanolol* blocks beta recep-

tors and the effect of epinephrine, leading to cardiac slowing and regulation of heartbeat.

TABLE 16.1 Effects of autonomic stimulation		
Organ affected	Parasympathetic effects	Sympathetic effects
Iris	Contraction of sphincter pupillae; pupil size decreases	Contraction of dilator pupillae; pupil size increases
Ciliary muscle	Contraction; accommodation for near vision	Relaxation; accommodation for distant vision
Lacrimal gland	Secretion	Excessive secretion
Salivary glands	Secretion of watery saliva in copious amounts	Scanty secretion of mucus rich saliva
Respiratory system:		
Conducting division	Contraction of smooth muscle; decreased diameters and volumes	Relaxation of smooth muscle; increased diameter and volumes
Respiratory division	Effect same as on conducting division	Effect same as on conducting division
Blood vessels	Constriction	Dilation
Heart:		
Stroke volume	Decreased	Increased
Stroke rate	Decreased	Increased
Cardiac output and blood pressure	Decreased	Increased
Coronary vessels	Constriction	Dilation
Peripheral blood vessels:		
Skeletal muscle	No innervation	Dilation
Skin	No innervation	Constriction
Visceral organs (except heart and lungs)	Dilation	Constriction
Organ affected	Parasympathetic effects	Sympathetic effects
Stomach:		
Wall	Increased motility	Decreased motility
Sphincters	Inhibited	Stimulated
Glands	Secretion stimulated	Secretion inhibited
Intestines:		
Wall	Increased motility	Decreased motility

Table 16.1 (continued)

Organ affected	Parasympathetic effects	Sympathetic effects
Sphincters:		
Pyloric, iliocecal	Inhibited	Stimulated
Internal anal	Inhibited	Stimulated
Liver	Promotes glycogenesis; promotes bile secretion	Promotes glycogenolysis; decreases bile secretion
Pancreas (exocrine and endocrine)	Stimulates secretion	Inhibits secretion
Spleen	No innervation	Contraction and emptying of stored blood into circulation
Adrenal medulla	No innervation	Epinephrine secretion
Urinary bladder	Stimulates wall, inhibits sphincter	Inhibits wall, stimulates sphincter
Uterus	Little effect	Inhibits motility of nonpregnant organ; stimulates pregnant organ
Sweat glands	No innervation	Stimulates secretion (produces "cold sweat" when combined with cutaneous vasoconstriction)

Higher autonomic centers

The discussion of autonomic function so far might suggest that control by this system is dependent only on reflex activity. This is not the case. The brain exercises the "right" to provide input to the systems, so that psychological changes can influence visceral activity. We are undoubtedly familiar with having a "stomach tied in knots" as a result of emotional upheaval, and ulcers are often associated with excessive autonomic activity.

The medulla of the brainstem contains the vital centers (vasomotor, cardiac, respiratory) that serve autonomic functions concerned with blood pressure and breathing. The cerebral cortex contains, in the frontal lobes, both parasympathetic and sympathetic areas. These regions may be tied into the limbic system, for visceral expression of rage and anger. The hypothalamus, in its control of temperature, gastric secretion, and emotional expression (see Chapter 15), also is involved in control of autonomic function.

Clinical considerations

Of the various conditions that might be expected to be associated with excessive autonomic activity, one is potentially life-threatening: HYPERTENSION or high blood pressure. Severe strain is placed on the heart and blood vessels as pressure rises with vasoconstriction and increased heart action; vessels may rupture or the heart may fail. In many cases of hypertension, excessive norepinephrine or epinephrine secretion may be the culprit.

Surgery and the autonomic system

The effect of sympathetic stimulation on all blood vessels, except those of heart and skeletal muscle, is to cause them to constrict. Blood pressure rises. A SYMPATHECTOMY that removes the sympathetic ganglia from T10-L2—and therefore the effect of sympathetic stimulation on blood vessels of the gut—may alleviate hypertension by causing dilation of these vessels. The individual who undergoes such an operation cannot normally respond to changes of posture or stress.

VAGOTOMY, or sectioning of the vagus (cranial nerve X) branches to the stomach, is sometimes employed to treat ulcers. The rationale behind this operation is that excessive stress may, by acting through the brain and vagus nerve, cause secretion of gastric juices into the stomach when there is no food to digest. The juice thus attacks the stomach wall.

Drugs and the autonomic system

An alternative to surgery is often the use of drugs to depress or block autonomic expression. Drugs that *imitate* (mimic) the effects of stimulation of a given autonomic division are said to be *parasympathomimetic* or *sympathomimetic*. Those that prevent synaptic transmission are called *ganglionic blocking agents.*

A chemical that produces the same effect as a *cutting* of parasympathetic nerves is termed a *parasympatholytic* drug. Most such drugs operate by blocking the effects of acetylcholine at nerve terminals on organs. Atropine and scopolamine are examples of these drugs. A common use for atropine or its derivatives is to dilate the pupil of the eye, for eye examinations.

A *sympatholytic agent* is one that blocks the effects of epinephrine or norepinephrine. Ergot, propanolol, phenoxybenzamine, and piperoxan are examples of sympatholytic drugs. Many of these drugs produce troublesome side effects. Table 16.2 lists some drugs employed to treat autonomic dysfunction.

TABLE 16.2 Drugs and the autonomic system			
Drug	How acts	Use	Comments
Reserpine	Depletes norepinephrine from postganglionic endings by increasing release. Causes decrease in heart action.	To alleviate hypertension	Is one of, and the most potent of a series of alkaloids derived from the Rauwolfia plant. Norepinephrine loss results in vasodilation and fall in blood pressure, and decrease in heart action.
Guanethidine	As above, but exerts no effect on heart	As above	Limited side effects make it a "clinically advantageous drug"
Methyldopa	Lowers brain and heart content of norepinephrine by interfering with synthesis.	As above	May produce toxicity of liver.

Table 16.2 (continued)

Drug	How acts	Use	Comments
Hydralazine	Depress vasoconstrictor center and inhibit sympathetic stimulation.	As above	Many side effects are common.
Veratrum (a series of plant alkaloids)	Slow heart action, stimulate vagus (parasympathetic) activity.	As above	Some side effects.
Hexamethonium	Blocks pre- to postganglionic transmission, Anticholinergic.	As above	
Acetylcholine	Increase or mimics parasympathetic stimulation. Causes some vasodilation.	As above	Not routinely employed. It is rapidly destroyed and affects many organs other than blood vessels.
Atropine	Inhibits acetylcholine and parasympathetic effects	Eye examinations; before general anaesthesia to dry respiratory secretions.	Sympathomimetic
Pilocarpine	Mimics parasympathetic stimulation.	Treatment of glaucoma.	Parasympathomimetic.

Summary

1. The autonomic nervous system consists of visceral afferent (sensory) and visceral efferent (motor) neurons. It controls the activity of smooth and cardiac muscle and glandular tissues.

2. Autonomic fibers use the cranial and spinal nerves, and, in the spinal cord, the gray and white communicating rami.

3. The efferent autonomic pathway usually consists of two neurons and one synapse. The neurons are designated as pre- and postganglionic. Transmission between them is by acetylcholine. Postganglionic fibers produce either acetylcholine or norepinephrine at their endings on effectors.

4. Autonomic ganglia and plexuses provide areas for synapse of pre- and postganglionic neurons.

5. Two divisions of the autonomic system are recognized.
   a. The parasympathetic (craniosacral) division is composed of cranial (cranial nerves 3, 7, 9, 10) and sacral (sacral spinal nerves 2–4) outflows. It conserves body resources.
   b. The sympathetic (thoracolumbar) division consists of all thoracic and lumbar outflows, and increases body activity and utilization of resources.

c. Most body viscera and glands receive fibers from both divisions, creating a dual innervation. Sweat glands, the adrenal medulla, spleen, and blood vessels of skin and skeletal muscles receive only sympathetic innervation.

d. Effects of both systems are tonic and continuous (*see* Table 16.1), and are usually antagonistic in those organs receiving a dual innervation.

6. Different effects of norepinephrine and epinephrine on body organs are explained by the hypothesized presence of alpha and beta receptors on cell membranes.

a. Alpha receptors respond to both chemicals and cause vasoconstriction and relaxation of smooth muscle of the gut.

b. Beta receptors respond only to epinephrine, and mediate vasodilation, increase of heart rate, and dilation of skeletal and heart blood vessels.

c. Drugs may block one or the other of these receptors and be useful in controlling hypertension.

7. Higher centers in brainstem, cerebral cortex, and hypothalamus control or initiate autonomic effects.

8. In treatment of autonomic disorders, surgery or drugs may be employed.

a. Sympathectomy removes sympathetic ganglia to cause (primarily) vasodilation and lowering of blood pressure.

b. Vagotomy reduces gastric secretion in ulcers.

c. Drugs may be employed to mimic or block the effects of autonomic nerves.

Questions

1. Compare the effects on the body of stimulation of the two divisions of the autonomic system.

2. Compare the anatomical differences between the two autonomic divisions.

3. How are the differential effects of stimulation of the two autonomic divisions explained?

4. What body organs receive a dual innervation by autonomic fibers? Which receive only a single innervation?

5. How is the activity of an organ that receives only a single autonomic innervation controlled?

6. What areas of the brain have autonomic effects?

7. How do drugs affect the autonomic system?

Readings

Di Cara, L. V. "Learning in the Autonomic Nervous System." *Sci. Amer.* 222:30, Jan. 1970.

Lefkowitz, Robert J. "β-adrenergic Receptors: Recognition and Regulation." *New Eng. J. Med.* 295:323, Aug. 5, 1976.

Pick, J. *The Autonomic Nervous System.* Lippincott. Philadelphia, 1970.

Chapter 17

Blood Supply of the Central Nervous System; Ventricles and Cerebrospinal Fluid

Objectives

After studying this chapter, the reader should be able to:

- Outline the blood supply to the brain.

- List the components of the Circle of Willis and state its importance to brain blood supply.

- Outline the blood supply to the spinal cord.

- Give the anatomical basis for the blood-brain barrier.

- Suggest some functions for the blood-brain barrier.

- Give a simplified outline of causes of vascular disorders of the brain.

- Describe the locations and shapes of the ventricular system of the brain.

- Describe cerebrospinal fluid regarding sites of production, composition, and circulation.

- Discuss some causes for excessive amounts of cerebrospinal fluid in the CNS.
- Suggest some treatments for excessive cerebrospinal fluid accumulation.

Blood supply

The central nervous system depends on a continual supply of blood perhaps more than any other body organ. The vessels of the brain and cord supply the glucose and oxygen that form the essential nutrients of these organs.

Brain (Fig. 17.1)

A pair of INTERNAL CAROTID ARTERIES, derived from the common carotid arteries of the neck, and the paired VERTEBRAL ARTERIES, arising from the subclavian arteries to the upper appendage, form the arterial supply of the brain. The carotids give rise to two pairs of *cerebral arteries* supplying the cerebrum. The vertebrals join to form the *basilar* artery, and the basilar gives rise to a third pair of cerebral arteries, and to vessels supplying the cerebellum and brain stem. The carotid and basilar arteries are joined by communicating vessels to form the CIRCLE OF WILLIS. The circle allows blood from one set of arteries to flow into the area supplied by the other in case of blockage or diminished flow. The veins draining the brain are largely unnamed except for the dural sinuses (Fig. 17.2).

Spinal cord

A single ANTERIOR SPINAL ARTERY (*see* Fig. 17.1) arises in a Y-shaped manner from the vertebral arteries. Paired POSTERIOR SPINAL ARTERIES arise separately from the vertebral or posterior cerebellar arteries. These three vessels run the length of the cord. SPINAL ARTERIES arise from the abdominal aorta and also supply blood to the cord.

The blood-brain barrier

Many substances that pass easily through the walls of capillaries elsewhere in the body are slowed or stopped in their passage through the walls of cerebral capillaries. Protein molecules, antibiotics, urea, chloride, and sucrose are some of the materials whose passage is slowed. Such observations suggest that there is some sort of *blood-brain barrier* that restricts solute passage from the cerebral capillaries. Electron microscope studies of cerebral capillaries show the anatomical features presented in Figure 17.3. Several points may be emphasized.

Cerebral capillaries seem to lack the "pores" found in other capillaries.

Endothelial cells overlap in cerebral capillaries rather than being placed "end-to-end." This creates four membranes that a solute must pass through, instead of two.

A continuous basement membrane runs around each cerebral capillary, not just beneath each endothelial cell.

About 85 percent of the outer cerebral capillary surface is covered by glia. A substance moving from the capillary to the neuron may therefore have to pass through two cells (endothelial and glial).

The barrier may protect the brain neurons from the entry of potentially harmful substances; it also makes antibiotic treatment of brain inflammation very difficult.

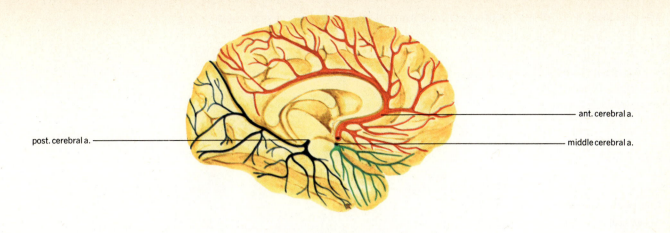

post. cerebral a.

ant. cerebral a.

middle cerebral a.

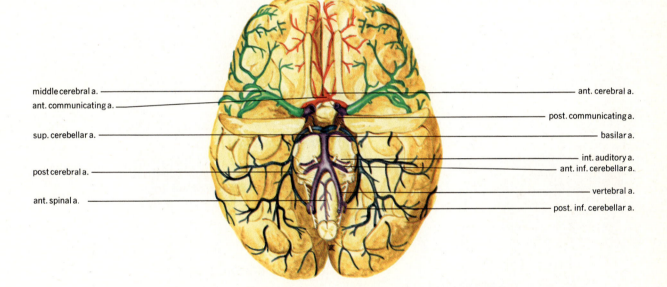

middle cerebral a.

ant. communicating a.

sup. cerebellar a.

post cerebral a.

ant. spinal a.

ant. cerebral a.

post. communicating a.

basilar a.

int. auditory a.

ant. inf. cerebellar a.

vertebral a.

post. inf. cerebellar a.

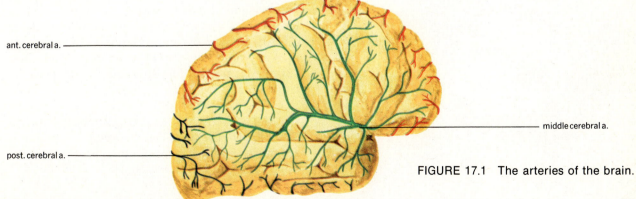

ant. cerebral a.

post. cerebral a.

middle cerebral a.

FIGURE 17.1   The arteries of the brain.

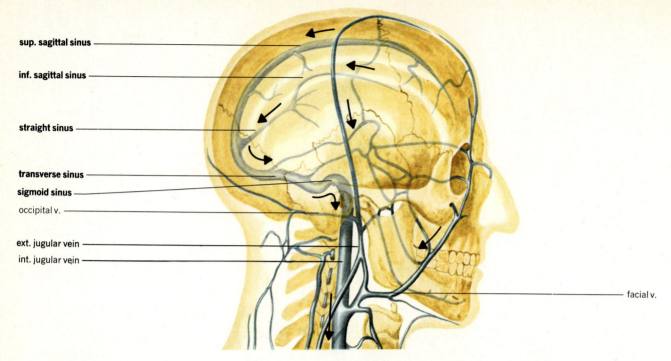

sup. sagittal sinus

inf. sagittal sinus

straight sinus

transverse sinus

sigmoid sinus

occipital v.

ext. jugular vein

int. jugular vein

facial v.

FIGURE 17.2 Veins of the skull and neck (dural sinuses in boldface type).

FIGURE 17.3 The anatomy of the blood-brain barrier. (Drawn from an electron micrograph.)

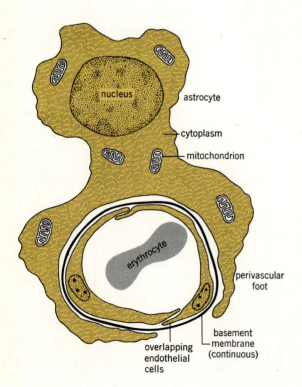

nucleus

astrocyte

cytoplasm

mitochondrion

erythrocyte

perivascular foot

overlapping endothelial cells

basement membrane (continuous)

Clinical considerations

No attempt is made to present all possible disorders involving blood vessels of the central nervous system. It may be stated that cerebral vascular lesions ["strokes," cerebral vascular accidents (CVA)] account for more neurological disorders than any other category of pathological processes. "Stroke" is the third leading cause of death in the United States.

CLASSIFICATION OF VASCULAR DISORDERS OF THE BRAIN. The vascular disorders of the brain are those in which normal nervous system function is impaired in two ways.

Failure of circulation to all or part of the brain, with death of nerve cells resulting.

Rupture of one or more blood vessels, with development of pressure or dissolution of brain tissue or both.

There are three major categories of cerebral vascular disorders, based on cause.

Narrowing or occlusive disorders can occur in any of the vessels of the brain circulation. Vessel diameter is reduced, flow decreases, and cells starve.

Embolic disorders are those in which brain vessels are blocked by masses of substance, such as fats, air bubbles, or clots that are floating in the bloodstream.

Weakening of vessel walls, usually followed by dilation (aneurysm) and possible rupture. If the vessel ruptures, an intracranial hemorrhage results.

Narrowing and occlusion usually occur as the result of deposition of fat in the walls of vessels and subsequent incorporation into the vessel wall (atherosclerosis) or hardening of the wall (arteriosclerosis). Infections (syphilis) may thicken the blood vessel walls and reduce diameter.

Embolism may result from sudden decompression and gas bubbles forming in the bloodstream (the "bends") or by entry of air into the bloodstream after cardiac, pulmonary, or cerebral surgery. Fatty deposits on vessel walls may tear loose and become floating hazards. Clots may form during left heart surgery, and if not removed, may break loose and lodge in the brain.

Weakness is usually the result of structural defects, such as failure to develop vessel coats, infection, toxic processes (poisons), and is secondary to the mechanical trauma of hypertension.

SYMPTOMS OF VASCULAR DISORDER. Symptoms of any vascular disorder affecting the brain are produced by gradual or sudden reduction of blood supply and are essentially the same. Any differences are due to the area of the brain whose blood supply is diminished. The symptoms common to most vascular disorders of the brain are:

Headache, usually vague and short-lived.

Dizziness and *faintness*, which is exaggerated by postural changes.

Nausea and *vomiting*.

Reflexes are unequal; for example, pupillary light reflex.

Muscular weakness, often one-sided, and on the opposite side from where the lesion has occurred.

Mental confusion and speaking difficulties.

Paralysis and/or *difficulty in walking* if motor areas are involved.

These symptoms suggest that a CVA has occurred, and do not lead to dignosis of specific disorders.

Ventricles and cerebrospinal fluid

The nervous system develops as a hollow tube. The central cavity of this tube is modified, as the system develops, into a series of brain cavities known as the *ventricles*. Within the ventricles are vascular structures that secrete the *cerebrospinal fluid*. The fluid fills the ventricles and surrounds the spinal cord.

The ventricles (Fig. 17.4)

Each cerebral hemisphere contains a *lateral ventricle* that communicates through an *interventricular foramen* with the *third ventricle*. The third ventricle is a single, narrow and slitlike structure lying between the two halves of the diencephalon. The *cerebral aqueduct* is a small channel running through the midbrain and pons to the *fourth ventricle* located in the medulla. *Three openings* are found in the roof of the fourth ventricle; two lateral foramina of Luschka, and a single midline foramen of Magendie. These foramina communicate with the *subarachnoid space* of the cord and brain.

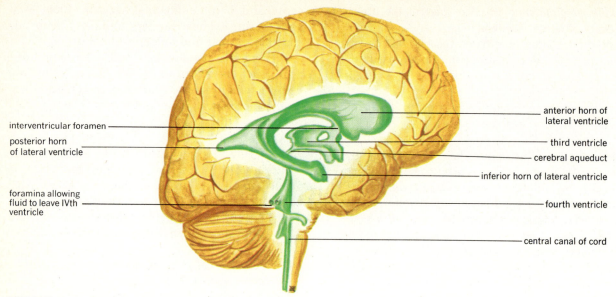

interventricular foramen

posterior horn
of lateral ventricle

foramina allowing
fluid to leave IVth
ventricle

anterior horn of
lateral ventricle

third ventricle

cerebral aqueduct

inferior horn of lateral ventricle

fourth ventricle

central canal of cord

FIGURE 17.4   The ventricles of the brain.

Cerebrospinal fluid

Vascular structures known as *choroid plexuses,* located in the lateral, third, and fourth ventricles, produce the CerebroSpinal Fluid (CSF) from the blood plasma. CSF differs from plasma in enough ways to suggest that it is formed by active processes and not filtration (Table 17.1).

The fluid circulates in a top to bottom direction — that is, from lateral ventricles to third ventricle, to aqueduct, to fourth ventricle, to subarachnoid space. Absorption of the fluid into blood vessels occurs through spinal venous vessels and subarachnoid villi (Fig. 17.5) in the skull.

The fluid acts as a shock absorber for the brain and cord, is easily removed by spinal puncture for analysis, and compensates for changes in blood volume, keeping cranial volume constant.

Clinical considerations

The term hydrocephalus (G. *ydor,* water, + *kephale,* head) means an increased accumulation of

cerebrospinal fluid in the ventricles of the brain. It may result from increased production of CSF, to decreased absorption, or blockage of the flow of CSF.

TABLE 17.1 Plasma and CSF compared for some major constituents		
Constituent or property	Plasma	CSF
Protein	6400–8400 mg%	15–40 mg%
Cholesterol	100–150 mg%	0.06–0.20 mg%
Urea	20–40 mg%	5–40 mg%
Glucose	70–120 mg%	40–80 mg%
NaCl	550–630 mg%	720–750 mg%
Magnesium	1–3 mg%	3–4 mg%
Bicarbonate (as vol % of CO_2)	40–60 mg%	40–60 mg%
pH	7.35–7.4	7.35–7.4
Volume	3–4 liters	200 ml
Pressure	0–130 mm Hg	110–175 mm CSF

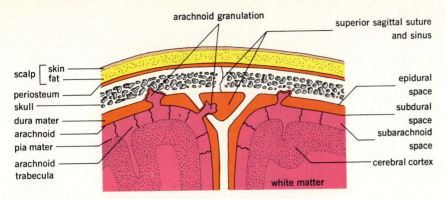

FIGURE 17.5   The subarachnoid villi (arachnoid granulations).

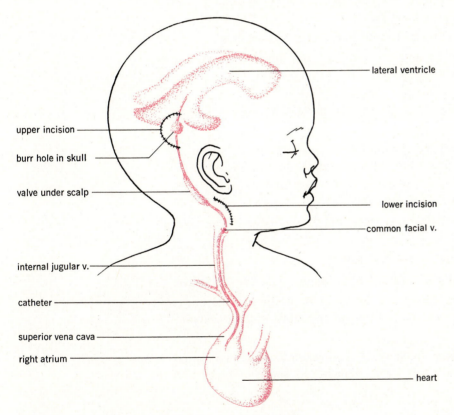

FIGURE 17.6   Surgical treatment of hydrocephalus.

Choroid plexus tumor may increase production of CSF. A pneumoencephalogram, in which the CSF in the ventricles is replaced by air usually demonstrates the presence of such a tumor.

Decreased absorption is usually due to failure of the subarachnoid villi (*see* Fig. 17.5) of the meninges to develop.

Blockage, by tumor or failure of the ventricular system to develop, accounts for about one-third of the cases. The cerebral aqueduct, because of its small size, is a common site of blockage of CSF flow.

As a result of excessive fluid filling the ventricle(s), ballooning of the ventricles may occur and the head enlarges because the sutures be-

tween the cranial bones have not yet grown together. Diagnosis of the condition is made in infants by repeated measurements of head circumference that demonstrate the abnormal rate of increase of head size.

Treatment is usually surgical and involves bypassing the obstruction (if present) with a tube and a one-way valve (Fig. 17.6). The CSF may be delivered to the right atrium of the heart or the peritoneal (abdominal) cavity.

Summary

1. The central nervous system has blood vessels designed to provide it with nutrients and to remove its wastes of activity.
   a. The brain is supplied by the internal carotid and vertebral arteries.
   b. The spinal cord is supplied by anterior and posterior spinal arteries, and by spinal arteries arising from the aorta.
   c. Capillaries in the cerebrum are less permeable to solute passage than elsewhere in the body. A blood-brain barrier exists. The barrier has an anatomical basis.

2. Vascular disorders of the central nervous system deprive nerve cells of their nutrients and lead to cell death. There are three types of disorders.
   a. Occlusive disorders result from narrowing of vessels and deprivation of neurons of oxygen and fuels.
   b. Embolic disorders result when floating masses or bubbles lodge in a cerebral or brain vessel.
   c. Weakening of vessel walls may result in rupture.

3. Symptoms of vascular disorders are basically the same regardless of cause.
   a. Headache, dizziness, nausea, unequal reflexes on both sides of the body, muscular weakness, and mental confusion are the most common symptoms.

4. A series of cavities or ventricles, filled with cerebrospinal fluid, are found within the brain. The fluid also fills the subarachnoid space of brain and cord.
   a. There are four ventricles.
   b. The fluid is formed from plasma in choroid plexuses. It is removed by absorption from the subarachnoid spaces.
   c. Excessive cerebrospinal fluid, from whatever cause, produces hydrocephalus.

Questions

1. What would be expected to happen to brain circulation if (as may happen) flow through a carotid artery is reduced? What portions of the cerebral circulation would be most greatly affected?

2. What is implied by the term blood-brain barrier? What is the structural basis of this barrier?

3. What are the major types of vascular disorders that may affect the central nervous system? What are the symptoms that suggest a vascular disorder?

4. Describe the flow of cerebrospinal fluid.

5. How may hydrocephalus develop?

6. What is the clinical importance of the cerebrospinal fluid?

Readings

Pappas, G. D. "Some Morphological Considerations of the Blood-Brain Barrier." *J. Neurol. Sci.* 10:241, 1970.

Rapoport, Stanley I. *Blood-Brain Barrier in Physiology and Medicine*. Raven. New York, 1976.

Chapter 18

Sensation

Objectives

After studying this chapter, the reader should be able to:

■ Define a receptor, a sensory unit, and a peripheral receptive field.

■ Describe what is meant by sensitivity, what it depends on, and give the properties that all receptors possess.

■ Construct a classification of receptors based on stimulus and location of receptor.

■ Discuss, for the following sensations, the adequate stimulus, neural pathway, and brain termination of the pathway: touch, pressure, heat, cold, pain, kinesthesia, "synthetic senses," visceral sensations.

■ List the several varieties of pain and their characteristics.

■ Describe the structure of the olfactory organ, the types of odors, and the neural pathways for smell.

- Describe the structure of taste buds, the varieties of taste, the type of chemical responsible for each variety of taste, and the neural pathways for taste.

- Describe the structure of the eye, including the extrinsic muscles, their innervations and actions, and the lacrimal apparatus.

- Show how images are normally focused on the retina and account for nearsightedness, farsightedness, and astigmatism, explaining how each may be corrected.

- List the changes occurring during eye accommodation and the causes of each.

- Explain the roles of the rods and cones in vision and list some other functions of the eye.

- Trace the visual pathways, and the effects of lesions in different parts of the pathways.

- Describe the structure of each part of the ear, giving functions for each part.

- Show how sound waves lead to the formation of impulses in the cochlear hair cells and trace the auditory pathways.

- Describe the equilibrial organs of the inner ear, their functions, and trace the neural pathways for equilibrium.

Homeostatic regulation implies the presence in the body of RECEPTORS that can detect alterations in the internal and/or external environments of the body. AFFERENT NERVES then make connections with the CENTRAL NERVOUS SYSTEM. Spinal tracts and various brain stem or diencephalic structures may be involved in the appreciation of sensations, and the cerebral cortex is involved in the final analysis of sensory input.

Receptors

Definition and organization

A receptor is considered to be the peripheral or outer end of an afferent neuron that is specialized to respond best to a particular type of stimulus. An afferent nerve, its branches, and the attached receptors (if any) constitute a *sensory unit*.

The area of the body supplied by that sensory unit is the unit's *peripheral receptive field*. Receptive fields may overlap, and density of receptors (numbers per area) may vary in different regions of the body. Thus, *sensitivity* to a particular stimulus is not necessarily the same in all body areas. Typically, three neurons, designated first order, second order, and third order, carry impulses from the periphery to the cerebral cortex.

Basic properties of receptors and their nerve fibers

All receptors follow certain principles of operation.

The *law of adequate stimulus*. A given receptor responds best, but not exclusively, to a particular stimulus, creating specificity on the part of the receptor. This principle is what accounts for the great variety of receptors the body contains, since there has to be a different receptor for each stimulus.

The *law of specific nerve energies*. All impulses in different sensory nerves are alike, regardless of the receptor involved. What we appreciate as a sensation is thus due to the central connections the fiber eventually makes, and the brain becomes the important organ for analysis of sensory input.

All *receptor transduce* (change) a stimulus into a nerve impulse. A particular stimulus produces a permeability change in the receptor, which results in the formation of a *generator potential*. This, in turn, depolarizes the nerve fiber to create the nerve impulse. This property enables a wide variety of different stimuli to be reduced to a form that the brain can interpret—that is, a nerve impulse.

Most *receptors show accommodation*. As a stimulus of a given strength is continued, the frequency of receptor discharge decreases, or may even cease. This "adjustment" constitutes accomodation, and enables familiar or trivial stimuli to be ignored. Thus we do not feel the clothes on our bodies, rings on our fingers, or continue to smell odors in a room.

Classification of receptors

Receptors are most meaningfully classified according to the stimuli that activate them, as shown in Table 18.1 Another way to classify receptors is according to their location. Thus, receptors at or near body surfaces are termed *exteroceptors* (sensitive to changes in the external environment); those within body viscera are termed *enteroceptors;* those in muscles and tendons are called *proprioceptors.*

TABLE 18.1 A classification of receptors	
Adequate stimulus	**Example(s)**
Mechanical (mechanoreceptors) pressure, bending, tension	Touch and pressure receptors in skin. Pressure or stretch sensitive receptors (baroreceptors) in blood vessels, lungs, gut. Equilibrium and balance receptors in inner ear (semicircular canals, maculae). Organ of hearing (hair cells of cochlea). Kinesthetic receptors in muscles, tendons, joints.
Chemical (chemoreceptors) substances in solution	Taste Smell
Stimulation by type or concentration of chemical	Carotid and aortic bodies (CO_2 and O_2 sensitive) Osmoreceptors in hypothalamus
Light (photoreceptors)	Eye
Thermal change (thermoreceptors)	Heat Cold
Extremes of most any stimuli (nociceptors)	Pain

Specific sensations: somesthesia

Somesthetic sensations are those resulting from stimulation of receptors other than the ones for sight, hearing, taste, and smell. They are, therefore, sensations originating in the skin, muscles, tendons, joints, and viscera. Sensations from the skin are the *cutaneous sensations,* and include touch, pressure, heat, cold, and some types of pain. Those from muscles, tendons, and joints are the *kinesthetic sensations.* Those from visceral organs are *visceral sensations.*

Touch and pressure

The receptors for touch and pressure are the MEISSNER'S CORPUSCLES of the skin and mucous membranes, and the PACINIAN CORPUSCLES of the skin, mesenteries, and pancreas (Fig. 18.1). Naked nerves, or nerve fibers without corpuscles on their endings around hair follicles, also convey impulses of touch when the hair is bent. The stimulus that causes a nerve impulse to form is mechanical change of shape of the receptor. Touch receptors are most numerous on the tongue and fingertips, are less numerous on the face, and least numerous in body areas such as the back. The first-order neurons carrying the senses of touch and pressure enter the CNS through the spinal nerves and the fifth cranial nerves. The gracile and cuneate tracts of the cord are the second-order neurons and carry the impulses to the brain stem. A band of fibers goes next to the thalamus. The thalamus sends third-order neuron relay fibers to the cortex (areas 3,1,2).

Thermal sensations: heat and cold

Discrete end organs for heat and cold cannot always be demonstrated in the skin. A group of ill-defined organs called SENSORY END BULBS, and that differ from other receptors in size, shape, and configuration of their nerve fibers have been described. The receptors for heat and cold belong to this group. Sensitivity to thermal change appears "spotlike," with cold spots outnumbering heat spots by 4–10 to 1. Cold thus appears to be a greater threat to body homeostasis than heat, if numbers of receptors indicate anything.

Warmth receptors respond best between 37 and 40° C (98.6 to 104° F). Cold receptors respond maximally at 15 to 20° C (59 to 68° F) and again strongly at 46 to 50° C (114 to 122° F). The latter range of discharge accounts for the *paradoxical cold,* or feelings of coldness that may be experienced when stepping into a hot shower.

The first-order neurons carrying thermal sensations center the CNS through spinal nerves and the fifth cranial nerves. The lateral spinothalamic tracts are the second-order neurons and end in the thalamus. Relay fibers pass from the thalamus to the cerebral cortex (areas 3,2,1), as the third-order neurons.

Pain

Pain is a protective sensation for the body, as it usually indicates damage or potential damage. It may result from overstimulation of any nerve fiber, and is served by naked nerve endings. There are three varieties of pain.

"Bright" or pricking pain is intense, usually short-lived, and is easily localized to the body part affected. Cutting a finger gives bright pain.

Burning pain, such as that experienced in burning a body part, is slow to develop, lasts longer, and is less easily localized.

Aching (visceral) pain is often nauseating, persistent, and may be referred, as shown in Table 18.2, to the skin of body areas other than that of its source.

Pain may be *referred* to a different area than that in which it originates, because impulses from the painful area enter the spinal cord and stimulate

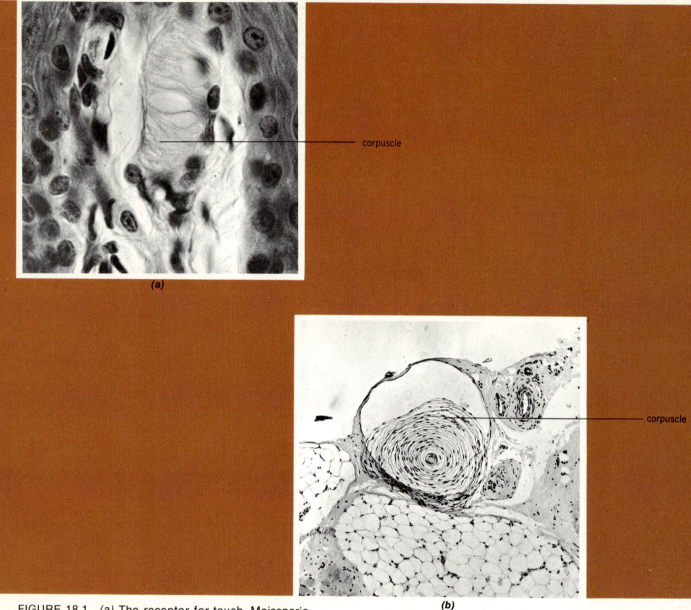

FIGURE 18.1   *(a)* The receptor for touch, Meissner's corpuscle. *(b)* The receptor for pressure, Pacinian corpuscle.

neighboring neurons receiving fibers from the skin. Impulses pass to the cortex and are localized as coming from the skin. Referred pain indicates visceral overstimulation and should be of the utmost concern to the physician and allied medical personnel. It may signal life-threatening disorders of visceral organs.

TABLE 18.2   Referred pain

True point of origin of pain	Common cause	Area to which pain is referred
Heart	Ischemia secondary to infarction	Base of neck, shoulders, pectoral area, arms
Esophagus	Spasm, dilation, chemical (acid) irritation	Pharynx, lower neck, arms, over heart ("heart burn")
Stomach	Inflammation, ulceration, chemicals	Below xiphoid process (epigastric area)
Gall bladder	Spasm, distention by stones	Epigastric area
Pancreas	Enzymatic destruction, inflammation	Back
Small intestine	Spasm, distention, chemical irritation, inflammation	Around umbilicus
Large intestine	As above	Between umbilicus and pubis
Kidneys and ureters	Stones, spasm of muscles	Directly behind organ or groin and testicles
Bladder	Stones, inflammation, spasm, distension	Directly over organ
Uterus, uterine tubes	Cramps (spasm)	Lower abdomen or lower back

HEADACHE. Headache is probably the most common pain complaint. Several things may contribute to its production.

Tension on blood vessels or meninges, as in brain tumor.

Increased intracranial pressure, from vasodilation. The increased amount of blood entering the rigid skull then puts pressure on the brain and its coverings.

Inflammation of the sinuses or any other pain-sensitive structure in the brain or cranium. Commonly, the pain of an inflamed sinus is felt in the forehead or cheek.

Spasm of cranial muscles. This produces the so-called "tension headaches." This type of headache is most commonly felt in the back of the head, or may be a "pain in the neck."

Eye disorders may result in pain in or behind the eyes as a result of eyestrain.

The conduction pathways for pain include the spinal nerves, the fifth cranial nerves and the spinothalamic tracts. The thalamus may interpret pain, and cortical areas 3,1,2 localize pain and determine its intensity.

Aspirin appears to be effective in treatment of pain by reducing nerve conduction and raising thresholds for pain in the thalamus-hypothalamus areas.

Kinesthetic sense

Sense of body position and movement is provided by receptors in muscles, tendons, and joints. Tension on, or shortening of, the receptor appears to be the adequate stimulus. *Conscious muscle sense,* or appreciation of body position, is carried over afferent spinal and cranial nerves to the gracile and cuneate tracts, to the thalamus, and then to the cerebrum; *unconscious muscle sense,* concerned with muscle tone, is carried over afferent spinal and cranial nerves to the spinocerebellar tracts, and then to the cerebellum.

The MUSCLE SPINDLE is an important sensory receptor of the muscles that provides information on the muscle's length and the rate of shortening when the muscle contracts. The spindle consists of several tiny skeletal muscle fibers known as the *intrafusal fibers,* around which are wrapped the *annulospiral sensory nerves.* Stretching of the fibers causes stimulation of these spiral endings and provides the information about muscle length. The amount of stretch required to cause stimulation of those endings is determined by the degree of contraction of the intrafusal fibers themselves. Logically, if the fibers are contracted, it will take more force to lengthen them, that is, a greater stretch on the muscle. The degree of contraction of the intrafusal fibers is determined by so-called *gamma-efferent nerves* that are believed to come from the basal ganglia of the brain. The spindles are important parts of the crossed-extensor and other spinal muscular reflexes.

Synthetic senses

Itch, tickle, and *vibrational sense* have no specific nerve endings to serve them. They appear to result from particular modes (methods of how a stimulus is delivered) of stimulation, or simultaneous stimulation of touch, pain, and pressure receptors. Itch appears to result from chemical stimulation of skin pain fibers. If a light stimulus moves across the skin, tickle results, through stimulation of touch receptors. Vibration sense appears to result from rhythmical and repetitive stimulation of pressure receptors; frequencies of 250 to 300 cycles per second are the most effective.

The reader is reminded of the roles of thalamus and hypothalamus in sensory interpretation. Those sections may be profitably reviewed at this time.

Visceral sensations

Visceral or *organic sensations* result from stimula-tion of internal organs, and include hunger and thirst. Other sensations (bladder and rectal fullness) will be discussed in appropriate chapters.

Hunger is associated with powerful contractions of the stomach, which may result from lowering of the blood glucose levels. The compression of the gastric nerves by the powerful contractions may create the unpleasant and often near-painful quality of the sensation.

Thirst occurs when the oral membranes dry as a result of decreased salivary secretion. The decreased secretion is, in turn, due to dehydration, and signals a need for fluid intake.

The somesthetic senses are summarized in Table 18.3.

Clinical considerations

Damage to peripheral nerves produces symptoms that depend on the location of the lesion and the function of the nerve. Sensory disturbances include changes in total sensation (anesthesia) or specific changes in position sense, vibratory sense, or pain, touch, and pressure.

ANESTHESIA. Complete cutting of a nerve produces loss of all sensation in the area the nerve supplies. Injuries from bumps, heat, friction rubs (shoe on the heel) are not felt, and severe damage to the skin may occur. Protection of the anesthetized area becomes the primary consideration.

PARESTHESIA. This term refers to abnormal sensations arising from damage to peripheral nerves or cord tracts. It usually involves pain being referred to the area served by the nerve, although no damage has occurred in that area.

HYPERESTHESIA. This term describes lowered pain thresholds and a greater sensitivity to pain.

It is again noted that peripheral nerves may regenerate if the separation or area of damage is not great, and if the neurilemmal sheath is intact.

Otherwise, the regenerating fiber may not make its proper connection as before, or it may form a tangled mass known as a *neuroma*, which is both painful and functionless.

TABLE 18.3 A summary of somesthetic senses

Sensation	Receptor	Adequate stimulus	First order neuron	Second order neuron	Third order neuron	Comments
Touch	Meissner's corpuscle	Mechanical pressure to skin	Spinal or cranial afferent	Gracile and cuneate tracts	Thalamus to cortex	Light pressure causes change in receptor shape
Pressure	Pacinian corpuscle	As above	As above	As above	As above	Heavier pressure causes change in receptor shape
Heat	Sensory ending	Rise of temperature, especially between 37–40°C	As above	Spinothalamic tracts	As above	—
Cold	Sensory ending	Fall of temperature, especially between 15–20°C	As above	As above	As above	—
Pain	Naked nerve endings	Overstimulation of any nerve; strong stimulation of naked nerves as by tension, pressure	As above	As above	As above	Three types of pain: bright, burning, visceral
Kinesthesis	Muscle, joint and tendon organs	Tension, stretch or movement	As above	Conscious: gracile and cuneate tracts.	Thalamus to cortex	Conscious is body position
				Unconscious: spinocerebellar tracts	Cord to cerebellum	Unconscious is tone
Synthetic (itch, tickle, vibration)	None specific	Chemical and mechanical stimulation	—	—	—	Itch: chemical stimulation of pain fibers Tickle: moving stimulation Vibration: touch organs
Visceral (hunger, thirst)	None specific	Mechanical and drying	—	—	—	Most visceral afferents are carried in the vagus nerve

Smell and taste

Smell and taste are chemical senses, in that the adequate stimuli are chemicals in solution in air or water. The senses reinforce one another; anyone with a cold and blocked nasal passages knows how "tasteless" food becomes. Smell has survival value for lower animals, and perhaps also for people, in terms of identification of enemies, territories, and food seeking. Taste may play a part in food selection if an animal is deficient in certain nutritional elements. Children have been shown over the long run to select a diet that is adequate if presented with a variety of foods in separate dishes. Food "fads" should not be a cause for concern unless they continue for long periods of time.

Smell

The receptors for the sense of smell are found in the OLFACTORY EPITHELIUM of the nasal cavities (Fig. 18.2). The olfactory cells are bipolar nerve cells. The outer portion of each cell is a dendrite modified into a cylindrical process that proceeds vertically and then turns parallel to the epithelial surface. On the dendritic process are "receptor sites" into which the molecules to be smelled fit.

Classification of odors appears to depend on molecular shape, and is highly subjective. *Seven basic odors* are described:

Camphoraceous, musky, floral, etheral, pungent, putrid, and pepperminty.

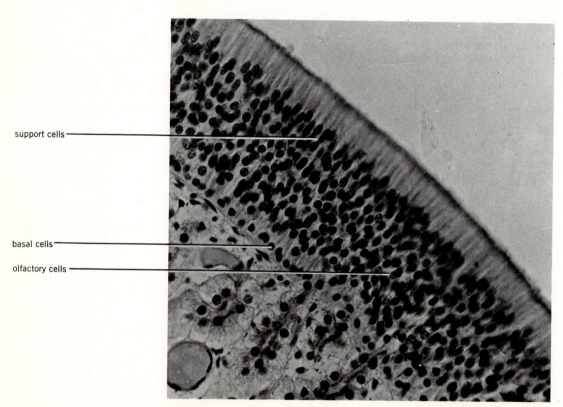

support cells

basal cells

olfactory cells

FIGURE 18.2   The olfactory epithelium.

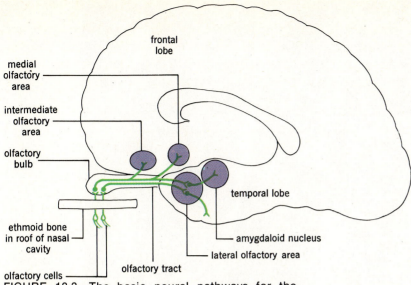

FIGURE 18.3 The basic neural pathways for the sense of smell.

In general, the sense of smell is much more acute than the sense of taste, about 25,000 times more sensitive, as judged by the concentrations of substances required to create a sensation. Low concentrations of substances cause maximum receptor response, followed by rapid adaptation. This fact suggests that the sense of smell is designed to detect the presence, rather than the amount, of an odor.

The neural pathways for smell (Fig. 18.3) pass from *olfactory cells* to the *olfactory bulb* of the brain, then over the *olfactory tracts* to the *olfactory areas* in the temporal and frontal lobes of the brain. Fibers also pass to autonomic centers in the hypothalamus and brain stem to produce salivation, sexual behavior, and emotional expression as a result of olfactory stimulation.

CLINICAL CONSIDERATIONS. Disorders of the olfactory system are rare, but have considerable importance when present. Tumors pressing on or invading any portion of the system may produce loss of smell. Temporal lobe lesions may cause olfactory hallucinations, usually of disagreeable odors. Such odors often precede an epileptic seizure.

Taste

The receptors for taste are the TASTE BUDS (Fig. 18.4), located primarily on the tongue. A few buds may be found on the soft palate, and in the walls of the pharynx and epiglottis. The nerve cells in the buds have a rodlike shape, and have "taste hairs" on their outer ends. The hairs are actually short dendritic processes. The material to be tasted must fit into receptor sites on the taste hairs to depolarize the nerve cell. *Four basic taste sensations* are described:

Sour tastes are produced by H^+ or acids.

Salty tastes are produced by cations of ionized salts, such as Na^+, NH_4^+, Ca^{2+}, Li^+, and K^+.

Sweet tastes are produced by the presence of compounds containing hydroxyl ($-OH$) groups. Alcohols, amino acids, sugars, ketones, and lead salts all taste sweet (children may eat paint for this reason and suffer lead poisoning).

Bitter tastes are produced by alkaloids such as quinine, strychnine, and caffeine and also by many long chain organic molecules.

The buds serving these sensations are not

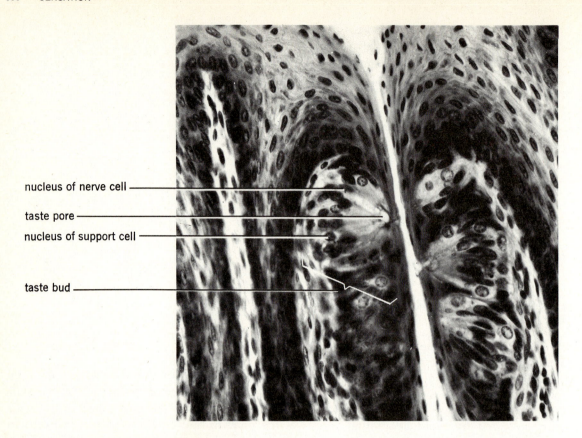

nucleus of nerve cell

taste pore

nucleus of support cell

taste bud

FIGURE 18.4   Taste buds.

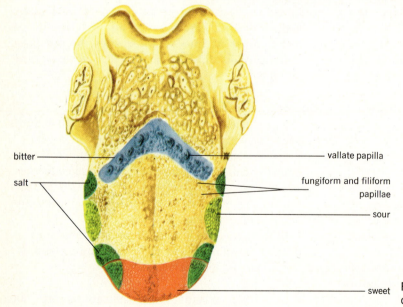

bitter

salt

vallate papilla

fungiform and filiform
papillae

sour

sweet

FIGURE 18.5   The distribution of the sense
of taste on the tongue.

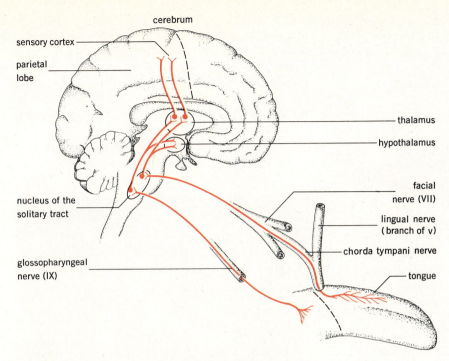

FIGURE 18.6 The basic neural pathways for the sense of taste.

uniformly distributed over the tongue (Fig. 18.5) and do not have the same thresholds. Greatest sensitivity is to bitter, followed by sour, salt, and sweet.

The afferent pathways for taste include the VII (facial) cranial nerve for buds on the anterior two-thirds of the tongue, the IX (glossopharyngeal) nerve for buds on the posterior one-third of the tongue, and the X (vagus) nerve for those elsewhere. The impulses pass to brain stem, thalamus, and to the parietal cortex (areas 3,1,2). These paths are shown in Figure 18.6.

Vision

The EYES are photoreceptors specialized to respond to light energy. They are located within the orbits of the skull and are moved by six extrinsic muscles (Fig. 18.7 and Table 18.4).

The orbits protect the eyes on all but the anterior one-sixth or so, and this latter portion is protected by the conjunctiva and the eye lids (Fig. 18.8). A two-lobed lacrimal gland lies on the upper outer surface of each eye and produces the tears that cleanse, lubricate, and destroy bacteria on the anterior surface of the eye. Tears are drained into the nose by a system of ducts (Fig. 18.9).

Structure of the eye (see Fig. 18.8)

The adult eyeball is an organ about 25mm (1 inch) in diameter. It is composed of three layers of tissue.

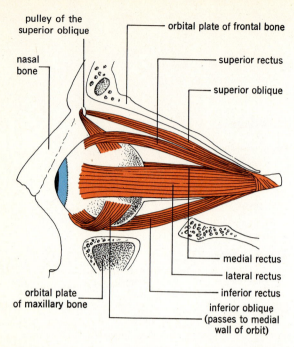

FIGURE 18.7 The six extrinsic muscles of the left eye, viewed from the lateral side. Notice how the pulley converts the direction of pull of the superior oblique.

The SCLERA is the outermost coat. Posteriorly, it is opaque and forms the "white of the eye." Anteriorly, it is transparent, and forms the *cornea*.

The UVEA forms the middle coat of the eyeball. Anteriorly, it consists of the *ciliary body,* including the *ciliary muscle, iris, lens,* and *lens ligaments.* Posteriorly, it is composed of the vascular *chorioid* (choroid).

The RETINA is the innermost layer. It extends only as far anteriorly as the posterior border of the iris.

TABLE 18.4	The extrinsic eye muscles	
Name	Innervation (cranial nerve)	Action on eyeball
Lateral rectus	VI Abducent	Out
Medial rectus	III Oculomotor	In
Superior rectus	III Oculomotor	Up and in
Inferior rectus	III Oculomotor	Down and in
Inferior oblique	III Oculomotor	Up and out
Superior oblique	IV Trochlear	Down and out

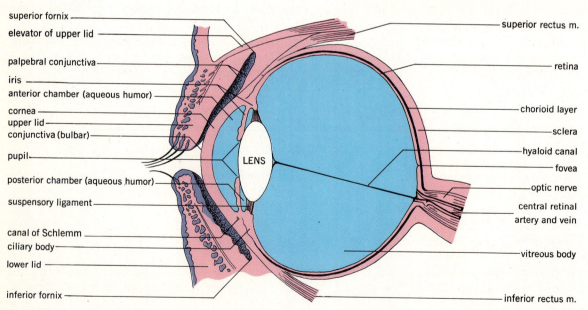

FIGURE 18.8   The eye and eyelids in section.

It contains the *rods* and *cones* that are the visual receptors of the eye. Its nerve fibers form the *optic nerves*, that exit from the eye.

The portion of the eye anterior to the lens is filled with the watery *aqueous humor,* secreted from the ciliary body. A ring of veins, known as the *canal of Schlemm,* lies at the junction of sclera and cornea (*see* Fig. 18.8), and drains the humor from the anterior chamber. Blockage of the canal raises intraocular pressure higher than its normal 20 mm Hg (millimeters of mercury), and *glaucoma* results. Glaucoma causes compression of blood vessels in the eye and compression of the optic nerve. Retinal cells die and the optic nerve may atrophy, producing blindness, Glaucoma is the leading cause of blindness in the world.

The remaining portion of the eye is filled with the gel-like *vitreous humor,* which serves mainly to maintain the shape of the eyeball.

The blood vessels that are responsible for most of the nourishment of the eye enter the eyeball from behind at the *optic disc*. The disc is where the optic nerve leaves the eye, and is lacking in visual receptors, hence the name "blind spot." Ophthalmoscopic examination of the eye is often performed to determine the status of the eye structures and vessels. What is seen is termed the fundus, and it normally appears as is shown in Figure 18.10.

Functioning of the eye

IMAGE FORMATION. Light waves entering the eye must be bent or *refracted* so that they form a sharp image on the retina. Ability to bend light waves depends on the curvature of the refracting surface and the media (air or fluid) on either side of the surface. Most of the refraction of light entering the eye occurs at the anterior corneal surface that might be said to act like a coarse adjustment on a microscope. The refracting power of the cornea is fixed by its anatomical curvature and cannot change. Final focusing occurs by changes in curvature of the lens to create sharp retinal images.

CLINICAL CONSIDERATIONS. *Errors of refraction* include *myopia, hypermetropia* (or *hyperopia*) *astigmatism,* and *presbyopia.*

The normal or *emmetropic eye* focuses an inverted or reversed image directly on the retina. The *myopic* (nearsighted) eye focuses images in front of the retina, because the eyeball is too long for the refracting power of the eye. The *hypermetropic* (farsighted) eye focuses images behind the retina because the eyeball is too short for the refracting power of the eye. These abnormalities are easily corrected by fitting lenses ahead of the cornea. These conditions and their correction are shown in Figure 18.11. *Astigmatism* results when the cornea or lens, instead of having the same curvature vertically and horizontally, is shaped like a teaspoon. The unequal curvatures cause two separate focal points and double vision. *Presbyopia* ("old eye") results from hardening of the lens with age, and causes an inability to accurately focus near objects.

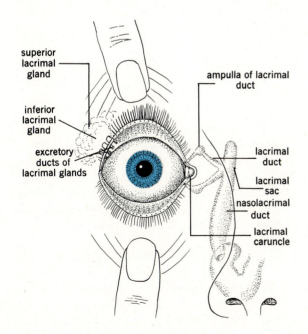

superior lacrimal gland

inferior lacrimal gland

excretory ducts of lacrimal glands

ampulla of lacrimal duct

lacrimal duct

lacrimal sac

nasolacrimal duct

lacrimal caruncle

FIGURE 18.9   The lacrimal apparatus.

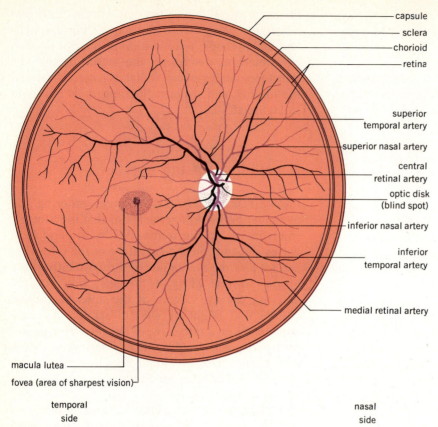

capsule
sclera
chorioid
retina

superior
temporal artery

superior nasal artery

central
retinal artery

optic disk
(blind spot)

inferior nasal artery

inferior
temporal artery

medial retinal artery

macula lutea

fovea (area of sharpest vision)

temporal
side

nasal
side

FIGURE 18.10   The fundus of the right eye as seen through an ophthalmoscope.

ACCOMMODATION. A series of three changes occur in the eye as images are formed: change in lens curvature for focusing, pupillary constriction, and convergence of the eyes.

Change in lens curvature occurs as the ciliary muscle contracts, tension is relaxed on the lens ligaments, and the lens becomes more rounded as a result of its own elastic capsule. Curvature is adjusted so that images are normally brought to a sharp focus on the retina.

Pupillary constriction blocks light rays that would pass through the edges of the lens and cannot be brought to sharp focus. Pupillary constriction also occurs in response to light shown into the eye, and is a protective response.

Convergence directs both eyes at the object being viewed. If one of the extrinsic eye muslces contracts more forcibly, or is shorter than the others, accurate directing of the eyes on the object may not occur, and double vision may result (strabismus).

CLINICAL CONSIDERATIONS. If the oculomotor nerve or its central connections are damaged, pupillary constriction to light may not occur. An Argyll-Robertson pupil results. Pupil size also changes with anesthesia, being dilated in light and deep stages, and constricted in the intermediate stage.

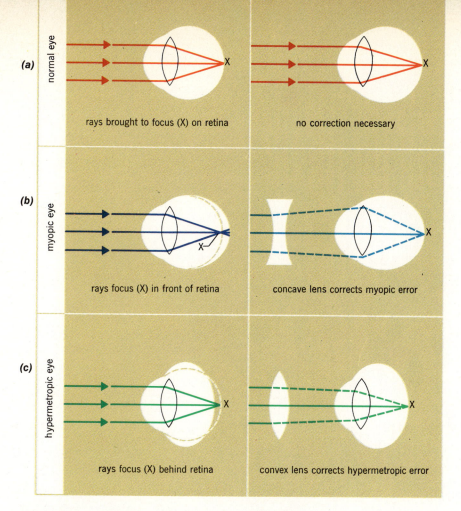

FIGURE 18.11 *(a)* Emmetropic, normal. *(b)* Myopic, nearsighted. *(c)* Hypermetropic, farsighted eyes. Diagram indicates type of lens required to correct abnormal eyes.

Strabismus (cross-eye) is the result of improper convergence. It is a condition that, if uncorrected, can lead to the brain ignoring one of the two images it receives from the eyes, and the subject may become functionally blind in one eye.

Cataract is lens opacity, and may result from aging, infection, or trauma of the eye, or as the result of diabetes mellitus. It results in failure to transmit light rays to the retina. The lens may be surgically removed if it is too opaque to transmit enough light to see by; glass lenses then correct the vision.

Corneal transplant may be employed in corneal damage. The tissue is not vascular, and rarely is rejected when transplanted.

RETINAL FUNCTION. The retina consists of 10 layers of neurons and fibers (Fig. 18.12). The second layer contains the rods and cones (Fig. 18.13).

The rods. Rods are sensitive to low-intensity light and are responsible for *scotopic vision* (night vision). Rods contain a visual pigment called

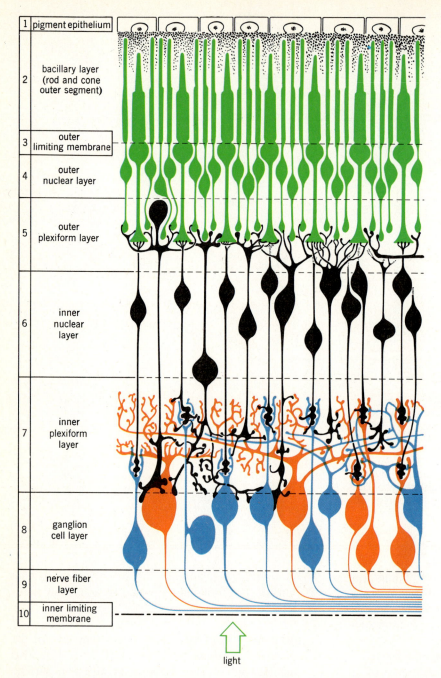

1	pigment epithelium
2	bacillary layer (rod and cone outer segment)
3	outer limiting membrane
4	outer nuclear layer
5	outer plexiform layer
6	inner nuclear layer
7	inner plexiform layer
8	ganglion cell layer
9	nerve fiber layer
10	inner limiting membrane

light

FIGURE 18.12   The organization of the retina.

rhodopsin. The pigment is thought to be split by light, and the products of the splitting cause depolarization of the rod cell. A nerve impulse is

formed. Resynthesis of the pigment requires a derivative of *vitamin A*. Thus, if the body is deficient in vitamin A, "night blindness" may

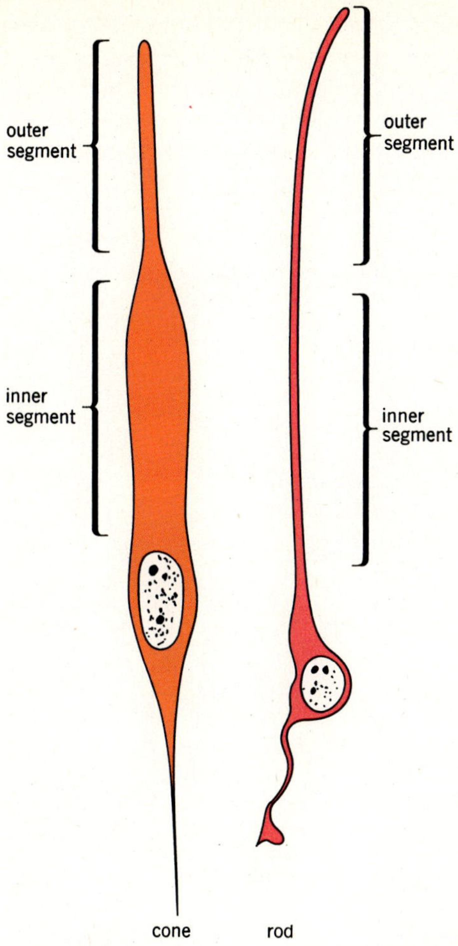
outer
segment
inner
segment
outer
segment
inner
segment
cone
rod

FIGURE 18.13 The structure of the rods and cones.

result. Yellow vegetables are sources of vitamin A.

The cones. There appear to be three classes of cones, each with a different visual pigment in it. There are cones that respond maximally to red, green, or blue light. It is known that all colors may be created by mixing these three basic colors. Differential (unequal) stimulation of the three cone types may be the basis for *color vision,* one of the functions of the cones. *Photopic vision* (daylight vision) is the other function of the cones. The basic mechanism of cone function is presumed to be the same as that for rods—that is, light causes pigment breakdown and depolarization of the cone.

CLINICAL CONSIDERATIONS. *Color blindness* is easily explained on the basis of the three cone types, by presuming that either a given type of cone is missing, or that the pigment is not synthesized. An individual is said to have *protanopia* (red blindness), *deuteranopia* (green blindness), or *tritanopia* (blue blindness). Most color blindness is inherited, and is sex-linked (defective gene is carried on sex chromosomes), so that males are more commonly affected than females.

OTHER RETINAL FUNCTIONS. *Visual acuity* is a function of the retina. This refers to sharpness of vision and depends on the "density of packing" of cones in the fovea. The fovea contains only cones that are narrower than in other parts of the retina. Acuity depends on the presence of an unstimulated cone between two stimulated ones. If the cones are narrower, an object may be moved farther away before its image falls on neighboring cones. How "far" we can see clearly is the most common measure of acuity.

The ability to *fuse* separate images into a continuous sequence (as in viewing a movie) depends on *visual persistence* in the retina. The products of destruction of visual pigments are not removed immediately and the "image lasts" a while. The next image then fuses with the first, and we see "continually," and not in "bits and pieces."

After images also result from persistence of visual pigment breakdown, and one may "see something" for several seconds after the stimulus has disappeared.

Visual pathways (Fig. 18.14)

Impulses originating in the retina pass over the *optic nerves* to the *optic chiasma,* and from there over the *optic tracts* to the *thalamus.* From the thalamus, fibers pass over the *optic radiation* (or *geniculocalcarine tract*) to the *occipital lobes* (area 17) of the cerebrum. Further interpretation of visual impulses is provided by the visual association areas (18, 19).

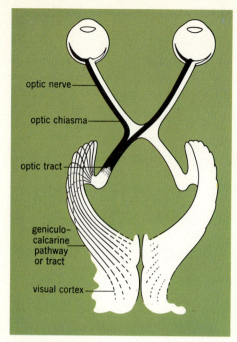

FIGURE 18.14 The neural pathways for the sense of sight.

CLINICAL CONSIDERATIONS. Figure 18.15 presents the types of visual loss that would result from damage to the visual pathways. One should note that there is a definite spatial relationship of the various retinal fibers, so that loss is characteristic for the site of injury.

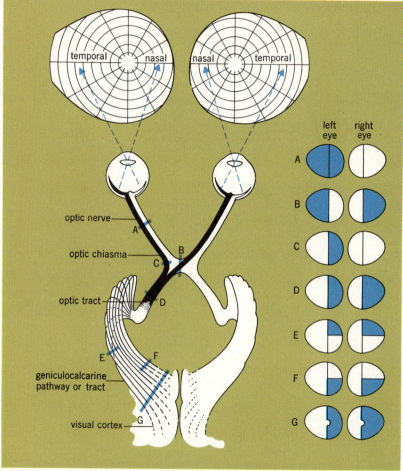

FIGURE 18.15 The types of visual loss resulting from lesions in various parts of the visual pathways. Blue areas indicate deficits of vision.

Hearing

The auditory mechanism responds to sound waves, and is divided into three portions; the *outer, middle, and inner ears* (Fig. 18.16).

The outer ear

The outer ear consists of the AURICLE or *pinna,* the fleshy flap attached to the side of the skull, and the EXTERNAL AUDITORY MEATUS or ear canal. The pinna may aid in collection and directing of sound waves to the middle ear. The canal is slightly S-shaped, and directs sound waves to the eardrum. The canal is guarded externally by large hairs and by glands (ceruminous glands) that produce the ear wax (cerumen). The wax is bitter to the taste, and is supposed to aid in keeping insects out of the ear canals.

CLINICAL CONSIDERATIONS. Accumulation of cerumen may block transmission of sound waves down the canal, and result in one type of *transmis-sion deafness.* This cause of the condition is more common in children.

The middle ear

The middle ear consists of the MIDDLE EAR CAVITY, and its contents. The cavity is an irregular, air-filled space that communicates with the throat by way of the *pharyngo-tympanic (Eustachian) tube.* The tube allows equalization of air pressure on the two sides of the eardrum. The tympanum or ear-drum forms the lateral wall of the cavity. Sound waves cause vibration of the drum. Three EAR OSSICLES—the *malleus, incus,* and *stapes*—are caused to move by the vibrations of the drum. The ossicles form a system of levers that increases the pressure and decreases the movement of the stapes on the oval window of the cochlea. Since the drum vibrates in air and the cochlea in fluid, more energy is required to vibrate the fluid. Two small muscles in the middle ear (tensor tympani

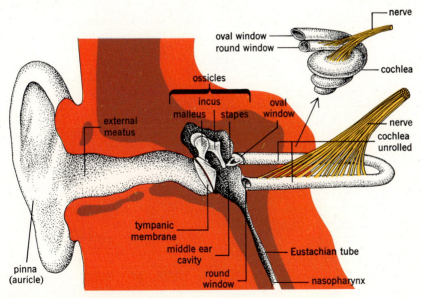

FIGURE 18.16   The ear. a, b, and c indicate the 1st, 2nd, and final turns of the cochlea.

and stapedius) contract when sounds are very loud and prevent excessive movement of the ossicles, which could damage the inner ear. The contraction of the muscles is reflexly controlled and is called the *auditory reflex.*

CLINICAL CONSIDERATIONS. Transmission (conduction) deafness occurs when sound waves are prevented from reaching the eardrum or by failure of the ossicles to transmit the sound waves to the cochlea. *Inflammation* of the middle ear (otitis media), *fusion* of the ossicles, or *fixation* of the stapes in the oval window (otosclerosis) may all result in loss of hearing. Bone conduction may be employed if the ossicles cannot be freed. This utilizes the fact that vibrations may reach the fluid of the cochlea by transmission through the skull bones, rather than by way of the ossicles. This forms the basis for many types of hearing aids.

The inner ear

The inner ear (Fig. 18.17) contains the cochlea, the organ of hearing, and organs concerned with balance and equilibrium (maculae and semicircular canals).

FIGURE 18.17 The membranous labyrinth of the inner ear.

THE COCHLEA AND HEARING. As the stapes moves in the oval window of the cochlea, shock waves are created that cause vibration of the BASILAR MEMBRANE (Fig. 18.18) of the ORGAN OF CORTI (Fig. 18.19). The basilar membrane is composed of 25,000 to 30,000 strands of tissue that are "tuned" to vibrate to particular pitches of sound (place or selective theory of hearing). The motion of the basilar membrane is translated into nerve impulses through bending of the HAIR CELLS that form the receptors of the organ of Corti. The hair cells are bent as the basilar membrane vibrates; this creates generator potentials that depolarize the hair cells to create a nerve impulse. So that the shock waves do not continue to "bounce around" in the cochlea and cause continuous nerve impulse formation, the vibrations are damped by movement of the round window at the base of the cochlea. As the stapes moves inward at the oval window, the round window bulges outward and tends to cushion the shock wave.

AUDITORY PATHWAYS. Impulses pass from the hair cells over the cochlear portion of the *eighth cranial nerve* to the *cochlear nuclei* in the brain stem. From the brain stem, impulses pass to the *thalamus,* then to the *auditory cortex* in the tem-

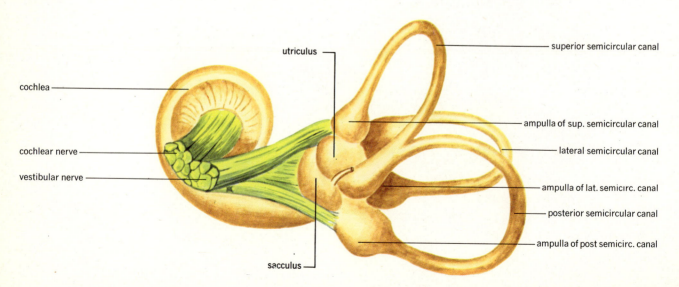

cochlea — cochlear nerve — vestibular nerve — sacculus — utriculus — superior semicircular canal — ampulla of sup. semicircular canal — lateral semicircular canal — ampulla of lat. semicirc. canal — posterior semicircular canal — ampulla of post semicirc. canal

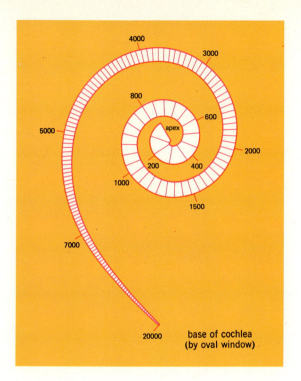

poral lobes (area 41,42). Both cochleas are represented in both temporal lobes, with high tones anteriorly, and low tones posteriorly. The temporal lobes determine localization of sounds, the quality of the sound, and its intensity. The temporal association areas are concerned with understanding of auditory input.

CLINICAL CONSIDERATIONS. *Nerve deafness* results from damage to the cochlea and/or the auditory pathways to the brain. The antibiotic streptomycin, if given to treat bacterial infections, may damage the eighth cranial nerve and may cause deafness and disturbances of equilibrium.

FIGURE 18.18 Pitch analysis by the basilar membrane of the cochlea. Numbers indicate cycles per second (Hz). Membrane averages 32 millimeters long, 0.04 millimeters wide at base, 0.5 millimeters wide at apex.

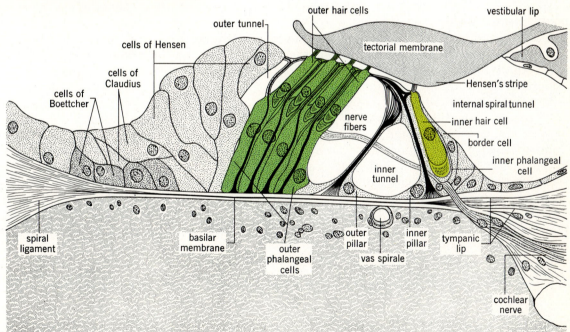

FIGURE 18.19 The organ of Corti. The sensory elements are the hair cells.

The inner ear and equilibrium and balance

The SEMICIRCULAR CANALS are three fluid-filled channels in each ear (*see* Fig. 18.17). An enlarged ampulla contains a CRISTA that responds, by being bent, to changes in acceleration and direction of motion. Bending occurs as the head moves and causes the fluid in the canals to move toward or away from the cristae. The three canals are placed in three mutually perpendicular planes to one another so that any head movement will cause excitation of one or more cristae. The MACULAE are organs of position, not movement. They consist of HAIR CELLS whose processes are set in a gelatinous mass containing tiny granules called otoliths (ear stones). The entire macula rests in a cavity in the vestibule of the inner ear. The maculae are activated by gravity pulling or pushing on the otoliths of the organ and bending the hair cells. Excitation of semicircular canals or maculae results in reflex responses (labyrinthine reflexes) tending to maintain the body properly in space. LABYRINTHINE REFLEXES are categorized into the two following groups.

Acceleratory reflexes, arising from stimulation of the semicircular canals, produce responses of eyes, neck, trunk, and appendages to "starting," "stopping," or "turning" motions. *Nystagmus*, a horizontal, vertical, or rotary movement of the eyes, occurs on angular acceleration. Limb responses are those of extension on the side away from an angular acceleration, and toward the force if linear. Thus, a person "leans into" a curve, leans forward on acceleration, and backward on deceleration, maintaining balance.

Positional reflexes, arising from stimulation of the maculae, involve righting reflexes and notification of head position. The classical statement that "a cat always lands on its feet" illustrates the operating of the righting reflexes. In this example the head is first turned into a proper relationship to the ground, and the body is then brought into relationship to the head. The pathway for communication of information from the canals and maculae is via the *vestibular portion* (vestibular nerve) of the eighth cranial nerve. The fibers pass to brain stem *vestibular nuclei*, located in the lower lateral pons, and thence to the *cerebellum*, where appropriate muscular response is integrated.

CLINICAL CONSIDERATIONS. Although not strictly a disorder, *motion sickness* involves repetitive changes in angular and linear acceleration affecting the receptors. *Nausea* and *vomiting* may result. Vomiting results from motion sickness as follows.

There is, in the medulla of the brain stem, a "trigger zone" that works with the vomiting center to cause vomiting. The zone is larger than the actual center. Impulses that arrive at this trigger zone from the semicircular canals, excite the zone; impulses pass from the zone to the vomiting center and cause the nausea and vomiting associated with motion sickness.

Infections of the vestibular nerve may result in loss of equilibrium as may trauma that injure the nerve. Indeed, any basic loss of equilibrium without change of muscle tone is indicative of altered inner ear equilibrium function. Testing equilibrium function typically involves spinning the subject in a Barany chair with the head in various positions to excite the different canals. If nystagmus, dizziness and loss of equilibrium are not produced, a lesion is presumed to exist somewhere in the system.

Summary

1. Receptors detect changes in internal and external environments, and lead to homeostasis-maintaining responses.

2. Receptors:
   a. Follow the law of adequate stimulus and respond best to one type of stimulus.
   b. Follow the law of specific nerve energies that states that all impulses in different nerves are the same, and the sensation we appreciate depends on where the nerve pathways terminate in the brain.
   c. Cause a stimulus to be changed into a nerve impulse.
   d. Show accommodation (decrease of discharge with continued stimulation).

3. Receptors are best classified by adequate stimulus (Table 18.1).

4. Touch and pressure sensations:
   a. Are initiated by a change in shape of the receptor.
   b. Are served by Meissner's and Pacinian corpuscles.
   c. Are conducted over gracile and cuneate tracts to the brain stem. Fibers then pass to the thalamus and then to the cerebrum.

5. Heat and cold receptors:
   a. Respond to changes in temperature.
   b. Send impulses over spinothalamic tracts to the thalamus and then fibers go to the cerebrum.

6. Pain:
   a. Results from overstimulation of any nerve.
   b. Is protective in nature.
   c. Is served by naked nerve endings.
   d. Is transmitted over spinothalamic tracts to thalamus and then to the cerebrum.
   e. Is of several types: bright pain, burning pain, and aching pain; the latter may be referred to a different body area than its point of origin. Referred pain indicates involvement of visceral organs.
   f. Headache results from tension on cranial structures, increased intracranial pressure, or inflammation, spasm, and eyestrain.

7. Kinesthetic sense originates in muscles, joints, and tendons.
   a. It maintains awareness of body position.
   b. It maintains muscle tone.
   c. It results from stimulation of specific end organs, muscle spindles, and tendon organs.

8. Synthetic senses (itch, tickle, vibration):

   a. Have no specific end organs.

   b. Result from stimulation of other end organs.

9. Visceral sensations (hunger, thirst):

   a. Originate in visceral organs.

   b. Are important in regulation of nutrition and water balance.

10. Loss or alteration of the above sensations is significant in determining what nerves have been damaged. Anesthesia, paresthesia, or hyperesthesia may occur when sensory nerves are damaged.

11. Smell:

   a. Has receptors in the nasal cavities.

   b. Has seven basic odors.

   c. Is more sensitive than taste.

   d. Is connected by olfactory bulbs and tracts to temporal and frontal lobes of the brain.

12. Taste:

   a. Has the taste bud as receptor.

   b. Has four basic sensations: sour, salt, sweet, and bitter.

   c. Uses VII and IX cranial nerves as pathways, and terminates in the parietal lobes.

13. Vision is served by the eyes.

   a. The eyes are located in the orbits.

   b. Each eye is moved by six muscles.

   c. Three coats form each eye.

      1) The sclera and cornea are outermost.

      2) The uvea is central and includes: chorioid, ciliary muscle, iris, lens, and lens ligaments.

      3) The retina, which responds to light.

   d. The eyes contain aqueous and vitreous humors.

      1) Blockage of aqueous humor drainage creates glaucoma.

14. Image formation by the eye:

   a. Requires action of cornea, lens, and iris.

   b. If defective, causes myopia, hyperopia, and astigmatism.

   c. Involves accommodation, which includes three things:

      1) Lens shape changes for focusing.

      2) Pupil constriction for correction of aberrations and light control.

      3) Convergence for directing eyes to object.

15. The retina:

  a. Contains rods and cones.

  b. Is responsive to low light levels (rods).

  c. Is responsive to high light levels and colors (cones).

  d. Is responsible for visual acuity.

  e. Fuses separate images into a continuum.

  f. Creates afterimages.

16. The visual pathways include the optic nerves, chiasm, tract, thalamus, and occipital lobes.

17. The ear is divided into outer, middle, and inner ears.

18. The outer ear:

  a. Gathers sound waves.

  b. Conducts sound waves to the eardrum.

19. The middle ear:

  a. Consists of middle ear cavity, ossicles, and Eustachian tube.

  b. Transmits eardrum vibrations to the cochlea.

  c. Equalizes pressure on the drum.

  d. Controls ossicle movement according to intensity of sound.

20. The inner ear:

  a. Contains the cochlea, with its organ of Corti and hair cells for hearing. Vibrations (sound waves) cause selective vibration of Corti's organ and hearing results. Auditory pathways pass from cochlea to brain stem to thalamus to temporal lobes.

  b. Contains organs for equilibrium. The maculae signal held position of the head. The semicircular canals detect motion and initiate motor responses to maintain posture.

Questions

1. What is a receptor? What is a peripheral receptive field?

2. What basic properties do all receptors share?

3. Give examples of chemoreceptors, baroreceptors, and thermoreceptors, and describe what the adequate stimulus is for each one listed.

4. What is somesthesia?

5. Name the receptor, spinal pathway, and interpretive area for the senses of touch, pain, and unconscious muscle sense.

6. What are the varieties of pain? How do they differ?

7. What are paresthesia and hyperesthesia? How may each be brought about?

8. What are the roles of the thalamus in sensation?

9. What structures are involved in image formation by the eye? What does each contribute to the formation of the image?

10. What changes occur during accommodation?

11. Describe the defects present that result in myopia. How is the condition corrected?

12. What evidence exists to suggest the presence of two types of receptors in the retina?

13. What functions are served by the structures located in or communicating with the middle ear?

14. Describe the structure and function of the cristae ampullaris.

15. What are the basic taste sensations and what are the adequate stimuli for each?

16. What are some basic odors? Compare sensitivity of the olfactory organ to that of taste.

Readings

Dunn-Rankin, Peter. "The Visual Characterization of Words." *Sci. Amer.* 238:122, Jan. 1978.

Goldbert, J. M., and Cesar Fernandez. "Vestibular Mechanisms." *Ann. Rev. Physiol.* Vol. 37. Palo Alto, Calif., 1975.

Goldstein, Avram. "Opioid Peptides (Endorphins) in Pituitary and Brain." *Science* 193:1081, 17 Sept. 1976.

Ross, John. "The Resources of Binocular Perception." *Sci. Amer.* 234:80, Mar. 1976.

Rushton, W. A. H. "Visual Pigments and Color Blindness." *Sci. Amer.* 232:64, Mar 1975.

Chapter 19
Body Fluids and Acid-base Balance

Objectives

After studying this chapter, the reader should be able to:

■ Define what is meant by, and give the functions of, the internal environment.

■ Define a fluid compartment.

■ Define what is meant by total body water, and give its values in percent of body weight and volume.

■ Explain what extracellular fluid is and list its subdivisions, with percents of body weight and volumes.

■ Explain intracellular fluid, giving its percent of body weight and volume.

■ List the solutes characteristic of extra- and intracellular fluids.

- Explain the processes involved in shifts of fluids and solutes between plasma, interstitial fluid, and cells.

- Explain where body stores of fluids and solutes come from, where they may be absorbed, and how much comes from each source.

- Show the routes of loss of fluids and solutes from the body, with values for each route.

- Explain the mechanisms available for regulation of body fluid and solute content.

- Explain the genesis of dehydration and edema, giving common symptoms of each.

- Define acids and bases.

- Explain the origin of hydrogen ion in the body.

- List, with examples, the several methods the body employs to keep hydrogen ion within normal limits in the body.

- Explain the genesis of, the results of, and the compensatory reactions the body exhibits in the following conditions: respiratory acidosis, metabolic acidosis, respiratory alkalosis, metabolic alkalosis.

Life is said to have originated in the sea. The sea provided rich stores of nutrients for the organisms that lived (and still do today) in it, and also supplied a means of carrying away from those organisms their often toxic wastes of metabolism. As organisms colonized the land, they carried with them a fluid in their bodies that served the same functions as the sea: provision of nutrients and removal of wastes. In humans, this "sea" is the body fluids that form the INTERNAL ENVIRONMENT for the body cells. All living cells are *surrounded* by fluid, though amounts vary according to the location of the cell, and cells *contain* fluid. The composition and characteristics of these fluids must be narrowly regulated to ensure survival of the living units and requires constant expenditure of energy by the cells themselves.

Fluid compartments

One may speak of total body water, and this refers to all the body water, whether around or within cells. There are various membrane-separated subdivisions in the total body water, and each constitutes what is called a FLUID COMPARTMENT.

Total body water

In the average adult, the total body water constitutes 55 to 60 percent of the body weight and some 40 to 45 liters in volume. There is individual variation in these figures because of age (older persons have less fluid volume) and amount of adipose or fat tissue in the body (fat tissue is nearly water-free, thus more fat means *relatively* less water). Total body water is often expressed as a percentage of lean (fat-free) body mass and, on this basis, is nearly constant for all adult individuals at about 72 percent of the lean body mass. This total body water contains many types of solutes, and these solutes—by passive and active processes—are distributed between the various body water compartments. Recalling that water tends to distribute itself osmotically according to solute concentrations, the solutes of the body water compartments will largely determine how much water will enter, leave, or remain within a given compartment.

Extracellular fluid

The term EXTRACELLULAR FLUID *(ECF)* refers to all the body water lying outside of body cells themselves. It amounts to about 37.5 percent (three-eighths) of the total body fluids, or about 17 liters in volume. This compartment represents the internal environment (mentioned in an earlier sec-

tion) that land animals have incorporated. It may be subdivided as follows.

The BLOOD PLASMA, or liquid portion of the blood, amounts to about 7 percent of the body weight, or about 3 liters in the adult. It is caused to circulate through the body's blood vessels by heart action and is the major source of fluid and solutes for all other compartments.

INTERSTITIAL *(tissue)* FLUID lies outside of blood vessels and immediately surrounds the cells. LYMPH is interstitial fluid that has entered the lymphatic vessels of the body. Fluid and solutes move slowly through these areas, but are in equilibrium with the plasma for major inorganic ions. Together, interstitial fluid and lymph comprise about 18 percent of the body water, or about 8 liters in the adult.

FLUIDS OF DENSE CONNECTIVE TISSUE AND BONE are actually a part of the interstitial compartment, but their water exchanges so slowly that it *behaves* like a separate compartment. These fluids account for about 10 percent of the body water, or about 4.5 liters in the adult.

TRANSCELLULAR FLUIDS are those fluids separated from other compartments by epithelial membranes. Cerebrospinal, synovial, ocular, pleural (around lungs), pericardial (around the heart), peritoneal (in

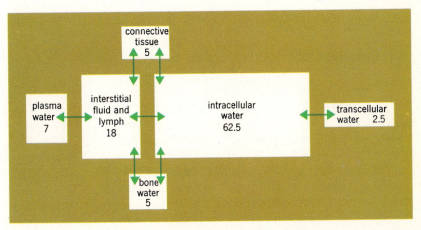

FIGURE 19.1 Relationships of body water compartments. Numbers represent percent each compartment comprises of total body water in an adult.

TABLE 19.1 Compartments of the body water		
Compartment	Approximate percentage of adult body weight	Approximate percentage of adult body fluid
Extracellular fluid	21.4	37.5
Plasma	5	7
Interstitial fluid	15	28
Transcellular fluid	1.4	2.5
Intracellular fluid	38.6	62.5
Total body water	60	100

abdominal cavity) fluids, and the fluids within hollow organs of the body (stomach, intestines, urogenital organs) are some of the components of this category. They differ markedly from other compartments in composition and characteristics, reflecting their origin by active processes and their specialized functions. These fluids constitute about 2.5 percent of the body water, or about 1.5 liters in the adult.

Intracellular fluid

The term INTRACELLULAR FLUID (ICF) refers to the fluids enclosed by the plasma membranes of the body cells. It amounts to about 62.5 percent (five-eighths) of the body fluids, or about 28 liters in the adult. It may differ in composition within different areas of a single cell, because of the membrane-separated organelles that have their own inner environments. In considering overall fluid balances, one ignores such variations and considers ICF as a single compartment.

Figure 19.1 and Table 19.1 summarize the relationships and compartments of the body fluids.

Composition of the various compartments

In general, the subdivisions of the ECF (transcellular fluids excepted) are nearly identical in ionic composition and in content of small molecules (glucose, urea, for example). The reason for this is that bulk flow, under the force of the blood pressure, pushes small solutes and water through capillary membranes as through a sieve. Plasma contains much more protein than interstitial fluid because the protein molecules pass with difficulty through capillary walls; they are simply too large. The primary ECF ions are sodium, bicarbonate, and chloride.

Intracellular fluid is rich in potassium and phosphate ions and protein, which is derived from the synthetic activities of the cell.

Recalling again that water distributes itself according to solute distributions, we may conclude that there *must* be mechanisms to control solute distribution outside and inside a cell, or the cell may take on or lose water with detrimental effects on its activities. Figure 19.2 and Table 19.2 summarize facts concerning the composition of the plasma, interstitial fluid, and ICF.

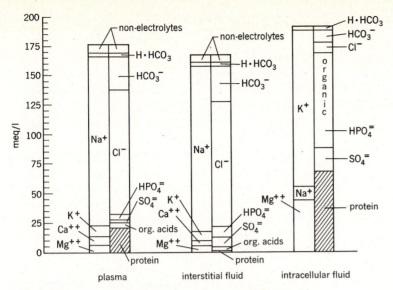

FIGURE 19.2 A comparison of the constituents of the three major compartments of the body water. (Reproduced, with permission, from J. L. Gamble, Jr., *Chemical Anatomy, Physiology, and Pathology of Extracellular Fluid.* 6th ed. Harvard University Press. Boston, 1954.)

Constituents and properties	Extracellular fluid		Intracellular fluid
	Plasma	Interstitial fluid	
Sodium	142 meq/L	145 meq/L	10 meq/L
Potassium	4 meq/L	4 meq/L	160 meq/L
Calcium	5 meq/L	5 meq/L	2 meq/L
Magnesium	2 meq/L	2 meq/L	26 meq/L
Chloride	101 meq/L	114 meq/L	3 meq/L
Sulfate	1 meq/L	1 meq/L	20 meq/L
Bicarbonate	27 meq/L	31 meq/L	10 meq/L
Phosphate	2 meq/L	2 meq/L	100 meq/L
Organic acids	6 meq/L	7 meq/L	—
Proteins	16 meq/L	1 meq/L	65 meq/L
Glucose (av)	90 mg %	90 mg %	0–20 mg %
Lipids (av)	0.5 gm %	—	—
pH	7.4	7.4	6.7–7.0[a]

TABLE 19.2 Composition of extracellular and intracellular fluid compartments

[a] Average value; difficult to measure.

Exchanges of fluid and solutes between compartments

That fluids and solutes do pass from one compartment to another is a consequence of filtration, diffusion, osmosis, and active mechanisms. Various combinations of these forces are responsible for exchange.

Plasma and interstitial fluid

Fluid and solutes within the blood vessels of the body are under a pressure determined by the heart action (the *"blood pressure"*). Of the various blood vessels that allow passage of substances through their walls, *capillaries* permit most of this passage. Blood arrives at the arterial end (the one closest to the heart) of a capillary under a hydrostatic or blood pressure of about 32 mm Hg. This pressure tends to force all molecules that are small enough through the walls of the capillaries in a

process of bulk flow (*see* Chapter 3). Not passing — or passing in only minute amounts through the capillary walls — are the large plasma protein molecules. They create an osmotic pressure within the capillaries of 28 mm Hg that tends to draw water back into the capillaries. However, at the arterial end of the capillaries, the blood pressure (32 mm Hg) *exceeds* the protein osmotic pressure (28 mm Hg), so that there will be a *net outward movement* of water and small solutes. As the blood passes through the length of the capillary, it loses hydrostatic pressure, to a value of 16 mm Hg. Osmotic pressure from proteins has not changed, remaining at 28 mm Hg. Thus, at the venous end of the capillary (the "other end"), the force causing osmotic movement of water *into* the capillary (28 mm Hg) exceeds the hydrostatic pressure, (16 mm Hg) and water osmoses and carries some solutes back into the capillary. About 90 percent of the water

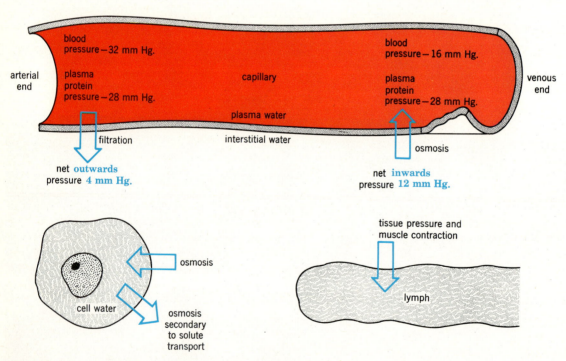

FIGURE 19.3   Some determinants of water movement between body water compartments.

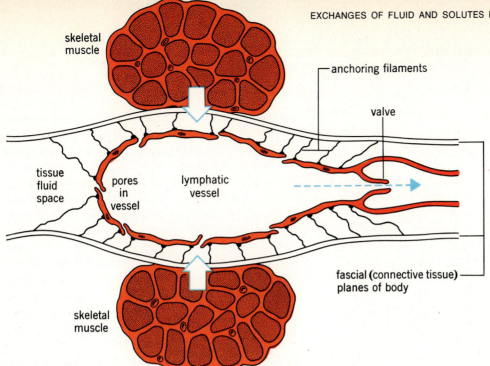

FIGURE 19.4 The relationships of lymphatic vessels to skeletal muscles and fascial planes of the body. The anchoring filaments tend to expand the vessel; muscle contraction tends to compress the vessels (↓). Lymph tends to move to the right (→) and is prevented from backflowing by the valve. A "bellows" or "pumping" action thus draws tissue fluid into the vessels and moves it onwards.

that is filtered at the arterial end of the capillary is returned to the venous end of the capillary. Filtered solutes may enter cells, return with water to the capillary, or enter lymphatic vessels between the cells. The 10 percent of filtered water and "extra solutes" not returned to blood capillaries pass into lymph vessels to be ultimately returned to the bloodstream.

Interstitial and intracellular fluid

Most animal cell membranes are considered to be freely permeable to water. ICF has a slightly greater solute concentration than interstitial fluid, and thus water tends to enter the cell by osmosis. Sodium and chloride ions tend to follow a con-

centration gradient *into* cells (diffusion). The cell contains much nondiffusible solute that, with diffusible solutes, nearly balances the osmotic pressure outside the cell. Nevertheless, the flow of water into the cell tends to slightly exceed the outflow *unless* the cell takes steps to eliminate this tendency. What the cell apparently does is to actively pump diffused sodium *out of* the cell to the point where the inflow and outflow of water are balanced. Movement of water and solutes into lymph vessels occurs by a "bellows-like" action of skeletal muscles alternately compressing and releasing the vessels. This same action also moves the lymph through the vessels. Figures 19.3 and 19.4 show the forces responsible for exchange between compartments and entry and movement of lymph.

Exchanges of fluids and solutes between the extracellular compartment and the external environment

Fluid and solute intake

Body water volume and solute composition tend to remain remarkable constant on a day-to-day basis in spite of variable water and solute intake. This reflects a balance between intake and loss of these substances. In the average adult, some 2500 ml of fluid, with variable amounts of solutes, is added per day to body stores. Of this amount of fluid, some 2300 ml is INGESTED in food and drink. The remainder is METABOLIC WATER that results from chemical reactions occurring within body cells. Of these two sources, ingested water is adjustable, while water of metabolism remains fairly constant. The THIRST MECHANISM is a basic device for governing the amount of ingested fluid. It is activated when hypothalamic cells are stimulated by any condition that causes a rise in osmotic pressure of the blood reaching the hypothalamus. For example, loss of fluid as in burns, heavy sweating, excessive salt intake or retention all result in increased blood osmolarity. Salivary secretion diminishes, the mouth becomes dry, and a sensation (called *thirst*) develops that normally leads to fluid intake.

Fluid and solute absorption

Ingested fluids and solutes enter the digestive tract (stomach, intestines). Most of the absorption of solutes occurs in the small intestine. Ions (Na^+, K^+, Cl^-) are actively removed (active transport) from the intestine, as are the small molecules (amino acids, simple sugars) resulting from di-

gestion of proteins and carbohydrates. Larger molecules (lipids, for example) are removed by pinocytosis into the epithelial cells, and from there to the interstitial fluid. All this combines to create a higher solute concentration in the interstitial fluids around the epithelial cells, and water moves passively by osmosis from the intestine to the interstitial area. From here, small molecules enter blood vessels, mainly by diffusion, and larger molecules enter the more permeable lymphatic vessels. Ultimately, all will enter the plasma compartment to be distributed to the entire body.

Fluid and solute loss

Loss of water and solutes occurs from the body by two general types of routes.

OBLIGATORY LOSS, or loss that cannot be controlled, occurs through the *lungs* (water loss), *perspiration* (mostly water loss; a few solutes), *feces* (both water and solutes), and *evaporation from the mouth* (water loss). About 300 ml of water loss occurs per day via the lungs, 500 ml per day in perspiration, and 200 ml of water, 3 meq of Na^+ and Cl^-, 10 meq of K^+, 45 meq of Ca^{2+}, and 20 meq of Mg^{2+} are lost per day via the feces.

FACULTATIVE LOSS refers to *controllable* (within limits) *loss*, with the *kidneys* as the organ involved. Loss of fluid via the urine amounts to about 1500 ml per day in the average adult, with the primary solute loss consisting of NaCl and urea. In general, fluid and solute loss via the URINE is inversely proportional to loss via the obligatory routes, so that, for example, if one sweats heavily, urine volume will decrease.

Regulation of volume and osmolarity of the extracellular fluid

REGULATION OF VOLUME AND OSMOLARITY of the ECF is accomplished by controlling the excretion of water *and* the excretion of sodium ion. The latter process is effective because sodium is the major ECF ion and because the retention of sodium ion must be associated with retention of enough water to keep the ECF isotonic (isosmotic) to the body cells. Usually, excretion of both occurs at the same time to maintain proper ratios of both in the ECF. Control is achieved by two hormone-dependent systems.

The hypothalamus contains cells that act as *osmoreceptors*, that is, they swell or shrink according to the osmotic pressure of the blood reaching them (Fig. 19.5). These same cells are neurosecre-

tory, producing a hormone designated as ANTI-DIURETIC HORMONE *(ADH)*. This hormone is passed over nerve fibers to be stored in and released from the posterior lobe of the pituitary gland. The effect of ADH is to make certain portions of the kidney tubules more permeable to water. Thus, if the blood reaching the hypothalamus contains too much solute relative to water (hypertonic or hyperosmotic), the hypothalamic cells will lose water to the bloodstream and shrink. The obvious need here is to dilute the blood, so ADH release is increased, more water passes from kidney tubules *back into* (ultimately) the bloodstream, and isotonicity is restored. Decrease of ADH secretion occurs with a

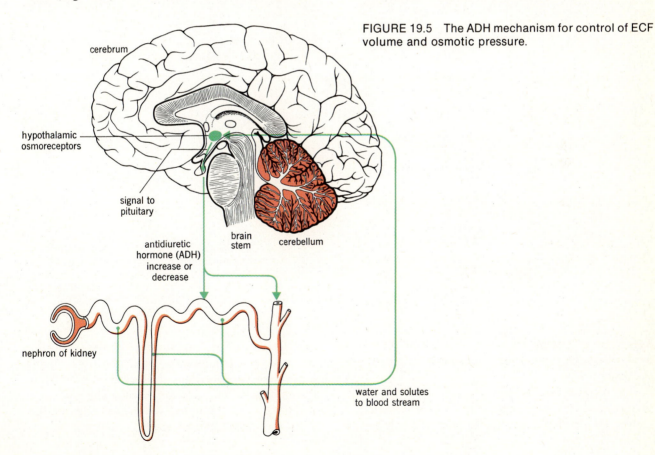

FIGURE 19.5   The ADH mechanism for control of ECF volume and osmotic pressure.

cerebrum

hypothalamic osmoreceptors

signal to pituitary

antidiuretic hormone (ADH) increase or decrease

brain stem

cerebellum

nephron of kidney

water and solutes to blood stream

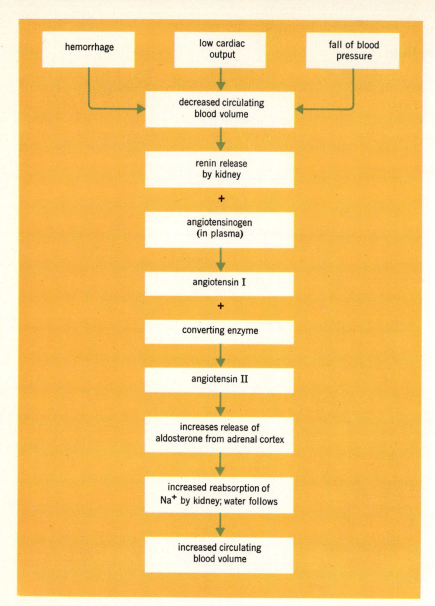

FIGURE 19.6 The relationships of angiotensin and aldosterone to the regulation of ECF volume.

hypotonic (hyposmotic) blood; the "extra" water passes through the kidney tubules and is lost via the urine.

The second mechanism involves a hormone called ALDOSTERONE, produced by the adrenal cortex. Its primary effect is to increase the re-absorption of Na^+ from kidney tubules. Aldosterone release is triggered by any condition causing a lowered blood volume, decreased blood flow, or lowered oxygen level to the kidney. The kidney, in response to such stimuli, releases an enzyme-like substance called RENIN. Then, renin + angiotensinogen (an inactive substance in the plasma) ⟶ angiotensin I (an active form of the inactive material). angiotensin I + plasma converting enzyme ⟶ angiotensin II.

The final compound, angiotensin II, is an extremely potent vasoconstrictor (causes narrowing of blood vessels) that raises blood pressure and stimulates aldosterone secretion. The kidney reabsorbs more Na^+, and water follows osmotically. Thus, blood volume is increased and blood pressure rises. The steps just described are summarized in Figure 19.6.

An additional mechanism suggested as being involved in control of sodium excretion (and therefore also of water) is a NATRIURETIC HORMONE. It is released whenever blood or ECF volume has been increased and causes increased excretion of Na^+ into the urine (by preventing its reabsorption by the kidney). The site of production of such a hormone is not known, and its release is believed to be controlled by nerve impulses arriving at the site of production from volume receptors thought to be located in the heart, thorax, liver, or brain.

This section of the chapter has indicated that regulation of solute and fluid intake and output, osmolarity, and volume of the ECF is essential to normal cellular function. This regulation is accomplished primarily by regulation of ECF sodium concentrations, and the kidney is an important organ in maintaining proper Na^+ levels. We thus see that there are hormone-dependent, self-regulating systems that maintain ECF volume nearly constant on a day-to-day basis.

Clinical considerations

Dehydration

ISOTONIC DEHYDRATION (equal loss of fluid and electrolytes) occurs during loss of whole blood from the body (hemorrhage) or during short-term vomiting or diarrhea. The total volume of the ECF changes, but composition and osmotic pressure remain within normal limits.

HYPERTONIC DEHYDRATION (excess fluid loss compared to electrolyte loss) occurs when there is a lack of water intake while loss continues through skin, lungs, and urine. It may also occur during excessive sweating, in burned patients, and during diabetes mellitus (sugar diabetes). The loss of relatively more water than solutes tends to raise the concentrations of sodium, potassium, and calcium in the body fluids, singly or in combination, and makes the ECF hypertonic to the cell. The cell tends to lose water. Excessive sodium levels tend to cause water retention by the kidney, and this is a beneficial effect. Excessive potassium levels lead to muscular weakness and eventual paralysis, to cardiac and respiratory irregularities, and to diminished urine formation. Excessive calcium levels cause relaxation of muscles, flank pain, and loss of reflexes.

HYPOTONIC DEHYDRATION (excess electrolyte loss compared to fluid loss) occurs in deficiency of certain adrenal cortical hormones (aldosterone), in kidney disease, during encephalitis (ADH secretion may be abnormal), during prolonged diarrhea or vomiting, and on a low sodium diet. The ECF tends to become hypotonic relative to the cell, and the cell tends to gain water. Loss of sodium leads to muscle cramps, convulsions, and a feeling of apprehension. Loss of potassium leads to relaxed muscles (the muscles feel like half-filled hot water bags), paralysis, weakness, and irregularities of heart beat. Lowered calcium levels lead to muscle cramps and tetany (muscle in a sustained contraction), and tingling of the fingers. Loss of magnesium causes disoriented behavior, muscle tremor, and exaggerated reflexes.

Edema

Edema is a condition caused by accumulation of excess fluid in the interstitial compartment. It results in swelling and "waterlogging" of the tissues. Some factors involved in the production of edema are the following.

Increased filtration of fluid and solutes from capillaries. If capillary permeability permits protein loss through the capillary wall, the osmotic force tending to draw water from tissues to capillary is decreased. Thus, water remains in the tissues.

Decreased heart action blocks blood flow through veins, and causes engorgement of tissues with blood. Venous pressure rises and opposes osmotic return of water from the tissues to the capillaries. Therefore more water remains in the tissues.

Salt retention or high salt diets cause water to be retained as well, and ECF volume increases. If ECF volume rises, cell volume may also increase by increased water movement into cells. The nervous system shows such diluting effects first, with the development of disoriented behavior, convulsions, and coma.

Although it is usually not possible to determine which particular solute has been altered and in which direction, without blood analysis, one can get an idea of whether one is dealing with an altered state of hydration or an alteration of solutes in general. Several PHYSIOLOGICAL OBSERVATIONS should be made on a patient suspected of having a hydration disorder or on one who is under treatment for such a disorder.

Temperature. In the absence of an obviously fever-producing condition, an increase in temperature indicates decreased water or increased sodium levels in the body.

Pulse. Rapid weak but regular pulses indicate hemorrhage or sodium loss. Weak and irregular pulses suggest potassium deficiency.

Respirations (breathing). Difficult breathing suggests excessive fluid volume and possible pulmonary edema. Shallow breathing may indicate potassium deficiency, while harsh, high-pitched sounds (stridor) during breathing suggests severe calcium deficiency.

Blood pressure. Increased pressures suggest excessive fluid volumes; decreased pressures suggest sodium and/or potassium deficit, or loss of fluid volume.

Skin condition. Does the skin appear dry, does it stay "tented" when pinched, or does it retain a depression when pushed? Such observations give a clue as to the state of skin hydration (water content) and thus of the body in general.

General behavior. Is the patient thirsty, does he or she talk and walk well, is he or she apprehensive or anxious?

Keep in mind that these symptoms are not exclusively the result of alterations of water and solute levels; they do, however, strongly suggest an alteration in fluid and/or electrolyte balance in the body.

Acid-base balance of body fluids

Metabolism produces large quantities of substances that release free hydrogen ions. Free hydrogen ions are potentially dangerous because they change the acidity of the body fluids and thus threaten many enzyme controlled chemical reactions in the body. The body normally maintains a pH value of 7.4 ± 0.02 in the ECF to ensure normal body function.

Acids and bases

Acids are materials that, in water, are capable of losing or giving up hydrogen ions to create free H^+; bases are substances capable of accepting free hydrogen ions. Strong acids release hydrogens easily, or may be said to have a high degree of dissociation ("come apart" in solution). A weak

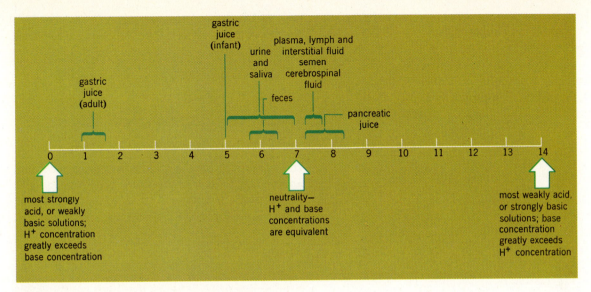

FIGURE 19.7   The range of pH values for several body fluids.

TABLE 19.3 Relationships between pH and concentration of hydrogen ion	
pH	Concentration of H$^+$ (gm/L)
1	0.1
2	0.01
3	0.001
4	0.0001
5[a]	0.00001

[a] Relationships continue to pH 14, with a tenfold decrease of [H$^+$] for each pH unit change.

acid prefers to remain associated, and gives up hydrogen ions grudgingly. It has a low degree of dissociation. Concentrations of hydrogen ion are usually described by the symbol pH. The pH values range from 0 to 14 (Fig. 19.7), or from strongly acid to weakly acid solutions. At a pH of 7, a solution is neither acidic (values below 7), nor basic (values above 7), and is said to be neutral, because acid and base concentrations are equal. A change of one pH unit represents a ten fold change in hydrogen ion concentration, as is shown in Table 19.3.

Because of the production of large quantities of acids during metabolism, most clinically important deviations in pH occur between pH 7.40 and 7.25, that is, to the acid side of normal (pH 7.4). In general, living processes cannot be maintained if pH falls below 6.8 or rises above 7.8.

Sources of hydrogen ion in the body

The COMPLETE COMBUSTION of carbon containing compounds, such as carbohydrates and fatty acids, produces CO_2 that reacts with water to form carbonic acid. The carbonic acid then dissociates to liberate H$^+$. These reactions are shown by the equation:

$$CO_2 + H_2O \rightleftharpoons \underset{\substack{\text{carbonic}\\\text{acid}}}{H_2CO_3} \rightleftharpoons \underset{\substack{\text{hydrogen}\\\text{ion}}}{H^+} + \underset{\substack{\text{bicarbonate}\\\text{ion}}}{HCO_3^-}$$

The reactions are reversible, and while carbonic acid is a weak acid, so much of it is produced from CO_2 that it forms a major source of H$^+$ for the body fluids.

The PRODUCTION OF ORGANIC ACIDS results from the incomplete combustion of carbon containing compounds in metabolic cycles such as glycolysis and the Krebs cycle. The hydrogens then dissociate from the carboxyl (−COOH) groups of the acids. The equation below shows dissociation of H^+ from pyruvic acid as an example.

$$CH_3-\overset{\overset{O}{\|}}{C}-COOH \rightleftharpoons H^+ + CH_3-\overset{\overset{O}{\|}}{C}-COO^-$$
pyruvic acid

The INGESTION OF ACID-producing substances may add H^+ to the body fluids. While one does not normally ingest acid, it is of interest to note that substances such as aspirin contain organic acids (salicylic acid) capable of dissociating H^+.

Methods of minimizing the effects of free hydrogen ion

The threat posed to body homeostasis by the production of free H^+ from such processes as described above, is that pH will fall and inhibit vital enzyme-controlled reactions that are pH dependent. The pH of either intracellular or extracellular fluid may be altered with detrimental effects on body function. Several methods are employed to keep hydrogen ion concentrations within normal limits. Most are aimed at maintaining normal ratios between acidic and basic substances in the body fluids.

BUFFERING. Buffering is a process that utilizes certain chemicals in the body fluids to react with and thus "trap" hydrogen ion in a more or less nondissociable compound. The reaction requires a weak acid, and the "conjugate base" (completely ionizable salt) of that acid. Both compounds will share a common anion (negatively charged ion), but will have different cations (positively charged ions). The anion of the base reacts with excess H^+ to form the weak acid. The combination of the weak acid and its conjugate

base is termed a *buffer pair* or *buffer system*. The system of *carbonic acid and sodium bicarbonate* is the major buffer of the extracellular fluid.

weak acid: $H_2CO_3 \rightleftharpoons H^+ + HCO_3^-$
           carbonic   hydrogen   bicarbonate
           acid   ion   ion

conjugate base: $NaHCO_3 \rightarrow Na^+ + HCO_3^-$
           sodium   sodium   bicarbonate
           bicarbonate   ion   ion

The conjugate base provides a reserve of anion (in this case, bicarbonate ion) which reacts with H ion, from whatever source, to form the weak acid. For example,

$H^+ + HCO_3^- \rightleftharpoons H_2CO_3$
hydrogen   bicarbonate   carbonic acid,
ion   ion   a weak acid

Normal body mechanisms maintain a ratio of one molecule of H_2CO_3 to 20 molecules of $NaHCO_3$. The extra HCO_3^- provides an alkali reserve, or base excess, to react with H^+ produced under most normal activity.

As H^+ reacts with HCO_3^- to form the weak acid, it would appear that the base cation Na^+ would tend to increase its concentration in the body fluids. What does the body do to prevent accumulation of excess Na^+ that can cause disturbances in excitability? Keeping in mind that the body's primary task is to maintain nearly constant ratios of acids and bases in the body regardless of absolute concentrations of these substances, it should become clear that if there is an increased amount of H_2CO_3 and Na^+ in the fluids, the body must add HCO_3^- to the fluids to keep ratios constant. This job may be taken care of by the kidney. The kidney reabsorbs HCO_3^- from the filtrate (fluid formed by the kidney). Alternatively, the breathing rate may be stimulated by a slight fall in pH, which results as excessive H^+ is released from the carbonic acid. This response causes a greater liberation of CO_2 from H_2CO_3 according to the equation:

$$H_2CO_3 \xrightleftharpoons{enzyme} H_2O + CO_2$$

The reaction is aided by an enzyme called *carbonic anhydrase*. Thus acidic and basic substances are again balanced. The excess water is then eliminated by the kidney. The responses the body makes to keep acid and base substances in proper ratio constitute compensation.

Although the H_2CO_3 : $NaHCO_3$ system is probably the most important buffer system of the extracellular fluid, it is not the only buffer system operating in the body. Other ECF buffer systems include

Phosphate systems
  weak acid: $NaH_2PO_4 \rightleftharpoons Na^+ + H_2PO_4^-$
  conjugate base: $Na_2HPO_4 \longrightarrow Na^+ + NaHPO_4^-$

Plasma proteins
  weak acid: $H \text{ Proteinate (Pr)} \rightleftharpoons H^+ + Pr^-$
  conjugate base: $Na \text{ Proteinate} \longrightarrow Na^+ + Pr^-$

Intracellular buffer systems include
Hemoglobin (Hgb) in red cells
  weak acid: $HHgb \rightleftharpoons H^+ + Hgb^-$
  conjugate base: $KHgb \longrightarrow K^+ + Hgb$

Phosphate systems
  weak acid: $KH_2PO_4 \rightleftharpoons K^+ + H_2PO_4^-$
  conjugate base: $K_2HPO_4 \longrightarrow K^+ + KHPO_4^-$

Other methods of minimizing free H^+ effects are:

DILUTION. Some hydrogen ion, as it is produced, is set free in the body fluids. It is diluted by these fluids and thus local accumulation is prevented. This method does not remove the H^+; it only prevents excessive local accumulation.

UPTAKE BY HYDROGEN ACCEPTORS. Associated with the basic metabolic cycles are molecules that take up hydrogen ions as they are released by those cycles. These acceptors then carry the hydrogens to oxidative phosphorylation or to a synthetic scheme that uses them. Three acceptors are recognized:

  Nicotinamide adenine dinucleotide (NAD)
  Nicotinamide adenine dinucleotide phosphate
    (NADP)
  Flavine adenine dinucleotide (FAD)

As long as the hydrogens are on the acceptor or in a cycle, they are not free to cause pH change. It is critical to the understanding of acid-base disturbances to recognize that both acids and bases are present in the body, and that pH can change either as a result of too much of one substance or too little of the other. A "scalelike" bal-

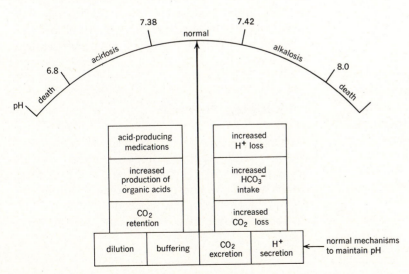

FIGURE 19.8   Factors affecting body fluid pH.

ance is thus achieved, which can be upset from either side (Fig. 19.8).

LUNG ELIMINATION OF CARBON DIOXIDE. The weak acid, carbonic acid, formed when carbon dioxide reacts with water is said to be a *volatile acid*. This is because the CO_2 that forms the acid is a gas and can be eliminated via the lungs as such, *if* the CO_2 can be liberated from the acid. Then, less acid will be available to dissociate hydrogen ion.

The cells lining the capillaries of the lungs contain an enzyme called CARBONIC ANHYDRASE that catalyzes the reaction

$$H_2CO_3 \longrightarrow CO_2 + H_2O$$

As H_2CO_3 is broken down into CO_2 and H_2O, hydrogen ion and bicarbonate ion from the blood plasma combine to form more H_2CO_3

$$H^+ + HCO_3^- \longrightarrow H_2CO_3 \longrightarrow CO_2 + H_2O$$

The CO_2 is exhaled and H^+ concentration of the ECF is reduced. This method lowers H^+ in body fluids very rapidly, since CO_2 elimination by the lungs is a direct function of rate and depth of breathing.

KIDNEY SECRETION OF HYDROGEN ION. The kidneys can adjust the amount of H^+ they eliminate in the urine. This process is slower than lung elimination of CO_2, and operates continually to rid the body of H^+. If for some reason lung elimination of CO_2 is decreased, this mechanism becomes of primary importance in regulating body fluid pH.

Clinical aspects of acid-base balance

Normally, mechanisms for maintaining ECF pH regulate it between 7.38 and 7.42. Alkalosis (a pH value greater than 7.42), or acidosis (a pH value less than 7.38) occur beyond these limits. Four basic types of disturbances may occur.

Respiratory acidosis

Respiratory acidosis occurs when elimination of CO_2 by the lungs is diminished. CO_2 is retained, more carbonic acid and hydrogen ion is formed, and pH falls. This disturbance occurs in many types of lung conditions. In mild respiratory acidosis, excess H^+ resulting from CO_2 retention stimulates breathing, causing an increased CO_2 elimination; the kidney increases H^+ secretion, and the condition is compensated.

Metabolic acidosis

Metabolic acidosis results from loss of base, as in diarrhea, which causes loss of HCO_3^-; by impaired secretion of hydrogen ion by the kidney; by excessive production of organic acids by abnormal or exaggerated metabolic processes (e.g., diabetes mellitus). Excessive liberation of hydrogen ion from these acids causes pH to fall. Fall of pH again stimulates breathing and eliminates CO_2 to reduce the source of H^+, and achieve compensation.

Respiratory alkalosis

Respiratory alkalosis results from excessive loss of CO_2, as in hyperventilation; excessive alkali accumulates, and pH rises. Slowing of breathing occurs when pH rises, and decreased kidney secretion of H^+ raises the concentration of H^+ in the body fluids. This compensates for the alkalosis.

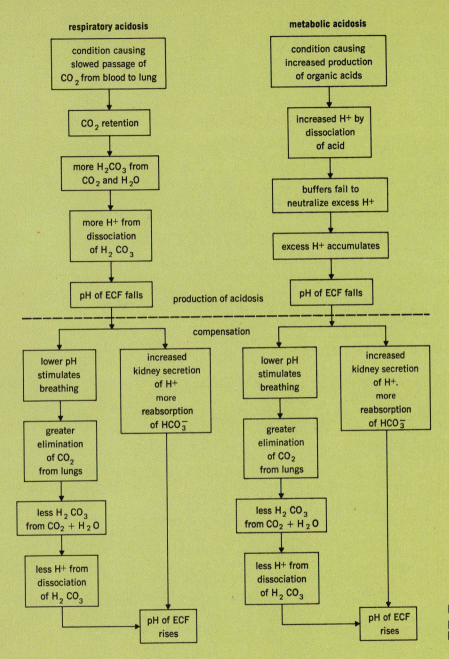

respiratory acidosis

condition causing slowed passage of CO_2 from blood to lung

CO_2 retention

more H_2CO_3 from CO_2 and H_2O

more H^+ from dissociation of H_2CO_3

pH of ECF falls

production of acidosis

compensation

lower pH stimulates breathing

greater elimination of CO_2 from lungs

less H_2CO_3 from $CO_2 + H_2O$

less H^+ from dissociation of H_2CO_3

increased kidney secretion of H^+ more reabsorption of HCO_3^-

pH of ECF rises

metabolic acidosis

condition causing increased production of organic acids

increased H^+ by dissociation of acid

buffers fail to neutralize excess H^+

excess H^+ accumulates

pH of ECF falls

lower pH stimulates breathing

greater elimination of CO_2 from lungs

less H_2CO_3 from $CO_2 + H_2O$

less H^+ from dissociation of H_2CO_3

increased kidney secretion of H^+. more reabsorption of HCO_3^-

pH of ECF rises

FIGURE 19.9 Production and compensation of respiratory and metabolic acidosis.

Metabolic alkalosis

Metabolic alkalosis results from excessive alkali intake (e.g., antacids) and by vomiting, which causes loss of hydrogen ion in stomach fluids. Again excessive alkali accumulates and pH rises. Compensation occurs as in respiratory alkalosis. These conditions and their compensations are

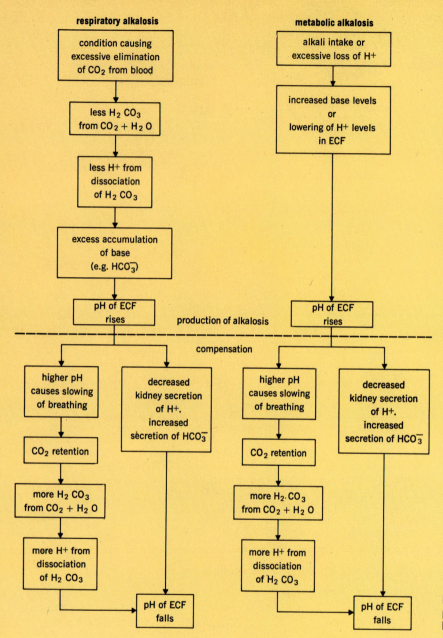

FIGURE 19.10 Production and compensation of respiratory and metabolic alkalosis.

summarized in the "flow sheets" below (Figs. 19.9 and 19.10).

In achieving recovery from acid-base disturbances, CO_2 elimination and secretion of hydrogen ion by the body are considered to be of critical importance. Metabolic acidosis and alkalosis can normally be compensated over several days to the extent of 50 to 75 percent by alteration of breathing. Respiratory acidosis and alkalosis can be nearly completely compensated, also in several days, by alteration of kidney secretion of hydrogen ion. If it seems paradoxical to alter a given condition by using the opposite organ (e.g., kidney to compensate for a respiratory acidosis), remember that initially something was wrong with the organ that caused the condition and that the body will seek another route to correct it.

Summary

1. The body fluids constitute an internal environment for the body cells, and compose 55 to 60 percent of the body weight.

   a. The fluids supply necessary nutrients for cellular activity and serve as the route for removal of wastes.

   b. Regulation of composition and characteristics of the body fluids is essential to life.

2. Two major divisions and several minor subdivisions may be made in total body fluids.

   a. Extracellular fluids (37.5%), fluid outside cells:

      1) Blood plasma (7%).
      2) Interstitial fluid (tissue fluid) and lymph (18%).
      3) Connective tissue fluids (10%).
      4) Transcellular fluids (2.5%): fluids in hollow organs (eye, digestive organs, lungs, joints, cavities of and surrounding the central nervous system).

   b. Intracellular fluid (62.5%), fluid within cells.

3. All extracellular fluids are nearly identical in electrolyte types and concentrations, but differ in protein concentration. Sodium and chloride are the primary extracellular electrolytes. Protein concentration is highest in the plasma compartment.

4. Intracellular fluid contains much more protein than extracellular fluid and includes potassium and phosphate as the primary electrolytes.

5. Most absorbed water enters the plasma.

   a. Plasma water is filtered into the interstitial fluid.

6. Water is returned to the plasma by osmosis due to plasma proteins.

7. Interstitial water passes osmotically into the plasma, is removed by lymph vessels, or enters cells.

8. Maintenance of water balance requires intake equivalent to output.

   a. Requirements for water are increased with increase of metabolism.

   b. Sources of water include:

1) Ingestion of food and drink.
2) Production by metabolism.

   c. Routes of loss of water include:

1) "Fixed loss"; lungs, perspiration, feces.
2) Variable loss, kidneys.

9. Regulation of volume and osmotic pressure of extracellular fluid is by:

   a. Thirst that increases intake.

   b. Osmoreceptors in the brain (hypothalamus) that determine release of anti-diuretic hormone.

   c. The adrenal hormone, aldosterone, that deals with sodium reabsorption.

   d. An hypothesized natriuretic hormone, increasing Na excretion.

10. Dehydration results when fluid and electrolyte loss exceeds intake.

    a. "Isotonic dehydration" involves equivalent loss of fluid and electrolytes.

    b. "Hypertonic dehydration" involves greater fluid than electrolyte loss. Cell water is lost.

    c. "Hypotonic dehydration" involves greater electrolyte than fluid loss. Water enters cells.

11. Edema results from accumulation of fluid in the interstitial compartment. It occurs when:

    a. Filtration from blood vessels exceeds osmotic return.

    b. The heart fails to pump blood effectively and venous pressure increases.

    c. There is salt and water retention.

    d. Lymph vessels are blocked.

12. Regulation of pH of body fluids insures continuance of enzymatic reactions in the body.

    a. Regulation of pH maintains pH at 7.4 ± 0.02.

13. Acids are substances that release hydrogen ions.

    a. Strong acids release much hydrogen ion.

    b. Weak acids release little hydrogen ion.

14. Bases are substances that accept free hydrogen ions and neutralize their effects.

15. Sources of hydrogen ion include:

    a. Reaction of CO_2 with H_2O to produce carbonic acid with release of hydrogen ion.

    b. Release of hydrogen ion from organic acids.

    c. Ingestion of acidic substances.

16. Methods of minimizing the effect of free H ion is by:

   a. Buffering. A weak acid and its corresponding completely ionized salt are a buffer pair. Hydrogen ion added to the salt forms the weak acid. Several buffer systems in ECF (bicarbonate, phosphate, protein) and in ICF (hemoglobin, phosphate) are described. Compensation is discussed as a method by which breathing and kidney secretion of H^+ are adjusted to maintain proper H^+ and base ratios.

   b. Dilution. Prevents local accumulation.

   c. Uptake by hydrogen acceptors.

   d. Lung elimination of CO_2.

   e. Kidney secretion of H^+.

17. The body has base in excess of normal need to neutralize hydrogen ion. The extra base is the alkali reserve or base excess.

18. Acid-base disturbances include acidosis (pH less than 7.38), and alkalosis (pH greater than 7.42).

   a. Respiratory acidosis results from CO_2 retention and is compensated by increased kidney secretion of H^+.

   b. Metabolic acidosis results from loss of alkali, excessive intake of acids, or accumulation of organic acids, and is compensated by increased breathing, and increased kidney secretion of H^+.

   c. Respiratory alkalosis results from excessive CO_2 loss, and is compensated by decreased breathing and decreased kidney secretion of H^+.

   d. Metabolic alkalosis results from loss of hydrogen ion, and is compensated in the same manner as respiratory alkalosis.

Questions

1. What are the three largest body water compartments? Give the appropriate percentage of body weight for each in the adult.

2. Which of the body water compartments does the kidneys regulate and, thus, ultimately regulate all the fluid compartments of the body?

3. What are the sources and routes of loss of body water?

4. What forces are responsible for water movement into and out of each fluid compartment?

5. Compare the compositions of intracellular and extracellular fluid for the predominant ions.

6. How is the volume of the ECF regulated?

7. What is the ADH mechanism for control of fluid osmotic pressure?

8. What is a weak acid?

9. Where does hydrogen ion come from in the body?

10. How does the body minimize the effect of hydrogen ion?

11. Discuss the "automatic mechanisms" that go into operation, when acidosis or alkalosis arises, to compensate for the excessive or deficient H^+ concentration.

12. Compare a metabolic and a respiratory acidosis as to cause and method of compensation.

Readings

Andersson, Bengt. "Regulation of Body Fluids." *Ann. Rev. Physiol. 39*:185, 1977.

Burke, S. R. *The Composition and Function of Body Fluids.* Mosby. St. Louis, 1972.

Christensen, H. N. *Body Fluids and Their Neutrality.* Oxford Univ. Press. New York, 1963.

Davenport, H. W. *The ABC of Acid-Base Chemistry.* Univ. of Chicago Press. Chicago, 1969.

Feig, Peter U., and Donna K. McCurdy. "The Hypertonic State." *New Eng. J. Med. 297*:1444, Dec. 29, 1977.

Kuntz, I. D., and A. Zipp. "Water in Biological Systems." *New Eng. J. Med. 297:* 262, Aug. 4, 1977.

O'Brien, Donough. "The Management of Fluid and Electrolyte Problems in Childhood." *Pediatric Basics.* No. 17. Gerber Products Co. Fremont, Mich., 1977.

Chapter 20

The Blood and Lymph

Objectives

After studying this chapter, the reader should be able to:

- List the major subdivisions of the blood.

- Outline the development of the blood.

- List the functions of the blood, with examples.

- Describe the plasma as to percent of the blood, its major constituents, and the functions of those constituents.

- List the type and amounts of plasma proteins and give their functions.

- Account for the origin of plasma constituents.

- Describe, for each formed element: its structure; numbers; site(s) of production; life history; functions.

- Describe hemoglobin and its functions.

- Define anemia, and give some causes for the condition.

- Describe the several types of leukemia.

- Define what hemostasis is, and describe the intrinsic and extrinsic schemes of clotting.

- Describe how clotting may be hastened or prevented.

- Describe the general plan of the lymphatic system.

- Explain how lymph is formed, its composition, how it moves through the lymph vessels, and its functions.

- Describe the structure and functions of lymph nodes, spleen, and tonsils.

Blood is classified as a connective tissue with a complex liquid intercellular material, the PLASMA, in which cells and cell-like structures, the FORMED ELEMENTS, are suspended. Blood plasma is one of the extracellular fluid compartments, one that despite its small volume (3L) is one of the most dynamic in terms of turnover of constituents. Blood constitutes about 7 percent of the body weight at any age and, in the average male adult, amounts to some 5 to 6 liters in volume. The adult female, because of a usually smaller body size, has 4.5 to 5.5 liters of blood. Blood is easily obtained from superficial veins (venipuncture) of the body and, because it circulates to and from body cells, its composition reflects cellular activity. Therefore, analysis of the blood gives a good idea of the status of body function.

Lymph is formed by passive processes from the blood and is tissue fluid that has entered lymphatic vessels. It is also an extracellular fluid, derived from the interstitial compartment. It is ultimately returned to the bloodstream by lymphatic vessels.

Development of the blood (Fig. 20.1)

Growth of the embryo depends on adequate supplies of oxygen and nutrients, and thus blood and blood vessels are among the earliest structures to develop. At about 2½ weeks of development—

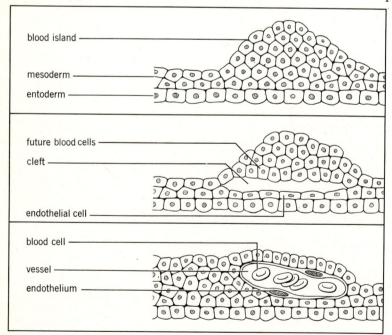

FIGURE 20.1   The development of blood islands, with vessel and blood cell formation.

when the embryo is only 1.5 millimeters long—dense masses of cells called BLOOD ISLANDS appear in the mesoderm of the embryo and its surrounding tissues (yolk sac, chorion). Cavities appear within the islands, and the cells around these cavities flatten to form the linings (endothelium) of what are to become primitive blood vessels. Some cells are shed *into* the cavities and they differentiate into primitive blood cells. The first

cells to be formed are nucleated red blood cells, capable of transporting oxygen and carbon dioxide. The liver, spleen, lymph glands, and bone marrow form blood cells beginning at about 6 weeks of development and continue until shortly before birth, when bone marrow and lymph organs remain as the only important sites of blood cell formation.

A general description of functions of blood

Functions of the blood center around two major activities: TRANSPORT and REGULATION OF HOMEOSTASIS.

Transport

Because of its liquid intercellular material, the blood can DISSOLVE and/or SUSPEND many materials, and CARRY them to and from the cells as the blood is caused to circulate by heart action.

Amino acids, lipids, carbohydrates, minerals, vitamins, water, and other materials are transported from their absorption sites in the digestive tract to the cells.

Red blood cells bind oxygen in the lungs and carry it to the cells; carbon dioxide, produced by metabolic activity, is transported to appropriate organs for excretion or elimination.

Heat is absorbed by the water of the plasma and is carried to skin and lungs for elimination.

Hormones are carried from their sites of production to body cells that respond to them.

Excess body water and inorganic materials are carried to appropriate organs of excretion, as are solid wastes of metabolism.

The blood is thus acting as a vehicle to carry substances, and, in the case of substances such as hormones, forms the *only* way these materials can reach *all* body cells.

Regulation of homeostasis

Regulation of homeostasis is concerned with activities such as regulation of extracellular fluid volume, regulation of body fluid pH, body temperature, and protection against infection and blood loss.

REGULATION OF EXTRACELLULAR FLUID VOLUME. Fluids in the interstitial compartment are derived by *filtration* (primarily) from the blood capillaries. Plasma proteins are generally too large to filter and create an osmotic force that causes *osmotic return* of water to the capillaries. The composition of the blood and its concentrations of osmotically active particles determines, in part, how much water will remain in the interstitial compartment.

REGULATION OF pH. Recall that the blood contains *buffer systems*, utilizing bicarbonates, phosphates, proteins, and hemoglobin, that remove free hydrogen ion from the fluids. Thus, maintenance of proper levels of these substances in the bloodstream aids in maintenance of acid-base balance.

REGULATION OF BODY TEMPERATURE. Water of the blood receives heat from the metabolic processes of cells and then delivers the heat to appropriate organs for elimination. A diminished blood volume is often associated with development of fever, as total heat carrying capacity is reduced.

PROTECTION AGAINST INFECTION. The blood contains chemicals known as ANTIBODIES that can neutralize foreign chemicals that may enter the body. Several of the white cells are good PHAGO-CYTES that can engulf and destroy microorganisms.

PROTECTION AGAINST BLOOD LOSS. As part of the mechanisms that operate to prevent loss of blood (hemorrhage) from the vessels of the body, the blood changes from a liquid to a semisolid state in the process of COAGULATION (*clotting*). Many of the chemicals necessary for this process are found in the plasma and formed elements.

Components of the blood

By drawing a sample of blood, adding an anticoagulant (to prevent clotting), and then letting the blood stand or centrifuging it in a tube, whole blood may be separated into two major fractions: FORMED ELEMENTS and PLASMA.

Being heavier than the plasma, the formed elements will settle or be centrifuged to the bottom of the tube, while a straw-colored fluid, the plasma, will remain on top. The formed elements normally constitute 43 to 45 percent of the volume of the blood, the plasma 55 to 57 percent.

Plasma

Plasma is an extraordinarily complex solution and suspension of chemical substances. It consists of 90 percent water, 1 percent inorganic substances, 6 to 8 percent protein, with the remainder glucose, lipids, nitrogenous wastes of metabolism (e.g., urea) gases, enzymes, hormones, and other substances. Table 20.1 summarizes some of the more important plasma constituents.

As may be recalled, the water of the plasma serves transport functions and the inorganic substances act as buffers and osmotically active particles and ensure proper excitability of cells.

THE PLASMA PROTEINS. The plasma proteins are characteristic for the plasma and serve the following general functions.

They are the primary agents responsible for the osmotic return of interstitial fluid to blood capillaries.

They give viscosity to the blood ("blood is thicker than water").

They create a suspension stability in the blood that aids in preventing materials (proteins, lipids, etc.) from "settling out" of the blood.

They serve as a reserve of amino acids for cell use, although they are not ordinarily utilized for this purpose unless the person is starving.

They aid in the buffering process.

Chemicals called antibodies, that aid in the defense against foreign chemicals, are proteins in the plasma.

By relatively simple techniques, such as addition of salts to the plasma ("salting out"), three main fractions of proteins may be separated from the plasma. They are designated as the ALBUMINS, GLOBULINS, and FIBRINOGEN.

Albumins are the most plentiful (55–64 percent) of the plasma proteins, and their concentration is 4 to 5 gm per 100 ml of blood. They are also the smallest (molecular weights 69,000–70,000) of the proteins. Since osmotic pressure depends on numbers of solute particles, the abundant albumins contribute *most* of the osmotic pressure that "pulls" water osmotically from the interstitial compartment back into the capillaries. Albumins also serve as "vehicles" that bind several substances for transport through the plasma (barbiturates; thyroxin, a hormone).

TABLE 20.1   Some important plasma constituents

Constituent/type	Amount/concentration	Functions/comments
Water	90% of plasma	Dissolves, suspends, ionizes electrolytes, carries heat.
Electrolytes	About 1% of plasma (Examples: Na^+, K^+, Ca^{++}, Mg^{++}, Cl^-, HCO_3^- PO_4	They buffer, establish osmotic gradients, are responsible for excitability of cells.
Proteins	6–8% of plasma	General functions: give viscosity to blood, clotting, antibodies, reserve of amino acids, establish colloid osmotic pressure of plasma.
Albumins	50–65% of total proteins	Are responsible for most of the plasma colloidal osmotic pressure.
Globulins (alpha, beta, gamma)	14.5–27% of total proteins	α and β serve general functions; γ are major source of antibodies.
Fibrinogen	2.5–5% of total proteins	Clotting factor
Other substances		
Glucose	about 0.1% of plasma	Nutrients or Wastes
Gases (O_2, CO_2)	variable (O_2 av 20 ml/100 ml blood) (CO_2 av 9 ml 100 ml blood)	
Lipids	variable	
Vitamins	variable	
Nitrogenous materials (urea)	variable	

Globulins constitute about 2 percent of the plasma proteins and are present to the extent of 2 to 3 gm per 100 ml of blood. Their molecular weights range from 150,000 to 900,000. Three fractions may be separated in the globulins by such procedures as electrophoresis (plasma is subjected to an electric current, and the proteins migrate at different rates according to size and electrical charge).

ALPHA GLOBULINS (molecular weights 150,000–160,000) serve the general functions of the plasma proteins and are the major proteins that bind other substances for transport.

BETA GLOBULINS are a bit larger (molecular weight 160,000–200,000), also serve the general functions of

plasma proteins, and include several of the clotting factors (e.g., prothrombin).

GAMMA GLOBULINS range in size from 150,000 to 900,000 molecular weight. This group contains the immunoglobulins (Ig) or antibodies. Five types of Ig, designated IgA, IgM, IgG, IgD, and IgE, have been isolated. Each Ig is produced in response to a particular type of antigenic challenge. As detailed in the following chapter, Ig's provide protection against chemical challenges in the body.

Fibrinogin is a soluble plasma protein with a molecular weight of about 200,000. It is present in the blood to the extent of 0.15 to 0.3 gm per 100 ml. It is converted to insoluble fibrin as the blood

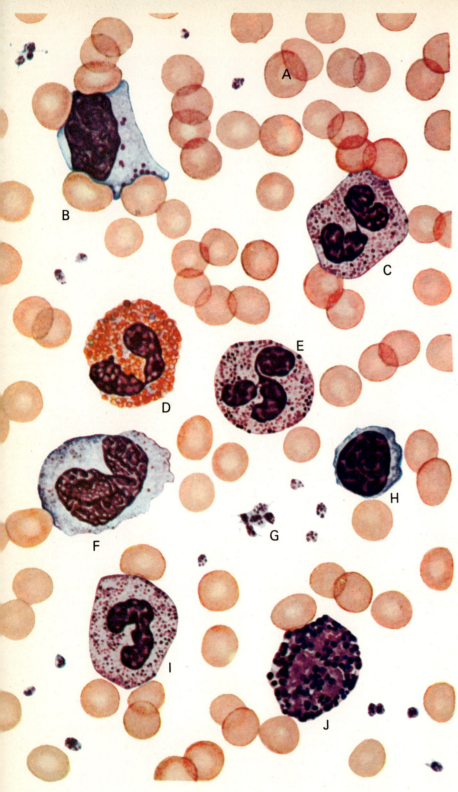

FIGURE 20.2 The morphology of the blood elements.

Legend key.

CELL TYPES FOUND IN SMEARS OF PE-RIPHERAL BLOOD FROM NORMAL INDI-VIDUALS. The arrangement is arbitrary and the number of leukocytes in relation to erythrocytes and thrombocytes is greater than what would occur in an actual microscopic field.

A Erythrocytes
B Large lymphocyte with azurophilic granules and deeply indented by adjacent erythrocytes
C Neutrophilic segmented
D Eosinophil
E Neutrophilic segmented
F Monocyte with blue gray cytoplasm, coarse linear chromatin, and blunt pseudopods
G Platelets (thrombocytes)
H Lymphocyte
I Neutrophilic band
J Basophil

coagulates, and forms a barrier tending to prevent blood loss from damaged vessels.

Concentrations of plasma proteins vary little in persons in good health, and the albumins and globulins normally maintain about a 2:1 ratio (A/G ratio). Levels of proteins decrease during starvation (they are metabolized), liver damage (production is diminished) and renal disease (they are excreted). A primary sign of plasma protein deficiency is the development of edema, as water filtered from blood vessels is not osmotically returned to the capillaries.

SOURCES OF PLASMA COMPONENTS. Water, inorganic substances, carbohydrates, lipids, and vitamins are chiefly derived by absorption from the gut, as food and drink are ingested. Proteins are manufactured primarily by the liver, with some coming from disintegration of white cells in the bloodstream. Gases are derived from the lungs (O_2) or cells (CO_2). One can thus understand why liver disease or starvation can lower plasma protein concentrations.

Formed elements (Fig. 20.2)

Three categories of formed elements occur in normal human peripheral blood. They are ERYTHROCYTES (red blood cells or corpuscles); LEUCOCYTES (white blood cells); and PLATELETS (thrombocytes).

ERYTHROCYTES. Mature erythrocytes, as they occur in the bloodstream, are biconcave, nonnucleated, flexible discs averaging 8.5 μm in diameter, 2 μm thick at their edges and 1 μm thick at their center. This shape (biconcave disc) provides the maximum surface area for the cell's volume, ensuring the greatest possible surface for diffusion of gases. Although they contain no nuclei, erythrocytes do consume small amounts of O_2, glucose, and ATP, and produce small quantities of CO_2. This indicates a low level of metabolism, probably concerned with the operation of active transport systems in the cell membrane. These systems maintain solute balance between the cell and its

environment so that net shifts of water do not occur to cause shrinkage or swelling of the cell.

Numbers. Erythrocytes are the most numerous of the formed elements, averaging about 5×10^6 units per cu mm in the adult. This number remains quite constant in good health, and reflects a balance between production and destruction. About 1 percent of the total erythrocyte number is regenerated each day, or about 3.5 million per kg of body weight per day.

Production. During intrauterine life, the yolk sac, spleen, liver, lymph nodes, and red bone marrow serve as sites of erythrocyte production. At or shortly before birth, all areas except the red bone marrow in such regions as the sternum, skull bones, vertebrae, ribs, and epiphyses of the long bones, cease production of red cells. The red cells originate from a stem cell common to all formed elements, and pass through a series of steps (Fig. 20.3) to form the mature element. At that stage known as the *rubricyte*, the first appearance of hemoglobin occurs. Condensation and loss of the nucleus occurs in the *metarubricyte* stage, giving more room for accumulation of hemoglobin.

Life history. Mature red cells circulate in the bloodstream for 90 to 120 days. As they age in the bloodstream, subtle chemical changes apparently occur in the cells. These changes render the cells more liable to phagocytosis by cells located primarily in the liver, spleen, and bone marrow. The phagocytes largely recycle the erythrocyte components for reuse by cells in general and in the synthesis of new cells and hemoglobin.

Requirements for production of erythrocytes. Many substances are required for the production of red cells. The more important ones, and their roles, are presented in Table 20.2. ERYTHROPOIETIN is a substance that stimulates erythrocyte production and is a major factor in achieving the balance necessary to maintain normal numbers of erythrocytes in the bloodstream. The kidney, in response to any condition that lowers the arterial oxygen levels (*hypoxemia*), produces *renal erythropoietic factor* (REF). REF acts on a plasma globulin precursor to cause the production of erythropoietin that stimulates mitosis of marrow cells and their

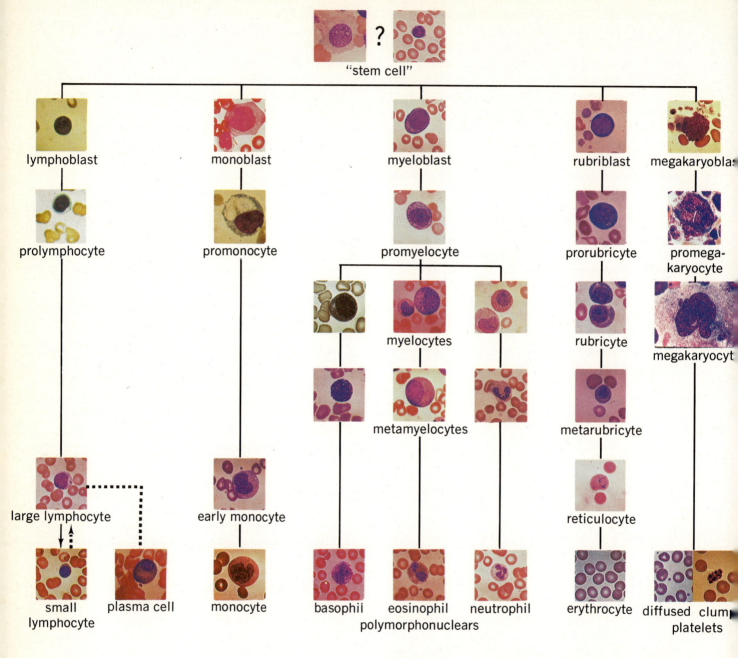

FIGURE 20.3   The formation of blood cells. The exact morphology of the stem cell is still open to question. Wright stain. 1200×. (Courtesy of American Society of Hematology National Slide Bank and Health Sciences Learning Resources Center, University of Washington. Used with permission.)

TABLE 20.2 Requirements for erythrocyte production	
Substance	Description and/or use
Lipid	Cholesterol and phospho-lipids; incorporated in membrane and stroma
Protein	Incorporated into cell membrane
Iron	Incorporated into hemoglobin
Amino acids	Incorporated into hemoglobin
Erythropoietin	A glycoprotein, it is released from the kidney with hypoxia, hemorrhage, and excessive androgen secretion, and stimulates production of erythrocytes
Vitamin B_{12}	Used in the formation of DNA in nuclear maturation
Intrinsic factor	A mucopolysaccharide produced by the stomach. It combines with vitamin B_{12} and ensures absorption of the vitamin from the gut
Pyridoxin	Increases the rate of cell division
Copper	Catalyst for hemoglobin formation
Cobalt	Aids synthesis of hemoglobin
Folic Acid	Promotes DNA synthesis

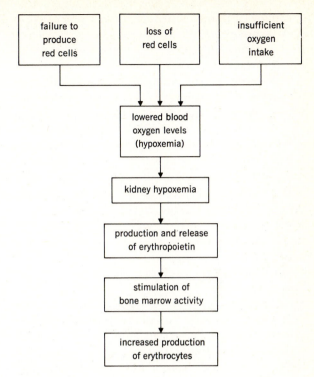

FIGURE 20.4   The homeostatic mechanism maintaining red cell production.

maturation to red cells. Erythropoietin can cause an increase of about 2 percent in the total circulating erythrocyte mass. We thus see that the major factor controlling red cell production is blood oxygen level. The relationships of erythropoietin to red cell production are summarized in Figure 20.4.

Hemoglobin. Hemoglobin is an iron-containing respiratory pigment found in the red cells. Four nonprotein red pigment molecules called *heme*, containing the iron, and four polypeptide chains, constituting the *globin* portion of the molecule, are combined to create the hemoglobin molecule. Each molecule can combine with eight atoms of oxygen in a reversible fashion.

The reaction between hemoglobin and oxygen may be represented by the equation

$$Hgb + O_2 \rightleftharpoons HgbO_2$$

reduced hemoglobin (bluish in color)   oxyhemoglobin (red in color)

The hemoglobin loads oxygen in the lungs by diffusion, with the oxygen attaching to the iron atoms of the heme, and releases it by diffusion to the cells where oxygen levels have been reduced by metabolism that utilizes the oxygen. Without hemoglobin, $1/3$ ml of oxygen could be carried by 100 ml of blood; with hemoglobin, 20 ml O_2 per 100 ml of blood may be transported, a 60 fold increase. Thirteen to 16 gm of hemoglobin per 100 ml of blood is considered to be a normal amount.

Hemoglobin may also transport carbon dioxide by attachment to the amine groups of amino acids in the globin molecules. The compound formed is called *carbaminohemoglobin.*

The amino acid sequences in the globin portions of the hemoglobin molecules are genetically determined. Mutations may cause the production of abnormal globin molecules and alter the ability of hemoglobin to function properly. For example, hemoglobin S, the characteristic hemoglobin of sickle-cell anemia, has a substitution of one amino acid for another in the globins, and precipitates under low blood oxygen levels. The red cells assume a crescent shape *(sickling).* Such cells are more rapidly phagocytosed and the subject may suffer *anemia.*

Functions of the erythrocytes. As may be deduced from the preceding discussion, erythrocytes are the agents for transport of some 95 percent of the body's oxygen in the bloodstream, and can carry about 25 percent of metabolically produced carbon dioxide. Hemoglobin is also a buffering substance, aiding the maintenance of proper pH of the blood.

Clinical considerations. In addition to the production of abnormal hemoglobins, faulty production of either hemoglobin or the cells themselves may create a variety of abnormalities in the red cells.

ANEMIA results when there is a reduced amount of hemoglobin in the bloodstream, either as a result of fewer cells or less hemoglobin per cell. Table 20.3 summarizes several types of anemia by cause, and gives characteristics of and suggested treatments for the various types of anemia.

POLYCYTHEMIA refers to excessive numbers of red cells in the bloodstream. It is usually a response to low blood oxygen levels, such as might result from ascent to high altitudes, or a condition that results in failure of O_2 to pass from lung to blood.

ANISOCYTOSIS refers to red cells that are not of normal size; *macrocytes* are larger than normal, *microcytes* smaller than normal.

POIKILOCYTOSIS refers to abnormally shaped red cells. The sickle cells of sickle cell anemia and spherocytes (ball-shaped cells) are examples.

Metabolism of hemoglobin. Destruction of the heme fraction of the hemoglobin molecule by phagocytic cells results in the production of a greenish pigment, BILIVERDIN. Biliverdin undergoes reduction to a yellowish pigment known as bilirubin that is released into the bloodstream. Liver cells absorb the bilirubin and conjugate it with a compound called glucuronic acid to form a *bile salt.* The salt is excreted into the small intestine in the bile and is acted on by intestinal bacteria to form UROBILINOGEN *(stercobilinogen).* Some of the urobilinogen remains in the intestine and is excreted as stercobilin, imparting the orange-brown color to the feces; the remainder is absorbed by the intestine, and is ultimately excreted by the kidney as urobilin, the amber coloring matter of the urine. The metabolism of heme is summarized in Figure 20.5.

LEUCOCYTES. The white blood cells are a heterogeneous population of nucleated cells that contain no hemoglobin *(see* Fig. 20.2). They range in size from 8 μm to about 25 μm.

Numbers. White cells are the least numerous of the formed elements. In the adult, they normally range from 5000 to 9000 per cu mm with an average of 7500 per cu mm.

Life history *(see* Fig. 20.3). According to where they are produced, two categories of white cells are distinguished: GRANULAR leucocytes (neutrophil, eosinophil, basophil) are produced by red bone marrow and are said to be of *myeloid* origin; NONGRANULAR leucocytes (lymphocytes, monocytes) are produced, after birth, in lymphoid tissue such as lymph nodes, tonsils, and spleen, and are said to be of *lymphoid* origin. Additionally, myeloid leucocytes have *lobed* or *segmented nuclei* (polymorphonuclear cells) and *characteristic granules* in their cytoplasm, while lymphoid leucocytes have a nonlobed nucleus and lack characteristic cytoplasmic granules; these are sometimes called mononuclear cells (unlobed nucleus).

TABLE 20.3   Some common anemias

	Type of anemia	Causes	Characteristics	Symptoms	Treatment
INCREASED LOSS	*Hemorrhagic* Acute	Trauma Stomach ulcers Bleeding from wounds	Cells normal	Shock	Transfusion
	Chronic (iron deficiency)	Stomach ulcers Excessive menses	Cells small (microcytes), deficient in hemoglobin content	None or fatigue	Iron administration
	Hemolytic	Defective cells Destruction by parasites, toxins, antibodies	Young cells (reticulocytes) very prominent serum haptoglobin[a] reduced. Morphology of cells may be abnormal when detected	None or fatigue	Dependent on cause
DECREASED FORMATION	*Deficiency states* Folic acid	Nutritional deficiency	Cells large (macrocytes), with normal hemoglobin content	None or fatigue	Folic acid
	Vitamin B$_{12}$	Lack of intrinsic factor in stomach (pernicious anemia)	Cells large with normal hemoglobin content		Administration of vitamin B$_{12}$
	Hypoplastic or *Aplastic*	Radiation Chemicals Medications	Bone marrow hypoplastic	None or fatigue	Transfusion Androgens Cortisone

[a] A protein strongly binding hemoglobin in plasma.

Granular leucocytes remain functional for about 10 hours to 3 days in the bloodstream; nongranular cells for 100 to 300 days. Leucocytes are lost by fragmentation in the bloodstream, by migration through epithelial surfaces, and by destruction during inflammation and infectious processes.

Production of leucocytes is influenced primarily by body levels of adrenal steroid hormones and by products of infection; both stimulate production.

Materials required for leucocyte production are those for cells in general, but particularly important is folic acid. This substance is essential for nucleic acid production as cells divide.

Functions of the leucocytes. All leucocytes are capable of ameboid motion and phagocytosis. They can thus pass through capillary walls (*diapedesis*) and move through body tissues where they provide defense against microorganisms and foreign chemicals and remove the products of cell

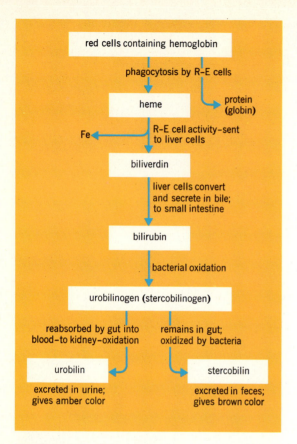

red cells containing hemoglobin

↓ phagocytosis by R-E cells

heme → protein (globin)

Fe ← R-E cell activity-sent to liver cells

biliverdin

↓ liver cells convert and secrete in bile; to small intestine

bilirubin

↓ bacterial oxidation

urobilinogen (stercobilinogen)

reabsorbed by gut into blood–to kidney–oxidation   remains in gut; oxidized by bacteria

urobilin   stercobilin

excreted in urine; gives amber color   excreted in feces; gives brown color

FIGURE 20.5   The metabolism of the heme portion of the hemoglobin molecule.

destruction and death. The protective relationships are explored more extensively in the next chapter.

Clinical considerations. LEUCOPENIA refers to a *decrease* in the numbers of leucocytes. It may result from increased destruction of cells, such as occurs during viral infections, or from treatment with certain drugs (sulfonamides, anticonvulsants). LEUCOCYTOSIS refers to an *increase* in numbers of leucocytes. A physiological leucocytosis (temporary increase) results from exercise, as cells are "washed out" of normally closed capillaries. Sustained elevations of white cell numbers suggest pathological processes in the body, such as leukemia.

LEUKEMIA is a neoplastic disorder of the blood-forming tissues, primarily those giving rise to the white blood cells. Viruses have been suggested to be the cause of the cell proliferation, but no proof exists for this cause in the human. The major symptoms that appear are the production of great numbers of white cells, and diminished production of erythrocytes. According to the type of cell that predominates in the bloodstream, and the maturity of those cells, four general types of leukemia are distinguished.

Acute lymphoblastic leukemia (ALL) is a disease primarily of children, and is characterized by increased numbers of immature lymphocytes. Anemia, due to decreased production of red cells, is found in 90 percent of patients with ALL.

Acute myeloblastic leukemia (AML) can occur at any age, and is characterized by increase in all of the granulocytes (neutrophil, eosinophil, basophil) and by increased numbers of immature granulocytes.

Chronic lymphocytic leukemia (CLL) occurs most commonly in the middle aged and elderly, and is characterized by greatly increased numbers of small, mature lymphocytes. Anemia is milder than in the acute leukemias.

Chronic myelocytic leukemia (CML) occurs more commonly in young adults and is characterized by increased numbers of all granular leucocytes.

Radiation and chemotherapy are the measures most frequently employed to treat leukemia. The acute forms of the disease are more difficult to arrest than the chronic forms; remissions of as much as 15 years have been achieved with adequate therapy that has been instituted early.

HODGKIN'S DISEASE is a chronic lymphoma of unknown cause characterized by enlargement of the lymph nodes, spleen, and liver. White cell numbers do not increase greatly in this disorder, but a giant cell with many-lobed or multinuclei (Reed-Sternberg cells) appear in the nodes. There is no certain cure for this disease, but radiation and chemotherapy are employed to lengthen life.

MONONUCLEOSIS is a disease thought to be due to a virus. It causes elevation of nongranular white cells, fatigue, swollen lymph nodes, fever, and

Element	Normal numbers	Origin (area of)	Diameter (μm)	Morphology	Function(s)
Erythrocytes	4.5–5.5 million /mm³	Myeloid (marrow)	8.5 (fresh) 7.5 (dry smear)	Biconcave, non-nucleated disc; flexible	Transports O_2 and CO_2 by presence of hemoglobin; buffering
Leucocytes	6000–9000/mm³		9–25		
Neutrophil	60–70% of total	Myeloid	12–14	Lobed nucleus, fine heterophilic specific granules	Phagocytosis of particles, wound healing. Granules contain peroxidase for destruction of microorganisms.
Eosinophil	2–4% of total	Myeloid	12	Lobed nucleus; large, shiny red or yellow specific granules	Detoxification of foreign proteins? Granules contain peroxidases, oxidases, trypsin, phosphatases. Numbers increase in autoimmune states, allergy, and in parasitic infection (schistosomiasis, trichinosis, strongyloidiasis).
Basophil	0.15% of total	Myeloid	9	Obscure nucleus; large, dull, purple specific granules	Control viscosity of connective tissue ground substance? Granules contain heparin (liquefies ground substance) serotonin (vasoconconstrictor), histamine (vasodilator)
Lymphocyte	20–25% of total	Lymphoid			
Small			9	Nearly round nucleus filling cell, cytoplasm clear staining	Phagocytosis of particles, globulin production
Large			12–14	Nucleus nearly round, more cytoplasm	
Monocyte	3–8% of total	Lymphoid	20–25	Nucleus kidney or horseshoe-shaped, cytoplasm looks dirty	Phagocytosis, globulin production
Platelets (Thrombocytes)	250,000–350,000 mm³	Myeloid	2–4	Chromomere and hyalomere	Clotting

TABLE 20.4 Summary of formed elements

sore throat. It lasts 10 days to 2 weeks, and is specifically diagnosed by the presence of abnormal lymphocytes and a specific antibody (Paul-Bunnel antibody) in the bloodstream.

PLATELETS (*see* Fig. 20.2). Platelets are nonnucleated structures 2 to 4 μm in size. They contain one or more dark-staining granules (*chromomeres*) in a lighter-staining area (*hyalomere*), and are found singly or in clumps among the other formed elements on a peripheral blood smear.

Life history (*see* Fig. 20.3). The red bone marrow contains giant cells known as *megakaryocytes*. As these cells mature, they develop in their cytoplasm many small granules (chromomeres). They then send pseudopods through capillary walls and "pinch off" a portion of the cytoplasm that contains one or more granules. The portion of cytoplasm pinched off is a platelet. Platelets are believed to circulate in the bloodstream for 10 to 12 days, and are there either phagocytosed or form plugs across tears in capillary walls.

Functions. Platelets contain one of the several factors required for blood clotting. They are "sticky," and aid in sealing the microscopic tears that occur in capillary walls as a normal consequence of life. Platelet deficiency is associated with many tiny pinpoint hemorrhages appearing in the capillaries (*purpura*) that are particularly evident in the skin.

Table 20.4 summarizes facts about the formed elements.

Hemostasis

When blood vessels are damaged, a series of reactions occur that aid in preventing blood loss through the wound. These reactions constitute HEMOSTASIS (G. *amia*, blood + *statikos*, standing). Three types of reactions occur.

There is a NARROWING OF THE VESSELS (vasoconstriction) in the area of the wound, which presumably causes some reduction of the blood flow in the injured area.

A PLATELET PLUG is formed across the tear in the vessel.

The blood changes from its normal fluid consistency to a gelated or semisolid state, in the process of COAGULATION or clotting.

Vascular responses and the platelet plug

VASOCONSTRICTION occurring when blood vessels are traumatized is first due to nerve impulses arriving at the vessels because of the pain associated with the injury. A reflex involving the cord is responsible, but this response lasts only a few minutes. Second, the vessel muscle undergoes an intense spasm because of the direct mechanical damage to the vessels. This reaction lasts for 20 to 30 minutes.

Injury to the vessels also creates a roughened surface to which the PLATELETS ADHERE. Several layers of platelets accumulate across a tear in a vessel and may completely seal the injury. This platelet mechanism operates continually to seal the ruptures in the small vessels of the body that occur as a normal consequence of living.

Coagulation

At least 12 factors or chemical substances are required for the blood to clot. These are present in the platelets and plasma or are produced during the clotting reactions. They are shown in Table 20.5.

In general outline, blood clots in a definite sequence of events as shown below:

Thromboplastin + prothrombin $\longrightarrow$ thrombin
Thrombin + fibrinogen $\longrightarrow$ fibrin (clot)

TABLE 20.5 Factors definitely implicated in blood coagulation

International committee designation	Synonyms	Origin	Location
Factor I	Fibrinogen	Liver	A plasma protein
Factor II	Prothrombin	Liver	A plasma protein
Factor III	Thromboplastin	By series of reactions in blood; also found, as such, in cells	Produced in the clotting process or released into fluids by injured cells
Factor IV	Calcium	Food and drink; from bones	As Ca^{2+} in plasma
Factor V	Labile factor (accelerator globulin)	Liver	Plasma protein
Factor VI[a]			
Factor VII	SPCA (serum prothrombin conversion accelerator)	Liver	Plasma
Factor VIII	AHF (antihemophilic factor) AHG (antihemophilic globulin)	Liver	Plasma
Factor IX	PTC (plasma thromboplastin component)	Liver	Plasma
Factor X	Stuart-Prower factor; develops full factor III power	Liver	Plasma
Factor XI	PTA (plasma thromboplastin antecedent)	Liver	Plasma
Factor XII	Hageman factor; contact factor; initiates reaction	?	Plasma
Factor XIII	Fibrin stabilizing factor; renders fibrin insoluble in urea. (Laki-Lorand factor)	?	Plasma
Platelet factor	Cephalin	Marrow	Platelets

[a] No longer considered a separate entity; considered to be identical to Factor V.

Production of thromboplastin requires the interaction of several other substances, and certain substances speed the reactions producing thromboplastin. Additionally, thromboplastin may be produced by two methods: an EXTRINSIC SCHEME occurs when cells are damaged; an INTRINSIC SCHEME occurs within the blood itself. One theory of how these chemicals and schemes fit together is presented on the next page.

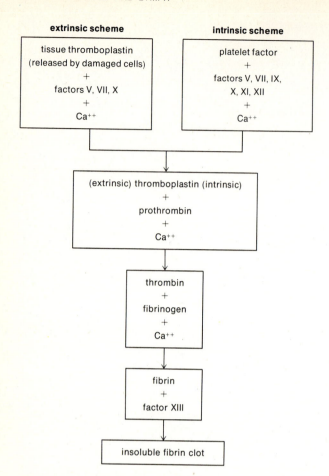

extrinsic scheme

tissue thromboplastin
(released by damaged cells)
+
factors V, VII, X
+
Ca++

intrinsic scheme

platelet factor
+
factors V, VII, IX,
X, XI, XII
+
Ca++

(extrinsic) thromboplastin (intrinsic)
+
prothrombin
+
Ca++

thrombin
+
fibrinogen
+
Ca++

fibrin
+
factor XIII

insoluble fibrin clot

These reactions are autocatalytic, that is, each step speeds the production of materials in the next step, so that, once begun, the clotting process proceeds rapidly. The clot is eventually removed by an enzyme, FIBRINOLYSIN (plasmin), produced by activation of a plasma precursor.

Procoagulation and anticoagulation

Clot formation may be speeded by heat or by spraying a wound with a clotting factor (e.g., fibrin). Prevention of clotting is essential when drawing blood for tests, transfusions, or storage, and is most easily accomplished by removing one of the materials in the clotting process; Ca^{2+} is the one most easily removed, by simply mixing the blood with substances (e.g., oxalates) that exchange their cation for Ca^{2+}. The blood is then said to be *decalcified. Heparin* is an organic substance that decreases thromboplastin production; *dicoumarol* is an organic substance that interferes with liver synthesis of prothrombin, and factors V and VII. Dicoumarol requires 36 to 48 hours for its effect to be manifested.

TABLE 20.6 Some genetically determined disorders of clotting			
Disorder or deficiency	Synonyms	How inherited	Treatment
Factor VIII deficiency (80% of all hemophilias)	Classical hemophilia; Hemophilia A	Sex linked (on X chromosome) recessive; males show disorder	Transfusion of *fresh* blood containing Factor VIII (VIII disappears with aging of blood)
Factor IX deficiency (15% of hemophilias)	PTC deficiency; Christmas disease; Hemophilia B	Sex linked recessive; males show disorder	Transfusion of stored blood is effective
Factor XI deficiency	PTA deficiency; Hemophilia C	Rare; autosomal dominant	Transfusion of stored blood is effective
Hypofibrinogenemia	Afibrinogenemia	Autosomal recessive	Transfusion of whole blood, or plasma

Clinical considerations

Since many of the major clotting factors are proteins (e.g., Factors I, II, VIII, IX, and XI), and since proteins are produced under the direction of DNA, alterations in genes (mutations) may result in defective production of clotting factors. Table 20.6 presents several genetically determined disorders of clotting, all of which are characterized by failure of the blood to clot when the body is wounded, bruised, or traumatized.

Disorders of platelet production leads to the development of THROMBOCYTIC PURPURA, characterized by easy bruising and bleeding from the nose, gums, gastrointestinal tract, and kidneys. Low platelet levels result in fragile capillaries and slowed clotting.

SPONTANEOUS INTRAVASCULAR CLOTTING (clotting in vessels in the absence of trauma) may create a THROMBUS (stationary clot) that blocks a vessel and causes tissue death beyond the block (e.g., coronary thrombosis).

Lymph and lymph organs

Lymph formation, composition, and flow

Lymph is tissue or interstitial fluid that has entered the lymph vessels (Fig. 20.6) of the body. It arises by a process of filtration of the blood from blood capillaries. Thus it has the same composition as cell free, low protein plasma.

Tissue fluid enters lymph vessels when they are expanded as muscle becomes active in the body ("suction effect"), and because of a small pressure gradient that exists from blood to tissue space to lymph vessel. Protein enters the lymph vessels easily because they are much more permeable than blood capillaries. The lymph moves through the vessels because of the massaging action of muscles as they contract (muscle pump), and because of breathing movements that alternately compress and release the lymphatics (respiratory pump). Valves assure flow toward the veins beneath the clavicles, where the large lymph ducts connect to the blood vascular system. Rate of flow is about 1.5 milliliters per minute.

FUNCTIONS. The lymph returns, to the blood vascular system, filtered protein and excess tissue water (10%) that is not osmotically returned (90%) to blood capillaries. The lymph of the intestine contains much absorbed end products of fat digestion because these end products enter the lymph vessels of the intestinal villi and not the blood vessels.

Lymph nodes

FORM AND FUNCTIONS. Lymph nodes (Fig. 20.7) are ovoid or rounded structures varying in size from 1 to 25 millimeters. They are situated along most medium-sized lymph vessels and form four major groups (see Fig. 20.6) in the *axilla, groin, neck,* and *abdomen.* The nodes are sites of production of nongranular white cells. Within the nodes are many fixed phagocytic cells (macrophages). Lymph passing through the nodes is FILTERED or cleansed of particulate matter (bacteria, dirt, and cell debris). The nodes may also trap cancer cells that have entered the lymph vessels from a neoplasm.

CLINICAL CONSIDERATIONS. Enlargement of the nodes is common in infections, due to an increase in the number of cells to trap the infectious agents. Nodes are generally removed in conditions where it is believed that cancer cells have metastasized. The status of the nodes is commonly determined

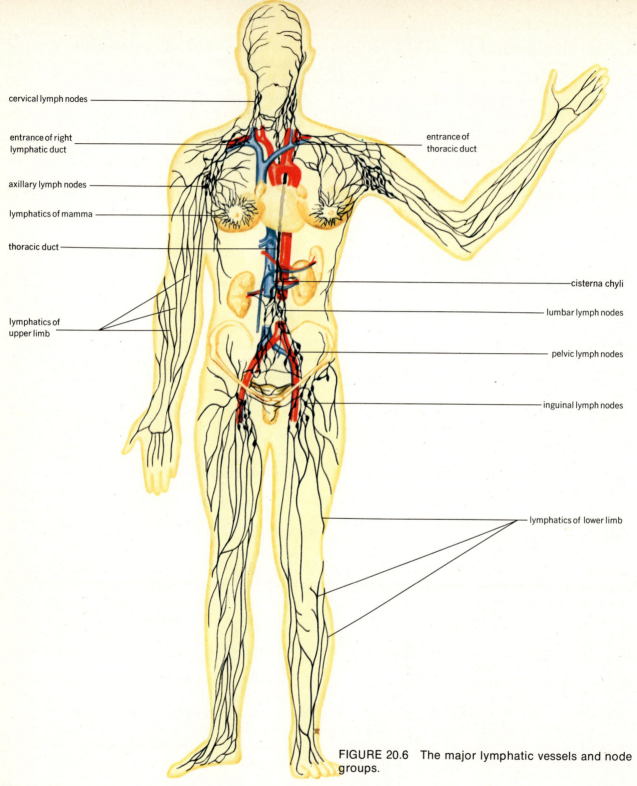

cervical lymph nodes

entrance of right
lymphatic duct

axillary lymph nodes

lymphatics of mamma

thoracic duct

lymphatics of
upper limb

entrance of
thoracic duct

cisterna chyli

lumbar lymph nodes

pelvic lymph nodes

inguinal lymph nodes

lymphatics of lower limb

FIGURE 20.6   The major lymphatic vessels and node
groups.

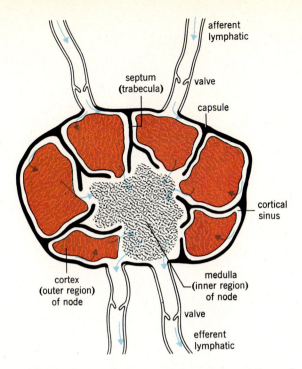

FIGURE 20.7 A diagrammatic representation of a lymph node. Cortical and medullary areas contain both lymphocytes and macrophages. Arrows indicate direction of lymph flow.

by feeling (palpation) in individuals presenting themselves for physical examination.

Other lymph organs

The TONSILS (Fig. 20.8) are three pairs of lymphoid organs located in the oral cavity and the upper part of the throat. The *palatine tonsils* are found in the side walls of the posterior part of the oral cavity and are the ones referred to when one "has the tonsils out." The *pharyngeal tonsils*, also called the adenoids, are located in the upper part of the throat or pharynx just behind the hard palate. The *lingual tonsils* are found embedded in the base of the tongue. All the tonsils consist of masses of lymphocytes and germinal centers in which lymphocytes are produced. The lymphocytes are shed into outgoing lymph vessels and eventually reach the bloodstream. The tonsils do not filter or cleanse the lymph because they do not have afferent or incoming lymph vessels.

The SPLEEN (Fig. 20.9) is a soft reddish-colored organ located behind the left side of the stomach. It lies between arteries and veins and contains areas where lymphocytes are produced that are

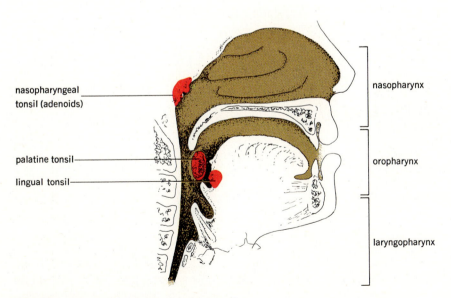

FIGURE 20.8   The location of the tonsils.

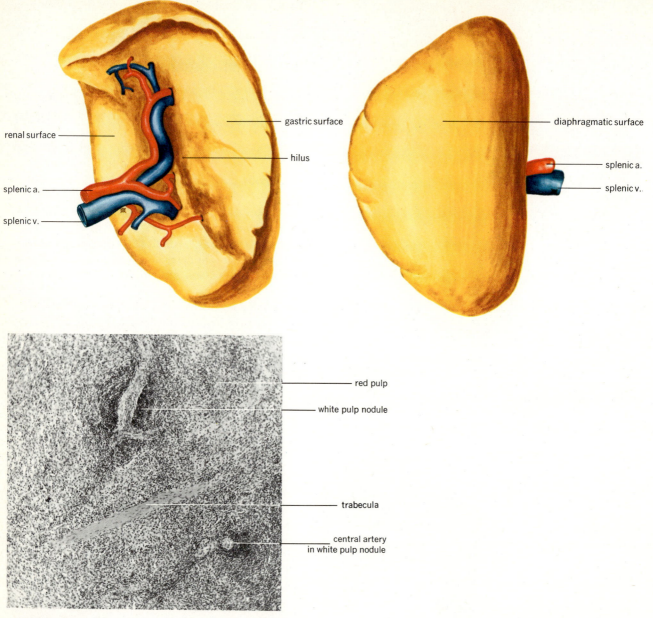

FIGURE 20.9   The gross and microscopic structure of the spleen.

then placed into outgoing lymphatic vessels. These areas are known as the *white pulp* of the spleen. The spleen is also an area where old red blood cells are phagocytosed and removed from the circulation, and where up to 200 milliliters of blood can be "stored" and injected into the circulation when the spleen contracts. The spleen does not cleanse the lymph, because it, like the tonsils, has only outgoing lymphatic vessels.

The functions of the THYMUS are considered in Chapter 21. Masses of lymphoid cells form NODULES in the walls of the digestive and respiratory systems and are considered in those chapters.

CLINICAL CONSIDERATIONS. The palatine, pharyngeal, and lingual tonsils form a ring (Waldeyer's ring) of lymphoid tissue that encircles the entrance of the digestive and respiratory systems into the body. The ring offers a means of combating microorganisms that may penetrate the mucous membranes of the upper respiratory and digestive systems. Removal of any part of the ring may thus interfere with the protective function the ring serves.

The components of Waldeyer's ring normally enlarge as the child grows, reaching peak development at puberty. It grows because it is exposed to a variety of new infectious agents. Mere size of one of the components should not be used as the primary reason for its surgical removal. Only if one of the tonsils is chronically infected, or has enlarged to the point where it obstructs the breathing or digestive passageways, should it be removed.

The spleen is not an organ essential to life, since it may be removed without harmful effects to the body. It may enlarge during infections (typhoid fever, malaria, syphilis, tuberculosis), or during conditions that stress its red-cell-destroying capacity (e.g., sickle cell anemia). Trauma may cause the spleen to rupture, and this usually requires surgical removal of the organ to avoid "bleeding to death."

Summary

1. Blood is a connective tissue that flows through the blood vessels. It constitutes about 7 percent of the body weight.

2. Blood develops from mesodermal blood islands at about 2½ weeks of embryonic life.

3. Blood functions in transport, and in regulation of several aspects of homeostasis (heat control, water balance, pH regulation). Its cells and antibodies protect against bacteria and their products.

4. Two main components compose blood.
   a. The liquid plasma exceeds one half the blood volume, and contains organic and inorganic components. It dissolves, suspends, and transports many materials.
   b. The formed elements are cells or cell-like structures, and constitute a little less than one half the blood volume.

5. Plasma proteins serve to maintain blood water levels, give viscosity to the blood, aid in clotting, and act as a reserve of amino acids.
   a. Albumins are osmotically active.
   b. Globulins contain antibodies.
   c. Fibrinogen is involved in clotting.

6. Erythrocytes or red cells:
   a. Are the most numerous of the formed elements (4.5–5 million/mm³); and are produced in the bone marrow.
   b. Are of small (8μm) size.
   c. Contain hemoglobin for O_2 transport.
   d. Last 90–120 days in the circulation.

7. Hemoglobin:
   a. Contains a pigment (heme) portion.
   b. Contains a protein (globin) portion.
   c. Requires many substances for its production.
   d. Acts as a buffer.
   e. Forms a loose combination with O_2.

8. Anemia:
   a. Results when hemoglobin levels are decreased.
   b. May be due to decreased production of cells or hemoglobin, to increased loss, or to shortage of building blocks.
   c. May be due to formation of abnormal hemoglobins due to genetic causes.

9. Leucocytes or white cells:
   a. Are all nucleated and contain no hemoglobin.
   b. Are the least numerous of the formed elements (7500/mm³).
   c. Are produced in marrow and lymph organs.
   d. Last from hours to a year in the body.
   e. Afford protection against disease organisms and their products.
   f. Alter their numbers in various types of diseases, such as leukemia and mononucleosis.

10. Platelets:
    a. Are fragments of large bone marrow cells.
    b. Are intermediate in number (250,000/mm³).
    c. Are important in blood clotting.

11. Blood flow and characteristics change when vessels are injured. The vessels constrict, a platelet plug is formed, and the blood clots.
    a. Many factors are required to form a clot.
    b. Thromboplastin, prothrombin, fibrinogen and, Ca^{2+} are essential for clotting.
    c. There are two schemes of clotting: the intrinsic and extrinsic.

12. Failure of the blood to clot involves lack or removal of an essential clotting factor.

   a. Removal of calcium may stop clotting.

   b. Hemophilias result from failure of DNA to cause production of essential clotting proteins.

13. Lymph:

   a. Is formed by filtration of blood.

   b. Is tissue fluid in lymph vessels.

   c. Flows by "pumping action" of muscles and breathing.

   d. Serves to return protein and water to the blood vessels and as a route of fat absorption from the gut.

14. Lymph nodes:

   a. Produce nongranular white cells.

   b. Filter and cleanse lymph.

   c. Trap cancer cells that have metastasized.

   d. Enlarge during infections.

   e. Are 1–25 millimeters in size, and are disposed in four main areas of the body.

15. Tonsils and spleen are accessory lymphoid organs and produce lymphocytes and monocytes, but do not filter the lymph. The tonsils form Waldeyer's ring around the entrance of the respiratory and digestive systems.

Questions

1. What are the functions of the blood?

2. What are the components of the blood and their percentages?

3. What are the major components of plasma?

4. What are the subdivisions of the plasma proteins and what are their functions?

5. Describe erythrocytes as to appearance, size, numbers, and function.

6. What is the importance of hemoglobin to the body?

7. What are the two major causes of anemia?

8. What are the properties and functions of leucocytes?

9. Outline the phases of the clotting of the blood.

10. Compare the extrinsic and intrinsic schemes of clotting as to materials required and results.

11. Compare the composition of plasma and lymph and account for any significant differences.

12. Describe the functions of the lymph nodes.

13. Compare the composition of plasma, lymph, and intracellular fluid.

14. What functions do lymph nodes and tonsils have?

Readings

Carmel, Ralph, and Cage S. Johnson. "Racial Patterns in Pernicious Anemia." *New Eng. J. Med. 298*:647, March 23, 1978.

Chaplin, Hugh, Jr. "Frozen Blood." *New Eng. J. Med. 298*:679, March 23, 1978.

Erslev, A. J. *Pathophysiology of Blood.* Saunders. Philadelphia, 1975.

Maines, M. D., and A. Kappas. "Metals as Regulators of Heme Metabolism." *Science 198*:1215, 23 Dec. 1977.

O'Reilly, Richard J., et al. "Successful Transplantation of Marrow from an Unrelated HLA-D-Compatible Donor." *New Eng. J. Med. 297*:1311, Dec. 15, 1977.

Science News. "Sickle Cell Anemia: Test Tube Treatment." *113*:134, March 4, 1978.

Tullis, James L. *Clot.* Thomas. Springfield, Ill., 1976.

Weiss, H. J. "Platelets: Physiology and Abnormality of Function." *New Eng. J. Med. 293*:531, Sept. 11, 1975; *293*:580, Sept. 18, 1975.

Chapter 21

Body Defenses against Disease; Mechanisms of Protection, Antigen Antibody Reactions, Immunity, Blood Groups

Objectives

After studying this chapter, the reader should be able to:

- Outline the various mechanisms the body employs to protect itself against disease.

- Give the purposes of the inflammatory response, the steps in the response, and how wound healing takes place.

- Define the immune system and what its functions are.

- Characterize antigens and antibodies, define an antigen-antibody reaction, and give its purpose(s).

- ■ Describe the roles of complement and properdin in the antigen-antibody reaction.

- ■ Discuss current theories of how antibodies are produced.

- ■ Explain the origin and functions of T- and B-lymphocytes.

- ■ List some diseases against which immunity may be acquired and methods of achieving that immunity.

- ■ Explain the relationships of interferon and transfer factor to the development of immunity.

- ■ Explain what allergy is and what the characteristics of the several types of allergic responses are.

- ■ Explain what the problems of transplantation are on an immunological basis.

- ■ Explain the genesis of autoimmunity and immunodeficiency.

- ■ Explain the origins of the ABO and Rh blood groups, and how blood type is determined.

- ■ Discuss the basis of antigen-antibody reactions that may occur with blood transfusion or pregnancy.

- ■ Outline the considerations involved in transfusion of blood or other fluids, and list some possible transfusion reactions.

The various mechanisms of protection

From within and without, our bodies constantly face factors that have great potential to harm us. Natural and artificial radiations assault our skin; microorganisms on our body surfaces and in systems that open on the body surfaces are, in many cases, potentially pathogenic; foreign substances enter our bodies in our food and drink and in the air we breathe; products of metabolism may be toxic and must be detoxified or eliminated from the body.

This chapter concentrates primarily on those devices that provide us with immunity to foreign chemicals and microorganisms. Defense against airborne materials is explored in greater detail in the respiratory chapter; protection against ingested materials is treated in the chapter on di-

gestive system; elimination of toxic substances is explored in the chapter on the urinary system.

Finally, because the blood groups represent examples of antigen-antibody systems, they are considered after the reader has an understanding of antigens and antibodies and how they interact.

The external body surfaces

The skin is the largest organ of the body, having a surface area of about 1.75 m² (3000 in²) in the adult, and it accounts for about 7 percent of the body weight.

Whole skin offers a MECHANICAL BARRIER that prevents entry of noxious materials into the body *and* prevents loss of vital substances from the body. A keratinized layer of cells provides most of the mechanical barrier, while an acidic (pH 4–6.8) SURFACE FILM, derived from breakdown of surface cells and secretions of the skin glands, acts as an antiseptic device that retards microorganism and fungal growth on the skin surface. Lacrimal fluid (tears) contains *lysozyme*, a bacteriostatic agent that protects the anterior surface of the eyeball. Melanocytes, pigment-containing cells in the skin, absorb solar radiation and increase in number (tanning) as the skin is exposed to sunlight.

The internal body organs

A variety of protective devices afford protection to internal body organs.

The mouth, anal canal, nasal cavities, and urogenital organs that open on a body surface are protected by MUCUS, lying on a *mucous membrane.* Mucus traps microorganisms and is then moved to areas where it is usually eliminated.

SALIVA is acidic (5.8–7.0) and retards microorganism growth in the mouth.

HYDROCHLORIC ACID, secreted by cells of the stomach, is an effective bacteriolytic agent that destroys ingested pathogens.

URINE is normally acidic, and also is a bacteriostatic agent.

The LIVER, by its ability to DETOXIFY potentially harmful chemicals within the body, forms a line of defense against accumulation of toxins.

These examples should give some idea of the range of defenses that the body possesses.

Inflammation and wound healing

If the skin or a mucous membrane is penetrated by some agent, the area around the site of injury becomes swollen, tender, warm to the touch, reddened (hyperemic), and painful. These symptoms are part of a response to injury known as the INFLAMMATORY REACTION. The purpose of such a reaction is to destroy or neutralize the agent responsible for the injury and to facilitate repair of the injury. The response is essentially the same regardless of the nature of the injuring agent and follows a typical pattern.

There is a short-lived VASOCONSTRICTION (narrowing of blood vessels) that is due to a nervous reflex associated with the pain of the injury. It tends to momentarily localize the agent and/or the products of the injury.

VASODILATION (opening of blood vessels) follows, as products of cell destruction reach the vessels and more blood, with white cells, enters the area. This increased flow of blood accounts for the reddening of the injured area and the warmth.

LYSOSOMES ARE RELEASED from injured cells and clear the area of the debris of cell death; they may also destroy microorganisms.

WHITE BLOOD CELLS PASS THROUGH CAPILLARY WALLS and phagocytose (engulf) microorganisms and cell debris.

HEALING by replacement of lost cells or by scar tissue formation occurs. Scar tissue consists mainly of collagenous fibers.

How long the inflammation lasts — and its eventual outcome — depends on the nature of the agent and how long it acted (longer acting generally means greater destruction), the tissue affected (more vascular tissues recover more rapidly), and the effectiveness of the defensive response.

As wounds heal, three general stages are recognizable.

The initial injury causes a flow of blood into the area, and a clot and scab is formed.

The inflammatory reaction constitutes the second step.

Healing occurs as the last step.

Figure 21.1 shows the sequence of events just described, with a cut on the skin as the example.

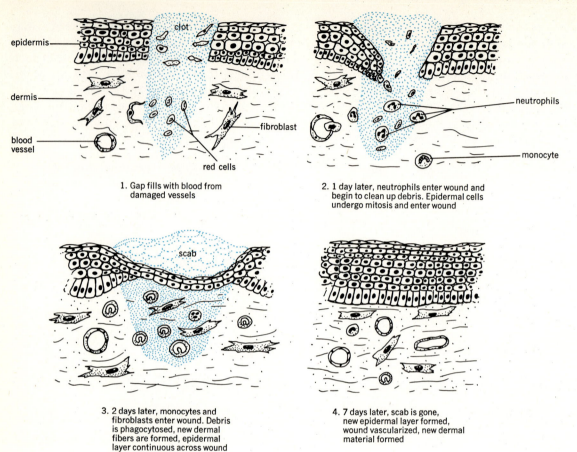

FIGURE 21.1 The basic stages in wound healing. (Redrawn from "Wound Healing," by Russel Ross. Copyright © June, 1969, by Scientific American, Inc. All rights reserved.)

Immune reactions

The IMMUNE SYSTEM is a series of ORGANS, including the *thymus gland, lymph nodes, spleen, tonsils,* and *lymphocytes* and *plasma cells* that produce ANTIBODIES against foreign chemicals. Its function, in general terms, is to PROTECT THE BODY AGAINST MACROMOLECULES that may enter it on viruses, bacteria, or cells (as in a transplant), or which may originate within the body. In most cases, the system can differentiate between molecules that are not part of the body (*nonself*) and those that are (*self*).

Antigens and antibodies and their interactions

Definitions and characteristics

An antigen is typically described as a large molecule having a molecular weight of 10,000 or more. It may be part of a cell or microorganism or a product thereof. Chemically, antigens are most commonly composed of proteins, polysaccharides, nucleic acids, or combinations of these. An antigen has a complicated shape, and contains, somewhere on its molecule a sequence of 3 to 10 amino acids called the EPITOPE. It is this epitope that an antibody "recognizes" and thus a reaction between the antigen and an antibody can occur.

An antibody is itself a protein molecule, belonging to the gamma globulins. Molecular weight is between 150,000 and 900,000, and antibodies are produced by modified lymphocytes called PLASMA CELLS. An antibody has a complicated shape (Fig. 21.2) made up of four polypeptide chains. The four chains consist of paired *light* and *heavy chains* that are identical in all antibodies, except for areas called *variable regions*. Here, amino acid sequences are different in different antibodies and constitute *combining sites* that react with the antigen's epitope and lead to an ANTIGEN-ANTIBODY REACTION. Thus, each antigen calls forth the production of a *specific* antibody.

In the antigen-antibody reaction, "matching" antigens and antibodies react in a manner that is designed to remove the antigen as a threat. It may take a variety of visible or nonvisible forms including: *neutralization*, in which the effect of the antigen is removed; *precipitation*, in which an aggregate is formed that is removed by phagocytosis; *agglutination*, in which a clumping together of cells occurs; or *lysis*, in which the cells are destroyed.

Complement and properdin

In order for lysis of cells to occur, a series of serum factors called COMPLEMENT must be present. In this system, an antibody attaches to one of several molecules in the complement system, and a series of chemical reactions produces enzymes that create holes in the wall of the affected cell. Ions and water flow into the cell and it swells and bursts. A similar series of reactions occurs in the PROPERDIN system, but in this system no antibody is required; bacterial polysaccharides may initiate the reactions that cause cell lysis.

Theories of antibody production

Two general categories of theories have been advanced to explain how antibodies are produced against the literally millions of antigens the body may encounter during its lifetime.

The INSTRUCTIVE *(template)* THEORY suggests that the antigen acts as a form around which a "standard" unfolded gamma globulin molecule is molded to give a shape that is the mirror image of the antigen. Once formed, the antibody molecule retains its shape by chemical bonds within the folds. The antibody is then supposed to release from its antigen template and could react with other antigens like the one that caused its formation. We see that antibodies are *not* present in the producing cell until arrival of an antigen; also, if this theory is correct, antibody producing cells should demonstrate antigens in them, which they don't.

The SELECTIVE *(modified clonal)* THEORY suggests that the capacity to synthesize different antibodies is already present, and that contact with an antigen "switches on" (induces) production of a corresponding antibody. Support for this theory includes the facts that: antibody-producing cells *do not* have demonstrable antigen content; antibodies have different amino acid sequences in their variable regions (which they wouldn't have if they all came from a "standard" molecule); mu-

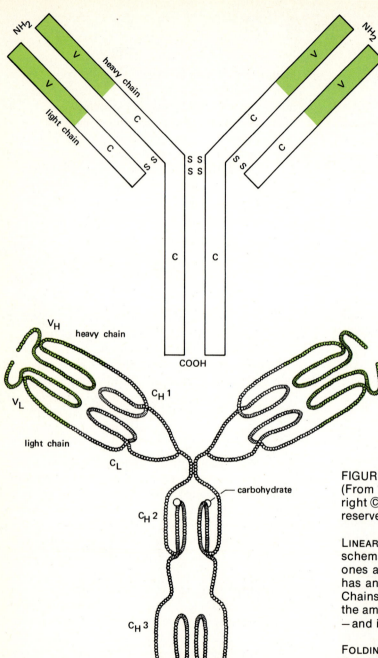

FIGURE 21.2   The structure of an antibody molecule. (From "The Immune System," by Niels K. Jerne. Copyright © July 1973 by Scientific American, Inc. All rights reserved.)

LINEAR STRUCTURE of an antibody molecule is shown schematically. The two heavy chains and two light ones are connected by disulfide bridges. Each chain has an amino end (NH_2) and a carboxyl end (COOH). Chains are divided into variable (V) regions—in which the amino acid sequence varies in different antibodies—and into constant (C) regions.

FOLDING OF THE FOUR CHAINS is suggested in this drawing based on a bead model of the antibody molecule made by Gerald M. Edelman and his colleagues. Each bead represents an amino acid, of which there are more than 1200. The variable regions are in color. (V_H, variable region of heavy chain; V_L, variable region of light chain; C_L, light chain; C_H1, C_H2, C_H3, folded regions of constant portion of heavy chains.)

tations can produce a loss of ability to synthesize *particular* antibodies. This theory accounts for tolerance to one's own chemicals by suggesting that an antigen-antibody reaction occurring between embryonically developed chemicals and the antibody producing cell leads to destruction of the cell. Thus, no response occurs to further production of that antigen.

The true explanation of antibody production will only be clarified by further research. Genetically based theories (selective) seem to have the edge at the moment, as further discussed in the section on immunodeficiency later in this chapter.

The immune response

Development of the immune system

Two distinct but interdependent systems of immunity protect the body from antigenic challenge. CELL-MEDIATED IMMUNE RESPONSES combat fungi, initial invasion by viruses, and are responsible for rejection of foreign tissues (e.g., transplanted organs). The lymphocytes involved in this type of response have antibody molecules attached to their surfaces and the whole cell must contact the antigen before a reaction can occur. In HUMORAL IMMUNITY, plasma cells synthesize antibodies and *release them into the bloodstream.* This type of immunity provides protection against bacteria and reinfection by viruses.

The basis for these two "lines of defense" lies in embryonic and fetal development of two different immunologically competent cell lines. One cell line develops from stem cells that originate in the yolk sac, liver, spleen, and bone marrow, and which migrate to the developing thymus gland. Here, they are endowed with the capacity (by certain cells or chemicals produced in the thymus) to become the cell-mediated line of defense. Such cells resemble lymphocytes and are designated as T-CELLS. Other cells originate in the same areas as T-cells, but do not respond to thymic stimulation. These cells also resemble lymphocytes and are termed B-CELLS; they differentiate into plasma cells to provide humoral immunity. Figure 21.3 summarizes the relationships of T- and B-cells.

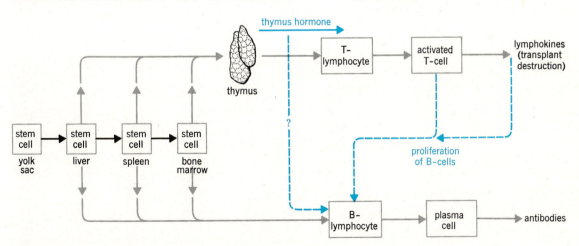

FIGURE 21.3   Development and interrelationships of B- and T-cells.

Types of immunity
and duration of protection

The term *immunity* is generally understood to mean the PROTECTION achieved against *second* exposure to antigens. To achieve immunity involves both cell-mediated and humoral response. To trigger those responses, one may "catch" the disease or may be VACCINATED to stimulate the body to provide protection. Table 21.1 shows some of the methods employed to stimulate the immune responses of the body. Catching the disease is called active immunity, while passive immunity is achieved when antibodies produced in another individual are placed into one's body. Table 21.2 summarizes the various "types" of immunity one has and Table 21.3 shows several diseases against which we may be protected and the method of achieving immunity most commonly employed. As you look at Table 21.3, see if you can pick out those diseases, for which protection against involves cell-mediated immunity and those involving humoral immunity. (*Clue:* humoral immunity depends on circulating antibodies that may pos-

TABLE 21.1   Methods of achieving immunity

Method	Example(s)	Comments
Live, weakened organism is given	Polio, measles	Organism stimulates antibody production, but is too weak to cause disease (usually).
Use of organism similar to disease organism	Use of cowpox for smallpox	Use is based on similarity of antigens on organism surface.
Use of killed organism	Polio, typhoid	Cannot produce disease, but stimulate antibody production.
Use of toxoid	Diphtheria, tetanus	Toxoid is a toxic substance of the infectious organism which has been treated to reduce its toxic properties but not its antibody stimulating properties.
Injection of gamma globulin		A "shotgun" approach to any disease; may decrease severity of the disease in the exposed subject.
Antitoxins	Used against snake venoms	Is antibody to that specific antigen.

TABLE 21.2   Types of immunity

Type of immunity	Examples	Comments
Tissue (cellular)	Measles	Antibody stays in cell; does not go to bloodstream, thus, less liable to destruction.
Humoral (chemical)	Diphtheria, tetanus	Antibody enters bloodstream, and is liable to destruction thus, "booster" required.
Active	Any viral or bacterial disease	Subject gets disease and acquires immunity to it.
Passive	Tetanus, poisonous snakebite	Antibody is given directly.
Passive-active	?	Antibody to protect; antigen to stimulate further antibody production.
Active-passive	Measles	Antibodies given to reduce disease severity; disease itself causes antibody production.

TABLE 21.3   Some diseases with methods and duration of protection

Disease	Method used	Duration of protection
Measles 　Rubella 　Rubeola	Attenuated organism	Expected to be life[a]
Poliomyelitis	Attenuated or killed organism	After a series of injections or oral administrations, boosters are required about every 10 years
Pertussis (whooping cough)	Killed organism	About 10 years
Smallpox	Similar organism toxoid	3–10 years
Diphtheria	Toxoid	About 10 years
Tetanus	Toxoid Antitoxin (for exposure)	To 10 years after first booster or if penetrating injury of skin
Mumps	Attenuated organism	Expected to be life[a]
Typhoid	Killed organism	3 years (minimum)
Influenza	Attenuated organism	1 year for a given strain

[a] Vaccines have been available too short a time to assess their protection for life; they presently produce a lasting effect.

sibly be metabolized or excreted.) In those diseases that do not create lifetime protection, periodic ''boosters'' may be given to restimulate antibody production and keep high blood levels of antibody.

Interferon and transfer factor

INTERFERON is a substance produced by cells that have been contacted or invaded by a virus. Within such a cell, interferon prevents the replication of virus particles, possibly by blocking RNA synthesis. Some interferon is released from an infected cell and ''coats'' uninfected cells, preventing viral attachment. Initially, after its discovery, it was hoped that interferon might provide protection against the common cold and flu, both viral-caused diseases. It has, however, been difficult to purify, one finds it difficult to obtain quantities sufficient for widespread use, and only human interferon is maximally effective in humans (species specificity).

TRANSFER FACTOR is an antibody-like material

that may be isolated from the white cells of individuals who have acquired immunity to a particular disease. When given to a person with the same disease as the first person who acquired immunity to it, transfer factor brings about dramatic lessening of symptoms and sometimes ''cures.'' Further research on this material may reveal possibilities for therapeutic use.

Allergy

Allergies represent an immune response that was abnormal or incomplete, so that immunity is not conferred against the challenging antigen. The reaction is also associated with a ''runny'' nose, hives, and edema. These symptoms appear to result from release of a chemical called HISTAMINE from cells damaged by the antigen-antibody reaction. ANTIHISTAMINES are substances that reduce the effect of histamine and bring relief of allergy symptoms. PRECIPITIN ALLERGIES are exemplified by the ''penicillin reaction.'' An antigen causes production of enough antibody to cause an antigen-antibody reaction. The reaction causes severe tissue damage with development of shock. DELAYED-REACTION ALLERGIES are exemplified by reactions to poison oak and poison ivy. A first exposure "sensitizes" the individual, and antibodies are produced that react with a second exposure to the antigen. ATOPIC ALLERGIES are exemplified by hay fever and asthma. Exposure to the antigen brings about itching, runny nose, and spasm of muscle in the small tubes of the lungs. This type of allergy may have an hereditary basis and is most commonly treated by giving the individual regular doses of the antigen in the skin to try to stimulate antibody production. This is called *desensitization*.

Transplantation and tissue rejection

Surgical replacement of diseased or worn-out organs is a relatively new field of medicine. Replacement organs (except in identical twins) constitute foreign antigens and call forth reactions to reject the transplant. Transplants of tissues or organs from one part of a person's body to another are called AUTOLOGOUS TRANSPLANTS. Skin, bone, and cartilage grafts are examples of autologous transplants or *autografts* (G. *autos*, self, + L. *graphium*, grafting knife). A transplant from one species to a member of the same species is called a HOMOLOGOUS TRANSPLANT. If the exchange of organs is between genetically identical individuals, for example, between identical twins, it is called an *isograft* or *homograft* (G. *isos*, equal, or G. *homo*, likeness). The term *allograft* (G. *allos*, other) describes an organ transplant between genetically different members of the same species and results in the introduction of foreign antigens into the recipient. The transplantations of hearts, kidneys, lungs, and livers between humans are examples of homologous transplants. HETEROLOGOUS TRANSPLANTS, also known as *heterografts* or *xenografts* (G. *eteros*, other, or G. *xeno*, foreign), involve transplantation of organs between different species, and are the least successful of all types of grafts because they introduce nonhuman or dissimilar antigens into the recipient's body. The use of chimpanzee livers to remove toxic wastes from the bloodstream of humans is an example of a heterologous transplant. Several methods may be employed to reduce the tendency of the body to reject the transplant, as is shown in Table 21.4.

TABLE 21.4 Methods of minimizing tissue transplant rejection	
Method	Comments
Culture of tissue (e.g., skin, adrenal glands, corneas)	Cells assume embryonic characteristics, and (presumably) do not produce antigens.
Radiation of the lymphoid tissues	Kills plasma cells; also kills other cells.
Tissue matching	Uses lymphocytes in a typing or matching test to get as close a genetic match as possible. Reduces severity of rejection reaction, but does not eliminate it.
Drug therapy	Reduces inflammatory response to transplant.
Antilymphocyte serum (ALS)	Take subject's lymphocytes, inject into animal, isolate antibodies and give to subject. "Kills" subject's lymphocytes; also reduces resistance to other diseases.
Remove lymphocytes from lymph	Lessens numbers of cells that can become plasma cells.

Autoimmunity

The production of antibody and an ensuing reaction between that antibody and *one of the body's own components acting as an antigen* is the basis for AUTOIMMUNITY ("reaction to self"). Because of some disease process, there may be a chemical alteration in a normal body constituent. The immune system recognizes this chemical as "foreign" or "new," and responds with antibody production. The antigen-antibody reaction that follows may severely damage a body organ.

Immunodeficiency

This term refers to genetic or drug-induced deficiencies in the function of the immune system. There may be no, or a defective, response to antigens, and diseases that most of us can successfully combat become life-threatening to the immunodeficient person. Causes would seem to be obvious, if we recall the cell-mediated and humoral lines of defense the body has. In other words: B- or T-cells are abnormal and cannot carry out their functions properly; plasma cells are abnormal and cannot produce circulating antibodies; the thymus is defective or fails to develop. Hope for ameliorating such disorders may be in thymus transplants or use of transfer factor. The only other treatment that preserves life is to place the patient in a "germ-free" environment, possibly isolating the person from normal contact with the world.

Blood groups

The blood groups are examples of genetically determined antigens that have corresponding antibodies. Two main groups are recognized: the ABO SYSTEM of antigens, and the RH SYSTEM of antigens. In both groups, the antigens are designated as ISOANTIGENS (isoagglutinogens, agglutinogens), and the antibodies are designated as ISOANTIBODIES (isoagglutinins, agglutinins).

The ABO system

Two basic isoantigens on the red cell surface are involved, called A and B. With two antigens, four possible combinations may exist: the cell may have one, the other, both, or neither antigen. The particular antigen involved determines the blood group. The four basic blood groups are of variable occurrence, and are summarized in Table 21.5.

Genetically, A and B inheritance is dominant to O, so that a given individual may be homozygous (have two like genes) or heterozygous (have two unlike genes) for his blood group.

Blood group	Genotype (genes present)
A	AA, Ai
B	BB, Bi
AB	AB
O	ii

where

A = group A ⎱ codominant, and
B = group B ⎰ dominant to group O
AB = group AB
i =   group O, recessive to both A and B

A child thus may have a blood group different from that of its parents, unless they are both group O.

The plasma contains isoantibodies corresponding to the isoantigens listed above. These are designated as **a,** *Anti-A,* or *alpha,* and **b,** *Anti-B,* or *beta. Isoantibodies are not present at birth.* They are gamma globulins that appear between 2 and 8 months after birth in response to A and/or B antigens ingested in foods (meats). In the blood

TABLE 21.5 Blood groups and frequency of occurrence

Antigen present	Blood group (same as antigen)	Frequency of population	
		Percentage of white	Percentage of black
A	A	40.8	27.2
B	B	10.0	19.8
AB	AB	3.7	7.2
Neither A nor B	O	45.5	45.8

of any one given individual, *the antibody present is always the reciprocal of the antigen.* This avoids an antigen-antibody reaction between corresponding materials. Thus, the setup of antigens and antibodies in the basic blood groups would be as shown below.

Blood group	Dominant Antigen(s)	Antibody
A	A	b
B	B	a
AB	AB	None
O	None	ab

The Rh system

This blood system is composed of three allelic genes for each group. The dominant genes are usually designated **CDE**, the recessives as **cde.** Eight genotypes, grouped into two major categories are recognized. These are shown below.

CDE ⎫		Cde ⎫	
	Rh positive		Rh negative
cDE	(85% of white	CdE	(15% of white
	Americans,		Americans,
cDe	88% of black	cdE	12% of black
	Americans)		Americans)
CDe ⎭		cde ⎭	

These antigens (CDE) are inherited independently of the ABO antigens, and any combination between the two systems may exist. The Rh positive individual has Rh antigen(s) on the red cells, but has no antibodies in the plasma; the Rh nega-

tive individual has no D antigen(s) on the red cells, *and* no antibody in the plasma. *The Rh negative individual can produce antibodies to Rh positive cells if they ever enter his bloodstream.*

Typing the blood

The blood may be typed for ABO and Rh antigens (and therefore blood group) by mixing commercially available antibody on a slide with the blood to be typed. The antibodies will then visibly react with antigen-laden red cells to CLUMP or AGGLUTINATE them. As the slide is inspected, one fact and one question should be kept in mind: fact— a reaction will occur only if corresponding antigens and antibodies are mixed; question—what had to be present in the unknown blood to give a reaction with known antibody? The chart below summarizes the possible reactions that may result (+ indicates a reaction; − indicates no reaction).

Antibody		Group and antigen present is:
a	b	
+	−	A
−	+	B
+	+	AB
−	−	O
Anti Rh		
+		Rh positive
−		Rh negative

TABLE 21.6   Transfusion fluids other than blood

Fluid	Use	Comments
Plasma	To preserve blood volume and pressure	Used when fluid and protein has been lost (e.g., burns). "Pulls" water from tissues since proteins are still present. May contain antibodies.
Colloidal solutions: 　Albumin (a protein)	As above	As above; also, may be antigenic.
Dextran (a poly-saccharide)	As above	A "plasma expander," it filters slowly and is not antigenic. Draws water from tissues by osmosis.
Crystalloid solutions (isotonic salt solutions)	As above	Particles are very small and filter rapidly. Short-lived effect.

Transfusions

The significance of the blood group antigens is apparent when it becomes necessary to transfuse blood from a DONOR into a RECIPIENT. If the mixed bloods contain corresponding antigens and antibodies, a reaction may occur with life-threatening results. It thus becomes necessary to perform tests to determine the COMPATIBILITY of the two bloods.

The blood is TYPED (as described above) and CROSS MATCHED. A cross match mixes donor's red cells with recipient's serum and vice versa. If no reaction occurs in each combination, the bloods are usually considered compatible and capable of being transfused without the development of reactions.

Whole blood is obviously the best transfusion fluid, for it provides all necessary substances and cells. If blood is not readily available, other substances may be employed to preserve blood volume and pressure. Table 21.6 describes some alternative substances.

Clinical considerations

Other than the transfusion or agglutination reactions mentioned above, the major clinical consideration concerned with the blood groups is HEMOLYTIC DISEASE OF THE NEWBORN (HDN) or, as it is sometimes still called, erythroblastosis.

If a child inherits from the father an antigen *not* possessed by the mother, and if that antigen crosses the placenta into the mother's bloodstream before birth, the mother will make an antibody to that antigen. The antibody may then recross the placenta to the child and damage its red cells. ABO and Rh antigens may cause this series of events, with Rh difficulties the more severe of the two. If Rh antibodies are discovered in the mother's bloodstream or if the infant is born with hemolytic disease, the antibodies may be neutralized by giving the mother, within 72 hours after birth of her child, massive doses of anti-Rh gamma globulin (RhoGAM* or Rho-Imune†). A product that may be excreted by the kidney is formed, and the mother suffers no harm. The infant may have to have replacement of his or her blood supply in an exchange transfusion if the hemolytic disease is severe. If it is not severe, transfusion or no treatment is employed.

* Ortho Chemical Co.

† Lederle Laboratories.

Summary

1. The body has a variety of defense mechanisms to protect it from radiation, microorganisms, and chemicals ingested or produced within the body.

   a. The skin provides a mechanical barrier to entry of materials and prevents loss of body constituents. It has an antiseptic surface film.

   b. Tears contain an enzyme to keep the eyeball clean.

   c. Mucous membranes, moistened by mucus, trap materials.

   d. Saliva cleanses the mouth.

   e. Acid secreted by the stomach destroys organisms.

   f. Urine is acid and washes materials through the tract.

   g. The liver detoxifies harmful chemicals.

2. The inflammatory reaction is a device that aids in destruction of noxious agents and speeds healing. It has several steps that are the same regardless of the agent causing it.

   a. Vasoconstriction localizes the damage.

   b. Vasodilation brings more white cells and blood to the area.

   c. Lysosome release clears the area of cells and debris.

   d. White cells phagocytose products.

   e. Healing occurs by replacement of cells and/or by scar tissue formation.

3. The immune system provides defense against chemicals that may enter the body on foreign tissues or on transplanted tissues or organs.

   a. It consists of several organs (thymus, lymph nodes, spleen, tonsils) and cells (lymphocytes, plasma cells), plus antibodies.

   b. It can discriminate between chemicals that are part of the body and those that are not.

4. Antigens are large molecules that are foreign to the body and trigger an immune response.

   a. They may be proteins, polysaccharides, or nucleic acids.

   b. They possess an epitope that enables an antibody to attach to it.

5. Antibodies are gamma globulins that are produced by specific antigenic challenge.

   a. They have combining sites that attach to antigen epitopes to neutralize or remove the antigen threat to the body.

   b. Antibodies are produced mainly by plasma cells.

6. An antigen-antibody reaction involves combination of an antigen and its specific antibody. The reaction may take several forms.

   a. Neutralization

    b. Precipitation

    c. Agglutination (clumping)

    d. Lysis (cell destruction)

7. Complement and properdin are required for cell lysis in an antigen-antibody reaction.

    a. A series of enzymes are formed that "punch a hole" in the membrane of the attached cell.

    b. Ions and water enter the cell and it swells and bursts.

8. There are two major types of theories by which antibody production is explained.

    a. The instructional theory suggests that *any* antigen causes a "standard" antibody molecule to be folded into a particular shape that enables it to react with a particular antigen. The antigen itself causes the folding, acting as a template.

    b. The selective theory suggests that there is a different antibody for each antigen and that an antigen "switches on" a preexisting mechanism in the antibody producing cell.

9. The immune response involves two "types" of responses and involves two types of cells.

    a. Cell-mediated response involves production of antibodies that remain on or in the cell of production. T-cells that require thymus action handle this type of response, and protect against fungi, initial virus contact, and reject transplants.

    b. Humoral response involves release of antibodies into the bloodstream and combats bacteria and reinfection by viruses. B-cells, which do not require thymus activity, handle this response.

10. Several methods of acquiring immunity and the several types of immunity are presented in Tables 21.1 to 21.3.

11. Additional protection against microorganisms is provided by interferon (combats mainly viruses) and transfer factor.

12. Allergy represents an incomplete response to antigenic challenge, and no lasting immunity is acquired.

    a. Precipitation allergies result in cell destruction and severe shock.

    b. Delayed-reaction allergies require a second antigen exposure to create symptoms.

    c. Atopic allergies are milder and may be treated by desensitization.

13. Organ transplants call forth antibody production because they constitute foreign antigens. The organ may be rejected.

14. The blood groups are examples of blood antigens and antibodies.

    a. The ABO group consists of two antigens (A and B) and two antibodies (a and b) usually present in reciprocal relationship. For example, Ab, Ba, AB−, and Oab.

    b. The Rh group consists of one antigen (Rh) and it may be present (Rh positive) or absent (Rh negative) in the blood.

    c. Transfusion of blood requires that the antigens are the same between donor and recipient bloods to avoid an antigen-antibody reaction.

    d. Erythroblastosis may result if a mother produces antibodies to her child's blood Rh antigens while the child is still *in utero* and the antibodies recross to the child through the placenta.

15. Replacement of lost blood may be made by transfusion of whole blood, plasma, colloids, or crystalloids. Whole blood provides all missing elements; the other solutions provide volume.

Questions

1. What mechanism(s) exist(s) to protect the body against the potential effects of each of the following?

    a. Solar radiation.

    b. Microorganisms on the external body surface.

    c. Bacterial or viral invasion of the body.

2. How is the inflammatory process of value to the body in combating a harmful agent? What are the steps in this process?

3. What constitutes the immune system and what are its functions?

4. What are the characteristics of antigens and antibodies and how do they relate to one another?

5. How is the complement system related to an antigen-antibody reaction? What does the activated system do to a cell?

6. Contrast the two major theories of antibody production.

7. What are the jobs of T- and B-lymphocytes? From where does each originate?

8. Contrast active and passive immunity as to how each is achieved.

9. How do allergies differ from immunity? What type of a reaction is a "penicillin reaction" and why is it so threatening to the body?

10. What considerations must be kept in mind if a person needed a kidney transplant? How can the rejection phenomenon be reduced?

11. What determines a person's blood type?

12. What are some of the reactions that can occur if bloods are mismatched for transfusion and then transfused?

13. In Rh disease between a child *in utero* and its mother, how can the severity of the reaction to that child and to subsequent children be reduced?

Readings

Allison, A. C., and J. Ferluga. "How Lymphocytes Kill Tumor Cells." *New Eng. J. Med.* 295:165, July 15, 1976.

Beer, Alan E., and Rupert E. Billingham. "The Embryo as a Transplant." *Sci. Amer.* 230:36, April 1974.

Burke, Derek C. "The Status of Interferon." *Sci. Amer.* 236:42, Apr. 1977.

Cooper, Max D., and Alexander R. Lawton III. "The Development of the Immune System." *Sci. Amer.* 231:58, Nov. 1974.

Mayer, Manfred M. "The Complement System." *Sci. Amer.* 229:54, Nov. 1973.

Merigan, Thomas C., et al. "Interferon for the Treatment of Herpes Zoster in Patients with Cancer." *New Eng. J. Med.* 298:981, May 4, 1978.

Raff, Martin C. "Cell-surface Immunology." *Sci. Amer.* 234:30, May 1976.

Science News. "Cystic Fibrosis: An Immune Disease?" *111*:232, Apr. 9, 1977.

Science News. "Finding an Immune Fighter in the Blood." *113*:246, Apr. 22, 1978.

Science News. "Molecule Suppresses Allergic Reactions." *113*:150, March 11, 1978.

Science News. "Self-Styling and the Thymus." *113*:186, March 25, 1978.

Science News. "Stress May Damage Cell Immunity." *113*:151, March 11, 1978.

Science News. "Tumors Not Typed to a T." *113*:278, April 29, 1978.

Scientific American Readings. *Immunology.* Freeman. San Francisco, 1976.

Siegal, Frederick P. "Suppressors in the Network of Immunity." *New Eng. J. Med.* 298:102, Jan. 12, 1978.

Chapter 22
The Heart

Objectives

After studying this chapter, the reader should be able to:

■ Describe, in general terms, the early development of the heart, the plan of the fetal circulation, and its differences from the adult's.

■ Locate the heart accurately within the thorax, describe its relationships to the pericardium and the structure of the heart wall.

■ Accurately label the parts of the heart as they would appear in a longitudinal section of the organ.

■ Describe the coronary circulation.

■ Name and locate the several masses of nodal tissue in the heart and give their functions.

■ Describe the anatomy of cardiac muscle, list the properties of cardiac muscle, and relate them to the tasks of the heart.

■ Define an arrhythmia, and some causes for it (them).

- Define what a cardiac cycle is and give time values for each phase of the cardiac cycle.

- Draw and accurately label the phases of a normal electrocardiogram, and correlate each phase with cardiac activity.

- Show how pressures change within the heart and its great vessels as activity proceeds and how these changes operate the valves.

- Explain the causes of heart sounds.

- List and explain the five factors involved in creation of blood pressure.

- Explain how each of the following affects cardiac output, and which component—stroke rate or stroke volume—is affected most: venous inflow, peripheral resistance, temperature, chemicals.

- Describe the cardioaccelerator and inhibitor nerves to the heart concerning origin, distribution, and effects.

- Define what heart failure is, and the steps in its development.

- Describe the energy sources of the heart.

The heart provides a source of pressure to cause the circulation of blood through the body. It is a hollow muscular pump whose contractions normally ensure circulation of enough blood to meet the varying demands of the body for nutrients, oxygen, and waste removal.

Development of the heart (Fig. 22.1)

As the embryo grows, cells tend to become removed from immediate environmental sources of nutrients.

The development of vessels to carry blood to these cells, along with a source of pressure to move the blood thus become early priorities in embryonic development.

At about 2 weeks of age, a mass of mesoderm called the CARDIOGENIC PLATE develops in the anterior part of the embryo. A cavity forms around the plate as the future pericardial cavity. By 3 weeks of age, paired HEART TUBES, actually blood vessels, have formed in the plate. By 4 weeks of age, the paired tubes have fused from anterior to posterior to create a single tube. Constrictions from anterior to posterior in the tube mark the positions of the BULBUS ARTERIOSUS, VENTRICLE, ATRIUM, and SINUS VENOSUS. It should be noted that, at this time, the chambers are reversed in position from the fetal or adult position. Rapid growth of the heart tube occurs next, and the ventricle and sinus venosus fold over the atrium and

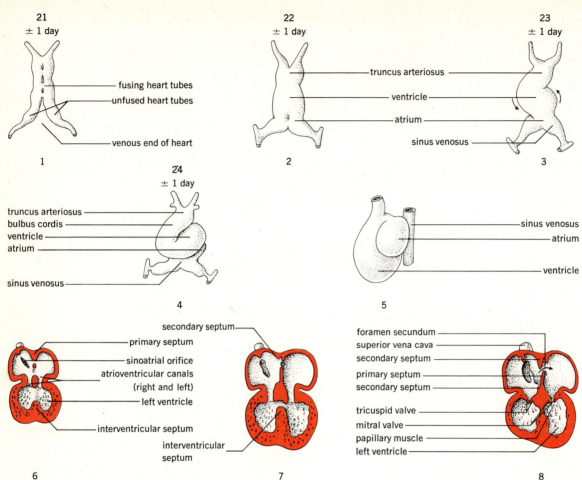

FIGURE 22.1   Stages in the development of the heart.
(Numbers 5 to 8 occur between 5 and 8 weeks.)

come to lie posterior to the atrium, which is the normal fetal position. Between 5 and 7 weeks, a series of changes occur that result in the formation of the typical fetal structures of the heart. These changes include the following.

The common atrium is divided into two chambers by a primary septum that is perforated almost immediately by an opening called an ostium. Later, a secondary septum develops just to the right of the first one, but it remains incomplete. The opening between

the two portions of the secondary septum is the FORAMEN OVALE. The foramen allows a large portion of the blood entering the right atrium of the heart to go to the left atrium and bypass the nonfunctional lungs of the fetus. The lower part of the primary septum overlaps the foramen, and seals it at birth to establish the adult type of flow through the heart.

The ventricle is partitioned into two chambers by an interventricular septum, which grows upward from the apex of the heart.

The bulbus arteriosus is subdivided into two ves-

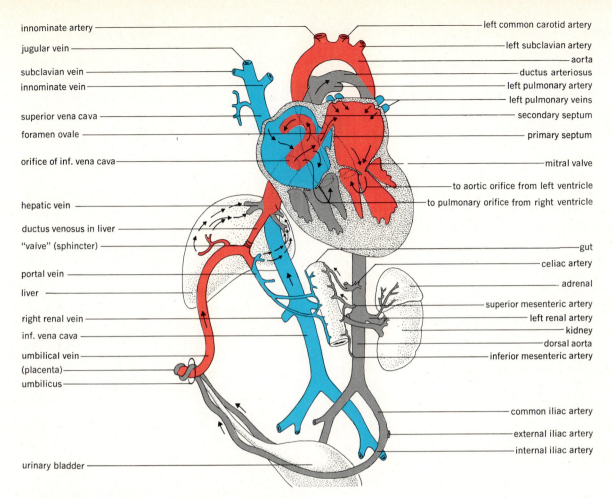

innominate artery
jugular vein
subclavian vein
innominate vein
superior vena cava
foramen ovale
orifice of inf. vena cava
hepatic vein
ductus venosus in liver
"valve" (sphincter)
portal vein
liver
right renal vein
inf. vena cava
umbilical vein
(placenta)
umbilicus
urinary bladder

left common carotid artery
left subclavian artery
aorta
ductus arteriosus
left pulmonary artery
left pulmonary veins
secondary septum
primary septum
mitral valve
to aortic orifice from left ventricle
to pulmonary orifice from right ventricle
gut
celiac artery
adrenal
superior mesenteric artery
left renal artery
kidney
dorsal aorta
inferior mesenteric artery
common iliac artery
external iliac artery
internal iliac artery

FIGURE 22.2   The plan of the fetal circulation. Colors indicate state of oxygenation of blood. (Blue, lowest; red, highest; gray, intermediate.)

sels, the aorta and pulmonary trunk, which leave the left and right ventricles, respectively.

The three layers of the heart wall are formed, as is the special nodal tissue that causes the heart to beat.

Simultaneously, connections with the placenta, for exchange of nutrients and wastes, are established. The heart and circulation show the features depicted in Figure 22.2.

A critical time period in these changes occurs when the atria and ventricles are being divided.

Interference with these processes, as by rubella or rubeola (measles) virus, may lead to ventricular septal defects (VSD), atrial septal defects (ASD), or defects in the arteries. At birth, the shunts in the fetal circulation are closed and the normal adult pattern of circulation is established. Two shunts are closed: the foramen ovale; and the ductus arteriosus, which allows any blood entering the pulmonary artery to go to the aorta, by-passing the lungs.

Basic relationships and structure of the heart

Size

The human heart approximates the size of the clenched fist of its owner. At birth, it measures about an inch (25 mm) in length, width, and depth, and weighs about 20 grams. In the adult, the heart is about 5 inches (12.5 cm) long, 2 inches (5 cm) deep, and 3½ inches (9 cm) wide, and weighs about 300 grams. There are two periods of greatest growth; between 8 and 12 years of age, and between 18 and 25 years of age.

Location and description (Fig. 22.3)

The adult human heart is located in the central portion of the thorax known as the MEDIASTINUM. The broad upper or atrial end of the heart is known as the BASE, and is directed toward the right shoulder. The tapering lower end of the heart, or APEX, is directed toward the left hip.

Thus, the AXIS of the heart (a line drawn from the center of the base to the apex) does not lie vertically within the chest. The heart is not centered within the chest; about one-third lies to the right of the midsternal line, and two-thirds lies to the left.

The heart is enclosed within a double-walled sac known as the PERICARDIUM (Fig. 22.4), which consists of two layers of tissue. An outer tough FIBROUS PERICARDIUM attaches to the bases of the pulmonary trunk and aorta, the diaphragm, and sternum. It aids in maintaining the heart's position, and protects it. The SEROUS PERICARDIUM is a thin delicate membrane that lines the fibrous sac and is reflected onto the surface of the heart at the bases of the arteries. The part of the serous pericardium lining the fibrous sac is termed the PARIETAL PERICARDIUM, and the portion on the heart is the VISCERAL PERICARDIUM or epicardium.

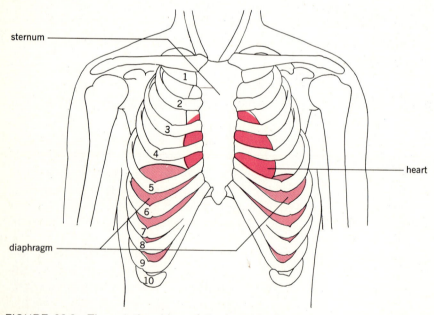

FIGURE 22.3   The relationships of the heart to thoracic structures.

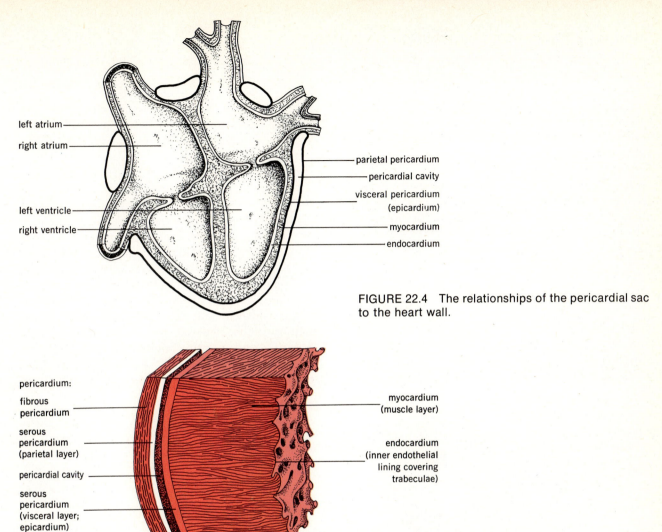

left atrium

right atrium

left ventricle

right ventricle

parietal pericardium

pericardial cavity

visceral pericardium
(epicardium)

myocardium

endocardium

FIGURE 22.4   The relationships of the pericardial sac to the heart wall.

pericardium:

fibrous
pericardium

serous
pericardium
(parietal layer)

pericardial cavity

serous
pericardium
(visceral layer;
epicardium)

myocardium
(muscle layer)

endocardium
(inner endothelial
lining covering
trabeculae)

FIGURE 22.5   The structure of the heart wall.

A space exists between the visceral and parietal layers, known as the PERICARDIAL CAVITY. A thin film of pericardial fluid lubricates the surfaces of the visceral and parietal layers and allows nearly frictionless movement of the heart within the pericardial sac. In PERICARDITIS, the serous membrane becomes inflamed and *adhesions* (sticking together of the membranes) may develop. In CARDIAC TAMPONADE, excessive pericardial fluid secretion or blood accumulation in the cavity may hinder the heart's action.

The heart wall (Fig. 22.5)

As development and differentiation of the heart occurs, its wall develops three basic layers of tissue. An inner ENDOCARDIUM is composed of a lining endothelial layer and of several layers of connective tissue and smooth muscle. The endocardium also forms the valves of the heart. In some parts of the heart, a *subendocardial layer* is present and carries the specialized nodal tissue that causes the contraction of the heart muscle.

A central MYOCARDIUM, composed of cardiac muscle makes up about 75 percent of the thickness of the heart wall. Many capillaries and small arteries and veins are found between the cardiac muscle fibers of the myocardium. The fibers are organized in circular or spiral fashion around the chambers of the heart, so as the heart muscle contracts, the chamber "wrings" the blood from it, rather than pushing it out. The apex twists as the beat occurs, and taps the chest wall to give the apex beat, or apical pulse. This pulse is often used to determine the effects of medication on the heart rate. The outer layer of the heart wall is the EPICARDIUM, and consists of the visceral layer of the serous pericardium described above.

Chambers and valves (Fig. 22.6)

The heart is actually two pumps in one. The right side of the heart acts as a pump to receive blood from the body generally and to send it to the lungs for oxygenation and carbon dioxide elimination.

This circuit constitutes the PULMONARY CIRCULATION. The left side of the heart receives blood from the lungs and pumps it to the body generally through the SYSTEMIC CIRCULATION.

The upper receiving chambers of the heart are called the ATRIA, and the lower pumping chambers are the VENTRICLES. The RIGHT ATRIUM is a thin-walled chamber that receives blood from the "great veins" or venae cavae. It pumps its blood through the right atrioventricular orifice into the RIGHT VENTRICLE. The orifice is guarded by the RIGHT ATRIOVENTRICULAR or TRICUSPID VALVE. This valve consists of three flaps (cusps) of tissue attached by their bases to a fibrous ring around the orifice. The free edges are attached to strands of fibrous tissue called CHORDAE TENDINAE. The chordae attach to PAPILLARY MUSCLES of the ventricular wall and prevent the valve from reversing as the ventricle contracts. The right ventricle has a thicker muscular wall than the right atrium and pumps blood into the pulmonary artery for oxygenation in the lungs. In the base of the pulmonary artery is the PULMONARY VALVE. It is

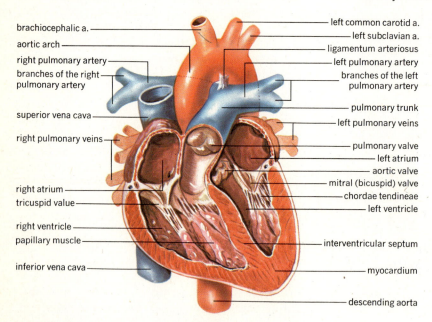

brachiocephalic a.
aortic arch
right pulmonary artery
branches of the right pulmonary artery
superior vena cava
right pulmonary veins
right atrium
tricuspid value
right ventricle
papillary muscle
inferior vena cava

left common carotid a.
left subclavian a.
ligamentum arteriosus
left pulmonary artery
branches of the left pulmonary artery
pulmonary trunk
left pulmonary veins
pulmonary valve
left atrium
aortic valve
mitral (bicuspid) valve
chordae tendineae
left ventricle
interventricular septum
myocardium
descending aorta

FIGURE 22.6   The heart in frontal section, with associated organs.

composed of three pockets of tissue, without supporting chordae; the valve prevents return of blood to the right ventricle. After being oxygenated in the lungs, the blood returns to the thin-walled LEFT ATRIUM through four pulmonary veins. No valves are present in these vessels. The left atrium sends blood through the left atrioventricular orifice into the LEFT VENTRICLE. This orifice is guarded by the LEFT ATRIOVENTRICULAR VALVE, also known as the MITRAL, or BICUSPID VALVE. It has a structure similar to that of the tricuspid valve, except it has only two flaps of tissue. The very thick-walled left ventricle pumps blood into the aorta for distribution to the entire body. Again, a valve, the AORTIC VALVE, is found in the base of the aorta. It is constructed like the pulmonary valve and prevents return of blood to the left ventricle.

Thickness of chamber wall reflects the amount of work each chamber must do. Atria receive blood and move it to the adjacent ventricles; thus, little force is required, and the atria have thin walls. The right ventricle pumps to the lungs, a task requiring more force; thus, it has a thicker wall. The left ventricle pumps to the entire body and has a wall about three times thicker than the wall of the right ventricle.

CLINICAL CONSIDERATIONS. Congenital anomalies of the heart include: failure of the ductus arteriosus to close at birth (patent ductus arteriosus); failure of the foramen ovale to close; and a variety of atrial and ventricular septal defects that may occur in various combinations. The most common of these congenital defects are shown in Figure 22.7. Short descriptions of each condition are provided under each picture. Valves may also not develop properly, or may be damaged by disease and trauma (as in hypertension). If a valve fails to open completely, it is said to be stenosed; the condition is called STENOSIS. If the fibrous rings supporting the valves are stretched, the orifice may not seal properly when the valve closes. Such a valve is said to be INCOMPETENT, and permits regurgitation of blood in a wrong-

way flow. Rheumatic fever, the result of streptococcal bacterial infection, may result in stiffening and erosion of the mitral cusps causing stenosis and incompetence.

Blood supply of the heart (Fig. 22.8)

Two CORONARY ARTERIES, a right and a left, arise from the aorta just behind the pockets of the aortic valve. This origin permits the heart to receive blood with the highest oxygen level and under the highest possible pressure. The right coronary artery passes to the posterior side of the heart, giving off branches to the left ventricle, right ventricle, and right atrium. The left coronary artery supplies the left ventricle and left atrium. *Anastomoses*, or communications between the smaller branches of both coronary arteries, are common. About 48 percent of the United States population has what is called a right coronary predominance, in which the right coronary artery supplies more than just the right side of the heart. About 18 percent have left coronary predominance. The remainder, about 33 percent, have a balanced coronary supply.

The arteries form capillary and sinusoidal (large irregular vessels) beds in the myocardium, and substances diffuse to and from these beds to the muscle and other layers of the heart wall. About 30 percent of the blood reaching the heart wall passes through the sinusoids directly into the ventricular cavities. The remaining 70 percent is collected by the CORONARY VEINS that, by way of the CORONARY SINUS, empty into the right atrium.

Blood flows through the system of coronary arteries, capillaries, and veins only when the myocardium is *not* contracted. Contraction compresses the small vessels of the heart and stops flow. Also, when the ventricles are ejecting blood, the open semilunar valve in the aorta blocks the openings of the coronary arteries. Thus, the only time when nutrients may be supplied to the heart is when the myocardium is relaxing or at rest, and when the aortic valve is closed.

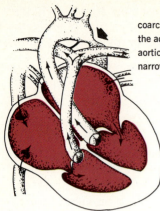

coarctation of
the aorta. the
aortic lumen is
narrowed.

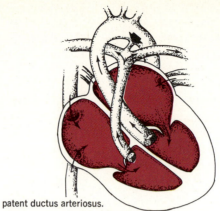

patent ductus arteriosus.
the shunt between the pulmonary
artery and aorta is open.
blood bypasses the lungs.

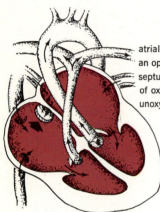

atrial septal defect.
an opening in the atrial
septum permits mixing
of oxygenated and
unoxygenated blood

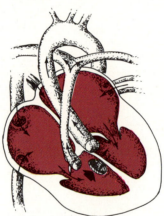

ventricular septal defect.
an abnormal opening between
the ventricles. blood passes
from left to right ventricle.

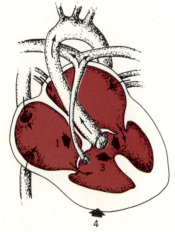

tetralogy of Fallot. four defects
occur simultaneously. (1) narrowing
of pulmonary artery or valve (2) ventricular septal defect, (3) aorta (overriding)
receiving blood from both ventricles
(4) hypertrophy of wall of right
ventricle.

FIGURE 22.7   Some of the most common congenital
anomalies of the heart.

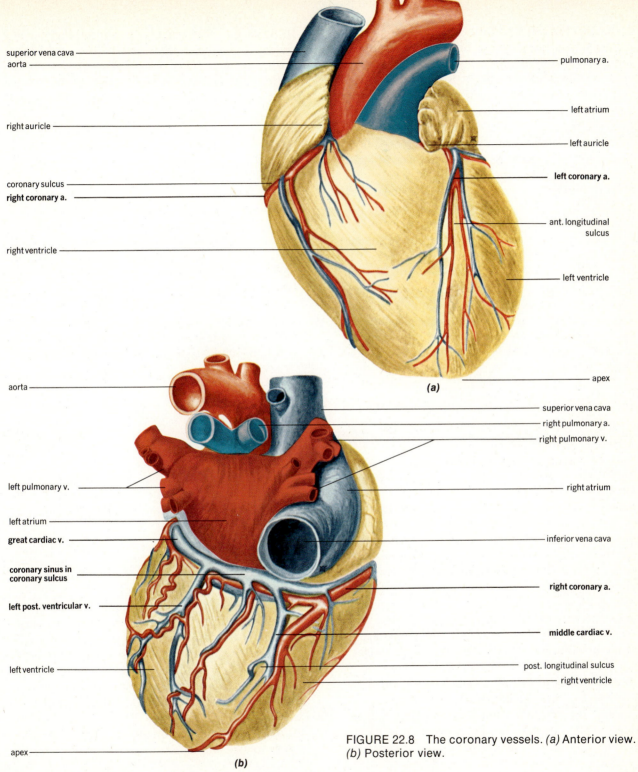

superior vena cava

aorta

right auricle

coronary sulcus

right coronary a.

right ventricle

pulmonary a.

left atrium

left auricle

left coronary a.

ant. longitudinal sulcus

left ventricle

apex

(a)

aorta

left pulmonary v.

left atrium

great cardiac v.

coronary sinus in coronary sulcus

left post. ventricular v.

left ventricle

apex

superior vena cava

right pulmonary a.

right pulmonary v.

right atrium

inferior vena cava

right coronary a.

middle cardiac v.

post. longitudinal sulcus

right ventricle

(b)

FIGURE 22.8   The coronary vessels. *(a)* Anterior view. *(b)* Posterior view.

CLINICAL CONSIDERATIONS. Blockage of the coronary vessels, as by a blood clot, results in coronary thrombosis; the myocardium is deprived of blood flow (ischemia) and death of tissues beyond the block (infarction) may occur. If the thrombosis results in incomplete blockage of the vessel, a reduced blood flow to the area supplied (ischemia) may be the only result. Decreased flow of blood to the heart is usually associated with the development of the pain of ANGINA. In this condition, pain associated with myocardial ischemia is referred to the left chest and arm. If the vessel plugged is relatively small, the anastomoses between coronary arteries may permit sufficient blood flow to compensate for the blockage. If a large vessel is plugged, the amount of tissue that dies may be too great to permit continued heart action. The *s*erum *g*lutamic *o*xaloacetic *t*ransaminase levels or SGOT (an enzyme involved in heart metabolism) bears a relationship to severity of cardiac muscle death following a thrombosis. Normally found in a concentration of about 10 to 40 units per milliliter of serum, the enzyme increases in amount according to the severity of cardiac damage. Thus, an SGOT of 90 to 100 units indicates mild heart muscle damage, and 200 or more units indicates severe damage. The SGOT may be used to indicate the severity of the initial attack, and to follow recovery from the attack.

Nerves of the heart

The heart receives MOTOR FIBERS from both divisions of the autonomic nervous system. Parasympathetic fibers are carried in the vagus nerves and act primarily to slow the beat. Sympathetic fibers, carried in the cardiac nerves, increase both heart rate and strength of beat. SENSORY FIBERS arise within the heart and are carried in the vagus nerves to the brain stem. These fibers carry impulses for the reflex control of heart rate and transport pain impulses to the central nervous system.

The physiology of the tissues composing the heart

The heart consists of several types of tissues that are responsible for its ability to contract. About half of the 300 gm weight of the adult human heart consists of connective tissue that forms the CARDIAC SKELETON. In general, the main mass of this connective tissue is disposed roughly in the form of a cross. The vertically oriented portion composes a support for the septae between atria and ventricles, and the "cross bar" forms the supporting structures for the atrioventricular valves. This "cross bar" also acts to electrically insulate the atrial muscle mass from that of the ventricles. NODAL TISSUE is developmentally derived from cardiac muscle, but has largely lost its ability to contract and has developed the powers of *spontaneous depolarization* and *conductivity* to a high degree. This tissue is responsible for the stimulation that causes the muscle of the heart to contract.

CARDIAC MUSCLE comprises about half the adult heart weight and its contraction creates the pressure necessary to circulate the blood through the body.

Nodal tissue

The presence of the nodal tissue in the heart gives AUTORHYTHMICITY to the organ. What the term implies is that the heart possesses, within itself, the ability to beat independently of any outside influence.

The nodal tissue exists as a series of discrete masses, bundles, and fibers within the subendocardial layer of the heart wall (Fig. 22.9). The SINOATRIAL (SA) NODE measures about 18 mm by 6 mm in size and is located at the junction of the

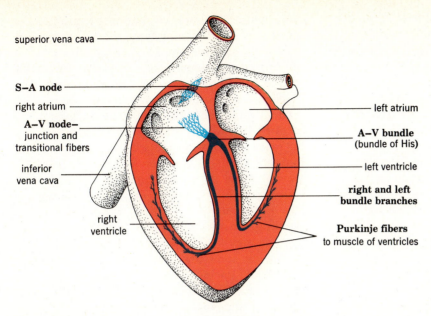

superior vena cava

S–A node

right atrium

A–V node–
junction and
transitional fibers

inferior
vena cava

right
ventricle

left atrium

A–V bundle
(bundle of His)

left ventricle

right and left
bundle branches

Purkinje fibers
to muscle of ventricles

FIGURE 22.9   The locations of the nodal tissue in the
human heart.

superior vena cava with the right atrium. This
mass of nodal tissue is designated as the PACE-
MAKER of the heartbeat, for it possesses the fastest
rate of spontaneous depolarization. Its basic
rhythm of depolarization is about 80 times per
minute, and the mechanisms of depolarization are
believed to result from a "leakage" of sodium ions
across the membranes of the nodal fibers into the
cells. As sodium movement occurs, the membrane potential decreases, and at some point depolarization occurs with formation of an action
potential. The depolarization wave spreads
through the atrial musculature at a rate of about
0.3 m per second, stimulating the muscle to contract. Note that the *only* way that the depolarization wave may reach the next segment of the nodal
system is by passing through the atrial muscles;
there is *no direct connection* between the SA node
and any other nodal tissue.

The depolarization wave eventually reaches the
ATRIOVENTRICULAR (*AV*) NODE, located in the right
atrium. Small, slow-conducting *junction* and *transition fibers* within the AV node momentarily im-
pede the progress of the wave through the node,
which allows the depolarization wave to excite
all the atrial muscle before the wave is transmitted
to the ventricles. The need for this *AV delay* will
become apparent by referring to Figure 22.9; note
that the distance from SA node to AV node is *less*
than from SA node to the farthest reaches of the
left atrium. Thus, if the AV node did not slow the
depolarization wave, the ventricles could contract
before the left atrium had completed contraction.

From the AV node, the depolarization wave is
transmitted very rapidly (2–4 m per sec) through
the AV BUNDLE (of *His*), the RIGHT and LEFT
BUNDLE BRANCHES (that pass through the ventricular septum), to the PURKINJE FIBERS that terminate on the innermost layer or two of myocardial cells. The basic function of this portion
of the nodal tissue is to achieve a nearly simultaneous stimulation of the ventricular muscle mass.
This part of the system becomes the only *normal*
way of distributing SA node influence to the
ventricles. As discussed later, the rate of SA node
depolarization and, to some extent, the rate of
conduction through the rest of the tissue may be
altered by outside influences (nerves, chemical

environment), but the *ability* to contract resides within the heart itself. Heart transplants could not be attempted unless contractile ability lay within the heart; to reconnect nerves and reestablish beat (if the contraction *did* depend on outside influences) would be impossible.

Because all parts of the nodal system possess the ability to spontaneously depolarize, if there is SA node failure, other parts of the system (e.g., AV node) may take over pacemaker duties. However, the farther down the "nodal line," the slower the rate of depolarization, and if rate falls to about 40 beats per minute, an inadequate cardiac output will result.

CLINICAL CONSIDERATIONS. ARRHYTHMIA is a term referring to alterations in the normal rhythm or sequence of events in heart excitation. Three basic conditions may be cited as causative factors in arrhythmias.

Alterations in rate of SA node activity
Interference with conduction
Presence of ectopic foci

Alterations in rate of SA node activity. BRADY-CARDIA (decreased or slow heart rate) or TACHY-CARDIA (elevated heart rate) are most often associated with chemicals released during stress, or which are ingested. Atropine, caffeine, epinephrine (adrenalin), and camphor raise heart rate; amyl nitrite, nitroglycerine, alcohol in small quantities, and acetylcholine slow the heart rate.

Interference with conduction. HEART BLOCK refers to alterations in ability of the nodal tissue to *conduct* impulses (thus involving the tissue from AV node onwards). Normally, atria and ventricles contract at the same rate, with AV node responding to each SA node output. If there is no dissociation of atrial and ventricular rates, but merely a slowing, a *first degree block* has occurred. *Second degree block* occurs when the ventricle fails to respond to all SA node impulses (e.g., atrial rate 75, ventricular rate 60). A *third degree block* sees the atria beating at a rate determined by the SA node, while the ventricles are responding to the activity of the other parts of the nodal system; one or the other of the bundle branches may be defective and stimulation is basically being delivered to half the ventricular mass. Third degree block usually results in a rate too low to adequately pump enough blood ("heart failure") to meet body needs and an *artificial pacemaker* may be required to correct this condition. The pacemaker delivers stimulation to the ventricles by way of leads or wires from an external power source and may provide continuous stimulation at a preset rate ("set" pacing), a variable rate depending on activity levels ("synchronous" pacing), or may "fire" only when normal heart rate falls below some preset rate of SA node discharge ("demand" pacing). Hazards of pacemaker installation include infection, clot formation, lead breakage, and power failure.

Presence of ectopic foci. Areas of excitability other than nodal tissue may develop in the heart. Depolarization leads to additional stimuli being delivered to the cardiac muscle and very rapid heart rates may occur *(fibrillation)*. Defibrillation causes all the muscle to assume the same electrical state simultaneously, and a normal rhythm may be established.

Cardiac muscle (Fig. 22.10)

Cardiac muscle, forming the contractile tissue of the heart, is a type of STRIATED (striped) muscle. As indicated in an earlier section, the muscle is disposed in two masses (one for the atria, one for the ventricles) that are electrically insulated from one another.

The cardiac fibers are anatomically separated from one another by INTERCALATED DISCS, and the branching fibers form a network within the heart. The discs apparently offer little resistance to the passage of a wave of depolarization, and thus a mass of muscle behaves as though it was one continuous cell. This situation is spoken of as a FUNCTIONAL SYNCYTIUM. This means that stimulation to any part of the cardiac muscle will result in excita-

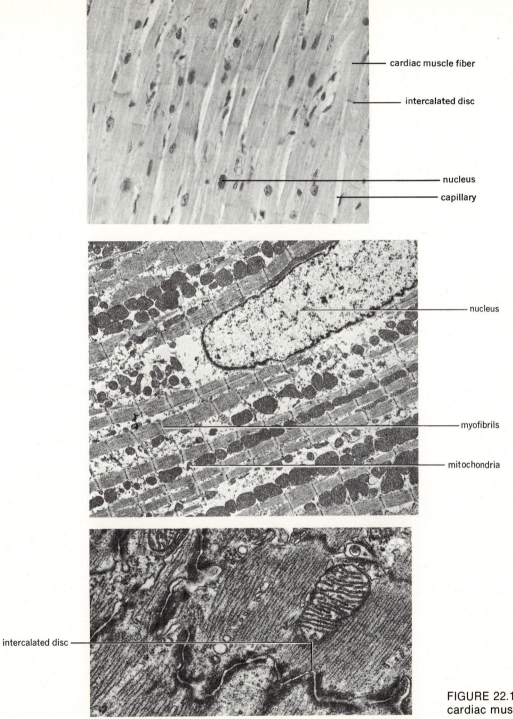

FIGURE 22.10A   The histology of cardiac muscle.

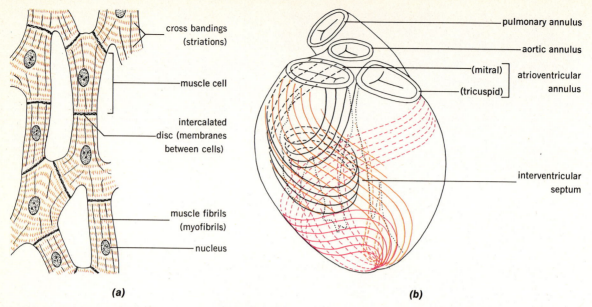

cross bandings
(striations)

muscle cell

intercalated
disc (membranes
between cells)

muscle fibrils
(myofibrils)

nucleus

pulmonary annulus

aortic annulus

(mitral)
(tricuspid)
atrioventricular
annulus

interventricular
septum

(a)

(b)

FIGURE 22.10B *(a)* Diagram showing the structure of cardiac muscle. *(b)* The courses of the major groups of ventricular muscle fibers.

tion of all parts, a condition necessary for near-simultaneous contraction and maximum pressure development. Cardiac muscle has a contraction time about three times longer than that of skeletal muscle (cardiac muscle, 0.3 sec). The muscle follows the ALL-OR-NONE LAW, which states that a stimulus capable of causing depolarization will give a maximum response in terms of contraction. *Strength* of that contraction will depend on the chemical and thermal environment of the muscle at the time of stimulation.

Cardiac muscle follows the LAW-OF-THE-HEART (Frank-Starling law). This law states that, to a point, greater stretch or tension placed on the cardiac muscle will cause a stronger contraction, and, of course, the reverse is true also. In the body, greater tension occurs as the ventricles are filled to a greater degree by activity or other causes, and provides a device by which *automatic* increases in amount of blood pumped help to meet the demands of that exercise.

After depolarization, cardiac muscle fibers re-

polarize more slowly than skeletal fibers, and so the REFRACTORY PERIOD IS LONG in cardiac muscle. In fact, the muscle is into its relaxation period before it becomes capable of responding to a second stimulus. Again, the value of this length of period is obvious if one remembers to pump blood, the heart must fill, and to fill, the muscle must at least partially relax. Thus, a long refractory period normally ensures a continual, though possibly reduced, circulation of blood.

EXTRASYSTOLES, or additional contractions, occur when a stimulus occurs after the refractory period is over but before the next regular SA node impulse is due (Fig. 22.11). The extrasystole is usually less strong than a regular contraction because the chamber is less full than normal (remember law-of-the-heart). When the next regular SA node stimulation arrives, the muscle is refractive from the extrasystole, and a *compensatory pause* occurs. During this time, the ventricle will fill to a greater degree than normal, and the next regular contraction will be stronger (law-of-the-heart).

The several properties of nodal tissue and cardiac muscle described in these sections illustrate that the combination of nodal tissue and cardiac muscle provide an inherent rhythmical beat that can be adjusted to body need for blood circulation. The properties of cardiac muscle might well be compared at this time to those of skeletal and smooth muscle presented in Chapter 10 to see how each type of nuscle is best fitted for its tasks in the body. Table 22.1 summarizes the basic facts about nodal tissue and cardiac muscle properties.

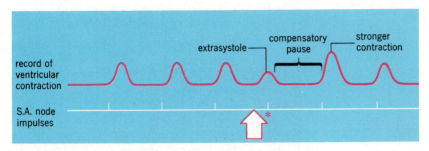

FIGURE 22.11   The genesis of an extrasystole and the compensatory pause. Stimulus applied at ↑ causes extrasystole; next normal SA node impulse arriving at * finds muscle refractory from extrasystole. Muscle does not respond again until SA node impulse arrives at normal time.

TABLE 22.1 A summary of properties of nodal tissue and cardiac muscle		
Tissue	Property	Comments
Nodal	Autorhythmicity	Allows a basic rate of beat (70–80/min) to be established. Not dependent on outside factors
	Conductivity	Fastest in A-V bundle onwards to excite ventricular muscle nearly simultaneously. Slowest in A-V node
Cardiac Muscle	Syncytial arrangement of fibers	Allows rapid spread of depolarization through a mass of muscle. Gives near simultaneous contraction for greatest pressure development
	Follows the all-or-none law	Maximal contraction occurs if stimulus is strong enough to cause depolarization. Strength varies according to environment and mechanical factors (tension)
	Follows the law-of-the-heart	Increased tension results in stronger contraction. Allows variation in amount of blood pumped. Adjustability
	Has long refractory period	No tetanus. Some filling before contraction to maintain pumping action

The cardiac cycle

A cardiac cycle is defined as one complete series of events occurring within the heart, and usually is said to begin with depolarization of the SA node. Each cycle requires a certain length of time for its completion, and pressure, volume, electrical, and sound changes occur during each cycle.

Timing of the cycle (Fig. 22.12)

At a rate of 75 beats per minute, considered to be a typical normal or resting rate, it takes 0.8 seconds for the completion of one cardiac cycle. Within this time span, atrial SYSTOLE (contraction) takes about 0.1 seconds, atrial DIASTOLE (relaxation) and DIASTASIS (rest) about 0.7 seconds. Ventricular systole takes about 0.3 seconds, and ventricular diastole and diastasis take 0.5 seconds. Each chamber, therefore, spends more time relaxing or resting than contracting. If heart rate increases, the periods of relaxation and rest become shorter. Relaxation and rest permit blood flow through the coronary vessels to occur and allow filling of the heart chambers with the blood to be pumped during the next cycle.

Electrical changes during the cycle

Formation and transmission of impulses through the nodal tissue and depolarization of the cardiac muscle are associated with electrical changes that may be recorded as an ELECTROCARDIOGRAM or ECG. A normal ECG is shown in Figure 22.13. The parts of the recording are explained as follows.

The *P wave* indicates atrial excitation.

The *P-R segment* or *interval* is the time required for the impulses to reach the ventricles.

The *QRS wave* represents ventricular excitation.

The *ST segment* or *interval* is the time between the end of ventricular depolarization and the start of repolarization.

The *T wave* indicates completion of ventricular repolarization.

The *U wave* is not present in all ECG's. If it shows up, it indicates either slow papillary muscle repolarization, or hypokalemia.

CLINICAL CONSIDERATIONS. The value of an ECG lies in the fact that any changes in the nodal tissue or cardiac muscle will be seen as alterations

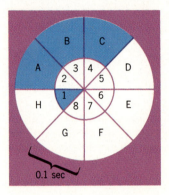

FIGURE 22.12   The timing of a cardiac cycle. Rate of beating is 70 to 72 beats per minute. 1, atrial systole; 2, atrial diastole; 3–8, atrial diastasis. A, B, C, ventricular systole; D, ventricular diastole; E, F, G, H, ventricular diastasis.

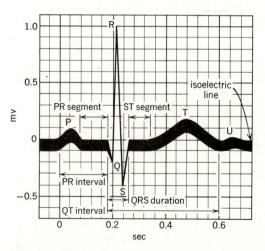

FIGURE 22.13   The normal electrocardiogram.

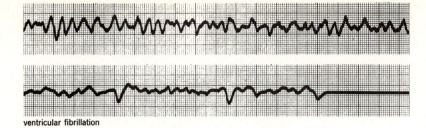

ventricular fibrillation

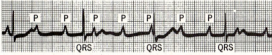

complete heart block (atrial rate, 107; ventricular rate, 43)

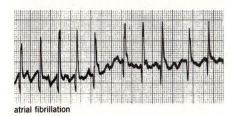

atrial fibrillation

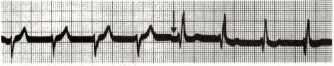

intermittent right bundle branch block

FIGURE 22.14   Abnormal electrocardiograms in various conditions.

of the normal ECG. Alterations of function are reflected as changes in direction of the waves, in their height, or in the length of the intervals. Damage of the nodal tissue or cardiac muscle following "heart attacks" may be determined; recovery may be followed; the severity of the damage may be assessed. Some abnormal ECG's are presented in Figure 22.14. Each should be examined to determine the specific changes that occur in the ECG in each type of abnormality.

Pressure and volume changes during the cycle (Fig. 22.15)

Since blood only moves through the body from an area of higher pressure to one of lower pressure, the creation of pressure "highs" by the heart is essential to blood circulation. Pressure changes also operate the valves in the heart and the valves in the arteries leaving the heart. Pressure is usually measured in mm Hg (millimeters of mercury).

Blood enters the right atrium in a steady stream from the venae cavae. Right atrium contraction creates a pressure of about 7 mm Hg, which aids in moving blood to the right ventricle. Right ventricular contraction raises the pressure in that chamber above that of the right atrium. The blood trying to return to the right atrium shuts the tricuspid valve. Pressure rises in the right ventricle until it reaches about 18 mm Hg, at which time the pulmonary valve is forced open and blood is ejected into the pulmonary

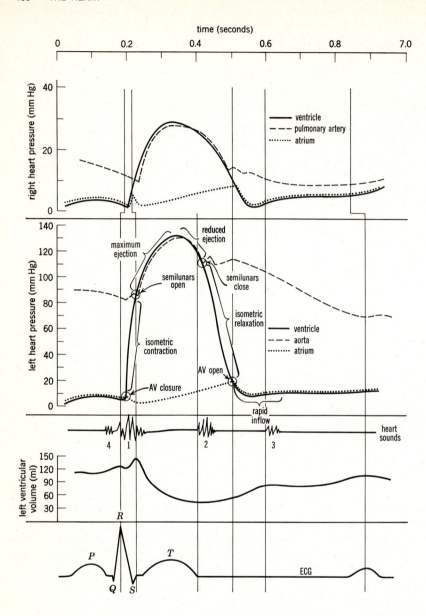

FIGURE 22.15 Composite diagram of heart activity.

circulation. Pressures rise to a maximum of about 30 mm Hg in the right ventricle. Relaxation of the right ventricle causes pressure to fall to less than that in the pulmonary artery and blood tries to return to the ventricle. This closes the pulmonary valve. Continued fall of right ventricular pressure causes, eventually, a fall of pressure to less than that in the right atrium, the tricuspid valve opens, and blood again fills the ventricle. The same series of events occurs in the left side of the heart as blood passes from left atrium, to left ventricle, to aorta. Several important differences may be noted in the action of the left side of the heart.

The left atrium achieves a peak pressure of about 10 mm Hg, as compared to 7 mm Hg in the right atrium.

The left ventricle must create a pressure of 80 mm Hg to open the aortic valve as compared to 18 mm Hg by the right ventricle, and a peak pressure of about 130 mm Hg is achieved by the left ventricle, as compared to 30 mm Hg by the right ventricle.

As the ventricles contract, their volume obviously decreases as blood is pumped into the pulmonary and systemic circulations. Each ventricle has a normal volume of 100 to 120 milliliters. At normal or resting rates and volumes of beat, each ventricle empties about two thirds of its contained blood into the two circulations. Thus, about 65 to 80 milliliters of blood enters the circulation at each heartbeat. Increased volumes of blood may be pumped into the vessels by either increasing the heart rate or the volume emptied per beat. Normally, each ventricle pumps an equal amount of blood, so that no "accumulation" of blood occurs in any part of the system.

Heart sounds during the cycle

As one listens (auscultation) to the heart with a stethoscope placed on the chest, it may be determined that the heart creates sounds as it beats. Most of these sounds occur as valves open or close, and as the myocardium contracts. With a stethoscope, two heart sounds are heard, designated the FIRST HEART SOUND, and the SECOND HEART SOUND. Sensitive microphones may pick up a third and fourth heart sound. The times of occurrence of each sound are shown in Figure 22.15, and their causes are presented in Table 22.2.

CLINICAL CONSIDERATIONS. Heart MURMURS are abnormal sounds produced during heart activity, or are modifications of normal sounds. They usually indicate abnormal valve action (stenosis or incompetence). FUNCTIONAL MURMURS are usually of no concern, are produced only when the heart is beating forcibly, and are usually outgrown. RESTING MURMURS are a cause for concern, for they usually indicate valve damage.

TABLE 22.2 Causes of the heart sounds		
Sound (S)	Cause(s)	Comments
First (S_1)	Closure of AV valves	Sound is low pitched and drawn out ("Lubb")
	Tensing of valves	
	Vibration of chordae	
	Contraction of ventricular muscle	
Second (S_2)	Closure of arterial valves	Sound is higher pitched and short duration ("Dup")
Third (S_3)	Rush of blood from atrium to ventricle	Occurs during or after atrial contraction; a sound of turbulent flow
Fourth (S_4)	Muscular contraction of atria	—

The heart in creation and maintenance of blood pressure

The discussion of heart action has emphasized the role of the heart in creating a pressure to eject blood from the heart into the pulmonary and systemic circulations. In order that blood flow be continuous despite intermittent heart action, and to ensure that the heart can develop a pressure on the blood, factors other than heart action enter the picture. The five factors responsible for the origin and maintenance of arterial blood pressure are presented here, so that an introduction to the concepts of blood pressure may be made; then the specific role of the heart in the creation and maintenance of arterial pressure is discussed in this chapter, and that of the blood vessels is considered in Chapter 23. The five factors responsible for creating and maintaining arterial blood pressure are as follows.

1. *The pumping action of the heart.* Ventricular contraction injects into a vascular system of a given volume, variable quantities of blood per minute. The greater the amount of blood injected into the arteries, the higher the pressure in the system. Without heart action, *no* pressure would exist. The term *cardiac output* refers to the volume of blood pumped by one ventricle per minute.

2. *Peripheral resistance.* Mainly a function of the diameter of blood vessels, this factor refers to the *ease of outflow* of blood through the small muscular arteries of the body. The smaller the vessel, the greater the pressure required to force blood through it.

3. *Elasticity of arteries* near the heart. Ventricular activity injects a quantity of blood into the large arteries near the heart. They expand to contain this volume and, in so doing, prevent excessive rise of systolic blood pressure. During ventricular relaxation, elastic recoil of these arteries prevents diastolic pressure from falling to low values, by moving the blood onward through the arteries.

4. *Viscosity of the blood.* A thicker (viscous) fluid such as the blood requires more pressure to circulate it; therefore, pressures are higher than if a fluid such as water was being pumped.

5. *Volume of blood* in the system. In a system of pipes with a given volume or capacity, the placing of more fluid into that system must raise the pressure.

Of these factors, *heart action* and *peripheral*

resistance are the most important in creating and maintaining arterial blood pressure and are the only ones that can be changed to meet the varying demands of the body for blood flow.

Cardiac output

As the ventricular muscle contracts (ventricular systole), pressure is created on the blood within those chambers and blood is moved into the pulmonary and systemic circulations. The volume of blood leaving *each* ventricle during one minute's time is termed the CARDIAC OUTPUT. Cardiac output *(CO)* is the product of two other factors: the rate of beat or STROKE RATE *(SR)* and the volume of blood ejected per beat, or STROKE VOLUME *(SV)*. Represented as a mathematical formula

$$\text{CO (ml/min)} = \text{SR (beats/min)} \times \text{SV (ml/beat)}$$

As one can see, beats cancel in the equation, and CO is expressed as ml/min.

Over a period of time, the normal heart will eject the same amount of blood from each ventricle, although minor variations may exist on a beat-to-beat basis.

Stroke rate may be easily determined by resting the fingers on a superficial artery of the body (e.g., radial at the wrist; carotid on the neck) and counting beats for a minute. Determining stroke volume requires finding the difference between *end systolic volume* (amount of blood remaining in a ventricle at the end of contraction), and *end diastolic volume* (the volume of blood in the ventricle after filling). To determine stroke volume accurately is a difficult task. For purposes of illustrating our point, let us assume an end systolic volume of 50 ml and our end diastolic volume of 110 ml when the heart is beating under normal resting conditions; stroke volume is thus 60 ml (110 minus 50). It may be noted that it is not necessary to empty the ventricle completely with each beat; the blood remaining forms part of the basis for increasing cardiac output during activity.

With normal resting SR and SV, cardiac output lies between 4.2 and 5.6 liters per minute.

SR = 70 to 80 beats/min; SV = 60 to 70 ml/beat

During exercise, heart rate may increase to 120 beats/min, and volume to 120 ml/beat. Thus, cardiac output becomes 14 liters per minute. Increase of SR beyond about 140 beats/min allows insufficient time for ventricular filling and SV drops drastically, reducing CO.

Factors controlling cardiac output

Obviously, to alter cardiac output, SR and SV may be altered together or separately, and in the same or opposite directions. Taken together, the following mechanisms control SR and SV.

VENOUS INFLOW. This term refers to the volume of blood that is returned to the right atrium from the body in general. It represents the amount of blood available to enter the right ventricle and, ultimately, the left ventricle (after passing through the lungs). In exercise, for example, a greater venous inflow occurs because of the muscles "massaging" a greater amount of blood through the veins and more enters the right atrium. As this blood passes into the ventricles, the law-of-the-heart is invoked, and a *greater stroke volume* ensues. This mechanism is presented in Figure 22.16.

PERIPHERAL RESISTANCE. Muscular arteries in the body have smooth muscle circularly disposed around them. Contraction of that muscle diminishes the size of the vessels (*vasoconstriction*), while relaxation increases their size (*vasodilation*). If constriction occurs, blood flow to tissues decreases, and pressure rises in the vessels between the heart and constricted area. The heart must increase its output, mainly by *increased stroke volume* to maintain circulation in the face of a higher pressure. The steps in this sequence of events are shown in Figure 22.17.

TEMPERATURE. Exercise is associated with utilization of increased amounts of fuels, and that utilization releases heat. The blood is warmed and

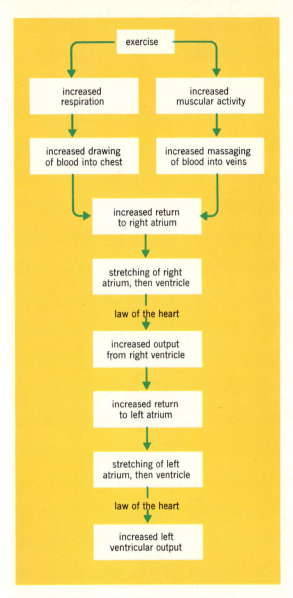

FIGURE 22.16 The effects of exercise on heart action.

this stimulates SA node activity. *Stroke rate increases.* Although of little importance physiologically, a 3 to 4 beat per minute increase in rate will occur. Rapid transfusion of large volumes of blood still cold from storage has been shown to be correlated with increased incidence of heart

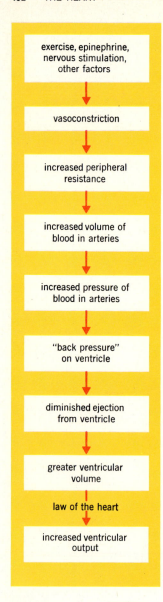

exercise, epinephrine, nervous stimulation, other factors

↓

vasoconstriction

↓

increased peripheral resistance

↓

increased volume of blood in arteries

↓

increased pressure of blood in arteries

↓

"back pressure" on ventricle

↓

diminished ejection from ventricle

↓

greater ventricular volume

law of the heart

↓

increased ventricular output

FIGURE 22.17 The effects of vasoconstriction upon heart action.

block, indicating that nodal tissue conductance is depressed by a fall of temperature of the blood.

CHEMICALS. The chemical environment of nodal tissue and cardiac muscle affects heart action. *Increase of sodium ion* has little effect on the normal heart until concentrations two to three times nor- mal are reached; then *heart rate is slowed* as the cardiac muscle becomes harder to depolarize. *Decrease of sodium ion* by 10 to 20 percent of normal also *slows rate* as less ion is available to pass through muscle membranes. *Increase of potassium ion* to two to three times normal *decreases conduction through nodal tissue,* and blocks, arrhythmias, and fibrillation (rates to 300–400 beats/min) occur. *Decrease of potassium ion decreases stroke rate* by causing slower depolarization. *Increase of calcium ion slows rate* by causing slowed depolarization and *increases stroke volume,* presumably by making more calcium available for the contractile process. *Decrease of calcium ion* causes *increased excitability* of cardiac muscle and ectopic foci may result.

Other chemicals affecting heart action include *epinephrine* (adrenalin), which increases both rate and strength of beat. It is sometimes used as a heart stimulant. *Digitalis* is a "cardiac glycoside" that increases strength of heart beat and raises stroke volume. It is a toxic substance that has many side effects including nausea, vomiting, and headache. Its use should be carefully monitored by medical personnel. Atropine and caffeine increase rates while nitroglycerine, acetylcholine, and small quantities of alcohol decrease rate. *Low oxygen levels* in the bloodstream (*hypoxemia*) cause weakened contractions and *reduce stroke volume* if the shortage is prolonged. *Excessive carbon dioxide levels* act as an anesthetic and cause depression of SA and AV node activity; *heart rate falls.* Fall of pH also lowers rate, but its rise increases heart rate.

NERVOUS FACTORS. Nervous factors provide the finest degree of control of stroke rate; they have little effect on stroke volume.

Both parts of the autonomic nervous system send fibers to the heart (Fig. 22.18). The *sympathetic supply* forms the CARDIAC NERVES that secrete norepinephrine at their terminations on SA and AV nodes and atrial and ventricular musculature. Stimulation of SA node rate of depolarization, increase of AV node conductance, and an

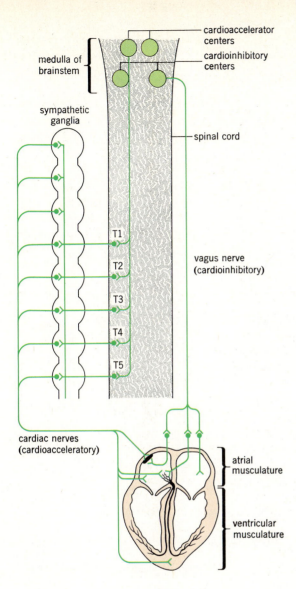

cardioaccelerator centers

cardioinhibitory centers

medulla of brainstem

sympathetic ganglia

spinal cord

T1

T2

vagus nerve (cardioinhibitory)

T3

T4

T5

cardiac nerves (cardioacceleratory)

atrial musculature

ventricular musculature

FIGURE 22.18 A diagram showing the arrangement of the autonomic nerves to the heart. Only one side of each set is represented for clarity (T = thoracic).

increased strength of contraction ensue. Cardiac output is thus increased by increase of both SR and SV. These nerves are termed CARDIOACCELERATOR NERVES. *Parasympathetic supply* to the heart is carried in the VAGUS NERVE. Acetylcholine is secreted by these nerves at their termination on SA and AV nodes and atrial musculature. Acetylcholine decreases heart rate and thus these nerves are termed CARDIOINHIBITORY NERVES. Both sets of nerves are *tonically* (continuously) active, and so the heart rate at any given moment reflects a *balance* in the effects of the two divisions.

The medullary neurons giving rise to the vagal fibers to the heart, and those influencing the sympathetic ganglia form the CARDIOINHIBITORY and CARDIOACCELERATOR CENTERS respectively. They serve as coordinating centers for reflex activity originating in many body areas. Figure 22.19

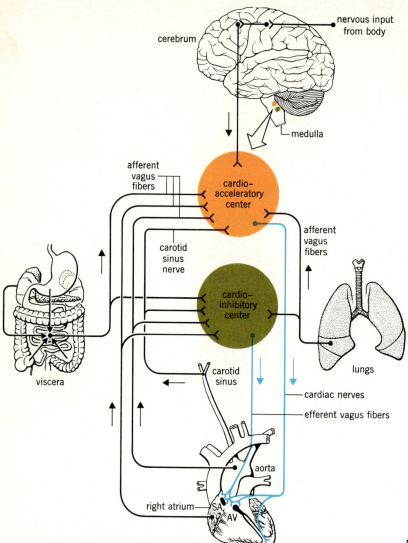

FIGURE 22.19   Cardiac reflex pathways.

and Table 22.3 show origins of this reflex activity and name the major cardiac reflexes with their effects. Note, as you study these reflexes, that control over cardiac output that matches body needs is provided.

Figure 22.20 provides an overall view of the several mechanisms described and their effects on cardiac output. It may also indicate the reserve the heart has for meeting varying needs.

Clinical considerations.

Inability of the cardiac output to keep pace with body demands for nutrients and waste removal constitutes HEART FAILURE. This generally occurs when output falls to 2 to 2.5 liters per minute. One ventricle or both may fail. If the right ventricle fails to adequately pump blood, the blood will accumulate in systemic veins and venous pressure

TABLE 22.3 Cardiac reflexes

Name of reflex	Input from	Stimulus triggering	*Effect on CIC[a]	*Effect on CAC[b]	*Effect on SR[c]	*Effect on SV[d]
Aortic depressor	Baroreceptor (sensitive to stretch) in aorta	Stretch of aorta	+	−	↓	0
Carotid sinus	Baroreceptor in carotid sinus	Stretch of sinus	+	−	↓	0
—	Baroreceptors in pulmonary arteries	Stretch of arteries	+	−	↓	0
Bainbridge reflex (existence disputed)	Vena cava or right atrium	Increased filling of atrium (stretching)	−	0/+	↑	0
—	Cerebrum— anger/fear	Stress, thoughts	−	+	↑	0/↑
—	Strong stimulation of any nerve	Pain	−	+	↑	↑

* + (stimulate); − (inhibit); 0 (no change); ↑ (increase); ↓ (decrease).
[a] CIC = Cardioinhibitor center.
[b] CAC = Cardioaccelerator center.
[c] SR = Stroke rate.
[d] SV = Stroke volume.

rises. This may cause fluid to be forced back into, or remain within the tissues, creating EDEMA of the appendages and abdomen (ascites). Left ventricular failure causes pulmonary edema as fluid accumulates in the lungs. The sequence of events in heart failure is present in Figure 22.21.

Energy sources for cardiac activity

As stated in an earlier section, nutrients and oxygen supply to the heart, and waste removal is handled by the coronary circulation (*see* Fig. 22.8). A pair of CORONARY ARTERIES leave the aorta just behind the semilunar valve flaps. These vessels supply the myocardium with capillary beds. CORONARY VEINS parallel the arteries and form the CORONARY SINUS that empties into the right atrium.

During ventricular contraction, the capillaries are squeezed shut and no flow occurs through the system. Flow is about 250 ml/min (at rest) and occurs only during relaxation and rest of the muscle.

Carbohydrates (glucose) or the products of its metabolism (pyruvic acid, lactic acid) account for about 35 percent of the energy supply for cardiac activity. About 60 percent is obtained from fatty acid breakdown, and 5 percent from metabolism of amino acids. Oxygen consumption is, at rest, about 9 ml/100 gm of tissue per minute.

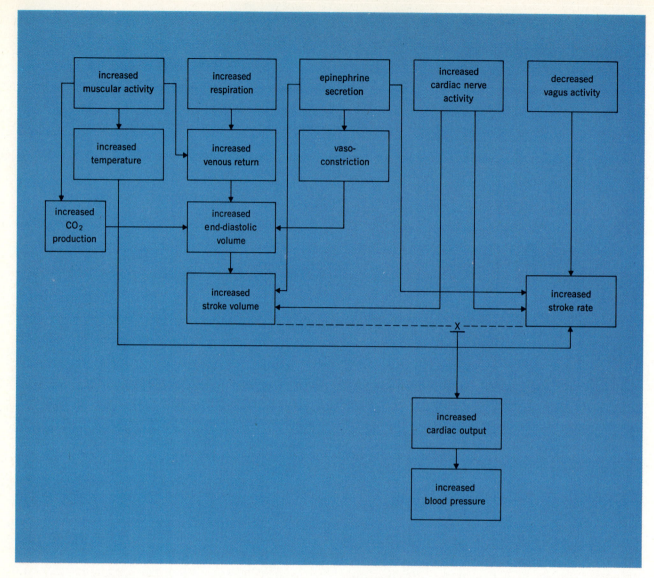

FIGURE 22.20   Factors controlling cardiac output.

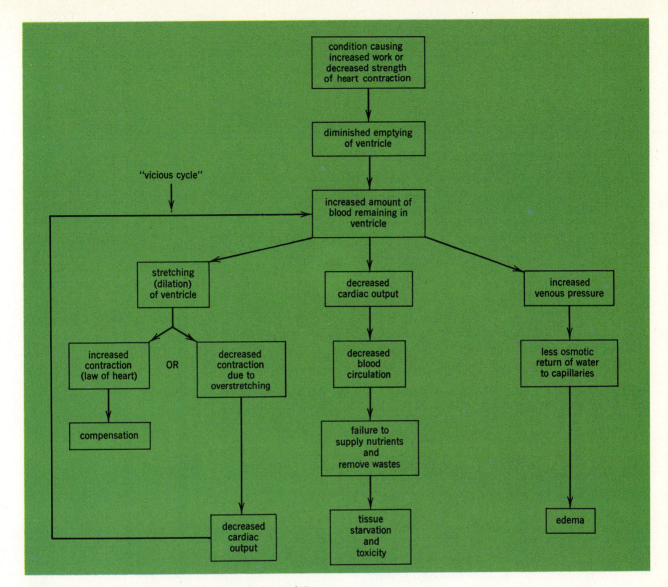

FIGURE 22.21   The sequence of events in heart failure. If the right ventricle fails, the symptoms generally appear in the systemic circulation; if the left ventricle fails, the symptoms generally appear in the pulmonary circulation.

Summary

1. The heart provides a pressure to circulate blood to the body to provide nutrients and remove wastes.

2. The heart develops early in embryonic life to assure substances necessary for cellular activity.

   a. At 2 weeks, a cardiogenic plate is present.

   b. At 3 weeks, paired heart tubes are formed.

   c. At 4 weeks, fusion of the tubes forms the heart.

   d. Between 5 and 7 weeks, the four chambers are established, the wall is developed, and a circulation of blood is established.

3. The heart has certain structural and locational characteristics.

   a. It is about the size of one's clenched fist (age dependent).

   b. It is located in the chest, about one-third to the right of the midline, two-thirds to the left.

   c. It is enclosed in a pericardial sac having fibrous and serous portions.

   d. The wall is composed of an inner endocardium, a middle myocardium, and an outer epicardium.

   e. It is four-chambered, with two atria and two ventricles.

   f. The right side of the heart pumps into the pulmonary circulation; the left to the systemic circulation.

   g. Valves are present between atria and ventricles, and in the great arteries, and control direction of blood flow.

4. Congenital anomalies may cause defects in formation of the heart; infections may damage it.

5. The heart is supplied with blood through the coronary arteries.

   a. Flow through the vessels occurs only when the heart muscle is not contracted.

   b. The coronary circulation provides nutrients to sustain heart activity, and ions to govern rate and strength of beat.

   c. Blockage of the arteries is associated with "heart attack," tissue death, and possible stoppage of life.

6. Nerves to the heart include:

   a. Parasympathetic nerve (vagus) to slow heart rate.

   b. Sympathetic nerves (cardiac nerves) to speed heart rate and strength of contraction.

7. Two types of tissue are most important in heart activity.

   a. Nodal tissue (SA node, AV node, AV bundle, bundle branches, Purkinje

fibers) originates and distributes the stimuli necessary for heart muscle contraction. The SA node is the pacemaker.

b. Cardiac muscle contracts and creates pressure to circulate the blood. It is in the form of a functional syncytium, is involuntary, inherently rhythmic, cannot be thrown into a sustained contraction if ionic concentrations are normal, follows the law of the heart, and the all-or-none law.

8. A cardiac cycle is one complete series of events during cardiac activity.

   a. It normally lasts about 0.8 seconds.

   b. It creates electrical changes (ECG) that reflect the status of the nodal and cardiac tissue. The ECG is used to determine injury to the heart and to follow recovery from injury.

   c. Pressure changes operate valves and move blood.

   d. Sounds are heard as the heart works (*see* Table 22.2).

   e. Murmurs during the cycle usually indicate improper valve functioning.

9. Cardiac output is the volume of blood pumped per minute by one ventricle.

   a. It is the product of stroke rate (beats/min) and stroke volume (ml of blood ejected per beat).

   b. Stroke volume is the difference between end systolic volume and end diastolic volume and is adjustable depending on force of contraction.

   c. Cardiac output is adjusted to body demands by changes in stroke rate and stroke volume.

10. Several factors determine stroke rate and stroke volume.

    a. Venous inflow varies according to exercise and determines degree of ventricular filling and stroke volume.

    b. Peripheral resistance depends on diameter of muscular blood vessels and determines the pressure against which the heart must work to pump blood, and also influence stroke volume.

    c. Temperature increase causes a small rate increase, and vice versa.

    d. A variety of chemicals influence both rate and strength of beat.

    e. Several ions influence heart action. Sodium, potassium, and calcium are the most important.

    f. Cardioaccelerator (sympathetic) and cardioinhibitor nerves (parasympathetic) control rate and strength of beat. In general, sympathetics increase rate and strength of beat; parasympathetics have the opposite effects.

    g. A variety of cardiac reflexes influence the medullary centers from which the nerves arise. Most act to depress heart action (*see* Table 22.3).

11. Heart failure occurs when cardiac output is insufficient to meet body demands.

    a. A cardiac output less than 2 to 2.5 liters indicates failure.

    b. Edema is a common result of heart failure.

12. The heart utilizes carbohydrates (35%), lipids (60%), and amino acids (5%) for its energy sources. It consumes, at rest, some 9 ml/100 gm tissue/min of oxygen.

Questions

1. What devices are available to the heart to enable increase in volume of blood pumped when demand for increased circulation occurs?

2. Explain the role of the nodal tissue in establishing the rhythmical nature of the heartbeat.

3. In what ways is cardiac muscle especially suited to its tasks of circulation of the blood, and adjustment of amount of blood pumped?

4. Describe the role of nerves in alteration of heart rate.

5. Discuss the role of cardiac reflexes in control of blood pressure.

6. Suppose that a kidney malfunction results in potassium retention and calcium loss. What would be expected in terms of derangement of heart action?

7. What defect is implied in "heart block"? Speculate as to the effect of the various degrees of block on maintenance of adequate cardiac output.

8. What is the value of an electrocardiogram? How does the recording indicate changes in conduction times? In strength of depolarization (and, therefore, condition of the cardiac muscle)?

9. Discuss the pressure changes occurring within the ventricles that lead to: ejection of blood; filling of the ventricles.

10. What are the causes of the first and second heart sounds? How are modifications of these sounds related to valvular defects?

11. How does an increased return of blood to the right atrium stimulate cardiac output? How is the increased venous return brought about?

Readings

Benditt, Earl P. "The Origin of Atherosclerosis." *Sci. Amer. 236*:74, Feb. 1977.

Blackburn, Henry. "Contrasting Professional Views on Atherosclerosis and Coronary Disease." *New Eng. J. Med. 292*:105, Jan. 9, 1975.

Braunwald, Eugene. "Current Concepts: Determinants of Cardiac Function." *New Eng. J. Med. 296*:86, Jan. 13, 1977.

...

Cannon, Paul J. "The Kidney in Heart Failure." *New Eng. J. Med. 296*:26, Jan. 6, 1977.

Kolata, Gina Bari. "Coronary Bypass Surgery: Debate Over Its Benefits." *Science 194*:1263, 17 Dec. 1976.

Kolata, Gina Bari. "Detection of Heart Disease: Promising New Methods." *Science 194*:1029, 3 Dec. 1976.

Kolata, Gina Beri, and Jean L. Marx. "Epidemiology of Heart Disease: Searches for Causes." *Science 194*:509, 29 Oct. 1976.

Marx, Jean L. "After the Heart Attack: Limiting the Damage." *Science 194*:1147, 10 Dec. 1976.

Marx, Jean L. "Cardiovascular Disease and the Forms It Takes." *Science 194*:511, 29 Oct. 1976.

Marx, Jean L. "Sudden Death: Strategies for Prevention." *Science 195*:39, 7 Jan. 1977.

Scheuer, James, and Charles M. Tipton. "Cardiovascular Adaptations to Physical Training." *Ann. Rev. Physiol. 39*:221, 1977.

Science News. "Heart Disease and Life Stress." *112*:166, Sept. 10, 1977.

Chapter 23

The Blood and Lymphatic Vessels: Regulation of Blood Pressure

Objectives

After studying this chapter, the reader should be able to:

■ Give a brief outline of development of blood and lymphatic vessels and state when each begins development.

■ List the three major types of blood vessels and how they are related to the heart and tissues.

■ Compare the wall structure of arteries, capillaries, and veins, and correlate structure with the functions of each type of vessel.

■ Relate sclerosis of vessels to changes occurring in their walls.

- Name the major arteries and veins of the body, and indicate the areas supplied or drained.

- Describe the special features of the pulmonary and portal circulations and name the vessels involved.

- State the principles that govern blood flow and pressure in blood vessels, and relate these to the actual pressure, velocity, and surface area of vessels in the body.

- Describe the origins and effects of the two types of nerves that innervate most blood vessels, some reflexes that control blood vessel size, and indicate the value of such reflexes to the body.

- Describe some nonnervous factors that influence blood vessel size.

- Give the symptoms and causes of shock and describe some of the shock-compensating devices the body has, with their effects.

- Define hypertension, some causes, its dangers, and the aim of treatment.

- Define what hypotension is, its dangers, and the aim of treatment.

- Describe the organization of the lymph vessels of the body.

- Give functions of the lymph vessels.

- Describe what causes flow of lymph through lymph vessels and rate of flow.

Development of vessels

Blood vessels develop (Fig. 23.1) from masses of mesoderm called BLOOD ISLANDS that form first in the yolk sac and then in the embryo. At about 2 weeks of embryonic development, cavities appear in these islands and some of the mesodermal cells are found within the cavities. The cells lining the cavities flatten to form an endothelial lining for the cavities. Growth and joining of these cavities create networks of blood channels or what may now be called vessels. The network is extended by "budding" from the original vessels. The mesodermal cells within the vessels differentiate to form, first, red blood cells and, somewhat later, the various types of white blood cells. Other mesodermal cells outside the primitive vessels will give rise to the muscular and connective tissue coats around all vessels except those that will become capillaries. By about 3 weeks of develop-

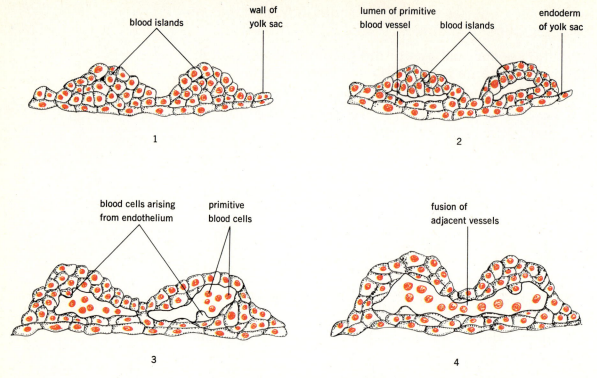

FIGURE 23.1   Early development of blood vessels.

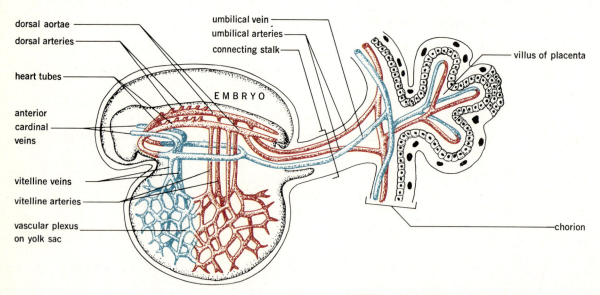

FIGURE 23.2   Later development of blood vessels. Degree of development shown occurs at about three weeks.

ment, the vessels indicated in Figure 23.2 have formed, and a rudimentary circulation is established between the embryo and the developing placenta.

Lymph vessels begin to develop in the embryo at about 5 weeks. They develop in a manner similar to that of the blood vessels and make connections with those blood vessels that will become veins. Initial development of vessels occurs from large areas known as LYMPH SACS. There are six sacs that initially form: two in the neck region (*jugular lymph sacs*); two in the pelvic area (*iliac lymph sacs*); one against the posterior abdominal wall (*retroperitoneal lymph sac*); one dorsal to the retroperitoneal sac (*cisterna chyli*). Lymph vessels (or lymphatics) grow from these sacs following the course of the major veins, to form networks in the developing body organs and muscles. Head, neck, upper trunk, and arms are supplied with lymphatics from the jugular sacs, the lower trunk and legs from the iliac sacs, and the gut and its derivatives from the retroperitoneal sac and cisterna chyli.

Further information on development of blood vessels and lymphatics may be found in the readings at the end of this chapter.

Types of blood vessels

There are three main types of blood vessels.

ARTERIES carry blood *away from the heart* and are subdivided according to size and structure into *large* or *elastic arteries*, and *medium sized* and *small* (arterioles) or *muscular arteries*. The latter two types of vessels have circularly disposed smooth muscle in their walls, and can actively change their size, influencing blood flow and blood pressure in the body.

CAPILLARIES are microscopic endothelial tubes that permeate the body tissues and organs. Their thin walls permit *exchange* of nutrients and wastes between body cells and bloodstream.

VEINS carry blood *toward the heart* and are divided into *small* (venules), *medium sized*, and *large* veins. These vessels have little muscle in their walls, and act mainly as volume vessels for transporting blood.

The general plan of these vessels is shown in Figure 23.3, with capillaries indicated by the white boxes.

The structure of blood vessels

Arteries

Arteries, regardless of size, have three layers of tissue in their walls (Fig. 23.4). An inner TUNICA INTIMA (*tunica interna*) is composed of endothelium and a thin layer of connective tissue. A TUNICA MEDIA is the middle layer and consists primarily of elastic connective tissue in large arteries and smooth muscle in medium and small arteries. It is the thickest layer in arteries. An outer TUNICA ADVENTITIA (*tunica externa*) is composed of connective tissue, and is one-third to one-half as thick as the media. It contains small vessels that nourish the tissue coats, and that are called the VASA VASORUM. Artery structure is summarized in Table 23.1.

CLINICAL CONSIDERATIONS. ARTERIOSCLEROSIS is a generalized term referring to any vascular condition that results in thickening and loss of elasticity and resiliency of the artery wall.

ATHEROSCLEROSIS refers specifically to the changes that occur as fats accumulate in the intima and subintimal layers of the vessel wall.

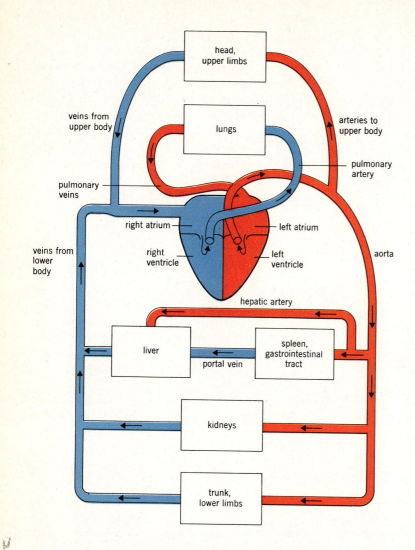

FIGURE 23.3   A schematic representation of the circulatory system.

Accumulation of fats in the walls of arteries leads to *narrowing* of their size, and to *decrease of blood flow* to the tissues supplied by the affected vessel.

Atherosclerosis is present in almost all animal species, and its changes can be detected at any age. It is considered to be an inevitable result of aging. Although the exact cause of the disorder is not known, circumstantial evidence suggests that the intake of a high-fat (cholesterol) diet, cigarette smoking, and lack of exercise accelerate the changes associated with the atherosclerotic process.

The changes that occur have been shown to occur in the following order.

Fatty streaks appear in the arteries in the first days, months, or years of life, and deposition continues into the second decade of life.

Fibrous and fatty plaques appear beginning with the second decade of life and increase until some

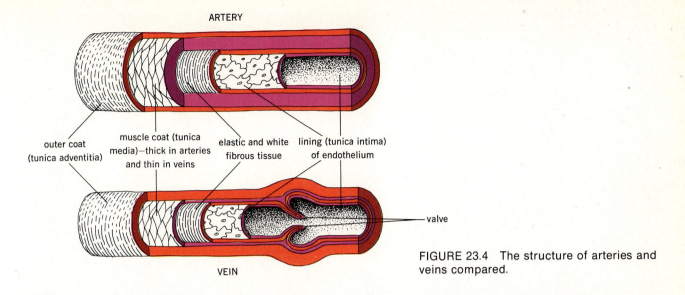

FIGURE 23.4   The structure of arteries and veins compared.

TABLE 23.1   A summary of artery structure

Type of vessel	Average size (mm)	Example(s) or location	Tissue in: Tunica intima	Tunica media	Tunica adventitia	Comments
Large artery	20–25	Aorta, pulmonary artery	Endothelium and c.t. (connective tissue)	Elastic c.t.	Loose c.t.	Elastic vessels expand to contain cardiac output; recoil to push blood onward.
Medium artery	2–10	Named vessels of arms, legs, viscera	As above	25–40 layers of smooth muscle	As above	Muscle makes them important in terms of changing size; not too numerous
Small artery (arteriole)	2 mm to about 20 μm	Unnamed; close to or in tissues and organs	As above	5–10 layers of smooth muscle	As above	Large numbers make these the primary controllers of blood flow and pressure.

individuals show the signs of vessel narrowing and diminished blood flow to body organs (cardiac infarction, stroke, gangrene).

Once a lesion is established, it acts as an irritant to the artery wall, causing an inflammatory reaction that further thickens and stiffens the vessel wall. Ulceration may then cause rupture of the wall, or ischemia may cause tissue death.

TABLE 23.2   A summary of vein structure

| Type of vessel | Average size (mm) | Example(s) or location | Tissue in: | | | Comments |
			Tunica intima	Tunica media	Tunica adventitia	
Large vein	20–30	Vena cava	Endothelium and c.t.	C.t. and scattered smooth muscle cells	Loose c.t.	Few in number, little change in size
Medium vein	2–20	Named veins of appendages and viscera	As above	As above	As above	As above
Small vein (venule)	<2 mm	In tissues and organs	As above	Collagenous c.t. and a few smooth muscle cells	Lacking	Most numerous; can significantly alter venous volume by change in size

Treatment of the disorder is centered about reducing dietary intake of lipids, which is thought to reduce the rate of accumulation of fats in the vessel walls. If an artery is of such a size and is in a favorable location, *thromboendarterectomy* may be performed. This procedure in effect "reams out" the vessel, removing its lipid deposits and increasing blood flow. Bypass grafting of blood vessels to renew the supply to an organ or tissue that has been deprived of blood is sometimes successful.

In some individuals, the tunics of the vessel walls do not develop properly, and the wall thins out. Under the pressure of the blood, the artery develops a thin blisterlike *aneurysm*, which may rupture, leading to hemorrhage, shock, and death.

Veins

Veins, like arteries, usually have three layers of tissue in their walls (*see* Fig. 23.4). The predominant layer in veins is the tunica adventitia, being two to five times thicker than the media. Small veins are two layered, lacking a layer comparable to an adventitia. Valves, controlling the flow of blood through veins toward the heart, are found in medium sized and large veins (except the venae cavae). A summary of vein structure is presented in Table 23.2.

CLINICAL CONSIDERATIONS. Superficial veins, having thinner walls and less muscle than arteries, and less supporting tissue around them, are more easily enlarged or stretched than deep veins.

VARICOSITIES or VARICOSE VEINS are superficial veins that have become abnormally elongated or dilated. The human's upright posture, combined with the effect of gravity, tends to cause pooling of blood in the superficial veins of the legs. Valves in these vessels and the massaging action of the skeletal muscles as they contract and relax tend to move the blood toward the heart. Accumulation of blood in these veins, accompanied by lack of muscular activity, may cause enlargement of the veins to a point where the valves become incompetent. Once the cycle of enlargement and valvular incompetence is established, it tends to cause the development of larger and larger varicosities. Support hose and elevation of the feet may aid the

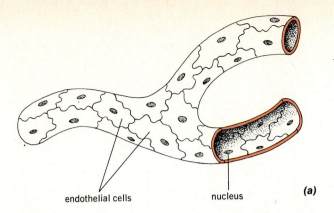

endothelial cells   nucleus   *(a)*

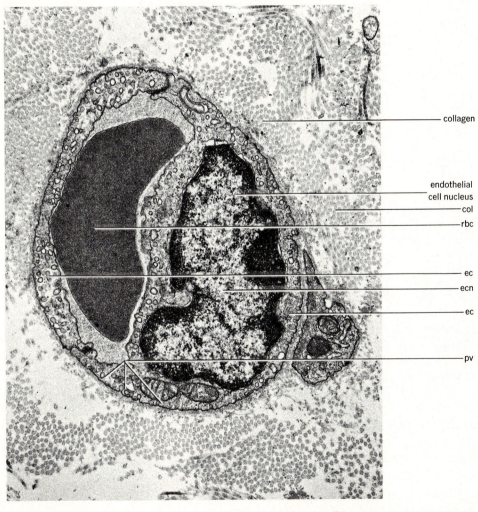

(b)

FIGURE 23.5 *(a)* The microscopic and *(b)* ultramicroscopic structure of a capillary (col. collagen; rbc, red blood cell; ec, endothelial cell; ecn, endothelial cell nucleus; pv, pinocytic vesicle).

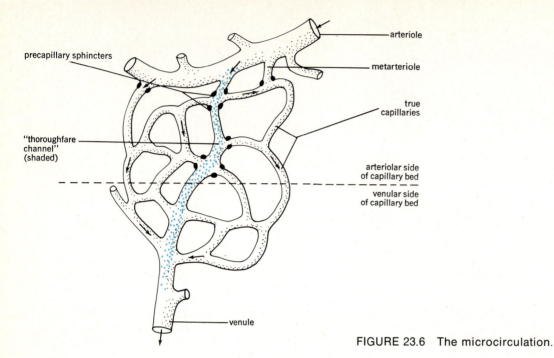

FIGURE 23.6   The microcirculation.

condition, or the veins may be removed surgically.

Capillaries (Fig. 23.5)

Capillaries, as the vessels where exchange of materials occurs, must have thin walls. Thus, all layers but the endothelium are missing. Typical capillaries are 7 to 9 micrometers in diameter, and permit the diffusion and filtration of all blood substances except cells and most of the plasma proteins. There are literally miles (one estimate says over 61,000 miles) of capillaries in the body, and their large surface area (800 × that of the aorta) provides not only a great surface for exchange, but also slows the speed of blood flow to allow time for exchange of materials to occur. Sinusoids are large irregular vessels similar to capillaries but they have a lining of phagocytic cells. They are found in the liver and spleen.

The microcirculation

Physiologists define a microcirculatory unit consisting of an *arteriole*, the *capillaries* rising from it, and the *venules* draining the capillaries as an important functional unit of the circulation (Fig. 23.6). The arterioles in these units can, by narrowing or increasing their diameter through activity of smooth muscle in their walls, control the amount of blood flowing through the capillary beds. The importance of the microcirculation is discussed in a later section of this chapter.

The names of the systemic vessels of the body

Arteries

The systemic arteries of the body include the AORTA and its BRANCHES. The major systemic arteries are shown in Figures 23.7 to 23.10 and are summarized in Table 23.3 to 23.6.

Veins

The major systemic veins are presented in Figures 23.11 through 23.14 and are summarized in Tables 23.7 and 23.8. Veins usually accompany their corresponding arteries and are generally named in similar fashion. In the appendages, a DEEP SET of veins courses alongside the corresponding arteries and are named as are the arteries; a SUPERFICIAL SET lies just beneath the skin. The superficial veins are often visible on the appendages, and the larger ones are used as site for *venipuncture* to obtain blood for analysis. In the upper appendages, the major superficial veins are the CEPHALIC, BASILIC, and ANTECUBITAL veins; in the lower appendages, the SAPHENOUS system is the major superficial set. The SUPERIOR and INFERIOR VENAE CAVAE are the vessels carrying blood to the right atrium of the heart.

Vessels in special body regions

Pulmonary vessels (Fig. 23.15)

The PULMONARY TRUNK arises from the right ventricle, and branches to form the RIGHT and LEFT PULMONARY ARTERIES. These then divide to form the arterioles and capillary networks of the lungs. Four PULMONARY VEINS are formed by the vessels leaving the lungs, and return blood to the left atrium. This system operates at a pressure about one-fourth that of the systemic system, and the vessels are fewer and thinner walled.

Portal circulation (Fig. 23.16)

The veins draining the abdominal digestive organs do not empty separately into the inferior vena cava. Instead, they come together to form the hepatic PORTAL VEIN to the liver. The liver thus gets "first choice" of the nutrients absorbed from the intestines. The portal vein empties into the liver sinusoids and its contents are there mixed with the oxygen-rich blood brought to the liver by the hepatic artery. The liver thus has two afferent blood supplies, both of which empty into the sinusoids. The HEPATIC VEINS are formed by the vessels draining the liver sinusoids and carry blood to the inferior vena cava.

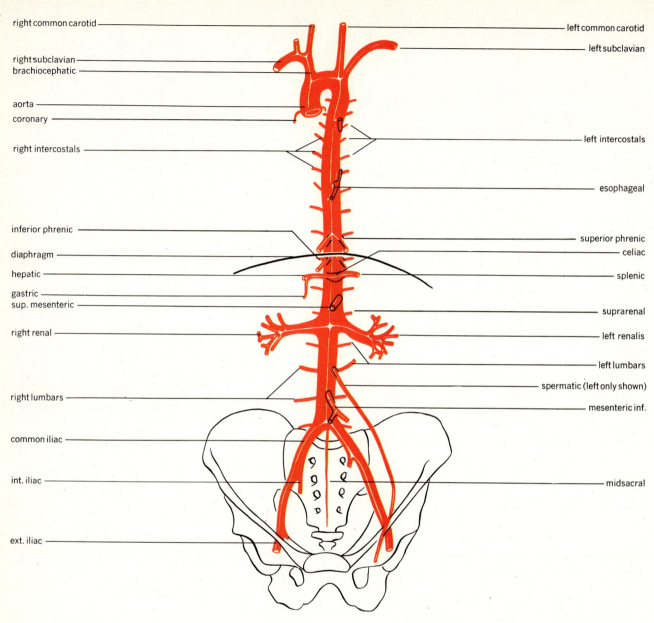

right common carotid

right subclavian
brachiocephatic

aorta
coronary

right intercostals

inferior phrenic

diaphragm

hepatic

gastric
sup. mesenteric

right renal

right lumbars

common iliac

int. iliac

ext. iliac

left common carotid
left subclavian

left intercostals

esophageal

superior phrenic
celiac
splenic

suprarenal
left renalis

left lumbars
spermatic (left only shown)
mesenteric inf.

midsacral

FIGURE 23.7   The major arteries originating
from the aorta.

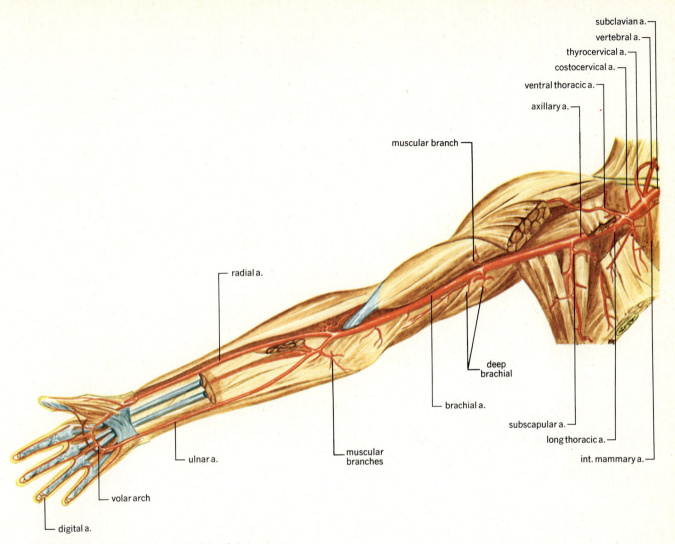

FIGURE 23.8   The major arteries of the upper appendage.

FIGURE 23.9   The major arteries of the lower appendage.

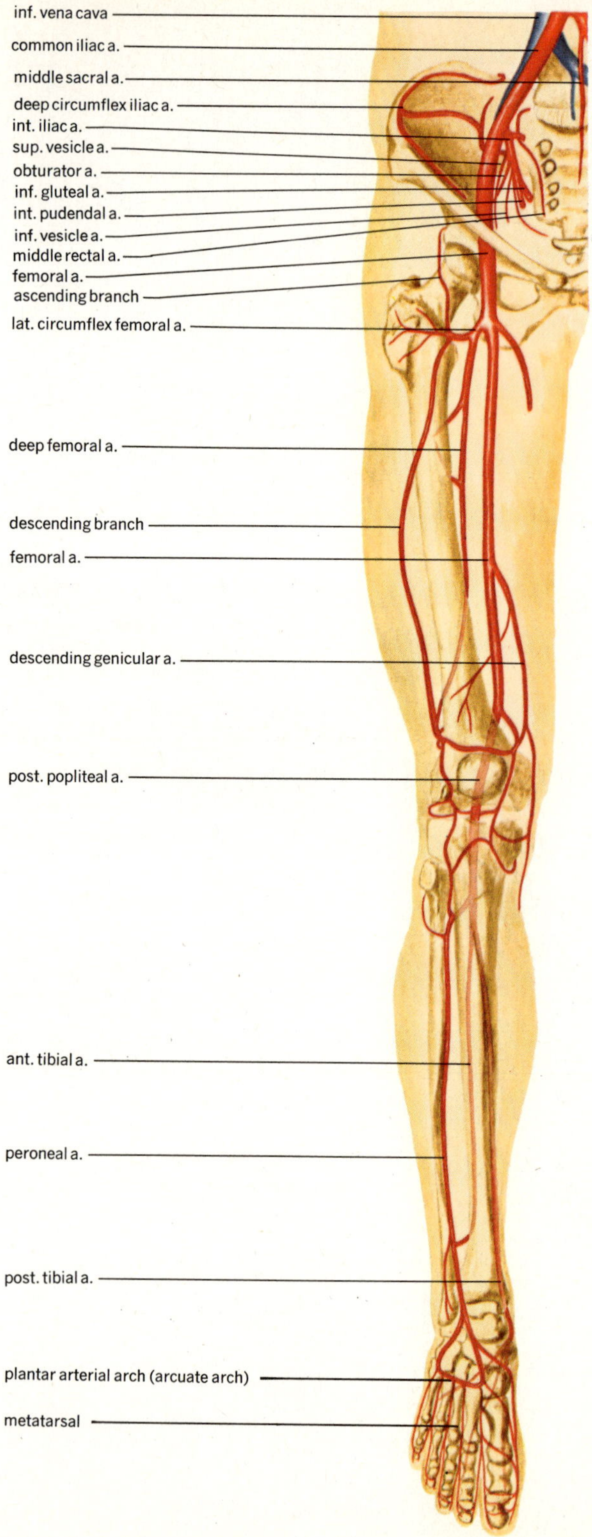

inf. vena cava

common iliac a.

middle sacral a.

deep circumflex iliac a.

int. iliac a.

sup. vesicle a.

obturator a.

inf. gluteal a.

int. pudendal a.

inf. vesicle a.

middle rectal a.

femoral a.

ascending branch

lat. circumflex femoral a.

deep femoral a.

descending branch

femoral a.

descending genicular a.

post. popliteal a.

ant. tibial a.

peroneal a.

post. tibial a.

plantar arterial arch (arcuate arch)

metatarsal

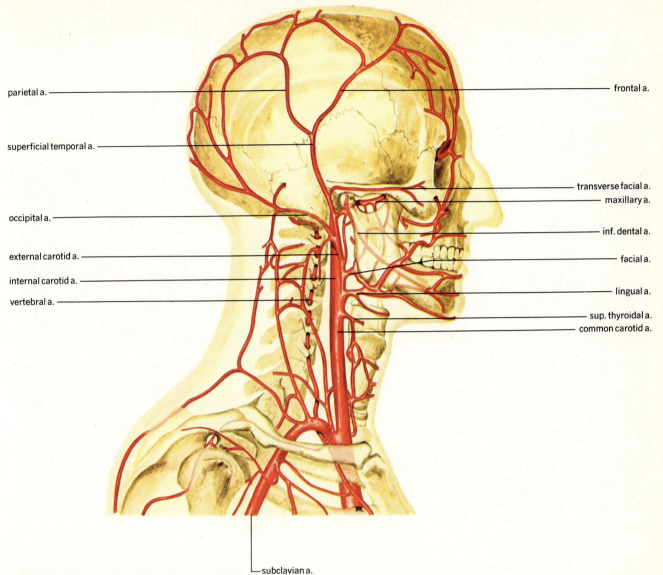

parietal a.

superficial temporal a.

occipital a.

external carotid a.

internal carotid a.

vertebral a.

frontal a.

transverse facial a.

maxillary a.

inf. dental a.

facial a.

lingual a.

sup. thyroidal a.

common carotid a.

subclavian a.

FIGURE 23.10   The major arteries of the
head and neck.

TABLE 23.3 Branches of the aorta and area(s) supplied (Fig. 23.7)

Arising from	Branch	Area supplied by branch
Ascending aorta	Coronaries	Heart
Arch of aorta	Brachiocephalic; gives rise to Right common carotid Right subclavian	Right side head and neck Right arm
	Left common carotid	Left side head and neck
	Left subclavian	Left arm
Descending aorta: Thoracic portion	Intercostals	Intercostal muscles, chest muscles, pleurae
	Superior phrenics	Posterior and superior surfaces of the diaphragm
	Bronchials	Bronchi of lungs
	Esophageals	Esophagus
	Inferior phrenics	Inferior surface of diaphragm
Abdominal portion	Celiac; gives rise to: Hepatic Left gastric Splenic	Liver Stomach and esophagus Spleen, part of pancreas and stomach
	Superior mesenteric	Small intestine, cecum, ascending and part of transverse colon
	Suprarenals	Adrenal glands
	Renals	Kidneys
	Spermatics (male) or ovarians (female)	Testes Ovaries
	Inferior mesenteric	Part of transverse colon, descending and sigmoid colon, most of rectum
	Common iliacs, which give rise to: External iliacs Internal iliacs Midsacral	Terminal branches of aorta Lower limbs Uterus, prostate gland, buttock muscles Coccyx

TABLE 23.4 Arteries of the upper appendage and area(s) supplied (Fig. 23.8)

Artery	Major side branches	Area(s) supplied by branches
Subclavian (beneath clavicle)	Vertebral	Brain and spinal cord
	Internal thoracic (mammary)	Mammary glands, diaphragm, pericardium
	Thyrocervical Costocervical	Muscles, organs, and skin of neck and upper chest
Axillary (armpit)	Long thoracic Ventral thoracic Subscapular	Muscles and skin of shoulder, chest and scapula

(TABLE 23.4 continued)

Brachial (upper arm)	Muscular	Biceps muscle
	Deep brachial	Triceps muscle. Both also supply skin and other tissues of upper arm
Radial, Ulnar (forearm)	Muscular Muscular	Muscles and skin of the forearm
Palmar (volar) arch	Metacarpals	Muscles and skin of the hand
Digitals	—	Muscles and skin of fingers

TABLE 23.5   Arteries of the lower appendage (Fig. 23.9)

Artery	Major side branches	Area(s) supplied by branches
Femoral (thigh)	Deep femoral, pudendals	Muscles and skin of thigh, lower abdomen and external genitalia
Popliteal (knee)	Cutaneous, muscular, genicular	Skin of back of leg, hamstrings, and structures of knee joint
Anterior tibial (anterior side of leg)	Muscular	Anterior crural muscles and skin of leg
Posterior tibial (posterior side of leg)	Muscular, fibular, cutaneous	Posterior crural muscles and skin of leg
Peroneal (lateral side of leg)	Muscular, cutaneous	Lateral crural muscles and skin of leg
Plantar and dorsis pedis Arcuate	Form arcuate arch Metatarsal	Muscles and skin of foot
Digitals		Muscles and skin of toes

TABLE 23.6   Arteries of the head and neck (Fig. 23.10; see Fig. 17.1)

Artery	Major side branches	Area(s) supplied by branches
Common carotid	Internal carotid	Brain
	External carotid: (superior thyroidal, lingual, facial, occipital, superficial temporal, maxillary)	Muscles and skin of cranium, face, teeth, and eyes
Basilar (formed from vertebrals)	Cerebellar arteries	Cerebellum
	Pontine	Pons and brain stem
	Auditory	Inner ear structures

Circle of Willis—not a vessel but a circular pathway on the base of the brain. It is composed of the anterior cerebral arteries (from internal carotid), the anterior communicating arteries, the internal carotid arteries, and the posterior communicating arteries.

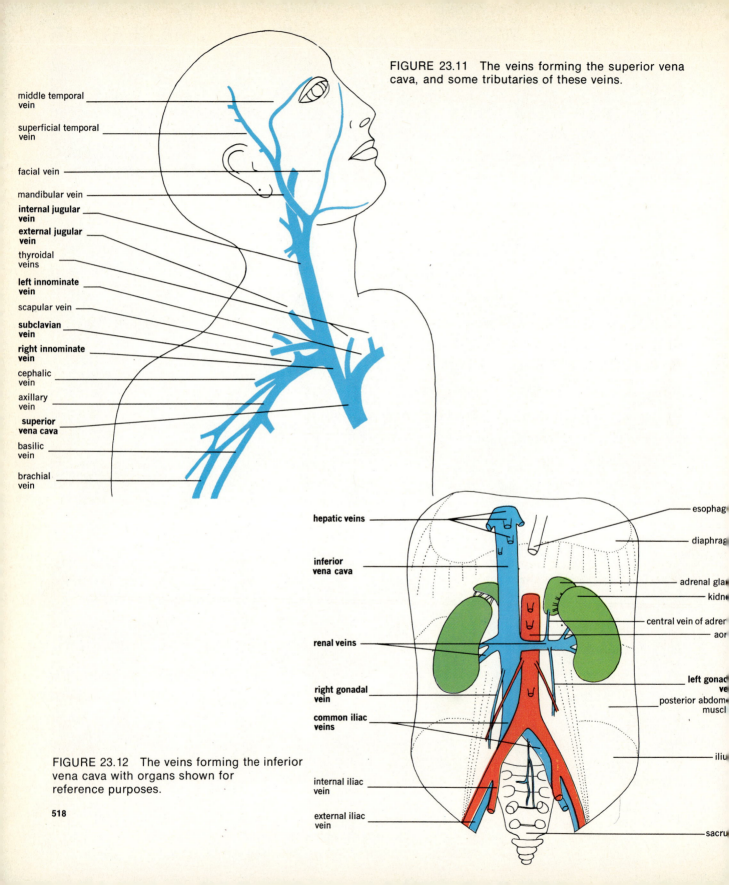

middle temporal
vein

superficial temporal
vein

facial vein

mandibular vein

**internal jugular
vein**

**external jugular
vein**

thyroidal
veins

**left innominate
vein**

scapular vein

**subclavian
vein**

**right innominate
vein**

cephalic
vein

axillary
vein

**superior
vena cava**

basilic
vein

brachial
vein

FIGURE 23.11   The veins forming the superior vena
cava, and some tributaries of these veins.

hepatic veins

**inferior
vena cava**

renal veins

**right gonadal
vein**

**common iliac
veins**

internal iliac
vein

external iliac
vein

esophag

diaphrag

adrenal gla

kidne

central vein of adrer

aor

**left gonad
ve**

posterior abdom
muscl

iliu

sacru

FIGURE 23.12   The veins forming the inferior
vena cava with organs shown for
reference purposes.

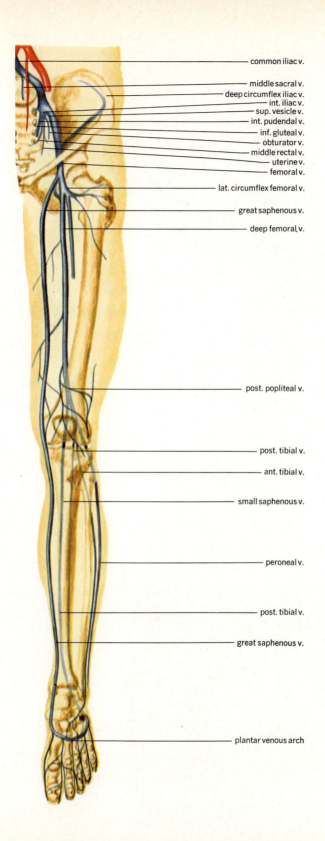

common iliac v.

middle sacral v.
deep circumflex iliac v.
int. iliac v.
sup. vesicle v.
int. pudendal v.
inf. gluteal v.
obturator v.
middle rectal v.
uterine v.
femoral v.

lat. circumflex femoral v.

great saphenous v.

deep femoral, v.

post. popliteal v.

post. tibial v.

ant. tibial v.

small saphenous v.

peroneal v.

post. tibial v.

great saphenous v.

plantar venous arch

FIGURE 23.13   The major veins of the lower limb.

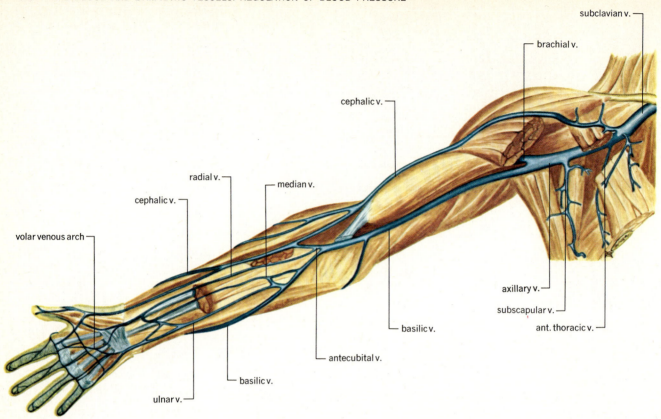

FIGURE 23.14   The major veins of the upper limb.

TABLE 23.7 Veins forming the superior vena cava (Fig. 23.11)		
Vein	Formed from	Area(s) drained
Internal jugular	Dural sinuses	Inside of skull and brain
External jugular	Veins of face	Muscles and skin of scalp and face
Subclavian	Axillary, caphalic*, basilic*, and their tributaries, scapular, and thoracic veins	Upper appendage, chest, mammary glands
Innominate (brachiocephalic)	Internal jugular, external jugular, and subclavian	
* These are the superficial veins of the extremities.		

TABLE 23.8   Veins forming the inferior vena cava (Fig. 23.12)

Vein	Formed from	Area(s) drained
Hepatics	Sinusoids of liver	Liver
Renals	Veins of kidney	Kidney
Gonadals	Veins of gonads	Gonads; testes, and ovaries
Common iliac	External iliac	Lower appendage
	Internal iliac	Organs of lower abdomen
	Saphenous*	Superficial structures of lower appendages

* These are the superficial veins of the extremities.

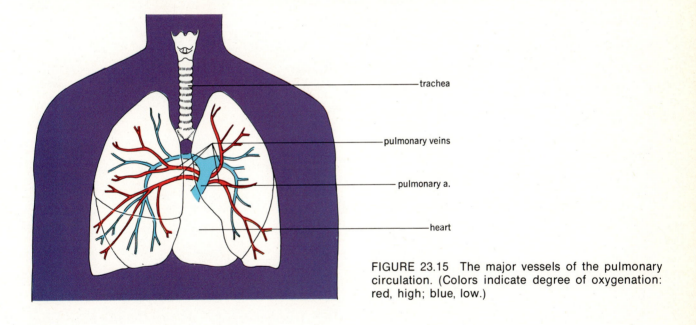

FIGURE 23.15 The major vessels of the pulmonary circulation. (Colors indicate degree of oxygenation: red, high; blue, low.)

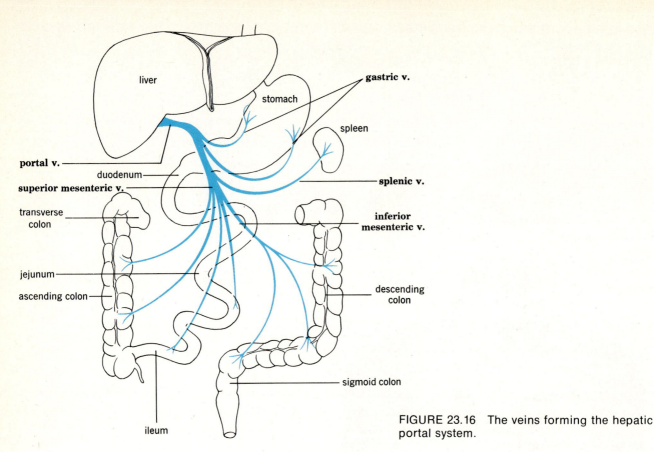

FIGURE 23.16   The veins forming the hepatic portal system.

The role of blood vessels in control of blood pressure and blood flow

Principles governing pressure and flow in tubes

The laws of hydrodynamics that govern flow of fluids through rigid tubes may be adapted to the flow of blood through blood vessels. Although the analogy is not perfect—because blood vessels are *not* rigid tubes and may actively change size, and because blood contains cells that change its behavior from that of a homogeneous fluid—the principles give a basis for understanding pressure and flow characteristics in the various vessels.

It may be recalled that *cardiac output*, by injecting a volume of blood into a system of tubes having a given volume, creates a basic pressure and flow in the vessels. Pressure and flow will rise with increased output or if the volume of the tubes decreases. *Peripheral resistance* refers to the ease or difficulty blood encounters in flowing through the small arteries of the vascular system. If these arteries constrict (become smaller), resistance is increased, and pressure rises proximal to the constricted area. A diversion of blood flow may also occur if arteries in one body area constrict, while those of another area simultaneously dilate (become larger). It should thus become clear that the muscular blood vessels can alter the basic pressure and flow patterns created by heart action.

The principles governing flow and pressure in tubes may be stated as follows.

Resistance to flow is directly proportional to the length of the tubes and inversely proportional to the total cross-sectional area presented by the tubes. Thus, a longer vessel presents a greater frictional surface, as does a greater cross-sectional area, and consequently a pressure drop will occur as blood flows through such vessels.

Flow is directly proportional to the fourth power of the radius of the tube. Thus, if the radius is doubled, flow (volume) of blood increases by 16 times (2^4).

Pressure is directly proportional to cardiac output, and inversely proportional to total cross-sectional area of the vessels. Pressure is thus highest close to the heart, and rises if output increases, and will decrease as more smaller vessels are traversed.

Velocity (speed) of flow is directly proportional to pressure, and inversely proportional to total cross-sectional area of the vessels. Flow is thus rapid in arteries, where pressure is high and surface less, is slowest where surface area is greatest (in the capillaries), and increases in veins, where pressure is low *but* surface area is less.

To aid the understanding of how these factors are interrelated, study Figure 23.17 and Table 23.9. Note especially how total cross-sectional area changes in passing from arteries to capillaries to veins. This is the primary factor in determining pressure and velocity of flow. Note also that *volume* of flow is equal in the several parts of the vascular system; otherwise blood would "pile up" somewhere in the body. An interesting clinical fact revolves around the statement that one pound of fat in the body adds one mile of capillaries to the circulation. The heart must work harder (increase its output) to force blood through the resistance imposed by these added vessels, and thus blood pressure is often higher in obese individuals, as is the chance for "heart attack."

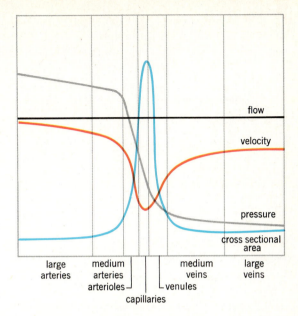

FIGURE 23.17   Relationships of cross-sectional area, blood pressure, velocity, and blood flow in the several areas of the vasculature.

Pressure and flow in arteries, capillaries, and veins

ARTERIES. As the previously stated principles suggest, pressure is highest in the aorta. SYSTOLIC PRESSURE (highest pressure generated by ventricular contraction) is about 125 mm Hg (range at rest of 100–150 mm Hg); DIASTOLIC PRESSURE (lowest pressure determined by ventricular relaxation and peripheral resistance) is about 75 mm Hg (range 70–90 mm Hg). Flow is greatest (5–6 L/min) and velocity is high (100–140 cm/sec), primarily because of the high pressure and low cross-sectional area (2.5 cm²). At the end of the aorta in the lower abdomen, the mean blood pressure, represented by

$$\frac{(\text{systolic pressure} + 2 \text{ diastolic pressure})}{3}$$

is about 90 mm Hg, compared with 100 mm Hg in the upper aorta. This is because about 25 cm of vessel has been traversed. Blood passes next into

TABLE 23.9 Surface area, pressure, flow, and velocity in various regions of the vascular system

Region	Estimated surface area (sq cm)	Average pressure High Low (mm Hg)		Flow (L/min)	Velocity of flow (cm/sec)
Large arteries (e.g., aorta)	2.5	125	75	5–6	100
Muscular arteries (medium and small arteries)	60	75	40	5–6	4–5
Capillaries	2500	40	10	5–6	0.1–0.15
Veins	325	10	0	5–6	0.8–1.1

the muscular arteries that have about 60 cm² of cross-sectional area. Mean pressure drops to about 40 mm Hg, and velocity has decreased.

CAPILLARIES. Blood next enters the microcirculatory unit described previously, where exchange of materials will occur. At the arteriolar ends of the capillaries, pressure is 32 to 40 mm Hg, while at the venous end of the capillaries, pressure is about 16 mm Hg (Fig. 23.18). A higher "entrance pressure" ensures filtration of materials *from* the capillaries; a lower "exit pressure" allows osmotic return of filtered water and transported solutes (bulk flow) since plasma proteins that are too large to filter create a colloidal osmotic pressure to "draw" fluid back into the capillary. Velocity of flow is slowest in the capillaries (0.3 cm/sec) because of their huge total cross-sectional area (estimated at 2500 cm²), and this allows *time* for exchanges to occur.

VEINS. As venules join to form the fewer and larger vessels leaving capillary beds, total cross-sectional area falls to about 325 cm² in all the body veins. Pressure was lost in traversing arterioles and capillaries, so it continues to fall in the veins as blood is returned to the heart (10 mm Hg in venules, 0 mm Hg at right atrium). However, velocity increases as cross-sectional area falls, to a value of about 60 cm/sec. Although pressure is low in the veins, flow is aided by the following factors (Fig. 23.19).

The MASSAGING ACTION OF SKELETAL MUSCLES (the "venous pump"). As muscles contract, they become thicker, compressing veins and forcing blood toward the heart (backflow is prevented by valves).

BREATHING MOVEMENTS (the "respiratory pump"). As inspiration occurs, the venae cavae are expanded, drawing blood into the vessels. Expiration compresses the vessels and forces blood toward the heart.

Control of blood vessel size

In previous sections, it was indicated that muscular blood vessels can actively change their size and that this will affect blood flow and pressure. Change in size is governed mainly by nervous and nonnervous (chiefly chemical) factors.

NERVOUS CONTROL. Two types of nerve fibers may be traced to muscular blood vessels. Collectively, they are called VASOMOTOR NERVES.

VASOCONSTRICTOR NERVES originate from or are controlled by the *vasoconstrictor centers* in the medulla of the brainstem. They cause constriction of muscular arteries. These nerves are part of the sympathetic nervous system.

VASODILATOR NERVES originate from or are controlled by the *vasodilator centers* of the medulla, and usually are carried in the fibers of the parasympathetic nervous system. They cause constriction of muscular blood vessels.

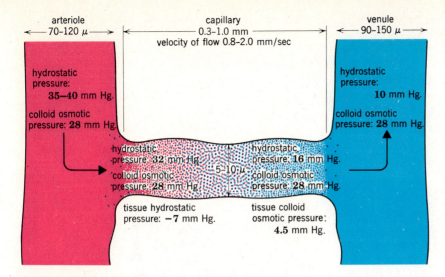

FIGURE 23.18 Capillary dynamics. Pressures favor filtration of fluids from arteriolar end of capillary and osmotic return at the venous end. Velocity of flow is slow because of the large number of capillaries and their large surface area.

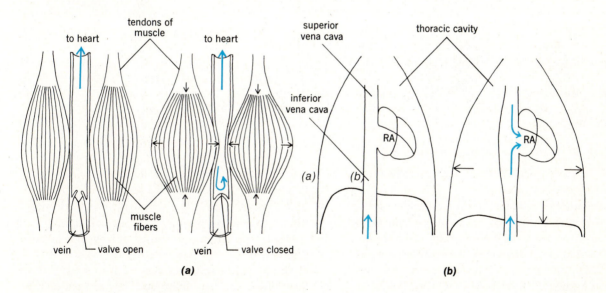

(a)

(b)

FIGURE 23.19 The skeletal and respiratory pumps. *(a)* The skeletal muscle pump. Relaxed muscles *(left)* allows blood to flow through open valve. Contracted muscles *(right)* become shorter and thicker, compressing the vein. Blood attempts to go away from the heart, closing the valve. Blood is thus massaged to-ward the heart. *(b)* The respiratory pump. Inspiration increases thoracic volume, decreases intrathoracic pressure, and expands the thin-walled venae cavae. Blood is drawn into the larger vessels and speeds the flow through the venous system, increasing blood flow to the right atrium (RA).

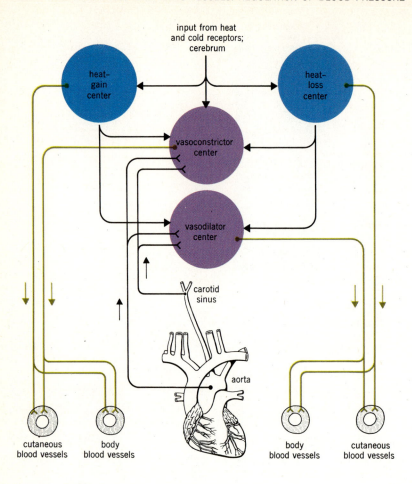

FIGURE 23.20   Vasomotor reflex pathways.

The vasomotor centers in the medulla are, in turn, controlled by impulses reaching them from receptors in many areas of the body. Thus, reflex control over blood vessel size is established. Several of these vasomotor reflex pathways are depicted in Figure 23.20 and are described below.

BARORECEPTORS (sensitive to pressure rise) are found in the *aortic arch, subclavian arteries, carotid arteries,* and *lungs.* In general, stimulation of these receptors inhibits vasoconstrictor activity and stimulates vasodilator activity. Thus, blood pressure falls. This is particularly important to the brain encased as it is in a rigid "box"; too much blood in the brain may lead to cell damage by compression, and possibility of vascular rupture ("stroke").

The hypothalamus and cerebrum may influence vasomotor activity. Increased loss of heat as a result of cutaneous vasodilation is important in regulating body temperature; the "blush" or "blanch" associated with emotional change, fear, or anger reflects vasodilation and constriction respectively.

NONNERVOUS CONTROL. Chemical, thermal, and mechanical factors may influence blood vessel size. EPINEPHRINE, a hormone of the adrenal medulla, dilates vessels in the heart, skeletal muscles, and lungs, while constricting vessels elsewhere. Because more vessels constrict than dilate, the net effect of epinephrine is to raise blood pressure. CARBON DIOXIDE causes local vasodilation, inhibiting contraction of smooth muscle. An active tissue, burning more fuel and liberating more CO_2, ensures itself a greater blood flow. Direct application of either heat or cold to denervated skin causes vasodilation. Vessel damage and release of a chemical called *histamine* accounts for this effect.

These various types of control devices ensure that blood flow is being maintained to active tissue; they also protect the vessels from damaging effects of excessively high pressure.

The term SHOCK refers to the state of circulatory collapse that may follow depression of cardiovascular function. It is associated with the development of certain characteristic signs, including:

Pallor
Cold extremities
Decreased body temperature
Depression of central nervous system activity
Rapid, weak pulse
Low blood pressure
Reduced cardiac output

The most threatening developments are the low cardiac output and low blood pressure that may result in inadequate renal and cerebral circulation. The body can to some degree compensate for these effects, as considered below.

Classification of shock

Shock is usually classified on the basis of cause.

HEMORRHAGIC SHOCK results from loss of whole blood or plasma. The loss reduces circulating blood volume and for this reason is sometimes called *hypovolemic shock* (*hypo-*, less; *volemic*, volume of blood).

CARDIOGENIC SHOCK results from cardiac failure, with resulting low cardiac output and low blood pressure. The cardiac muscle may be weak or fails to be stimulated normally.

VASCULAR SHOCK results from vasodilation because of chemicals or nervous influences.

Compensation of shock

The body automatically adjusts — to the best of its ability — to offset the effects of shock. Some of the adjustments made include:

Restoration of lost fluid volume by movement of fluid from tissues to blood vessels.

Increased secretion of hormones such as antidiuretic hormone (ADH) that increases fluid reabsorption from kidney tubules, and aldosterone that causes sodium chloride retention by the kidney, along with water to keep the solution isotonic.

Vasoconstriction that reduces the capacity of the vessels, and adjusts capacity to the remaining fluid volume.

Release of epinephrine (adrenalin) from the adrenal medulla. The hormone increases both stroke rate and volume and causes vasoconstriction. The net result will be a rise of blood pressure.

The compensatory reactions are efficient enough to compensate for a blood loss of 10 to 15 percent of the blood volume. Above 15 percent loss, compensation becomes increasingly difficult, and with a loss greater than 25 to 30 percent, death will ensue unless the volume is replaced with some type of transfusion fluid.

Hypertension

Systolic blood pressures of 140 to 150 mm Hg and diastolic blood pressures of 90 to 100 mm Hg are generally regarded as the upper limits of normal. Sustained elevations of one, the other, or both pressures constitutes HYPERTENSION or *high blood pressure*. Hypertension puts a strain on the heart to circulate the blood against high pressures in the system and can cause rupture of weak vessels. Sustained elevation of the diastolic pressure appears to be particularly dangerous, for to combat a sustained pressure, the artery walls may become thickened and hardened (sclerosis). This in turn may cause narrowing of the size of the involved vessels, diminution of blood flow, and tissue death. A common denominator of hypertension appears to be generalized peripheral vasoconstriction. Several hypotheses have been advanced to explain this vasoconstriction.

Reduction of blood flow or oxygen levels to the kidney causes production by specialized cells in the organ of a substance called RENIN. Renin acts as an enzyme to convert a substance in the plasma to a powerful vasoconstrictor agent called ANGIOTENSIN. Removal of an offending kidney—one that

has perhaps developed a poor blood supply—may ameliorate the condition.

Excessive sympathetic nervous system discharge, causing generalized peripheral vasoconstriction, will raise blood pressure. Anxiety, stress, and psychological disorders are often associated with hypertension. Drugs that block sympathetic effects (sympatholytic drugs) may successfully treat the condition. A *sympathectomy* is a surgical procedure that removes sympathetic ganglia from T10 to L2. This procedure deprives visceral blood vessels of vasoconstrictor impulses and lowers blood pressure, but blood may now "pool" in the dilated vessels.

Excessive intake of salt (NaCl) requires retention of sufficient water to keep the solution isotonic to cells. This excess water increases blood volume and raises blood pressure. Often, restriction of salt in the diet will significantly reduce blood pressure, as the kidney eliminates the water that is no longer required to "balance" the salt.

Table 23.10 presents additional information on hypertension.

TABLE 23.10	Some types, causes, and effects of hypertension		
Type	Definition	Cause(s)	Effect(s)
Benign (essential)	Hypertension of slow onset	No apparent cause	Usually without symptoms
Malignant	Severe hypertension with degenerative changes in vessel wall	The common denominator in all these types is an increase in peripheral resistance due to loss of vessel elasticity and/or narrowing of size. Constriction may be due to vascular spasm, or emotional disturbance, or release of vasoconstrictive chemicals. Retention of Na+ and water also increases pressures.	Reduced blood flow and tissue death
Systolic	Elevation of systolic pressure		Vascular congestion and danger of rupture. High cardiac output; vessels often sclerose
Diastolic	Elevation of diastolic pressure		Reduce blood flow and tissue death
Renal	Hypertension from kidney disease or ischemia		Results in Na+ and water retention. Ischemia causes renin release[a]

[a] An ischemic kidney releases the enzyme renin. Renin converts plasma materials ultimately to angiotensin II, the most powerful vasoconstrictive agent known.

Hypotension

HYPOTENSION refers to chronically *low* blood pressure, usually of a degree that normal perfusion of organs cannot be maintained. Fainting may result if the brain fails to receive enough flow, or retention of toxins occurs if the kidney does not have sufficient filtration pressure to operate properly. Causes are not definitely known, but may be associated with damage to the vasoconstrictor centers from mechanical or chemical (drug) insults, or by a blood loss that cannot be compensated. Treatment is aimed at restoring sympathetic vasoconstrictor activity, increasing strength of heart contraction, or restoring blood volume.

Blood pressures toward the lower limits of the normal range are to be desired if there is no lack of perfusion of body organs. A lower blood pressure means less work for the heart and less strain on blood vessels.

Lymphatic vessels

Plan, structure, and functions

The lymphatic vessels of the body (Fig. 23.21) carry lymph or tissue fluid from the body tissues and organs to the blood vascular system. Filtered water, protein, and white blood cells are therefore returned or added to the bloodstream, maintaining water and protein homeostasis.

The LYMPH CAPILLARIES are tiny vessels that are found in all body areas except the brain, spinal cord, and eyes. They have the same basic structure as blood capillaries but are larger and irregular in size. Dense networks of capillaries form *lymphatic plexuses* in most organs and collect tissue fluid from the interstitial fluid compartment. Lymph capillaries are much more permeable to large solutes than blood capillaries, and thus filtered proteins, or large molecules such as enzymes that are produced by cells, easily enter the small lymph vessels. Capillaries form larger lymphatics known as COLLECTING VESSELS along which the lymph nodes are placed. The collecting vessels resemble small or medium-sized veins in structure, having three thin tunics or coats named as in blood vessels. Valves are numerous in the collecting vessels and give a "beaded" appearance to the vessel as they control movement only toward the neck.

The LYMPH DUCTS are formed from the collecting vessels, and resemble large thin-walled veins in structure. They too have three tunics and valves. The *thoracic duct* collects lymph from all but the upper chest, right arm, the right side of the neck and head, and empties into the left subclavian vein at its junction with the left internal jugular vein. Its lower end is called the *cisterna chyli*, although its does not form an enlarged bulb as it does in other animal species. The *right lymphatic duct* collects lymph from the areas listed above and empties into the right innominate (brachiocephalic) vein.

Flow through the vessels

The contraction and relaxation of skeletal muscles, breathing, and the expansion of lymph vessels by radially arranged elastic fibers around them contribute to the entry and movement of lymph in the vessels. Flow averages about 1.5 milliliters per minute from the cut end of a medium-sized lymph vessel.

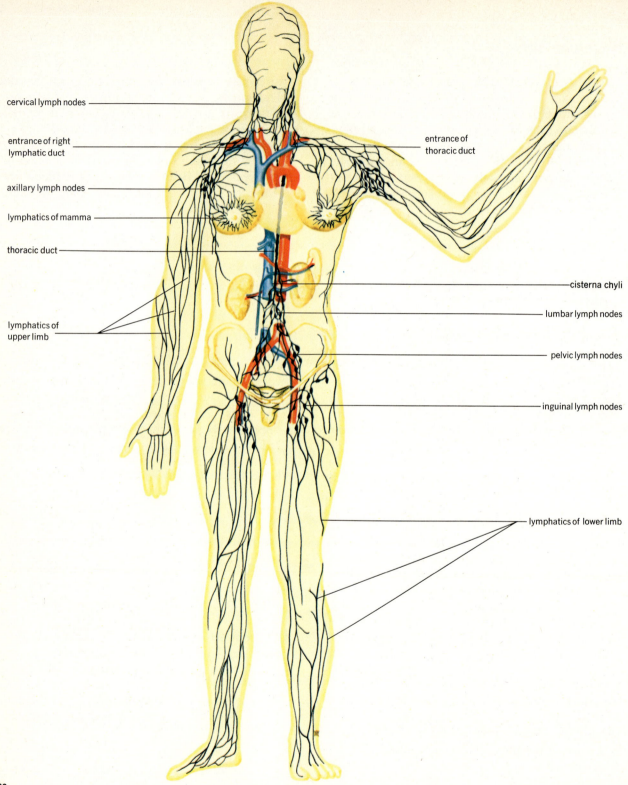

cervical lymph nodes

entrance of right
lymphatic duct

axillary lymph nodes

lymphatics of mamma

thoracic duct

lymphatics of
upper limb

entrance of
thoracic duct

cisterna chyli

lumbar lymph nodes

pelvic lymph nodes

inguinal lymph nodes

lymphatics of lower limb

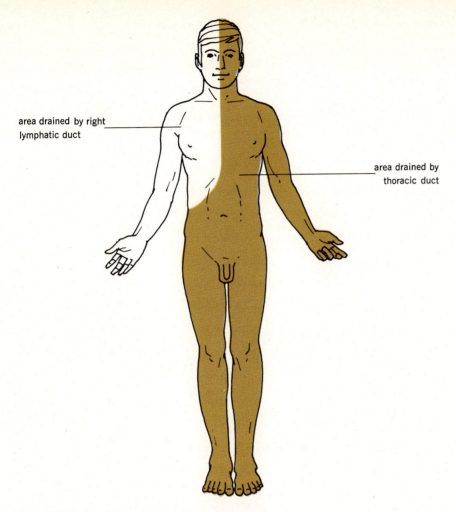

area drained by right
lymphatic duct

area drained by
thoracic duct

FIGURE 23.21   The major lymph vessels of the body
(left) and their areas of drainage *(right)*.

Clinical considerations

Since the lymphatics return about one-tenth of the filtered plasma water to the blood vessels, blockage of a major lymph vessel is associated with the development of EDEMA in the area served by the vessel.

ELEPHANTIASIS is the result of blockage of lymph vessels by a parasite, and used to be common in Africa. Blockage of lymphatics by cancer cells that have metastasized and been trapped may cause mild edema in the body part served by the plugged vessel.

Summary

1. Blood and lymph vessels develop as cavities in masses of mesoderm.

   a. Blood vessels are formed at about 2 weeks of development and by 3 weeks form a circulation in the embryo and placenta.

   b. Lymph vessels develop at about 5 weeks from several areas known as lymph sacs, invade developing tissues and organs, and ultimately connect to the subclavian veins in the upper thorax.

2. There are three major types of blood vessels in the body.

   a. Arteries carry blood away from the heart and have elastic tissue and/or smooth muscle in their walls.

   b. Capillaries are one cell thick, are found in the body's tissues and organs, and serve as areas of exchange.

   c. Veins carry blood toward the heart, and have mainly connective tissue in their walls.

   d. The microcirculation is a unit composed of an arteriole, its associated capillaries, and the venules draining the capillaries.

3. Blood vessels—except capillaries—have a layered wall structure.

   a. Arteries have three layers in their walls: intima, media, and adventitia. The media is thickest and contains elastic tissue and/or smooth muscle. The muscle enables active size change to occur.

   b. Arteriosclerosis and atherosclerosis involve thickening of inner and middle layers, loss of elasticity, and reduced blood flow.

   c. Veins usually have three layers in their walls: intima, media, and adventitia. The adventitia is thickest.

   d. Varicosities develop when veins are stretched.

   e. Capillaries have only an endothelium.

4. The major named systemic arteries and veins of the body are named and pictured.

5. Some special features of the pulmonary and portal circulation are presented.

   a. Pulmonary vessels are shorter, wider, and require less pressure to perfuse them.

   b. The portal circulation involves vessels from spleen, stomach, and intestines fusing to form the portal vein to the liver.

6. Blood vessels control pressure and flow of blood in the body.

   a. The principles governing flow and pressure in tubes are presented. In general, total cross-sectional area of the vessels determines pressure and velocity, assuming a basic pressure created by heart action, and distribution of blood depends on size of vessels in different body areas.

   b. In arteries, pressure, velocity, and flow are greatest; in capillaries pressure and velocity are low; in veins, pressure is lowest, but velocity increases.

   c. Flow of blood through veins is aided by skeletal muscle contraction and breathing.

7. Diameter of muscular blood vessels is under nervous and chemical control.

   a. Vasoconstrictor and vasodilator nerves pass to most muscular blood vessels.

   b. The nerves are derived from or are controlled by medullary vasoconstrictor and vasodilator centers.

   c. The centers are controlled by reflexes originating in peripheral pressure receptors and areas in hypothalamus and cerebrum.

   d. Chemicals produced by active tissues (CO_2), hormones (epinephrine), and products of injury (histamine) influence vessel size.

8. Shock refers to circulatory collapse, with low cardiac output and low blood pressure.

   a. Shock is often classified according to cause; for example, hemorrhagic, cardiogenic, vascular.

   b. The body can compensate for shock that is not too severe by fluid shifts, hormone secretion, vasoconstriction, and hormone release. These devices restore blood volume and raise blood pressure.

9. Hypertension refers to high blood pressure.

   a. It increases the work of the heart and is potentially damaging to blood vessels.

   b. It may result from low blood flow to the kidney that ultimately causes formation of angiotensin, to excessive nervous vasoconstriction, or to salt retention.

   c. Treatment, by drugs or surgery, is aimed at causing vasodilation, or eliminating salt and water from the body.

10. Hypotension refers to excessively low blood pressure.

   a. It may result in blood flow, to vital organs, too low to maintain normal function.

   c. Treatment is aimed at stimulating vasoconstriction and heart action.

11. Lymphatic vessels compare to the venous portion of the blood vascular system. There are lymph capillaries, collecting vessels, and two major ducts that connect to the subclavian veins.

   a. The vessels return tissue fluid and protein to the blood veins.

   b. Edema may develop if a vessel is blocked.

Questions

1. Construct a table comparing velocity, pressure, and surface area in the three main subdivisions of the vascular system (arteries, capillaries, veins).

2. What are the physical principles related to the changes described in question one?

3. Compare and contrast the factors causing blood flow through arteries, veins, and lymphatics.

4. What are the common denominators in the development of shock, regardless of cause?

5. What are some of the causes of hypertension? What are the common denominators of the disorder?

6. What is the relationship between pressure within a blood vessel and the type and thickness of tissues in its wall?

7. Compare arterial and venous systems in terms of total surface area, number of pathways to and from the tissues or organs, and pressure within the systems.

8. How are arteriolar diameter and cardiac output integrated to achieve alterations in blood pressure?

9. It has been said that the whole circulatory system exists to serve the capillaries. Discuss this statement.

10. Trace a drop of lymph from the left leg to the blood vascular system.

Readings

Baez, Silvio. "Microcirculation." *Ann. Rev. Physio. 39*:391, 1977.

Gavras, Haralambos, et al. "Antihypertensive Effect of the Oral Angiotensin Converting-Enzyme Inhibitor SQ14225 in Man." *New Eng. J. Med. 298*:991, May 4, 1978.

Gutstein, William H., John Harrison, Fritz Parl, George Kiu, and Matt Avitable. "Neural Factors Contribute to Atherogenesis." *Science 199*:449, 27 Jan. 1978.

Iwatsuki, K., G. J. Cardinale, S. Spector, and S. Udenfriend. "Hypertension: Increase of Collagen Biosynthesis in Arteries but Not in Veins." *Science 198:* 403, 28 Oct. 1977.

Marx, Jean L. "Stress: Role in Hypertension Debated." *Science 198*:905, 2 Dec. 1977.

Nicoll, Paul A., and Aubrey E. Taylor. "Lymph Formation and Flow." *Ann. Rev. Physiol.* *39*:73, 1977.

Reisin, Efrain, et al. "Effect of Weight Loss Without Salt Restriction on the Reduction of Blood Pressure." *New Eng. J. Med.* *298*:1, Jan. 5, 1978.

Science News. "Arteriosclerosis: Virus-induced?" *113*:58, Jan. 28, 1978.

Science News. "Lowering Blood Pressure." *111*:347, May 28, 1977.

Chapter 24

The Respiratory System

Objectives

After studying this chapter, the reader should be able to:

■ Outline the early development of the respiratory system.

■ Name the organs of the system, divide them into conducting and respiratory divisions, and describe their structure and functions.

■ Explain the mechanisms responsible for inspiration and expiration of air in the lungs.

■ Show how, and explain why, pressures in the thorax and lungs change during breathing.

■ Explain what surfactant is and explain its role in respiration.

■ Name and give volumes for the several lung volumes and capacities.

■ Give values for gas composition in the atmosphere and the body areas involved in gas supply and removal, and explain how and why oxygen and carbon dioxide take the pathways they do.

■ Explain how, and to what extent, oxygen and carbon dioxide are transported by the bloodstream.

■ List factors influencing oxygen transport.

■ Give some functions of the lungs other than gas exchange.

■ Describe some protective devices present in the respiratory system.

■ Name and give the roles of the respiratory centers in breathing.

■ Discuss the roles of reflexes in control of breathing.

■ Contrast lung and placenta as respiratory organs in terms of rate and degree of gas diffusion.

■ Describe changes in respiration and circulation occurring at birth.

Development of the system

The nose and nasal cavities are formed at 3½ to 4 weeks of embryonic life, as the face develops (Fig. 24.1). An OLFACTORY (*nasal*) PLACODE, or area of thickened ectoderm, appears on the front and lower part of the head. An OLFACTORY PIT develops in the placode, and grows posteriorly to fuse with the front part of the primitive gut (foregut). This latter structure forms the pharynx. The pit extends above the mouth cavity, and is separated from it by a thin membrane. This membrane ruptures and creates a single large nasal-oral cavity. At about 7 to 8 weeks of age, division of the cavity into an upper nasal cavity and a lower oral cavity is begun by plates that grow horizontally across the cavity. At about the same time, a vertical plate grows downward from the roof of the nasal cavity. These plates meet and fuse by 3 months of development. The horizontal plate becomes the HARD PALATE, and the vertical plate the NASAL SEPTUM.

If the horizontal processes fail to meet in the midline, a CLEFT PALATE will result. Cleft palate makes it very difficult for the infant to swallow and create suction to nurse.

The development of the rest of the system (Fig. 24.2) begins as a LARYNGOTRACHEAL GROOVE in the floor of the foregut. The groove deepens, and the walls come together to form a tube. The lower end of the tube forms a lung bud that undergoes growth and repeated branching. The various tubes of the "respiratory tree" and the air sacs are derived from these branchings. In all, some 24 divisions of the original tube occur, to create the vast number and great surface of the respiratory organs.

Until about 26 weeks of age the lungs do not contain alveoli. After this time, there are usually enough alveoli developed to sustain life if the child is born prematurely.

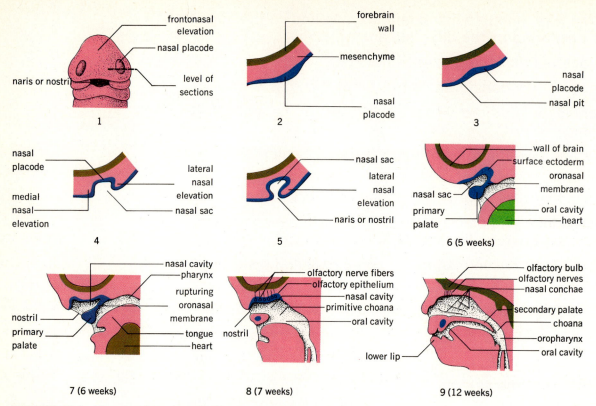

FIGURE 24.1   The development of the nasal region from fourth to twelfth week. Numbers 1, 2, 3, 4, 5 all occur in the *fourth week*.

The organs of the system

The organs of the respiratory system (Fig. 24.3) provide a means of bringing air into contact with a surface through which oxygen may diffuse *to* the bloodstream; it also provides a diffusing surface for carbon dioxide to pass *from* the bloodstream and be eliminated.

Accordingly, the organs of the respiratory system may be conveniently divided into those whose function it is to merely transport air and those that permit gas exchange with the bloodstream. The name CONDUCTING DIVISION is given to the tubes that transport air; the name RESPIRA-TORY DIVISION to the organs that permit gas diffusion.

The conducting division

This division is too thick walled to permit gas exchange with the bloodstream, and is composed of the NASAL CAVITIES and their associated PARANASAL SINUSES, the PHARYNX, LARYNX, TRACHEA, BRONCHI, BRONCHIOLES, and TERMINAL BRONCHIOLES. All parts beyond the bronchi are contained within the lungs.

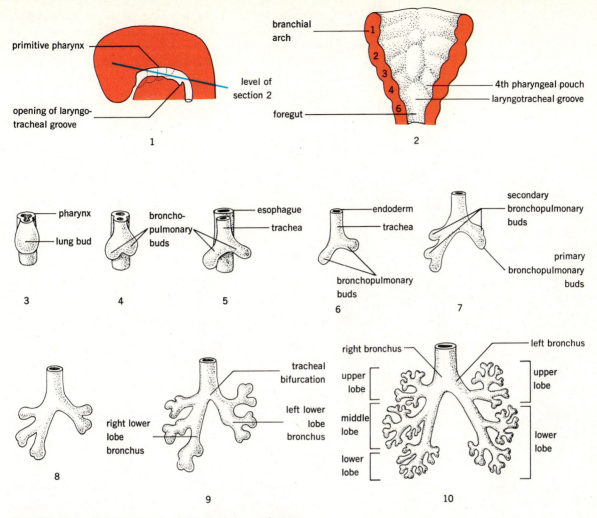

FIGURE 24.2 Development of the lower respiratory system. Numbers 1 and 2, *3½ weeks;* 3, 4, 5, 6, *4 weeks;* 7 and 8, *5 weeks;* 9, *6 weeks;* 10, *8 weeks.*

THE NASAL CAVITIES (Fig. 24.4)

Structure. The anterior openings into the nasal cavities are the NOSTRILS. The posterior openings are the CHOANAE. The floor is formed by the hard palate, and the side walls by the maxillary, inferior concha, ethmoid, and vomer bones. The lateral walls are irregular due to the CONCHAE or TURBINATES, which project into the cavities "like sagging shelves." These "shelves" increase the surface area of the cavities. The epithelium is ciliated, and covered with mucus secreted by mucous cells and glands in the lining tissue. Many large thin-walled veins are found in the connective tissue beneath the epithelium. The paranasal sinuses (frontal, maxillary, ethmoid, sphenoid) open into spaces or MEATUSES beneath the conchae. The tear or nasolacrimal duct from the eye also opens into the nasal cavities. The

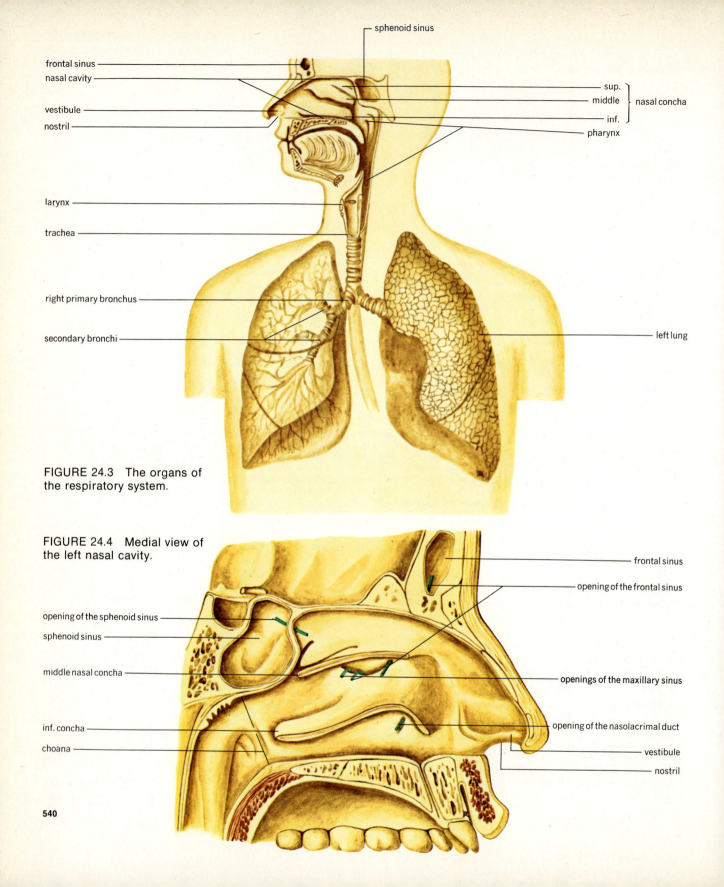

sphenoid sinus

frontal sinus
nasal cavity

vestibule
nostril

sup.
middle } nasal concha
inf.
pharynx

larynx

trachea

right primary bronchus

secondary bronchi

left lung

FIGURE 24.3   The organs of
the respiratory system.

FIGURE 24.4   Medial view of
the left nasal cavity.

frontal sinus

opening of the frontal sinus

opening of the sphenoid sinus

sphenoid sinus

middle nasal concha

openings of the maxillary sinus

inf. concha

opening of the nasolacrimal duct

choana

vestibule

nostril

olfactory epithelium is found in the roof of each nasal cavity.

Function. The nasal cavities warm, moisten, and partially cleanse the incoming air. Heat radiates from the blood vessels, moisture is supplied from the secretions, and the sticky mucus traps dust, pollens, and other solids. Also, immune globulin (IgA) is secreted in the nasal cavities. Ciliary action moves the mucus to the pharynx (or throat) where it is swallowed.

Clinical considerations. The most common condition affecting the nose is, of course, the "common cold." It is a viral infection of the linings of the nose, and produces *rhinitis* (G. *rhin*, nose, + *itis*, inflammation). Rhinitis is associated with swelling of the linings ("plugged nose"), over-secretion by the glands ("runny nose"), fever, and general tiredness. It is hoped that, within the next few years, vaccines may be made against the viruses causing colds, thus giving protection against the organisms.

SINUSITIS refers to the inflammation of the para-nasal sinuses. It may be caused by microorganisms or allergic reactions. Swelling of the membranes that line the openings of the sinuses into the nasal cavities blocks drainage of secretions from the sinuses and fever, headache, dizziness, and tenderness when pressing on the affected sinus may develop. Sinusitis may be acute or chronic, and treatment is directed toward opening the exits from the sinuses to promote drainage and relieve pressure.

EPISTAXIS refers to hemorrhage or bleeding from the nose. Trauma, nose picking, fractures of the skull bones, and foreign bodies in the nose are the most common causes of nasal bleeding. Systemic disorders such as leukemia, low platelet levels, and hemophilia are also associated with epistaxis.

THE PHARYNX

Structure. The pharynx (*see* Fig. 24.3) is a tube common to both the respiratory and digestive systems. It extends from behind the nasal cavities to the level of the larynx ("voice box"). It has three parts.

The *nasal pharynx* lies behind the nasal cavities. The Eustachian tubes from the middle ear cavities open into its lateral walls. Posteriorly, it contains the pharyngeal tonsil or adenoid, a mass of lymphoid tissue. Ciliated and mucus-covered epithelium lines this part.

The *oral pharynx* lies behind the oral cavity. It carries both food and air. It is lined with a tough stratified squamous epithelium that resists the wear and tear of foods hitting its walls.

The *laryngeal pharynx* lies behind the larynx. In its lower end, the tubes for food (esophagus) and air (larynx) separate. It too has a stratified squamous epithelial lining.

Function. The pharynx conducts either air and/or foods.

Clinical considerations. Enlargement of the adenoids from inflammation may interfere with breathing, and may convert an individual to "mouth breathing" with great loss of the warming, moistening, and cleansing functions the nasal cavities serve. Surgical removal of the enlarged organ is usually indicated. Since the body has many other lymphoid organs, loss of the adenoids is not severely missed, although a portion of Waldeyer's ring has been removed.

THE LARYNX (Fig. 24.5)

Structure. The larynx is a cartilagenous box forming the opening into the lower respiratory tract. It is composed of three major and three pairs of minor cartilages. The cartilages are connected by ligaments.

The *thyroid cartilage* is the largest cartilage of the larynx, and forms the "Adam's apple."

The *cricoid cartilage* is "signet-ring" shaped with the widest part of the ring placed posteriorly. It is located beneath the thyroid cartilage.

The *epiglottis* is a leaf-shaped cartilage located above the thyroid cartilage. When one swallows, the epiglottis seals the opening (glottis) into the respiratory system, and prevents entry of food and secretions into the larynx.

The *arytenoid, cuneiform,* and *corniculate carti-*

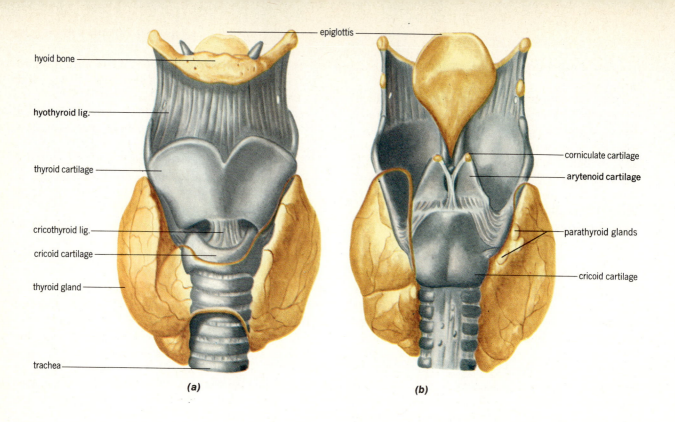

epiglottis

hyoid bone

hyothyroid lig.

thyroid cartilage

cricothyroid lig.

cricoid cartilage

thyroid gland

trachea

(a)

corniculate cartilage

arytenoid cartilage

parathyroid glands

cricoid cartilage

(b)

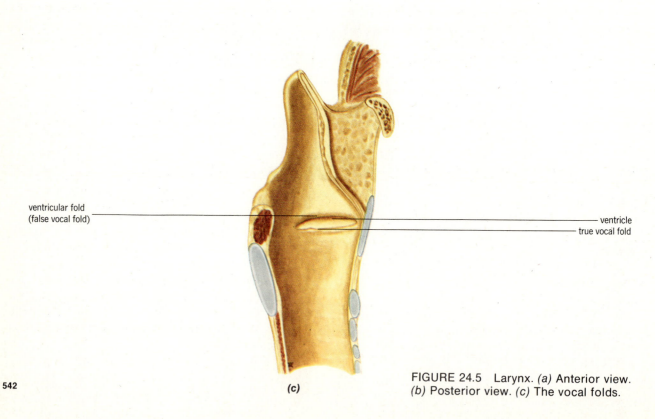

ventricular fold
(false vocal fold)

ventricle

true vocal fold

(c)

FIGURE 24.5   Larynx. *(a)* Anterior view.
(b) Posterior view. *(c)* The vocal folds.

lages are the paired cartilages that support the vocal cords.

Function. Besides conducting air, the larynx creates the sounds that tongue and mouth shape into speech. The vocal cords—fibrous bands suspended across the larynx—are caused to vibrate by air forced across them. According to the tension on and length of the cords, the pitch of the voice is changed. Tension on the cords is controlled by skeletal muscles, which attach to the cartilages supporting the cords.

During swallowing, the larynx rises, due to muscular action, the epiglottis is depressed, and the opening into the lower respiratory tract is sealed. Food and liquids are thus not aspirated nor drawn into the respiratory system.

Clinical considerations. Laryngitis, or inflammation of the larynx, may result in swelling of the vocal folds and cords to where the air passageway is obstructed and breathing becomes difficult or impossible. A hole may be made in the trachea (tracheotomy) below the larynx, to permit air movement to occur. LARYNGECTOMY, or removal of the larynx, is sometimes necessary in cancer of the larynx. Speech is still possible, even though the larynx is removed, by belching air through the esophagus and shaping it into speech (esophageal speech), or by applying a special vibrator to the neck, and shaping the sound into speech. LARYNGOSCOPY is a term referring to visual examination of the larynx through an instrument called a LARYNGOSCOPE.

THE TRACHEA, BRONCHI, AND BRONCHIOLES
(Fig. 24.6)

Structure. The TRACHEA extends from the larynx about 11 centimeters (4½ inches) into the chest. It divides into the RIGHT and LEFT BRONCHI or primary bronchi that, in turn, branch to form SECONDARY BRONCHI and BRONCHIOLES in some 14 to 15 generations of branchings. The most characteristic feature of the trachea and bronchi is the presence in their walls of C- or Y-shaped rings of hyaline cartilage. These rings support the tubes, and keep them open as breathing causes pressure changes in the system. In the bronchioles, cartilage rings become scattered plates of cartilage, and finally disappear altogether in the last portion of the conducting division, the terminal bronchioles. As cartilage disappears, smooth muscle increases, to give support to the tube wall. Diameter de-

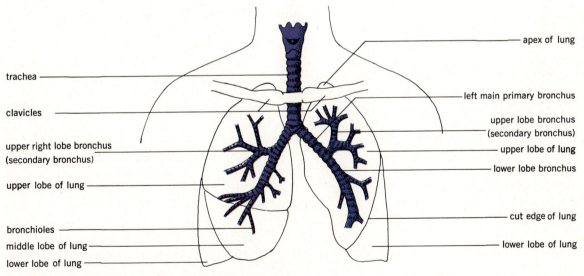

FIGURE 24.6   The trachea, bronchi, and bronchioles. (Sternum removed.)

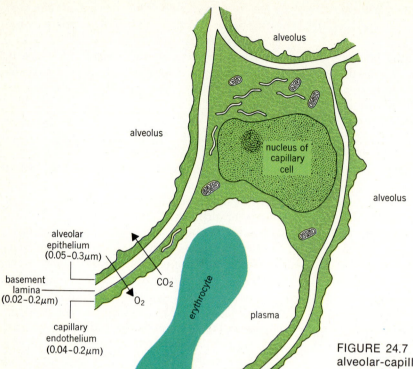

alveolus

alveolus

nucleus of capillary cell

alveolus

alveolar epithelium (0.05–0.3μm)

basement lamina (0.02–0.2μm)

capillary endothelium (0.04–0.2μm)

CO_2

O_2

erythrocyte

plasma

FIGURE 24.7 A diagrammatic representation of the alveolar-capillary membrane as drawn from an electron micrograph at 20,000×.

creases as branching occurs, so that an increasing number of smaller tubes are produced.

Function. These tubes conduct and cleanse the incoming air.

Clinical considerations. Since the terminal bronchioles have a wall that has no cartilage, and mostly smooth muscle, contraction of that muscle may severely obstruct air flow through the tubes. IN ASTHMA, spasm of the muscle due to irritants may cause extreme difficulty in filling and emptying the alveoli. Any foreign body or tumor that interferes with air flow through these tubes also leads to labored breathing. As one of its many effects, EMPHYSEMA causes a collapse and kinking of small and terminal bronchioles, thus interfering with air flow through the tubes.

The respiratory division

In the respiratory division, epithelial linings become thinner, until a very delicate diffusion membrane is formed (Fig. 24.7). Also, supporting or surrounding tissue such as connective tissue and muscle virtually disappears. The object here is to get as thin a membrane as possible to allow maximum diffusion of respiratory gases. The organs composing this division are the RESPIRA-TORY BRONCHIOLES, ALVEOLAR DUCTS, and ALVE-OLAR SACS (Fig. 24.8). About nine generations of branchings occur in this division.

STRUCTURE. It is best to consider the changes that occur in the respiratory divisions as one proceeds to the end of the system, rather than to state the structure of each specific organ. Among the changes occurring are the following.

The epithelium changes from a tall to a flat single layer of cells. This allows for increasing diffusion of O_2 and CO_2.

There appear in the walls of the tube, thin-walled

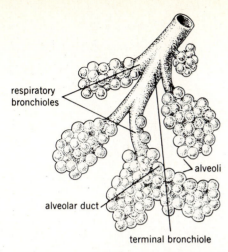

respiratory
bronchioles

alveoli

alveolar duct

terminal bronchiole

FIGURE 24.8   The terminal portions of the respiratory
system.

TABLE 24.1 Some characteristics of the respiratory system				
Organ	Generation of branching	Number of organs	Diameter (mm)	Total cross-sectional area (cm²)
Trachea	0	1	18–25	2.5
Bronchus	1	2	12	2.3
Lobe bronchi	2	4–5	8	2.1
Small bronchi	5–10	1024	1.3	13.4
Terminal bronchioles	14–15	32,768	0.7	113.0
Respiratory bronchioles	16–18	262,000	0.5	534
Alveolar ducts	19–22	4.2 million	0.4	5880
Alveoli	23–24	300 million	0.2	(50–70m²)

ALVEOLI (air sacs) that have many capillaries in their walls. Thus, the alveoli are the areas where gas exchange occurs.

FUNCTION. The respiratory division, because it contains alveoli and dense networks of capillaries derived from the pulmonary arteries, serves as the area for O_2 and CO_2 exchange between the lungs and the bloodstream.

Some interesting facts about the two divisions are presented in Table 24.1.

CLINICAL CONSIDERATIONS. The respiratory division, because of its alveoli, is subject to a number of conditions that interfere with gas exchange. Some of the more important are summarized below.

In ATELECTASIS, there is *collapse* of alveoli, or *incomplete expansion* in the newborn. Because alveolar collapse reduces the surface through which gas exchange may occur, symptoms of O_2 lack (cyanosis,

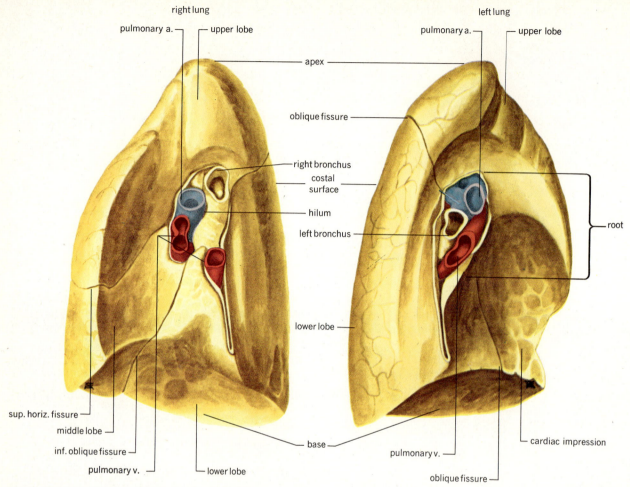

FIGURE 24.9   The gross anatomy of the lungs.

labored breathing), or CO$_2$ retention (hyperventilation, acidosis) may occur.

In PNEUMONIA, production of fluid fills alveoli with secretions that reduce diffusion. Basically the same symptoms occur as in atelectasis.

In EMPHYSEMA, the walls between alveoli disappear, creating large cavities that reduce diffusing surface.

One may rightly conclude that the most important conditions affecting the respiratory division are those that reduce gas diffusion between alveoli and blood.

The lungs (Fig. 24.9)

STRUCTURE. All parts of the respiratory system beyond the bronchi are contained within the lungs. The paired lungs lie within the two lateral PLEURAL CAVITIES of the thorax or chest. The cavities are lined by PARIETAL PLEURA, which is reflected onto the lung surface at the root of the lung as the VISCERAL PLEURA. In life, the lungs fill the cavities and thus the pleural cavities are potential, not actual, spaces. Other features of the lungs are shown in Figure 24.9. Notice that the

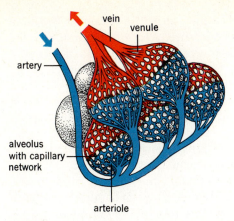

FIGURE 24.10   The capillary beds around lung alveoli.

shape of the lungs corresponds to the shape of the cavities in which they lie, and that the right lung has three subdivisions or LOBES while the left has only two. The anatomical unit of the lung is the SECONDARY LOBULE, consisting of several terminal bronchioles and all their branches. The lungs are composed of many of these lobules separated by thin connective tissue partitions. The functional unit is the PRIMARY LOBULE, consisting of one respiratory bronchiole and its branches. The parts making up these lobules are shown in Figure 24.8.

Blood supply. The branches of the conducting division from the trachea onward are nourished by blood from the BRONCHIAL ARTERIES. These arise from the aorta and the upper intercostal arteries. The respiratory division is nourished by PULMONARY ARTERY blood, which is being oxygenated during its passage through the lungs.

The branches of the pulmonary artery form rich CAPILLARY BEDS around the alveoli (Fig. 24.10) for gas exchange. The PULMONARY VEINS collect blood from these beds and return it to the left atrium.

Nerves. The VAGUS NERVE is the most important nerve serving the lungs. It provides afferent or sensory nerves that serve a number of respiratory reflexes. Vasomotor nerves supply the arteries and veins of the lung but have little effect on blood flow through the lungs.

CLINICAL CONSIDERATIONS. Total or partial LUNG COLLAPSE may follow wounds or disease that cause communication of a pleural cavity with the outside (pneumothorax) or with the lumina of the system's tubes. CANCERS usually require that a part of a lung be removed; if so, a lobe is usually removed because it is more or less a separate unit and chances of bleeding are reduced. If a whole lung requires removal, the procedure is called a *pneumonectomy*.

The effects of smoking on the respiratory system have been described in detail in the Surgeon General's report on smoking and health. In general, it may be stated that there are changes in the epithelium and loss of cilia that reduce the cleansing and protective functions of the system; there is loss of elastic tissue in the lungs and the respiratory tree that causes collapse of tubes and inability to adequately ventilate the lungs; there is loss of walls between alveoli that results in decrease of diffusing surface. All in all, smoking appears to alter the protective, ventilatory, and diffusing capabilities of the system.

Physiology of respiration

As we breathe, air is brought into and emptied from the lungs. Gases diffuse between lung and blood and, after transport by the bloodstream, diffuse between bloodstream and tissues or cells. These various aspects of respiration are often given names.

PULMONARY and ALVEOLAR VENTILATION refers to the series of events that lead to getting air into (inspiration), and out of (expiration) the lungs. Inspiration fills both the conducting *and* respiratory divisions of the system; *pulmonary ventilation* strictly refers to filling the conducting division, while *alveolar ventila-*

tion means actually filling the alveoli with air.

EXTERNAL RESPIRATION refers to the exchange of gases occurring between bloodstream and lungs.

INTERNAL RESPIRATION involves the exchange of gases occurring between bloodstream and tissues or cells.

TRANSPORT OF GASES refers to the several methods by which the gases are carried in and by the bloodstream.

All these activities require CONTROL, provided by nervous and chemical factors.

This section examines these activities in the general order indicated in the preceding list.

Pulmonary and alveolar ventilation

Of any given volume of inspired air, about 30 percent remains within the structures of the conducting division and is unavailable for exchange through these thick-walled organs. This volume is termed DEAD AIR. The remaining 70 percent normally reaches alveoli and is available for exchange through the thin-walled alveoli. It is this latter fraction that constitutes alveolar ventilation.

RESPIRATORY MOVEMENTS. INSPIRATION, or drawing of air into the lungs, occurs through muscular activity that increases the volume of the thorax. The external intercostal muscles, contracting from nerve impulses reaching them over the intercostal nerves, elevate the ribs. As the ribs are elevated, they swing outwards as well as upwards, and the front-to-back and side-to-side dimensions of the thorax are increased. The diaphragm, stimulated through the phrenic nerves, contracts, its central tendon is pulled downwards, and the vertical dimension of the thorax is increased (Fig. 24.11). During resting breathing, the average adult achieves an increase in thoracic volume of about 500 ml by the combined activity of these muscles.

As the volume of the chest cavity is increased, there are alterations in the pressure within this cavity. This pressure is designated the INTRA-

THORACIC PRESSURE, and it is normally less than atmospheric pressure. During inspiration, the pressure falls to even less than it was during rest. This occurs because the chest (thoracic) cavity does not communicate with the atmosphere, and to increase the volume of such a cavity makes it larger without allowing more air molecules to enter it. Thus, since pressure depends on molecules of air striking a surface per unit of time, fewer molecules mean a lower pressure. The problem now is to transmit this lowered intrathoracic pressure to the lungs. They *do* communicate with the atmosphere by way of the conducting division, and if their pressure can be lowered, air will rush from atmosphere into the lungs.

The lungs cohere or "stick to" the walls of the thorax because of fluid films between lungs and chest walls, and thus a movement of the chest will cause the lungs to follow. (Try to separate two microscope slides that have a film of water between them—you cannot *pull* them apart, but you can *slide* them apart.) As the chest size increases, the lungs are expanded as they follow the chest increase, and this expansion lowers the pressure in the lungs relative to the atmosphere. Air from the atmosphere will then flow to the low pressure area—that is, *into* the lungs.

The pressure within the lungs is termed the INTRAPULMONIC PRESSURE, and it falls as the lungs are expanded, then equalizes as the air inflow equalizes the difference in pressure between lungs and atmosphere.

EXPIRATION is normally a passive process, one that at rest requires no muscular activity to occur. As the lungs are expanded during inspiration, elastic tissue within them is stretched. Relaxation of the external intercostal muscles and diaphragm allows the thorax to return to its original size and volume. The lung size decreases and the elastic recoil of the previously stretched tissue creates a *higher* than atmospheric pressure within the lungs. Air is driven *from* the lungs. Intrapulmonic pressure again equalizes with atmospheric pressure as outward airflow ceases. Expiration may become active as the abdominal muscles contract and force the viscera against the diaphragm,

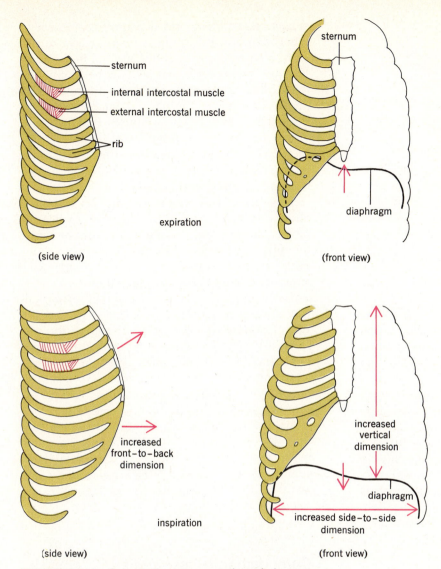

FIGURE 24.11 Schematic representation of changes in thoracic dimensions on expiration and inspiration.

hastening its return to a normal resting position.

Alterations of the pressures within chest and lung during breathing are depicted in Figure 24.12. Note the cyclical nature of the fluctuation in intrapulmonic pressure, above and below atmospheric pressure, while intrathoracic pressure remains less than atmospheric.

SURFACTANT. Alveoli behave like bubbles; that is, they try to assume the smallest diameter for their volume. This means that they will tend to collapse as the lung deflates. SURFACE TENSION is the force that draws the bubble or the alveolus to its smallest size. If the material on the alveolar wall in contact with the air in the alveolus is water,

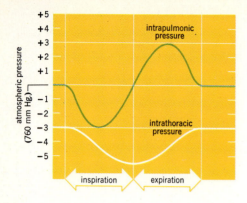

FIGURE 24.12 Changes in intrathoracic and intra-pulmonic pressures during breathing.

surface tension forces are very high and there would be a very strong tendency for alveolar collapse. But there is a substance on the alveolar wall called SURFACTANT. It is a phospholipid, and lowers the alveolar surface tension by 7 to 14 times that expected of an air–water junction. Thus, as alveoli deflate, they have a much lesser tendency to collapse. If surfactant is diminished or absent in the alveoli, the alveoli are very difficult to expand (as at birth) or they collapse easily. In the *respiratory distress syndrome* (RDS) called HYA-LINE MEMBRANE DISEASE (HMD), surfactant is greatly diminished, alveoli cannot be expanded, and the infant suffers oxygen lack because of a diminished surface for gas exchange.

Exchange of air

VOLUMES AND CAPACITIES. The amount of air that is exchanged with each breath—regardless of the depth of breathing—is called the TIDAL VOLUME. Breathing normally, the average adult exchanges about 500 ml of air (this is a figure that represents both an inspiratory *or* an expiratory volume). If, at the end of a normal inspiration, a forced inspiration is made, an additional volume of air may be inspired. It is called RESERVE IN-

SPIRATORY VOLUME, and is a quantity drawn on when depth of breathing is increased, as during acitvity. It amounts to about 3000 ml in the adult. RESERVE EXPIRATORY VOLUME may be forcibly exhaled after a normal expiration and amounts to about 1100 ml in the adult. Thus, we see that normal breathing utilizes a volume that lies somewhere between "full lungs" and "empty lungs" (Fig. 24.13). RESIDUAL VOLUME is the air remaining in the lungs even after the most forcible expiration, and reflects the fact that the lungs do not collapse each time we breathe out. Its volume is about 1200 ml in the adult. If the lung *is* collapsed, part of the residual volume will be driven out of the lung, but there will still be air left in the lung. This volume is called the MINIMAL VOLUME, and amounts to about 600 ml in the adult.

If we add these volumes together in various ways, we arrive at what are called LUNG CAPACITIES. The sum of tidal, reserve inspiratory, and reserve expiratory volumes is called the VITAL CAPACITY, and is a measure of a given individual's total "movable" air. The FUNCTIONAL RESIDUAL CAPACITY is the sum of residual and reserve expiratory volumes. It is the air remaining in the lungs after a normal expiration from which continued oxygenation of the blood will occur between inspirations. TOTAL LUNG CAPACITY is the sum of all volumes except minimal volume, and gives an idea of lung size. These capacities are also indicated in Figure 24.13.

Composition of respired air; external and internal respiration

The air we breathe (atmospheric air) contains about 78 percent nitorgen and 20 percent oxygen. Small amounts of carbon dioxide are present as well as variable amounts of water vapor (Table 24.2). In reporting concentrations of gases in the body, one uses the expression PARTIAL PRESSURE. This is a figure, expressed in mm Hg, obtained by multiplying the percent of a gas in a mixture times the total pressure of the mixture. Assuming

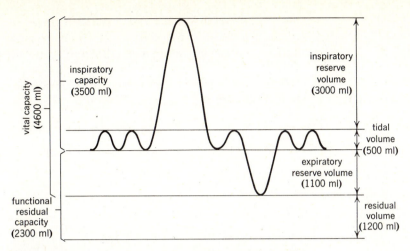

FIGURE 24.13   Respiratory volumes and capacities for an adult.

Gas	Atmosphere	Trachea; mixture of "in" and "out" air	Alveoli	Arterial blood	Venous blood	Tissues
TABLE 24.2 Partial pressures in mm Hg of gases in various areas of the body						
Nitrogen	596	564	573	573	573	573
Oxygen	158	149	100	95	40	40
Carbon dioxide	0.3	0.3	40	40	46	46
Water vapor	5.7 av.	47	47	47	47	47
Total	760	760	760	755	706	706

standard, corrected, barometric pressure (760 mm Hg), the partial pressure expresses the pressure of one gas as though it was present alone. The value is represented as Po_2, Pco_2, PH_2O, with the figure following the symbol.

As atmospheric air is inhaled, it is mixed with air remaining in the system from the previous exhalation. This air contains less oxygen and more carbon dioxide and water vapor than atmospheric air. The term TRACHEAL AIR is applied to this mixture, and as inhaled air mixes with it, the Po_2, and PN_2 are lowered in the inhaled air, while Pco_2, and PH_2O are elevated (see Table 24.2). A forced expiration delivers air from the alveoli into the trachea, and, if a sample from the trachea is

obtained under these conditions, its composition is taken to represent the partial pressures of gases in the alveoli. This sample is called ALVEOLAR AIR and it has the composition shown in Table 24.2. It is relatively easy to obtain sample of ARTERIAL and VENOUS BLOOD and to determine their gas concentrations. Figures shown in Table 24.2 present these values. Finally, gas pressures in the TISSUES may be determined; they also are shown in Table 24.2.

Study of the table discloses several interesting and important facts.

Po_2 decreases from atmosphere to alveoli to blood to tissue.

P_{CO_2} decreases in the opposite direction, that is, from tissues to blood to alveoli to atmosphere.

P_{H_2O} is constant within the several areas.

P_{N_2} is constant once it is "inside" the body.

The force responsible for exchange of gases between alveoli and bloodstream (*external respiration*) is DIFFUSION, a passive process dependent on a diffusion gradient. Since the gradient for oxygen is "inwards," oxygen will pass from alveoli to bloodstream, and from bloodstream to tissues (*internal respiration*). Oxygen will diffuse from alveoli to bloodstream until values are nearly equal between alveoli and arterial blood, a figure of about 95 mm Hg. Equilibrium with the alveoli (100 mm Hg) is not obtained because of the rapid movement of blood through lung capillaries (about 1 sec) and because oxygen does not easily dissolve in the plasma prior to becoming attached to hemoglobin in the red cells. At the tissues, about 30 percent of the oxygen will diffuse from the hemoglobin, under resting conditions. Thus, venous blood leaving the tissues still has much oxygen in it. This "extra" oxygen may be given up to the tissues during activity, as tissue levels of oxygen fall during accelerated aerobic metabolism.

Carbon dioxide, a product of cellular metabolism, is highest in the tissues. It diffuses into the bloodstream and is carried to the lungs where it diffuses "outwards" into the alveoli. Equilibrium is attained in the case of CO_2 because it is much more soluble in plasma and because its release from the plasma is aided by an enzyme, CARBONIC ANHYDRASE.

Thus, we see that the gases follow their concentration gradients, which are *in opposite directions*.

Transport of gases

OXYGEN. NINETY-FIVE PERCENT of the oxygen of the bloodstream is carried attached to the iron atoms of the HEMOGLOBIN molecules. FIVE PERCENT is carried in simple SOLUTION in the plasma water.

As oxygen loads on the four iron atoms present in a molecule of hemoglobin, it combines in steps, each governed by its own equilibrium constant (K). The reactions may be represented as below [Hgb = hemoglobin; numbers indicate iron atoms (e.g., Hgb_4) or oxygen atoms (e.g., O_2)].

$$Hgb_4 \quad + O_2 \overset{K_1}{\rightleftharpoons} Hgb_4O_2$$
$$Hgb_4O_2 + O_2 \overset{K_2}{\rightleftharpoons} Hgb_4O_4$$
$$Hgb_4O_4 + O_2 \overset{K_3}{\rightleftharpoons} Hgb_4O_6$$
$$Hgb_4O_6 + O_2 \overset{K_4}{\rightleftharpoons} Hgb_4O_8 \text{ (saturated)}$$

Some molecules of hemoglobin take up or release oxygen more readily than others, as determined by K.

Several factors determine how much oxygen will be taken up by the hemoglobin.

P_{O_2}. Obviously, if more oxygen is available for pickup, more will become attached to the hemoglobin, up to its capacity (100 percent saturation). The relationship of P_{O_2} to percent saturation takes the shape of a sigmoid curve (Fig. 24.14). It is called an OXYGEN DISSOCIATION CURVE.

pH. The lower the pH of the plasma, the less oxygen will be taken up by hemoglobin or the more oxygen will be driven off of it. This effect is called the BOHR EFFECT, or a shift of the dissociation curve to the right.

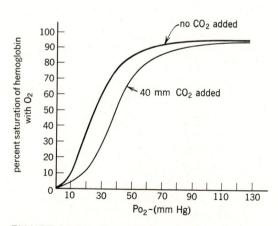

FIGURE 24.14   Oxygen dissociation curve.

What primarily determines plasma pH is the amount of CO_2 in it, by the reaction

$$CO_2 + H_2O \rightarrow H_2CO_3 \rightarrow H^+ + HCO_3^-$$

The H^+ determines pH. The Bohr effect automatically ensures that an active tissue, producing more CO_2, will have an increased oxygen supply as pH of the plasma falls.

Temperature. Increased temperature decreases the amount of oxygen the hemoglobin can hold. Thus, an active, warmer tissue can be assured of an increased supply of oxygen.

DPG levels. DPG is an abbreviation for 2, 3 *di*phosphoglycerate. Red cells produce the chemical, and it binds in the center of the hemoglobin molecule, reducing affinity of the hemoglobin for oxygen. In fetuses, DPG levels are low, and the fetal hemoglobin loads more oxygen at the relatively low Po_2 of the maternal blood. In an adult, DPG levels are higher, and with higher Po_2 in the alveoli than in the placenta, oxygen loads less rapidly but achieves a higher percent saturation on the hemoglobin.

CARBON DIOXIDE. Carbon dioxide is carried in several ways in the bloodstream.

Nearly two-thirds (64 percent) of the carbon dioxide is carried in the form of BICARBONATE ION in plasma and red cell. The bicarbonate is produced by the reaction of carbon dioxide and water

$$CO_2 + H_2O \rightarrow H_2CO_3 \rightarrow H^+ + HCO_3^-$$

Bicarbonate ion builds up in the red cells and begins to diffuse out of the cell. H^+ is attached to hemoglobin and is *buffered.* Chloride ion moves into the cells to offset loss of negatively charged bicarbonate ion.

Up to 27 percent of the carbon dioxide is carried attached to the amine ($-NH_2$) groups of the amino acids of the hemoglobin molecules. The combination is called CARBAMINOHEMOGLOBIN.

About 9 PERCENT is carried in SOLUTION in the plasma.

The series of reactions that occur as gases are picked up or released between lungs and bloodstream and bloodstream and tissues are shown in Figures 24.15 and 24.16. Note particularly the reactions that attach oxygen to hemoglobin and buffer the H^+ resulting from the reaction of carbon dioxide and water. The following explanation describes the changes.

At the tissues, carbon dioxide dissolves in plasma water or enters the erythrocytes to react with cellular water. In either case, carbonic acid is formed, which dissociates into hydrogen ion and bicarbonate ion. The hydrogen ions produced

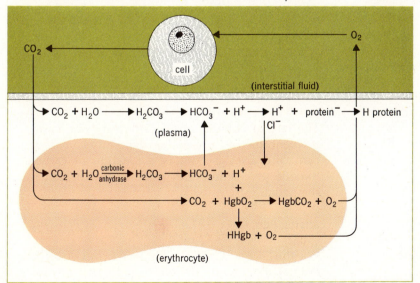

FIGURE 24.15 Gas exchange between the blood and tissues.

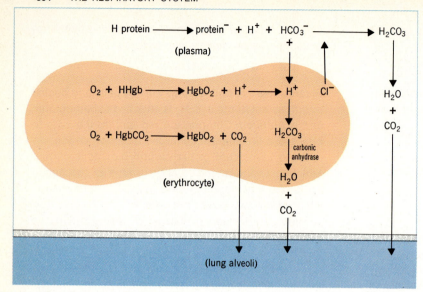

FIGURE 24.16   Gas exchange between the blood and the lungs.

in the plasma are buffered by reacting with plasma proteins; the hydrogen ions produced within the erythrocytes do not pass through the cell membrane and are buffered by hemoglobin. Bicarbonate ion accumulates within the erythrocytes and soon begins to diffuse out of the cell into the plasma. A loss of negative charge results and, to return to electrical neutrality, chloride ion moves from plasma into the erythrocytes (chloride shift). Some carbon dioxide displaces O_2 from the hemoglobin, and the excess H^+ in the erythrocyte drops pH and also drives O_2 from the hemoglobin.

At the lungs, the action of carbonic anhydrase liberates CO_2 from carbonic acid within the erythrocytes and plasma. Bicarbonate ion moves into the erythrocyte to form carbonic acid with the H^+ inside the cell and an excess of negative charges accumulates; chloride ion moves back into the plasma to maintain electrical neutrality. The high oxygen levels in the lung region also cause displacement of CO_2 from the hemoglobin as the formation of oxyhemoglobin occurs. Hydrogen ion is released from its combination with protein as bicarbonate enters the erythrocyte, and the protein is made available to buffer more H^+.

Acid-base regulation by the lungs

The lung is the most important body organ involved in the regulation of acid-base balance. The production of carbon dioxide by body cells results in the addition of 13,000 to 20,000 meq. per day of H^+ to the body fluids. The reaction involved in the formation of H^+ from carbon dioxide is

$$CO_2 + H_2O \rightarrow H_2CO_3 \rightarrow \mathbf{H}^+ + HCO_3$$

This reaction is reversed in the lungs by the elimination of carbon dioxide, which shifts the chemical equilibria of the reaction toward the formation of carbonic acid:

$$H^+ + HCO_3^- \rightarrow H_2CO_3 \rightarrow H_2O + \underset{\text{(eliminated)}}{CO_2}$$

The rate of elimination of carbon dioxide is dependent on the process of ventilation, and, in turn, the arterial P_{CO_2} is a function of carbon dioxide production and ventilation according to the equation:

$$\text{arterial } P_{CO_2} = \frac{P_{CO_2} \text{ production}}{\text{alveolar ventilation}}$$

We thus have a system for SELF-REGULATION of arterial P_{CO_2} and, therefore, pH. The respiratory centers controlling respiration are stimulated by increase in P_{CO_2} and/or a decrease of pH, and elimination of carbon dioxide increases. Conversely, decreases of P_{CO_2} and/or rise of pH result in decreased stimulation of the centers, and carbon dioxide elimination is decreased. The entire system is thus controlled within very narrow limits by a feedback mechanism that monitors the P_{CO_2} as determined by the balance between carbon dioxide production and elimination. Adjustments are made rapidly and accurately.

CLINICAL CONSIDERATIONS. The reader should review the sections on respiratory acidosis and alkalosis in Chapter 19.

Other functions of the lungs

The lungs serve functions other than merely acting as the areas for gas exchange between atmosphere and bloodstream. Some of their additional functions include:

SECRETION OF SURFACTANT by specialized cuboidal-shaped cells in the alveolar lining called *septal cells* (alveolar cells, great alveolar cells, pneumonocytes).
PRODUCTION OF: KININS, vasodilating compounds; KALLIKREINS, compounds that convert inactive kininogens to kinins; KININASES that destroy kinins.
INACTIVATION OF GASTRIC HORMONES.
FORMATION OF ANGIOTENSIN after renin has initiated the process.
METABOLISM OF INSULIN.
SYNTHESIS OF PROSTAGLANDINS.
PRODUCTION OF ENZYME INACTIVATORS such as antitrypsin. Trypsin is a digestive enzyme that could digest lung tissue; the antitrypsin destroys or blocks trypsin effects.

All-in-all, the lung appears to be a protective and highly active organ, a role not easily appreciated if one only looks at the thin alveolar lining cells.

Protective mechanisms of the respiratory system

Because the respiratory system transports gases from the atmosphere to the lungs, a wide variety of potentially dangerous microorganisms, antigens, and particulate matter may be inhaled. To aid in the understanding of the operation of the protective mechanisms, recall that the greater part of the respiratory system is lined with a mucus coating secreted by mucous glands in the subepithelial connective tissue and the epithelial goblet cells. Also recall that the epithelium is ciliated. There are lymphatics, alveolar macrophages, and reflex mechanisms that aid in dislodging or combatting foreign materials in the system. Several specific mechanisms may be cited as providing protection.

THE MUCOCILIARY ESCALATOR. The presence of a sticky mucus layer on the epithelium of the system allows trapping of particles as incoming air contacts the surface. Coordinated ciliary activity moves the mucus layer posteriorly in the nasal cavities to the throat where the mass is usually swallowed. A similar mechanism operates from the terminal bronchioles upward through the conducting division. Continuous cleansing of the system above the terminal bronchioles is provided. In normal individuals, the rate of mucus production and removal by ciliary action is nicely balanced, and mucus rarely comes to our conscious attention. One of the demonstrated effects of tobacco smoke is to slow the ciliary action and change the nature of the mucus, usually making it more viscous. Such mucus is harder to move and may result in retention of potentially harmful substances.

ALVEOLAR MACROPHAGES. Since both cilia and goblet cells are not found beyond the terminal bronchioles, a cleansing mechanism is provided for the respiratory division. Phagocytic cells (alveolar macrophages) are found in the alveoli of this division. These cells engulf and destroy bacteria, particles of dust, foreign antigens, and other harmful substances. Their numbers are partially dependent on the levels of contamination of inhaled air, increasing as the load of pollutants increases. Their phagocytic ability may also vary, becoming less as the amounts of chemical contaminants increase. These cells thus provide an extremely important line of defense.

FILTERING. The presence of the large hairs around the nostrils tends to restrict the entry of large objects into the system.

SECRETION OF IMMUNE GLOBULINS. The presence of immunoglobulin A (Ig A) has been demonstrated in the secretions of the glands in all parts of the respiratory tree. The substance is a nonspecific antibody whose production is apparently triggered by a wide variety of antigenic challenges. The amount produced is directly proportional to the degree of antigenic challenge and forms an important defense mechanism against foreign chemicals.

LYMPHATICS. Lymphatics are present in somewhat greater numbers in the respiratory and digestive systems than elsewhere in the body. The combination of lymphocytes in the mucous membranes, and lymphatic vessels carrying lymph to the lymph nodes, aids removal of matter that has entered the tissues themselves.

REFLEX PROTECTIVE MECHANISMS. These include sneezing and coughing. The impulses responsible for initiating a sneeze originate from irritation of trigeminal nerve endings in the nasal cavities, while those initiating a cough are the result of irritation of glossopharyngeal and vagus endings in the pharynx, larynx, trachea, bronchi, and alveoli. Both reflexes involve a deep inspiration brought about by stimulation of the inspiratory centers. A violent expiration, aided by powerful contractions of the abdominal muscles, produces air flow velocities that may approach the speed of sound. In a sneeze, the uvula and soft palate are positioned to allow air to escape both through the nose and mouth; in a cough, the uvula blocks the nasal cavities and the full force of the expiration is directed through the mouth.

Control of breathing

CENTRAL INFLUENCES. Located within or adjacent to the medulla of the brainstem are groupings of nerve cells that constitute the RESPIRATORY CENTERS (Fig. 24.17). Included in these groupings are the paired *inspiratory centers* that send nerves to the diaphragm and external intercostal muscles. When stimulated, these centers lead to a maximal inspiratory effort. Posterior to the inspiratory centers are the paired *expiratory centers* that provide part of the influence required to achieve expiration of air from the lungs. In the pons are the paired *pneumotaxic centers* that provide activity also designed to cause expiration. In the reticular formation of the brainstem are paired areas that "drive" the inspiratory centers; they are called the *apneustic* centers. The nuclei of the *vagus* (Xth cranial) *nerves* receive impulses from baroreceptors of the lungs and provide inhibition to inspiratory activity. Lastly, unnamed centers in the region of the *fourth ventricle* monitor pH of the cerebrospinal fluid, and, if pH falls, provide stimulation of rate and depth of breathing. This latter effect results in increased elimination of CO_2, lowering pH toward normal values.

Integration of these centers may be described as follows.

The inspiratory centers have an inherent rhythm that causes the muscular contraction drawing air into the lungs.

As lung tissue is stretched during inspiration, stretch receptors send impulses over vagal afferents to inhibit the inspiratory center and to activate the expiratory center.

Simultaneously, impulses pass from inspiratory centers to the pneumotaxic centers, stimulating them

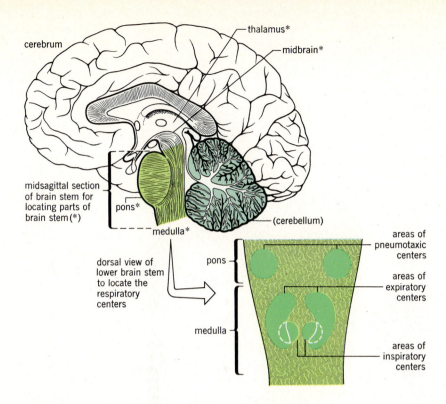

cerebrum

thalamus*

midbrain*

midsagittal section
of brain stem for
locating parts of
brain stem (*)

pons*

medulla*

(cerebellum)

dorsal view of
lower brain stem
to locate the
respiratory
centers

pons

medulla

areas of
pneumotaxic
centers

areas of
expiratory
centers

areas of
inspiratory
centers

FIGURE 24.17   The location of the
respiratory centers.

to inhibit the activity of the apneustic center, and the "drive" to the inspiratory center is reduced.

All these influences combine to interrupt the activity of the inspiratory center, the respiratory muscles relax, and expiration occurs. These events are depicted in Figure 24.18.

THE STIMULUS TO BREATHING. The source of the inherent rhythm that the inspiratory centers exhibit appears to lie in the effect of the P_{CO_2} of the blood reaching the inspiratory centers. A 10 percent increase of P_{CO_2} will cause an eightfold increase in ventilation. Perhaps directly, as CO_2, or through the H^+ liberated from the reaction of CO_2 with water, the cells of the inspiratory center are triggered to form impulses. Of less importance is P_{O_2} of the blood; fall of 50 percent in P_{O_2} is required to even stimulate breathing. Carbon dioxide is perhaps a greater threat to the body because of its relationship to pH, and to tie

ventilation to it not only assures more oxygen when ventilation is stimulated but regulates acid-base balance as well.

PERIPHERAL INFLUENCES. There are a variety of REFLEX MECHANISMS that influence rate and depth of breathing. Stretching the aorta, lungs, and carotid sinus stimulates baroreceptors that slow the rate and depth of breathing. The HERING-BREUER REFLEX refers to the reflex originating in the lungs. CHEMORECEPTORS in the aorta and carotid sinus monitor P_{O_2}, P_{CO_2}, and pH of the blood reaching them. Decrease of P_{O_2}, increase of P_{CO_2}, or fall of pH produce increased rate and depth of breathing that supplies more oxygen, eliminates more CO_2, and lowers pH. The reflex mechanisms involved in regulation of breathing are shown in Figure 24.19.

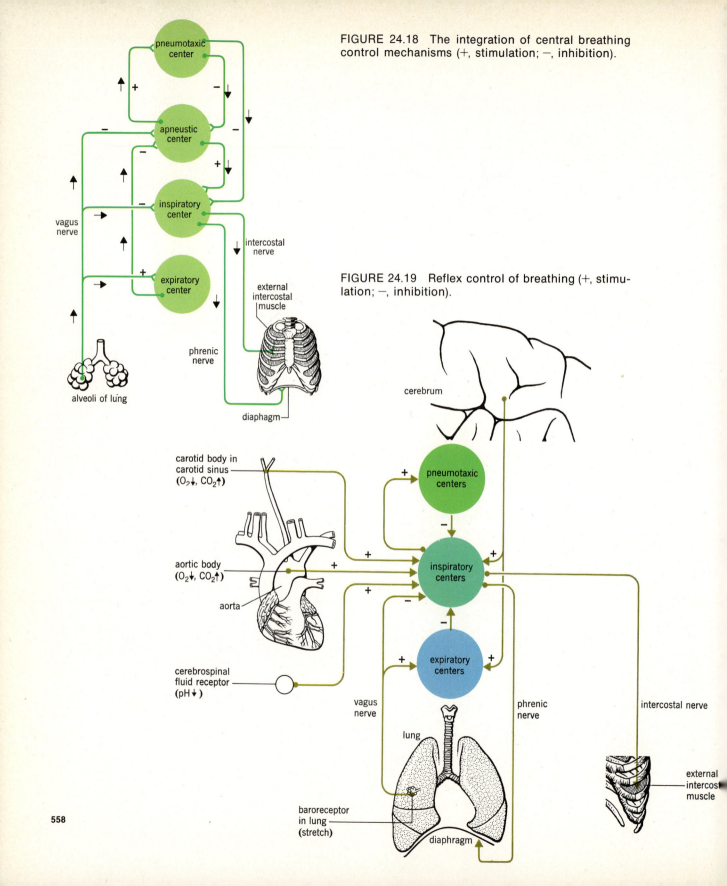

FIGURE 24.18 The integration of central breathing control mechanisms (+, stimulation; −, inhibition).

FIGURE 24.19 Reflex control of breathing (+, stimulation; −, inhibition).

pneumotaxic center

apneustic center

inspiratory center

expiratory center

vagus nerve

intercostal nerve

external intercostal muscle

phrenic nerve

alveoli of lung

diaphagm

cerebrum

carotid body in carotid sinus (O_2↓, CO_2↑)

aortic body (O_2↓, CO_2↑)

aorta

cerebrospinal fluid receptor (pH↓)

pneumotaxic centers

inspiratory centers

expiratory centers

vagus nerve

phrenic nerve

intercostal nerve

lung

baroreceptor in lung (stretch)

diaphragm

external intercostal muscle

The fetal respiratory system

The fetus does not use its lungs to acquire oxygen or eliminate carbon dioxide; this function is served by the placenta. The transition from placenta to lung breathing is one of the most dramatic events in nature. A few words on this change may provide insight into the drama.

The placenta as a "lung"

Exchange of gases across the placental membranes is slower than that across the alveolar membranes primarily because the placental membranes are much thicker than alveolar walls (3.5 μm versus 0.7 μm or less). The diffusing surface area of the placenta is about one-fourth that of the adult lung and thus the volume of gas diffusing per unit of time will be less. Circulation is slower through the placenta than through the fetal body, and a somewhat longer time for gas diffusion is thereby provided, somewhat offsetting the slowing of diffusion resulting from thick membranes. The fetus operates in a gas environment that is lower in oxygen and higher in carbon dioxide than that of adults. P_{O_2} is about 44 mm Hg in maternal placenta blood, and the umbilical vein blood leaves the placenta with a P_{O_2} of about 30 mm Hg. P_{CO_2} in fetal umbilical artery blood is about 48 mm Hg, that of the mother 40 to 45 mm Hg. Thus, oxygen diffuses from maternal blood to fetus, and carbon dioxide in the opposite direction. The fetal circulation (*see* Fig. 22.2) also contains devices to shunt the blood returning from the placenta past the nonfunctional lung. These devices include the *foramen ovale* that passes blood from right atrium to left atrium and the *ductus arteriosus* that connects pulmonary artery directly to the aorta.

Changes at birth

Birth by the normal route (through the vagina) compresses the fetal chest that expels amniotic fluid from the fetal nostrils, mouth, and trachea. As the chest is delivered, it expands and partially fills the system with air. Some air probably reaches the lungs. Diaphragmatic contraction follows to further expand the lungs and draw air into the alveoli. The diaphragmatic contraction is termed the "first true breath."

The stimuli that cause the initial contraction of the diaphragm are basically unknown. Among those hypothesized to be involved are: auditory, tactile, and visual stimuli as the baby is delivered; exposure to the less-than-body-temperature environment as delivery occurs; high P_{CO_2} in the bloodstream stimulating the inspiratory center (P_{CO_2} rises as the placenta separates); low P_{O_2} in the bloodstream. In any event, lung inflation occurs, and any fluid in the lungs is rapidly absorbed by blood vessels and lymphatics in the lung, so that diffusion of gases may increase across the alveolar membranes. Another interesting thing that occurs as the child is delivered is that uterine contractions "squeeze" 80 to 100 ml of placental blood into the umbilical veins and from there to the fetus. The higher blood pressure that results speeds removal of wastes and CO_2 from the bloodstream of the infant.

In response to the higher P_{O_2} occasioned by infant breathing, the ductus arteriosus and umbilical arteries close down; the pressure changes in the heart cause a *functional* closure of the foramen ovale within minutes after birth; eventually the flaps grow together to form a permanent closure. Thus, the transition from water (amniotic fluid) to air breathing is assured.

Abnormalities of respiration

RESPIRATORY INSUFFICIENCY develops when the respiratory system cannot meet tissue demands for oxygen or carbon dioxide removal. Basically, the problem is the result of insufficient alveolar ventilation caused by: narcosis, from drugs that depress breathing; respiratory diseases (emphysema that reduces diffusing surface, pneumonia that fills the alveoli with fluid, atelectasis or alveolar collapse); or insensitivity of the inspiratory center to carbon dioxide. *Dyspnea* (labored breathing) and *hypoxemia* (low arterial P_{O_2}) inevitably result. RESUSCITATION or artificial ventilation of the lungs may be required. Mouth-to-mouth resuscitation (Figure 24.20) may provide immediate relief, or respirators that force air into the lungs under pressure may be employed.

FIGURE 24.20   Mouth-to-mouth resuscitation. (Courtesy American Red Cross.)

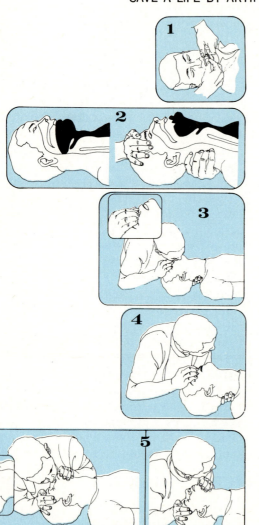

WHEN BREATHING STOPS SECONDS COUNT
SAVE A LIFE BY ARTIFICIAL RESPIRATION

MOUTH-TO-MOUTH METHOD

1. If foreign matter is visible in the mouth, wipe it out quickly with your fingers, wrapped in a cloth, if possible.

2. Tilt the victim's head backward so that his chin is pointing upward. This is accomplished by placing one hand under the victim's neck and lifting, while the other hand is placed on his forehead and pressing. This procedure should provide an open airway by moving the tongue away from the back of the throat.

3. Maintain the backward head-tilt position and, to prevent leakage of air, pinch the victim's nostrils with the fingers of the hand that is pressing on the forehead.

   Open your mouth wide; take a deep breath; and seal your mouth tightly around the victim's mouth with a wide-open circle and blow into his mouth. If the airway is clear, only moderate resistance to the blowing effort is felt.

   If you are not getting air exchange, check to see if there is a foreign body in the back of the mouth obstructing the air passages. Reposition the head and resume the blowing effort.

4. Watch the victim's chest, and when you see it rise, stop inflation, raise your mouth, turn your head to the side, and listen for exhalation. Watch the chest to see that it falls.

   When his exhalation is finished, repeat the blowing cycle. Volume is important. You should start at a high rate and then provide at least one breath every 5 seconds for adults (or 12 per minute).

   When mouth-to-mouth and/or mouth-to-nose resuscitation is administered to small children or infants, the backward head-tilt should not be as extensive as that for adults or large children.

   The mouth and nose of the infant or small child should be sealed by your mouth. Blow into the mouth and/or nose every 3 seconds (or 20 breaths per minute) with less pressure and volume than for adults, the amount determined by the size of the child.

   If vomiting occurs, quickly turn the victim on his side, wipe out the mouth, and then reposition him.

MOUTH-TO-NOSE METHOD

5. For the mouth-to-nose method, maintain the backward head-tilt position by placing the heel of the hand on the forehead. Use the other hand to close the mouth. Blow into the victim's nose. On the exhalation phase, open the victim's mouth to allow air to escape.

RELATED INFORMATION

6. If a foreign body is prohibiting ventilation, as a last resort, turn the victim on his side and administer sharp blows between the shoulder blades to jar the material free.

7. A child may be suspended momentarily by the ankles or turned upside down over one arm and given two or three sharp pats between the shoulder blades. Clear the mouth again, reposition, and repeat the blowing effort.

8. Air may be blown into the victim's stomach, particularly when the air passage is obstructed or the inflation pressure is excessive. Although inflation of the stomach is not dangerous, it may make lung ventilation more difficult and increase the likelihood of vomiting. When the victim's stomach is bulging, always turn the victim's head to one side and be prepared to clear his mouth before pressing your hand briefly over the stomach. This will force air out of the stomach but may cause vomiting.

When a victim is revived, keep him as quiet as possible until he is breathing regularly. Keep him from becoming chilled and otherwise treat him for shock. Continue artificial respiration until the victim begins to breathe for himself or a physician pronounces him dead or he appears to be dead beyond any doubt.

Because respiratory and other disturbances may develop as an aftermath, a doctor's care is necessary during the recovery period.

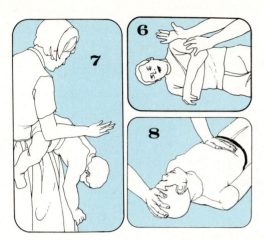

THE AMERICAN NATIONAL RED CROSS

Summary

1. The respiratory system develops between 3½ and 12 weeks of age.

   a. Nasal cavities and bronchial tubes develop prior to 12 weeks.

   b. Alveoli develop after 26 weeks of age.

2. The organs of the system provide a means of conducting air to the alveoli, and the alveoli provide a means of allowing gas exchange between lungs and bloodstream.

   a. The conducting division includes nasal cavities, pharynx, larynx, trachea, bronchi, and bronchioles. It is too thick walled to allow gas exchange between lungs and blood.

   b. The respiratory division includes the respiratory bronchioles, alveolar ducts and sacs, and has alveoli that allow gas exchange between lungs and blood.

   c. Each part of the system has special functions including sound production, warming, moistening and cleansing of inhaled air, and secretion of immune globulins.

3. The nasal cavities warm, moisten, and cleanse the inhaled air.

4. The pharynx is a tube common to both respiratory and digestive systems and conducts air. It has nasal, oral, and laryngeal subdivisions.

5. The larynx is a cartilagenous structure that conducts air and creates sound that is shaped into speech by the tongue and mouth.

6. The trachea, bronchi, and bronchioles conduct air and form the branching system of tubes commonly referred to as the "respiratory tree."

7. The respiratory division allows gas exchange between lung and blood because of alveoli.

8. The respiratory system below the throat is nourished by blood from bronchial and pulmonary arteries and receives nerve fibers primarily from the vagus nerve.

9. The physiology of respiration involves intake and output of air, diffusion between lungs and blood and tissues, transport of gases, and control of breathing.

10. Inspiration of air occurs by active muscular contractions of the external intercostals and diaphragm.
    a. Contraction elevates ribs and increases the vertical dimension of the thorax, increasing chest volume.
    b. Increase of volume lowers intrathoracic and then intrapulmonic pressures; air enters the lungs.
    c. Elastic tissue in the lung is stretched.

11. Expiration is normally passive.
    a. Relaxation of muscles diminishes thoracic volume.
    b. Intrapulmonic pressure is elevated as elastic lung recoil occurs.
    c. Air is driven out of the lungs.

12. A phospholipid designated as surfactant lowers alveolar surface tension and prevents alveolar collapse as expiration occurs. Lack of the material creates respiratory distress syndrome.

13. There are several volumes and capacities related to the air of the lungs.
    a. Tidal volume is air breathed in and out. 500 ml, at rest, in the adult.
    b. Reserve inspiratory volume may be *inhaled* above tidal volume. 3000 ml in the adult.
    c. Reserve expiratory volume may be *exhaled* beyond tidal volume. 1100 ml in the adult.
    d. Residual volume remains in the lungs after forced expiration. 1200 ml in the adult.
    e. Minimal volume remains in the lungs after lung collapse. 600 ml in the adult.

f. Vital capacity is the sum of tidal, reserve inspiratory, and reserve expiratory volumes.

g. Functional residual capacity is the sum of expiratory reserve and residual volumes.

h. All volumes added comprise total lung capacity.

14. Oxygen enters the bloodstream by diffusion and is transported to the cells where it diffuses from the bloodstream. The gradient for oxygen is *into* the body.

15. Carbon dioxide leaves cells and enters the bloodstream by diffusion and is carried to the lungs where it diffuses to the alveoli. The gradient for oxygen is *out of* the body.

16. Gas transport occurs in and by the bloodstream.

a. Oxygen is carried on hemoglobin (95%) and in solution in the plasma (5%).

1) An oxygen dissociation curve relates amount of oxygen carried to P_{O_2}.

2) A decrease of pH results in a lowered hemoglobin O_2 saturation.

3) Increased temperature lessens hemoglobin O_2 saturation.

4) Increased DPG levels decrease hemoglobin O_2 saturation.

b. Carbon dioxide is carried in three basic ways.

1) 64 percent is carried as bicarbonate ion.

2) 27 percent is carried on hemoglobin.

3) 9 percent is carried in solution in the plasma.

17. The lungs are involved in acid-base regulation by their ability to eliminate carbon dioxide.

18. The lungs produce a variety of substances required for body function and metabolize substances that might exert unwanted effects on the body.

19. Protective devices the lung possesses include the mucociliary escalator, phagocytic cells, immunoglobulin secretion, filtering mechanisms, and reflexes intended to remove irritants (sneezing, coughing).

20. Breathing is initiated and controlled by central and peripheral influences.

a. Respiratory centers (inspiratory, expiratory, pneumotaxic, apneustic) provide for inspiration and expiration.

b. Blood P_{CO_2} or $[H^+]$ is believed to be the stimulus initiating inspiration.

c. Peripheral influences are reflexes that originate in baroreceptors and chemoreceptors and alter the rate and depth of breathing.

21. The fetus obtains oxygen and eliminates carbon dioxide via the placenta.

a. Exchange of gases is slower than in the adult lung.

* *See* Table 24.2 for actual values of P_{O_2} and P_{CO_2} in various body areas.

$b.$ P_{O_2} is lower, and P_{CO_2} higher than after birth; the hemoglobin is less saturated with oxygen.

$c.$ The fetal circulation is constructed so as to bypass the nonfunctional lungs (shunts are present).

22. At birth, lung breathing begins.

$a.$ Stimuli to initiate breathing may include visual and tactile stimuli, high blood P_{CO_2}, and lowered environmental temperature.

$b.$ The fetal circulation assumes the normal adult pattern with closure of the shunts.

23. Respiratory insufficiency occurs when the system cannot adequately supply O_2 or remove CO_2.

$a.$ Techniques of resuscitation may be employed to artificially ventilate the lungs in respiratory insufficiency, regardless of cause.

Questions

1. What is the functional significance of the changes in epithelial type that occur in passing from the nasal cavities to the alveoli?

2. Compare the blood supplies of conducting and respiratory divisions and comment on the effect on the tissues that a decreased blood flow in each system might have.

3. Compare lobes, and primary and secondary lobules of the lungs in terms of anatomical extent and functional significance.

4. Name the volumes and capacities of the lung and give a volume for each.

5. What is the role of surfactant in pulmonary physiology?

6. What ensures an "outside to inside" movement of oxygen? An "inside to outside" movement of carbon dioxide?

7. How, and to what extent, are oxygen and carbon dioxide carried by the blood?

8. What factors determine the extent of oxygenation of the blood?

9. How is the lung involved in acid-base regulation?

10. What mechanisms exist to protect the body from inhaled toxins?

11. How is breathing controlled? What supplies the stimulus that leads to a rhythmical pattern of breathing?

12. How does the fetus acquire O_2 and eliminate CO_2? Are there "special" features of its blood that ensure adequate oxygenation; if so, what are they?

13. How is the fetal circulation adapted to bypass the nonfunctional lung?

14. How do respiratory and circulatory systems change at or after birth? What may cause "the first breath"?

Readings

Avery, Mary E. et al. "The Lung of the Newborn Infant." *Sci. Amer. 228*, Apr. 1973.

Comroe, Julius H. *Physiology of Respiration.* Year Book Medical Pubs. Chicago, 1965.

Comroe, Julius H. "The Lung." *Sci. Amer. 214*:57, Feb. 1966.

Fraser, R. G., and J. A. P. Pare. *Structure and Function of the Lung.* Saunders. Philadelphia, 1971.

Heinemann, H. O, and A. P. Fishman. "Nonrespiratory Functions of the Mammalian Lung." *Physiol. Rev. 49*:1, 1969.

Hochachka, P. W., G. C. Liggins, J. Qvist, R. Schneider, M. Y. Snider, T. R. Wonders, and W. M. Zapol. "Pulmonary Metabolism During Diving: Conditioning Blood for the Brain." *Science 198*:831, 25 Nov. 1977.

Hock, Raymond J. "The Physiology of High Altitude." *Sci. Amer. 222*:52, Feb. 1970.

Macklem, Peter T. "Respiratory Mechanics." *Ann. Rev. Physiol. 40*:157, 1978.

Newhouse, M., J. Sanchis, and J. Bienenstock. "Lung Defense Mechanisms." *New Eng. J. Med. 295*:990, Oct. 28, 1976; *295*:1045, Nov. 4, 1976.

Science News. "Sudden Infant Death." *113*:234, April 15, 1978.

Smith, J. Robert, and Stephen A. Landaw. "Smokers' Polycythemia." *New Eng. J. Med. 298*:6, Jan. 5, 1978.

Smith, Kline, and French Laboratories. *Emphysema and Related Diseases.* Cliggott Publishing Co. Hackensack, N.J., 1967.

Strang, L. B. "Growth and Development of the Lung: Fetal and Postnatal." *Ann. Rev. Physiol. 39*:253, 1977.

Wyman, Robert J. "Neural Generation of the Breathing Rhythm." *Ann. Rev. Physiol. 39*:417, 1977.

Chapter 25
The Digestive System

Objectives

After studying this chapter, the reader should be able to:

- Outline the early development of the digestive system and describe some of its congenital anomalies, with causes.

- Diagram and label a cross section of the gut to illustrate the tissue plan of the wall.

- Describe the mouth as to limits, the organs contained within it, and its role in the digestive process.

- Describe the structure of the tongue and its role in digestion.

- List the types of teeth in the mouth, their numbers, and explain the structure of a tooth.

- Name the salivary glands, describe their cellular makeup, explain the role of saliva in digestion and control of salivary secretion.

- Explain the structure and function of the esophagus.

- Describe the membranes lining the abdominopelvic cavity and how they suspend organs within that cavity.

- Describe the structure of the stomach, its role in digestion, and control of its secretion.

- Describe the structure of the small intestine, with special attention to devices for increasing its surface area.

- Name the enzymes and other substances that are secreted into, or produced by, the small intestine, and what each contributes to the digestion of foodstuffs.

- Explain how secretion of the pancreas, gall bladder, and intestinal wall are controlled.

- Describe the structure and functions of the large intestine.

- Describe the gross and microscopic structures of the pancreas and liver, and give their functions.

- Describe the types of movements that occur in the alimentary tract, their purposes, and describe some reflexes that are important in control of motility in the tract.

The digestive system is composed of the tubelike ALIMENTARY TRACT and the ACCESSORY ORGANS that lie inside and outside of the tract. It performs the essential tasks of RECEIVING FOODS and nutrients (ingestion), PREPARATION of foodstuffs for absorption and utilization by cells (digestion), ABSORPTION of end products of digestion, and ELIMINATION (egestion) of unusable residues of the digestive process.

Development of the system

Formation of the gut

Development of the embryo in the first 3 weeks produces a cavity lined with endoderm that is the primitive gut. Growth of the embryo results in the formation of a gut tube which, by 3.5 weeks, may be subdivided into an anterior FOREGUT, a middle MIDGUT, and a posterior HINDGUT (Fig. 25.1). Each section is supplied by a major branch of the aorta that will form its blood supply throughout life. The *celiac artery* supplies the foregut, the *superior mesenteric artery* supplies the midgut, and the *inferior mesenteric artery* supplies the hindgut. Derivatives of each portion are given in Table 25.1, with indication of when each feature becomes obvious. Formation of the mouth and pharynx was outlined in Chapter 24.

Clinical considerations

As the major changes mentioned above take place, several other processes are occurring. For example, the open cavities of the gut tube are obliterated by proliferation of the lining epithelium; later, recanalization occurs. The membranes separating the mouth and anal cavities from the outside normally rupture to provide a tube open at both ends. Nerve cells, muscular and connective tissue coats are forming around the tube. Alterations in these processes create a wide variety of congenital disorders. Several of the more common disorders are presented in Table 25.2 and are shown in Figure 25.2.

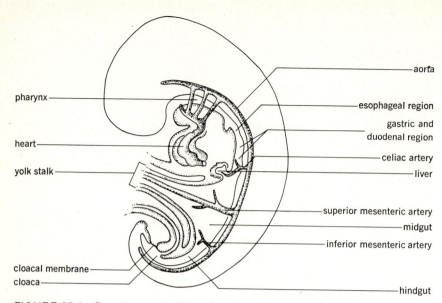

pharynx

aorta

esophageal region

gastric and
duodenal region

heart

celiac artery

liver

yolk stalk

superior mesenteric artery

midgut

inferior mesenteric artery

cloacal membrane

cloaca

hindgut

FIGURE 25.1   Development of the digestive system at
3½ weeks.

TABLE 25.1	Derivatives of the gut tube	
Part of tube	Derivatives	Comments
Foregut	Esophagus	Recognizable at 4 weeks.
	Stomach	Recognizable at 4 weeks, becomes baglike at about 8–10 weeks, and assumes typical form at about 12 weeks.
	Duodenum to entrance of bile duct	Recognizable at 4 weeks.
	Liver Pancreas	Develop as outgrowths of gut at about 4 weeks; liver lobes form by 6 weeks; pancreas complete by 10 weeks.
	Bile ducts and gall bladder	Connection retained to duodenum forms ducts. Bladder is an outgrowth of duct.
Midgut	Rest of small intestine	Recognizable at 4 weeks; elongates and coils at 5 weeks; villi at 8 weeks.
	Cecum, appendix, one half of large intestine	Separated at 5 weeks; completed by 8 weeks.
Hindgut	Remainder of large intestine, rectum, anal canal	Formed by 4 weeks; completed by 7 weeks.

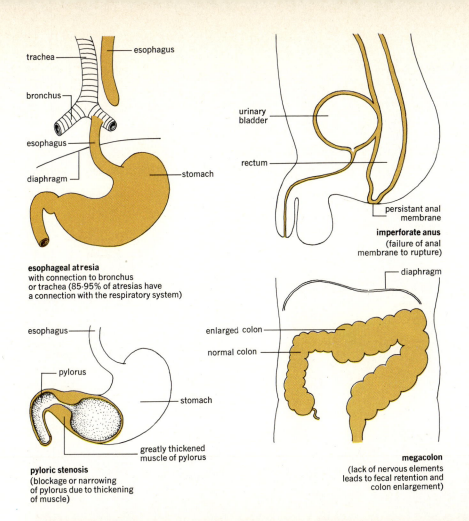

esophageal atresia
with connection to bronchus
or trachea (85-95% of atresias have
a connection with the respiratory system)

pyloric stenosis
(blockage or narrowing
of pylorus due to thickening
of muscle)

imperforate anus
(failure of anal
membrane to rupture)

megacolon
(lack of nervous elements
leads to fecal retention and
colon enlargement)

FIGURE 25.2 Some common congenital disorders of the digestive system.

TABLE 25.2 Some common congenital disorders of the digestive system			
Disorder	Frequency (no. per births)	Cause of disorder	Comments
Esophageal atresia	1/2500–3000	No recanalization of esophagus, improper separation from respiratory system. No passage to stomach.	Infant shows excess of saliva and poor nutrition since foods cannot reach stomach.
Pyloric stenosis	1/200 male 1/1000 female	Excessive development of muscle fibers of distal end of stomach	Blockage of stomach exit causes vomiting of feedings, weight loss, and dehydration.
Imperforate anus	1/5000	Failure of anal membrane to rupture. No anus formed.	Surgery necessary.
Megacolon	1/25,000	Failure of nerve cells to innervate a section of the colon.	Part without nerves does not move contents onward, accumulates feces and dilates.

The organs of the digestive system

The term ALIMENTARY TRACT is used to refer to the tubular organs of the system, including the mouth, pharynx, esophagus, stomach, small and large intestines, rectum, and anal canal. The ACCESSORY ORGANS of the system include those that develop from the tube, but which (usually) lie outside the tube and communicate with it by ducts. The salivary glands, liver, and pancreas are the major organs in this category. Also included as accessory organs are the tongue and teeth. The major organs of the system are shown in Figure 25.3.

FIGURE 25.3   The organs of the digestive system.

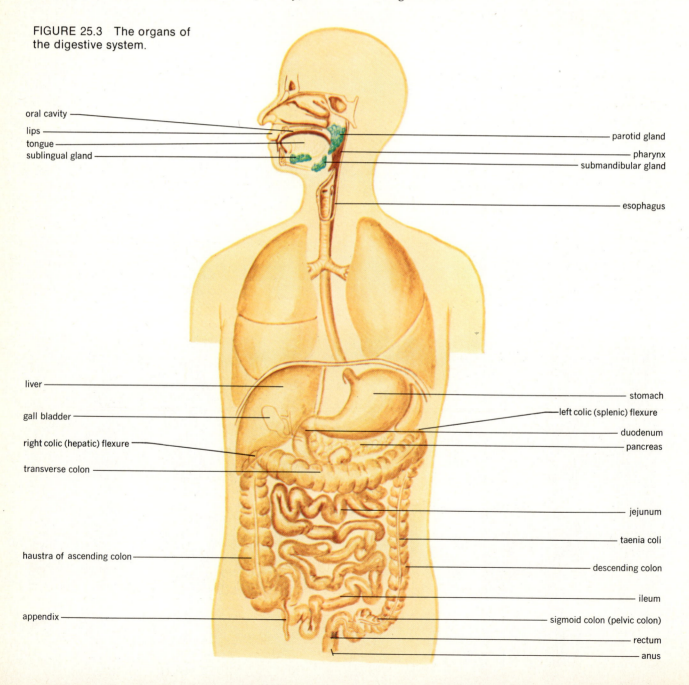

The structure and functions of the organs of the system

Structural layers of the alimentary tract

The organs of the alimentary tract, particularly those from esophagus to anus, have a basic organization of tissues in their walls (Fig. 25.4). Four main layers are described from inside outward.

   1. The *mucosa* (mucous membrane) consists of an epithelial lining of a type that varies according to the jobs each organ is carrying out. The epithelium is underlain by connective tissue that is rich in glands. These glands produce most of the substances necessary for digestion of foods. Absorption also occurs through this layer.

   2. The *submucosa* is a layer of vascular connective tissue serving to nourish the tissues of the tube wall; it also holds the mucosa to the muscular layers. Nerve fibers and cells form the *submucosal plexus* (of Meissner) in the connective tissue of this layer. The

nerve cells of this plexus supply the smooth muscle of the villi with impulses.

   3. The *muscularis externa* typically consists of two layers of smooth muscle that are responsible for the movements that occur in the tract. The inner layer is circularly arranged (actually a tight spiral), and the outer layer is longitudinally arranged (actually a very loose spiral). The *myenteric plexus* (of Auerbach) is formed by nerve fibers and cells lying between the two layers of muscle, and controls the activity of the muscularis externa.

   4. An outer *serosa* or *fibrosa* completes the tube wall. The term serosa is used to describe the layer of connective tissue and epithelium that forms the outer covering of most of the organs located in the abdominopelvic cavity; a fibrosa is a layer of connective tissue that is not separated from the surrounding tissues. Fibrosas occur on the organs lying in the thoracic cavity, such as the trachea and the esophagus.

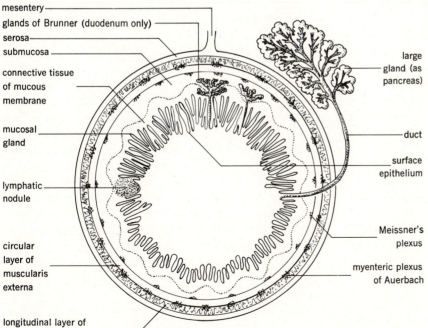

FIGURE 25.4   A diagram of a cross section of the intestinal tract to show its tissue layers.

mesentery

glands of Brunner (duodenum only)

serosa

submucosa

connective tissue of mucous membrane

mucosal gland

lymphatic nodule

circular layer of muscularis externa

longitudinal layer of muscularis externa

large gland (as pancreas)

duct

surface epithelium

Meissner's plexus

myenteric plexus of Auerbach

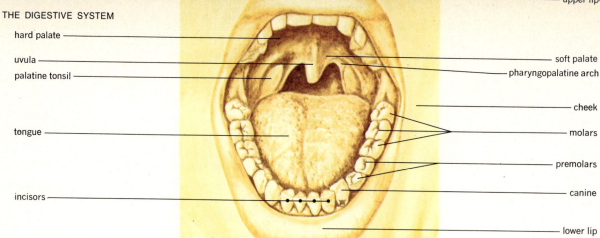

FIGURE 25.5   The mouth and oral cavity.

The mouth and oral cavity (Fig. 25.5)

STRUCTURE. The MOUTH is that part of the digestive system bounded by the lips, cheeks, hard palate, tongue, and soft palate. The ORAL CAVITY is that part of the mouth enclosed by the teeth and gums. The VESTIBULE lies between the teeth and the cheeks and lips. The lips and cheeks are important in speech and in guiding food between the teeth during chewing. The epithelium in this area is stratified squamous to withstand the wear-and-tear the mouth must take, and is translucent, allowing the blood vessels beneath to show their red color.

FUNCTION. The mouth receives food, and mechanical and chemical digestion is begun here.

CLINICAL CONSIDERATIONS. Many types of lesions may occur in the mouth lining. *Koplik's spots* are reddish spots with blue-white centers that appear on the side walls of the mouth before the rash of measles appears. *Canker sores* are viral lesions of the mouth that resemble tiny volcanoes; that is, they are elevated with an ulcerlike depres-

sion in their center. *Epstein's pearls* are whitish plugs of keratin in the roof of the mouth of newborns. They have no clinical significance.

The tongue (Fig. 25.6)

STRUCTURE. The tongue is primarily a muscular organ, composed of skeletal muscle covered with mucous membrane. The mucosa on the upper surface is formed into a variety of folds or PAPILLAE. FILIFORM PAPILLAE are more or less pointed projections, and roughen the surface of the tongue to enable it to more efficiently guide foods during chewing and swallowing. They also contain nerves for touch sensations. FUNGIFORM PAPILLAE are rounded projections and also serve to roughen the tongue surface. Most also contain taste buds, serving the sense of taste. VALLATE (circumvallate) PAPILLAE are large papillae forming a V-shaped line on the posterior one third of the tongue. They are sunken into the tongue and contain taste buds.

FUNCTION. As indicated above, the tongue serves in swallowing, manipulating food for chewing,

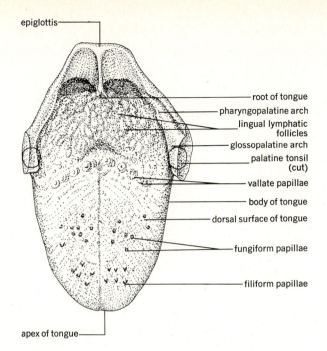

epiglottis

root of tongue
pharyngopalatine arch
lingual lymphatic follicles
glossopalatine arch
palatine tonsil (cut)
vallate papillae
body of tongue
dorsal surface of tongue
fungiform papillae
filiform papillae

apex of tongue

FIGURE 25.6   The superior surface of the tongue.

and in taste. It is also an important organ in the articulation of speech. It is anchored anteriorly to the floor of the oral cavity by the FRENULUM.

CLINICAL CONSIDERATIONS.   An abnormally short frenulum restricts tongue movements and may interfere with speech *(tongue-tie)*. Inflammation of the tongue is GLOSSITIS. "Coatings" on the tongue are most commonly associated with digestive upsets and smoking. The appearance of the tongue, its size, and mobility are often used to aid the diagnosis of pernicious anemia, endocrine disorders, and nerve defects.

In PERNICIOUS ANEMIA, the tongue is often sore, appears beefy red in color, and may develop patchy white vesicles on its surface. Loss of papillae and the development of a slick smooth surface is a later development in this disease.

In ACROMEGALY, characterized by excessive growth hormone production in an adult, the tongue becomes enlarged. In HYPOTHYROIDISM, the tongue also becomes enlarged.

DAMAGE TO THE XIITH CRANIAL NERVE (hypoglossal) may be reflected by the tongue deviating to one side or the other when its owner is told to "stick out your tongue."

The teeth

Each person receives two sets of teeth during his or her lifetime. The DECIDUOUS or milk teeth consist of 20 teeth that usually begin to erupt about 6 months of age, with one appearing about each month thereafter until the full set has erupted. These teeth include incisors, canines and premolars. The permanent teeth are 32 in number, include

    8 incisors
    4 canines
    8 premolars
    12 molars

and appear between 6 and 17 years of age. The last 3 teeth in each half of each jaw come in only once; all other teeth are replaced.

There are several types of teeth *(see* Fig. 25.5).

Incisors are chisel-shaped and exert a scissorslike action useful in biting.

Canines are conical, and are most useful in tearing or shredding food.

Premolars (bicuspids), and *molars* (tricuspids) are specialized for grinding foods and are the most important teeth in subdividing food (mechanical digestion).

STRUCTURE. Each tooth has a CROWN, a NECK, and a ROOT. A section of a tooth shows these and other features (Fig. 25.7). The CLINICAL CROWN is the visible part, the ANATOMICAL CROWN extends to the neck and is covered with ENAMEL, the hardest substance in the body. The root is covered with CEMENTUM, a bonelike material. Where enamel and cementum meet is the NECK *(cervix)* of the tooth. The main mass of the tooth is com-

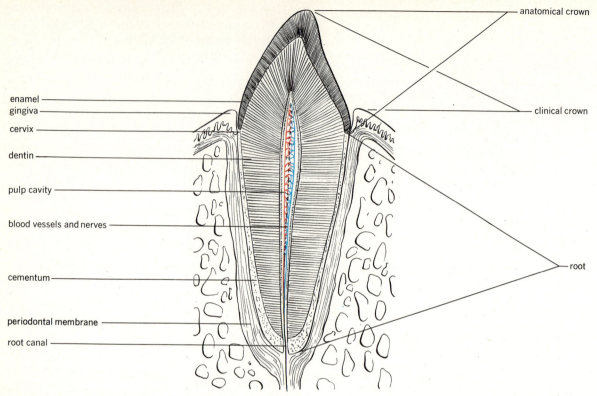

enamel
gingiva
cervix
dentin
pulp cavity
blood vessels and nerves
cementum
periodontal membrane
root canal

anatomical crown
clinical crown
root

FIGURE 25.7   Longitudinal section of a tooth to show its structure.

posed of DENTINE, a soft, yellowish, bonelike substance. The roots of the teeth fit in the sockets (alveoli) in the jaw bones. A PERIODONTAL MEMBRANE holds the tooth firmly but not rigidly, in its socket.

FUNCTION.   The teeth mechanically subdivide the food for easier digestion.

CLINICAL CONSIDERATIONS.   Cavity formation (caries) in the teeth appears to be one of the chief penalties of civilization and the consumption of soft, sweet foods. Acid production from bacterial action on foods in the mouth is regarded as the agent that dissolves tooth substance and leads to cavity formation. Adequate brushing within five minutes after the intake of food or drink is a

major aid in the prevention of cavities. Brushing should begin when the first tooth erupts. Coating the teeth with fluoride containing compounds and various types of plastic coatings appears to increase the resistance of the tooth to acid action.

The term PERIODONTAL DISEASE is used to refer to inflammation or degeneration or both of any of the tissues surrounding the teeth. These tissues include the gums (gingivae), bone, periodontal membrane, and the cementum of the teeth themselves. The diseases are characterized by loosening of the teeth, resorption of bone, and shrinking of the gums.

GINGIVITIS refers to inflammation, swelling, bleeding, and redness of the gums. It may be caused by local factors such as infection, foods jammed between the teeth and gums, and maloc-

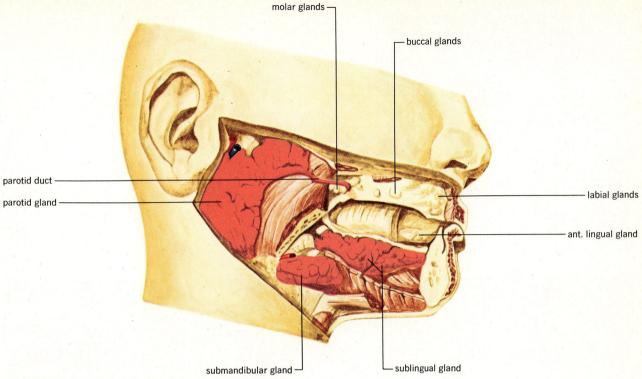

FIGURE 25.8 The locations of the salivary glands.

clusion, or by generalized factors such as vitamin C deficiency, allergy, diabetes, and leukemia.

PERIODONTITIS develops when gingivitis progresses to the point where it results in bone destruction. Most periodontal disease can be reduced by attention to oral hygiene and regular professional care.

The salivary glands (Figs. 25.8 and 25.9)

STRUCTURE. Three pairs of salivary glands lie outside the mouth and empty their secretions into the mouth. Two types of cells are found in various combinations in the glands: *serous cells* are small and granular, and produce a watery secretion rich in digestive enzyme; *mucous cells* are large and pale staining and produce a slimy mucus that is used mainly for purposes of lubrication. Table 25.3 summarizes the important facts concerning the structure of the salivary glands.

FUNCTION. The composite fluid of all three pairs of salivary glands is the SALIVA. Some of its characteristics and constituents are presented in Table 25.4.

Saliva functions to MOISTEN and SOFTEN ingested foods, to LUBRICATE them for swallowing, to CLEANSE the teeth and mouth, and acts as a route of EXCRETION for materials such as urea and uric acid.

The digestive function of the saliva centers around its content of SALIVARY AMYLASE (ptyalin) an enzyme that starts carbohydrate digestion according to the equation:

$$\text{Starch} \xrightarrow{\text{amylase}} \underset{(95-97\%)}{\text{dextrins}} \xrightarrow{\text{amylase}} \underset{(3-5\%)}{\text{disaccharides}}$$

Dextrins are polysaccharides containing fewer simple sugar units than starches. Foods normally remain in the mouth for only a short time, so that

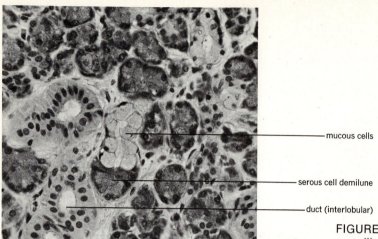

mucous cells

serous cell demilune

duct (interlobular)

FIGURE 25.9   Photomicrograph of a section of a sub-mandibular salivary gland.

TABLE 25.3 The salivary glands					
Name of gland	Location	Cellular composition	Name of duct	Entry of duct into mouth	Secretion contains
Parotid	Side of mandible in front of ear	All serous	Stensen's	Lateral to upper second molar	Water, salts, enzyme
Submandibular	Beneath the base of the tongue	Mostly serous, some mucous	Wharton's	Papilla lateral to frenulum	Water, salts, enzyme, some mucus
Sublingual	Anterior to submandibular under tongue	Mostly mucous, some serous	Rivinus'	With duct of submandibular	Mostly mucus, a little water, salt, and enzyme

TABLE 25.4 Some characteristics and constituents of human saliva	
Characteristic or constituent (examples)	Value or amount
Average daily volume	1000–1500 ml
Water	99.5%
Solids Inorganic salts (NaCl, KCl, NaHCO$_3$, KHCO$_3$, Na$_3$PO$_4$, K$_3$PO$_4$) Organic substances (urea, uric acid, proteins, salivary amylase, mucus)	0.5%
pH	5.8–7.1

breakdown to disaccharides (double sugars) occurs only to a slight degree.

CONTROL OF SALIVARY SECRETION (Fig. 25.10). The secretion of saliva is controlled by nervous reflexes, with the taste buds as receptors, cranial nerves VII and IX as afferent and efferent pathways, the salivary nuclei as centers, and the glands themselves as the effectors. The nature as well as quantity of the saliva may be altered by this mechanism. For example, dry foods stimulate both serous and mucus secretion; soft moist foods stimulate less secretion of mucus. Cerebral in-

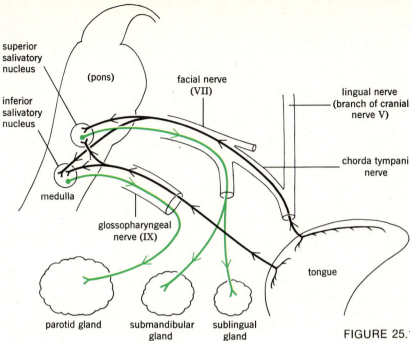

superior
salivary
nucleus

(pons)

facial nerve
(VII)

lingual nerve
(branch of cranial
nerve V)

inferior
salivary
nucleus

chorda tympani
nerve

medulla

glossopharyngeal
nerve (IX)

tongue

parotid gland

submandibular
gland

sublingual
gland

FIGURE 25.10   The control of salivary secretion.

fluences also cause salivary secretion as evidenced by "watering of the mouth" on sight, smell, or sound of food in preparation.

CLINICAL CONSIDERATIONS. PAROTITIS refers to inflammation of the parotid gland, as by mumps virus.

The pharynx (*see* Fig. 24.3)

The basic structure of the pharynx has been considered in Chapter 24.

The esophagus

STRUCTURE. The esophagus is a muscular tube about 25 centimeters (10 inches) in length that connects the pharynx to the stomach. Its wall structure follows the plan described previously. The epithelium is stratified squamous to resist abrasion created by swallowing food. A thick-

ened circular layer of muscle forms a functional sphincter at the junction of esophagus and stomach and tends to prevent regurgitation of stomach contents into the esophagus. The thickened muscle is named the *gastroesophageal* (cardiac) *constrictor*.

FUNCTION. The esophagus conducts food to the stomach; it has no digestive function.

CLINICAL CONSIDERATIONS. ACHALASIA refers to failure of the gastroesophageal constrictor to relax and permit substances to enter the stomach. Enlargement of the esophagus (MEGAESOPHAGUS) may occur as materials accumulate in the lower portion. If gastric contents do regurgitate into the lower portion of the esophagus, the acid within it may irritate the esophagus; HEARTBURN results.

Mesenteries and omenta (Fig. 25.11)

The abdominopelvic (peritoneal) cavity, in which

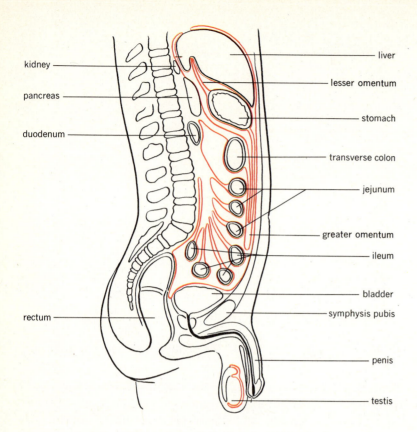

kidney

pancreas

duodenum

rectum

liver

lesser omentum

stomach

transverse colon

jejunum

greater omentum

ileum

bladder

symphysis pubis

penis

testis

FIGURE 25.11   The mesenteries and omenta of the abdomen.

most of the organs of the alimentary tract lie, has a serous membrane lining known as the PERITONEUM. Organs developing in tissues of the wall of the cavity may grow into the cavity, pushing the lining ahead of them. They may then lose their connections with the wall and become suspended by a double-layered fold of the peritoneum. The term PARIETAL PERITONEUM is given to the lining membrane of the cavity; VISCERAL PERITONEUM is the name given to the peritoneum on the organ surface; the double-layered suspending structure is known as a MESENTERY. Specific mesenteries are often named according to the organs they suspend, for example: mesogastrium (stomach), mesocolon (large intestine).

OMENTA (sing.: omentum) are mesenteries that lie between two organs, such as the liver and stomach, and stomach and duodenum. They arise when two organs develop one behind another

and move into the cavity. They do not suspend organs. Adipose tissue is commonly stored in the omenta.

Some organs, such as pancreas and kidney, remain against the body wall, and are not suspended by mesenteries in the cavity. Such organs lie behind (retro) the peritoneum and are said to be RETROPERITONEAL in position.

CLINICAL CONSIDERATIONS. PERITONITIS is inflammation of the peritoneum due to bacterial invasion of the cavity. It may follow rupture of a hollow organ, or organisms may enter through the female genital tract, bloodstream, lymph stream, or after surgery within the cavity. The continuous nature of the lining and covering membranes allows rapid spread of the inflammation and may result in involvement of most of the organs in the cavity.

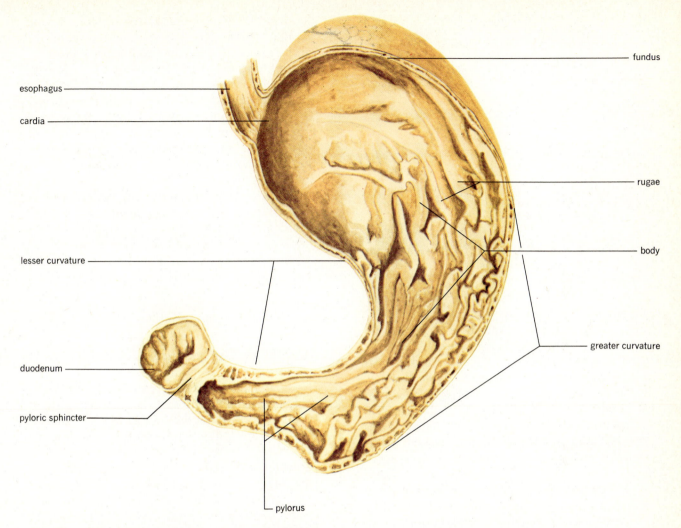

FIGURE 25.12   The gross anatomy of the stomach.

The stomach

STRUCTURE. The stomach is a J-shaped, baglike organ lying under the left side of the diaphragm. Its gross anatomy is illustrated in Figure 25.12. Wall structure follows the typical plan, with three layers of smooth muscle in the externa. The epithelium is simple columnar, with the outer portions of the cells filled with mucus. The mucus normally prevents the destruction of the stomach by the enzymes and acid within it. Some 35 million gastric glands lie in the mucosa, and these produce the gastric juice.

FUNCTION. The stomach RECEIVES FOOD, STORES IT while the initial stages of digestion are occurring, and COMMENCES THE DIGESTION OF PROTEINS. Gastric juice is a watery solution of HYDROCHLORIC ACID (HCl), containing PEPSIN, and small quantities of lipase. Its pH is between 0.9 and 2.0. Steps in the formation of active pepsin are shown in the following equations:

In cells of the gastric glands

$$CO_2 + H_2O + NaCl \longrightarrow HCl + NaHCO_3$$

Secretion of pepsinogen (an inactive precursor of pepsin)

Then

$$HCl + pepsinogen \longrightarrow pepsin$$

Pepsin attacks large protein molecules and breaks them into units known as *proteoses* and *peptones.* These units contain 4 to 12 amino acids. Gastric lipase is of little value in the human; it is destroyed by the low stomach pH.

CONTROL OF GASTRIC SECRETION. Gastric secretion occurs in three phases.

The CEPHALIC PHASE occurs when food is seen, smelled, or tasted. Impulses pass from brain to stomach over the vagus nerves, and about 50 to 150 milliliters of juice is produced by this phase.

The GASTRIC PHASE is controlled by two mechanisms. When food enters and stretches the stomach, and when the products of protein digestion act on the pyloric portion of the stomach, a hormonelike substance called GASTRIN is produced. Gastrin is distributed through the bloodstream to the entire stomach and causes production of 600 to 750 milliliters of gastric juice.

The INTESTINAL PHASE adds additional small quantities of juice to the previous phases. It occurs when the digesting food mass (chyme) enters the duodenum of the small intestine. This produces a hormone, unnamed at present, that stimulates gastric secretion.

The intestinal mucosa also produces a hormone called ENTEROGASTRONE that reaches the stomach via the bloodstream. This material slows the muscular activity of the stomach and allows a longer time for gastric digestion of proteins to occur.

CLINICAL CONSIDERATIONS. PEPTIC ULCERS may occur in the stomach if the mucus lining is destroyed or is not secreted normally. The acid and pepsin quite literally digest the stomach wall.

The cause of peptic ulcer of the stomach is not definitely known, but excessive secretion of gastric juice when there is no food in the stomach is an important factor in the production of the ulcer. Such production of acid commonly occurs during stress, and has been related to the emotional pressures generated by a competitive modern society. The "executive ulcer" is a classical example of this type of disorder. The ulceration usually extends only into the mucosa, but occasionally penetrates the muscularis externa and perforates into the abdominal cavity.

GASTRITIS is inflammation and irritation of the stomach lining produced by irritant foods or drink. Avoidance of a food creating the irritation is the only known help. LAVAGE means "to wash," and refers to flushing toxins out of the stomach ("having the stomach pumped").

The small intestine

STRUCTURE. The small intestine extends from the pyloric valve of the stomach some 3.4 to 4 meters (10–12 feet*) to the large intestine. It is divided into three parts on the basis of microscopic structure: the DUODENUM, the first 25 to 30 centimeters (10–12 inches); the JEJUNUM, the next 1 to 1.5 meters (3–4 feet); and the ILEUM, the last 2 to 2.5 meters (6–7 feet). All parts have a wall structure following the typical plan described earlier. VILLI, and PLICAE (Fig. 25.13) increase surface area of the intestine for production of digestive enzymes and absorption of end products of digestion. The villi contain networks of arteries, capillaries, and veins, and a lymph capillary known as a lacteal (Fig. 25.14).

FUNCTION. The small intestine receives the secretions of the pancreas, liver (bile), and its own glands, and digestion of all three basic foodstuff groups is completed here. Most of the absorption of nutrients occurs from the small intestine. The

* Measurement in the living person; postmortem measurements may reach 20 to 21 feet.

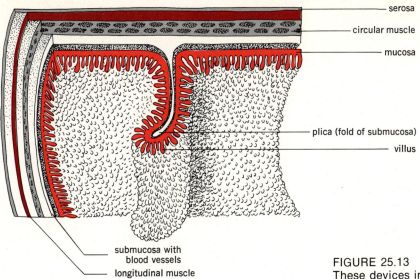

serosa

circular muscle

mucosa

plica (fold of submucosa)

villus

submucosa with
blood vessels

longitudinal muscle

FIGURE 25.13   Villi and plicae of the small intestine.
These devices increase surface area.

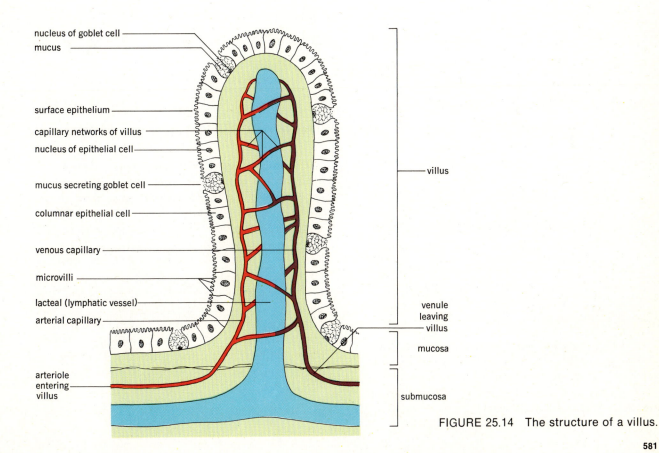

nucleus of goblet cell

mucus

surface epithelium

capillary networks of villus

nucleus of epithelial cell

mucus secreting goblet cell

columnar epithelial cell

venous capillary

microvilli

lacteal (lymphatic vessel)

arterial capillary

arteriole
entering
villus

villus

venule
leaving
villus

mucosa

submucosa

FIGURE 25.14   The structure of a villus.

digestive juices secreted into the small intestine include the following.

Pancreatic juice. Pancreatic juice is an alkaline fluid (pH 7.1–8.2) that stops pepsin action, and creates the proper environment (pH 7–8) for the action of all enzymes operating in the intestine. Enzymes in pancreatic juice act on all three main foodstuff groups and include:

1. *Trypsin*, secreted as trypsinogen and activated by enterokinase, an intestinal enzyme.

2. *Chymotrypsin*, secreted as chymotrypsinogen and activated by trypsin.

3. *Carboxypeptidase*, secreted as procarboxypeptidase and activated by trypsin.

These three enzymes further the digestion of proteins by converting proteoses and peptones to *dipeptides*, units containing two amino acids.

4. *Pancreatic amylase* (amylopsin) converts the dextrins of salvary digestion to *disaccharides*, such as maltose, lactose, and sucrose.

5. *Pancreatic lipase* (steapsin) splits triglycerides into three *fatty acids* and *glycerol*, products that are ready for absorption.

Bile. Bile, secreted by the liver, aids lipase in its action by preventing the coming together of the small fat droplets in the intestine and thus allows more rapid digestion of the fats. This action is called the *emulsifying action* of bile. The bile salts also combine with the fatty acids and render them more easily absorbed. This is called the *hydrotropic action* of the bile.

Intestinal juice *(succus entericus)*. Intestinal juice is produced by the glands of the intestine itself. It completes the digestion of proteins and carbohydrates. Its enzymes include:

Erepsin. Erepsin is a name for a group of enzymes that split dipeptides into their individual amino acids.

Specific amylases attack a particular disaccharide and reduce it to simple sugars.

Maltase acts on maltose and releases two glucose units.

Sucrase acts on sucrose and releases glucose and fructose.

Lactase acts on lactose and releases glucose and galactose.

Foodstuff digestion has been completed, and the end products, amino acids, simple sugars, glycerol, and fatty acids, are ready for absorption. The digestion of the three main foodstuff groups is summarized below.

$$\text{Starch} \xrightarrow{\substack{\text{salivary}\\ \text{amylase}}} \text{dextrins} \xrightarrow{\substack{\text{pancreatic}\\ \text{amylase}}} \text{disaccharides}$$

intestinal amylases:

$$\begin{array}{l}\text{Maltase}\\ \text{Lactase}\\ \underline{\text{Sucrase}}\end{array} \longrightarrow \text{monosaccharides.}$$

$$\text{Proteins} \xrightarrow{\substack{\text{pepsin}\\ \text{(stomach)}}} \substack{\text{proteoses \&}\\ \text{peptones}} \xrightarrow{\substack{\text{trypsin}\\ \text{carboxypeptidase}\\ \text{chymotrypsin}\\ \text{(pancreas)}}}$$

$$\text{dipeptides} \xrightarrow{\substack{\text{erepsin}\\ \text{(small}\\ \text{intestine)}}} \text{amino acids.}$$

$$\text{Triglycerides} \xrightarrow{\substack{\text{lipase}\\ \text{(pancreas)}}} \text{3 fatty acids + glycerol.}$$

Control of secretion of pancreas, liver, intestinal wall. Control of secretion of the three organs is carried out primarily by hormonal materials that are produced by the intestine and are carried by the blood to the organ controlled. The control mechanisms are summarized in Table 25.5.

CLINICAL CONSIDERATIONS. DUODENAL ULCER may result from the action of stomach acid and pepsin on the wall of the duodenum. A variety of irritants may cause inflammation of various parts of the intestine (duodenitis, jejunitis, ileitis). Severe inflammation may require removal of the affected portion and joining of the cut ends.

TABLE 25.5 Control mechanisms for secretion of pancreas, liver, and intestine

Organ	Stimulus	Hormone(s) produced	Effect of hormone
Pancreas	HCl in intestine (also water, meat juices, fats, alcohol)	Secretin	Stimulates secretion of watery, buffered fluid
	Fats	Cholecystokinin	Stimulates secretion of enzymes
Liver and gall bladder	Fats in intestine	Secretin	Stimulates liver secretion of bile
	as above	Cholecystokinin	Stimulates contraction of gall bladder and emptying of stored bile
Intestine	Stroking, stretching, flooding of intestine	Enterocrinin	Stimulates secretion of intestinal glands

The large intestine and associated structures (Fig. 25.15)

STRUCTURE. The LARGE INTESTINE is about 1½ meters (5 feet) long, and consists of the CECUM, APPENDIX, COLON, RECTUM, and ANAL CANAL. It is larger in diameter than the small intestine and throughout most of its length is easily recognized by its HAUSTRA or sacculations. The APPENDIX arises from the blind pouch of the cecum. The COLON consists of *ascending, transverse, descending,* and *sigmoid* portions, and has two *colic flexures* (*see* Fig. 25.3). The rectum (Fig. 25.16) lacks haustra, and has longitudinal RECTAL COLUMNS containing hemorrhoidal blood vessels. The anus, or opening to the exterior, is surrounded by an INTERNAL SPHINCTER of smooth muscle and an EXTERNAL SPHINCTER of voluntarily controlled skeletal muscle.

Microscopically, the large intestine and its parts are distinguished by lack of villi, and by the great numbers of goblet (mucus secreting) cells in the epithelium. The mucus lubricates the fecal material as water is absorbed from it. Additionally, the outer longitudinal layer of the muscularis externa is in the form of three bands known as the TAENIA COLI.

FUNCTION. The large intestine has no digestive function. It ABSORBS some 400 milliliters of water and much inorganic salt per day. It contains a large flora of microorganisms that produce vitamins K and B, and amino acids. These products are absorbed by the host. FECES ARE FORMED in the colon, and consist of 60 percent solid matter (bacteria, food residues, digestive secretions), and 40 percent water.

CLINICAL CONSIDERATIONS. COLITIS refers to inflammation of the colon. APPENDICITIS occurs when infectious agents enter an appendix that does not or cannot empty itself. The organ may require surgical removal before it ruptures and causes peritonitis. DIVERTICULOSIS describes a saclike protrusion of the colon wall where arteries enter the organ and weaken its wall. They may appear on X-rays, but usually do not cause symptoms. If the outpocketings accumulate fecal material and become infected, DIVERTICULITIS occurs. Diverticulitis produces pain in the lower left part of the abdomen, a "left-sided appendicitis."

HEMORRHOIDS are enlarged hemorrhoidal veins. They may bleed profusely and often require surgery.

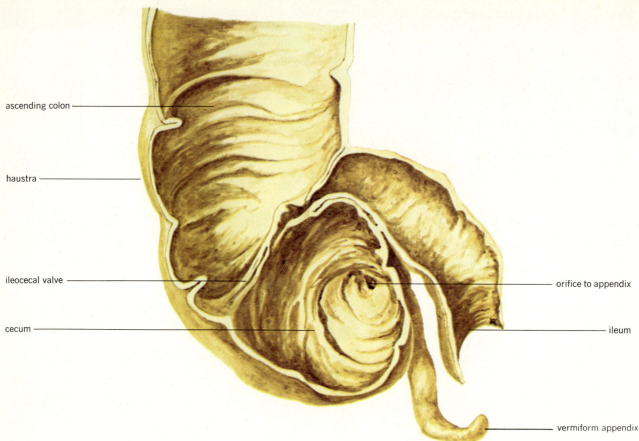

ascending colon

haustra

ileocecal valve

cecum

orifice to appendix

ileum

vermiform appendix

FIGURE 25.15   The ileocecal region.

The pancreas (Fig. 25.17)

STRUCTURE.   The pancreas is a carrot-shaped gland located along the greater curvature of the stomach. It has a HEAD, BODY, and tapering TAIL, is about 20 centimeters (8 inches) long, and weighs from 65 to 160 grams. It is a double gland (Fig. 25.18) having an EXOCRINE PORTION producing digestive enzymes, and an ENDOCRINE PORTION (islets of Langerhans) producing hormones.

FUNCTION. The digestive function of the pancreas has already been considered. The endocrine function includes the production of INSULIN, a hormone increasing the formation of glycogen and increasing cellular uptake of glucose, and of

GLUCAGON, a hormone causing glycogen breakdown to glucose and rise of blood sugar levels.

CLINICAL CONSIDERATIONS. CYSTIC FIBROSIS is a genetically transmitted abnormality that affects many exocrine glands, including the pancreas. Fibrous tissue invades the glands, reducing secretory capacity. Pancreatic insufficiency and failure to properly digest foods may result, with failure to thrive. Disorders of the endocrine portion are discussed in Chapter 29.

The liver (Figs. 25.19 and 25.20)

STRUCTURE. The liver is the largest gland in the body, weighing about 1500 grams (about 3 pounds)

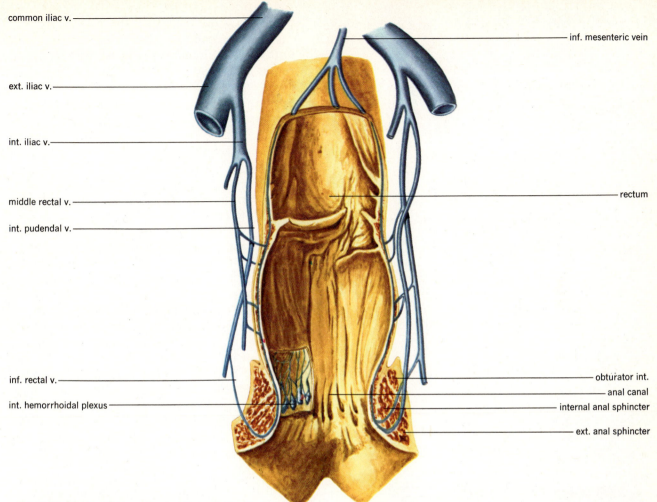

common iliac v.

ext. iliac v.

int. iliac v.

middle rectal v.

int. pudendal v.

inf. mesenteric vein

rectum

inf. rectal v.

int. hemorrhoidal plexus

obturator int.

anal canal

internal anal sphincter

ext. anal sphincter

FIGURE 25.16   The rectum and anal canal.

in the adult. It has four major LOBES, and is fixed in position by several ligaments. The unit of structure is the LOBULE, which consists of plates or cords of HEPATIC CELLS radiating from a CENTRAL VEIN, with SINUSOIDS lying between the plates of cells. The blood supply to the liver is double; the HEPATIC ARTERIES bring O_2 rich blood to the sinusoids, and the PORTAL VEIN brings nutrient-laden blood from the intestines to the sinusoids. Outgoing blood empties via the HEPATIC VEINS into the inferior vena cava.

A system of bile ducts begins in the lobules and ultimately carries bile to the gall bladder and small intestine.

FUNCTIONS. The liver produces an exocrine secretion (bile) whose functions have already been considered. Bile is stored in the gall bladder until required for digestion. Other functions of the liver include:

1. *Synthetic reactions.*
   a. Synthesis of certain amino acids, plasma proteins, prothrombin, antibodies, urea, creatine, cholesterol.
   b. Glycogen and glucose (gluconeogenesis).
   c. Phospholipids (for cell membranes, etc.).
   d. Fatty acids.
   e. Ketone bodies.

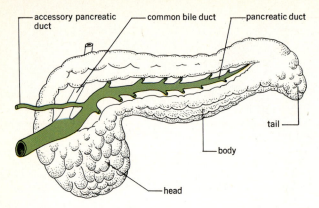

FIGURE 25.17   The gross anatomy of the pancreas.

2. *Metabolic reactions.*

a. Breakdown of glucose, glycogen, fatty acids, glycerol, amino acids, metabolism of hormones (insulin, aldosterone, testosterone, estrogens, and thyroid hormone).

b. Interconversion of certain amino acids to simple sugars and vice versa.

c. ATP formation.

3. *Embryonic formation of blood cells.*

4. *Destruction of aged red cells.*

5. *Storage.*

a. Glycogen.

b. Amino acids.

c. Fats.

d. Vitamins (A, B complex, D).

e. Iron and copper.

6. *Detoxification* (conversion of toxic substances to harmless compounds).

Specific metabolic functions of the liver are considered in Chapter 26.

CLINICAL CONSIDERATIONS. Because of the many and essential functions carried out by the liver, it can be considered one of the vital body organs. Interestingly, there is such a great reserve of liver tissue that a person can get along with only about one-fifth of the tissue functional. CIRRHOSIS of the liver involves replacement of cells by fibrous tissue when cells are destroyed by infection, parasites, alcohol, or other causes. HEPATITIS refers to inflammation of the liver. It is most commonly viral in origin, with the virus transmitted from one person to another by transfusion, needles, or other instruments, or by an oral route. The gall bladder not only stores bile, but concentrates it by absorption of water. If the solution in the organ becomes saturated, precipitation may occur with formation of GALLSTONES. The stones may block the bile ducts (Fig. 25.21) and cause JAUNDICE.

FIGURE 25.18   Photomicrograph of the exocrine and endocrine portions of the pancreas.

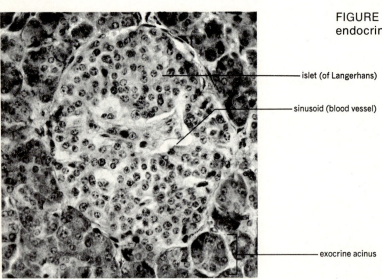

islet (of Langerhans)

sinusoid (blood vessel)

exocrine acinus

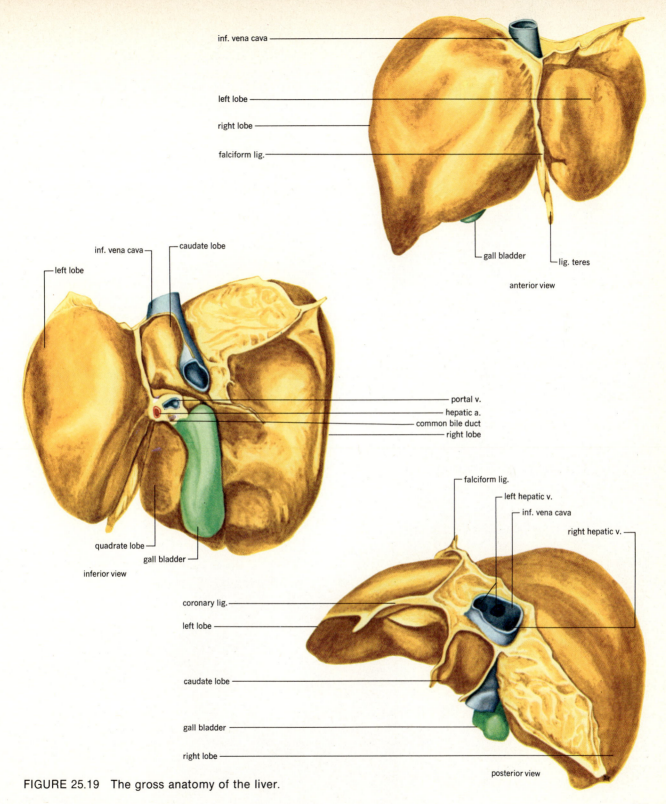

inf. vena cava

left lobe

right lobe

falciform lig.

gall bladder

lig. teres

anterior view

inf. vena cava — caudate lobe

left lobe

portal v.
hepatic a.
common bile duct
right lobe

quadrate lobe
gall bladder

inferior view

falciform lig.

left hepatic v.

inf. vena cava

right hepatic v.

coronary lig.

left lobe

caudate lobe

gall bladder

right lobe

posterior view

FIGURE 25.19   The gross anatomy of the liver.

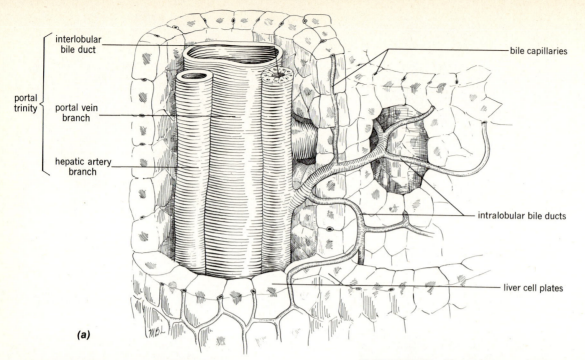

(a)

portal
trinity
- interlobular bile duct
- portal vein branch
- hepatic artery branch

bile capillaries

intralobular bile ducts

liver cell plates

FIGURE 25.20   The microscopic structure of the liver. *(a)* The relationship of the blood vessels and bile ducts to the liver cells. *(b)* A low-power micrograph of a liver lobule. *(c)* The portal canal area between lobules, showing vessels and ducts.

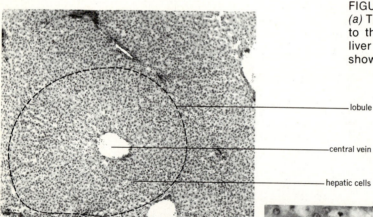

lobule

central vein

hepatic cells

(b)

portal canal

interlobular branch of portal vein

interlobular connective tissue

interlobular branch of hepatic artery

interlobular bile duct

sinusoid

hepatic cord

(c)

588

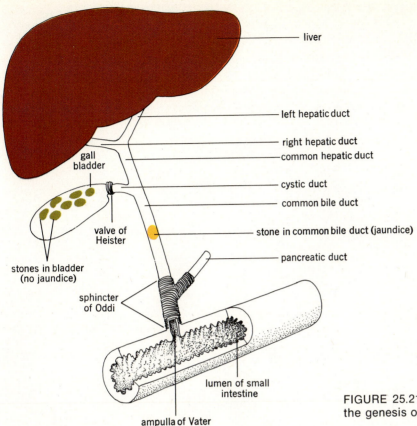

liver

left hepatic duct

right hepatic duct

common hepatic duct

gall bladder

cystic duct

common bile duct

valve of Heister

stone in common bile duct (jaundice)

stones in bladder (no jaundice)

pancreatic duct

sphincter of Oddi

lumen of small intestine

ampulla of Vater

FIGURE 25.21   The relationship of the bile ducts to the genesis of obstructive jaundice.

Motility in the tract

Movements in the digestive system, caused by muscular tissue, are essential to mechanical subdivision of foods and to transfer of foods along the tube. The main types of movement are briefly described below, and are summarized in Table 25.6.

CHEWING is started voluntarily by opening the mouth to receive food and closing it to start chewing. Rhythmical continuation of the activity is controlled by the brain.

SWALLOWING starts with collection of a mass of food (bolus) on the tongue, and movement of the bolus into the throat by the tongue. A positive pressure must be created by sealing the nasal and oral cavities to swallow. (A cleft palate prevents creation of a positive pressure.)

PERISTALSIS is the most characteristic type of movement occurring in the tract. It is a wavelike constriction of the gut that pushes ahead of it a mass of chyme. It is used mainly for transport of chyme along the length of the tube, and normally occurs only in one direction, from mouth to anus.

TONIC CONTRACTIONS and SEGMENTING CONTRACTIONS are local contractions of gut muscle that serve to mix and churn the contents of the organ.

The stomach shows a RECEPTIVE RELAXATION as food enters it, and thus it expands to contain a

TABLE 25.6   A summary of motility in the alimentary tract

Area	Type of motility	Frequency	Control mechanism	Result
Mouth	Chewing	Variable	Initiated voluntarily, proceeds reflexly	Subdivision, mixing with saliva
Pharynx	Swallowing	Maximum 20 per min	Initiated voluntarily, reflexly controlled by swallowing center	Clears mouth of food
Esophagus	Peristalsis	Depends on frequency of swallowing	Initiated by swallowing	Transport through esophagus
Stomach	Receptive relaxation	Matches frequency of swallowing	Unknown	Allows filling of stomach
	Tonic contraction	15–20 per min	Inherent by plexuses	Mix and churn
	Peristalsis	1–2 per min	Inherent	Evacuation of stomach
	"Hunger contractions"	3 per min	Low blood sugar level	"Feeding"
Small intestine	Peristalsis	17–18 per min	Inherent	Transfer through intestine
	Segmenting	12–16 per min	Inherent	Mixing
	Pendular	Variable	Inherent	Mixing
	Villus movements shortening and waving	Variable	Villikinin	Facilitates absorption
Colon	Peristalsis	3–12 per min	Inherent	Transport
	Mass movement	3–4 per day	Stretch	Fills pelvic colon
	Tonic	3–12 per min	Inherent	Mixing
	Segmenting	3–12 per min	Inherent	Mixing
	Defecation	Variable 1 per day-3 per week	Reflex triggered by rectal distension	Evacuation of rectum

meal. The stomach also exhibits "HUNGER CON-TRACTIONS," very strong peristaltic waves that usually signal a need for food intake.

The small intestine also shows PENDULAR MOVE-MENTS, a to-and-fro type of movement up and down short segments of the tube. They mix and churn intestinal contents. MOVEMENTS OF THE VILLI aid in absorption of substances by keeping "fresh" material next to the villus.

In the colon, DEFECATION is a strong peristaltic type of movement caused by stretching of the rectum by fecal material. Material is moved into the rectum by MASS MOVEMENT. Expulsion of material is aided by "straining," or using abdominal muscles to "push."

Control of motility

A variety of nervous, chemical, and physical factors control the strength and duration of motility in various parts of the tract.

In the esophagus, movement is controlled primarily by the VAGUS NERVE, which supplies the esophageal muscle.

Motility of the stomach depends on: the amount of fat it contains (more fat = slower activity); products of protein digestion (a nervous reflex called the ENTEROGASTRIC REFLEX is triggered by proteoses and peptones); osmotic pressure of stomach contents (hyper- and hypotonic solutions decrease motility).

In the intestine, the MYENTERIC REFLEX is a contraction caused by stretching the gut.

In the colon, the GASTROCOLIC REFLEX stimulates colon evacuation when the stomach is stretched.

The MUCOSAL REFLEX results in an increase of secretion and stimulation of muscular activity anywhere in the tract when it is irritated. The reactions tend to dilute and remove the irritating agent.

The PERITONEAL REFLEX causes an inhibition of motility in the tract when the peritoneum is traumatized, as after an abdominal operation.

Summary

1. The digestive system takes in food, digests it to absorbable end products, absorbs those products, and eliminates unusable materials.

2. The system begins its development during the third week of embryonic life with the formation of a fore-, mid-, and hindgut. Each part develops certain portions of the gut and its accessory structures (*see* Table 25.1).
   a. Abnormal development may cause relatively common congenital disorders of the system, including atresia, stenosis, and imperforate anus.

3. The organs of the system are divided into two groups.
   a. The tubelike alimentary tract includes the mouth, pharynx, esophagus, stomach, small and large intestines, appendix, rectum, and anal canal.
   b. The accessory organs include tongue, teeth, salivary glands, liver, pancreas, and gall bladder.

4. The alimentary tract has a typical layering of tissues in its wall that includes four layers.
   a. The mucosa is internal, glandular, and absorbs or protects.
   b. The submucosa is vascular connective tissue, and nourishes.
   c. The muscularis externa is muscular and creates the motility in the tract.
   d. An outer serous or fibrous layer surrounds and protects the tube.

5. The mouth receives food, and shows many lesions helpful in diagnosing disease.

6. The tongue is a muscular organ surfaced with a variety of papillae for roughening and taste. The organ manipulates food in the mouth and articulates speech.

7. The teeth subdivide food, occur in two sets (deciduous and permanent), are of four types (incisors, canines, premolars, and molars) and have a typical structure.

8. There are three pairs of salivary glands that secrete saliva.

   a. Saliva moistens and softens foods, cleanses the mouth, and excretes substances.

   b. Salivary amylase (ptyalin) splits starches to dextrins.

   c. Control of salivary secretion is by nerves.

9. The esophagus is a muscular tube conducting food to the stomach.

10. The peritoneum lines the abdominopelvic cavity, and covers organs. Mesenteries suspend organs within the cavity; omenta lie between organs and may store fat.

11. The stomach stores food, and secretes gastric juice.

   a. Gastric juice contains HCl, and pepsin.

   b. Pepsin breaks proteins to proteoses and peptones.

   c. Gastric secretion is controlled in three phases, cephalic, gastric, and intestinal. The first is under the control of the vagus nerve, the last two are hormonally controlled by gastrin and an unnamed hormone.

12. The small intestine completes the digestion of foods and absorbs the end products of digestion.

   a. It has a duodenum, jejunum, and ileum.

   b. Villi and plicae increase surface area for secretion and absorption.

   c. Pancreatic juice, emptied into the small intestine, contains three protein digesting enzymes called trypsin, chymotrypsin, and carboxypeptidase, that create dipeptides; amylase to create disaccharides; lipase to break down fats to glycerol and fatty acids.

   d. Bile, secreted by the liver and released from the gall bladder, emulsifies fats and renders fatty acids water soluble.

   e. Intestinal juice, secreted by the intestine wall, contains amylases that liberate simple sugars from double sugars; and erepsin for liberating amino acids from dipeptides.

13. Control of pancreatic, liver, and intestinal secretion is by hormones.

   a. Secretin and pancreozymin control pancreatic secretion.

   b. Secretin stimulates liver production of bile, and cholecystokinin releases bile from the gall bladder.

   c. Enterocrinin stimulates secretion by intestinal glands.

14. The large intestine has haustra, taenia coli, lacks villi, and consists of appendix, cecum, colon, rectum, and anal canal. The colon consists of

acending, transverse, descending, and sigmoid portions. It absorbs water, salts, products of bacterial activity, and forms feces.

15. The pancreas is carrot-shaped, and produces digestive enzymes and true hormones.

16. The liver is the largest gland of the body, has lobes and lobules, and carries on a variety of functions, including synthesis, metabolism, interconversion of substances, formation and destruction of blood cells, and storage of nutrients.

17. Movements (motility) of the tract subdivide foods, mix and churn food with enzymes, and move materials through and out of the tract (*see* Table 25.6 for types and details.
    a. Control of motility is by nerves, chemicals, and physical factors (osmotic pressure, stretch).

Questions

1. Now that you have studied the entire digestive system, explain the concept that the inside of the digestive organs is a part of the outside environment.

2. What is indicated by the phrase that, in its lumen each digestive organ had its "own peculiar ecology"?

3. Name the types of teeth in the human mouth and describe the anatomy of a tooth.

4. Name the tunics in the walls of a typical hollow, tubular, organ of the digestive system, and their tissue components.

5. List some of the ways by which the absorptive surface of the small intestine is increased.

6. What are some of the movements involved in the small intestine during digestion and absorption? How is each controlled?

7. What is the function of bile in the digestive system?

8. Describe the functions of the small intestine.

9. Describe the functions of the large intestine.

10. Name and locate the lobes of the liver.

11. Describe a liver lobule and its relationship to bile and blood channels.

12. Explain the statement: "The liver has a double blood supply."

13. The pancreas is a double gland with a twofold function. Explain this statement.

14. Trace, in proper order, the enzymes acting upon starch; list the products formed at each step. Do the same for protein.

Readings

Davenport, H. W. *Physiology of the Digestive Tract.* 3rd ed. Yearbook Medical Publishers. Chicago, 1971.

Davenport, H. W. "Why the Stomach Does Not Digest Itself." *Sci. Amer. 226:86,* Jan. 1972.

Gray, G. M. "Carbohydrate Digestion and Absorption: Role of the Small Intestine." *New Eng. J. Med. 292:1225,* June 5, 1975.

Haggerty, Robert J. "Sore Throats and Tonsillectomy." *New Eng. J. Med. 298:453,* Feb. 23, 1978.

Javitt, Norman B. "Hepatic Bile Formation." *New Eng. J. Med. 295:1464,* Dec. 23, 1976; *295:1511,* Dec. 30, 1976.

Johnson, Leonard R. "Gastrointestinal Hormones and Their Functions." *Ann. Rev. Physiol. 39:135,* 1977.

Kappas, Attallah, and Alvito P. Alvares. "How the Liver Metabolizes Foreign Substances." *Sci. Amer. 232:22,* June 1975.

Rabinowitz, Melba. "Why Didn't Anyone Tell Me About Bottle Mouth Cavities?" *Children Today,* vol. 3, no. 2, pp. 18–20. Children's Bureau, Office of Human Development. Washington, D.C., Mar–Apr 1974.

Rayford, Phillip L. et al. "Secretin, Cholecystokinin and Newer G-I Hormones." *New Eng. J. Med. 294:1093,* May 13, 1976; *294:1157,* May 20, 1976.

Science News. "Smoking and Gum Disease." *105:240,* April 13, 1974.

Thompson, D. A., and R. G. Campbell. "Hunger in Humans Induced by 2-Deoxy-D-Glucose: Glucoprivic Control of Taste Preference and Food Intake." *Science 198:1065,* 9 Dec. 1977.

Chapter 26

Absorption of Foodstuffs, Metabolism, and Nutrition

Objectives

After studying this chapter, the reader should be able to:

■ Explain the meaning of the following terms: absorption, metabolism, intermediary metabolism, and nutrition.

■ Explain what is absorbed, and how, from the mouth, stomach, and small and large intestines.

■ Outline the possible fates a glucose molecule has in the body.

■ State what glycolysis starts with, and the end products of the scheme.

■ Explain what glycogenesis and glycogenolysis involve.

■ Define what beta-oxidation is, and give its starting materials and products.

■ State how cholesterol is formed.

■ List the several types of lipoproteins and explain their relationship to atherosclerosis.

■ Outline what happens to amino acids in the body.

■ Show how lipids, carbohydrates, and proteins may be interconverted.

- Define basal metabolic rate, and give some factors influencing it.
- Explain how heat is gained, and lost, from the body, and how a nearly constant body temperature is maintained.
- State what a normal diet should contain in terms of basic foodstuffs.
- State sources of, requirements for, and uses of common vitamins in the body.
- State sources of, requirements for, and use of inorganic substances in the body.

As indicated in the previous chapter, the process of digestion has as its major aim the production of absorbable end products from more complex foodstuffs. Until these end products pass through the wall of the alimentary tract, they are not truly in the body and are unavailable for cellular use. ABSORPTION involves the TRANSFER OF MATERIALS through the mucosa of the alimentary tract into blood and lymph vessels. Transfer may be active, as by active transport and pinocytosis, or passive, as by diffusion and osmosis.

METABOLISM is defined as the sum total of the chemical reactions occurring in the body. It may be recalled that metabolism may be constructive or result in synthesis of new substances (anabolism), or may be energy releasing and result in the destruction of a compound (catabolism). The absorbed end products of digestion form the basis for anabolic and catabolic reactions. INTERMEDIARY METABOLISM refers to the specific steps a cell employs in metabolizing a given compound.

NUTRITION emphasizes the role of foodstuffs in meeting the energy and specific requirements of the body for continuation of cellular activity. It also studies the mechanisms that determine the rate at which foodstuffs are metabolized, and the use to which the products are put.

Absorption of foodstuffs

From the mouth

There is no absorption of foods from the mouth, because the molecules are too large to pass through the epithelium and the stratified squamous epithelium is rather impervious to most molecules. Certain drugs, in the form of lozenges, may be absorbed through the mouth lining particularly when held under the tongue.

From the stomach

Absorption from the stomach is slow and is limited to small molecules and inorganic salts. Water, alcohol, and compounds of Na and K apparently can pass through the stomach wall. Absorption of salts seems to occur by active transport, that of water and alcohol by passive processes.

From the small intestine

The bulk of nutrient absorption occurs from the small intestine. In this organ, foodstuffs have been reduced to a size where they can pass through the epithelium, the epithelial cells pos-

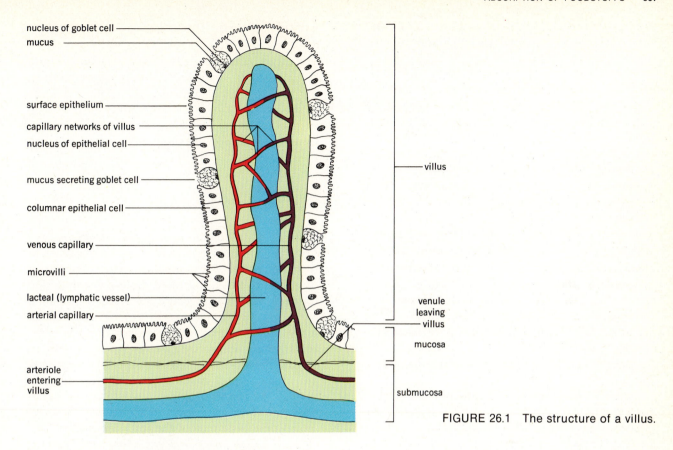

nucleus of goblet cell
mucus

surface epithelium

capillary networks of villus

nucleus of epithelial cell

mucus secreting goblet cell

columnar epithelial cell

venous capillary

microvilli

lacteal (lymphatic vessel)

arterial capillary

arteriole entering villus

villus

venule leaving villus

mucosa

submucosa

FIGURE 26.1   The structure of a villus.

sess transport systems to move materials, the villi and plicae (Fig. 26.1) provide a large surface through which substances may pass, and the intestine is relatively long, providing time for absorption to occur.

ABSORPTION OF SIMPLE SUGARS. Absorption of glucose and galactose occurs by active transport and places the sugars into the blood vessels of the villi for transport to the liver. Other sugars, including pentoses (e.g., ribose) are absorbed by carrier-facilitated diffusion or diffusion alone. Glucose is absorbed at a maximum rate of about 120 gms/hour; galactose passes about 10 percent faster than glucose.

ABSORPTION OF AMINO ACIDS. Most amino acids are actively transported from the intestine to the mucosal cells, and then diffuse into the blood vessels of the villi. Three separate transport systems appear to exist: one for amino acids that contain one carboxyl (—COOH) and one amine (—NH$_2$) group; one for basic amino acids containing two amine groups; one for acids containing ring structures. In infants, permeability of the intestine appears to be greater than in later life, and some large protein molecules such as antibodies in the mother's milk and protein allergens pass by diffusion through the wall. Pinocytosis appears to account for the absorption of these large molecules in children and adults.

ABSORPTION OF LIPIDS. Lipids are absorbed by diffusion into the epithelial cells of the mucosa.

Fatty acids containing less than 10 to 12 carbons enter the blood vessels of the villus and are carried to the liver as *free fatty acids* (FFA). Those containing more than 10 to 12 carbons are recombined with glycerol in the mucosal cells and are coated with protein, cholesterol, or phospholipid to form tiny globules called *chylomicrons*. These chylomicrons then leave the cells and enter the lymphatics of the villi (lacteal). Most of the absorption of fatty acids and glycerol occurs in the upper portion of the intestine; that of cholesterol occurs mainly in the lower intestine.

ABSORPTION OF VITAMINS, WATER, AND MINERALS. Pinocytosis appears to account for the absorption of vitamins from the intestine. Fat soluble vitamins (A, D, E, K) follow the large lipids into the lacteal, while water soluble vitamins (C, B complex) go into blood vessels. Many vitamins then form parts of enzymes necessary for metabolic processes.

Water moves by osmosis secondarily to removal of solutes from the intestine.

Sodium, potassium, and calcium are transported actively from the intestine. Since they are cations, anions (e.g., Cl^-, HCO_3^-) follow by electrostatic attraction (+ draws −).

From the large intestine

The large intestine absorbs salts, mostly sodium (60 meq/day), and water, about 400 milliliters per day, follows osmotically. Absorption of some of the products of bacterial activity in the colon such as vitamins B and K and amino acids occurs from this organ by the same processes that absorb them elsewhere.

Clinical considerations

Defects in absorption of foods result mainly from genetic abnormalities that cause malformation of the villi of the intestine (celiac disease), or from failure to synthesize digestive enzymes and transport systems because of disease processes or genetic abnormalities. In any event, two primary things occur: foods are not brought to an absorbable state; surface area of the gut is reduced so that absorption, even if normal, is very slow and malnutrition results. Interestingly, in gross obesity, where an individual is fat because of absorption of vast amounts of consumed food, removal of a large part of the intestine may be carried out as a device to reduce absorption when the subject cannot or will not control food intake.

Intermediary metabolism of foods

The main reason that foods are consumed is to degrade them to supply energy for the synthesis of adenosine triphosphate (ATP). ATP may be characterized as the universal biological energy carrier that forms the immediate source of energy for cellular activity such as contraction, secretion, and active transport. Thus, intermediary metabolism of foodstuffs becomes a series of catabolic reactions that provide energy for the anabolic reactions leading to formation of ATP. The following discussion presents the basic steps involved in intermediary metabolism. For those interested, complete metabolic cycles are presented in the Appendix.

Carbohydrates

Glucose forms the primary source of energy for synthesis of ATP. Depending on the economy of the body at any given moment, the fate of glucose entering the body follows one or the other of two main pathways.

It may be *catabolized* for energy and the synthesis of ATP.

It may be *converted* into a storage form from which it may be recovered as needed.

The basic schemes involved in each choice are

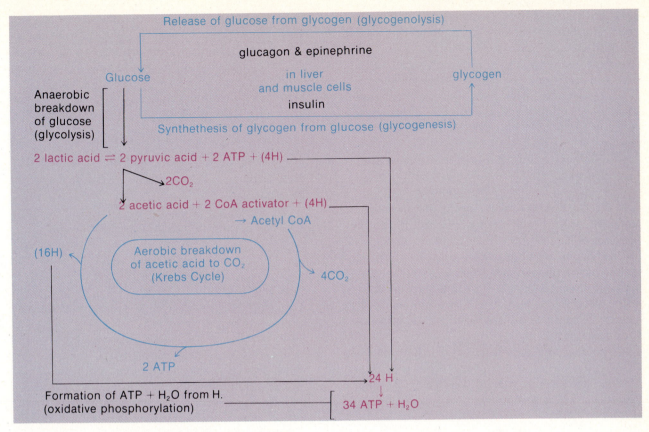

FIGURE 26.2   Interrelationships of the metabolic cycles involved in carbohydrate metabolism.

summarized in Figure 26.2. The catabolic scheme shows the fate of a single glucose molecule.

The presence of glucose within the cells in quantities greater than needed to run cellular machinery results in storage of the glucose as the polysaccharide glycogen in the process called GLYCOGENESIS. It occurs mostly in liver and muscle cells. *Insulin*, a hormone from the pancreas, aids entry of glucose into most body cells and speeds its conversion into glycogen. When blood levels of glucose are diminished, glucose may be recovered from glycogen in the process known as GLYCOGENOLYSIS. This process is speeded by a pancreatic hormone *glucagon*, and by *epinephrine* from the adrenal medulla.

Initial degradation of glucose is handled by GLYCOLYSIS, an anaerobic metabolic scheme that

produces 2 three-carbon units called pyruvic acid from the six-carbon glucose molecule. The scheme also releases nearly 60 kilocalories* of energy and results in the production of 2 new ATP molecules for cellular energy stores. Further degradation of pyruvic acid requires the presence of oxygen. The pyruvic acid has a CO_2 removed from it to form acetic acid, and then enters the KREBS CYCLE that produces 2 more CO_2, and 1 new ATP for each acetic acid molecule combusted. Per mol of acetic acid combusted, nearly 630 kilocalories of heat are also produced. Hydro-

---

* A calorie is the amount of heat required to raise the temperature of 1 gram of water 1°C.
A kilocalorie is 1000 times larger than a calorie, and raises the temperature of 1 kilogram of water 1°C.

gens are released by glycolysis and the Krebs cycle. They are collected by *hydrogen acceptors* and are carried to OXIDATIVE PHOSPHORYLATION. This scheme converts the hydrogen into water in the presence of oxygen, and releases enough energy to produce 34 new ATP molecules for body use. Therefore, 1 mol of glucose may be degraded to produce a total of 38 new ATP molecules for body use.

We emphasize several things about these interrelationships.

1. Glycogen synthesis will occur only if there is more glucose available than is required to run the body machinery.

2. Initial stages of glycogen breakdown do not require oxygen (anaerobic); later stages do (aerobic). If oxygen is not available to assure aerobic metabolism, lactic acid is formed as a means of temporary storage to conserve the energy in pyruvic acid until O_2 *is* available.

3. The *complete* degradation of glucose produces CO_2 and H_2O, with ATP formation. The majority of the ATP formation comes from oxidative phosphorylation that converts H to ATP.

Lipids

FATTY ACIDS AND GLYCEROL. Fatty acids are metabolized mainly in liver and muscle by removal of two carbon units in the process known as BETA-OXIDATION. The two carbon units are acetic acid that, when combined with Coenzyme A, enter the Krebs cycle for further oxidation. The Krebs cycle is, therefore, a sort of final common path for biological oxidation. Glycerol is metabolized by conversion to a compound found in glycolysis. It may then follow glycolysis and the Krebs cycle as does glucose. If body metabolism does not require breakdown of these lipids, they may be recombined into triglycerides and stored in the adipose cells of the body. The adipose tissue thus formed is like a savings account; it may be withdrawn when needed, but can accumulate (with interest!) if food intake exceeds utilization.

Unrestricted or excessive catabolism of fatty acids may produce more Acetyl CoA than can be combusted in the Krebs cycle. Excess AcCoA forms *acetone* and a compound called *beta-hydroxybutyric acid.* These are the KETONE BODIES, and their accumulation causes KETOSIS. Some of the interrelationships of fatty acid and glycerol metabolism are shown below.

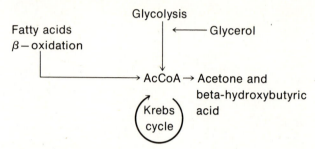

CHOLESTEROL. Cholesterol is a complex derived lipid containing the "sterol nucleus" It is the basic material for the synthesis of bile acids, many steroid hormones, and has been widely discussed as to its involvement in the production of atherosclerosis in the blood vessels. Cholesterol is synthesized from AcCoA, mainly by the liver. Restriction of dietary intake of cholesterol has been suggested to reduce the development of sclerosis of blood vessels, but no definitive evidence is available to prove the suggestion.

LIPOPROTEINS. Lipoproteins have attracted considerable attention recently because of their involvement in atherosclerosis and coronary heart disease. Four major groups of lipoproteins have been separated and identified by electrophoresis and ultracentrifugation. These are commonly named according to their densities, as measured in the ultracentrifuge, and are referred to as chylomicrons (lowest density), very low density lipoproteins (VLDL), low density lipoproteins (LDL), and high density lipoproteins (HDL). Increase in density is associated with an increase in protein content of the lipoprotein.

VLDL are formed primarily in the liver, and

chylomicrons and some VLDL are formed in intestinal mucosa cells, perhaps in response to the fat load being absorbed. LDL and HDL synthesis areas are not known, but they may be metabolic products of chylomicrons and VLDL. The term *apolipoprotein* (apoproteins) is applied to the proteins that combine with the lipids. They are relatively few in number, and differ from one another in their terminal amino acids.

If lipoprotein levels are low in the bloodstream, a HYPOLIPOPROTEINEMIA develops. In general, this is associated with elevated levels of free fatty acids (FFA) or triglycerides in the blood, and with depressed cholesterol levels in the blood. High FFA levels may contribute to the development of atherosclerosis, a disease associated with lipid deposition in arterial walls and development of an abnormal smooth muscle cell.

The *hyper*lipidemias and *hyper*lipoproteinemias are a greater threat because they indicate a greater fat load in the bloodstream. In these conditions, blood cholesterol and triglyceride levels are elevated with a potentially greater possibility for deposition in blood vessel walls. Cause may be genetic, or secondary to hypothyroidism, excessive intake of lipids in the diet, alcohol consumption, and use of contraceptive "pills." Normal values for triglycerides, cholesterol, and lipoproteins in the blood may be given as:

duces the effect of the first two factors, and minimizes the operation of the third.

Atherosclerosis cannot be guaranteed not to occur if blood lipid levels are kept low; neither is there an upper limit beyond which it is inevitable. The changes characteristic of atherosclerosis begin early in life with "fatty streaks" accumulating within the inner layer of a blood vessel wall. This accumulation is universal, but is reversible. Continued accumulation of fats in the vessel wall leads to the formation of a fatty plaque that encroaches on the vessel lumen, reducing blood flow. Progression to plaque formation can be slowed or stopped by dietary measures. An additional factor in plaque development is proliferation of smooth muscle cells in the vessel wall. Smooth muscle cells accumulate LDL, forming what are called "foam cells." In tissue culture, such cells are stimulated to proliferate by LDL, and to secrete elastic fiber protein, collagen and protein-polysaccharides that further contribute to plaque formation.

ESSENTIAL FATTY ACIDS. Several fatty acids that contain unsaturated carbon bonds (—C=C—) are necessary for normal lipid metabolism and to maintain growth and are not synthesized in the body. Among these "essential fatty acids" are linoleic, linolenic, and arachidonic acids.

	Triglycerides (mg/dl)	Cholesterol (mg/dl)	Lipoproteins (actual amounts vary, but relative % of total is fairly constant) — relative %			
			Chylomicrons	HDL	LDL	VLDL
Adult	10–150	130–250	0–2	10–38	38–74	1–37
Child	10–140	120–230	Levels in children are about $\frac{1}{3}$ that of the adult, and			
Infant	10–140	45–170	reach adult values about 13 yrs. of age.			

The suggestion has been made that hyperlipidemia results where there is obesity, a high dietary intake of lipids, and a genetic predisposition for elevated blood lipid levels. Dietary restriction of both lipid and carbohydrate re-

Proteins

Proteins, by their content of amino acids, provide the basis for synthesis of enzymes, many hormones, and the structural material (e.g., collagen, elastin) of the body. In general, proteins and amino acids are not oxidized for their energy content to any great degree, unless glucose and fatty acids have virtually disappeared from the body.

The amino acids composing proteins are of two general types: ESSENTIAL AMINO ACIDS* must be taken in the diet either because they are not synthesized in the body, or are synthesized in amounts too small to be significant; NONESSENTIAL AMINO ACIDS† may be synthesized by the liver usually from nonprotein precursors.

Essential to the formation of amino acids is the process of TRANSAMINATION, by which amine (—NH₂) groups are transferred from an amine donor to an organic acid (a keto acid) to form an amino acid. Conversely, DEAMINATION removes amine groups from amino acids and creates compounds (keto acids) that may be utilized in the Krebs cycle or glycolysis. Ammonia may be one product of deamination. It is very toxic and is converted in the liver to urea by the ORNITHINE CYCLE, and is excreted in that form. Some of these interrelationships are shown below.

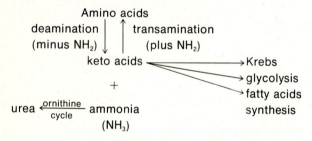

Interconversion of compounds

Synthesis of nonessential amino acids is an example of "creating" one substance from another. In general, it may be stated that lipids, carbo-

hydrates, and amino acids may be converted into one another *if* a compound common to the metabolism of two or more materials is present and *if* the metabolic cycles involved are reversible. Most cycles are reversible, and "crossover compounds" do exist. The body cells (again mainly the liver) therefore can assure themselves of their required building blocks even though the diet may not contain enough to supply their needs. Some of the interrelationships between the basic foodstuffs are shown in Figure 26.3, with the cycles involved named and crossover compounds indicated in italic type. This scheme of interrelationships is sometimes called the "metabolic mill" (Fig. 26.3). These schemes explain how one may become fat by eating starches, or how the nonessential amino acids may be produced from carbohydrates.

Control of metabolism

The term metabolic rate is used to describe how fast the body is catabolizing foodstuffs. The term basal metabolic rate (BMR) describes how many Calories the body liberates when carrying on only those activities essential to life, with no extra activity involved. BMR is influenced by several factors, including:

Size, or surface area. The skin serves as a surface through which heat is lost. In general, the larger the skin surface area, the higher the BMR, since more heat is lost through the greater surface and must be replaced by foodstuff catabolism.

Sex. Males oxidize foods about 7 percent faster than females, presumably due to their great amount of muscular tissue that calls for glucose to fuel its activities.

Age. In general, BMR declines as one ages because of loss of muscular tissue and general overall slowdown of activity.

Hormones. Thyroid hormone appears to "set the thermostat" of combustion in the body. Excessive hormone raises the BMR greatly.

---

* Threonine, Methionine, Valine, Leucine, Isoleucine, Lysine, Arginine, Phenylalanine, Tryptophane, and Histidine.

† Glycine, Alanine, Serine, Cysteine, Aspartic Acid, Glutamic Acid, Hydroxylysine, Cystine, Tyrosine, Diiodotyrosine Thyroxine, Proline, and Hydroxyproline.

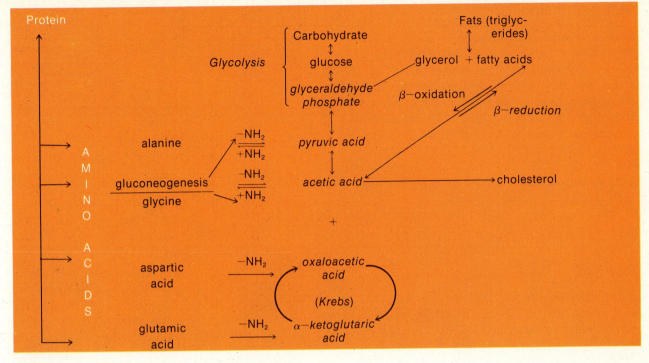

FIGURE 26.3 The "metabolic mill" showing inter-relationships of carbohydrate, lipid, and protein metabolism.

Drugs. Benzedrine is a drug raising BMR by stimulating CNS activity and increasing alertness and feelings of exhilaration.

Other factors. Pregnancy raises the BMR of the pregnant female; emotions, such as anger, also raise BMR.

We thus conclude that genetic, chemical, and environmental factors combine to govern the "intensity with which the fires of life burn."

The body temperature

Degradation of foodstuffs by glycolysis, the Krebs cycle, oxidative phosphorylation, and B-oxidation, releases energy. Only about 30 percent of the energy releases from foods is channeled into the synthesis of ATP. The rest is released as heat that primarily enables the human to maintain body temperature within the narrow limits necessary for continuation of *normal* cellular function. These limits are usually given as $98.6 \pm 1.8°F$, or $37 \pm 1°C$, when the temperature is taken orally. Rectal temperatures are 0.5 to 1°F higher, and internal organ temperatures (e.g., liver) may be as high as 105°F. Since heat is continually being produced by metabolic activity, there must be heat loss to maintain a near constant body temperature.

Heat production

Basal metabolism produces, in the average adult, about 1700 Calories of heat per day. Muscular activity, ingestion of foods, emotions, and other factors raise this output to higher values. Ingestion of foods is associated with secretion of materials, digestion, and absorption. These reactions are energy releasing and raise metabolic rate. This effect on metabolic rate is called the *specific dynamic effect* (SDE) of foods. Moderate work may cause an output of 3000 Calories; heavy work, 7000 Calories. Involuntary muscular activity, such as shivering, occurs when body temperature is lowered.

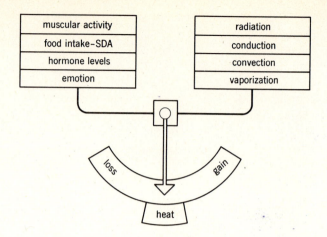

Heat loss

Heat loss occurs mainly through the skin. Small amounts of heat are lost by breathing warm air out through the lungs and by loss in urine and feces.

RADIATION is transfer of heat from a warmer to a cooler medium. It is effective only when there is a considerable difference (about 10°F) between the body and its environment. As environmental temperature approaches that of the body, radiation diminishes.

CONDUCTION is transfer of heat from a warmer to a cooler area when they are in contact. Thus, in a swimming pool, heat is transferred to the water from the body.

CONVECTION is heat transfer to a moving medium, such as moving air.

VAPORIZATION (evaporation) of water from the skin surface carries away heat. This method assumes increasing importance as a cooling device when environmental temperature approaches and exceeds that of the body.

The small "scale" shown below summarizes the methods of heat production and loss.

Control of body temperature

The hypothalamus of the brain controls body temperature. It contains heat gain and heat loss centers. Impulses reach these centers over nerves from the heat and cold receptors of the body, and the centers also monitor the temperature of the blood reaching them. Efferent connections of the centers are made to skeletal muscles, sweat glands, and skin blood vessels. If body temperature needs to be raised, muscles tense and may shiver (increase production of heat), and skin blood vessels constrict (heat conservation). If temperature must be lowered, muscles relax, skin vessels dilate, and sweating occurs. These processes decrease heat production and increase heat loss.

Alterations of body temperature

FEVER is a term used to describe a body temperature that is above the usual limits of normal. It may result from damage to the brain, from chemicals that affect the hypothalamic thermostat, from failure of the normal cooling mechanisms of the body (heat stroke), or from dehydration. In many diseases caused by bacteria or viruses, the toxic products produced by the microorganism, or the products of cell destruction, "reset" the thermostat to a higher value. Chemicals that have this effect are called PYROGENS (*pyro* = fire, *gen* = to produce). Brain tumors may affect the thermostat. Dehydration results in a smaller volume of body water to get rid of metabolic heat, and temperature rises. Elevated body temperature stimulates chemical reactions, more heat is produced, and a "vicious circle" is established. Heat stroke most commonly occurs when heavy work is performed in a hot, humid environment. To prevent this condition, one must cool off by whatever means are available. The upper limits of temperature tolerance appear to be about 43°C (109.4°F) rectally. Death is common from heat stroke at this temperature. A 41°C (106°F) temperature may cause convulsions and brain damage.

HYPOTHERMIA (low body temperature) is often created artificially during surgical operations (e.g., heart surgery). The combination of an anesthetic and cooling of the body can reduce core temperature to 70 to 75°F without permanent ill effects. Such low core temperatures result in less demand for nutrients by the brain, and promote the ability of the patient to tolerate the surgery. Core temperatures cannot be reduced below about 70°F without running the danger of damaging the hypothalamic centers.

Nutrition

To supply basic caloric and nutritional requirements per day, one should select a diet that contains all three basic foodstuffs. Recommendations suggest that 10 to 15 percent of the daily adult caloric intake be in the form of protein, 55 to 70 percent as carbohydrate, and 20 to 30 percent as fat. These percentages are roughly equal to 33 to 42 grams of protein, 250 to 500 grams of CH_2O, and 66 to 83 grams of fat. This type of intake may be achieved by ensuring that the diet includes the foods listed below. Such a diet will also guarantee adequate intake of essential minerals and vitamins.

Milk and milk products. Three or more glasses per day for children; four or more for teen-agers; two or more for adults. Cheese, ice cream, yogurt, cottage cheese, and other milk products may substitute for part of the milk requirement.

Meat group. Two or more servings daily of meat, fish, eggs, or cheese. Nuts, dry beans, and peas may substitute for one of the meats.

Vegetables and fruits. Four or more servings per day to include green and yellow vegetables, and citrus fruit or tomatoes. One serving per day should be raw.

Bread and cereals. Four or more servings per day. Servings should be composed of whole grain or enriched products.

Intake of specific foods

CARBOHYDRATES. Fruits, vegetables, and whole grains (in breakfast foods, bread) provide the major sources of carbohydrate after weaning. Sufficient carbohydrate should be provided in the diet to assure normal weight gain in terms of the activity level of the individual. It may be noted that until about 6 months to 1 year of age, the child has little salivary amylase and has low levels of pancreatic amylase. Thus it should not be fed a diet containing a large percentage of polysaccharides. The child does possess the ability to efficiently utilize lactose, and its diet should emphasize intake of this disaccharide.

PROTEIN. Protein is regarded as the mainstay of life because of its incorporation into enzymes, hormones, contractile tissues, and other body components. Egg and milk proteins appear to be the best sources of essential and nonessential amino acids. These sources are closely followed by animal proteins such as those of meat, fish, and poultry. Plant proteins are deficient in certain amino acids, or contain too low an amount to sustain optimum nutrition. Inadequate protein intake is usually reflected in muscular wastage, and in edema due to decreased plasma protein levels. Children need more protein to sustain their growth and maturation than do adults. A good measure of whether protein intake is adequate is to determine the nitrogen balance of the individual; that is, is nitrogen intake greater than, equal to, or less than nitrogen excretion? In general, greater nitrogen intake than output is associated with states of growth, while the reverse situation generally is associated with malnutrition and starvation.

FATS. Fats are important as energy sources and as the vehicle for absorption of fat soluble vitamins (A, D, E, K). Intake must include the essential fatty acids, which are necessary to combust other lipids. Fats are synthesized or are converted into a wide variety of compounds essential to the structure of cell membranes and myelin sheaths around nerves. Evidence is accumulating that the feeding of large amounts of fats early in life results in the development of excess numbers of adipose cells. These cells "like to be filled" and tend to accumulate fat, which may lead to fat babies who are not necessarily healthy babies. Adults who are obese tend to have been obese as children, possibly from having more adipose cells "crying for fat."

VITAMINS. Vitamins are classed as trace substances or accessory food factors that are required in very small amounts for normal cellular metabolism. None of these substances can be synthesized by the animal body, which must therefore depend mainly on dietary intake of them. Supplementation may occur through synthesis by intestinal flora and subsequent absorption through the gut (colon) wall. Some 16 organic substances are recognized as essential to normal body functions; 13 are designated as vitamins.

Two groups of vitamins are recognized: the fat soluble vitamins include A, D, E, and K; the water soluble vitamins are those of the B complex (thiamine or B_1, riboflavin or B_2, niacin, pyridoxine or B_6), vitamin C, vitamin B_{12}, folic acid, pantothenic acid, and biotin. Choline, inositol, and para-aminobenzoic acid may be essential to

body function, but are not usually designated as vitamins.

FAT SOLUBLE VITAMINS are ingested and absorbed along with dietary fats. Thus, any condition affecting fat digestion and absorption affects the intake of fat soluble vitamins. The use of oils as laxatives may cause dissolving of fat soluble vitamins in the oil and their excretion in the feces. Fat soluble vitamins are not present in all dietary fat sources. Liver tends to concentrate fat soluble vitamins and is thus an excellent source of these vitamins. Cod-liver oil has for years been a standard source of vitamins A and D. Vitamin K is widely distributed in all foods and is essential for prothrombin formation; vitamin E is present in eggs and grain oils. Vitamin E is an antioxidant, and may be necessary to maintain reduced states of hydrogen acceptors and electron transport systems.

WATER SOLUBLE VITAMINS are widely distributed in all foodstuffs, but many are destroyed by heat (cooking), sunlight, and are easily oxidized, as by standing in a vitamin pill bottle. Many water soluble vitamins are lost from foods if they are soaked in water or are cooked in large volumes of water. The water soluble vitamins are, to a large extent, incorporated into enzymes or cofactors for enzymes, and are utilized in the basic metabolic schemes of cells. Table 26.1 lists the vitamins, some of their sources, and minimum daily requirements (where known).

INORGANIC SUBSTANCES. Fourteen of the elements appearing in the periodic table have been shown to be essential for health. They are: calcium, phosphorus, magnesium, sodium, potassium, sulfur, chlorine, iron, copper, cobalt, iodine, manganese, zinc, and fluorine. Table 26.2 presents facts relative to acquisition, requirements, and uses of these inorganic materials.

HYPERALIMENTATION AND OBESITY. Malnutrition is usually associated with undernutrition. A greater problem in many parts of the world is overnutrition, or the intake of calories in excess

(Text continued on page 610.)

TABLE 26.1 The vitamins

Designation letter and name	Major properties	Requirement per day	Major sources	Metabolism	Function	Deficiency symptoms
A — Carotene	Fat soluble yellow crystals, easily oxidized	5000 I.U.	Egg yolk, green or yellow vegetables and fruits	Absorbed from gut; bile aids, in liver	Formation of visual pigments; maintenance of normal epithelial structure	Night blindness, skin lesions
D$_3$ — Calciferol	Fat soluble needlelike crystals, very stable	400 I.U. much made through irradiation of precursors in skin	Fish oils, liver	Absorbed from gut; little storage	Increase Ca absorption from gut; important in bone and tooth formation	Rickets (defective bone formation)
E — Tocopherol	Fat soluble yellow oil, easily oxidized	Not known for humans	Green leafy vegetables	Absorbed from gut; stored in adipose and muscle tissue	Humans — maintain resistance of red cells to hemolysis	Increased RBC fragility
					Animals — maintain normal course of pregnancy	Abortion, muscular wastage
K — Naphthoquinone	Fat soluble yellow oil, stable	Unknown	Synthesis by intestinal flora; liver	Absorbed from gut; little storage; excreted in feces	Enables prothrombin synthesis by liver	Failure of coagulation
B$_1$ — Thiamine	Water soluble white powder, not oxidized	1.5 mg	Brain, liver, kidney, heart; whole grains	Absorbed from gut; stored in liver, brain, kidney, heart; excreted in urine	Formation of cocarboxylase enzyme involved in decarboxylation (Krebs cycle)	Stoppage of CH$_2$O metabolism at pyruvate, beri-beri, neuritis, heart failure, mental disturbance
B$_2$ — Riboflavin	Water soluble; orange-yellow powder; stable except to light and alkalies	1.5–2.0 mg	Milk, eggs, liver, whole cereals	Absorbed from gut; stored in kidney, liver, heart; excreted in urine	Flavoproteins in oxidative phosphorylation (hydrogen transport)	Photophobia; fissuring of skin
Niacin	Water soluble; colorless needles; very stable	17–20 mg	Whole grains	Absorbed from gut; distributed to all tissues; 40% excreted in urine	Coenzyme in H transport, (NAD, NADP)	Pellagra; skin lesions; digestive disturbances, dementia

TABLE 26.1 (continued)

Designation letter and name	Major properties	Requirement per day	Major sources	Metabolism	Function	Deficiency symptoms
B_{12}—Cyanocobalamin	Water soluble; red crystals; stable except in acids and alkalies	2–5 mg	Liver, kidney, brain. Bacterial synthesis in gut	Absorbed from gut; stored in liver, kidney, brain; excreted in feces and urine	Nucleoprotein synthesis (RNA); prevents pernicious anemia	Pernicious anemia; malformed erythrocytes
Folic acid (Vitamin B_c, M, pteroyl glutamate)	Slightly soluble in water; yellow crystals; deteriorates easily	500 micrograms or less	Meats	Absorbed from gut; utilized as taken in	Nucleoprotein synthesis; formation of erythrocytes	Failure of erythrocytes to mature; anemia
Pyridoxine (B_6)	Soluble in water; white crystals; stable except to light	1–2 mg	Whole grains	Absorbed from gut; one-half appears in urine	Coenzyme for amino acid metabolism, fatty acid metabolism	Dermatitis; nervous disorders
Pantothenic acid	Water soluble; yellow oil; stable in neutral solutions	8.5–10 mg	?	Absorbed from gut; stored in all tissues; excreted in urine	Forms part of coenzyme A (CoA)	Neuromotor disorders, cardiovascular disorders; GI distress
Biotin	Water soluble; colorless needles; stable except to oxidation	150–300 mg	Egg white; synthesis by flora of GI tract	Absorbed from gut; excreted in urine and feces	Concerned with protein synthesis, CO_2 fixation and transamination	Scaly dermatitis; muscle pains, weakness
Choline (maybe not a vitamin)	Soluble in water; colorless liquid; unstable to alkalies	500 mg	?	Absorbed from gut; not stored	Concerned with fat transport; aids in fat oxidation	Fatty liver; inadequate fat absorption
Inositol	Water soluble; white crystals	No recommended allowance	?	Absorbed from gut; metabolized	Aids in fat metabolism; prevents fatty liver	Fatty liver
Para-amino benzoic acid (PABA)	Slightly water soluble; white crystals	No evidence for requirement	?	Absorbed from gut; little storage; excreted in urine	Essential nutrient for bacteria; aids in folic acid synthesis	No symptoms established for humans
Ascorbic acid (Vitamin C)	Water soluble; white crystals; oxidizable	75 mg/day	Citrus	Absorbed from gut; stored; excreted in urine	Vital to collagen and ground substance	Scurvy—failure to form c.t. fibers

TABLE 26.2 Inorganic substances

Substance	Requirements per day	High level sources	Where absorbed	Where found in body	Functions	Effects of Excess	Deficiency
Calcium	About 1 gm	Dairy products, eggs, fish, soybeans	Small intestine	Bones, teeth, nerve, bloodstream, muscle	Bone structure, blood clotting, muscle contraction, excitability, synapses	None	Tetany of muscles, loss of bone minerals
Phosphorus	About 55 mg/kg	Dairy products, meat, beans, grains	Small intestine	Bones, teeth, nerve, bloodstream, muscle, ATP	Bone structure, intermediary metabolism, buffers, membranes	None	Unknown; related to rickets, loss of bone mineral
Magnesium	Estimated 13 mg	Green vegetables, milk, meat	Small intestine	Bone, enzymes, nerve, muscle	Bone structure, factors with enzymes, regulation of nerve and muscle action	None	Tetany
Sodium	Newborn 0.25 gm Infant 1 gm Child 3 gm Others 6 gm	All foods, table salt	Stomach, small and large intestine	Extracellular fluids	Ionic equilibrium, osmotic gradients, excitability in all cells	Edema hypertension	Dehydration, muscle cramps, renal shutdown
Potassium	1–2 gms	All foods, meats, vegetables, milk	Stomach, small and large intestine	Intracellular fluids	Buffering, muscle and nerve function	Heart block (>10 meq/L)	Changes in ECG, alteration in muscle contraction
Sulfur	0.5–1 gm ?	All protein containing foods	Small intestine, as amino acids primarily	Amino acids, bile acids, hormones, nerve	Structural as amino acids are made into proteins	Unknown	Unknown
Chlorine	2–3 gm	All foods, table salt	Stomach, small and large intestine	Extracellular fluids	Acid-base balance, osmotic equilibria	Edema	Alkalosis, muscle cramps
Iron	Infant 0.4–1 mg/kg Child 0.4 mg/kg Adult 16 mg	Liver, eggs, red meat, beans, nuts, raisins	Small intestine	Respiratory proteins (hemoglobin, myoglobin, cytochromes)	O_2 and electron transport	May be toxic	Anemia (insufficient hemoglobin in red cells)
Copper	Infant and Child 0.1 mg/kg Adult 2 mg	Liver, meats	Small intestine	Bone marrow	Necessary for hemoglobin formation	None	Anemia
Cobalt	Unknown	Meats	Small intestine	Liver (Vitamin B_{12})	Essential to hemoglobin formation	None	Pernicious anemia
Iodine	Children 40–100 micrograms. Adult 100–200 micrograms	Iodized table salt, fish	Small intestine	Thyroid hormone	Synthesis of thyroid hormone	None	Goiter, cretinism
Manganese	Unknown	Bananas, bran, beans, leafy vegetables, whole grains	Small intestine	Bone marrow, enzymes	Formation of hemoglobin, activation of enzymes	Muscular weakness, nervous system disturbance	Subnormal tissue respiration
Zinc	Unknown	Meat, eggs, legumes, milk, green vegetables	Small intestine	Enzymes, insulin	Part of carbonic anhydrase, insulin, enzymes	Unknown	Unknown
Fluorine	0.7 part/million in water is optimum	Fluoridated water, dentifrices, milk	Small intestine	Bones, teeth	Hardens bones and teeth, suppresses bacterial action in mouth	Mottling of teeth	Tendency to dental caries

of body needs (hyperalimentation) leading to the development of obesity. Obesity is a more threatening form of mal (poor) nutrition than deficiency of caloric intake. A definition of obesity is that the person is 20 percent over his or her "ideal weight." In the United States, it has been estimated that 15 million persons are 20 percent over their ideal weight, and are thus obese. What causes obesity? Overeating because of boredom, unhappiness, or other emotional problems accounts for 95 percent of obesity. Genetic causes account for less than 5 percent. The penalties one pays for obesity include a higher mortality (death rate) from heart disease, the development of diabetes mellitus, digestive disorders, cerebral hemorrhage, and a higher morbidity (sickness) rate.

Basically, the control of obesity involves motivation to reduce and a caloric intake less than that required for activity and general living. One must have an appreciation of the caloric costs of activity and must adjust intake accordingly, if one is serious about losing weight. A loss of a pound per week requires about a 500 Calorie reduction in intake, or an increase in activity sufficient to increase caloric consumption by 500 Calories per day. All in all, it seems easier to reduce the caloric intake than to increase the exercise level. For example, dietary omission of an ice cream sundae or a piece of pie à la mode, eliminates the necessary calories; to burn up an equivalent amount of calories would involve running for an hour, walking (at 4 miles per hour for nearly 2 hours, or washing dishes for more than 7 hours.

The use of diet pills, usually amphetamines, is effective for about 2 weeks. Such drugs diminish appetite and elevate the spirits. Beyond 2 weeks time, the body adjusts to the drug, and appetite tends to return to previous levels. To get the effect again, a user may take more drug, or may take it more often, leading to a dependency on it.

Summary

1. Absorption involves transfer of foodstuffs across the linings of the alimentary tract.
   a. The mouth, pharynx, and esophagus absorb no food.
   b. Salts, water, and alcohol are absorbed from the stomach.
   c. All basic foodstuffs are absorbed by the small intestine.
   d. Salts, water, and bacterial products are absorbed by the colon.

2. Absorption of lipids, water, and certain sugars is by diffusion. Absorption of salts, amino acids, glucose, galactose, vitamins, and certain proteins is active, involving active transport and pinocytosis. Absorbed substances pass into blood vessels and lymphatics of the tract.

3. Intermediary metabolism of foods describes the changes cells make in absorbed products of digestion.
   a. Carbohydrates (glucose) are catabolized by glycolysis, the Krebs cycle, and oxidative phosphorylation. They are synthesized into glycogen (glycogenesis) from which they may be recovered (glycolysis).

b. Fatty acids are oxidized by beta-oxidation, which produces acetic acid that goes through the Krebs cycle. Glycerol is converted into a product found in glycolysis. Cholesterol is synthesized by the liver and forms bile acids and steroid hormones. Essential fatty acids must be included in the diet.

c. Proteins are composed of essential and nonessential amino acids, undergo conversion to compounds of glycolysis and Krebs cycle, and are oxidized for energy or incorporated into body structure.

4. Cholesterol and lipoproteins may be involved in the generation of atherosclerosis.

5. The basic building blocks of the three major foodstuffs are, to a great degree, interconvertible.

6. Control of metabolism is by genetic, chemical, and environmental factors.

7. Body temperature is maintained nearly constant in humans.

a. Heat production occurs through activity, food intake, and emotions.

b. Heat loss is by radiation, conduction, convection, and vaporization.

c. Control of body temperature is exerted by the hypothalamus and its connections to muscles, sweat glands, and blood vessels.

8. Nutrition is concerned with intake of foods necessary to maintain body function.

a. Protein should make 10–15 percent of the diet, carbohydrate 55–70 percent, fats 20–30 percent.

b. The diet should include milk and milk products, meats, vegetables, and fruits, and breads and cereals.

c. Tables summarizing vitamin and inorganic salt needs and uses are included.

9. Malnutrition includes excessive food intake as well as undernutrition.

a. Obesity exists when weight is 20 percent over ideal weight.

b. Obesity is caused primarily by overeating.

c. Obesity causes an increase in morbidity and mortality.

d. Diet control, involving reduction of caloric intake, is the only effective means of reducing weight.

Questions

1. Define absorption, metabolism, intermediary metabolism, and nutrition.

2. List the several parts of the alimentary tract and describe what is absorbed, by what method, and where the absorbed substance goes.

3. How is (are) glucose metabolized? Fatty acids? Amino acids?

4. How can one get fat by eating potatoes that are composed of carbohydrate?

5. What factors are involved in control of metabolism?

6. What are the normal limits of body temperature and how is it maintained?

7. What is fever? Hypothermia? What may cause each?

8. What constitutes an adequate diet?

9. To what uses are vitamins and minerals put in the body?

10. Why is obesity a threat to health, and how does it develop?

Readings

Goldstein, Joseph L., and Michael S. Brown. "The Low-Density Lipoprotein Pathway and Its Relation to Atherosclerosis." *Ann. Rev. Biochem. 46*:897, 1977.

Haussler, Mark R., and Toni A. McCain. "Vitamin D. Metabolism and Action." *New Eng. J. Med. 297*:974, Nov. 3, 1977; *297*:1041, Nov. 10, 1977.

Jacobson, Howard N. "Current Concepts: Diet in Pregnancy." *New Eng. J. Med. 297*:1051, Nov. 10, 1977.

Kolata, Gina Bari. "Obesity, A Growing Problem." *Science 198*:905, 2 Dec. 1977.

Masoro, E. J. "Lipids and Lipid Metabolism." *Ann. Rev. Physiol. 39*:301, 1977.

Science News. "Lipoproteins: A Delicate Balance." *113*:244, April 22, 1978.

Science News. "The Fat American." *113*:188, March 25, 1978.

Small, Donald M. "Cellular Mechanisms for Lipid Deposition in Atherosclerosis." *New Eng. J. Med. 297*:873, Oct. 20, 1977; *297*:924, Oct. 27, 1977.

Stacpoole, Peter W., George W. Moore, and David M. Kornhauser. "Metabolic Effects of Dichloroacetate in Diabetes and Hyperlipoproteinemia." *New Eng. J. Med. 298*:526, March 9, 1978.

Stricker, Edward M. "Hyperphagia." *New Eng. J. Med. 298*:1010, May 4, 1978.

Van Itallie, Theodore B., and Mei-Uih Yang. "Current Concepts in Nutrition: Diet and Weight Loss." *New Eng. J. Med. 297*:1158, Nov. 24, 1977.

Wirtshafter, David, and John D. Davis. "Body Weight: Reduction by Long-Term Glycerol Treatment." *Science 198*:1271, 23 Dec. 1977.

Chapter 27

The Urinary System

Objectives

After studying this chapter, the reader should be able to:

■ Describe some of the pathways the body utilizes as excretory routes, and the substances eliminated by each route.

■ List the "three kidneys" that develop in the human.

■ Name the organs of the urinary system.

■ Describe the features that would be visible upon making a frontal section of a human kidney.

■ Describe the structure of a nephron.

■ Explain the blood supply of the kidney.

■ Describe the processes involved in urine formation by the kidney including: glomerular filtration, tubular reabsorption and secretion, acidification, and urine concentration.

- Discuss the roles of antidiuretic hormone and aldosterone in regulation of urine volume.
- Describe the composition and properties of normal urine and compare it with plasma to show differences.
- Describe the structure and innervation of the ureters and urinary bladder and account for some disorders resulting from denervation of the bladder.
- Discuss the differences between the male and female urethras.

Pathways of excretion

Metabolism of substances by body cells produces a variety of waste products that must be removed from the body. Carbon dioxide, nitrogen containing substances such as urea, ammonia, and uric acid, heat, and excess water are the chief metabolic wastes. The lungs account for the elimination of the greater part of the carbon dioxide and small quantities of water and heat. The alimentary tract eliminates some carbon dioxide, water, and heat, and rids the body of unabsorbed ingested material and digestive secretions. The skin plays the major role in elimination of heat, but eliminates only small quantities of solids in the sweat.

Also treated as wastes—in that they will ultimately be eliminated from the body—are substances present in greater amounts than are needed for normal cellular function. Water, inorganic salts, H^+, sulfates, and phosphates are examples of such materials. Their levels in the extracellular fluid must be carefully controlled to assure the presence of only those amounts necessary for maintenance of homeostasis. Excretion of detoxified materials, antibiotics, and drugs also occurs. The organs of the urinary system, especially the kidney, are of primary importance in the regulation of composition of the extracellular fluid, and in the elimination of excess, detoxified and waste materials. The organs serving excretory functions in the body and some of the substances they deal with are summarized in Table 27.1.

Development of the system (Fig. 27.1)

The first signs of development of the urinary system are seen at about $3\frac{1}{2}$ weeks of embryonic growth. A ridge of mesoderm, the UROGENITAL RIDGE, located in the dorsal part of the celom (primitive peritoneal cavity) gives rise to a series of tubular structures. This series of tubular structures constitutes the first of three kidneys that will form during development, and is called the PRONEPHROS. At 4 weeks, the pronephros degenerates and is replaced by a MESONEPHROS. This in turn degenerates at about 7 weeks and is replaced by the METANEPHROS or permanent kidney, which contains the functional units of that organ.

As the pronephros develops, it sends a tubule, the PRONEPHRIC DUCT, to join the cloaca (the primitive common terminal cavity for digestive, urinary, and reproductive organs). The distal parts of the pronephric duct also serve the mesonephros as the mesonephric duct, and eventually connect to the gonads to form part of their duct system.

TABLE 27.1   Excretory organs of the body and substances dealt with

Organ	Substance(s)	Comments
Kidney	Water	Regulates body hydration
	Nitrogen containing wastes of metabolism (NH_3, urea, uric acid)	Originate from protein metabolism. Toxic if retained.
	Inorganic salts (Na^+, Cl^-, K^+, $PO_4^{\equiv}$, $SO_4^=$), H^+, HCO_3^-	Regulates osmotic pressure, pH of ECF; maintains excitability of cells.
	Drugs	
	Detoxified substances	Detoxification occurs in the liver; kidney ribs body of end product.
Lungs	Carbon dioxide	Regulation of acid-base balance
	Water	Fixed loss
	Heat	Fixed loss
Skin	Heat	Regulates temperature
	Water	Cooling
Alimentary tract	Digestive wastes	—
	Salts (Ca^{2+})	—

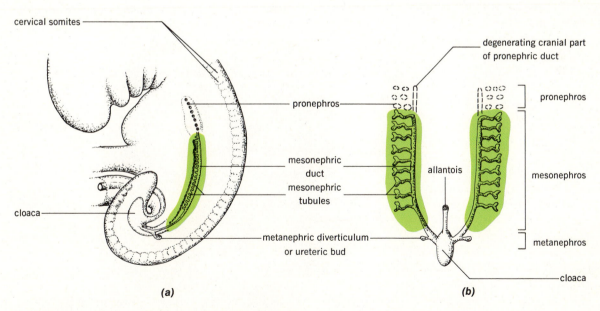

FIGURE 27.1   The development of the urinary system.
(a) Lateral view. *(b)* Ventral view.

TABLE 27.2 Some congenital anomalies of the kidney			
Disorder	Frequency	Cause of disorder	Comments
Agenesis Bilateral	1/6000 autopsies	Failure of both kidneys to develop	Fetus makes it to term, is using placenta to excrete. Dies several days after birth of uremia.
Unilateral	1/1200 autopsies	Failure of one kidney to develop	One kidney suffices to maintain life.
Aberrant blood vessels	1/105 autopsies	Extra blood vessels develop to supply kidney	Most common disorder. Not serious.
Horseshoe kidney	1/1200 individuals	Lower poles of kidney fused across midline	No symptoms.
Bifid (two) pelves	10% of individuals	2 pelves develop on a single kidney	No symptoms.

The portion of the duct connecting with the cloaca will form an outgrowth that will comprise the bladder, ureters, pelvis, and the collecting tubules of the kidney.

CLINICAL CONSIDERATIONS. Anomalies of the kidney are relatively common, but most are not severe enough to cause symptoms. Table 27.2 presents several congenital anomalities of the kidney.

The organs of the urinary system

The organs of the system include paired KIDNEYS, paired URETERS, and a single URINARY BLADDER and URETHRA (Fig. 27.2).

FIGURE 27.2 The organs of the urinary system and associated structures.

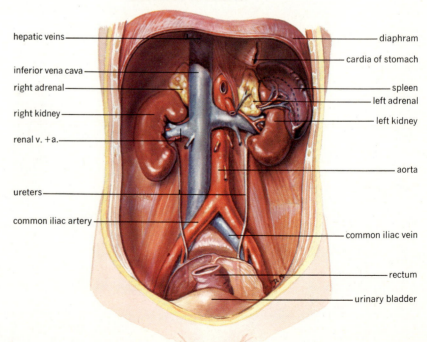

hepatic veins — diaphram
— cardia of stomach
inferior vena cava —
right adrenal — spleen
— left adrenal
right kidney — left kidney
renal v. +a. —
— aorta
ureters —
common iliac artery — common iliac vein
— rectum
— urinary bladder

The kidneys

Size, location, attachments

Living kidneys are reddish, bean-shaped organs, located behind the peritoneum (retroperitoneal) at the level of the twelfth thoracic to third lumbar vertebral bodies. They are about one half covered by the eleventh and twelfth ribs. They measure about 11 centimeters long, 6 centimeters wide, and 2.5 centimeters thick (about $4 \times 2 \times 1$ inches). The organs are attached to the body wall by adipose tissue and fibrous tissue that form the ADIPOSE CAPSULE and FIBROUS CAPSULE respectively.

Gross anatomy (Fig. 27.3)

The kidneys have a medially directed indentation, the HILUS (hilum), that marks the entry and exit of all blood vessels, nerves, and the excretory tube of the organ. A longitudinal (frontal) section of the kidney shows the RENAL PELVIS, its branches, the MAJOR and MINOR CALYCES (sing.: calyx), the RENAL SINUS, and the CORTEX and MEDULLA of the kidney. The medulla is composed of 8 to 18 MEDULLARY PYRAMIDS, whose tips contain duct openings that carry urine into the calyces for eventual elimination. RENAL COLUMNS are cortical substance lying between the pyramids.

Microscopic anatomy

The cortical and medullary portions of each kidney contain an estimated 1 to $1\frac{1}{2}$ million microscopic units known as NEPHRONS (Fig. 27.4). Each nephron contains a CAPSULE, PROXIMAL CONVOLUTED TUBULE, LOOP OF HENLE, and DISTAL CONVOLUTED

FIGURE 27.3   Frontal section of the kidney.

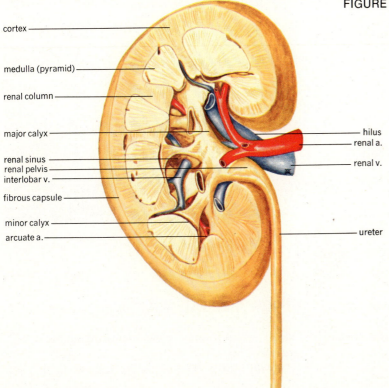

cortex
medulla (pyramid)
renal column
major calyx
renal sinus
renal pelvis
interlobar v.
fibrous capsule
minor calyx
arcuate a.

hilus
renal a.
renal v.
ureter

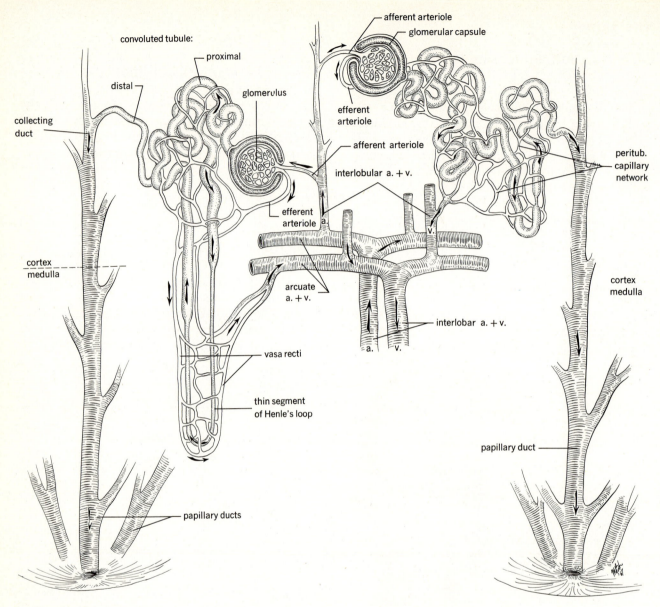

FIGURE 27.4   Two nephrons of the kidney and their associated blood vessels.

TUBULE, all of which are derivatives of the mesoderm, and a COLLECTING TUBULE and PAPILLARY DUCT, derivatives of the cloacal outgrowth. The nephrons are the units that control composition of the ECF and excrete solid wastes.

Blood supply (Fig. 27.5)

The aorta normally gives rise to a single RENAL ARTERY supplying each kidney. They carry about 1300 milliliters of blood per minute. Each renal

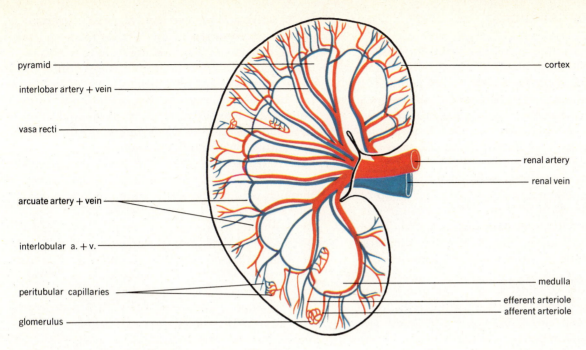

pyramid

interlobar artery + vein

vasa recti

arcuate artery + vein

interlobular a. + v.

peritubular capillaries

glomerulus

cortex

renal artery

renal vein

medulla

efferent arteriole

afferent arteriole

FIGURE 27.5   The blood vessels of the kidney.

artery branches to form, in order, INTERLOBAR ARTERIES, ARCUATE (arched) ARTERIES, and INTERLOBULAR ARTERIES. AFFERENT ARTERIOLES (*see* Fig. 27.4) lead to the GLOMERULI that lie in the nephron capsules. EFFERENT ARTERIOLES leave the glomeruli and form PERITUBULAR CAPILLARY NETWORKS or looped vessels (vasa recti) around the kidney tubules. INTERLOBULAR VEINS drain these capillary beds, and form ARCUATE VEINS, INTERLOBAR VEINS and usually, coalesce into a single RENAL VEIN draining each kidney. The attachment between vessels and nephron at the glomerulus is the critical one for urine formation. Also, the afferent arteriole, just before connecting with the glomerulus, and the adjacent part of the distal convoluted tubule, bear a curious mass of modified smooth muscle and tubule cells in their walls called the JUXTAGLOMERULAR APPARATUS or JG. These cells produce a substance that can ultimately result in hypertension.

Nerves to the kidneys are vasomotor in function and are derived from the renal plexus of the autonomic nervous system.

The formation of urine

Urine formation, and regulation of ECF composition involves several processes that are carried on by the nephron in a more or less stepwise fashion. These processes are:

Glomerular filtration
Tubular reabsorption and secretion
Acidification
Concentration

Glomerular filtration

Blood reaches the glomerulus and its capsule under a pressure of about 75 mm Hg. The membranes act like filters or sieves and permit the passage through their walls of any substance small enough to go through the "sieve pores." A fluid called the filtrate is formed, and it resembles plasma, except for a lack of blood cells and low protein content.

The blood pressure is opposed by an osmotic "back pressure," contributed by the plasma proteins that do not filter. It amounts to about 30 mm Hg. Additional forces opposing filtration (renal pressure) amount to about 20 mm Hg. Thus, the actual filtering pressure is about 25 mm Hg [75 − (30 + 20)].

The filtrate contains not only the nitrogenous wastes of metabolism, formed mainly in the liver, and excess substances not required for body use, but also substances (salts, water, glucose, amino acids) that are still useful for body function. Therefore, these useful materials must not be lost from the body.

Tubular reabsorption

By utilizing active and passive processes, the tubules remove substances from the filtrate and place them in the outgoing blood flow for return to the body generally. Since active reabsorption can be controlled, it is sometimes called facultative reabsorption. Obligatory ("have to") reabsorption occurs by passive processes and cannot be controlled.

Active removal of Na^+, K^+, amino acids, glucose, vitamins, and the small amounts of filtered protein returns these substances to the body. Negatively charged ions such as Cl^-, HCO_3^- follow the positive ions out of the tubules by electrostatic attraction. About 80 to 90 percent removal or reabsorption of such materials occurs in the proximal convoluted tubule. Glucose is reabsorbed completely if present in the filtrate in normal blood concentration.

As active and passive removal of solutes occurs, water concentration in the tubules tends to rise. Thus, osmosis of water occurs from tubule to the surrounding interstitial fluid in proportion to solute removal and the filtrate remains isotonic to the plasma. From the interstitial fluid, water enters blood vessels and returns to the circulation. Of about 190 liters of water filtered per day in the glomeruli, only about 1½ liters is excreted; the rest is reabsorbed. As water leaves the tubules, nitrogenous wastes accumulate in the tubule. Enough accumulates so that there is diffusion of these materials back into the circulation. Thus, both filtration and reabsorption of wastes are going on at the same time; fortunately, reabsorption of these wastes is less than filtration, and so the blood concentrations of these substances are kept below toxic levels. It should *not* be assumed that all of a waste is removed during a trip through the kidney. It is not; rather it is removed to the extent that we are not poisoned by it, but it is still there.

Tubular secretion

Active transport can move materials from the bloodstream into the tubules, as well as passing them out of the tubules. Secretion describes the movement of substances into the tubules. Both proximal and distal convoluted tubules secrete materials such as H^+, K^+, organic acids, and bases. One of the results of secretion is the acidification of the filtrate.

Acidification

Acid-producing foods predominate over alkali-producing foods in a normal diet. The body thus faces a real threat to its acid-base balance. Secretion of H^+ by the kidney provides a means of completely compensating a metabolic acidosis, given several days for the compensation to occur. The proximal and distal tubules are the primary sites of H^+ secretion. The ions are secreted in exchange for Na^+ or K^+ in the filtrate to maintain electrical neutrality of the cells. Some of the mechanisms of acidification are shown in Figure 27.6.

These mechanisms provide a means of controlling body pH. The reader should review Chapter 19 at this time.

Concentration of the filtrate

As the filtrate moves through the proximal tubule, it loses solute and water in equal ratios, so that its volume is reduced but the osmotic pressure is not changed. As the fluid enters the loop of Henle, it is subjected, in the ascending limb of the loop, to a selective active transport of sodium out of the fluid, *without* water following. The transported sodium is retained in the ECF around the loop, and tends to increase in concentration the deeper one goes toward the interior of the kidney. The mechanism responsible for this sodium concentrating effect is termed the COUNTERCURRENT MULTIPLIER, because sodium concentration is increased in the ECF by the mechanism. The multiplier ultimately results in the sodium concentrations shown in Figure 27.7. The fluid passing from the loop into the distal convoluted tubule is thus poor in solute (sodium) and rich in water, and is hypotonic to the plasma. The fluid now passes into the collecting tubules and papillary ducts that run through the extracellular fluid that has the increasing concentration of sodium ion.

An osmotic gradient for water is thus created between the tubule and the surrounding fluids. The height of the gradient is 3 to 4 times the osmotic pressure of the blood.

Therefore, water has a tendency to leave the tubule, *but* cannot do so unless a particular hormone is present.

ANTIDIURETIC HORMONE (ADH) is released from the posterior pituitary gland and allows water to pass through the membranes of the cells of the collecting tubules and papillary ducts. This hormone has the effect of matching need for reabsorption of water to the body requirements for water and is the single most important device for

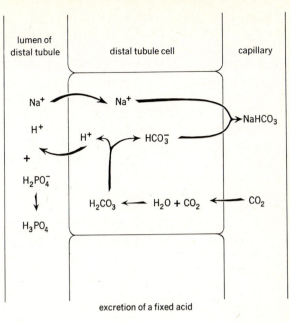

excretion of a fixed acid

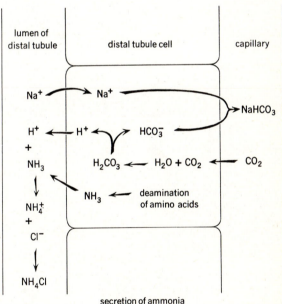

secretion of ammonia

FIGURE 27.6 Two mechanisms of acidification of the urine.

controlling blood-water concentration. The center controlling how much ADH is released is in the hypothalamus, and it monitors the osmotic pressure of the blood reaching it. If osmotic pressure

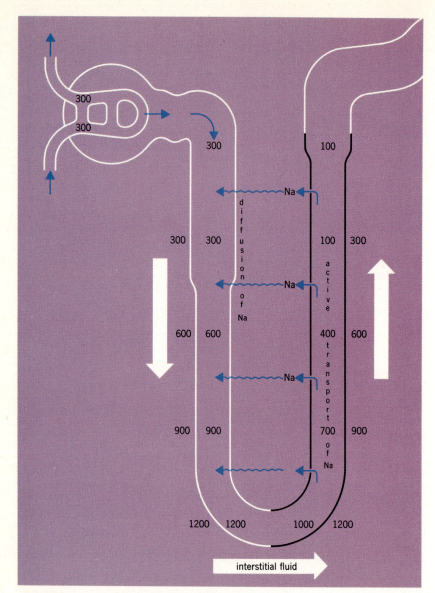

FIGURE 27.7 The result of the operation of the countercurrent multiplier. The heavy black lines indicate the portion of the loop of Henle that is not permeable to water.

of the blood increases, more ADH is released, more water is reabsorbed, and the blood is diluted. The reverse situation also occurs to concentrate the blood when it becomes too dilute.

The material leaving the papillary ducts has no further changes made in its composition and is now properly called URINE. A summary of the processes occurring in urine formation is provided in Table 27.3.

TABLE 27.3 Summary of the processes occurring in urine formation			
Process	Where occurring	Force responsible	Result
Filtration	Glomerulus	Blood pressure, opposed by osmotic, interstitial, and intra-tubular pressures	Formation of fluid having no formed elements and low protein concentration
Reabsorption	Proximal tubule Distal tubule Loop of Henle	Active transport	Return to bloodstream of physiologically important solutes
Secretion	Proximal tubule Distal tubule	Active transport	Excretion of materials Acidification of urine
Acidification (acid and base regulation)	Distal tubule	Active transport and exchange of alkali for acid	Excretion of excess H^+ Conservation of base (Na^+ and HCO_3^-)
Concentration of filtrate	Collecting tubule and papillary ducts	Osmosis of water under permissive action of ADH	Formation of hypertonic urine

Urine

Characteristics and composition

Normal urine differs from plasma, as is shown in Table 27.4, and contains a variety of substances, as is shown in Table 27.5. It is an amber or yellow, transparent, usually acidic fluid, and has a solute concentration about three times that of the plasma. Its volume is determined by several factors including the following.

The height of the blood pressure. Filtration is proportional to blood pressure, so that a higher blood pressure causes the formation of a larger volume of filtrate, from which relatively less water is reabsorbed; more water is therefore eliminated.

A high level of filtered solutes. High solute concentrations in the filtrate tend to cause water to remain in the tubules by "osmotic attraction." Thus, a greater volume is eliminated.

Amount of ADH present. This has been explained above.

Loss by other routes. Kidney loss of water is really the only route of loss that can be altered. If one sweats

heavily, or loses water by vomiting or diarrhea, less is excreted by the kidney to offset losses by other means.

Presence of diuretic substances. A diuretic is a substance promoting water loss by the kidney. It may act by slowing solute removal from the filtrate or by increasing solute secretion. Thus, the volume of water in the filtrate is increased, and more water is eliminated.

Control of urine formation

At least two hormones and one enzyme are involved in regulating both the volume and concentration of urine formation.

ADH has been discussed above.

ALDOSTERONE, a hormone of the adrenal cortex, controls the rate of Na^+ reabsorption and thus influences not only Na^+ reabsorption but also that of Cl^-, HCO_3^-, and water. Anions follow cations by electrostatic attraction; water follows solute removal.

RENIN, an enzyme, is secreted by the JG apparatus when the kidney receives an inadequate flow of blood. It acts as follows:

Renin + angiotensinogen → angiotensin I
(enzyme) (plasma substrate) (in plasma)

$$+$$

plasma enzyme

↓

angiotensin II
(potent vasocon-
strictor)

Angiotensin II causes a generalized vasoconstriction that raises blood pressure to the kidneys and assures adequate filtration. It also stimulates adrenal cortex secretion of aldosterone, increasing Na^+ and water reabsorption.

TABLE 27.4 Comparison of plasma and urine for certain constituents[a]

Property or Substance	Plasma	Urine
Osmolarity (mOs)	300	600–1200
pH	7.4 ± 0.02	4.6–8.2
Specific gravity	1.008–1.010	1.015–1.025
Sodium	0.3	0.35
Potassium	0.02	0.15
Chloride	0.37	0.60
Ammonia	0.001	0.04
Urea	0.03	2.0
Sulfate	0.002	0.18
Creatinine	0.001	0.075
Glucose	0.1	0
Protein	7–9	0

[a] Values, unless otherwise indicated, are gm %

TABLE 27.5 Some normal constituents, origins, and amounts excreted per day in normal urine

Constituent	Origin	Amount per day
Water	Diet and metabolism	1200–1500 ml
Urea	Ornithine cycle	30 g
Uric acid (purine)	Catabolism of nucleic acids	0.7 g
Hippuric acid	Liver detoxification of benzoic acid	Trace
Creatinine	Destruction of intracellular creatine phosphate of muscle	1–2 g
Ammonia	Deamination of amino acids	0.45 g
Chloride (as NaCl)	Diet	12.5 g
Phosphate	Diet and metabolism of phosphate containing compounds	3 g
Sulfate	Diet, metabolism of sulfate containing compounds, formation of H_2SO_4 in kidney tubules	2.5 g
Calcium	Diet	200 mg

The ureters

Gross anatomy and relationships

The ureters extend from the hilus of the kidney to the urinary bladder, a distance of 28 to 35 centimeters (about 11–14 inches). Their diameter is about 6 mm (¼ inch). They are retroperitoneal in placement and have an increasing diameter as they course toward the bladder.

Microscopic anatomy

Three coats of tissue form the wall of the ureter. A MUCOSA, composed of transitional epithelium and connective tissue, forms the inner layer. The central MUSCULARIS is composed of inner longitudinal and outer circular layers of smooth muscle throughout most of the length of the ureter. On the lower one-third of the organ, a third layer of muscle (outer longitudinal) is added. The outer FIBROUS COAT (adventitia) blends without demarcation into the surrounding fascia.

Innervation

The ureters in their upper part receive sympathetic fibers from the renal plexus; in the middle part they receive fibers from the ovarian or spermatic plexus; and near the bladder they receive fibers from the hypogastric nerves (Fig. 27.8). These fibers exert primarily a motor effect. They cause the ureters to exhibit rhythmical peristaltic contractions traveling at a speed of 20 to 35 milli-

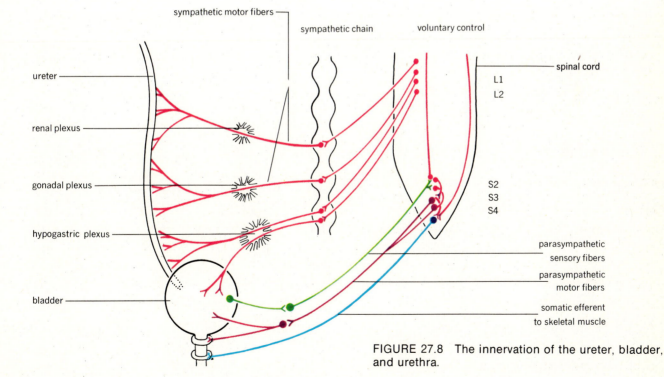

FIGURE 27.8 The innervation of the ureter, bladder, and urethra.

meters per second and at a frequency of 1 to 5 per minute. The urine, therefore, enters the bladder, not in a continuous stream, but in separate squirts synchronous with the arrival of each peristaltic wave.

The urinary bladder

Gross anatomy and relationships

The urinary bladder serves as a reservoir for the urine until it is voided. The organ is located superior to the symphysis pubis and is separated from the symphysis by a prevesicular space. The space, filled with loose connective tissue, allows for expansion of the filling bladder. Internally, three openings may be found in the bladder wall: the two ureters, and the urethra. An imaginary line drawn to connect these three openings outlines the *trigone*.

Microscopic anatomy

Although similar to the ureter in structure, the bladder has more cell layers in the transitional epithelial lining, a submucous layer of loose tissue between mucosa and muscularis, and three heavy layers of smooth muscle in the muscularis, disposed longitudinally, circularly, and longitudinally. Around the urethral opening, a dense mass of circularly oriented smooth muscle forms the internal sphincter of the bladder. An external sphincter is formed by the skeletal muscle of the

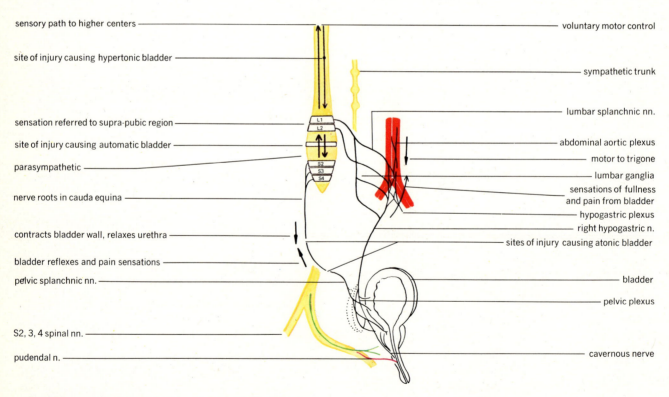

FIGURE 27.9 Nerves of the bladder illustrating the origin of bladder dysfunction.

pelvic floor. It is under voluntary control. A serous layer is formed by the peritoneum over the superior surface of the organ.

Innervation

The efferent nerves to the bladder and urethra (Fig. 27.9) are from both sympathetic and parasympathetic divisions. The sympathetic fibers furnish inhibitory fibers to the muscle of the bladder; they furnish motor fibers to the trigone and internal sphincter and the muscle of the upper part of the urethra. These fibers arise in the lumbar spinal segments and pass to the bladder via the inferior hypogastric plexus. The parasympathetic nerves supply motor fibers to the detrusor muscle (the muscle of the bladder) and inhibitory fibers to the internal sphincter. The desire to urinate occurs when a volume of 200 to 300 milliliters of urine has accumulated in the bladder. Stretch on the muscle of the bladder, brought about by filling, evokes an afferent impulse in the pelvic nerves. This impulse ascends through the spinal cord to a center in the hindbrain. An efferent discharge of motor impulses to the muscle of the bladder is accomplished through descending pathways in the cord and through the pelvic motor nerves. The same efferent or motor fibers bring about a simultaneous relaxation of the internal sphincter so that urine may be emptied from the bladder into the urethra (micturition). Urination may occur by reflex action not involving the hindbrain center. Filling evokes a reflex contraction of the detrusor and relaxation of the sphincter, which is served by lower cord segments only. The reflex is seen in infants, where the voluntary control over sphincters has not yet been achieved.

The urethra

The female urethra is about 4 centimeters (1¾ inches) in length, is completely separate from the reproductive system, and bears no regional differences. A mucosa is present, consisting of transitional epithelium near the bladder and stratified squamous elsewhere, and is underlain by connective tissue. A muscularis is present, consisting of circularly arranged fibers of smooth muscle. The female urethra opens just anterior to the vaginal orifice.

The male urethra is about 20 centimeters (8 inches) in length, serves as a common tube for the terminal portions of both urinary and reproductive systems, and does show regional variations. The PROSTATIC URETHRA is the first 3 centimeters of the organ; it is surrounded by the prostate gland and is lined with transitional epithelium. The MEMBRANOUS URETHRA is 1 to 2 centimeters long; it penetrates the pelvic floor, is very thin, and has a pseudostratified epithelium lining. The CAVERNOUS URETHRA is about 15 centimeters in length and lies within the penis.

Clinical considerations involving the urinary system

Kidney

Glomerulonephritis, nephrotic syndrome, and pyelonephritis are probably the most frequently encountered kidney disorders.

GLOMERULONEPHRITIS is almost always associated with a previous infection by a beta-hemolytic streptococcus bacterium elsewhere in the body (usually the pharynx) and is generally regarded as an "autoimmune disease." One of the components of the basement membrane of the glomerular capsule, sialic acid, is thought to be hydrolyzed by the bacterial toxin. The body then recognizes the hydrolyzed acid as a "new" or "foreign" substance and produces antibodies to neutralize it. In the ensuing antigen-antibody reaction the basement membrane is injured and allows red cells and more protein to leak into the filtrate. Edema is also a common occurrence because the osmotically active protein molecules are diminished in the blood, resulting in failure to return water to the circulation.

The NEPHROTIC SYNDROME also involves an increased glomerular permeability, but appears to involve the tubules to a greater degree than the glomeruli. The chief difference from glomerulonephritis lies in the fact that a massive loss of protein from the blood occurs, with cast formation in the tubules and a near shutdown of tubular function.

PYELONEPHRITIS is a bacterial infection of the kidney that usually begins in the lower tract and spreads upward to involve the kidney. Females are more susceptible to this disease as a consequence of the short urethra and the greater possibility of bladder infection occurring. It is the most common disease affecting the urinary system. Chills, fever, flank pain, nausea, and vomiting characterize the infection. Treatment is with antibiotics to which the organism cultured from the urine is sensitive.

Bladder and urethra

Nerve injuries produce three types of bladder dysfunction (*see* Fig. 27.9).

ATONIC BLADDER. Interruption of sensory supply results in loss of bladder tone, and the organ may become extremely distended with no development of an urge to urinate.

HYPERTONIC BLADDER. Interruption of the voluntary pathways results in excessive tone, and very small distentions create an uncontrolled desire to urinate.

AUTOMATIC BLADDER. Complete section of the cord above the first sacral nerve exit (S1) produces automatic emptying in response to filling, by the described cord reflex.

Other disorders associated with the lower urinary tract include cystitis, urethritis, and obstruction of the urethra.

CYSTITIS, or inflammation of the bladder, is usually secondary to infection of the prostate, kidney, or urethra.

INFECTION of the male urethra is most commonly associated with gonorrheal infections; in the female, almost any infection of the perineum may invade the urethra.

BLOCKAGE of the urethra is most commonly associated with calculi (stones) in the bladder. The stone may be voided in the urine, if small enough, or may be crushed by use of a cystoscope inserted through the urethra.

Stones may be found anywhere in the tract and are caused by infections, or metabolic disorders that cause excretion of large amounts of organic and inorganic substances. As the kidney concentrates the filtrate, the substances in the fluid may reach high enough levels to cause precipitation of the substance in the form of a tiny granule or stone. This granule then serves as a nucleus for further precipitation of materials to form the large stones. Stones are often described according to shape (e.g., staghorn stone in renal pelvis), or by what they contain (cystine, uric acid, calcium stones).

Individuals that form only an occasional stone may require only a large water intake (to 3 liters per day) to reduce stone formation; chronic stone formers usually are treated with substances to render the urine alkaline and to reduce the tendency for precipitation to occur.

Summary

1. Several pathways are utilized to rid the body of wastes of metabolism, excess materials, and undigestible substances.

   a. The lungs eliminate CO_2.

   b. The alimentary tract eliminates CO_2, water, undigested materials, and secretions.

   c. The skin eliminates heat.

   d. The kidney eliminates water, salts, nitrogenous wastes, and controls ECF fluid composition.

2. The system develops from mesoderm and endoderm, beginning at about $3\frac{1}{2}$ weeks.

   a. The kidney is mesodermal in origin.

   b. The terminal ducts are endodermal in origin.

   c. The kidney is a common organ for developmental anomalies. Most produce no symptoms.

3. The kidneys

   a. Are reddish, bean shaped, and are high against the back wall of the peritoneal cavity.

   b. Measure $4 \times 2 \times 1$ inches.

   c. Have cortex, medulla, and several branches of the ureter in them.

   d. Contain microscopic nephrons as the functional units.

   e. Have a "separate circulation" designed to supply large volumes of blood for cleansing by the kidney.

4. Urine is formed by several processes.

   a. Glomerular filtration removes all but the largest substances or cells from the blood.

   b. Tubular reabsorption actively removes needed solutes from the filtrate, and passively removes water and some wastes.

   c. Tubular secretion places into the tubules certain organic acids, bases, and ions.

   d. Acidification, by secretion of H^+, changes the filtrate to an acidic state.

   e. Concentration of the urine occurs by water reabsorption in the presence of antidiuretic hormone (ADH).

5. Urine differs from plasma and filtrate in both substances present, and amounts present.

   a. Its volume is determined by blood pressure, solute load, ADH, loss of water by other routes, and diuretics.

   b. ADH, aldosterone, and renin, control composition.

6. The ureters are 11–14 inches long and convey urine to the bladder.

  a. They have a three-layered wall.

  b. Their muscularis moves urine to the bladder by peristalsis.

  c. They receive nerves from lumbar and sacral spinal cord segments.

7. The bladder stores urine for voiding.

  a. Structure is similar to that of the ureters, but thicker.

  b. Nerves supply the bladder from lumbar and sacral cord, and contain both sensory and motor fibers.

  c. Micturition (urination) occurs reflexly, but may be voluntarily inhibited.

8. The urethra carries urine to the exterior. It differs in structure in male and female.

9. A variety of diseases and disorders involving the organs of the system are discussed.

Questions

1. Describe the location of the kidneys and their supporting structures.

2. What structures enter or leave the hilus of the kidney?

3. Describe the distribution of blood vessels within the kidney.

4. What is a nephron and what are its functions?

5. What are the processes the nephron uses in forming urine? Describe the main result of the operation of each process.

6. List five factors controlling urine volume and explain how each works.

7. Compare plasma, filtrate, and urine for the following properties: specific gravity, water content, glucose content, protein content, pH. What accounts for any differences?

8. Compare the structure of ureters and bladder.

9. Describe the mechanism involved in emptying of the bladder.

10. Describe differences between the male and female urethra.

11. Describe two disorders of the kidney, their cause, and their effect on normal nephron functioning.

12. What contribution do the tubules make to the formation of urine?

Readings

Brenner, Barry M., Thomas H. Hostetter, and H. David Humes. "Molecular Basis of Proteinuria of Glomerular Origin." *New Eng. J. Med. 298*:826, April 13, 1978.

Jamison, Rex L., and Roy H. Maffly. "The Urinary Concentrating Mechanism." *New Eng. J. Med. 295*:1059, Nov. 4, 1976.

Katz, Adrian I., and Marshall D. Lindheimer. "Actions of Hormones on the Kidney." *Ann. Rev. Physiol. 39*:97, 1977.

Lassiter, William E. "Kidney." *Ann. Rev. Physiol. 37*:371, 1975.

Nicoll, P. A., and T. A. Cortese, Jr. "The Physiology of Skin." *Ann. Rev. Physiol. 34*:177, 1972.

Reid, Ian A., Brian J. Morris, and William F. Ganong. "The Renin-Angiotensin System." *Ann. Rev. Physiol. 40*:377, 1978.

Chapter 28

The Reproductive Systems

Objectives

After studying this chapter, the reader should be able to:

- Outline the early development of the reproductive systems.

- List the organs of the male reproductive system, their locations, structure, and functions.

- List the organs of the female reproductive system, their locations, structure, and functions.

- Describe the stages in the development of an ovum.

- Describe the stages of the menstrual cycle, including hormonal relationships.

- Describe the mammary glands as to location, structure, and functions.

- List some venereal diseases, their effects, and potential for treatment.

- Cite reasons for the epidemic proportion of venereal diseases.

The male and female reproductive systems provide a means of creation of a new human individual. Certain of the organs are responsible for the production of sex cells, others transport and

632

nourish these cells, or provide a haven for development of a new individual. Some of the organs produce hormones that are essential for the process of reproduction of the species, and which may also affect the functions of the body generally.

Development of the systems

Development of the reproductive systems is evident after about 5 weeks of embryonic life. They originate from the urogenital ridge that gave rise to the kidney. At about 5 weeks, the structures shown in Figure 28.1 are present. The tubules shown in this figure were originally those developed in association with the kidney and become portions of the duct system of the reproductive organs. At about 8 weeks, the testes and ovaries assume an internal structure characteristic for each organ, with primitive sex cells visible. The testes also secrete small amounts of male sex hormone (testosterone) about this time.

Until about 8 weeks of development, the external genitalia are in an undifferentiated state, and are not characteristic of a particular sex. Genetic influences assure development of female external genitalia in the absense of hormones. Male external genitalia development requires testosterone production by the primitive testis to assure normal male differentiation. The changes occurring in the external genitalia from the undifferentiated state are shown in Figure 28.2. Organs in the two sexes having the same origin (homologous organs) are shown in the same color.

The male reproductive system

Organs

The organs of the male reproductive system (Fig. 28.3) may be grouped under three headings:

The *essential organs*, or *testes*, produce male sex cells and secrete the hormone *testosterone.*

The *excretory ducts* store or convey sperm, and include the straight tubules, rete tubules, efferent ductules, epididymis, ductus (vas) deferens, ejaculatory duct, and urethra.

The *accessory glands* include the seminal vesicles, prostate gland, and bulbourethral (Cowper's) glands.

The testes

LOCATION. After about 8 months of intrauterine life, the testes descend from their region of development in the abdomen into the SCROTUM. The latter structure is an outpocketing of skin and connective tissue of the lower anterior abdominal wall. It is divided internally into two lateral compartments, each of which houses a testis and its associated structures. Each scrotal compartment communicates with the abdominopelvic cavity through the inguinal canal. Descent of the testes into the scrotum ensures a temperature low enough to allow sperm cell development at puberty. Failure of descent results in *cryptorchidism* and sterility.

STRUCTURE (Fig. 28.4). Each adult testis is ovoid in shape, measuring 4 to 5 centimeters ($1\frac{3}{4}$–2 inches) long, and 2.5 centimeters (1 inch) in diameter. Several tunics or coats of tissue form coverings for the testis. The most obvious tunic is one made of fibrous tissue, and is called the TUNICA

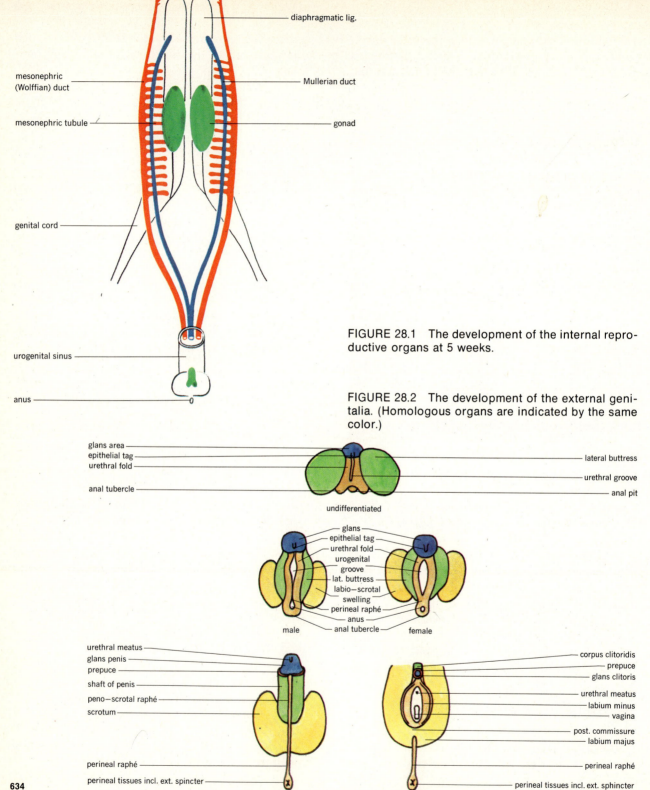

diaphragmatic lig.

mesonephric
(Wolffian) duct

Mullerian duct

mesonephric tubule

gonad

genital cord

urogenital sinus

anus

FIGURE 28.1   The development of the internal reproductive organs at 5 weeks.

FIGURE 28.2   The development of the external genitalia. (Homologous organs are indicated by the same color.)

glans area
epithelial tag
urethral fold

anal tubercle

lateral buttress

urethral groove

anal pit

undifferentiated

glans
epithelial tag
urethral fold
urogenital
groove
lat. buttress
labio—scrotal
swelling
perineal raphé
anus
anal tubercle

male

female

urethral meatus
glans penis
prepuce
shaft of penis
peno—scrotal raphé
scrotum

perineal raphé
perineal tissues incl. ext. spincter

male

corpus clitoridis
prepuce
glans clitoris
urethral meatus
labium minus
vagina
post. commissure
labium majus

perineal raphé

perineal tissues incl. ext. sphincter

female

634

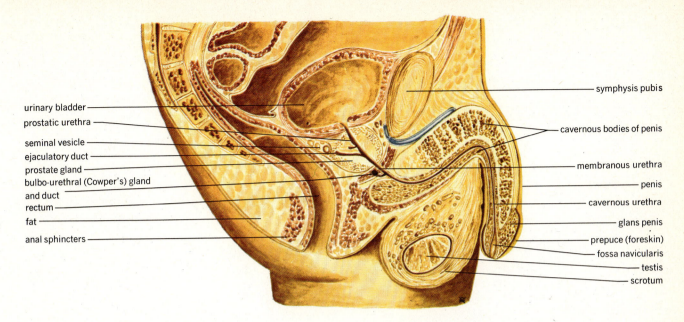

urinary bladder
prostatic urethra
seminal vesicle
ejaculatory duct
prostate gland
bulbo-urethral (Cowper's) gland
and duct
rectum
fat
anal sphincters

symphysis pubis
cavernous bodies of penis
membranous urethra
penis
cavernous urethra
glans penis
prepuce (foreskin)
fossa navicularis
testis
scrotum

FIGURE 28.3   The male reproductive organs and associated structures.

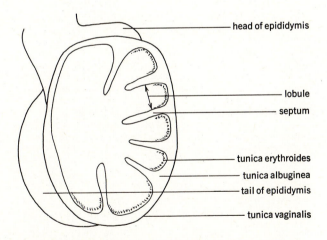

head of epididymis
lobule
septum
tunica erythroides
tunica albuginea
tail of epididymis
tunica vaginalis

FIGURE 28.4   The internal structure of the testis.

ALBUGINEA. It acts as a capsule for the testis, sends partitions or SEPTAE into the organ, dividing it into about 250 LOBULES, and is thickened on the medial side of the testis to form the MEDIASTINUM (TESTES).

FUNCTION. Each lobule contains 1 to 4 highly FOLDED SEMINIFEROUS TUBULES (Fig. 28.5), each of which is lined with a germinal epithelium (Fig. 28.6). At puberty, the germinal epithelium begins production of sperm cells, and shows several stages in the development of these sperm cells (see Fig. 28.6). The more primitive cells are located at the periphery of the tubules and maturation is more advanced toward the center of the tubules. Sperm are highly differentiated cells (Fig. 28.7) that carry a haploid chromosome number.

The hormone-producing cells of the testis are located in the connective tissue between seminiferous tubules and are called INTERSTITIAL CELLS (Fig. 28.8). Their product, testosterone, is necessary for development and growth of the external genitalia and development of male secondary sex characteristics.

At puberty, pituitary secretion of FSH (follicle stimulating hormone) initiates maturation of spermatozoa, and ICSH (interstitial cell stimulating hormone, also called LH, (luteinizing hormone) stimulates secretion of the interstitial cells. Testosterone secretion then causes develop-

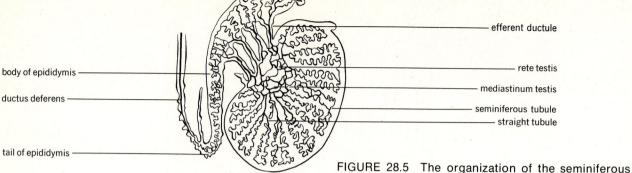

head of epididymis

duct of epididymis

efferent ductule

body of epididymis

rete testis

ductus deferens

mediastinum testis

seminiferous tubule

straight tubule

tail of epididymis

FIGURE 28.5 The organization of the seminiferous tubules and ducts of the testes.

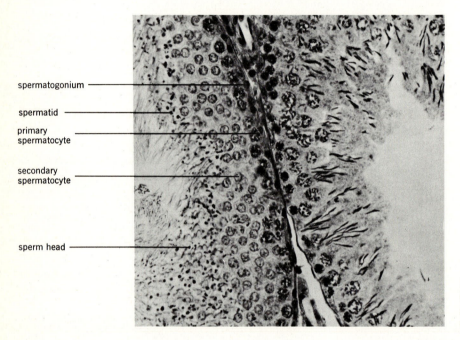

spermatogonium

spermatid

primary spermatocyte

secondary spermatocyte

sperm head

FIGURE 28.6 Cross section of a seminiferous tubule in an adult showing stages of sperm development.

ment of the genital ducts, external genitalia, and secondary sex characteristics.

The duct system
(Fig. 28.9; *see also* Fig. 28.5)

STRUCTURE. The STRAIGHT TUBULES connect the seminiferous tubules to the RETE TUBULES in the mediastinum. Only these two parts of the duct system lie within the testis. There are 12 to 16

EFFERENT DUCTULES that carry sperm to the EPI-DIDYMIS where they are stored until released. The epididymis is a single coiled tube about 20 feet in length. It is held by connective tissue to the testis. The DUCTUS (*vas*) DEFERENS is a muscular tube arising from the lower end of the epididymis and passes out of the scrotum, through the inguinal canal, to behind the bladder. It conveys sperm from the scrotum with much force during ejaculation. As the deferens passes out of the scrotum, it is bound by connective tissue along

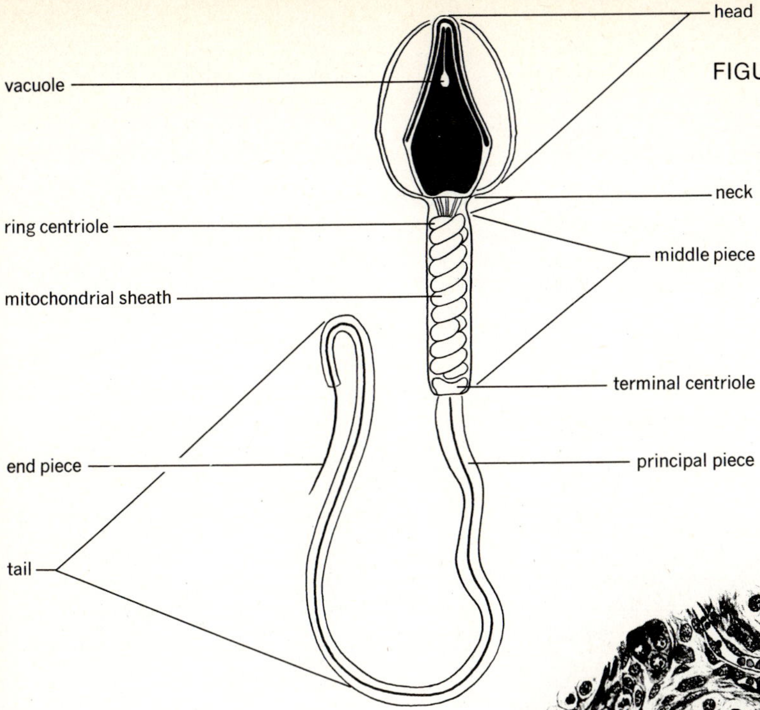

head

vacuole

neck

ring centriole

middle piece

mitochondrial sheath

terminal centriole

end piece

principal piece

tail

FIGURE 28.7 A spermatozoan.

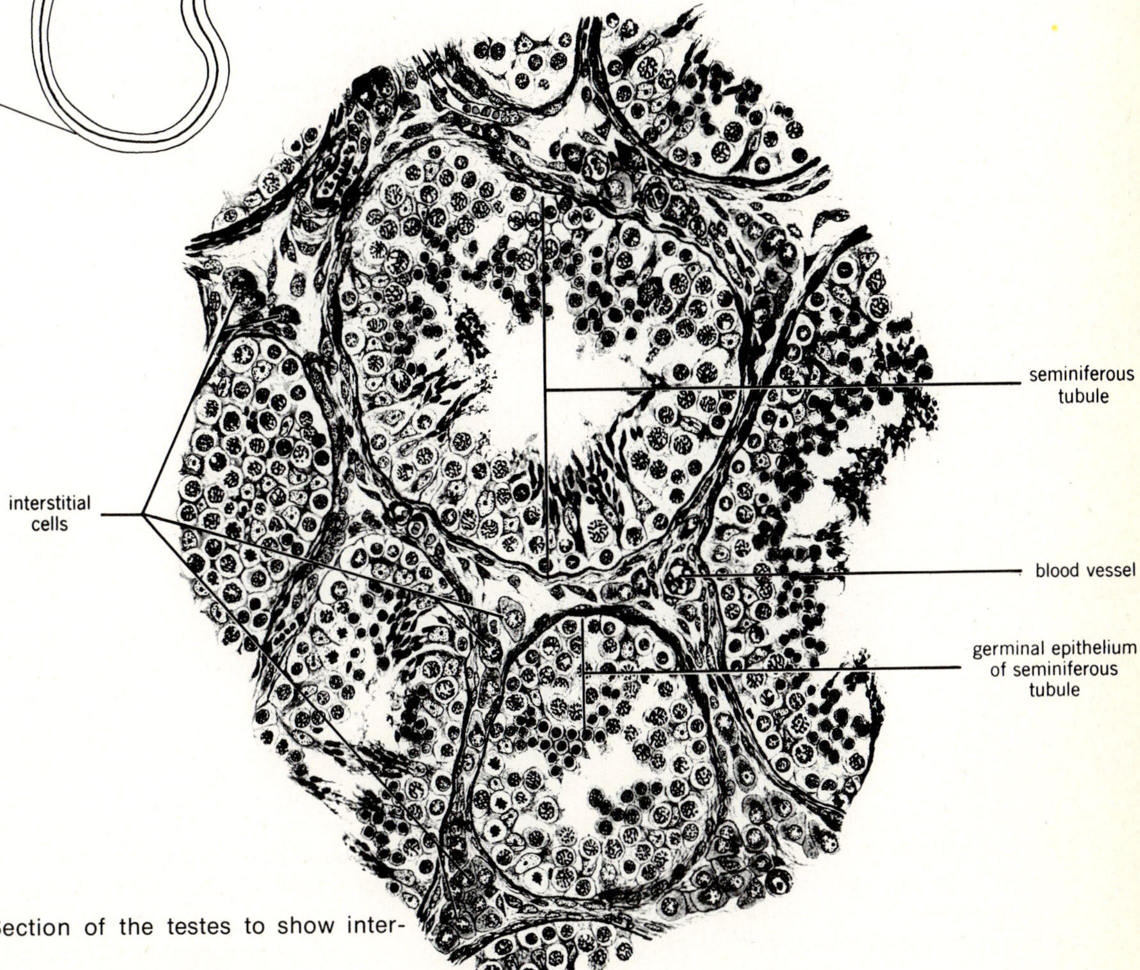

interstitial cells

seminiferous tubule

blood vessel

germinal epithelium of seminiferous tubule

FIGURE 28.8 Section of the testes to show interstitial cells.

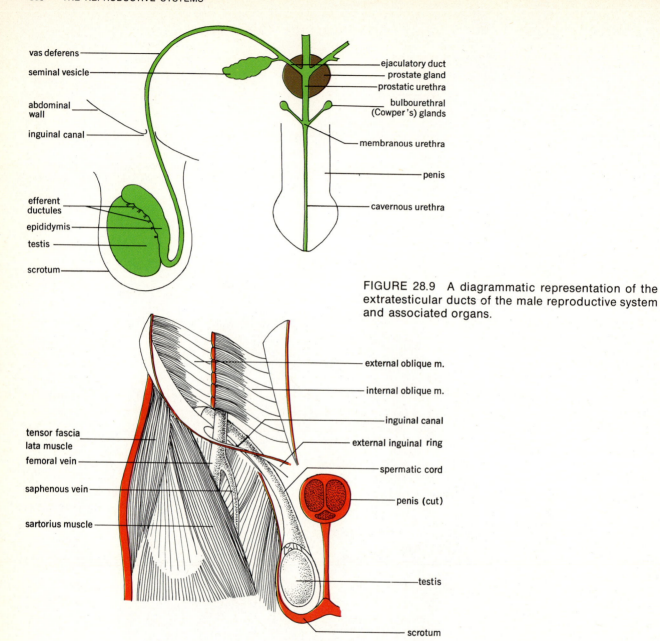

vas deferens

seminal vesicle

abdominal wall

inguinal canal

efferent ductules

epididymis

testis

scrotum

ejaculatory duct

prostate gland

prostatic urethra

bulbourethral (Cowper's) glands

membranous urethra

penis

cavernous urethra

FIGURE 28.9   A diagrammatic representation of the extratesticular ducts of the male reproductive system and associated organs.

tensor fascia lata muscle

femoral vein

saphenous vein

sartorius muscle

external oblique m.

internal oblique m.

inguinal canal

external inguinal ring

spermatic cord

penis (cut)

testis

scrotum

FIGURE 28.10   The spermatic cord is shown passing from the testis through the inguinal ring.

with nerves, arteries, veins, and lymphatics into a structure known as the SPERMATIC CORD (Fig. 28.10). The components of the cord separate within the abdominopelvic cavity. Behind the

bladder, the deferens is joined by the duct of the seminal vesicle, and the common tube, or EJACU- LATORY DUCT, joins the urethra. The structure of the urethra has been described in Chapter 27.

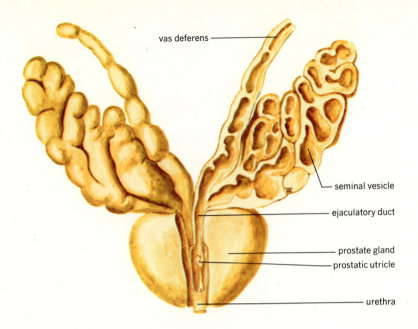

vas deferens

seminal vesicle

ejaculatory duct

prostate gland
prostatic utricle

urethra

FIGURE 28.11 The prostate, seminal vesicles, and associated structures.

FUNCTION. The ducts have smooth muscle in their walls and serve primarily to move sperm eventually to the exterior. The epididymis is so long that the sperm within it, though moving, are considered to be "in storage" as they pass slowly through. The deferens is a propulsive tube; the urethra is a conduction tube that directs sperm (via the penis) to the female reproductive system or to the exterior.

The accessory glands (Fig. 28.11)

LOCATION AND STRUCTURE. The SEMINAL VESI-CLES are paired tortuous sacs lying on the posterior aspect of the urinary bladder. The PROSTATE GLAND surrounds the upper urethra, just below the neck of the bladder. It is about the size of a walnut and consists of 30 to 50 separate glands all bound together by connective tissue. The paired BULBOURETHRAL (*Cowper's*) GLANDS are pea-sized, and empty into the membranous urethra.

FUNCTION. The secretion of the accessory glands, together with the sperm, constitutes the SEMEN.

The ejaculate is the total volume of semen expelled during *ejaculation*. Vesicle secretion is alkaline, contains much fructose and vitamin C, and constitutes about 60 percent of the 3 to 5 milliliters of the ejaculate. This secretion suspends and nourishes the sperm and aids in neutralizing acidity in the vagina. Prostatic secretion is rich in cholesterol, buffering salts, and phospholipids, makes up 40 percent of the ejaculate, and is thought to activate the sperm to make them motile. It also aids in neutralizing vaginal acidity. Some 300 to 400 million sperm are found in the semen, and are very sensitive to acid pH's. The secretion of the bulbourethral glands is an alkaline mucus that neutralizes any urine remaining in the male urethra. The bulbourethral glands secrete their material before ejaculation occurs.

The penis (Figs. 28.12 and 28.13)

The penis serves as a copulatory organ to introduce sperm into the vagina of the female. It is formed by three cylindrical masses of spongy

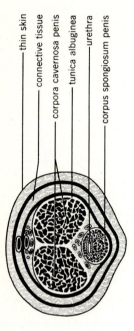

thin skin
connective tissue
corpora cavernosa penis
tunica albuginea
urethra
corpus spongiosum penis

FIGURE 28.12   Cross section of the penis.

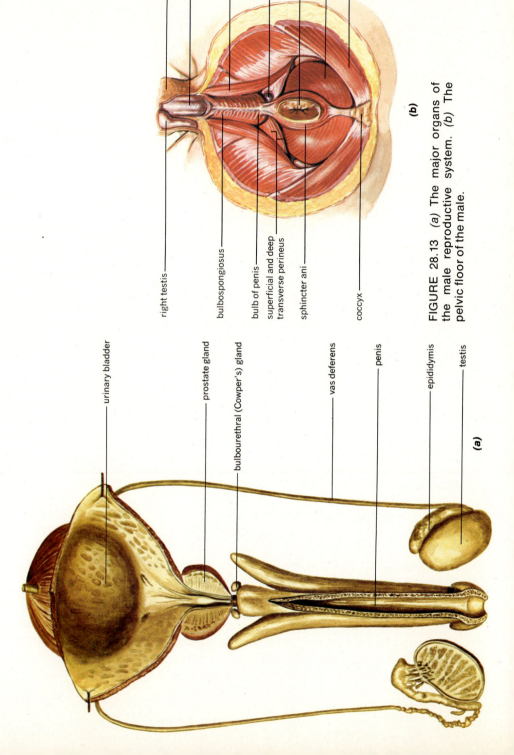

scrotal skin
shaft of penis
ischiocavernosus
bulbourethral (Cowper's) gland
anus
levator ani
gluteus maximus

right testis
bulbospongiosus
bulb of penis
superficial and deep transverse perineus
sphincter ani
coccyx

(b)

urinary bladder
prostate gland
bulbourethral (Cowper's) gland
vas deferens
penis
epididymis
testis

(a)

FIGURE 28.13   (a) The major organs of the male reproductive system. (b) The pelvic floor of the male.

tissue known as the cavernous bodies. Two of the bodies are dorsally placed in the penis and are known as the CORPORA CAVERNOSA (*spongiosa*) PENIS; a smaller, single, ventrally placed CORPUS SPONGIOSUM (*cavernosum*) URETHRAE carries the urethra. The cavernous bodies contain large venous sinuses that, when filled with blood, *erect* or stiffen the penis and make it an effective copulatory organ.

Clinical considerations

MALE INFERTILITY is defined as inability to produce children. Congenital causes, such as maldevelopment of the reproductive organs may result in infertility, as may diseases (e.g., venereal diseases, mumps) that attack the testes and destroy or reduce their ability to produce sperm.

Among the most common INFECTIONS affecting the male reproductive system are gonorrhea and syphilis. The microorganisms causing these diseases may be transmitted by sexual intercourse, body-to-body, or oral contact with an infected person. Recent evidence indicates that live gonorrhea organisms may be isolated from contaminated toilets, silverware, or doorknobs up to 8 hours after deposition of the bacteria.

IMPOTENCE, or an inability to create and sustain an erection may result in inability of the male to copulate successfully with the female. Impotence appears to be an increasingly common affliction of the male in "civilized" societies.

Sperm counts less than about 50 million per milliliter of semen make it difficult for the male to fertilize a normal female because not enough sperm are provided to:

Resist the loss of numbers due to acids in the vagina.

Survive the 6 inch trip through the uterus and uterine tubes to fertilize the ovum.

Provide sufficient hyaluronidase, an enzyme that liquefies the intercellular cement holding several layers of follicular cells to the released ovum.

Sperm are capable of surviving up to 3 days within the female reproductive tract, so that pregnancy can occur even though copulation has not occurred for several days. If ovulation occurs and sperm are present, pregnancy may result.

Enlargement of the prostate gland is a common occurrence with aging. If the gland enlarges greatly, it may compress the urethra to where both reproductive function and urination become difficult or impossible. Surgery is usually indicated to remove the enlarged organ. Cancer of the prostate is the second most common neoplasm in the male, and is the sixth leading cause of cancer death in the male.

The female reproductive system

Organs

The organs of the female reproductive system (Figs. 28.14; 28.15) may be grouped under two headings:

The *internal organs* are located within the abdominopelvic cavity and include the ovaries, uterine tubes, uterus, and vagina.

The *external organs* include the external genitalia (clitoris, labia majora, and minora) and the mammary glands. The latter are actually highly modified sweat glands, but are considered with this sytem because of their close functional relationship to the reproductive system.

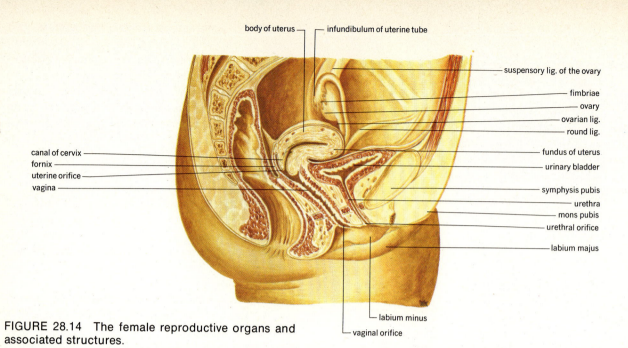

FIGURE 28.14 The female reproductive organs and associated structures.

FIGURE 28.15 The internal organs of the female reproductive system.

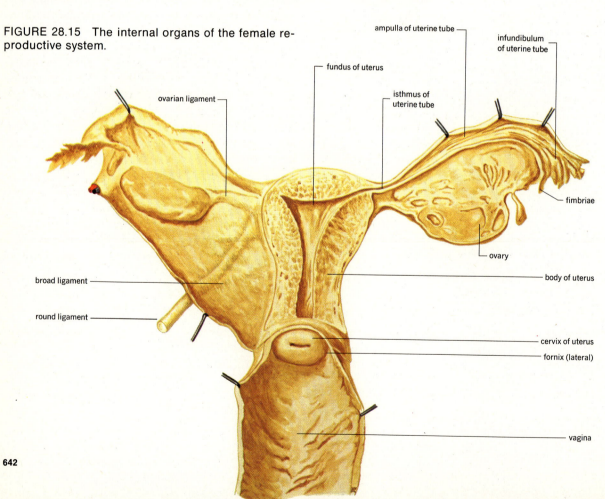

The ovaries (Fig. 28.16)

LOCATION AND STRUCTURE. The ovaries are paired, ovoid, organs lying within the pelvic portion of the abdominopelvic cavity and measure about 3 centimeters (1½ inches) long, 2 centimeters (¾ inches) wide, and 1 centimeter (½ inch) deep. Each ovary has an indented HILUS that marks the entry and exit of ovarian blood vessels, nerves, and lymphatics. The ovaries are supported by a mesentery, the MESOVARIUM, and OVARIAN and SUSPENSORY LIGAMENTS.

Microscopically (Fig. 28.17), each ovary is seen to be covered with a layer of cuboidal cells called the GERMINAL EPITHELIUM. This is a misnomer, inasmuch as the cells do not give rise to sex cells, but are merely a lining or covering tissue. A capsulelike TUNICA ALBUGINEA of fibrous tissue forms the major tunic of the ovary, and surrounds a connective tissue STROMA. The stroma is further subdivided into an outer CORTEX, containing OVARIAN FOLLICLES, and an inner vascular MEDULLA, which provides for the nutrition of the organ.

FUNCTION. The ovary, after puberty is the site of maturation of sex cells or OVA, and secretes several HORMONES necessary for development of the other internal organs, and of the female secondary sex characteristics.

The ovarian follicles (Fig. 28.18)

At birth, approximately 400,000 PRIMORDIAL FOLLICLES are present in the ovaries. These are composed of a small primitive ovum, surrounded by a single layer of flattened stromal cells. They will remain in this form until puberty, when pituitary hormones cause their further development. Of these 400,000 original ova, about 400 will mature during the reproductive life of the female. The rest undergo a degenerative process known as ATRESIA.

At puberty, follicular growth begins, caused by a pituitary hormone, FSH (follicle stimulating hormone). PRIMARY FOLLICLES are larger structures surrounded by several layers of cells that develop from the primitive stromal cells. Cavities known as ANTRA (sing.: antrum) next develop within these layers of cells, and the structure is known as a SECONDARY FOLLICLE. The separate cavities fuse to form a single large cavity and, at this stage, the structure is called a VESICULAR OR MATURE FOLLICLE. It may ultimately reach a size of ½ inch. Maturation requires 8 to 15 days, and is associated with production of a hormone known as ESTRADIOL. This hormone affects the lining of the uterus (causes proliferation of cells), and develops the ducts of the mammary glands. It also causes the appearance of the female secondary sex characteristics and the menses.

OVULATION, or release of the mature ovum and its surrounding coat of several layers of follicle cells, occurs next under the influence of pituitary luteinizing hormone. The tissues remaining in the ovary turn into a CORPUS LUTEUM (yellow body) also under the influence of LH or luteinizing hormone. The corpus luteum produces PROGESTERONE, a hormone that prepares the uterus to receive an egg if it is fertilized, and that develops secretory tissue in the breasts. The luteum will last about 2 weeks if fertilization does not occur; it will last about 3 months of a pregnancy if fertilization does occur. In either case, it ultimately degenerates, is replaced by scar tissue, and forms a CORPUS ALBICANS (white body).

The uterine tubes (oviducts, Fallopian tubes; see Figs. 28.14, 28.15)

STRUCTURE. The paired uterine tubes are about 4 inches in length, and connect to the uterus at their proximal ends. An outer funnel-shaped INFUNDIBULUM bears many fingerlike FIMBRAE. An intermediate AMPULLA lies between the infundibulum and the ISTHMUS that passes through the uterine wall. The tubes are lined (Fig. 28.19) by a ciliated and secretory epithelium and have considerable smooth muscle in their walls.

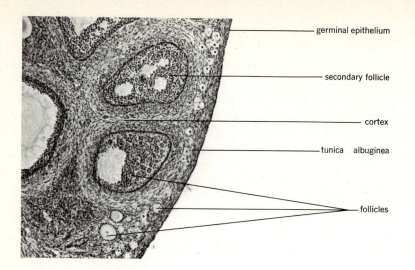

germinal epithelium

secondary follicle

cortex

tunica albuginea

follicles

FIGURE 28.16   Microscopic structure of the ovary, general view.

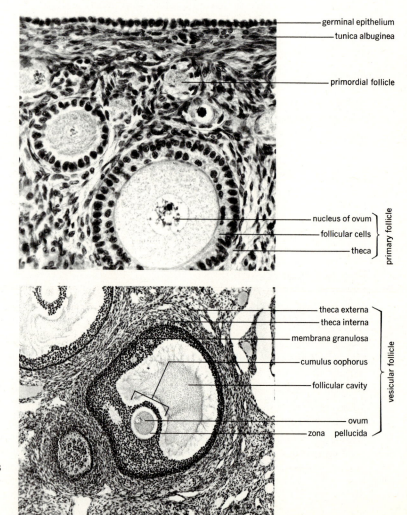

germinal epithelium
tunica albuginea

primordial follicle

nucleus of ovum ⎫
follicular cells ⎬ primary follicle
theca ⎭

theca externa ⎫
theca interna ⎪
membrana granulosa ⎪
cumulus oophorus ⎬ vesicular follicle
follicular cavity ⎪
ovum ⎪
zona   pellucida ⎭

FIGURE 28.17   Photomicrograph of the ovaries showing the development of follicles.

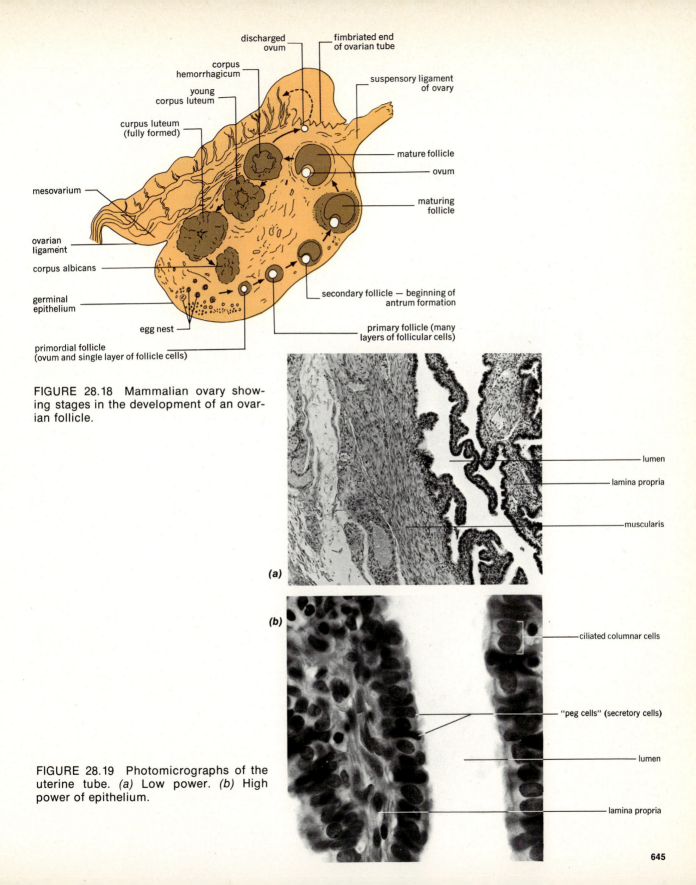

discharged ovum

fimbriated end of ovarian tube

corpus hemorrhagicum

young corpus luteum

suspensory ligament of ovary

curpus luteum (fully formed)

mature follicle

ovum

mesovarium

maturing follicle

ovarian ligament

corpus albicans

germinal epithelium

egg nest

secondary follicle — beginning of antrum formation

primary follicle (many layers of follicular cells)

primordial follicle (ovum and single layer of follicle cells)

FIGURE 28.18 Mammalian ovary showing stages in the development of an ovarian follicle.

lumen

lamina propria

muscularis

(a)

(b)

ciliated columnar cells

"peg cells" (secretory cells)

lumen

FIGURE 28.19 Photomicrographs of the uterine tube. (a) Low power. (b) High power of epithelium.

lamina propria

FUNCTION. The uterine tubes, by ciliary action of columnar ciliated cells and peristaltic movements of the smooth muscle, propel the egg to the uterus in 3 to 4 days. A released egg will degenerate if it is not fertilized within 18 to 24 hours after its release. Therefore, if fertilization is to occur, it *must* occur in the distal third of the uterine tubes. During passage of the ovum through the tube, PEG CELLS secrete nutrients to sustain the ovum.

The uterus
(*see* Figs. 28.14 and 28.15)

LOCATION AND STRUCTURE. The single uterus is, in the female who has never been pregnant, a pear-shaped organ about 7.5 centimeters in length, 5 centimeters wide, and $2\frac{1}{2}$ centimeters deep ($2\frac{1}{2} \times 2 \times 1$ inches). It lies behind and above the urinary bladder, at an angle of about 90 degrees to the vagina. It is supported by eight ligaments.

The *broad ligaments* run laterally from the sides of the uterus.

The *round ligaments* run anteriorly from the uterus to anterior abdominal wall.

Four *uterosacral ligaments* run from the uterus to the posterior body wall.

The uterus itself has an upper FUNDUS, a tapering BODY, and a NECK OR CERVIX that projects into the upper portion of the vagina.

The wall of the uterus is composed of three main layers of tissue (Fig. 28.20). An outer PERIMETRIUM consists of a thin mesothelial and connective tissue layer. A middle MYOMETRIUM contains smooth muscle and forms most of the thickness of the wall. An inner ENDOMETRIUM undergoes cyclical hormone-controlled changes after puberty (Fig. 28.21). The cyclical changes occurring in the endometrium constitute the menstrual cycles.

FUNCTIONS. The uterus is an organ designed to retain a fertilized egg, and to sustain and nourish it as it develops into a new individual during the approximate 280 days of a pregnancy. It expels the child and the placenta during and after birth.

The menstrual cycle

The menstrual cycle is hormonally controlled, and depends on the ovarian production of estradiol and progesterone. The cycle is divided into four phases:

The *menstrual phase* is characterized by a bloody discharge from the uterus. It occupies the first 3 to 5 days of a cycle, and is due to a lack of hormonal effect on the uterus. A small band of endometrial tissue remains in the uterus after menstruation has occurred.

The *proliferative (follicular* or *preovulatory) phase* is characterized by regrowth of endometrial tissue from the narrow band remaining in the uterus. It is under the control of estradiol from the maturing follicle, and causes about a 2-millimeter thick endometrial lining consisting of connective tissue, blood vessels, and glands. It occupies the next 7 to 15 days of the cycle.

The *secretory (luteal* or *postovulatory) phase* is the next 14 to 15 days of the cycle, and is characterized by an increase in glands and blood supply to the endometrium, in preparation for nourishing a fertilized ovum. If no fertilization occurs, the corpus luteum, which produces the progesterone responsible for this phase, degenerates.

The *premenstrual phase* is characterized by regression of blood vessels and glands, and by degeneration of tissue, as progesterone levels decrease. This phase terminates with the first external show of blood.

The length of a cycle varies from 24 to 35 days, depending on individual differences. It should not be assumed that all cycles are 28 days, or that ovulation occurs 14 days after the external show of blood. The most variable period is the length of the proliferative phase. Misinterpretation of these times can lead to accidental or unwanted pregnancy.

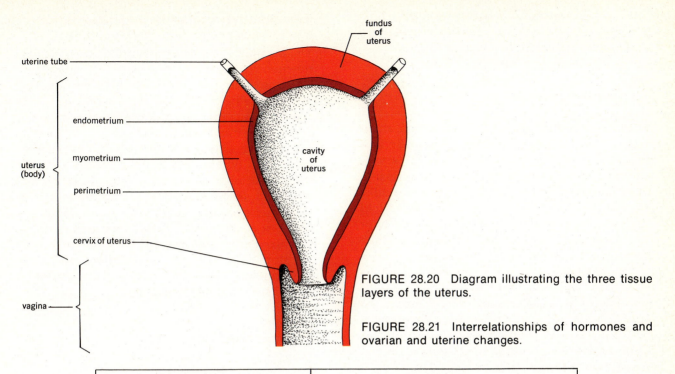

uterine tube

endometrium

uterus
(body)

myometrium

perimetrium

cervix of uterus

vagina

fundus
of
uterus

cavity
of
uterus

FIGURE 28.20 Diagram illustrating the three tissue layers of the uterus.

FIGURE 28.21 Interrelationships of hormones and ovarian and uterine changes.

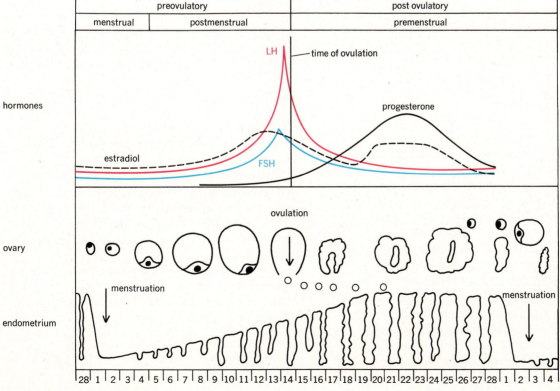

Some relationships between pituitary hormones, and ovarian and uterine changes are shown in Figure 28.21.

The vagina (*see* Fig. 28.14)

STRUCTURE. The vagina is a tubular organ, about 8 centimeters in length, which is normally in a collapsed state; that is, its cavity (lumen) is usually not open. The cervix of the uterus projects into the upper portion of the vagina, and the moatlike FORNICES surround the cervix. A membranous fold of tissue, known as the HYMEN, may partially or completely seal the distal end of the vagina. Three layers of tissue compose the wall of the organ: an inner MUCOSA is lined with stratified squamous epithelium containing much glycogen in the cells; a middle MUSCULARIS contains smooth muscle; an outer CONNECTIVE TISSUE LAYER fixes the organ in position. The glycogen is broken down into organic acids that create an acid environment to retard microorganism growth.

FUNCTION. The vagina receives the penis during sexual intercourse and serves as the birth canal during childbirth.

The external genitalia (Fig. 28.22)

Collectively, the external genitalia are known as the VULVA or PUDENDUM. The MONS PUBIS is a rounded pad of fat covering the pubic symphysis. The LABIA MAJORA are two fat-filled and (after puberty) hair-covered folds of tissue extending from the mons pubis toward the anus. The LABIA MINORA are smaller folds of tissue, lacking hair and fat, that closely surround the clitoris and vaginal opening. The CLITORIS is a single midline organ lying anterior to the urethral opening. The term OBSTETRIC PERINEUM refers to the area of the pelvic floor (perineum) between the vagina and anus. It may be incised (cut) to permit easier passage of an infant during birth, and to avoid tearing the perineum, in what is termed an EPISIOTOMY.

The mammary glands (Fig. 28.23)

The mammary glands are modified sweat glands, functionally related to the reproductive system. They produce milk after childbirth for nourishment of the newborn.

Each gland consists of a comma-shaped mass of connective and glandular tissue, located mainly between the second and sixth ribs on the anterior chest wall. Within each gland, there are 15 to 20 individual lobes of glandular tissue, each served by a milk (lactiferous) duct. These ducts form 3 to 5 larger ducts that empty to the exterior through the apex of the NIPPLE. Each larger duct contains an enlargement known as the *lactiferous sinus* that can store secreted milk until the infant removes it by suckling. The areola is the circular, pigmented region surrounding the nipple.

Clinical considerations

CARCINOMA (cancer) of the breast is the most common form of cancer in the female, and is the leading cause of death from cancer in the female. Cancer of the uterine cervix is the third most common type of cancer in the female. Breast cancer may be easily discovered by routine self-examination of the breasts, and the Papanicolaou (Pap) test can reveal many cervical cancers before they become life threatening. Chronic irritation, trauma, and viral infections appear to be the most common causes of cervical cancer.

AMENORRHEA refers to failure of a female to menstruate by 18 years of age or cessation of menstruation in a menstruating female, and calls for a complete investigation of the functional status of the hypothalamus, pituitary, ovaries, and uterus.

DYSMENORRHEA refers to painful menstruation. Its cause is not known, and is associated with cramps and pain 24 to 48 hours before menstrual flow.

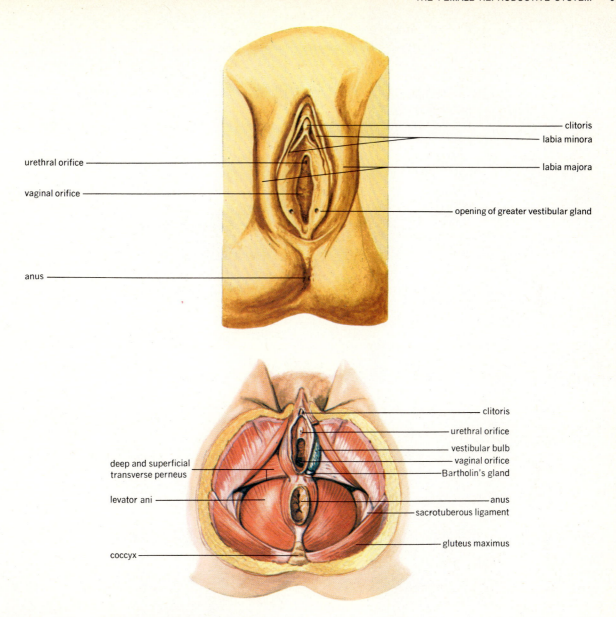

FIGURE 28.22   The external organs of the female reproductive system and associated structures.

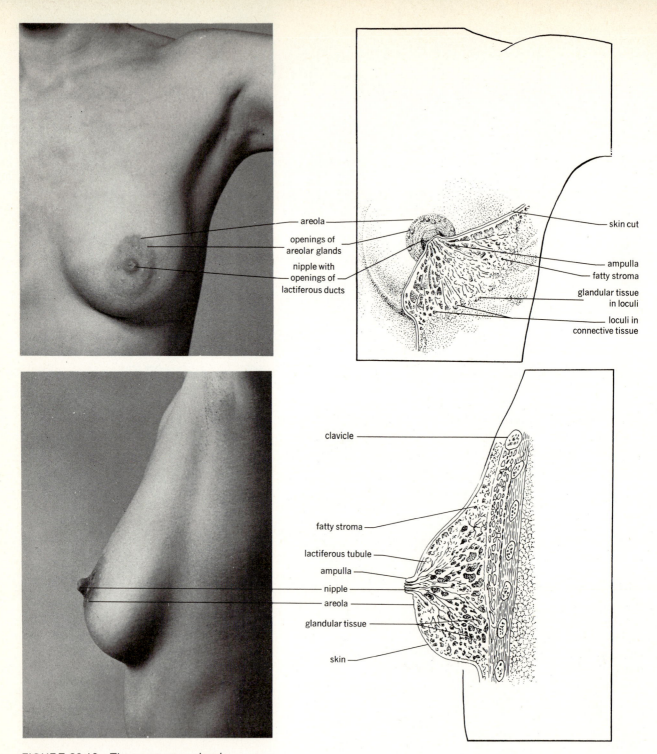

areola

openings of
areolar glands

nipple with
openings of
lactiferous ducts

skin cut

ampulla
fatty stroma

glandular tissue
in loculi

loculi in
connective tissue

clavicle

fatty stroma

lactiferous tubule

ampulla

nipple

areola

glandular tissue

skin

FIGURE 28.13   The mammary gland.

Venereal diseases

The VENEREAL DISEASES (VD) are the number one communicable diseases in the United States today. Explanations for the continuing increase in incidence of the diseases may point to several basic causes.

1. There have been and will continue to be changes in attitudes toward sexual intercourse that separate it from its reproductive function. Increased sexual contact has therefore increased the possibility of spread of the diseases among the population.

2. The increasing number of younger people in the population has increased the incidence of sexual contact with greater possibility of spread of the diseases. (The highest incidence is presently in the 15 to 30 age group.)

3. A large part of the social stigma associated with the diseases has disappeared and reporting of cases has increased. It may be hoped that increased reporting of cases has been an incentive for the establishment or increase in funding of programs designed for case finding and treatment of venereal diseases.

4. The discontinuance of indiscriminate use of antibiotics for "colds" and other infections that do not warrant their use has resulted in the removal of a "brake" on the spread of VD, in that many previously undiagnosed cases are no longer being treated inadvertently. Use of antibiotics can and has resulted in the development of antibiotic resistant strains of the VD organisms.

5. Better epidemiological studies have resulted in the discovery of reservoirs of infection that may not have been brought to light previously.

6. The belief of many segments of the public that a "shot" of antibiotic will cure the disease(s) has perhaps led to the development of complacency, as well as a false sense of security. All patients must be rechecked at a certain interval following treatment to ensure that the tests for organisms are negative. Also, after an initial set of symptoms, several venereal diseases may enter an asymptomatic period that may persist for years, leading to the belief that the disease has disappeared. During this time, damage may continue to be wrought in various body systems although no overt changes become apparent.

7. Reinfection is common since no immunity is developed to the organisms. Contacts must be examined and adequately treated if they are infected to prevent further spread. Sexual partners, diagnosed as infected, must be treated at the same time or they will be playing "ping-pong" with the organism. Persons having a gonococcal infection must also be checked for syphilis as both infections may be present at the same time but the amount and type of treatment may differ.

8. Antibiotic-resistant strains of disease-causing organisms are developing. "Old standby" antibiotics (e.g., penicillin) may no longer be effective, or require a higher dose to control the organisms.

Gonorrhea ("clap") and syphilis ("bad blood") are presently in epidemic proportions throughout the United States and some other countries. These diseases have no respect for race, creed, or color, and one cannot tell an infected person by just looking at him or her.

All personal information regarding cases, contacts, and suspects of VD is legally CONFIDENTIAL, which should encourage infected persons to seek early treatment of the disease(s).

Free treatment is available through your local health department. Other clinic facilities or private physicians or both are available in most communities. Information on your state's law regarding the treatment of minors may be obtained through your health department. Some states are now treating minors without the parents' consent, since the child's future welfare is at stake, and as an effort to cut down the VD incidence in the sexually active teenagers.

Some characteristics and effects of gonorrhea and syphilis are presented in Table 28.1.

The term venereal diseases today includes sexually or orally transmitted diseases other than gonorrhea and syphilis.

NONSPECIFIC URETHRITIS is caused by an unknown

agent, or may be secondary to or associated with parasitic infections, fungus infections, or trauma. That there is an organism involved is demonstrated by transmissibility of the infection from one person to another.

GRANULOMA VENEREUM is caused by a gram-negative rod, *Donovania granulomatis.* Evidence of infection includes a skin lesion, usually on the glans penis, typically followed by ulceration. The lesions heal slowly and tend to spread. Antibiotic therapy will cure the disease.

LYMPHOGRANULOMA VENEREUM is caused by *Bedsonia sp.,* and is associated with ulcerative lesions on the genitals. The organism then invades nearby lymph nodes, causing them to enlarge to form a *bubo,* an enlarged matted mass of nodes. Again, appropriate antibiotics will cure the condition.

HERPEX SIMPLEX VIRUS, type 2, produces vaginal, genital, and uterine cervix lesions. This virus may be transmitted to the fetus, either *in utero* or during birth. Herpes simplex encephalitis may occur that has a high mortality and common recurrences. "Treatment" is by acquisition of resistance or partial resistance through immune response; antibiotics are ineffective on viruses.

Recently, a compound designated Ara-A (for adenine arabinoside) has been shown to be effective against the herpes simplex organism causing herpes encephalitis. Perhaps the substance has promise in the therapy of venereal herpes infections.

TABLE 28.1   Some characteristics and effects of gonorrhea and syphilis

	Gonorrhea	Syphilis
Causative organism	Neisseria gonorrhoeae	Treponema pallidum
Incubation period	2–14 days (usually 3 days)	7–90 days (usually 3 weeks)
Method of transmission	Sexual, oral, or physical contact with an infected person. Contaminated object up to 8 hours after organisms deposited. *Infants*—during birth through vagina.	Sexual, oral, or physical contact with an infected person. Blood transfusion. On contaminated objects, dies quickly by drying. *Infants*—may acquire during birth, or through the placenta if mother not treated before third trimester and adequately.
Contact examination *(all sex contacts exposed within the following time periods)*	2 weeks (male) 1 month (female)	*Primary.* 3 months (+ duration of symptoms). *Secondary.* 6 months (+ duration of symptoms). *Early latent.* 1 year. *All syphilis.* "Family" contacts as indicated.

TABLE 28.1 (continued)

	Gonorrhea	Syphilis
Clinical characteristics	Discharge; burning, pain, swelling of genitals and glands. **Male.** Purulent urethral discharge, hematuria, chordee. Urethritis, prostatitis, seminal vesiculitis, epididymitis, occasional involvement of testes. If untreated, symptoms disappear in about 6 weeks, and organism persists in the prostate gland (Gc carrier). **Female.** *Child,* vaginitis. Leukorrhea, tubal abscess, urethritis, cervicitis, pelvic inflammation (Peritonitis). Tubal stricture, possible sterility due to closure of tubes. In both male and female, healing is by scar tissue formation. Strictures and closing of tubular structures may result. Arthritis, endocarditis, meningitis may occur.	**Primary.** Chancre present, solitary, nonpainful ulcer on genital or mucous membranes. **Secondary.** Rashes or mucous patches. Macules or papules on hands, feet, oral cavity, genitoanal area, trunk, extremities. **Tertiary:** *Latency.* No symptoms, positive serology. Profound changes are produced in the skin, mucous membranes, skeleton, GI tract, kidney, brain; heart and blood vessels show destructive changes (abscesses, scarring, tissue destruction). Tabes dorsalis in spinal cord destroys dorsal columns. Paresis and psychosis may result. *Relapse.* Recurrence of infectious lesions after disappearance of secondary lesions. *Late.* Cardiovascular, central nervous system, gummata. Obvious systemic damage appears.
Diagnostic procedures	Culture of discharge. Smear. Fluorescent antibody test (FAT). Currettage to get tissue containing cocci. Several cultures may be necessary as not all tests consistently show cocci. History, clinical and contacts. (Serologic test for syphilis).	Darkfield examination — microscopic for spirochete. Serologic (blood) test for antibody (reagin) to organism. [(Wasserman, VDRL — Venereal Disease Research Lab) False positive tests may be reported by smallpox antibodies, hepatitis, mononucleosis, and high fevers.] Spinal fluid test. X rays of long bones of infants. History, clinical and contacts.
Treatment	Penicillin Broad spectrum antibiotics (e.g. sulfonamides, streptomycin). Organism must be sensitive to the drug of choice.	Penicillin Broad spectrum antibiotics (e.g., Erythromycin, Tetracycline). Organism must be sensitive to drug of choice. Reexamination at 6 months and 1 year to evaluate treatment results.

Summary

1. The male and female reproductive organs provide for creation, development, and nourishment of new individuals. Several hormones are produced by certain of the organs and affect the whole body as well as the organs of reproduction.

2. The system develops from mesoderm at about 5 weeks. The organs go through an undifferentiated state and, at about 8 weeks, assume the appearance of male or female.

3. The male reproductive organs include the testes, a system of ducts, and accessory glands.

 a. The testes are found in the scrotum, are ovoid, are surrounded by tunics, and produce sperm cells in their seminiferous tubules, and a hormone (testosterone) in the interstitial cells.

 b. The ducts (straight and rete tubules, efferent ductules, epididymis, and vas deferens) transport and store sperm.

 c. The accessory glands (seminal vesicles, prostate, bulbourethral glands) produce the semen in which sperm are suspended. Semen nourishes, and activates the sperm, and neutralizes acid in male urethra and vagina.

 d. The penis serves as a copulatory organ, and is composed of three cavernous bodies.

4. The female reproductive organs include internal organs (ovaries, uterine tubes, uterus, and vagina), the external genitalia (labia majora and minora, and the clitoris), and the mammary glands.

 a. The ovaries are found in the pelvic cavity, are ovoid, are covered with germinal epithelium, a fibrous tunic, and contain a stroma (cortex and medulla) and follicles. They produce ova and two main hormones.

 b. The follicles are present in a primordial state at birth. At puberty they develop through primary, secondary, and vesicular follicles. The vesicular follicle produces estrogen, primarily responsible for female sexual maturity. Ovulation is followed by corpus luteum formation and progesterone production; progesterone ensures uterine wall development and mammary development for milk secretion. A scar is finally formed by the follicle.

 c. The uterine tubes are about 4 inches long and pass eggs to the uterus. Fertilization occurs in the tube.

 d. The uterus is an organ in which new individuals develop. It is supported by eight ligaments, has three parts (fundus, body, and cervix), and three layers in its wall (perimetrium, myometrium, and endometrium). The inner endometrium undergoes cyclical changes after puberty in the menstrual cycles.

 (1) Each cycle has four stages: menstrual (bleeding), proliferative (regrowth), secretory (development of many glands), and premenstrual (degeneration of blood vessels).

(2) The proliferative and secretory stages are controlled by estrogen and progesterone respectively.

e. The vagina is a tubular organ serving as the receptacle for the penis and as the birth canal.

f. The external genitalia consist of labia majora and minora and the clitoris.

g. The mammary glands are lobed structures that produce milk for nourishment of offspring.

5. The characteristics and effects of venereal disease are summarized in Table 28.1.

Questions

1. Compare ovaries and testes as to structure, function, and hormone production.

2. Trace an ovum's development in the ovary.

3. Describe the changes occurring in the uterus during a menstrual cycle.

4. How is maturation of eggs and sperm controlled in the two sexes?

5. With the continuing increase in venereal diseases, what do you think could be done to reduce the incidence of the diseases, or to discover undiagnosed cases?

Readings

Chan, Lawrence, and Bert O'Malley. "Mechanism of Action of the Sex Steroid Hormones." *New Eng. J. Med. 294*:1322, June 10, 1976; *294*:1372, June 17, 1976.

Fielding, Jonathan E. "Smoking and Pregnancy." *New Eng. J. Med. 298*:337, Feb. 9, 1978.

Kunin, Calvin M. "Sexual Intercourse and Urinary Infections." *New Eng. J. Med. 298*:336, Feb. 9, 1978.

Science News. "Sperm Inhibitors as Contraceptives." *113*:150, March 11, 1978.

Chapter 29

The Endocrines

Objectives

After studying this chapter, the reader should be able to:

■ Give a general review of the origins of the various endocrine organs.

■ List the criteria used to determine endocrine status.

■ Comment generally on the "life history" of hormones (production, utilization, fate).

■ List the methods by which endocrine secretion is controlled.

■ Describe, for each endocrine, the following:
The hormone(s) produced
The effect of those hormone(s) on body physiology

The method of control of hormone secretion
A major disorder of over- and underproduction, with obvious
  symptomology

■ Describe the involvement of adrenal cortical hormones in
  inflammation and stress.

■ Discuss the status of the pineal gland as an endocrine organ,
  with evidence to support or refute that status.

■ Describe prostaglandins as to sites of production, effects on
  the body, and possible therapeutic uses.

Working with the nervous system, the endocrine system provides a means of controlling many body functions. Endocrine control is exerted by chemical substances that are secreted into and carried by the bloodstream, and usually requires a somewhat longer time to exert an effect than does the nervous control. Additionally, endocrines control more widespread processes in the body, instead of controlling organs, which is more true of nervous control. For example, endocrines control processes such as metabolism, growth, and development.

Development of the system

Many endocrine glands are derivatives of the mouth, pharynx, and gut. Others come from the ectoderm that forms the nervous system, or from mesoderm. Table 29.1 indicates when the listed endocrines appear and their area or germ layer of origin.

TABLE 29.1 Development of some of the endocrines				
Endocrine	Age of first appearance (weeks)	Age of typical structure (weeks)	Area(s) formed in or from	Germ layer of origin
Thyroid gland	3½	12	Pharynx floor	Endoderm
Pituitary	4	16	Roof of mouth and brain	Ectoderm
Adrenal cortex	5	7	Dorsal celom	Mesoderm
Adrenal medulla	5	7	Neural crest by spinal cord	Ectoderm
Pancreas islets	12	14	Foregut	Endoderm
Parathyroids	7	9	Pharynx	Endoderm
Testes and ovaries	6	8	Pelvic cavity	Mesoderm
Pineal gland	7	11	Brain stem	Ectoderm

Organs of the system and general principles of endocrinology

Organs: Criteria of endocrine status

Although not a connected series of organs forming a system such as we have studied in previous chapters, all endocrines have several things in common: they have no ducts; they secrete specific chemicals; removal of an endocrine causes clearly defined alterations in function that may be restored to normal by administration of an extract of the suspected endocrine; they are extremely vascular and consist of well-defined groups of cells. The recognized endocrines are shown in Fig. 29.1.

Hormones

The specific chemical(s) secreted by an endocrine gland is a hormone. Strictly speaking, chemicals such as gastrin, secretin, cholecystokinin, and other substances are not true hormones, since their source is not definitely known; nevertheless, such substances are often called hormones.

Chemically, hormones fall into three classes: proteins or polypeptides, small amines, or steroids. In general, protein hormones are secreted by endodermal endocrines, amines by ectodermal, and steroids by mesodermal endocrines.

Control of secretion

Hormones are secreted according to body need for them. This obviously implies the presence of control mechanisms to govern production and release. Control is supplied by several methods:

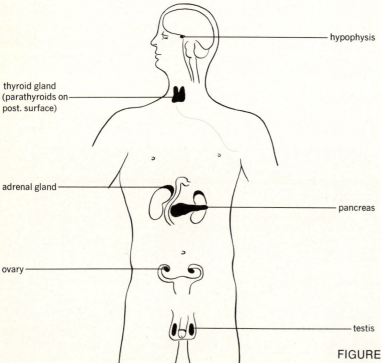

thyroid gland
(parathyroids on
post. surface)

adrenal gland

ovary

hypophysis

pancreas

testis

FIGURE 29.1 Locations and names of endocrine glands.

BLOOD LEVELS of substances other than hormones. Blood sugar (glucose) levels control pancreas hormone secretion; blood calcium levels control parathyroid secretion; blood levels of total solutes (osmotic pressure) control release of certain posterior pituitary hormones.

A summary of these generalities is presented in Table 29.2.

NEGATIVE FEEDBACK INVOLVING TWO HORMONES. One endocrine secretes a hormone, it acts on another endocrine, and the hormone of the second endocrine inhibits secretion by the first endocrine.

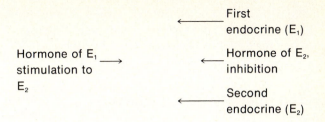

The thyroid, adrenals, and gonads are controlled in this manner.

NERVES. The posterior part of the pituitary and the adrenal medulla release hormones upon nervous stimulation. Response is thus very rapid.

TABLE 29.2 General summary of endocrine glands	
Item	Comments
Functions of endocrines	To integrate, correlate, and control body processes by chemical means.
Criteria for establishing function as endocrine	Cells are morphologically distinct.
	Cells produce specific chemicals not produced elsewhere.
	Chemicals exert specific effects; effect is lost if gland removed, restored if chemical administered.
Hormones	Hormones are produced by cells qualifying as endocrines.
Chemical nature of hormones	Steroid—mesodermal origin (e.g.: adrenal cortex, gonads).
	Polypeptides—endodermal origin (e.g.: pancreas).
	Small MW amines—ectodermal origin (e.g.: posterior pituitary, adrenal medulla).
Methods of control available	
1. Neurohumor	Nerve cells produce a chemical, it goes to endocrine and controls secretion. Stimulus is chemical.
2. Nerves	Nerve fibers pass to endocrine. Stimulus is electrical (nerve impulse).
3. Feedback	Target organ hormone influences secretion of another endocrine which stimulated target organ.
4. Nonhormonal organic substances in blood	Glucose acting on pancreatic islets. Rise causes increased secretion (usually).
5. Total osmolarity of blood	Requires nervous system to detect it. Nervous system then signals endocrine.
6. Inorganic substances in blood	Ca^{++} on parathyroid. Effect usually direct, not inverse.

PRODUCTION OF NEUROHUMORS. Chemicals produced by nerve cells are called neurohumors. They may pass over nerve fibers or be placed into blood vessels to control an endocrine. The hypothalamus controls the anterior part of the pituitary in this manner.

The pituitary (hypophysis)

Location and description

The pituitary gland is about the size of a large pea (about 10 millimeters in diameter), weighs ½ gram, and lies in the sella turcica of the sphenoid bone just beneath the hypothalamus of the brain. Part of it, the ADENOHYPOPHYSIS, is derived from the roof of the mouth, while the remainder, the NEUROHYPOPHYSIS, is a downgrowth of the hypothalamus. The adenohypophysis is, in turn, composed of a large ANTERIOR LOBE, and a narrow INTERMEDIATE LOBE. The neurohypophysis includes the STALK that attaches the pituitary to the brain, and the POSTERIOR LOBE. These divisions are shown in Figure 29.2.

The anterior lobe is composed of acidophil (red staining), basophil (blue staining), and chromophobe cells (little or no staining). Of these, the first two are regarded as being the sources of the anterior lobe hormones. The anterior lobe is connected to the hypothalamus by a system of capillaries called the PITUITARY PORTAL SYSTEM. The intermediate lobe is a narrow band of red- and blue-staining cells.

The posterior lobe is composed of modified nerve cells called PITUICYTES, and many nerve fibers. The posterior lobe is connected to the hypothalamus by these fibers, whose cell bodies are in the hypothalamus itself. These relationships are shown in Figure 29.3.

Hormones of the anterior lobe and their effects

At least six hormones are known to be secreted by the acidophils and basophils of the anterior lobe.

Growth hormone (somatotropin, Fig. 29.4) is a giant molecule composed of 188 amino acids. It increases protein synthesis in all body organs and thus aids in determining their rate of enlargement; it controls cartilage production in the ends of the long bones of

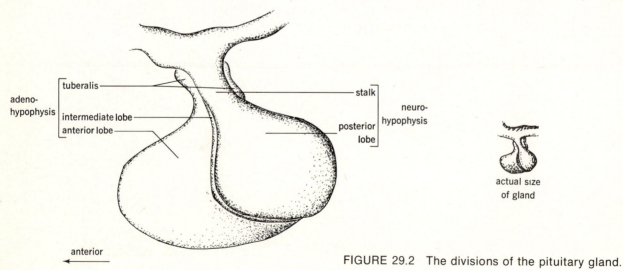

FIGURE 29.2   The divisions of the pituitary gland.

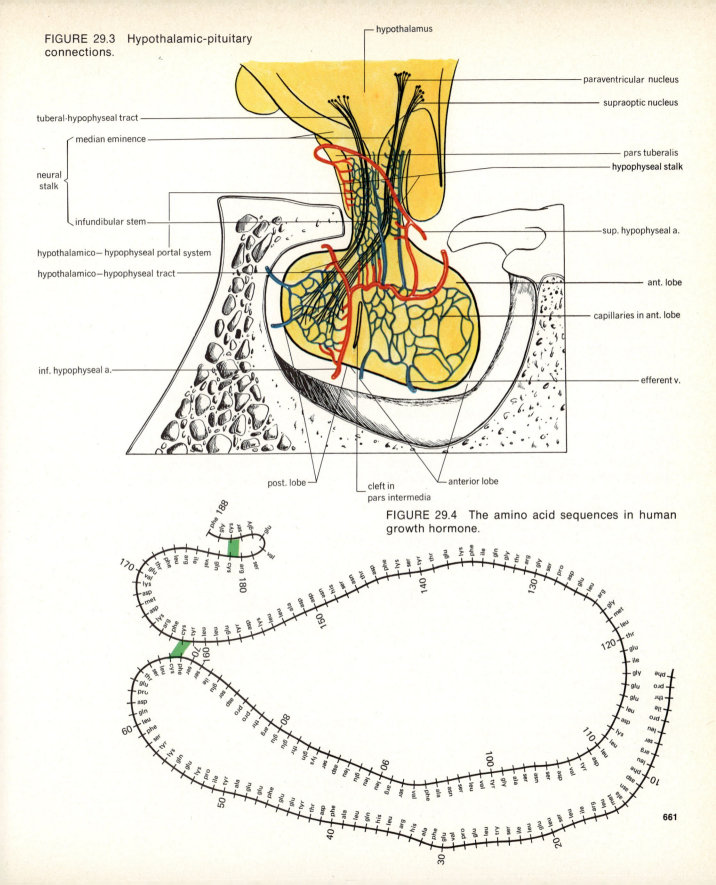

FIGURE 29.3 Hypothalamic-pituitary connections.

hypothalamus

paraventricular nucleus

supraoptic nucleus

tuberal-hypophyseal tract

median eminence

neural stalk

pars tuberalis

hypophyseal stalk

infundibular stem

sup. hypophyseal a.

hypothalamico—hypophyseal portal system

hypothalamico—hypophyseal tract

ant. lobe

capillaries in ant. lobe

inf. hypophyseal a.

efferent v.

post. lobe

cleft in pars intermedia

anterior lobe

FIGURE 29.4 The amino acid sequences in human growth hormone.

661

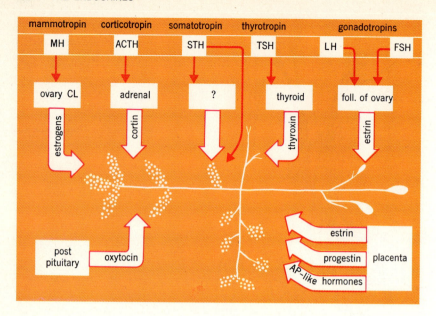

FIGURE 29.5 Hormones involved in lactation.

the body and thus controls bone growth in length; it promotes fat metabolism and decreases use of carbohydrates.

Thyroid stimulating hormone (TSH, thyrotropin) is concerned with controlling all phases of thyroid gland activity, including cell division, synthesis, and release of thyroid hormones.

Adrenocorticotropic hormone (ACTH, corticotropin) controls the growth and secretory activity of the adrenal cortex. It also provides resistance to stress imposed on the body by stimulating adrenal cortical secretion of steroids.

Two *gonadotropic hormones*, known as *follicle stimulating hormone* (FSH) and *luteinizing hormone* (LH, luteotropin) also known in the male as *interstitial cell stimulating hormone* (ICSH), are produced by the anterior lobe. FSH stimulates maturation of ovarian follicles, and sperm in the testes. LH is responsible for ovulation, corpus luteum formation and, as ICSH, secretion of testicular interstitial cells.

Prolactin (lactogenic hormone) aids in maintenance of the corpus luteum of pregnancy and in maintaining milk secretion, if the mammary glands have been brought to a "ready state" by several other hormones (Fig. 29.5).

Control of anterior lobe secretion

Primary control of anterior lobe secretion is afforded by—at least—eight neurohumors that are produced in the hypothalamus in response to nervous, hormonal, and other influences. These neurohumors are secreted into the pituitary portal system and are delivered to the anterior lobe to exert their effects. The pituitary secretion of TSH, ACTH, FSH, and LH (ICSH) may also be directly influenced by blood levels of thyroid, adrenal, ovarian, and testicular hormones in a negative feedback mechanism. Table 29.3 summarizes the known hypothalamic neurohumors.

The intermediate and "posterior lobe hormones" and their effects

MELANOCYTE STIMULATING HORMONE (MSH) is produced by the intermediate lobe. Its significance in the human has not been established. In lower animals, the hormone is secreted in response to changes in light intensity and causes contraction of certain pigment cells of the skin,

TABLE 29.3 Hypothalamic neurohumors[a] controlling the pituitary anterior lobe

Factor	Controls secretion of (hormone)	Effect on secretion	Comments
Growth hormone releasing factor (GHRF)	GH	Stimulates	GHRF is released by low blood sugar, increased amino acid intake, stress, sleep.
Growth hormone inhibiting factor (GIF)	GH	Inhibits	
Prolactin releasing factor (PRF)	Prolactin	Stimulates	Produced after childbirth; allows lactation to occur.
Prolactin inhibiting factor (PIF)	Prolactin	Inhibits	Present continually in nonpregnant and antenatal pregnant female.
Corticotropin releasing factor (CRF)	ACTH	Stimulates	Factor released by direct effect of hormones of target glands or via hypothalamus.
Thyrotropin releasing factor (TRF)	TSH	Stimulates	
Luteinizing hormone releasing factor (LRF)	LH; ICSH	Stimulates	
Follicle stimulating hormone releasing factor (FSHRF)	FSH	Stimulates	

[a] These factors were formerly called *releasing factors* (RF). Since it is now known that these factors may both stimulate or inhibit secretion, it is appropriate to call them *regulatory factors* when speaking collectively of the chemicals.

enabling camouflaging of the organism for survival purposes.

The posterior lobe produces no hormones. OXYTOCIN and ANTIDIURETIC HORMONE (ADH) are secreted by nerve cells in the hypothalamus and pass over nerve fibers to the posterior lobe where they are stored in the ends of the axons of the nerve cells that produced them. Release results from nerve impulses arriving over these same axons. Oxytocin is a stimulant to uterine contraction during childbirth and is important as an agent in causing uterine contraction after childbirth. This latter effect prevents hemorrhage. ADH is released according to the osmotic pressure of the hypothalamic blood, and promotes water reabsorption by the kidney.

In cases where a pregnant female is long overdue in delivery of her child or has been in pro-

longed labor without delivering, her uterus is sometimes caused to start contractions (induction) or to contract more strongly by injection of *oxytocics*. These are substances that act in the same manner on the uterus as does oxytocin. *Posterior pituitary injection* is an extract of the posterior lobe of domestic animals and contains both hormones. *Oxytocin* itself, or *synthetic oxytocin* may also be employed.

Table 29.4 summarizes the pituitary hormones and their effects.

Clinical considerations

In theory at least, one or several pituitary hormones may be affected by pathological conditions

TABLE 29.4   A summary of pituitary hormones and their effects

Hormone	Site of production	Effects
Growth hormone	Anterior lobe	Controls growth of body; influences fat and sugar metabolism
Thyroid stimulating hormone	Anterior lobe	Controls thyroid activity
Adrenocorticotropic hormone	Anterior lobe	Controls adrenal cortex activity
Follicle stimulating hormone	Anterior lobe	Causes ovarian follicle and sperm maturation
Luteinizing hormone	Anterior lobe	Causes ovulation and corpus luteum formation
Interstitial cell stimulating hormone	Anterior lobe	Causes secretion of interstitial cells of testes
Prolactin	Anterior lobe	Sustains lactation
Melanocyte stimulating hormone	Intermediate lobe	Contracts melanocytes
Oxytocin	Hypothalamus	Stimulates uterine contraction
Antidiuretic hormone	Hypothalamus	Permits kidney to reabsorb water

TABLE 29.5   Some disorders of the pituitary

Condition	Cause	Hormone(s) involved	Secretion is		Characteristics	Comments
			Excess	Deficient		
Giantism	Anterior lobe tumor before maturity	Growth hormone (GH)	X		Large body size	Proportioned but large
Acromegaly	Anterior lobe tumor after maturity	Growth hormone (GH)	X		Misshapened bones, hands, face, feet	Disproportional growth of face, and extremities
Hypophyseal infantilism	Destruction of anterior lobe cells by disease, accident, etc.	Growth hormone (GH)		X	Juvenile appearance, sexually and in height	Body properly proportioned
Water retention + demonstration of high ADH blood levels	Hypothalamic tumor	Antidiuretic hormone	X		Dilution of body fluids, weight gain	Must differentiate from H_2O retention due to other causes
Diabetes insipidus	Hypothalamic damage	Antidiuretic hormone		X	Excessive urine excretion	No threat to life

that cause the cells to either oversecrete or undersecrete hormones. Table 29.5 and Figures 29.6 to 29.8 present the essential features of several pituitary disorders.

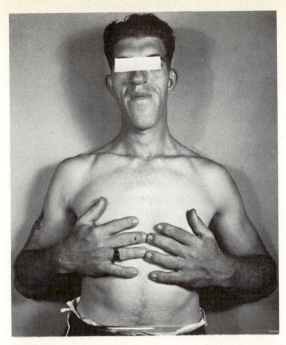

FIGURE 29.6 Acromegaly in an adult. Note the coarseness of the facial features, the enlarged mandible, and the large hands with thick, blunt fingers. (Armed Forces Institute of Pathology.)

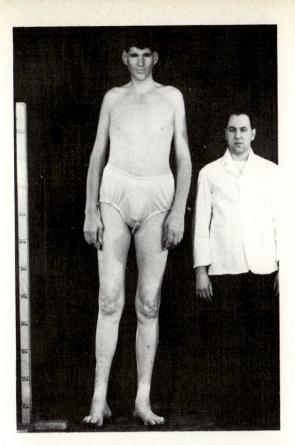

FIGURE 29.7 Giantism in a 42-year-old man. He is 7 feet, 6 inches tall. His companion is normal. The stick is 6 feet tall. (From the teaching collection of the late Dr. Fuller Albright. Courtesy Endocrine Unit and Department of Medicine, Massachusetts General Hospital.)

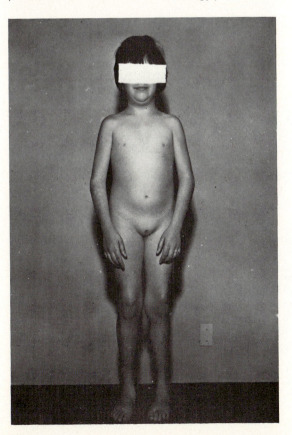

FIGURE 29.8 Hypophyseal infantilism in a 15-year-old female. Notice shortness of stature (4 feet, 3¾ inches tall), and failure to develop sexually. (Armed Forces Institute of Pathology.)

The thyroid gland

Location and description

The thyroid gland (Fig. 29.9) is a two-lobed organ lying on the anterior upper trachea and lower larynx. The two LATERAL LOBES are connected across the midline by the ISTHMUS, which often shows a PYRAMIDAL LOBE. It weighs about 20 grams in the adult, and receives a great blood supply (80–120 milliliters per minute).

The structural and functional units of the gland are spherical hollow units known as THYROID FOLLICLES. They are lined with an epithelium that synthesizes the thyroid hormones. The follicles contain a pink-staining material known as the THYROID COLLOID that is composed of a substance called THYROGLOBULIN. Thyroglobin is a storage form for thyroid hormone. Between the follicles, and scattered in between the follicular epithelial cells, are PARAFOLLICULAR CELLS.

Hormones and effect

The thyroid follicular cells synthesize THYROXIN (T_4) and TRIIODOTHYRONINE (T_3) (Fig. 29.10) by iodination (adding iodine) to the amino acid tyrosine in the thyroglobulin of the colloid. The hormones, when needed, are released from the colloid by an enzyme and enter the bloodstream. T_4 represents about 95 percent of the hormone secretion, T_3 represents about 5 percent. Both hormones govern the overall rate of biological oxidation in the body (metabolic rate), are essential for normal growth and development, especially of the brain, and exert effects on the metabolism of foods, such as increasing cell uptake of glucose, increasing glycogenolysis, and promoting protein synthesis.

CALCITONIN (thyrocalcitonin) is believed to be produced in the human by the parafollicular cells.

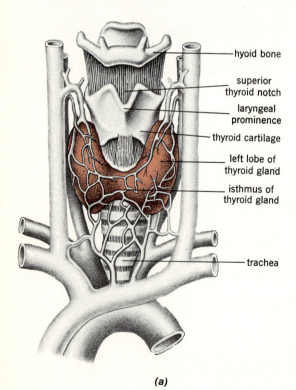

(a)

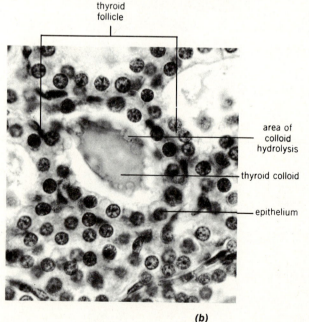

(b)

FIGURE 29.9   Thyroid gland. (a) Gross anatomy and location. (b) Microscopic anatomy.

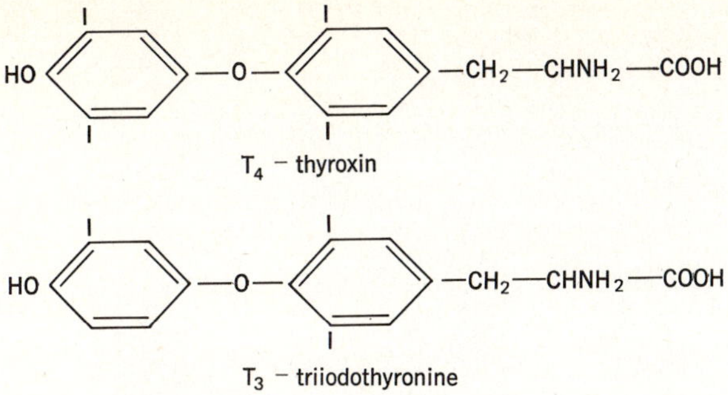

$$HO-\bigcirc-O-\bigcirc-CH_2-CHNH_2-COOH$$

T_4 – thyroxin

$$HO-\bigcirc-O-\bigcirc-CH_2-CHNH_2-COOH$$

T_3 – triiodothyronine

FIGURE 29.10   The formulas of thyroxin (T_4) and tri-iodothyronine (T_3).

It is involved in the metabolism of calcium and phosphate in the body. It causes a fall of blood Ca^{2+} and PO^{3-} levels, probably by increasing their incorporation into bony tissue.

Control of secretion

All phases of thyroid activity, including the production and release of T_4 and T_3, are controlled by pituitary TSH. T_4 and T_3, in turn, exert a negative feedback effect on TSH secretion. Calcitonin secretion is triggered by a rise of blood Ca^{2+} levels.

Clinical considerations

If dietary iodine intake is insufficient, the thyroid may enlarge and forms a SIMPLE GOITER (Fig. 29.11). Using iodized salt will prevent this disorder.

HYPOTHYROIDISM (Fig. 29.12), or failure of the gland to produce enough T_3 and T_4 to meet body needs, causes a lowering of metabolic rate, the development of dry flaky skin, and lack of resistance to cold temperatures. Myxedema is a term sometimes used to describe hypothyroidism occurring after birth. If hypothyroidism occurs in a child before birth, CRETINISM results, with dwarfing and mental retardation common.

HYPERTHYROIDISM (Fig. 29.13) results from overstimulation by TSH, or may be due to a tumor in

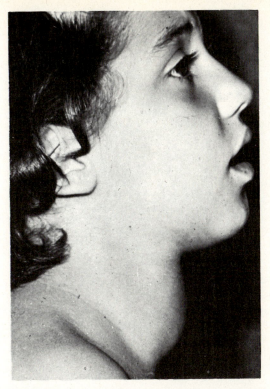

FIGURE 29.11   Goiter (enlargement of the thyroid gland) in an adolescent female. Notice the rounded bulge in the front part of the neck representing the enlarged thyroid gland. (Lester V. Bergman & Associates.)

the gland. In this case, metabolic rate is elevated, heart action is very rapid, blood pressure increases, there is weight loss, and the gland may enlarge (goiter). The eyes may bulge (exophthalmos) due to increase in mass and water content of the tissues behind the eyes.

STATUS OF THYROID FUNCTION is commonly determined by analyzing a blood sample for its content of *protein bound iodine* (PBI). T_4 and T_3, when secreted into the bloodstream, bind to plasma proteins; thus, a PBI determination is an indirect measure of iodine containing hormones. A normal PBI is 4 to 8 micrograms per 100 milliliters of plasma. To make hormones, the gland must accumulate iodine as one of its building blocks. Radio-

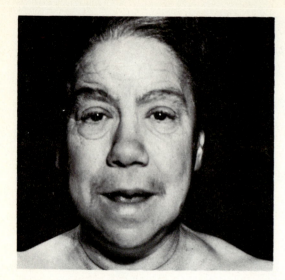

FIGURE 29.12 Myxedema in a 55-year-old female. Her PBI was less than 1 microgram per 100 milliliter (normal 4–8), and metabolic rate was 51 percent below normal. Notice puffy appearance of face and generalized fatigued appearance. (Armed Forces Institute of Pathology.)

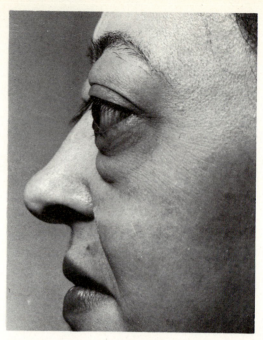

FIGURE 29.13 Hyperthyroidism. Note the protrusion of the eyeball (exophthalmos) that may accompany the disease.

TABLE 29.6 Summary of the thyroid	
Item	Comments
Location and parts; weight	Two-lobed, plus connecting isthmus; located on lower larynx and upper trachea; 20 grams in weight
Requirements for hormone synthesis	The amino acid tyrosine; iodide in diet; TSH to control all steps in synthesis
Hormones produced and effects: Thyroxin Thyrocalcitonin	Main controller of catabolic metabolism Lowers blood Ca^{++} and $PO_4^{\equiv}$ levels
Control of hormone secretion: Thyroxin Thyrocalcitonin	By thyroid stimulating hormone By blood Ca^{++} level
Disorders: Hypersecretion Hyposecretion	Creates hyperthyroidism: Elevated BMR, elevated heart action, exophthalamus, goiter. Creates: Goiter-enlarged gland, hypothyroidism (cretinism, myxedema) Low BMR, low heart rate, blood pressure, body temperature

active iodine may be given, and its rate of uptake by the gland determined by a counting device. Normally, 25 to 50 percent of a given dose will be picked up by the gland in 24 hours.

Table 29.6 summarizes facts about the thyroid gland.

The parathyroid glands

Location and description (Fig. 29.14)

There are usually four parathyroid glands, located on the posterior aspect of the thyroid lobes. Each measures about 5 millimeters in diameter, and all four weigh about 120 milligrams.

Microscopically, the gland shows tightly packed cells (principle cells) that produce a hormone.

Hormone and effects

The parathyroid glands produce PARATHYROID HORMONE (PTH), also called parathormone. The hormone controls, with calcitonin, the metabolism of calcium and phosphate in the body. Specifically, PTH increases absorption of these minerals from the gut, increases the reabsorption of calcium by the kidney, and governs blood calcium and phosphate levels by controlling the destruction of bony tissue.

Control of secretion

A fall of blood calcium level increases PTH secretion. More minerals are released from bone, and the blood calcium level rises.

Clinical considerations

Insufficient production of PTH produces HYPO-PARATHYROIDISM or TETANY. Because of the low blood calcium levels, muscle excitability is dis-

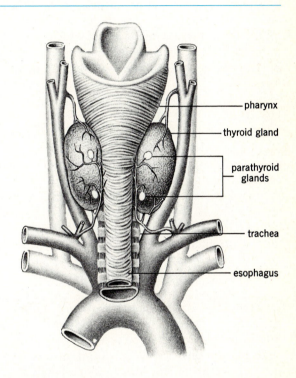

FIGURE 29.14   The location of the parathyroid glands on the posterior aspect of the thyroid gland.

turbed, and the skeletal muscles undergo painful cramps (tetany). Convulsions may also occur. Treatment requires raising the blood calcium level by intravenous injection of calcium salts.

HYPERPARATHYROIDISM is rare, and is usually due to a tumor of the glands. Excessive PTH secretion causes bone destruction, formation of kidney stones from the high levels of calcium filtered from the bloodstream, and may lead to calcium rigor (paralysis) in the heart and skeletal muscles.

The adrenal glands

Location and description

The adrenal glands are paired, hat-shaped organs, located above the kidneys. Each gland weighs 7 to 8 grams, and is composed of an outer CORTEX and an inner MEDULLA. The cortex is further subdivided into three zones, each of which produces hormones with different physiological effects. The cortex is essential for life, the medulla is not.

Hormones and effects

MEDULLA. Two hormones, EPINEPHRINE (*adrenalin*) and NOREPINEPHRINE are produced by the adrenal medulla. Both are sympathomimetic, that is, they have the same effects on the body as does stimulation of the sympathetic nervous system. For example, the hormones increase heart rate and output, stimulate breathing, raise blood pressure by causing vasoconstriction, and epinephrine alone causes acceleration of glycogenolysis.

CORTEX. The cortex produces steroid hormones generally known as CORTICOIDS. These may be grouped into three general types.

MINERALOCORTICOIDS, produced primarily by the outer zone of the cortex, are essential to the fluid and electrolyte balance of the body. ALDOSTERONE is the most potent mineralocorticoid, and deals with the transport of Na+ in the kidney. Recall that both negatively charged ions and water follow active transport of sodium from the kidney tubules.

GLUCOCORTICOIDS are secreted by the middle zone of the cortex and deal with metabolism of carbohydrate and proteins. For example, CORTISOL and CORTISONE, the major glucocorticoids, increase synthesis of glucose from amino acids (gluconeogenesis), increase glycogen formation in the liver, and stimulate ATP formation. Glucocorticoids are increased when the body is stressed, and confer resistance to stress of any sort. However, contin-

ued high levels of glucocorticoids in the body may cause ulcer formation, loss of resistance to disease, and high blood pressure. Glucocorticoids also have an *anti-inflammatory effect*, and are used in the treatment of diseases such as rheumatoid arthritis.

CORTICAL SEX HORMONES are produced by the inner zone of the cortex. Both male- and female-type hormones are produced. Their role in body function has not been established.

Control of secretion

MEDULLA. The medulla receives nerve fibers from the sympathetic nervous system that stimulate release of medullary hormones very rapidly. The hormones are rapidly destroyed by the liver so that their effect, though great, is short-lived.

CORTEX. The secretion of aldosterone is increased by ACTH, angiotensin II, fall of blood sodium levels, and hemorrhage. Increased reabsorption of salts and water tends to restore blood volume and pressure.

Glucocorticoid secretion is controlled by ACTH, and secreted cortisol exerts a negative feedback effect on ACTH secretion.

Cortical sex hormone secretion is also controlled by ACTH, with the sex hormones exerting a negative feedback effect.

Clinical considerations

MEDULLA. Since the medulla is not essential for life, there is no disorder associated with hyposecretion. A tumor of the medulla called a *pheochromocytoma*, liberates vast amounts of hormones into the blood and can cause circulatory damage if not removed.

CORTEX. ADDISON'S DISEASE is the main disorder of cortical deficiency, and usually involves all

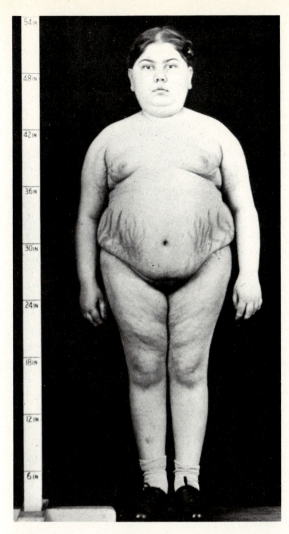

FIGURE 29.15   Cushing's Syndrome in a 12-year-old female. Notice the moon face, pendulous abdomen, abdominal striae, and concentration of adiposity on the trunk. (From F. Albright and E. C. Reifenstein, *The Parathyroid Glands and Metabolic Bone Diseases*, Williams & Wilkins, 1948. Courtesy Massachusetts General Hospital.)

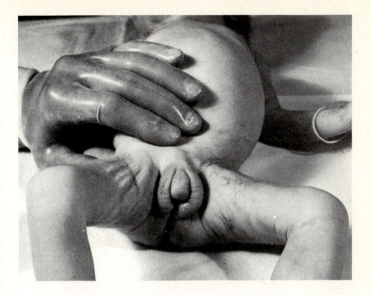

FIGURE 29.16   Adrenogenital syndrome in a 24-day-old female. Notice masculinization, with enlarged clitoris (center of genitalia) and scrotumlike development of the labia. (Armed Forces Institute of Pathology.)

zones of the cortex. Deficiency of glucocorticoids causes anemia, weakness, alimentary tract disturbances, and mineralocorticoid deficiency causes low blood sodium levels. Administration of the proper hormone corrects the deficiency. CUSHING'S SYNDROME (Fig. 29.15) results from excess cortisol secretion and causes a moonlike face with red cheeks, a "buffalo-hump" obesity, and stripes on the abdomen. ADRENOGENITAL SYNDROME (Fig. 29.16) results from overproduction of cortical sex hormones of male type. If the excess secretion occurs in a female, she will be virilized (develop male-type secondary sex characteristics). A male will have his masculine development accelerated.

The islets of the pancreas

Location and description

Small islets (a small mass of one type of tissue in another) of endocrine tissue are found in the body and tail of the pancreas. Their numbers have been estimated to lie between 500,000 and two million. Each islet is well supplied with blood vessels and consists of two main types of cells called ALPHA CELLS and BETA CELLS.

Hormones and effects

The alpha cells produce a hormone called GLUCAGON. It is a substance that speeds the conversion of glycogen to glucose and thus raises the blood glucose level (*hyperglycemia*).

The beta cells produce INSULIN, a hormone that increases cell uptake of glucose, and that speeds synthesis of glycogen in the liver. Since both these actions draw sugar from the blood and lower blood sugar levels, insulin may be said to have a *hypoglycemic* effect.

Control of secretion

The secretion of both hormones is controlled primarily by the blood glucose level. A fall of blood sugar causes glucagon secretion; a rise causes insulin secretion. Thus blood sugar is maintained within normal limits (80–100 mg/100 ml).

Clinical considerations

No named disease is associated with disorders of glucagon secretion. However, the literature contains reports of patients with chronic low blood sugar levels associated with lack or deficiency of alpha cells in the islets. The disorder appears to be hereditary.

Failure to produce insulin causes DIABETES MELLITUS. While primarily a disorder of carbohydrate uptake and metabolism, it affects the metabolism of all three basic foodstuffs. The following list of symptoms shows how metabolism is affected.

Insulin deficiency causes:

Hyperglycemia, due to failure of liver glycogen synthesis and cellular uptake of glucose.

Glucosuria, or sugar in the urine, caused by loss through the kidney. The tubules reabsorb only so much, the excess is lost. Cells, unable to take in glucose for metabolic activity, increase metabolism of fatty acids.

Ketosis and *acidosis* result from the production of ketone bodies derived from the excess acetyl CoA released during accelerated fatty acid oxidation.

Polyuria, or the secretion of large quantities of urine, results from osmotic attraction for water exerted by the large amounts of glucose in the kidney tubules.

Polyphagia, or excessive food intake, results from the caloric loss represented by loss of glucose.

Polydipsia, or excessive thirst, is the consequence of water loss.

Atherosclerosis may be accelerated by the high blood levels of fats, and the accumulation of fats inside blood vessels may cause a decrease of blood flow to organs and tissues. If the tissues die, gangrene may set in.

COMA may result if the hyperglycemia is not corrected. The coma is believed to result from dehydration of brain cells and by the increased H^+ concentration occurring in diabetes (diabetic acidosis). Insulin administration by injection may be required to control the disease. If there is deficiency, but not lack of insulin, diet control and orally administered drugs (e.g., Diabinese, Orinase, Tolinase, Dymelar) may control the disorder. The drugs stimulate secretion of remaining beta cells, increase cellular uptake of glucose, and inhibit liver glycogen breakdown.

Excessive secretion of insulin or insulin over-dose produces a hypoglycemia and INSULIN SHOCK. Since the adult brain depends entirely on glucose for its metabolism, the main symptoms of hypo-glycemia are nervous in origin and include: dis-turbances in walking, mental confusion, respira-tory disturbances, depression of body tempera-ture, and unconsciousness which may be fatal. Intake of glucose, either orally or intravenously, is required to control the condition.

The testis and ovary (gonads)

The anatomy of the testes and ovaries was de-scribed in Chapter 28. Hormones of both organs influence the same types of processes. For ex-ample:

An individual's maleness or femaleness depends on the chromosomal constitution, reinforced by hor-mones.

The secondary sex characteristics of fat distribu-tion, hair patterns, muscular development, and voice changes are dependent on gonadal hormones.

The ovarian hormones are responsible for the changes in the uterus during menstrual cycles.

The ovary: hormones and effects

Estradiol and estriol are steroid hormones pro-duced by the vesicular follicle. They are respon-sible for the proliferative phase of the menstrual cycle, increase mammary growth, stimulate con-traction of the uterus, cause and maintain the development of the accessory sex organs, and de-velop the female secondary sex characteristics. PROGESTERONE (*progestin*) is produced by the cor-pus luteum after ovulation has occurred. It is re-sponsible for the secretory phase of the menstrual cycle, causes milk production (but not secretion) by the mammary glands, causes ovulation, and is required for placenta formation.

Control of ovarian secretion

Four hormones are involved in a complex rela-tionship in control of ovarian secretion. FSH from the pituitary causes follicular development and estradiol and estriol secretion. These hormones, in turn, inhibit FSH secretion in a negative feedback mechanism. This relationship prevents other fol-licles from developing until the fate of the first one has been determined. LH from the pituitary causes ovulation, corpus luteum formation, and progesterone secretion. Progesterone, in turn, in-hibits both FSH and LH production. The recip-rocal relationships between these hormones are shown in Figure 29.17.

Oral contraception

The fact that estradiol and progesterone inhibit FSH secretion, and thus maturation of follicles, is the basis for "the Pill" to achieve contracep-tion. The first substances utilized in this manner resembled progesterone but created many un-pleasant side effects including nausea, water re-tention, and uncomfortable mammary swelling. Newer compounds are estrogenlike or are com-binations of estrogens and progesterone in low doses to imitate the natural secretion of hormones while preventing ovulation and ovum develop-ment. Tables 29.7 and 29.8 present the Pill and other methods of contraception that are available or are under investigation.

Clinical considerations: ovary
(Fig. 29.18)

Failure to produce sufficient estrogen results in female EUNUCHOIDISM, in which sexual develop-

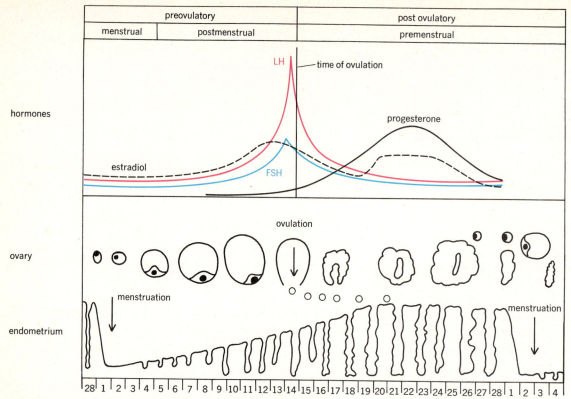

FIGURE 29.17 Interrelationships of hormones and ovarian and uterine changes.

ment remains infantile. Secondary characteristics develop minimally if at all. Excessive secretion of estrogen, usually from tumors, may create PRE-COCIOUS PUBERTY and adult sexual function at tender ages. The youngest person ever to become pregnant was a sufferer from precocious puberty, and conceived at 5 1/5 years of age.

TABLE 29.7 Methods of contraception currently available								
Method	Mode of action	Effectiveness if used correctly	Action needed at time of coitus	Requires resupply of materials used	Requires instruction in use	Requires services of physician	Suitable for menstrually irregular women	Side effects
Oral pill 21 day administration	Prevents follicle maturation and ovulation	Highest	None	Yes	Yes—timing	Yes—prescription	Yes	Early—some water retention, breast tenderness. Late—possible embolism, hypertension
Intrauterine device (coil, loop)	Prevents implantation	High	None	No	No	Yes—to insert	Yes	Some do not retain device. Some have menstrual discomfort

TABLE 29-7 (continued)

Method	Mode of action	Effectiveness if used correctly	Action needed at time of coitus	Requires resupply of materials used	Requires instruction in use	Requires services of physician	Suitable for menstrually irregular women	Side effects
Diaphragm with jelly	Prevents sperm from entering uterus, plus jelly spermicidal	High	Previous to coitus	Yes	Yes—must be inserted correctly each time	Yes—for sizing and instruction on use	Yes	None
Comdom (worn by male)	Prevents sperm entry into vagina	High	Yes	Yes	Not usually	No	—	Some deadening of sensation in male
Temperature rhythm	Determines ovulation time by noting body temperature at ovulation. T ↑	Medium	No	No	Definitely. Must learn to interpret chart correctly	No. Physician should advise	Yes, if are skilled in reading graph	None. Requires abstinence during part of cycle
Calendar rhythm	Abstinence during part of cycle	Medium to low	No—no coitus	No	Definitely. Must know when to abstain	No. Physician should advise	No!	None (pregnancy?)
Vaginal foams	Spermicidal	Medium to low	Yes. Requires application before coitus	Yes	No	No	Yes	None usually. May irritate
Withdrawal	Remove penis before ejaculation	Low	Yes. Withdrawal	No	No	No	Yes	Frustration in some
Douche	Wash out sperm	Lowest	Yes. Immediately after	No	No	No	Yes	None
Vasectomy (tying off of deferens)	Presents ejeculation of sperm	Highest	None	No	No	Yes—for surgery	—	May cause necrosis of testis, usually reversible
Tubal ligation (tying off of uterine tubes)	Prevents ova from reaching uterus	Highest	None	No	No	Yes—for surgery	—	Generally not reversible

The testis: hormone and effects

The interstitial cells of the testes produce the steroid TESTOSTERONE. Testosterone is responsible for the development of the male secondary sex characteristics, the external genitalia, and the accessory organs of the system. Testosterone is also anabolic in that it stimulates protein synthesis in muscle. Certain athletes (both male and female) have taken such hormones in the belief that they will increase muscular development and strength and thus will improve athletic performance.

Control of interstitial cell secretion

ICSH (interstitial cell stimulating hormone) is the male equivalent of LH. It stimulates secretion of testosterone, and testosterone inhibits ICSH secretion.

Clinical considerations: testis

Males may suffer EUNUCHOIDISM and PRECOCIOUS PUBERTY as described for females. Symptoms include failure of sexual development and accel-

TABLE 29.8   Newer methods of contraception under investigation

Method	Mode of action	Effectiveness if used correctly	Action needed at time of coitus	Requires resupply of materials used	Requires instruction in use	Requires services of physician	Side effects
"Mini-pill"— very low content of progesterone (1/4 mg.)	Inhibits follicle development	High	No	Yes	Yes	Yes— prescription	Irregular cycles and bleeding (25%)
"Morning-after pill"	Arrests pregnancy probably by preventing implantation. 50 × Normal dose of estrogen	By currently available data, high	No. For 1–5 days after coitus	Yes	No	Yes	Breast swelling, nausea, water retention
Vaginal ring— inserted in vagina; contains progesteroid in it	"Leaks" progesteroid into bloodstream through vagina at constant rate. Thereby inhibits follicle maturation	Studies are "promising"	No	Yes. Perhaps at yearly intervals	Yes	Yes	Spotting, some discomfort
Once-a-month pill	Injected in oil base into muscle. Slow passage of birth control drug into circulation inhibits follicle maturation	Said to be 100 percent	No	Yes. On monthly basis	No	Yes	Similar to oral pill
Depo-Provera (DMPA) 3 month injection	Injected. Inhibits follicle development	By currently available data, high	No	Yes. On 3 month basis	Yes	Yes	Similar to oral pill
Anti pregnancy vaccine	Antibodies to HCG cause pregnant female to menstruate	Said to be high	No	No	No	Yes	Unknown
Male "pill"	Inhibition of ICSH secretion by drug (denezol) given orally	85%	No	Yes	Yes	Yes	Unknown

eration of development and appearance of secondary characteristics. Figures 29.19 and 29.20 show the two conditions.

Climacterics

Males and females, as they age, undergo a cessation or diminution of gonadal secretion of hormones. This period is known as the climacteric or, in the female, as the menopause. Both sexes may show:

Psychic symptoms: nervousness, irritability, spells of emotional upheaval, and some loss of mental "sharpness."

Vasomotor symptoms: sweating, hot and cold "flashes," and headache.

Constitutional symptoms: fatigue, muscular weakness, and "lack of ambition."

Sexual symptoms: decrease of sexual drive, interest, and (in the female) ultimate sterility.

In the female, the symptoms appear to be related to decrease of progesterone secretion. In the male, about 15 percent of men show interstitial cell atrophy beginning about 40 years of age, which may be restored by testosterone administration. The remainder seem to retain nearly full interest and potency as they age, with only small gradual decrease of testosterone production.

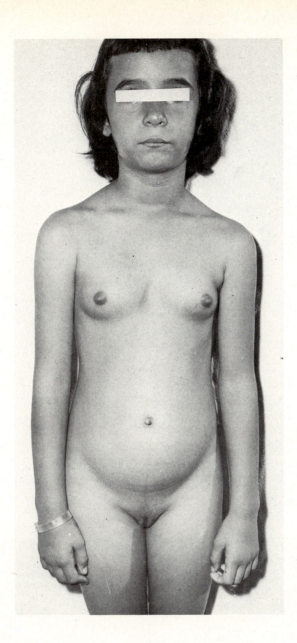

FIGURE 29.18   Precocious puberty in a 7-year-old female. Height was 4 feet 3¼ inches. Notice development of pubic hair and mammary glands. (Lester V. Bergman & Associates.)

The placenta

The placenta serves not only as the organ for exchange of nutrients and wastes between the fetal and maternal circulations, but produces several hormones that assure maintenance of pregnancy.

CHORIONIC GONADOTROPIN (CG) resembles LH in structure and effects. It maintains the corpus luteum of early pregnancy, and assures its production of progesterone.

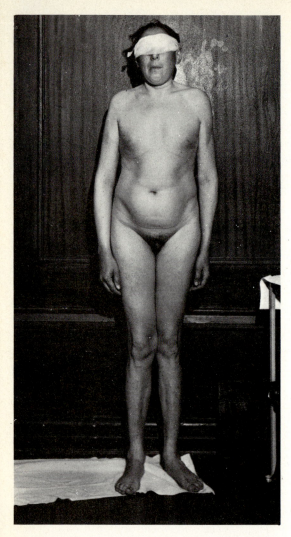

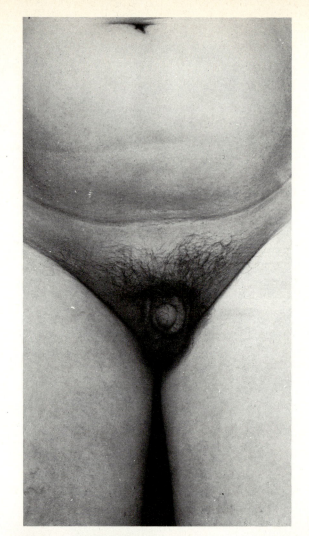

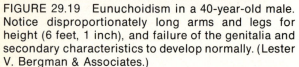

FIGURE 29.19 Eunuchoidism in a 40-year-old male. Notice disproportionately long arms and legs for height (6 feet, 1 inch), and failure of the genitalia and secondary characteristics to develop normally. (Lester V. Bergman & Associates.)

FIGURE 29.20 Precocious puberty in a 3-year-old male. Notice development of genitalia equivalent to that of an adolescent, and appearance of pubic hair. (Armed Forces Institute of Pathology.)

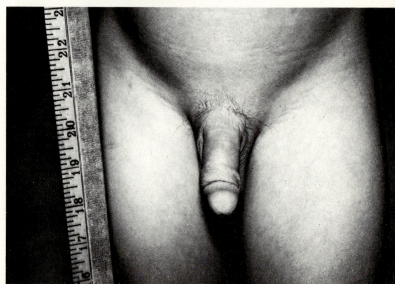

CHORIONIC ESTROGEN and PROGESTERONE are secreted in sufficient quantities, after 3 months of fetal life, to sustain a pregnancy even though the ovaries (and thus the corpus luteum) are removed.

CHORIONIC GROWTH-HORMONE-PROLACTIN (CGP) has lactogenic and growth-stimulating activity.

Finally, a hormone with TSH-like activity has been isolated from the placenta.

The placenta thus acts as insurance that the pregnancy will not suffer deficiencies of hormones that are necessary for continuation of development.

The pineal gland

The pineal gland (Fig. 29.21) has been variously regarded throughout recorded history as the residence of the soul, a valve to control movement of "spirits" through the body, and as an endocrine. In recent years, the pineal has acquired new status as an endocrine structure. Several known hormones or chemical substances have been shown to be synthesized in the pineal. Of these, MELATONIN is the only one that is synthesized only in the pineal. Melatonin causes contraction of pigment cells in amphibian skin. The eye apparently controls, by detecting changes in light intensity, the secretion of melatonin, and "matches" the skin color of the animal to light intensity. In this sense, the mechanism may have survival value. In humans, effects of the hormone are uncertain and its exact role in human physiology remains to be determined. It has been suggested that melatonin may be involved in the timing of human adolescence.

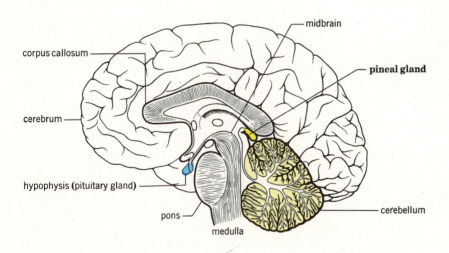

corpus callosum

midbrain

pineal gland

cerebrum

hypophysis (pituitary gland)

pons

medulla

cerebellum

FIGURE 29.21   The location of the pineal gland, as shown on a mid-sagittal section of the brain.

The prostaglandins

The prostaglandins (PGs) are a series of cyclic, oxygenated derivatives of prostanoic acid (Fig. 29.22). They are formed by the action of a prostaglandin synthetase enzyme system located on the microsomal (ER fragments) fraction of the cellular organelles of a wide variety of body cells (e.g., lungs, liver, muscle, reproductive organs). A long-chain polyunsaturated fatty acid is caused to assume a cyclic form under the influence of the synthetase.

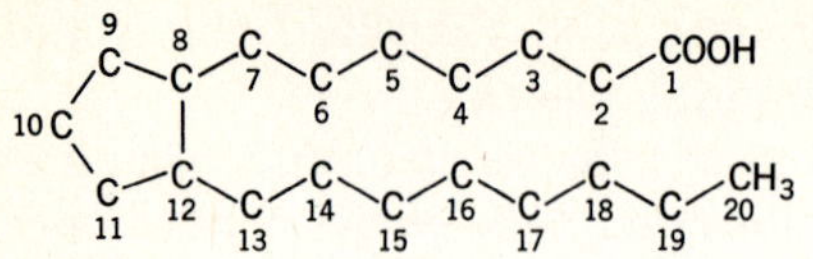

FIGURE 29.22 Prostanoic acid, the precursor of PG's.

With the discovery of the presence of cyclic nucleotides and their role in control of enzymatic reactions, PGs have been investigated as controllers of the adenylcyclase or guanylcyclase enzyme systems, and therefore the amount of cyclic nucleotides in cells. Evidence to date indicates that PG effect is indeed exerted via the cyclase enzymes, and may be either stimulatory or inhibitory in effect. The inhibitory effect may additionally be exerted through the mechanism of blocking the receptor site for a hormone that normally increases cyclic nucleotide formation. All this suggests that PGs may act as "intracellular switches" to turn chemical reactions off or on.

Release of PGs from tissues or cells is caused by a wide variety of stimuli, including chemicals, mechanical touching or distension, inflammation, and nervous stimuli, and may be associated with disturbances of any cell membrane by normal or pathological processes. Cells may release PGs to resist change, as when tension is applied to smooth muscle or when inflammation occurs. A primary result appears to be to resist tearing or rupture of the cell membrane.

Effects of PGs are nearly as widespread as the tissues in which they are found:

The list presented below indicates some, but certainly not all, the effects of PGs on the body.

Smooth muscle. Uterine smooth muscle is stimulated to increase its contractions by PGs. Vascular smooth muscle is relaxed or contracted, depending on type of PG.

Sodium excretion. PGs are natriuretic and can lower blood pressure by decreasing blood volume via Na and water excretion. The effect appears to be mediated via the kidney.

Ion and water exchange. PGs increases calcium release from binding sites and inhibits the ADH effect on toad bladder and perhaps also on kidney tubules.

Skin. The highest concentrations of PGs are found in the epidermis, where they are postulated to be involved in the synthesis of the waterproofing substances of the upper epidermal cell layers. If deficient in PG, these cells have been shown to become more permeable to water.

Neurons. PGs have been shown to depress synaptic transmission and elevate discharge rates of neurons. It is not a transmitter itself, but influences those processes that synthesize or degrade the normal transmitters.

Therapeutic uses of PGs will probably revolve around their effects on smooth muscle and the ovary. The use of PGs to CONTROL HYPERTENSION (through vasodilating effects), nasal congestion and asthma (through vasoconstrictive and smooth muscle relaxing effects respectively), ULCERS (through decreased gastric secretion), and as ABORTIFACIENTS (through luteolysis) has reached the experimental stage in humans. Minimal side effects appear to be produced, and the PGs may, in the future, achieve the fame accorded the antibiotics at their appearance.

Summary

1. The endocrine system provides control over metabolism, growth and development.

2. The organs of the system develop from the alimentary tract, the nervous system, or mesoderm of the body.

3. The organs or glands of the system include: the pituitary, thyroid, parathyroids, adrenals, pancreatic islets, ovaries and testes, and pineal gland.

   a. All endocrines are vascular, ductless, hormone-secreting structures whose removal disturbs body function or development.

   b. Hormones are the products of endocrine glands. They are proteins, steroids, or small amines.

4. Control of endocrine secretion is by nerves, neurohumors, negative feedback, or blood levels of specific or all solutes.

5. The pituitary gland lies on the undersurface of the brain. It has anterior, intermediate, and posterior lobes.

   a. The anterior lobe produces six hormones controlling growth (GH), thyroid activity (TSH), adrenal cortex activity (ACTH), gonadal activity (FSH, LH, or ICSH), and milk secretion (prolactin).

   b. Secretion of anterior lobe hormones is controlled by hypothalamic regulating hormones.

   c. The intermediate and posterior lobes release three hormones. Oxytocin and ADH are actually produced in the hypothalamus and are stored in and released from the posterior lobe by nerve impulses. Oxytocin stimulates uterine contraction; ADH insures water reabsorption by the kidney. MSH (intermediate lobe) has no known function in humans.

   d. Disorders of pituitary function include disturbances in growth, and water metabolism (*see* Table 29.5).

6. The thyroid gland is located in the neck. It has follicles as its functional unit.

   a. The thyroid produces thyroxin (T_4), triiodothyronine (T_3), and calcitonin as its hormones.

   b. T_4 and T_3 control body metabolism and growth.

   c. Calcitonin aids in Ca^{2+} and PO_4^{3-} deposition in bones.

   d. Goiter is an enlarged thyroid gland; cretinism and myxedema indicate hyposecretion; hyperthyroidism is excess T_4 and T_3 secretion.

   e. Control of T_4 and T_3 secretion is by TSH; calcitonin by blood calcium level.

   f. The PBI and iodine uptake tests are good tests for determining thyroid function.

7. The parathyroid glands lie on the thyroid. They contain principle cells that secrete a hormone.

   a. The parathyroid hormone (PTH) is concerned with increasing absorption, retention, and metabolism of calcium and phosphate.

   b. Secretion is controlled by blood calcium level.

   c. Hypothyroidism (tetany) is due to disturbed blood calcium levels and muscular paralysis. Excess PTH secretion causes bone destruction.

8. The adrenal glands lie above the kidneys and have an inner medulla and an outer cortex.

   a. The medulla produces epinephrine and norepinephrine that elevate body functions.

   b. The cortex produces cortical steroids. Mineralocorticoids deal with sodium (and thus Cl^-, HCO_3^-, and H_2O) metabolism. Glucocorticoids govern glucose, fat, and protein metabolism, and exert stress resistance and anti-inflammatory effects. Cortical sex hormones have no known effects in normal people.

   c. The medulla is controlled by nerves; the cortex is controlled by several chemicals, including ACTH.

   d. Deficient medulla secretion is no threat to life; deficient cortical secretion is associated with weakness; excess cortical secretion is associated with disturbances of metabolism.

9. The islets of the pancreas contain alpha and beta cells.

   a. Alpha cells produce glucagon, which raises blood sugar levels; beta cells produce insulin that lowers blood sugar levels, and controls carbohydrate, fat, and protein metabolism.

   b. Secretion of both hormones is controlled by blood glucose levels; fall increases glucagon secretion, rise increases insulin secretion.

   c. Diabetes (mellitus) is the result of insufficient insulin secretion. It is associated with disturbances of foodstuff metabolism, appetite, and acid-base balance. Excessive insulin creates disturbances in the brain due to lack of glucose for brain metabolism.

10. The testes and ovaries secrete hormones that control secondary sex characteristics and development of the sex organs.

    a. The ovaries produce estradiol and estriol that develops organs and characteristics; and progesterone, which insures ovulation, placenta formation, and milk production.

    b. The ovaries are controlled by pituitary FSH and LH.

    c. Oral contraception is based on the effects ovarian hormones have on the pituitary.

    d. Deficient ovarian hormone secretion results in failure of sexual maturity; excess causes early sexual development and function.

    e. The testes produce testosterone that controls male characteristics and organ development.

    f. The testes are controlled by pituitary ICSH.

    g. Clinical considerations are the same as for the ovary.

    h. Climacterics (menopause in the female) are associated with decrease of gonadal hormone secretion. Disturbing sexual, body, and neural symptoms may develop.

11. The pineal secretes melatonin, which alters skin color in "lower animals."

12. Prostaglandins are derivatives of fatty acids.

 a. They are widely distributed in animal cells.

 b. They appear to act to "turn off or on" certain chemical reactions in the body, notably cAMP production.

Questions

1. The hypophysis has sometimes been designated as "the master gland" of the endocrine system. What, in your mind, justifies this exalted position?

2. What hormones are required for normal body growth and development? Give the contribution of each to these processes.

3. What hormones are concerned with carbohydrate metabolism and how is each involved?

4. What endocrines exhibit a negative feedback mechanism of control? Which are controlled by blood levels of substances other than hormones?

5. What hormones enable the body to meet situations that are stressful or to meet sudden demands for accelerated activity?

6. What hormones are concerned with bone and tooth formation? Where are they produced and how are they involved?

7. Describe hypothalamic control of the hypophysis.

8. What justifies the statement that nervous and endocrine function are related?

9. What hormones ensure proper sexual development? Discuss contribution of each.

10. What roles do prostaglandins have in the body?

Readings

Chick, William L., et al. "Artificial Pancreas Using Living Beta Cells: Effects on Glucose in Rats." *Science* 197:780, 19 August 1977.

Frantz, Andrew G. "Prolactin." *New Eng. J. Med.* 298:201, Jan. 26, 1978.

George, J. M. "Immunoreactive Vasopressin and Oxytocin: Concentration in Individual Hypothalamic Nuclei." *Science* 200:342, 21 April 1978.

Greiner, A. C., and S. C. Chan. "Melatonin Content of the Human Pineal Gland." *Science* 199:83, 6 Jan. 1978.

Knodel, John. "Breast-feeding and Population Growth." *Science 198:*1111, 16 Dec. 1977.

Pacold, S. T., L. Kirsteins, S. Hojvat, and A. M. Lawrence. "Biologically Active Pituitary Hormones in the Rat Brain Amygdaloid Nucleus." *Science 199:*804, 17 Feb. 1978.

Patel, Y. C., D. P. Cameron, Y. Stefan, F. Malaisse-Lagae, and L. Orci. "Somatostatin: Widespread Abnormality in Tissues of Spontaneously Diabetic Mice." *Science 198:*930, 2 Dec. 1977.

Science News. "ACTH and Endorphins: A Common Origin?" *113:*102, Feb. 18, 1978.

Sterling, Kenneth, and John H. Lazarus. "The Thyroid and Its Control." *Ann Rev. Physiol. 39:*349, 1977.

Unger, R. H., R. E. Dobbs, and L. Orci. "Insulin, Glucagon, and Somatostatin Secretion in the Regulation of Metabolism." *Ann. Rev. Physiol. 40:*307, 1978.

Vale, Wylie, and Catherine Rivier. "Regulatory Peptides of the Hypothalamus." *Ann Rev. Physiol. 39:*473, 1977.

Epilogue

This book has concentrated primarily on the structure and function of the normal adult. Development of organs and systems has been considered. Few introductory physiology books acknowledge there is a life beyond "maturity," into old age and senescence. Inasmuch as we will all arrive, sooner or later, at this terminal part of our life span, it is appropriate to conclude this volume with a few remarks about what lies ahead.

Senescence

Physiological processes reach a peak during maturity (about 30 years of age), and thereafter undergo gradual declines. At some point, one or more functions decline below the point necessary to sustain life and death ensues. Heredity, environment, diet, and "speed of living" all influence the rate of decline and time of death.

Three general types of changes contribute to the aging process: SECULAR CHANGES are the result of natural wear and tear; SENESCENT CHANGES are the result of aging of tissues and organs, especially those with low mitotic rate; PATHOLOGIC COMPLICATIONS are those resulting from disease processes developing in the aging organs.

Secular changes

As one grows older, tissues appear to demand more metabolic support than the body can supply. In short, secular changes appear to be the result of an imbalance between vascular supply and tissue demand. Thus, the amount of active tissue declines in proportion to its vascularity, and body weight tends to be reduced. Little attempt appears to be made by the body to restore the balance. The changes are usually seen first in endocrine-dependent structures such as breast, prostate, internal reproductive organs, and thyroid.

Senescent changes

All body systems share in these changes. Some of the more noteworthy changes in systems are described below. Changes are gradual and may not be noticed by the individual.

Aging of the integument. The epidermis thins, becomes more translucent and dry, and the dermis becomes dehydrated and suffers loss of elastic fibers. The skin thus tends to "sag" on the body.

There is usually loss or thinning of hair as a result of lowered sex hormone levels, and lowered synthesis of proteins. Sweat gland secretion diminishes, making the older individual more susceptible to the effects of high environmental temperature. Skin lesions (cancer, ruptured blood vessels, etc.) are more common in the aged. Diabetic vasculopathy is also more common in the aged.

Aging of the eye is reflected by a high incidence of cataracts, occlusion of retinal vessels with subsequent retinal changes, glaucoma, and changes in the transparency of the cornea. About one person in six in the over-65 age group exhibits some ocular pathology.

Aging of the ear is evidenced by sclerotic alterations in the ear drum and ossicles that may result in loss of hearing. Vascular blockage may result in sudden hearing loss or loss of ability to maintain body equilibrium.

Neurological disorders are most commonly associated with change in vascular supply, either gradually (Parkinson's disease) or suddenly ("stroke"). Peripheral loss of sensation acuity, neuritis, and neuraliga appear to be more common in the aged.

Circulatory disorders may include hardening of arteries (arteriosclerosis) and deposition of lipids in the vessel walls (atherosclerosis). If deposition occurs in coronary vessels, the individual becomes a candidate for "heart attack." Heart action lessens, with fall in cardiac output. Shortness of breath may develop as a result of low cardiac output and low pulmonary perfusion. Hypertension is common.

At the cellular level, there is *diminution of DNA and RNA synthesis*, with failure to replace worn-out body proteins.

The *lungs* lose elasticity and dyspnea may develop. Diffusing capacity of gases diminishes, due primarily to loss of lung capillaries.

The *alimentary tract* undergoes changes in function to a greater degree than changes in structure. More than one half (56%) of aged persons show functional disorders such as heartburn, belching, nausea, diarrhea, constipation, and flatus. Many of these complaints have an emotional basis, based on fear of death and disease, and loss of contact with offspring. An organic basis may be demonstrated in some cerebral arteriosclerosis patients. Malignancy of the lower tract is more common in elderly persons (11%).

Renal disorders include lowered filtration rates, reduced reabsorption, and renal hypertension. Pyelonephritis (infection) is the most common renal disease. Calculi (stones) in the kidney and ureter are also more common in the aged.

Gonadal function decreases. The ovary atrophies, producing menopause in females; testicular function declines more slowly. Internal organs atrophy and often prolapse ("drop down") as, for example, the uterus entering the upper vagina.

Loss of the inorganic component of the *skeleton* produces osteoporosis and greater liability to fracture. The *muscles* become smaller and weaker and are prone to cramps and effects of altered electrolyte balance (hypokalemia, low blood potassium).

Articulations are affected by arthritis. *Osteoarthritis* is a noninflammatory condition in which joints undergo degenerative changes in cartilage and synovial membranes. *Rheumatoid arthritis* is typically inflammatory in nature. *Gouty arthritis* is a metabolic disorder in which uric acid crystals accumulate in the joints.

Endocrines may increase activity (as in acromegaly) or may decrease activity (diabetes, hypothyroidism). Gonadotropins appear to show the widest secretion ranges.

The *central nervous system* shows the greatest change in the brain, where neuron loss (estimated at 30 cells per minute from 30 years onward) may cause increased reaction times, loss of memory, and personality alterations.

Pathologic changes

The diseases of old age are too numerous to consider here. Suffice it to say that the aged person is more susceptible to the diseases we all are heir to, and that the presence of a disease process tends to accelerate the aging changes described previously.

Theories of aging

There are two general groups of theories as to why the body ages.

External theories

The external theories of aging suggest that factors acting on the body from outside determine the health of our cells and the length of time that they will continue to function. Environment plays a role in that air, food, or noise pollution may cause cell death. Nutrition exerts an effect in that diets slightly limited (caloric, not in essential foods) create an underweight but longer-living organism. Bacteria and viruses or their products or both may accelerate the aging process. Radiation may contribute to chemical changes within cells and to their inability to function normally.

Internal theories

The internal theories of aging suggest that the forces causing aging originate within the organism itself. Direct genetic programming may cause vital chemical reactions to cease, accelerate, or to escape control at definite times. If this theory is correct, there would have to be an "aging gene" present, which is turned on or off. Such a gene has not been shown to exist to this time. Mutations of existing genes might result in the production of abnormal proteins that would stop some vital chemical reaction(s). The production of specific youth or aging chemicals has also been investigated. No evidence exists to suggest that aging is associated with the disappearance from the body of some "youth potion," or the accumulation in the body of an "aging substance." Defective immune responses, the result of inability of plasma cells to react to antigens has been suggested to contribute to cell death. Establishment of cross-linking in collagen molecules has been shown to occur as aging processes. This implies that links which did not exist before may be detrimental to cell health. Thus far, no relationship between this phenomenon and cell death has been established.

If there is a common denominator in all these theories, it is that they all lead to an inability to adapt or adjust to environmental changes, and perhaps, in the last analysis, this is why we die.

Death

While old age itself does not cause death, organs and systems age at different rates. If function of a vital organ drops below the critical level for maintenance of life, life ceases. The operation of bodily systems is interrelated so that if one vital organ fails, all fail—"the chain is only as strong as its weakest link." Loss of recovery ability is obvious, so that alterations that would be rapidly returned to normal by a young person become life-threatening to the aged.

Definitions of death are extremely varied, and a strict definition is almost impossible. Among the definitions considered are:

Total irreversible cessation of cerebral function (as evidenced by a "flat" electroencephalogram) for at least 48 hours.
No spontaneous heart beat or respiration.
Permanent cessation of all vital functions.

At one time, stoppage of kidney function, cardiac standstill, or stoppage of respirations may have been considered signs of death. The use of the artificial kidney, respirators, and heart-lung machines has caused a reevaluation of the definition of death, as has the problem of securing organs from a donor for transplant. At present, spontaneous brain activity appears to be the most valid criterion of death, although no specific legal definition has as yet been advanced.

A philosophical statement

To be sure, there are demonstrable changes that occur in all systems of the body as life progresses. Some of these, such as deposition of fats in blood vessel walls, may be demonstrated in infants. Others may be seen to begin in adolescence, such as decline in auditory function. It thus becomes a difficult and highly individual matter at to when one "begins to age." Certain agencies suggest that 65 years is the time of beginning of "old age." Perhaps the best criterion is the old adage that "you're as young as you feel." Certainly, blanket application of some specific criterion that forces a person suddenly to quit as a productive member of society is not justified.

Research in gerontology may eventually result in great extensions of the human life span. But, without simultaneous efforts to increase the quality as well as the quantity of life, no effort will be deemed worthwhile. We must, in the last analysis, "add years to life," and "life to years," regardless of our age.

Similarly, it is becoming clear that the care of our bodies must begin with conception and terminate only with the death of a given individual. Thus, genetics, nutrition, environment, and the mental and physical health of the individual set the foundations for the next generation. We cannot expect to improve or maintain the quality of human resources without an all-out commitment to prevention of disease and disorders as the primary measure for promotion of optimum health throughout life.

Readings

Burnet, F. M. *Immunology, Aging, and Cancer*. Freeman. San Francisco, 1976.

Hayflick, Leonard. "The Cell Biology of Human Aging." *New Eng. J. Med. 295*:1302, Dec. 2, 1976.

Holliday, R., L. I. Huschtscha, G. M. Tarrant, and T. B. L. Kirkwood. "Testing the Commitment Theory of Cellular Aging." *Science 198*:366, 28 Oct. 1977.

Rabkin, Judith G., and Elmer L. Struening. "Life Events, Stress, and Illness." *Science 194*:1013, 3 Dec. 1976.

Rowe, John W. "Clinical Research on Aging: Strategies and Directions." *New Eng. J. Med. 297*:1332, Dec. 15, 1977.

Science News. "Senility: More Than Growing Old." *112*:218, Oct. 1, 1977.

General References

Allen, Garland. *Life Sciences in the Twentieth Century*. Wiley. New York, 1975.

Arey, L. B. *Developmental Anatomy*. 7th ed. Saunders. Philadelphia, 1974.

Barr, Murray L. *The Human Nervous System*. Harper & Row. New York, 1974.

Beeson, Paul B., and Walsh McDermott (eds.). *Textbook of Medicine*. 14th ed. Saunders. Philadelphia, 1975.

Bergersen, Betty S., and Andres Goth. *Pharmacology in Nursing*. 13th ed. C. V. Mosby Co. St. Louis, Mo., 1976.

Bloom, William, and Don W. Fawcett. *A Textbook of Histology*. 10th ed. Saunders. Philadelphia, 1976.

Cheek, Donald B. *Fetal and Postnatal Cellular Growth*. Wiley. New York, 1975.

"Children Are Different—Relation of Age to Physiologic Functions." Ross Laboratories. Columbus, Ohio, October, 1972.

Diem, K., and C. Lentner (eds.). *Scientific Tables*. 7th ed. J. R. Geigy. Basel, Switzerland, 1970.

Gray, Henry. *Anatomy of the Human Body*. C. M. Goss (ed.). 29th ed. Lea and Febiger. Philadelphia, 1973.

Guyton, Arthur C. *Textbook of Medical Physiology*. 5th ed. Saunders. Philadelphia, 1976.

Lockhart, R. D. *Living Anatomy*. Faber & Faber. New York, 1970.

Moore, Keith L. *Before We Are Born: Basic Embryology and Birth Defects*. Saunders. Philadelphia, 1974.

Moore, Keith L. *The Developing Human: Clinically Oriented Embryology*. Saunders. Philadelphia, 1973.

Mountcastle, V. B. (ed.). *Medical Physiology*. Vols. I and II. Mosby. St. Louis, Mo., 1974.

Netter, Frank H. *Ciba Collection of Medical Illustrations*. Ciba Pharmaceutical Products. Newark, N. J.
  Vol. I *Nervous System*.
  Vol. II. *Reproductive Systems*.
  Vol. III (Parts I, II, III). *Digestive System*.
  Vol. IV. *Endocrine System and Selected Metabolic Diseases*.
  Vol. V. *Heart*.

Nilsson, Lennart. *Behold Man*. Little, Brown and Co. Boston, 1973.

Normal Reference Laboratory Values. *New Eng. J. Med*. 298:34–45. Jan. 5, 1978.

Ruch, Theodore C., and Harry D. Patton. *Physiology and Biophysics*. Saunders. Philadelphia.
  Vol. I *The Nervous System*. 1977.
  Vol. II. *Circulation, Respiration, and Fluid Balance*. 1974.
  Vol. III. *Digestion, Metabolism, Endocrine Function, and Reproduction*. 1973.

Rubin, Alan (ed.). *Handbook of Congenital Malformations*. Saunders. Philadelphia, 1967.

Science vol 200. 26 May 1978 (Entire issue on medicine and public health).

Scientific American—Readings. *Human Physiology and the Environment in Health and Disease*. Freeman. San Francisco, 1976.

Smith, Clement A., and Nicholas M. Nelson. *The Physiology of the Newborn Infant*. Thomas. Springfield, Ill., 1976.

Thompson, James S., and M. W. Thompson. *Genetics in Medicine*. 2nd ed. Saunders. Philadelphia, 1973.

Timiras, Paola S. *Developmental Physiology and Aging*. Macmillan. New York, 1972.

Vaughn, Victor C., and R. James McKay. *Nelson Textbook of Pediatrics*. Saunders. Philadelphia, 1975.

Appendix

Glycolysis, the anaerobic degradation of glucose and fructose. Compound formulas are presented, along with enzymes required for the individual steps. $\triangle$ G expresses the energy change at each step. A negative $\triangle$ G indicates a reaction in which the products have a lower energy content than the reactants (the reverse for a $+ \triangle$ G). Boxed numbers indicate steps.

Phosphorylation of sugars as they pass through cell membranes. ATP is used and the energy level of the sugar is raised, enabling it to undergo further metabolism. Enzymes are required to carry out the phosphorylation.

The formation of glycogen (glycogenesis) and breakdown of glycogen (glycogenolysis). Both reactions require enzymes and are aided by hormones. Liver cells carry out these reactions, as do skeletal muscle cells.

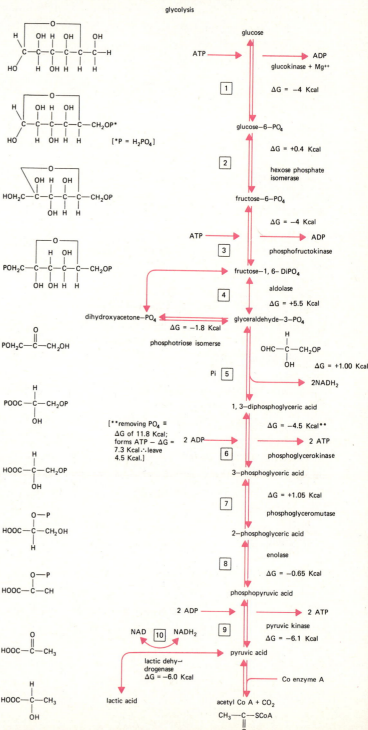

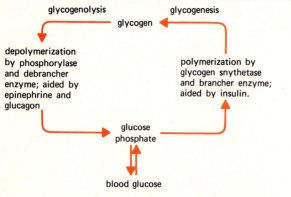

The formation of lactic acid from pyruvic acid by addition of 2H. The reaction is reversible, by removing 2H (see glycolysis).

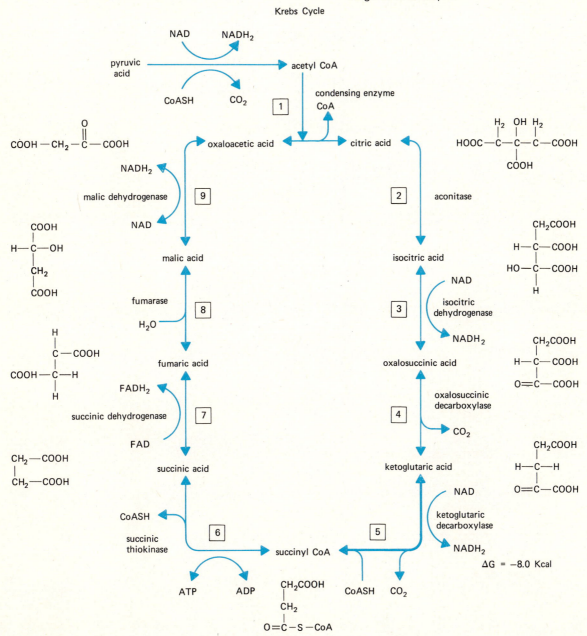

The Krebs cycle, for aerobic degradation of acetic acid. The cycle usually runs clockwise, owing to the large $\triangle$ G at step 5.

Oxidative phosphorylation. Hydrogens are converted
to water, with release of energy to synthesize ATP.

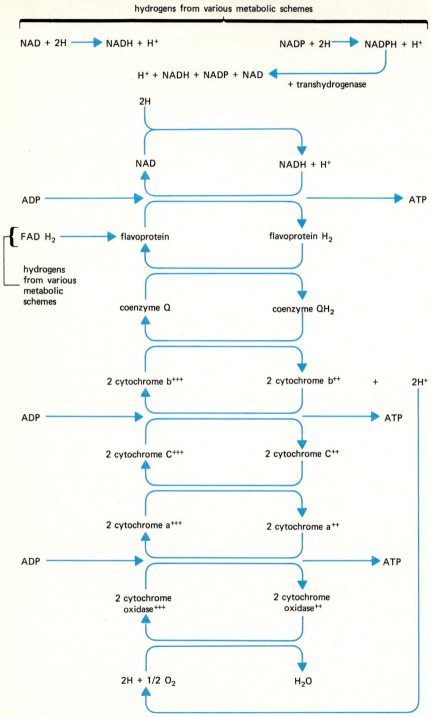

hydrogens from various metabolic schemes

NAD + 2H → NADH + H$^+$ NADP + 2H → NADPH + H$^+$

H$^+$ + NADH + NADP + NAD ←
+ transhydrogenase

2H

NAD   NADH + H$^+$

ADP →   → ATP

FAD H$_2$ →   flavoprotein   flavoprotein H$_2$

hydrogens
from various
metabolic
schemes

coenzyme Q   coenzyme QH$_2$

2 cytochrome b^{+++} 2 cytochrome b^{++} +   2H$^+$

ADP →   → ATP

2 cytochrome C^{+++} 2 cytochrome C^{++}

2 cytochrome a^{+++} 2 cytochrome a^{++}

ADP →   → ATP

2 cytochrome   2 cytochrome
oxidase^{+++} oxidase^{++}

2H + 1/2 O$_2$ H$_2$O

Net reaction:

2H + 1/2 O$_2$ + 3 ADP ←→ H$_2$O + 3 ATP

Beta-oxidation, the degradation of a fatty acid.

1. $R-CH_2-CH_2-\overset{\overset{O}{\|}}{C}-OH + CoA-SH \xrightarrow[\text{Mg}^{++}]{\text{ATP + thiokinase}} R-(CH_2)_2-\overset{\overset{O}{\|}}{C}\sim SCoA + AMP + PP^*$

*pyrophosphate

2. $R-(CH_2)_2-\overset{\overset{O}{\|}}{C}\sim SCoA \xrightarrow[\text{FAD} \longrightarrow \text{FADH}_2]{\substack{\text{acyl} \\ \text{dehydrogenase}}} R-CH=CH-\overset{\overset{O}{\|}}{C}\sim SCoA$

3. $R-CH=CH-\overset{\overset{O}{\|}}{C}\sim SCoA \underset{\text{crotonase}}{\rightleftharpoons} R-\overset{\overset{OH}{|}}{CH}-CH_2-\overset{\overset{O}{\|}}{C}\sim SCoA$

4. $R-\overset{\overset{OH}{|}}{CH}-CH_2-\overset{\overset{O}{\|}}{C}\sim SCoA \xrightarrow[\text{NAD} \longrightarrow \text{NADH}_2]{\text{dehydrogenase}} R-\overset{\overset{O}{\|}}{C}-CH_2-\overset{\overset{O}{\|}}{C}\sim SCoA$

beta-keto fatty acid

5. $R-\overset{\overset{O}{\|}}{C}-CH_2-\overset{\overset{O}{\|}}{C}\sim SCoA + CoA-SH \xrightarrow{\text{thiolase}} \boxed{R-\overset{\overset{O}{\|}}{C}-SCoA} + CH_3-\overset{\overset{O}{\|}}{C}\sim SCoA$

acetyl CoA

$CH_3-\overset{\overset{O}{\|}}{C}-SCoA + CH_3-\overset{\overset{O}{\|}}{C}-SCoA \underset{\text{thiolase}}{\rightleftharpoons} CH_3-\overset{\overset{O}{\|}}{C}-CH_2-\overset{\overset{O}{\|}}{C}-SCoA + CoA-SH$

acetoacetyl CoA

$CH_3-\overset{\overset{O}{\|}}{C}-CH_2-\overset{\overset{O}{\|}}{C}-SCoA \xrightarrow[\text{liver only}]{\text{deacylase in}} CH_3-\overset{\overset{O}{\|}}{C}-CH_2-\overset{\overset{O}{\|}}{\underset{\underset{OH}{|}}{C}} + CoA-SH$

acetoacetic acid

$CH_3-\overset{\overset{O}{\|}}{C}-CH_2-\overset{\overset{O}{\|}}{\underset{\underset{OH}{|}}{C}}$ $\xrightarrow{-CO_2}$ $CH_3-\overset{\overset{O}{\|}}{C}-CH_3$ (acetone)

$\xrightarrow{+2H}$ $CH_3-CHOH-CH_2-COOH$
(beta-OH butyric acid)

The formation of the ketone bodies.

The names and formulas of the essential fatty acids (they are unsaturated fatty acids).

Palmitoleic acid
$CH_3(CH_2)_5CH=CH(CH_2)_7COOH$

Oleic acid
$CH_3(CH_2)_7CH=CH(CH_2)_7COOH$

Linolenic acid
$CH_3CH_2CH=CHCH_2CH=CHCH_2CH=$
$CH(CH_2)_7COOH$

Linoleic acid
$CH_3(CH_2)_4CH=CHCH_2CH=$
$CH(CH_2)_7COOH$

Arachidonic acid
$CH_3(CH_2)_4(CH=CHCH_2)_4(CH_2)_2COOH$

Deamination of an amino acid.

$CH_3-\overset{\overset{NH_2}{|}}{\underset{\underset{H}{|}}{C}}-COOH + \tfrac{1}{2}O_2 \xrightarrow{\substack{\text{amino acid} \\ \text{oxidase}}} NH_3 + CH_3-\overset{\overset{O}{\|}}{C}-COOH$

alanine   ammonia   pyruvic acid
(an α keto acid)

Transamination of an α-keto acid, to form an amino acid.

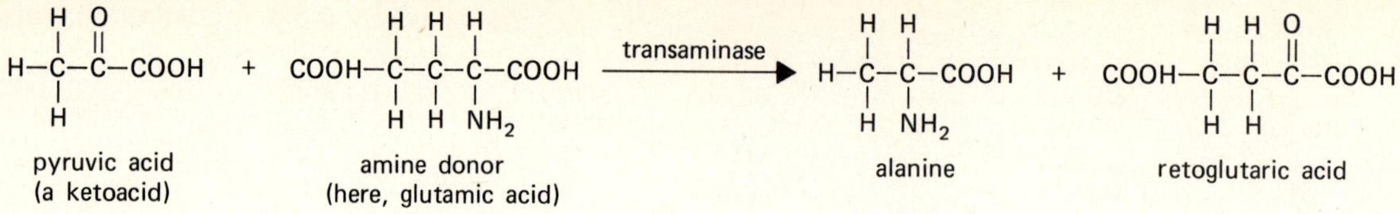

pyruvic acid
(a ketoacid)

amine donor
(here, glutamic acid)

transaminase

alanine

retoglutaric acid

the ornithine cycle.

ornithine

$+ CO_2 + NH_3$

ornithine
transcarbamylase

$+ H_2O$

citrulline

+

aspartic acid

arginase

urea

arginine

condensing
enzyme

arginosuccinase

arginosuccinic acid

fumaric acid

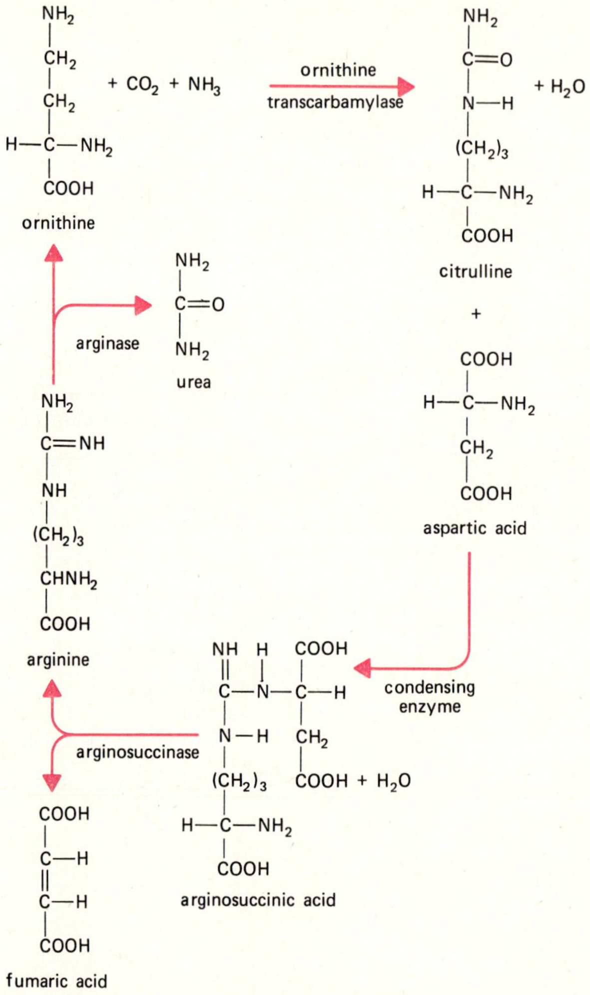

The ornithine cycle that produces urea from ammonia
(NH_3) and carbon dioxide (CO_2).

Glossary

Abbreviations

abbr.	abbreviation
A.S.	Anglo-Saxon
Colloq.	Colloquial
e.g.	for example
esp.	especially
Fr.	French
G.	Greek
Ger.	German
i.e.	that is
L.	Latin
M.E.	Middle English
pert.	pertaining
pl.	plural
sing.	singular

Pronunciation

Pronunciations may be only approximately indicated unless all the markings in Webster's *New International Dictionary* are used, which is not practical in a glossary.

Accent. Indicates the stress on certain syllables.

Diacritics. Marks over the vowels to indicate the pronunciations. Only two diacritics are used: The *macron,* showing the long or name sound of the vowels and the *breve,* showing the short vowel sounds, as

a in rāte	**a** in căt
e in ēat	**e** in ĕver
ee in sēēd	
i in īsle	**i** in ĭt
o in ōver	**o** in nŏt
oo in mōōn	**oo** in bŏok
u in ūnit	**u** in cŭt

a-, an- [G. *alpha*]. Prefix: without, away from, not.

ab- [L.]. Prefix: from, away from, negative, absent.

abdomen (ab-dō′men) [L.]. The part of the trunk between the chest and the pelvis.

absorb [L. *absorbere,* to suck in]. To take in.

absorption (ab-sorp′shun) Taking into body fluids or tissues, usually across a
membrane.

abducens (ab-dū′senz) [L. drawing away from]. The sixth cranial nerve; innervates the lateral rectus muscle of the eye that moves the eyeball outward or away from the median line of the body.

abduction (ab-dukt′shun) [L. *abducere*, to lead away]. Lateral movement away from the median plane of the body.

acceleration (ak-sel′e-rā-shun) [L. *acceleratus*, to hasten]. Increasing the speed of pulse, respiration, rate or motion.

accept′or [L. *accipere*, to accept]. Compounds taking up or chemically binding other substances, e.g., hydrogen acceptor.

accom′modate [L. *accomodare*, to suit]. To adjust or adapt; a decrease in nerve impulse discharge with continued stimulation; the adjustment of the eye for various distances by changing the curvature of the lens.

acetabulum (as-ē-tab′ū-lum) [L. a little saucer for vinegar]. The rounded cavity of the os coxae that receives the head of the femur.

acetic acid (a-sē′tik a′cid) [L. *acetum*, vinegar, + *acidus*, sour]. The acid of vinegar. CH_3COOH.

acetone (as′ē-tōn). A chemical [$(CH_3)_2CO$] produced when fats are not properly oxidized in the body. It is one of several substances called *ketone bodies* or *acetone bodies*.

acetylcholine (ă-sĕt-il-kō′lēn). An ester of choline released from certain nerve endings; thought to play a role in transmission of nerve impulses at synapses and myoneural junctions. Easily destroyed by the enzyme, *cholinesterase*.

Achilles tendon (ăh-kil′ēz). Heel tendon of the gastrocnemius and soleus muscles of the leg.

acid [L. *acidus*, sour]. Any substance releasing H ion and neutralizing basic substances; a pH less than 7.00.

acid-base balance. State of equilibrium between acids and alkalies so that the H^+ concentration of the arterial blood is maintained between pH 7.35–7.45.

acid′ification [L. *acidium*, acid, + *factus*, to make]. Process of making acidic; becoming sour.

acidosis (as-ĭ-dō′sis) [″ + G. -*osis*, condition]. Decrease of alkali in proportion to acid, or lowering of pH of extracellular fluid, caused by excess acid or loss of base.

acinus (as′i-nus) (pl. *acini*) [L. grape]. Smallest division of a gland; a saclike group of secretory cells surrounding a cavity.

acne (ak′nē) [G. *acme*, point]. Chronic inflammatory disease of the sebaceous glands and hair follicles of the skin.

acoustic (a-koos′tik) [G. *akoustikōs*, hearing]. Pert. to sound or to the sense of hearing. The cochlear portion of the eighth cranial nerve.

acromegaly (ak-ro-meg′ă-lĭ [G. *akros*, extreme, + *megas*, big]. Enlargement of the extremities (hands, feet) and jaw because of hypersecretion of the somatotrophic or growth hormone from the anterior pituitary, after full growth has been achieved.

actin (ak′tĭn). A muscle protein responsible for the shortening of the muscle when it contracts. Forms a combination with the protein *myosin* when muscle contracts.

ac′tion potential (pō-těn′shăl) [L. *actio*, from *agere*, to do + *potentia*, power]. The measurable electrical changes associated with conduction of a nerve impulse or contraction of a muscle.

ac′tivator. Substances converting inactive materials to active ones, or causing activity.

ac′tive transport′ [L. *trans*, across, + *porta*, to carry]. Using carriers, energy (ATP), and enzymes of cell to cause a substance to cross a membrane.

acu′ity [L. *acuere*, to sharpen]. Clearness or sharpness, usually in reference to vision or hearing.

adap′tation [L. *adaptare*, to adjust]. The ability of an organism to adjust to environmental change; the ability of the eye to adjust to various intensities of light by change in size of the pupil.

adduction (ăd-dŭk′shŭn) [L. *ad*, toward, + *ducere*, to bring]. Movement toward the median plane of the body.

adenine (ad′ĕ-nēn). A nitrogenous base found in DNA and RNA (nucleic acids).

adenohypophysis (ad-ē-no-hi-poph′i-sis) [G. *adēn*, gland + *ypo*, under, + *physis*, growth]. The glandular lobes of the hypophysis or pituitary gland. Derived from the mouth, and it includes the anterior lobe *(pars distalis)* and intermediate lobe *(pars intermedia)*.

adenoid (ad′ĕ-noid) [G. *adenoeides*, glandular]. Lymphoid tissue; the pharyngeal tonsil.

adenosine triphosphate (ad′ĕ-nō-sen). The major source of cellular energy; found in all cells. Composed of a nitrogenous base *(adenine)*, ribose sugar, and 3 phosphoric acid radicals. Abbr: **ATP.**

adhe′sive [Fr. *adhesif*]. Sticky; that which causes two materials to adhere.

ad′ipose [L. *adeps*, fat]. Refers to a type of connective tissue composed of fat containing cells; fatty.

adolescence (ad-o-les′ens) [L. *adolescere*, to grow up]. The period from the beginning of puberty until maturity; variable among individuals.

adrenal (ăd-rē′năl) [L. *ad*, to, + *ren*, kidney]. An internally secreting gland located above the kidney. Outer *cortex* produces hormones essential to life; inner *medulla* not essential to life.

adrenalin (ă-dren′ă-lin) [" + " + *in*, within]. Proprietary name for *epinephrine*, the main hormone of the adrenal medulla.

adrenergic (ad-ren-er′jik) [" + " + G. *ergon*, work]. Term used to describe nerve fibers that release *norepinephrine* (noradrenalin or sympathin) at their endings when stimulated.

adsorption [" + *sorbere*, to suck in]. "Sticking together" of a gas, liquid, or dissolved substance on a surface. No bonding.

adventitia (ad′ven-tish′ē-ă) [L. *adventitius*, coming from abroad]. The outermost covering of an organ or structure.

aerobic (ā-er-ō′bik) [G. *aer*, air + *bios*, life]. Requiring the presence of oxygen for life or function.

af′ferent [L. *ad*, to, + *ferre*, to bear]. Carrying toward a center or toward some specific reference point, e.g., afferent glomerular vessels, or afferent nerves to the brain.

agglutinin (ă-glū′tĭ-nĭn) [L. *agglutinare,* to glue a thing]. An antibody in the red blood plasma capable of causing clumping of specific antigens on red blood cells.

agglutinogen (ă-glū-tĭn′ō-gĕn) [″ + G. *gennan,* to produce]. An antigen on red blood cells stimulating the production of a specific agglutinin.

ag′onist [G. *agōn,* a contest]. The contracting muscle; a muscle whose contraction is permitted by relaxation of a muscle (antagonist) having the opposite action to the agonist.

albumin (al-bū′min) [L. *albus,* white; *albumen,* coagulated egg white]. One of a group of simple proteins in both animal and vegetable cells or fluids.

aldehyde (al′de-hīd) [*al,* abbr. alcohol, + *dehyd,* abbr. dehydrogenatum, alcohol deprived of H^+]. A compound containing the CHO group; derived by oxidation of a primary alcohol.

aldosterone (al-dos′tĕr′ōn). The most active mineralocorticoid secreted by the cortex of the adrenal gland; regulates metabolism of sodium, potassium, and chloride.

-algia (al-je-a) [G. *algeis,* sense of pain]. Suffix: denoting pain.

alimen′tary [L. *alimentum,* nourishment]. Pert. to nutrition.

alkali [Arabic *al-qili,* ashes of salt wort]. A substance that can neutralize acids; combines with acid to form soaps; basic in reaction (pH).

alkali(ne) reserve. The amount of base in the blood, mainly bicarbonates, available to neutralize acids.

alkalo′sis [″ + G. *-osis,* condition of]. Excessive amount of alkali, or rise of pH of extracellular fluid caused by excess base or loss of acid.

allantois (a-lan′tō-ĭs) [G. *allanto,* sausage, + *eidos,* resemblance]. A kind of long bladder located between the chorion and amnion of the fetus; its blood vessels contribute to the formation of the umbilical vessels.

allele (ă-lēl′) [G. *allēlōn,* reciprocally]. One of two or more genes occurring at the same position on a specific pair of chromosomes and controlling the expression of a given characteristic.

al′lergy [G. *allos,* other, + *ergeia,* work]. A condition characterized by reaction of body tissues to specific antigens without production of immunity.

allograft (al′ō-graft) [″ + L. *graphium,* grafting knife]. Grafts or transplants between genetically dissimilar members of the same species.

alveolus (al-vē′ō-lus) (pl. *alve′oli*) [L. small hollow or cavity]. A little hollow; an air sac of the lungs; a socket of a tooth.

ame′boid [G. *amoibē,* change, + *eidos,* resemblance]. Like an ameba; ameboid movements imply cellular motion by use of *pseudopods* (false feet or cytoplasmic outflows).

amenorrhea (a-men-ō-rē-ă) [G. *a-,* without, + *mēn,* month, + *rein,* to flow]. Absence or suppression of menstruation.

amine (ă-mēn′). A nitrogen containing organic compound with one or more hydrogens of ammonia (NH_3) replaced by organic radicals.

amino acid (ă-mē′no). Unit of structure of proteins; a compound containing both an amine group ($-NH_2$), and a carboxyl group (-COOH). Two groups: *essential,* not synthesized in body or not produced in amounts necessary for normal function; *nonessential,* synthesized in body.

am'nion (am'nĭ-on) [G. little lamb]. The inner fetal membrane that holds the fetus suspended in amniotic fluid. By 8 weeks it fuses with the *chorion* to form the *bag of waters* or *caul.*

amorphous (a-mor'fus) [G. *a-*, without, + *morphē*, form]. Without definite shape or structure.

amphiarthrosis (am-fi-ar-thrō'sis) [G. *amphi*, on both sides, + *arthrosis*, joint]. A form of articulation in which the mobility in all directions is slight.

ampulla (ăm-pŭl'a) [L. little jar]. A saclike dilation of a duct or canal.

anabolism (an-ab'ō-lizm) [G. *anabolē*, a building up]. Synthetic or constructive chemical reactions.

anaerobic (an-ā-er-ōb'ic) [G. *an-*, without, + *aer*, air, + *bios*, life]. Not requiring oxygen for life or function.

anaphase (an'a-fāz) [G. *ana*, up, + *phainein*, to appear]. A stage in mitosis characterized by separation and movement of chromosomes toward poles of the dividing cell.

anastamosis (a-nas-to-mō'sis) [G. opening]. An opening of one vessel into another, or the joining of parts to form a passage between any two spaces or organs.

androgen (ăn'drō-jĕn) [G. *anēr, andros*, man, + *gen*, to produce]. A substance stimulating or producing male characteristics, as testosterone, the male sex hormone.

anemia (an-e'mĭ-a) [G. *an*, without, + *aima*, blood]. A condition where there is reduction of circulating red blood cells or of hemoglobin.

anesthesia (an-es-the'zĭ-ă) [G. *an-*, without, + *aisthesis*, sensation]. Partial or complete loss of sensation with or without loss of consciousness.

aneurysm (an'ū-rism) [G. *aneurysma*, a widening]. Abnormal dilation of a blood vessel due to weakness of the vessel wall or to a congenital defect.

angina (ănjī'na) [L. *angere*, to choke]. Any disease characterized by suffocation or choking. *Angina pectoris* occurs when the heart muscle is deprived of blood flow and includes severe pain referred to the chest and left arm.

angiotensin (an-jĭ-ō-ten'sin) [G. *angeion*, vessel, + L. *tensio*, tension]. A pressor substance that increases arterial muscular tone, producing vasoconstriction and a rise of blood pressure. (Formerly called *angiotonin* or *hypertensin*.)

Angström unit (ong'strum) [Anders J. Angstrom, Swedish physicist, 1814-1874]. An international unit of wave length measurement. $\frac{1}{10}$ millionth of a millimeter, or $\frac{1}{250}$ millionth of an inch. Abbr: Å or A.

anion (an'ī-on) [G. *ana*, up, + *ion*, going]. A negatively charged ion or radical. In electrolysis, anions go toward the anode.

anisocytosis (an-ĭ-sō-sī-tō'sis) [G. *anisos*, unequal, + *kytos*, cell, + *-osis*, condition]. An abnormal condition with inequality in size of cells, esp. erythrocytes.

ann'ulus [L]. a fibrous ring surrounding an opening.

anoxemia (an-oks-ē'mĭ-ă) [" + oxygen + "]. Deficiency of oxygen in the arterial blood. Also, *hypoxemia.*

anoxia (an-ox'ĭ-ă) [" + oxia, oxygen]. Deficiency of oxygen in body tissues. Also, *hypoxia.*

antag'onist [G. *antagōnizesthai*, to struggle against]. That which counteracts or has the opposite action of something else *(agonist)*; esp. muscular contraction.

antenatal (ăn-tē-nā′tal) [L. *ante*, before, + *natus*, birth]. Occurring before birth.

ante′rior [L.] Before, or in front of, or the ventral (belly) portion.

antibiotic (ăn-tĭ-bī-ŏt′ik) [G. *anti*, against, + *bios*, life]. A substance produced by bacteria, molds, and other fungi, that has power to inhibit growth of or destroy other organisms, esp. bacteria.

antibody (an′tĭ-bod-ē). A protein (usually a globulin) developed against a specific substance (antigen), whether the antigen is foreign or not.

antigen (an′tĭ-jēn) [″ + *gennan*, to produce]. A substance causing production of antibodies.

antitox′in [″ + G. *toxikon*, poison]. An antibody capable of neutralizing a specific chemical (toxin) produced by microorganisms.

an′trum (pl. *antra*) [G. *antrom*, cavity]. Any nearly closed chamber or cavity, e.g., the antrum of the stomach *(pylorus)*.

anus (ā′nŭs) [L.]. The outlet of the rectum lying in the fold between the buttocks.

aorta [G. *aorte*, aorta]. The main trunk of the arterial system of the body arising from the left ventricle of the heart. It has a thoracic portion (ascending, arch, and descending subdivisions) and an abdominal portion (part of descending aorta). Branches from these supply the body with arterial (oxygenated) blood.

aperture (ap′er-tūr) [L. *apertura*, opening]. An opening or orifice.

apex (ā′peks) (pl. *apices*) [L. apex, tip]. Summit or peak; the narrow portion or tip of an organ shaped in a pyramidal form.

Apgar score [Virgina Apgar, M.D.]. System of scoring an infant's physiological condition at 1- and 5-minute intervals after birth.

aphasia (ă-fā′zĭ-ă) [G. *a-*, not, + *phasis*, speaking]. Loss of verbal comprehension, or inability to express oneself properly through speech.

apnea (ap′nē-a) [G. ″ + *pnoē*, breath]. Temporary stoppage or cessation of breathing.

apocrine (ap′ō-krēn, -krĭn) [G. *apo*, from, + *krinein*, to separate]. A method of production of a secretion by an externally secreting gland in which the cells lose part of their cytoplasm in forming the secretion, e.g., mammary gland, apocrine sweat glands.

aponeurosis (ăp-ō-nū-rō′sis) [G. *apo*, from, + *neuron*, sinew]. A white, flat, fibrous sheet of connective tissue that attaches muscle to bone or other tissues at their origin or insertion.

appendage (ă-pĕn′dĭj) (pl. *appendices*) [L. *appendere*, hang to]. Any part attached or appended to a larger part, as a limb or the appendix.

apposition (ap-ō-zĭ-shun) [L. *ad*, to, + *ponere*, to place]. A fitting together or contact of two substances, usually at a surface, e.g., appositional growth of cartilage by addition of new cartilage on the surface.

aqueduct (ak′we-dukt) [L. *aqua*, water, + *ductus*, duct]. A passage or canal to convey fluids; cerebrospinal fluid flows from the third to fourth ventricle through the cerebral aqueduct (of Sylvius).

aqueous (ā′kwē-ŭs) [L. *aqua*, water]. Watery; having the nature of water. Aqueous humor.

a. humor [L. fluid]. Watery fluid in the anterior and posterior chambers of the eye.

arachnoid (ă-rak'noid) [G. *arachne*, spider, + *eidos*, form]. The central membrane covering the brain and spinal cord, having the appearance of a spider web.

arborization (ar-bor-ĭ-zā'shun) [L. *arbor*, tree]. A treelike structure; terminations of nerve fibers and arterioles.

ar'bor vi'tae [" + *vita*, life]. In the cerebellum the treelike outline of the internal white matter; a series of branching ridges on the uterine cervix.

areola (ă-rē'ō-lă) (pl. areolae) [L. a small space]. A small cavity in a tissue; around the nipple, the ringlike pigmented or red area; the part of the iris enclosing the pupil.

arrhythmia (ă-rith'mĭ-ă) [G. *a-*, not, + *rhythmos*, rhythm]. Lack of rhythm; irregularity.

arteriole (ar-tē'rĭ-ōl) [L. *arteriola*, small artery]. A small or tiny artery; its distal end leads into a capillary or capillary network.

arteriosclerosis [" + G. *sklērōsis*, hardening]. General term referring to many conditions where there is thickening, hardening, and loss of elasticity of the walls of blood vessels.

artery (ar'ter-ĭ) (pl. arteries) [G. *arteria*, windpipe]. Any vessel carrying blood away from the heart to the tissues. Usually refers to the larger vessels.

arthritis (ar-thrī'tis) (pl. *arthritides*) [G. *arthron*, joint, + *itis*, inflammation]. Inflammation of a joint, usually accompanied by pain and frequently structural changes.

arthrosis [" + *-osis*, increased]. A degenerative condition of a joint; a joint.

artic'ulate [L. *articulatus*, jointed]. To join together as in a joint; to speak clearly.

artic'ulation [L.]. A connection between bones; a joint.

-ase. Suffix used in forming the name of an enzyme; added to the part of or name of substance on which it acts.

asphyxia (ăs-fĭk'sĭ-ă) [" + *sphyxis*, pulse]. Decreased amount of O_2 and an increased amount of CO_2 in the body as a result of some interference with respiration.

associa'tion. Uniting or joining together; coordination with another structure, idea; a neuron lying between and joining an afferent and efferent neuron.

asthenia (as-thē'nĭ-ă) [" + *sthenos*, strength]. Loss or lack of strength; weakness in muscular or cerebellar disease.

athero- (ath-er-ō) [G. *athērē*, porridge]. Prefix: referring to lipid (fatty) substances.

atherosclerosis. Arteriosclerosis characterized by lipid deposits in the inner layer of blood vessel walls (chiefly arteries).

asthma (as'mă) [G. panting]. Intermittent severe difficulty of breathing accompanied by wheezing and cough; due to spasms or swelling of the bronchioles.

astigmatism (a-stig'mă-tism) [" + *stigma*, point, + *ismos*, condition of]. Refractive surfaces of the eyeball are different in one or more planes; usually due to unequal curvature of the cornea and lens in vertical and horizontal planes.

astrocyte (as'trō-sīt) [G. *astron*, star, + *kytos*, cell]. A star-shaped neuroglial cell possessing many branching processes.

ataxia (ă-taks'ĭ-ă) [G. lack of order]. Muscular incoordination when voluntary muscular movements are attempted.

atlas. The first cervical vertebra by which the head articulates with the occipital bone. Named for Atlas who was supposed to have supported the world on his shoulders.

atom [G. *atomos,* indivisible]. The smallest particle of an element that can take part in a chemical change, keeping its identity, and which cannot be further divided without changing its structure; composed of protons, neutrons, electrons.

atre'sia [" + *trēsis,* a perforation]. Congenital absence or pathological closure of a normal opening; degeneration of ovarian follicles.

ATP. Abbr. for *adenosine triphosphate.*

atrium (ā'tri-um) (pl. *atria*) [L. corridor]. A cavity; one of the two upper chambers of the heart; tympanic cavity of the ear; a space that opens into the air sacs of the lungs.

at'rophy [G. " + *trophe,* nourishment]. Wasting or decrease in size of a part resulting from lack of nutrition, loss of nerve supply, or failure to use the part.

aud'itory [L. *auditiō,* hearing]. Pert. to the sense of hearing.

auditory nerve. Eighth cranial nerve, sensory. Also, *vestibulocochlear* n.

auricle (aw'rĭ-kl) [L. *auricula,* the ear]. The outer ear flap on the side of the head; a small conical portion of the upper chambers of the heart, *NOT* synonomous with atrium.

auscultation (aws-kul-tā'shun) [L. *auscultāre,* listen to]. Listening for sounds in body cavities, esp. chest and abdomen, to detect or judge some abnormal condition.

autograft [G. *autos,* self, + L. *graphium,* grafting knife]. A graft or transplant taken from one part of the body to fill in another part on the same person.

autoimmune [" + L. *immunis,* safe]. A condition in which antibodies are produced against a person's own tissue.

autonomic (aw-tō-nom'ik) [" + *nomus,* law]. Self-controlling; refers to the involuntary or self-regulating portion of the peripheral nervous system.

autosome (aw'to-sōm) [" + *soma,* body]. Any of the paired chromosomes other than the sex (X and Y) chromosomes.

axial (aks'ĭ-al) [L. *axis,* lever]. Pert. to or situated in an axis.

axil'la [L. little pivot]. The armpit.

ax'is (pl. *axes*). A line running through the center of a body, or about which a part revolves. The second cervical vertebrae that has the dens *(odontoid process)* about which the atlas rotates.

ax'on [G. axis]. A neuron process that conducts impulses away from *(efferent)* the cell body.

balance. State of equilibirum; state in which the intake and output or concentrations of substances are approximately equal (homeostasis).

baroreceptors (baro-rē-sĕp'tor) [G. *baros,* weight + L. a receiver]. Receptor organs sensitive to pressure or stretching.

bar'rier. An obstacle or impediment.

bā'sal [G. *basis,* base]. The base of something; of primary importance; the lowest level to maintain function.

ba′sal metab′olism [G. *basis,* base, + *metabole,* change]. The amount of energy needed for maintenance of life when the subject is at physical, emotional, and digestive rest.

ba′sal metab′olic rate [G.]. The metabolic rate as measured under so-called *basal conditions,* expressed in large Calories/M² of body surface/hr. Abbr: **BMR.**

bāse [G. *basis,* base]. The lower or broader part of anything; an alkali, any substance that will neutralize an acid.

benign (bē-nīn′) [L. *benignus,* mild]. Mild; not malignant.

beta- (bā′tă) [G. β, Second letter of alphabet]. Prefix: to note isomeric position or variety in compounds of substituted groups.

beta-oxidation (ŏk′sĭ-dā′shŭn) [″ + *oxys,* sour, pungent]. A metabolic scheme that catabolizes fatty acids by removal of two carbon units.

biceps (bī′sĕps) [L. *bis,* two, + *caput,* head]. A muscle with two heads.

bifid (bī′fid) [″ + *findere,* to cleave]. Cleft or split into two parts.

bilirubin (bil-ĭ′ru-bin) [L. *bilis,* bile, + *ruber,* red]. The orange or yellow pigment in bile; derived from breakdown of heme ($C_{33}H_{36}O_6N_4$).

biliverdin (bil-ı-ver′din. [″ + G. *viridis,* green]. The green pigment of bile formed by oxidation of bilirubin ($C_{33}H_{34}O_6N_4$).

bio- [G. *bios,* life]. Prefix: life.

-blast [G. *blastos,* germ or bud]. Suffix: immature or primitive.

blastocele (blas′tō-sēl) [″ + *koilos,* hollow]. An embryonic stage of development, the cavity of the blastula.

blastocyst (blas′tō-sist) [″ + *kystis,* bag]. Stage in embryonic development that follows the morula; also called the *blastodermic vesicle,* it is a hollow mass of cells.

blastomere (blas′to-mēr) [″ + *meros,* a part]. One of the cells resulting from the segmentation or cleavage of a fertilized ovum; the cells composing the *morula.*

blastula (blas′tu-lă). A stage in the early development of an ovum that has a hollow sphere of cells enclosing the *blastocele.* Also, *blastocyst.*

blind spot [A. S. *blind,* unable to see]. Area in the retina in which there are no light receptor cells (no rods nor cones); point where optic nerve enters the eye *(optic disk).*

blood-brain bar′rier [A. S. *blōd,* blood, + *braegen,* cranial portion CNS, + *barrier,* an obstacle or impediment]. Special mechanism that prevents the passage of certain substances, such as dyes or toxins, from the blood to the cerebrospinal fluid.

blood pressure. The pressures measured on a peripheral artery during contraction and relaxation of the heart.

    b.p., diastolic. The pressure remaining in the arteries during relaxation of the left ventricle.

    b.p., systolic. The peak pressure created by contraction of the left ventricle of the heart, and the resistance to flow offered by the blood vessels.

bo′lus [G. a mass]. A mass of food ready for swallowing.

bone marrow [A. S. *bōn,* bone, + *mearh,* marrow]. Soft tissue within the hollow of certain bones of the skeleton; responsible for production of certain blood cells.

booster (boos'ter) [colloq. *bouse*, to haul up]. An additional dose of an immunizing agent given to maintain or increase the protection or immunity originally conferred.

-borne (born, [*bore*, to carry]. Refers to types of methods of transmission of communicable diseases, e.g., airborne, waterborne, foodborne, etc.

Bowman's capsule [Sir Wm. Bowman, Eng. phys., 1816–1892, + L. *capsula*, small box]. The capsule containing the glomeruli of the kidney; the cuplike depressions on the expanded ends of the renal tubules or nephrons that surround the capillary tufts *(glomeruli)*.

brachium (brā'kĭ-um) [G. *brachiōn*, arm]. The upper arm from shoulder to elbow.

brady- [G. *bradys*, slow]. Prefix: slow, as in *bradycardia* or slow heart rate.

breathe (brēth) [A. S. *braeth*, odor]. To take in and release air from the lungs and respiratory system; *respire*.

bronchi (bron'kī) (sing. bronchus) [G. *bronchos*, windpipe]. The first two divisions of the trachea that penetrate the lungs and terminate in the bronchioles; air tubes to and from the lungs.

bronchiectasis (bron-kĭ-ek'tă-sis) [" + *ektasis*, dilatation]. Condition characterized by dilatation of the bronchus(i) with secretion of large amounts of foul substance.

bronchiole (bron'kĭ-ōl) [L. *bronchiolus*, air passage]. A smaller division of the bronchi that lead to the alveolar ducts to the alveoli.

bucca (buk'a) [L. mouth, cheek]. The mouth; hollow part of the cheek.

buffer (bŭf'ĕr) [Fr. *buffe*, blow]. A substance preserving pH upon addition of acids or bases; reduces the change in H⁺ concentration. *b. pair, b. system*, the combination of a weak acid and its conjugate base, which work together to maintain pH nearly constant.

bul'b [G. *bolbos*, bulb-shaped root]. An expansion of an organ, canal, or vessel.

bulk [M. E. *bulke*, a heap]. Size, mass, or volume; the main body of something.

bun'dle of His [M. E. *bundel*, bind, + W. *His*, Gr. anat., 1863–1934]. A bundle of fibers of the impulse-conducting system that initiates and controls contraction of the heart muscle. It extends from the A-V node and becomes continuous with the Purkinje fibers of the ventricles.

bur'sa [G. a leather sac]. A sac or cavity filled with a fluid that reduces friction between tendon and ligament or bone, or other structures where friction may occur.

buttocks (but'uks) [M. E. *butte*, thick end]. The gluteal muscle prominence, the "rump" or "seat."

-calcemia (kal-sē'mĭ-ă) [L. *calcarius*, pert. to calcium or lime, + G. *amia*, blood]. The calcium level of the blood.

calcification (kal-sif'ĭk-a shun) [L. *calx*, lime, + *ferre*, to make]. Deposit of lime salts in bones and other tissues.

calcitonin (kal-sĭ-tōn'ĭn) [G. *calci*, pert. to calcium + *tonos*, tone]. A hormone secreted by the thyroid gland that aids in regulating calcium-phosphorus metabolism.

calculus (kal′kū-lus) (pl. calculi) [L. pebble]. A "stone" formed within the body usually composed of mineral salts.

cal′lus [L. hard]. Circumscribed area of the horny layer of the skin that has thickened; osseous material formed between the ends of a fractured bone.

calorie, small (kăl′ŏ-rē) [L. *calor,* heat]. A unit of heat; amount of heat required to raise temperature of 1 gram of water 1°C. Abbr: **cal.**

Calorie, large [L.]. A large calorie or kilocalorie; 1000 times as large as a calorie; amount of heat required to raise temperature of 1 kilogram of water 1°C. (distinguished from small calorie by a capital C). Abbr: **Cal.,** or **kcal.**

calyx (kā′lix) (pl. *calyces*) [G. *kalix,* cup]. Cuplike divisions of the kidney pelvis.

canal (ka-nal) [L. *canā′lis,* canal]. A narrow channel, tube, duct, or groove.

canaliculus (kan-ă-lĭk′ū-lus). A tiny canal or channel.

can′cellous [L. *cancellus,* a grating]. Having a latticework or reticular structure, as the spongy tissue of bone.

cancer (kan′ser) [L. a crab; ulcer]. A malignant tumor or neoplasm (new growth); a carcinoma or sarcoma.

capillary (kap′ĭ-lā-rē) (pl. *capillaries*) [L. *capillaris,* hairlike]. Minute blood vessels that connect the arterioles with the venules to carry blood; minute lymphatic ducts. Capillary *networks* allow passage of oxygen and nutrients from the blood to the tissues, and of wastes from the tissues into the blood.

capitulum (ka-pĭt′ū-lum) [L. *caput,* head]. A small rounded end of a bone that articulates with another bone.

ca′put [L.]. The head; upper part of an organ.

carbaminohemoglobin (karb′ă-mē-nō-hē′mō-glō-bin). The compound formed by the combination of carbon dioxide with the amine groups of the globin molecule of hemoglobin.

carbohyd′rates (kar-bō-hĭd′ratz) [L. *carbo,* carbon, + G. *ydōr′* water]. A class of organic compounds composed of C, H, and O, with H and O usually in the ratio of 2:1; sugars, starches and cellulose; the *monosaccharides, disaccharides,* and *polysaccharides.*

car′bon [L. *carbo,* carbon or coal]. Constituent of organic compounds, found in all living things; combines with H, O, N to make life possible.

carbon dioxide. A colorless gas, heavier than air; produced by complete combustion, fermentation, or decomposition of carbon (compounds) found in air, and exhaled by all animals. CO_2.

carbonic anhydrase. An enzyme catalyzing the reaction of CO_2 and water to form *carbonic acid* (H_2CO_3) and vice versa; present primarily in red blood cells.

carcinogens (kar′sĭ-nō-jens′) [G. *karkinos,* cancer, + *genere,* to produce]. Substances known to cause cancer or neoplasms.

carcinoma (kar-sĭ-nō′mă) [″ + *oma,* tumor]. A malignant cancer or neoplasm of epithelial tissues.

cardiac cycle (sī′kle) [G. *kardia,* heart, + Fr. *cycle,* circle]. The period from the beginning of one heart beat to the next beat, includes systole and diastole.

cardiac output. Amount of blood ejected from one ventricle per minute.

caries (ka′rēz) [L. rottenness]. Tooth or bone decay.

carotene (car′ō-ten) [G. *karōton,* carrot]. A yellow crystalline pigment obtained from yellow vegetables; stored and converted to *vitamin A* in the liver.

carpus (kar′pus) [G. *karpos,* wrist]. The eight bones of a wrist. Also, *carpals.*

carrier [Fr. *carier,* to bear]. A large molecule transporting substances across cell membranes in *active transport.*

carotid (kă-rot′ĭd) [G. *karōtides,* from *karos,* heavy with sleep]. Arteries providing the main blood supply to the head and neck; pressure on them may produce unconsciousness.

cartilage (kar′til-ăj) [L. *cartilagō,* gristle]. A form of connective tissue usually with no blood nor nerve supply of its own; it is firm, elastic, and a semiopaque bluish-white or gray; cells lie in cavities called *lacunae.*

catabolism (kă-tab′ō-lism) [G. *katabolē,* a casting down]. The destructive phase of metabolism; breaking down of complex chemical compounds into simpler ones, usually with the release of energy.

catalyst (kat′ă-list) [G. *katalysis,* dissolution]. A substance that alters or speeds up the rates of chemical reactions without being itself altered in the process.

cat′aract [G. *katarraktēs,* a rushing down]. Opacity or loss of transparency of the lens of eye, its capsule, or both.

cation (kat′ĭ-on) [G. *katiōn,* descending]. A positively charged ion or radical.

cauda equina (kaw′dă e-kwin′a) [L. *cauda,* tail, + *equus,* horse]. The terminal portion of the roots of the spinal nerves and spinal cord below the first lumbar nerve: resembles a horse's tail.

cavernous body (kăv′ĕr-nŭs) [L. *caverna,* a hollow, + *corpus,* body]. One of several columns of tissue containing blood spaces in the dorsum of the clitoris or penis that aids in the erection of these organs.

cavity (kav′ĭ-tĭ) [L. *cavitas,* hollow]. A hollow space, such as in a body organ or a decayed tooth.

cecum (sēkŭm) [L. *caecum*]. The blind pouch that forms the first portion of the large intestine.

-cele [G. *hernia,* tumor]. Suffix: a swelling.

celiac (se′lĭ-ak) [G. *koilia,* belly]. Pert. to the abdominal region.

celom (sē′lom) [G. *koilōma,* a hollow]. The embryonic cavity between the layers of mesoderm that develops into the peritoneal, pleural, and pericardial cavities. Also **coelom.**

center (sen′ter) [G. *kentron,* middle]. Midpoint of a body; usually refers to a group of nerve cells in the brain or spinal cord; controls some specific activity.

centimeter [L. *centum,* a hundred, + *metron,* measure]. Abbr. **cm.** 2.5 cm equals 1 inch; 1 cm equals 1/100 meter, or about 2/5 inch (0.3937 inch).

centriole (sĕn′trĭ-ōl). A cell organelle that is involved in cell division; forms a spindle.

centromere (sen′trō-mēr). The structure at the junction of the two arms (chromatids) of a chromosome.

centrosome (sen′trō-sōm) [G. *kentron,* center, + *soma,* body]. An area of the cell cytoplasm lying near the nucleus that contains one or two *centrioles;* active during cell division.

cephalic (sef-ăl-ic) [G. *kephalē,* head]. Cranial, pert. to the head; superior in position.

cerebellum (ser-ē-bel′um) [L.]. A dorsal portion of the brain that is involved in coordinating activity of skeletal muscles, and in the coordination of voluntary muscular movements.

cerebrum (ser′ĕ-brum) [L.]. The upper largest part of the brain; contains the motor, and sensory areas, and the association areas that are concerned with the higher mental faculties.

cerumen (se-rū′men) [L. *cera*, wax]. The waxlike secretion in the external ear canal.

cervical (ser′vĭ-kal) [L. *cervicalis*, pert. to neck]. The neck region, or neck of an organ, e.g., **cervix** or neck of the uterus. The first 8 spinal nerves; the first 4 cervical spinal nerves forms the **c. plexus.** The first 7 vertebrae.

chambers (cham′ber) [G. *kamara*, vault]. A compartment or closed space, as the chambers of the heart.

chem′oreceptor. A sense organ or sensory nerve ending that is stimulated by a chemical substance or change, as in the taste buds.

chemotactic (kem-o-tak′tĭk) [G. *chēmeia*, chemistry, + *taxikos*, arranging]. Responding to a chemical stimulus by being attracted or repelled.

chiasm, chiasma (ki′azm) [G. *chiazen*, to mark with letter X]. A crossing or decussation; optic chiasm; the point of crossing of the optic nerve fibers.

chemotherapy (kem-o-ther′a-pi) [″ + *therapeia*, treatment]. Treatment of disease by use of chemical agents that have a specific effect on the causative microorganism or agent, usually without producing toxic effects on the person.

chloride shift (klō′rĭd) [G. *chlōrus*, green, + A. S. *sciftan*, to divide]. The movement of chloride ions into red blood cells as bicarbonate ions are formed and leave the cell during the reaction of CO_2 and water in the cell.

choana (kō′ă-nă) (pl. choanae) [G. *choane*, funnel]. A funnel-shaped opening; the communicating passageways between the nasal cavities and the pharynx; the posterior nares or nostrils.

cholecystokinin (ko-le-sĭs-tō-kī′nin) [G. *cholē*, bile, + *kystis*, cyst, *kinein*, to move]. A hormonelike chemical secreted by the duodenal mucosa that stimulates contraction of the gallbladder.

cholesterol (ko-les′ter′ol) [″ + *stereos*, solid]. A sterol (monohydric alcohol, $C_{27}H_{45}OH$) forming the basis of many lipid hormones in the body, e.g., sex hormones, adrenal corticoids; synthesized in the liver; a constituent in the bile.

cholinergic (kō′len-er-gik). Term applied to nerve endings that liberate acetylcholine.

chondroblast (kŏn′drō-blăst) [G. *chondros*, cartilage, + *blastos*, germ]. A primitive cell that forms cartilage.

chondrocyte (kŏn′drō-sīt) [G.]. A cartilage cell.

cholinesterase (kō-lĭn-ĕs′ter′ăs). Any enzyme that catalyzes the hydrolysis of choline esters (organic acid + an alcohol), e.g., *acetylcholinesterase* catalyzes the breakdown of *acetylcholine*.

chorion (kō′rĭ-ŏn) [G.]. Outer embryonic membrane of the blastocyst from which *villi* connect with the endometrium to give rise to the placenta.

chordae tendinae (kor′da tĕn′dĭn-ē) [G. *chorde*, cord, + L. *tendinōsus*, like a tendon]. Small tendinous cords that connect the edge of an atrioventricular valve to a papillary muscle, and prevent valve reversal.

chord'ee (kor-dē') [Fr. corded]. Downward painful curvature of the penis on erection as in *gonorrhea.*

chorea (ko-re'ā [G. *choreia,* dance]. Rapid, irregular, and wormlike movements of body parts; a nervous affliction.

chorioid (kō'rĭ-oid) [G. *chorioeidēs,* skinlike]. The middle vascular layer of the eyeball between the sclera and retina; a part of the *uvea.* Also, **choroid.**

choroid plexus (kō'roid plĕk'sŭs) [G. *chorioeidēs,* skinlike, + L. *plexus,* interwoven]. Vascular structures in the roofs of the four ventricles of the brain that produce *cerebrospinal fluid.*

chromatid (kro'mă-tĭd) [G. *chrōma,* color, + *tīd,* time]. One of the two bodies resulting from longitudinal separation of duplicated chromosomes, each goes to a different pole of the dividing cell.

chromatin (krō'ma-tin) [" + *in,* within]. Darkly staining substance within the nucleus of most cells; consists mainly of *DNA,* and is considered to be the physical basis of *heredity;* forms *chromosomes* during mitosis and meiosis.

chromosome (kro'mō-sōm) [" + *soma,* body]. A microscopic J- or V-shaped body developing from chromatin during cell division. Contains the gene or hereditary determinants of body characteristics. In human, total **46** (23 pairs).

chronaxie (kro'nak-sĭ) [G. *chronos,* time, + *axia,* value]. A value expressing sensitivity of nerve fibers to stimulation; the time that a current that is twice threshold value must last to cause depolarization.

chronic [G. *chronos,* duration]. Long drawn out with slow progress; a disease that is not acute.

chyle (kīl) [G. *chylos,* milklike]. The contents of intestinal lymph vessels. Consists mainly of absorbed products of fat digestion.

chyme (kīm) [G. *chymos,* juice]. The mixture of partially digested foods and digestive juices found in the stomach and small intestine.

cilia (sil'ĭ-ā) [L. *ciliaris,* pert. to eyelashes]. Motile hairlike projections from epithelial cells; eyelashes.

circadian (sir-kā'dĭ-en) [L. *circa,* about, + *dies,* day]. Describes events that repeat in a length of time approximating a 24-hour cycle.

circumduction (sir-kum-duk'shun) [L. *circum,* around + *ducere,* to lead]. Movement of a part in a circular direction, involves flexion, extension, abduction, adduction, and rotation.

cirrhosis (sĭ-rō'sĭs) [G. *kirros,* yellow, + *osis,* infection]. A chronic liver disease characterized by degenerative changes in the liver structure.

cister'na [L. a vessel]. A cavity or enlargement in a tube or vessel.

cleavage (kle'vej) [A. S. *cleofian,* to adhere]. Mitotic cell division of a fertilized egg; splitting a complex molecule into two or more simpler ones.

cleft [M. E. *clyft,* crevice]. A fissure; divided or split.

cleido (kli'do) [G. *kleis,* clavicle]. Prefix: pert. to the clavicle (collarbone).

clitoris (klĭ'to-rĭs, klit'ō-ris) [G. *kleitoris,* clitoris]. Female erectile genital organ that is homologous to the penis of the male.

clone (klōn) [G. *klon,* a cutting used for propagation]. A group of cells descended from a single cell; something reproduced exactly like its predecessor.

clot (klŏt) [A. S. *clott*]. The semisolid mass of fibrin threads plus trapped blood cells, which forms when the blood coagulates; to coagulate.

Co-A A dinucleotide that activates many substances in metabolic cycles.

coagulation [L. *coagulatiō*]. The process of clotting by which liquid blood is changed into a gel to aid in preventing blood loss through injured vessels.

coccygeal (kok-sij′ē-al) [G. *kokkyx*, coccyx]. Pert. to the coccyx, tailbone, last four fused spinal bones.

cochlea (kok′lē-ă) [G. *kochliās*, a spiral]. A cone-shaped winding tube that forms the portion of the inner ear that contains the organ of Corti, the receptor for hearing.

codon. The sequence of three nitrogenous bases of messenger RNA that specifies a given amino acid and its position in a protein.

coen′zyme [L. *co*, together, + G. *en*, with, + *zymē* leaven]. A nonprotein substance that activates enzymes or chemical compounds.

cofactor A substance essential to or necessary for the operation of another material such as an enzyme.

coitus (ko′i-tus) [L. a uniting]. Sexual intercourse between man and woman. Also, *coition, copulation.*

collagen (kol′ă-jen) [G. *kolla*, glue, + *gennan*, to produce]. A fibrous protein forming the main organic structures of connective tissues.

collateral (ko-lat′er-al) [L. *con*, together, + *lateralis*, pert. to a side]. Side branch of a nerve axon or blood vessel.

colloid (kol′oid) [G. *kollōdēs*, glutinous]. A system formed by large particles that remain suspended and do not settle. Size of particles lies between 1 and 100 nanometers.

colon [G. *kōlon*]. The portion of the large intestine from the end of the ileum to the rectum. It consists of the ascending, transverse, descending, and sigmoid portions.

co′ma [G. *kōma*, a deep sleep]. An abnormal deep stupor from which the person cannot be aroused by external stimuli.

combust′ [L. *comburere*, to burn up]. To burn or to oxidize.

compact′ bone [L. *compactus*, joined together]. The hard or dense part of bones containing subunits of structure called osteons or Haversian systems.

compensate [L. *cum*, with + *pensdāre*, to weigh]. To make up for a defect or deficiency.

compensatory. Serving to balance or offset; returning to a normal state.

compliance In the lung, ease of expanding.

component A constituent part.

compound [L. *componere*, to place together]. A substance composed of two or more parts and having properties different from its parts.

commissure (kŏm′ĭ-shūr) [L. *commissura*, a joining together]. Band of nerve fibers passing transversly over the midline in the central nervous system.

concave (kon′kāv) [L. *con*, with, + *cavus*, hollow]. Having a rounded hollow or depressed surface.

concentration (kon-sen-tra′shun) [L. *contratiō*, in the center]. Amount of a solute per volume of solvent, increase in strength of a fluid by evaporation or removal of water by other means.

concep′tion [L. *conceptiō*, a conceiving]. The point at which when a sperm unites with an ovum to initiate formation of a new individual; *fertilization.*

concha (kong'kǎ) [G. *kogchē,* shell]. A shell-shaped structure, as the nasal turbinates.

conduc'tion [L. *conducere,* to lead]. The transmission of a nerve impulse by exciting progressive segments of a nerve fiber; the transfer of electrons, ions, heat, or sound waves through a conducting media.

congenital (kon-jen'ĭt-al) [L. *congenitus,* born together]. Present at birth.

cone (kōn) [G. *kōnos,* cone]. A receptor cell in the retina concerned with color vision.

conjugate (kon'jū-gāt) [L. *con,* with, + *jugum,* yoke]. To pair or to join something with something else.

conjunctiva (con-junk-tī'vǎ) [L. " + *jungere,* to join]. Mucous membrane lining of the eyelid that is reflected onto the eyeball.

conscious (kon'shus) [L. *conscius,* aware]. Awake, being aware and having perception.

consolidate (kon-sal-ĭ-dā'shun) [L. *consolidāre,* to make firm]. To become solid or firm.

constric'tion [" + *stringere,* to draw]. The narrowing or becoming smaller in diameter as of a pupil of the eye or blood vessel.

commu'nicable disease. A disease that may be transmitted directly or indirectly from one person to another; due to an infectious agent, or toxic products produced by it.

continuity (kon-tĭ-nū'ĭ-tĭ) [L. *continuus,* continued]. The state of being intimately united, continuous, or held together.

contraception (kon-tra-sep'shun) [L. *contra,* against, + *conceptiō,* a conceiving]. The prevention of conception, impregnation, or implantation.

contrac'tion [L. *contractio,* a drawing up]. Shortening or tightening as of a muscle, or a reduction in size.

convection (kon-vek'shun) [L. *convehere,* to convey]. Transference of heat in liquids or gases by means of currents.

convergent (kon-ver'jent) [" + *vergere,* to incline]. Tending toward a common point; coming together.

conver'sion [L. *convertere,* to turn around]. Change from one state to another.

con'vex [L. *convexus,* vaulted, arched]. Having an outward curvature or bulge.

convul'sion [L. *convulsio,* a pulling together]. Paroxysms or involuntary muscular contractions and relaxations.

coordination (ko-or-din-a'shun) [L. *co,* together + *ordināre,* to arrange]. The working together of different body systems in a given process; the working together of muscles to produce a certain movement.

cope (kōp). The ability to effectively deal with and handle the stresses to which one is subjected.

copulation (kop-u-la'shun) [L. *copulātiō*]. Sexual intercourse.

cor'nea [L. *corneus,* horny]. The transparent anterior portion of the fibrous coat of the eye, continuous with the sclera.

corium (ko'rĭ-um) [G. *chorion,* skin]. The dermis or true skin that lies immediately under the epidermis; contains nerve endings, capillaries, and lymphatics, and is composed of connective tissue.

coronary (kor′o-na-rĭ) [L. *coronarius*, pert. to a circle or crown]. Blood vessels of the heart that supply blood to its walls.

cor′pus (pl. *corpora*) [L. body]. Any mass or body; the principal part of any organ.

cor′puscle [L. *corpusculum*, little body]. An encapsulated sensory nerve ending; any small rounded body; old term for blood cell, erythrocyte or leukocyte.

cor′pus luteum (lū′tē-um) [″ + *luteus*, yellow]. Yellow body of cells that fills the ovarian follicle in the stage following ovulation.

coro′na [G. *korōnē*, crown]. Any structure resembling a crown; **coronal** plane or section, same as *frontal*.

cor′tex (pl. *cortices*) [L. rind]. An outer portion or layer of an organ or structure.

cor′ticoid. One of many adrenal cortical steroid hormones. Also, *corticosteroid*.

corticospi′nal [″ + *spina*, thorn]. Pert. to cerebral cortex and spinal cord.

cos′tal [L. *costa*, rib]. Pert. to a rib.

countercurrent (kown′ter-ker′rent) [L. *contra*, against, + *currere*, to run]. When flow of gases or fluids in opposite directions in two closely spaced limbs of a bent tube.

crā′nium [L. from G. *kranion*, skull]. That portion of the skull that encloses the brain; consists of eight bones.

creatine (krē′ă-tĭn) [G. *kreas*, flesh]. A crystalline substance found in organs and body fluids; combines with phosphate and serves as a source of high-energy phosphate released in the anaerobic phase of muscle contraction.

crest [L. *crista*, tuft]. A ridge or long prominence, as on bone.

crista galli [″ + *galea*, helmet]. A crest or shelf on the ethmoid bone that attaches the *falx cerebri.*

criterion (pl. *criteria*) [G. *krites*, judge]. A standard or means of judging a condition or establishing a diagnosis.

crura (krōō′ra) (sing. *crus*) [L. legs]. The legs; a pair of long bands or masses resembling legs.

cross linkage. Chemical bonds linking fibrous molecules.

crypt (kript) [G. *kryptein*, to hide]. A small cavity or sac extending into an epithelial surface; a tubular gland.

crys′talloid [G. *krystallos*, clear ice, + *eidos*, form]. A substance that is capable of forming crystals, and that in solution diffuses rapidly through membranes; less than 1 nanometer in diameter; smaller than a *colloid.*

currettage (ku-ret′aj) [Fr. *curette*, a cleanser instrument]. Scraping of a cavity.

cuta′neous [L. *cutis*, skin]. Pert. to the skin.

cyanosis (sī-an-o′sis [G. *kyanos*, dark blue, + *-osis*, state of]. A condition of blue color being given to skin and mucous membranes by excessive amounts of reduced hemoglobin in the bloodstream.

cybernetics (sī′ber-net′iks) [G. *kybernetes*, helmsman]. The study of self-monitoring and regulating mechanisms for maintenance of near constant values of a function.

cycle (sī′kl) [G. *kyklos*, circle]. A series of events; something that has a predictable period of repetition; a sequence.

cystoscope (sīst′o-skōp) [G. *kystis*, bladder, sac, + *skopein*, to examine or view]. An instrument for examining or viewing the interior of the bladder.

-cyte (sīt) [G. *kytos,* cell]. Suffix denoting cell.

cyto- [G.]. Prefix indicating the cell.

cytochrome (si′to-krōm [″ + *chrōma,* color]. A yellow, iron-containing protein pigment that functions to transport electrons and/or hydrogens in cellular respiration.

cytokine′sis [″ + *kinēsis,* movement]. Division of cytoplasm in latter stages of mitosis.

cytoplasm (sī′to-plazm) [″ + *plasma,* matter]. The cellular substance between the membrane and nucleus of a cell.

damping. Steady decrease in amplitude of successive vibrations, as in the cochlea.

de-. Prefix; down; from.

deamination (dē-am-ĭ-na′shun). Enzymatic removal of an amine group (-NH₂) from an amino acid to form ammonia.

death (deth) [A. S. *death*]. Permanent cessation of all vital functions. Absence of life. A flat EEG for 48 hours is taken as a sign of loss of spontaneous brain activity and death.

decompression [″ + *compressio,* a squeezing together]. Release or removal of pressure from an organ or organism.

decussate (de-kus′āt) [L. *decussāre,* to cross, as an X]. To undergo crossing, or crossed.

defecation (def-ĕ-kā′shun) [L. *defaecāre,* to remove the dregs]. Evacuation or emptying fecal matter from the large bowel.

de′fect. A flow or imperfection.

deficit (def′e-sit) [L. *deficere,* to be lacking]. Lack of or lacking in; to be less than.

degeneration (di-jen′er-ā-shun) [L. *degenerāre,* to degenerate]. To "go bad" or deteriorate; to lose quality.

degrade (di-grāde) [L. *degradare,* to reduce in rank]. To tear down or break down, as to degrade a chemical compound.

dehydration (de-hī-drā′shun) [″ + G. *ydōr,* water]. Removal of water from something; the result of water removal.

den′drite [G. *dendritēs,* pert. to a tree]. A branched process of a neuron that conducts impulses to the cell body; form synapses.

dener′vate [L. *dē,* from, + G. *neuron,* nerve]. To block, remove, or cut the nerve supply to a structure.

de novo (dē-nō′vō) [L.]. Anew; again; once more.

dense [L. *densus,* compact]. Packed tightly together; in a bone, the outer layers of bony tissue that do not contain osteons.

deoxyribonucleic acid (dē-ok-sĭ-ri-bō-nu′klē-ik). Composed of nitrogenous bases, deoxyribose sugar, and a phosphate group. It is believed to be the site of determination of the body's hereditary characteristics. **DNA.**

depolarization (dē-pō-lar-ĭ-zā′shun) [″ + *polus,* pole]. Loss of polarity or the polarized state.

depressed (de-prest′). To push down; flattened or hollow; lowered in intensity; low in spirits.

depression (de-presh'un) [L. *depressiō*, a pressing down]. A lowered or hollowed region; lowering of a vital function; a mental state of dejection, lack of hope or absence of cheerfulness.

derivative (dĕ-riv'a-tiv) [L. *derivare*, to turn a stream from its channel]. A substance or structure originating from another substance or structure and having different properties than its source.

-derm- (durm) [L. *derma*, skin or covering]. A word used as a prefix or suffix referring to the skin or a covering.

derma (dermis). Pert. to the skin; the connective tissue layer of the skin under the epidermis or covering layer.

desquamate (des-kwam-āte) [" + L. *squamāre*, to scale off]. Shedding of the epidermal surface cells.

detoxify (de-toks'ĭ-fī) [" + G. *toxikon*, poison, + L. *facere*, to make]. To remove the toxic or poisonous quality of a substance.

development (Fr. *de'velopper*, to unwrap). Progress from embryonic to adult status; change from a less specialized to a more specialized state.

dextrose (deks'trōs) [L. *dexter*, right + *ose*, sugar]. Another name for glucose, a simple sugar (monosaccharide). Also, *grape sugar*.

di- [G.]. Prefix: twice or double.

dialysis (di-al'ĭ-sis) [" + *lysis*, loosening]. Passage of a diffusible solute through a membrane that restricts passage of colloids; separation of solutes by their rates of diffusion.

diameter (di-am'ĕ-ter) [" + *metron*, a measure]. A distance from any point on the periphery of a surface to the opposite point.

diapedesis (di-ă-ped-e'sis) [" + *pēdan*, to leap]. Movement of white blood cells through undamaged capillary walls.

diaphragm (di'ă-fram) [" + *phragma*, wall]. Any thin membrane; a rubber or plastic cup placed over the cervix of the uterus for contraceptive purposes; the muscle and connective tissue partition between chest and abdomen (muscle of respiration).

diaphysis (di-af'i-sis) [" + *plassein*, to grow]. The shaft or cylindrical part of a long bone.

diarrhea (di-ă-re'ă) [G. *dia*, through, + *rein*, to flow]. Passage of frequent watery stools: a symptom of gastrointestinal disturbance.

diastasis (di-as'tă-sis) [G. a separation]. In the cardiac cycle, the time when there is little change in length of muscle fibers and filling of the chamber is very slow. It is followed by contraction of the muscle of the chamber.

diastole (di-as'tō-le) [G. *diastellein*, to expand]. The relaxation phase of cardiac activity, when fibers are elongating and filling of a chamber is rapid.

diencephalon (dī-en-sef'ă-lon) [" + *egkephalos*, brain]. The second portion of the brain between telencephalon and mesencephalon; includes thalamus and hypothalamus.

differentiation (dif'er-en'shi-a'shun) [L. *dis*, apart, + *ferre*, to bear]. Acquiring a structure or function different from those originally present.

diffuse (dī-fūs') [" + *fundere*, to pour]. To spread, scatter, or move apart; usually caused by movement of molecules in a solution or suspension.

diffusion (dĭ-fu′zhun). Eventual even mixing of solutes and solvents as a result of motion of molecules.

digest [″ + *gerere*, to carry]. To break down foods using chemical or mechanical means; to convert foods to an absorbable state.

digitalis (dij-ĭ-tal′is) [L. *digitus*, finger]. The dried leaves of the purple foxglove plant; used in powdered form as a heart stimulant.

dilate (di′lāt) [L. *dilatāre*, to expand]. To become larger in size or diameter.

dilute (di-loot′) [L. *dilutus*, to wash away]. To thin down or weaken, by addition of water or other liquids.

dimension (dĭ-men′shun) [Fr. *dimensio*, a measuring]. A measurable extent, such as length, width; extent or size; importance.

diopter (dī-op′ter) [G. *dioptron*, something that can be seen through]. The refractive power of a lens with a focal distance of 1 meter.

dip′loe [G. a fold]. The spongy bone between the inner and outer tables (layers) of dense bony tissue in the skull bones.

diploid (dĭp′loyd). Cell having twice the number of chromosomes present in the egg or sperm of a given species.

disaccharide (dī-sak′i-rid) [G. *dis*, two, + *sakcharon*, sugar]. A sugar composed of two simple sugar molecules.

dissect (dis-sekt′) [L. *dissecare*, to cut up]. To separate tissues or organs of a cadaver for study.

dissociate (dis-so′si-āt) [″ + *sociatiō*, union]. To separate into components or parts.

dissolve (dĭ-zolv) [L. *dissolvere*, to dissolve]. To make liquid, liquefy or melt by placing in water or other liquids.

distal (dis′tal) [L. *distāre*, to be distant]. Fartherest from the center of the body or point of attachment; opposite of proximal.

diuresis (di-u-re′sis) [G. *dia*, through, + *ourein*, to urinate]. Secretion and passage of abnormally large amounts of urine.

diuretic (di-u-ret′ik). A substance causing diuresis.

diverticulum (di-ver-tik′u-lum) [L. *diverticulāre*, to turn aside]. An outpocketing, sac, or pouch in the wall of a canal or hollow organ.

dominance (dom′i-nense) [L. *dominare*, to rule]. In inheritance of characteristics, the effects of one allele overshadow the effects of the other and the character determined by the dominant (stronger) gene prevails.

donor (dō′ner) [Fr. *doneor*, to give]. One who gives or donates (such as blood or organs).

dor′sal [L. *dorsum*, back]. Pert. to the upper or back portion of the body.

drug [Fr. *drogue*]. A medicinal substance used in the treatment of disorder or disease.

duct [L. *ducere*, to lead]. A tubular vessel or channel carrying secretions from a gland onto an epithelial surface.

ductus. A channel from one organ, or part of an organ, to another.

duplication (doo′plĭ-cā-shun) [L. *duplicare*, to double]. To double something; to form a copy.

dura (du′ră) [L. *durus*, hard]. Hard or tough; *dura mater*, the tough outer membrane around the brain and spinal cord.

dwarf (M. E. *dwerf*, little). An abnormally small, short, or disproportioned person.

dysmenorrhea (dis-men-o-re′a) [L. *dys*, bad, difficult, or painful, + *men*, mouth + *rein*, to flow]. Painful or difficult menstruation.

dyspnea (disp-ne′a) [″ + *pnoē*, breathing]. Labored or difficult breathing.

dystrophy (dis′tro-fĭ) [″ + G. *trephein*, to nourish]. Degeneration of an organ resulting from poor nutrition, abnormal development, infection, or unknown causes.

ecchymosis (ek-ĭ-mo′sis) [G. *ek*, out, + *chymos*, juice, + *osis*, caused by, state of]. A blue-black, greenish-brown, or yellow area of hemorrhage in the skin.

ectoderm (ek′tō-derm) [G. *ekto*, outside, + *derma*, skin]. The outer layer of cells in the embryo, from which develops the nervous system, special senses, and certain endocrines.

-ectomy (ek′tō-my) [G. *ektomē*, a cutting out]. Suffix pert. to surgical removal of an organ or gland.

ectopic (ek-top′ik) [G. *ek*, out, + *topia*, place]. In an abnormal place or position.

eczema (ek′zĕ-mă) [G. *ekzein*, to boil out]. A chronic or acute inflammation of the skin characterized by the development of skin lesions.

edema (e-de′ma) [G. *oidema*, swelling]. Swelling due to increase of extracellular fluid volume; dropsy.

effect′or [L. *effectus*, accomplishing]. An organ that responds to stimulation; a gland or muscle cell.

ef′ferent [L. *ex*, out + *ferre*, to carry]. Carrying away from a center or specific point of reference.

egest (ē-jest′) [L. *egestus*, to discharge]. To rid the body of something. Also, *eliminate*.

ejaculation (ē-jak-u-la′shun) [L. *ejaculāri*, to throw out]. Ejection of semen from the male urethra during sexual excitement, or the discharge of secretions from vaginal glands.

elastance (ē-last′ance) [G. *elasticos*, elastic]. The power of elastic recoil after being stretched. Also, *elasticity*.

electro- (ē-lec′trō) [G. *elektron*, amber]. Prefix pert. to electrical phenomena of one sort or another.

electrolyte (e-lek′trō-līt) [″ + *lytos*, solution]. A substance that dissociates into electrically charged ions; a solution of conducting electricity because of the presence of ions.

elec′tron [G.]. A minute body or charge of negative electricity; a component of atoms.

el′ement [L. *elementum*, a rudiment]. A substance that cannot be separated into its components by conventional chemical means.

eliminate (e-lim′ĭ-nāte) [L. *ē*, out, + *limen*, threshold]. To rid the body of wastes by emptying of a hollow-organ; to expel.

em′bolus [G. plug]. A floating undissolved mass or bubble of gas brought into a blood vessel by the fluid flow.

embryo (em′brĭ-o) [G.]. The developing human from the *third* to *eighth weeks* after conception.

embryonic disc. The primitive two-layered structure formed about two weeks after fertilization, from which the embryo develops.

emeiocytosis (ēm-ē-ōsi-tō′sis) [G. *emein*, to vomit, + *cyte*, cell, + *osis*, state of]. Elimination of materials from a cellular vacuole by fusion of the vacuole with the cell membrane and release of contents to the outside of the cell.

emesis (em′ĕ-sis) [G. *emein*, to vomit]. Vomiting.

-emia (ē′mi-a) [G. *aima*, blood]. Suffix: blood.

emmetropic (em-me-trō′pik) [G. *emmetros*, in due measure, + *opsis*, sight]. Normal in vision.

emphysema (em-fi-se′mă) [G. *emphysan*, to inflate]. Condition resulting from rupture of expansion of alveoli of the lungs.

en- [G. *en*, in]. Prefix: in.

encephalo- (en-sef′a-lō) [G. *egkephalos*, brain]. Prefix pert. to brain or cerebrum.

en′dō- [G. *endon*, within]. Prefix: within; inner.

endocrine (ĕn′do-krīn, -krĭn) [″ + *krinein*, secrete]. internally secreting into the bloodstream.

endoderm (en′do-derm) [″ + derma, skin]. Inner layers of cells of the embryo; gives rise to *linings* of gut and all outpocketings of the gut (e.g., lungs, pancreas, liver).

endometrium (en-do-me′trĭ-um) [″ + *mētra*, uterus]. The mucous membrane lining the inner surface of the uterus.

endoplasmic reticulum (en-dō-plas′mik re-tĭk′ū-lum). The series of tubular structures found in the cytoplasm of cells; transports substances through cells.

endothelium (en-dō-the′lĭ-um) [″ + *thēlē*, nipple]. The simple squamous epithelium forming the inner lining of blood and lymphatic vessels, and the heart.

end product. The final product or waste of a digestive or metabolic process.

energy (en′er-ji) [″ + *ergon*, work]. Capacity to do work; heat or chemical bond capacity for work. Energy is seen as motion, heat, sound, etc.

engorged (en-gorjd′) [Fr. *engorger*, to obstruct or devour]. Swollen or distended as with blood.

engulf (in-gulf′) [Fr. *engolfer*, to swallow up]. To surround, take in, or swallow.

en′tero- [G. *enteron*, intestine]. Prefix pert. to the intestines.

enteroceptive (en-ter-ō-sep′tĭv) [″ + L. *capere*, to take]. Originating within body viscera.

environment (en-vī′rŏn-ment) [M. E. *environen*, about, + ment]. or [L. *in*, in, + *virer*, to turn]. The surroundings of a cell, organ, or organism.

enzyme (en-zīm) [L. *en*, in + *zymē*, leaven]. An organic catalyst causing alterations in rates of chemical reactions without being consumed in the reaction.

eosin (ē′o-sin) [G. *ēōs*, dawn]. A rose-colored dye used for staining cells and tissues. Something exhibiting a preference for the dye is called *eosinophilic*.

epi- [G.]. Prefix: upon, at, outside of, in addition to.

epididymis (ep-ĭ-did′ĭ-mis) [″ + *didymos*, testes]. A long convoluted tubule or duct resting on the testis and conveying sperm to the vas deferens.

epiglottis (ep-ĭ-glot′is) [″ + *glōttis*, glottis]. A leaf-shaped structure located over the superior end of the larynx. Aids in closing the larynx during swallowing.

epileptic (ep-ĭ-lep′tik) [G. *epilēptikos*, pert. to a seizure]. Pert. to a disturbance of consciousness occurring during epilepsy; an individual suffering from the disorder of epilepsy.

epiphysis (ĕ-pif′i-sĭs) [G. *epiphysis,* a growing upon]. One of the two ends of a long bone, separated from the diaphysis or shaft by a growth line.

epistaxis (ĕ-pis-tax′is) [G. *epistaxein,* to bleed from the nose]. Hemorrhage from the nose.

epithelial (ep-ĭ-thē′lĭ-al). Pert. to epithelia, the covering or lining tissues of the body.

equilibrium (e-kwil-ib′rĭ-um) [L. *aequus,* equal, + *libra,* balance]. A state of balance or rest, in which opposing forces are equal.

eruption (e-rup′shun) [L. *eruptio,* a breaking out]. Becoming visible, such as a lesion or rash appearing on the skin.

erythropoietin. A substance produced by the kidney which stimulates production of red blood cells.

erythropoiesis (ē-rith-ro-poy-e′sis) [G. *erythema,* redness, + *poiesis,* making]. The formation of erythrocytes or red blood cells.

eschar (es′kar) [G. *eschara,* scab]. A slough (dead matter or tissue); *debris* developing following a burn.

esthesia (es-the′zĭ-ā) [G. *aisthēsis,* sensation or feeling]. Perception, sensation or feeling.

estrogen (es′trō-jen) [G. *oistros,* mad desire, + *gennan,* to produce]. A substance producing or stimulating female characteristics; the follicular hormone.

eu- (ū) [G. *eu,* well]. Prefix: normal, well or good.

euphoria (u-fo′rĭ-ā) [″ + *pherein,* to bear]. An exaggerated feeling of well-being.

evacuate (e-vak′ū-āte) [L. *evacuāre,* to empty]. To empty or discharge the contents of, especially the bowels.

evaporate (ē-va′por-āte) [L. *ē,* out, + *vaporāre,* to steam]. To change from liquid to gaseous form usually by addition of heat, as to evaporate water by boiling, or on the skin.

evoke (e-vōk) [Fr. *ēvoquer,* to call out]. To cause to happen, to call forth.

ex- [L.]. Prefix: out, away from, completely.

excitable (ĕk-si′tă-bl) [L. *excitāre,* to rouse]. The property of being capable of responding to stimuli by alterations in ion separation.

excitatory [L.]. Something that causes changes in state of excitability; something that stimulates a function.

excrete (eks-krēt′) [L. *excernere,* to separate]. To separate and expel useless substances from the body.

exocrine (eks′ō-krēn) [″ + *krinein,* to separate]. External secretion by a gland, through ducts, upon an epithelial surface.

exogenous (eks-oj′ĕ-nus) [″ + *gennan,* to produce]. Originating outside the cell or organism.

expiration (eks-pi-rā′shun) [L. *ex,* out, + *spirare,* to breathe]. Expulsion of air from the lungs. Death.

exponential (ek-spō-nen′shul) [L. *exponere,* to put forth]. Change in a function according to multiples of powers (e.g., square or cube), rather than numerically (e.g., 1 time, 2 times, etc.).

external (ek-sturn′al) [L. *externus,* outside]. Outside or toward the surface.

exteroceptive (eks-ter-o-sep′tiv) [″ + *ceptus,* to take]. Sensations or reflex acts originating from stimulation of receptors at or near a body surface.

extracellular [L. *extra,* outside of, + cell]. Fluids outside of the cells. Abbr. **ECF.**

extraneous (eks-tra′ne-us) [L. *extraneus,* external]. Outside of or unrelated to something.

extrinsic (eks-trĭn′sik) [L. *extrinsecus,* outside]. From without or coming from without; separate or outside of an organ or cell.

exudate (eks′u-dāt) [″ + *sudāre,* to sweat]. Accumulation of fluid in a body cavity or on a body surface. Pus, serum, or the passing of the same.

facial (fā′shul) [L. *facies,* the face]. Pert. to the face.

facilitation (fah-sĭl-ĭ-tā′shun) [L. *ficilis,* to make easier]. An increased ease of passage of a nerve impulse across a synapse.

FAD. Abbr. for *flavine adenine dinucleotide,* one of several hydrogen acceptors.

fascia (fash′ĭ-ă) [L. *fascia,* a band]. Fibrous membranes covering, supporting, and separating muscles. *Superficial f.* is the hypodermis; *deep f.* is around muscles.

fascicle (fas′ik-l) [L. *fasciculus,* a little bundle]. An arrangement resembling a bundle of rods. Applied to bundles of nerve and muscle fibers.

fasciculation (fas-ĭck-u-lā′shun) A localized contraction of one or a few motor units in a skeletal muscle.

fatigue (fă-tēg) [L. *fatigare,* to tire]. A feeling of tiredness; the state of an organ, synapse, or tissue in which it no longer responds to stimulation.

feces (fē′sēz) [L. *faeces,* dregs]. The waste matter expelled from the bowel through the anus.

feedback. Detection of the nature of an output and using that to control the process producing the output. May be negative *(inhibitory)* or positive *(stimulating).*

fe′male [L. *femella,* woman]. The designation given to the sex that produces ova and bears offspring; a girl-child or woman.

fertilization (fur-tĭl-ĭ-zā′shun) [L. *fertilis,* to bear]. The impregnation of an ovum by a sperm.

fetus (fē′tus) [L. *fetus,* progeny]. The name given to a developing human after it assumes *clearly human form;* the period from 6 to 8 weeks after conception to birth.

fiber (fī′ber) [L. *fibra,* thread]. A larger threadlike or ribbonlike structure; a muscle cell; usually composed of smaller units.

fibril (fī′brĭl) [L. *fibrilla,* a little fiber]. A very small, threadlike structure, often the component of a muscle or nerve cell, or a connective tissue fiber.

fibrillate (fī′bril-āte) [L. *fibrilla,* a little fiber]. Spontaneous uncoordinated quivering of a muscle fiber, as in cardiac muscle.

fibrin (fī-brĭn) [L. *fibra,* a fiber]. A white or yellowish insoluble fibrous protein formed when blood clots.

fibrinogen (fī-brĭn′ō-gen). A soluble plasma protein that is acted upon to produce fibrin as blood or lymph clots.

fibrosis (fi-brō′sis). Abnormal deposition or increase in fibrous tissue in a body part, organ, or tissue.

fil′iform [L. *filum,* thread, + *form,* form]. Having a threadlike or hairlike form.

fil′ter [L. *filtare,* to strain through]. To strain, or separate on the basis of size; a device to strain liquids.

filtrate (fĭl′trāt). The name given to the fluid that has passed through a filter.

filtration (fĭl-trā′shun). The process of forming a filtrate by passing a fluid, under pressure, through a filter or selective membrane.

flaccid (flak′sid) [L. *flaccidus,* flabby]. Relaxed, flabby, having no tone, as in a muscle.

flatus (fla′tus) [L. *flatulentiā,* a blowing]. The gas in the digestive tract.

flora (flŏ′ra) [L. *flos.* flower]. Plant life; the microorganisms of the bowels, adapted to live in that organ.

follicle (fŏl′ĭ-cul) [L. *folliculus,* a little bag]. A small hollow structure containing cells or secretion.

foodstuff (fōōd′stŭf) [M. E. *fode,* to eat]. Any substance made into or used as a food; nutritive substances providing heat and energy to the body when metabolized.

fornix (for′nĭks) [L. *fornix,* arch]. Any structure having an arched or vaultlike shape; a band connecting lobes of a cerebral hemisphere; a cleftlike space of the vagina around the neck of the uterus.

free energy. Energy released from metabolic processes that is capable of doing work.

frequency (frē′qwĕn-si) [L. *frequens,* often or constant]. Number of repetitions of something in a given time period (e.g., cycles per second, *cps*).

fulcrum (fŭl′krŭm) [L. *fulcire,* to prop or support]. The point about which a lever turns or pivots.

function (fŭnk′shŭn) [L. *functia,* to perform]. The action performed by living matter (cell, tissue, organ, system, or organism); to carry on an action.

fundus (fŭn′dŭs) [L. *fundus,* sling or base]. The larger, usually blind end of an organ or gland.

fu′siform [L. *fusus,* spindle, + *forma,* shape]. Having a shape tapered at both ends.

fusion (fū′shun) [L. *fio,* to meet]. Coming together; meeting or joining.

gait (gāt) [M. E. *gaite,* street]. The manner of walking.

galactose (gă-lak′tōs) [G. *gala,* milk, + *ose,* sugar]. Milk sugar; a simple hexose sugar found in milk. $C_6H_{12}O_6$.

gamete (gam′ēt) [G. *gamēte,* a wife or spouse]. A male or female reproductive cell, that is, a sperm or egg (ovum).

gametogenesis (gam-ē-tō-jen′ĕ-sis) [″ + *genesis,* birth or origin]. The formation of gametes; spermatogenesis or oögenesis.

gamma (găm′ŭh) [G. letter **g** of alphabet]. The third item of a series; a microgram (one millionth gram).

gamma globulin. A plasma protein carrying most of the blood-borne antibodies.

ganglion (gang′lĭ-ŏn) [G. *gagglion,* a tumor or swelling]. A mass of nerve cell bodies outside of the brain and spinal cord; a cystic tumor developing on a tendon.

gangrene (gan′grēn) [G. *gaggraina,* an eating sore]. Decay of tissue when blood supply is obstructed.

gas′tric [G. *gaster,* stomach]. Pert. to the stomach.

gene (gēn) [G. *gennan*, to produce]. The unit of heredity responsible for transmission of a characteristic to the offspring; believed to be a part of a nucleic acid chain.

genetic (jen-ĕt′ik) [G. *genesis*, origin]. Pert. to genesis or origin of something.

genital (jen′ĭ-tal) [L. *genitalis*, genital]. Pert. to the organs of reproduction; *external genitalia*, the external organs of reproduction.

genotype (jēn′ō-tīp) [G. *gennan*, to produce, + *typos*, type or kind]. The hereditary makeup of an individual as determined by the genes.

germinal (jer′mĭn-ăl) [L. *germen*, microbe]. Pert. to a reproductive cell, or a "germ" (microorganism).

germ layer. One of the three basic tissue layers (ectoderm, endoderm, mesoderm) formed in the embryo that give rise to all body tissues and organs.

gerontology (jer-on-tol′ō-ji) [G. *geron*, old man, + *logos*, study of]. The study of the phenomena of old age.

gestation (jes-ta′shun) [L. *gestāre*, to bear]. The period of intrauterine development of a new organism; the time of pregnancy.

gingiva (jin′jĭ-vă) [L. *gingiva*, gum]. The gums, or tissues surrounding.

girdle (gĭr′dŭl) [A. S. *gyrdel*, to encircle]. A belt, zone, or a structure resembling a circular belt or band.

gland [L. *glans*, a kernel]. A secretory organ or structure.

glia (glē′ă) [G. *glia*, glue]. The nonnervous tissues of the nervous system; supporting and nutritive cells of various types.

globular (glob′ū-lar) [L. *globus*, a globe]. Having a rounded or spherical shape.

glomerulus (glō-mer′ū-lus) [L. a little skein, a tangle of thread or yarn]. A small round mass of cells or blood vessels.

glottis (glŏt′ĭs) [G. *glottis*, back of tongue]. The opening between the vocal cords in the larynx (voice box).

-glyc- (glik) [G. *glykus*, sweet]. Pert. to glucose.

glyceride (glĭs′ĕr-īd) [G. *glykus*, sweet]. Glycerin (an alcohol) together with one or more fatty acids.

glycogen (glī′ko-jen) [G. *glykos*, sweet, + "]. A complex polysaccharide; "animal starch"; stored in liver and muscle.

goblet cell. A one-celled mucus secreting gland found in epithelia of the respiratory and digestive system.

goiter (goi′tĕr) [L. *guttur*, throat]. An enlargement of the thyroid gland, irrespective of cause.

gon′ads [G. *gonē*, a seed]. A general name for an organ producing sex cells; the name of the embryonic sex gland before differentiation into ovary or testis.

gradient (grā′dĭ-ent). A slope or grade; a curve representing increase or decrease of a function.

graft [L. *graphium*, grafting knife]. A tissue or organ inserted into a similar substance or area to correct loss or absence of that structure; a transplant; to carry out the procedure of grafting or transplantation.

-gram. Combining form pert. to the record drawn by a machine (e.g., cardiogram, telegram).

granulation (grăn-ū-lā′shun) [L. *granulum*, little grain]. To form granules; tissue appearing in the course of wound healing, characterized by its large numbers of blood vessels.

gran′ule [L.]. Any minute or tiny grain in a cell.

grav′ity [L. *gravitās*, weight]. The force of the earth's gravitation attraction that gives objects weight.

gray matter. The tissue of the central nervous system consisting mainly of nerve cell bodies.

gristle (grĭs′l) [A. S.]. Cartilage.

gross (grōs) [L. *grossus*, thick]. Large; anatomy seen with the naked eye.

ground substance. The fluid or semifluid material occupying the intercellular spaces in connective tissues.

growth [A. S. *grōwan*, to grow]. Progressive increase in size of a cell, tissue, or whole organism.

gut [A. S]. The primitive or embryonic digestive tube (fore-, mid-, hindgut); Colloq. for intestine.

gyrus (jī′rus) (pl. *gyri*) [G. *gyros*, circle]. An upfold of the cerebrum.

[H⁺]. Symbol for hydrogen ion concentration.

hallucination (hă-lu-sĭ-nā′shun) [L. *alucinari*, to wander in mind]. A false perception having no basis in reality; may be auditory, visual, olfactory, etc.

hap′loid [G. *aploos*, simple]. Having one half the normal number of chromosomes characteristic of the species; sperm and ova are haploid cells.

hap′toglobin [G. *aptein*, to seize]. A protein in the plasma that combines specifically with hemoglobin released from red blood cells.

haustra (haws′tra) [L. *haurīre*, to draw water]. The sacculations or pouches of the colon.

heat (hēēt) [G. *heito*, fever]. High temperature; being hot; a form of energy; to make hot.

helix (pl. *helices*) (hē′liks) [G. *helix*, a coil]. A coil or spiral.

hematocrit (he-mat′ō-krit) [G. *aima*, blood, + *krinein*, to separate]. The volume of red blood cells in a tube after centrifugation.

heme (hēm) [G.]. The iron containing red pigment that, with globin, forms hemoglobin.

hemiplegia (hem-ĭ-plē′jĭ-ă) [G. *hemi*, half, + *plēgē*, a stroke]. Paralysis of one half of the body in a right-left direction.

hemoglobin (hē-mō-glō′bin) [G. *aima*, blood + *globus*, globe]. The respiratory pigment of red blood cells that combines with O_2, CO_2, and acts as a buffer.

hemolysis (he-mol′ĭ-sis) [″ + *lysis*, dissolution]. Destruction of red blood cells with release of hemoglobin into the plasma.

hemorrhage (hem′o-rij) [″ + *rēgnunai*, to burst forth]. Abnormal loss of blood from the blood vessels, either internally or externally.

hemorrhoids (hem′o-royds) [G. *aimorrois*, a vein liable to discharge blood]. Dilated vein in the rectal columns.

hemostasis (he-mō-stā′sis) [″ + *stasis*, stopping]. Stoppage of blood loss through a wound due to vascular constriction and coagulation.

he′par [G. *ēpar*, liver]. The liver.

hepatic (he-pat′ik). Pert. to the liver.

hereditary (hĕ-red′ĭ-tar-ē) [L. *hereditas*, heir]. Passed or transmitted (as a genetic characteristic) from parent to offspring; handed down.

hernia (her′nĭ-ă) [G. *ernos*, a young shoot]. Protrusion or projection of an abdominal organ through the wall that normally contains it.

heterogeneous (het-er-ō-je′nē-us) [G. *eteros*, other, + *gennos*, type]. Composed of things different or contrasting in type or nature.

heterograft [″ + *graphium*, grafting knife]. Grafting of tissues or organs between different species.

hex′ose [G. *hex*, six, + *ose*, sugar]. A six-carbon sugar (e.g., glucose, fructose).

hi′lus [L. *hilus*, trifle]. A depression or recess on the side of an organ; vessels and nerves usually enter and/or leave at this point.

histio-, histo- [G. *histos*, tissue]. Prefix; tissue.

holocrine (hōl-o-krĭn) [G. *olos*, whole + *krinein*, to secrete]. A manner of production of a secretion by a gland in which a cell is the product or in which the whole cell enters the secretion.

homeostasis (hō-mē-ō-stā′sis) [G. *omoios*, like, + *stasis*, a standing]. The state of near constancy of body composition and function.

homogenous (hō-moj′ĕn-ŭs) [G. *homo*, likeness, + *genos*, kind]. Like or uniform in structure or composition.

homograft [″ + L. ″]. Use of tissues or organs of the same species for grafting purposes. *Isograft.*

homologous (hō-mol′o-gus) [″ + *logos*, relation]. Similar in origin and structure.

hormone (hor′mōn) [G. *ormanein*, to excite]. The chemical produced by an endocrine gland.

humor (hū′mŭr) [L. *humor*, fluid, moisture]. Any fluid or semifluid substance in the body.

hyaline (hī′ă-lĭn) [G. *hyalinos*, glassy]. Clear, translucent, or glassy.

hyaluronidase (hi-ă-lur-on′ĭ-dās). An enzyme that liquefies the intercellular cement holding cells together.

hydrate (hī′drāt) [G. *ydōr*, water]. To cause water to combine with a compound; the compound after water has been added.

hydrogen (hi′drō-jen) [″ + *gennan*, to produce]. An inflammable gaseous element, the lightest of all known elements.

hydrostatic pressure [″ + *statikos*, standing]. Pressure exerted by liquids.

hyperemia (hī-per-ē′mĭ-ă) [G. *yper*, above, + *aima*, blood]. An extra amount of blood in an area.

hyperesthesia (hī-per-es-the′zē-ă) [″ + *aisthēsis*, sensation]. Excessive sensitivity to sensory stimulation.

hypermetropic (hi-per-mē-trō′pik) [″ + *metron*, measure + *ōps*, eye]. Pert. to farsightedness.

hyperphagia (hī-per-fā-jē-ă) [″ + G. *phagein*, to eat]. Excessive intake of food.

hyperplasia (hī-per-plā′zē-ă) [″ + *plassein*, to form]. Increase in size because of increased numbers of cells derived by division.

hyperpnea (hī-perp′nē-ă) [″ + *pnoe*, breath]. An increase in depth of breathing.

hypertension (hī-pĕr-tĕn′shun) [″ + *tensio*, tension]. High blood pressure.

hypertonic (hī-per-tŏn′ik) [″ + *tonos*, tension]. A solution containing a greater solute concentration and thus osmotic pressure than the blood, or having an osmotic pressure greater than another reference solution.

hypertrophy (hi-per′trō-fē) [″ + *trophē*, nourishment]. Increase in size by adding substance and *not* by increasing numbers of cells.

hyperventila′tion [″ + *ventiatio*, ventilate]. Increased exchange of air by increasing both rate and depth of breathing.

hypnotic (hip-nŏ′tik) [G. *ypnos*, sleep]. An agent producing sleep or depression of the senses.

hypokalemia (hī-pō-kăl-ē′mē-ă) [G. *ypo*, under, + L. *kali*, potash, potassium, + *aima*, blood]. Low blood potassium levels.

hypoten′sion [G.]. Low blood pressure.

hypothesis (hī-poth′ĕ-sis) [″ + *thesis*, a placing]. An assumption made to explain something. It is unproved and is made to enable testing of its soundness.

hypotonic (hi-pō-tŏn′ik) [″]. A solution containing less solute, and thus having a lower osmotic pressure, than the blood, or a solution having a lower osmotic pressure than some reference solution.

hypoxemia (hī-poks-ē′mē-ă) [″ + *oxys*, acid, + ″]. Lowered blood oxygen levels.

hypoxia (hī-poks′ē-ă) [″]. Low oxygen levels in inspired air or reduced tension in the tissues.

icteric (ik-tĕr′ik) [G. *ikteros*, jaundice]. Pert. to jaundice; excessive accumulation of bile pigments in the body tissues.

idiopathic (id-ĭ-ō-path′ik) [G. *idio*, own, + *pathos*, disease]. Any condition arising without a clear-cut cause; of spontaneous origin.

immune (ĭm-ūn′) [L. *immunis*, safe]. Protected from getting a given disease.

impermeable (im-pĕrm′ē-ă-bl) [L. *in*, not, + *permeāre*, to pass through]. Not allowing passage.

implantation (im-plan-tā′shun) [L. *in*, into, + *plantāre*, to plant]. Embedding of the blastocyst in the uterine lining; inserting something into a body organ.

impulse (im′puls) [L. *impellere*, to drive out]. A physicochemical or electrical change transmitted along nerve fibers or membranes.

inactivate (ĭn-ăk′tĭ-vāt) [L. *in*, not, + *activus*, acting]. To make inactive or inert.

inclusion (ĭn-klŭ′zhun) [L. *inclusus*, enclosed]. A lifeless, usually temporary constituent of a cell's cytoplasm.

independent irritability. Capable of reacting to external stimuli, or those delivered by other than the normal route.

indigestion (in-dī-jes′chŭn) [L. *in*, not + *digerere*, separated]. Inability to digest food; incomplete digestion; dyspepsia. Usually accompanied by pain, nausea, gas (belching), and heartburn.

indice (in-dūs) [L. *inducere*, to lead in]. To cause an effect; induction may be used to express the way that a gene causes an effect.

infarct (in′farkt) [L. *infacire*, to stuff into]. An area of an organ that dies following blockage of blood supply.

infection (in-fek′shun) [L. *inficere*, to taint]. The state when the body has been invaded by a pathogenic (disease-producing) agent.

infectious (in-fek′shus) [″]. Capable of being transmitted with or without contact. Usually used in connection with microorganism caused diseases.

inflame (in-flām) [L. *inflammare,* to set on fire]. To cause to become warm, swollen, red, sore, and feverish.

infundibulum (in-fun-dib'ū-lum) [L. *infundibulum,* funnel]. A funnel-shaped passage.

infusion (in-fū'zhun) [L. *infusiō,* from, + *fundere,* to pour]. Introduction of liquid into a vein.

ingest (in-jest') [L. *ingerere,* to carry in]. To take foods into the body.

inguinal (in'gwi-nal) [L. *inguinalis,* the groin]. Pert. to the groin.

inherit (in-her'ĭt) [L. *inhereditare,* to appoint as heir]. To receive from one's ancestors.

inhibit (in-hĭ'bĭt) [L. *inhibere,* to restrain]. To repress or slow down.

innervate (ĭn-nŭr'vāt) [L. *in,* in, + *nervus,* nerve]. To supply with nerves; to stimulate to action.

inorgan'ic [L. *in,* not, + G. *organon,* an organ]. A chemical compound not containing both hydrogen and carbon; not associated with living things.

input (in'po͝ot). That which is put into something, as in the passage of nerve impulses to the brain for processing.

insertion (in-sŭr'shen) [L. *insertio,* to join into]. The addition of something by "setting it into" something else, as insertion of parts of a chromosome into another chromosome.

in situ [L.]. In position.

inspiration (in-spī-rā'shun) [L. *inspiratio,* to breath in]. Taking air into the lungs.

integrate (in'tē-grāt) [L. *intergratus,* to make whole]. To blend, put together or unify.

integument (in-teg'ū-ment) [L. *integumentum,* a covering]. The outer covering of the body; skin.

intensity (in-ten'sĭ-tĭ) [L. *intensus,* tight]. Degree of strength, force, loudness, or activity.

intermediary (in-ter-me'dĭ-a-rĭ) [L. *inter,* between, + *mediāre,* to divide]. Something occurring between two time periods. In metabolism, the compounds formed as foodstuffs are utilized.

internuncial (in-tĕr-nun'sē-al) [" + *nuncius,* messenger]. A connector; between two other items, as neurons.

interphase. The "resting state" of a cell, when it is not dividing.

interstitial (in-ter-stish'al) [L. *interstitium,* thing standing between]. Lying between, as, interstitial fluid lies between vessels and cells or between cells.

in'tima [L. innermost]. The inner coat of a blood vessel.

intracellular (in-tra-sel'ū-lar) [L. *intra,* within, + *cellula,* cell]. Within cells.

intravenous (in-tră-ve'nus) [" + *vena,* vein]. Within or into a vein.

intrin'sic [L. *intrinsicus,* on the inside]. Located or originating within a structure.

in utero [L.]. Within the uterus.

inversion (in-ver'shun) [L. *in,* into, + *versiō,* a turning]. Turning, as upside down, or inside out; reversal.

in vitro [L.]. In glass, as in a test tube; *not* within the body.

in vivo [L.]. Within the body.

involuntary (in-vol'un-tĕr-i) [L. *involuntarius,* without act of will]. Occurring without an act of will.

involution (in-vō-lū'shun) [L. *in*, into, + *volvere*, to roll]. Change in a backward or diminishing direction.

i'on. [G. *iōn*, going]. An atom or group of atoms (radical) carrying an electric charge.

ipsilateral (ip-sĭ-lăt'er-al) [L. *ipse*, same, + *latus*, side]. On the same side; opposite of contralateral (*contra*, opposite).

irritability (ĭ-rĭ-tă-bĭl'ĭ-tē) [L. *irritare*, to tease]. Excitability, or the ability to respond to a stimulus.

ischemia (ĭs-kē'mĭ-ă) [G. *ischein*, to hold back, + *aima*, blood]. Local temporary reduction of blood flow to a body organ or part.

i'so- [G. *isos*, equal]. Prefix: equal to.

i'sograft [G. *isos*, equal, + *graphium*, grafting knife]. A graft or transplant from another animal of the same species. Also, *homograft*.

isometric (ī-sō-mĕ'trik) [G. *isos*, equal, + *metron*, measure]. Refers to no change of length, as an isometric muscle contraction.

isoton'ic [" + *tonos*, tension]. A solution having the same osmotic pressure as the blood or a reference solution; a muscular contraction in which shortening is allowed.

isthmus (ĭs-mus) [G. *isthmos*, narrow passage]. A narrow structure connecting two other structures.

itch [A. S. *giccan*, to itch]. An irritation of the skin, causing a desire to scratch. Also, *pruritus*.

-itis (ī'tĭs) [G.]. Suffix: inflammation of.

jaundice (jawn'dis) [M. E. *jaundis*, yellow]. A condition in which the skin is yellowed due to excessive bile pigment (bilirubin) in the blood.

joint [L. *junctura*, a joining]. An articulation or junction between bones or bone and cartilage.

jug'ular [L. *jugulum*, neck]. Large veins in the neck; pert. to the neck.

juice (jūs) [L. *jus*, broth]. Liquid secreted or expressed from an organism or any of its parts; esp., one of the digestive secretions.

ju'venile [L. *juvenilis*, young]. Pert. to youth or childhood; an immature white cell (metamyelocyte).

juxta- [L. near to]. Close or near to; in close proximity, as in juxtaglomerular.

karyoplasm (kar'ĭ-ōplăs-m) [G. *karyon*, nucleus, + *plasma*, a thing formed].

karyotype (kar'i-ō-tīp) [" + *typos*, form]. A grouping or arrangement of the 46 human chromosomes based on the size of the individual chromosomes.

keratin (ker'ă-tĭn) [G. *keras*, horn]. A protein found in the superficial epidermal cells.

keto (kē'tō). Prefix denoting the presence in an organic compound of the carbonyl or "keto" group.

ke'tone. Any substance (e.g., acetone) having the carbonyl group in its molecule. Also, *acetone* or *ketone bodies*.

ketosis (kē-tō'sis) [ketone + G. *-osis*, disease]. Accumulation of ketone bodies in blood or body.

kilocalorie (kĭl-ō-căl-ŏ-rē) [G. *chilioi*, one thousand, + *calor*, heat]. One thousand small calories, or one large Calorie.

kil′ogram [″ + *gramma*, a weight]. A metric measurement of weight; equals 1000 gm or 2.2 lb.

kinesthesia (kin-es-thē′zĭ-ă) [G. *kinesis*, motion, + *aisthēsis*, sensation]. The sensation concerned with appreciation of movement and body position.

kinetic (kĭ-net′ĭk) [G. *kinēsis*, motion]. Pert. to movement or work.

-kinin. Suffix denoting causing motion or action.

kyphosis (ki-fō′sis) [G. *kyphosis*, humpback]. An exaggeration of the normally anteriorly concave thoracic spinal curvature.

labium (lā′bĭ-um) [L. *labium*, lip]. A lip or liplike structure.

lacrimal (lak′rĭm-al) [L. *lacrima*, tears]. Pert. to the tears or any structure involved in producing or carrying tears.

lact- [L. *lac*, milk]. Prefix: milk.

lactation (lak-tā′shun) [L. *lactatio*, a suckling]. The function of secreting milk.

lactose (lak′tōse) [″ + ose, sugar]. Milk sugar, a double sugar (disaccharide).

lacuna (lă-kū′nă) [L. *lacuna*, a pit]. A hollow space in the matrix of cartilage and bone in which lie characteristic cells.

lamella (lam-el′ă) [L. *lamella*, a little plate or leaf]. Circularly arranged layers, as in the elastic layers in a large artery; a ring of bony tissue around a Haversian canal.

la′tent [L. *latēre*, to be hidden]. Not active, quiet.

leg [M. E.]. Specifically, the part of the lower limb between the knee and ankle; Colloq.: a lower appendage.

lemniscus (lĕm-nis′kŭs) [G. *lemniskōs*, a filet]. A bundle of sensory nerve fibers in the brain stem.

lesion (lē′zhun) [L. *laesio*, a wound]. A wound, injury, or infected area.

leuco-, leuko- (lu′ko) [G. *leukos*, white]. Prefix: white.

lever (lĕv′ĕr, lē′vĕr) [M. E. *lever*, to raise]. A rigid elongated structure used to change direction, force, or movement.

ligament (lĭg′ă-mĕnt) [L. *ligamentum*, a band]. A strong band or sheet of fibrous connective tissue commonly used to connect bones together; a structure formed by degeneration of a fetal blood vessel.

lim′bic system [L. *limbus*, border]. A group of nervous structures in the cerebrum and diencephalon serving emotional expression.

liminal (lĭm′ĭ-nal) [L. *limen*, threshold]. Threshold; just perceptible.

linear (lĭn′ē-ar) [L. *linea*, line]. Pert. to a line or lines; extended in a line.

linkage (link′ĭj) [M. E. *linke*, akin to]. In genetics, when two or more genes on a chromosome tend to remain together during sex cell formation.

lipid (lĭp′id) [G. *lipos*, fat]. Fats or fatlike substances that are not soluble in water.

lipoprotein (lī-pō-prō′tēēn) [″ + *proteios*, major, of first importance]. A combination between a lipid and a simple protein.

liter (lē′tĕr) [G. *litra*, a pound]. Metric measurement of fluid volume. Equals 1000 milliliters or 1.06 quarts or 61 cubic inches.

-lith- [G. *lithos*, stone]. Stone; presence of stones, *lithiasis*.

lobe (lōb) [G. *lobos*, part or section]. A section or part of an organ separated by clear boundaries from other sections or parts.

localize (lō′kăl-īz) [L. *locus*, place]. To restrict to a small area.

locomotion (lō-kō-mō′shun) [L. *locus*, place, + *motus*, moving]. To move from one place to another; movement.

locus (lō′kŭs) ["]. A place; in genetics the location or position of a gene in a chromosome; a place in the brain or heart where abnormal electric discharge may originate.

logarithmic (lŏg-ă-rĭth′mĭk) [G. *logos*, a proportion, + *arithmos*, number]. Progression of a function of powers (e.g., square or cube) and not by individual numbers.

loin (loyn) [Fr. *loigne*, long part]. The sides and back of the trunk between ribs and pelvis.

lozenge (loz-ĕnj) [Fr. *lozenge*, diamond-shaped]. A solid medicine to be held in the mouth until dissolved.

lucid (lū′sĭd) [L. *lucidus*, clear]. Clear, as of thought, mind, or speech.

lumen (lū′mĕn) (pl. *lumina)* [L. *lumen*, light]. The cavity of any hollow organ; a unit of light.

luteo- (lū′tē-ō) [L. *luteus*, yellow]. Prefix: yellow.

lympho- (lim′fō) [L. *lympha*, lymph]. Prefix: lymph or lymphatic system.

-lysis (lī′sĭs) [G. *lysis*, dissolution]. To destroy or break down.

lysosome (lī′sō-som) [" + *soma*, body]. Cell organelle concerned with digestion of large molecules.

L-tubule. Longitudinally arranged tubules of the sarcoplasmic reticulum in muscle cells.

macro- (mak′rō) [G.]. Prefix: large or long.

mal- (mahl) [L. *malum*, an evil]. Ill, bad, or poor.

male (māl) [L. *masculus*, man]. The designation given to the sex that produces sperm, fertilizes ova, and begets offspring; a boy-child or man.

malignancy (mă-lĭg′năn-sĭ) [L. *malignus*, of bad kind]. A severe form of something; tending to grow worse.

malnutrition (mal-nū-trĭ′shun). Literally, "poor nutrition"; most often used to refer to absence of essential foodstuffs in the diet.

maltose (mawl′tōs) [A. S. *mealt*, grain]. Malt sugar, a double sugar (disaccharide) found in malt and seeds. $C_{12}H_{22}O_{11}$.

mammary (mam′ă-rĭ) [L. *mamma*, breast]. Pert. to breast.

mammillary (mam′ĭl-lar-ĭ) [L. *mammilla*, nipple]. Resembling a nipple.

mastication (măs-tĭ-kā′shŭn) [L. *masticāre*, to chew]. Chewing of food in the mouth.

matrix (mā′triks) [L. *matrix*, mother or womb]. Intercellular substance of cartilage; formative portion of a tooth or nail; the uterus.

mature (ma-tūr) [L. *maturus*, ripe]. Fully developed or ripened.

maximal (maks′ĭ-mal) [L. *maximum*, greatest]. Highest; greatest possible.

mean (mēn) [L. *medius*, in the middle]. The average (sum of numbers divided by the number of numbers).

meatus (mē-ā'tŭs) [L. *meatus,* opening]. A passage or opening.

mediastinum (mē-dĭ-ăs-tī'nŭm) [L. in the middle]. A septum or cavity between two parts of an organ or between two other cavities.

medulla (mĕ-dul'lă) [L. marrow]. Inner or central portion of an organ; the medulla oblongata of the brain stem.

medullated (med'ū-lāt-ĕd). ['']. Having a myelin sheath.

mega- [G]. Large; one million.

meiosis (mī-ō'sĭs) [L. mēiosis, to make smaller]. A form of cell division that reduces chromosome number to haploid. Occurs in formation of sex cells.

melanophores (mel-ăn'ō-fōr) [G. *melas,* black, + *phoros,* a bearer]. A cell carrying dark pigment(s).

membrane (mĕm'brān) [L. *membrana,* membrane]. A thin soft layer of tissue lining a tube or cavity, or covering or separating one part from another.

meninges (mĕn-ĭn'jēz) [G. *mēnigx,* membrane]. The membranes around the central nervous system; *dura mater, arachnoid, pia mater.*

meniscus (men-is'kus) [G. *mēniskos,* crescent]. The crescent-shaped cartilages of the knee joint.

menstrual (men'strū-al) [L. *menstruāre,* to discharge menses]. Pert. to menstruation or the sloughing of the uterine endometrium.

merocrine (mer'ō-krĭn) [G. *meros,* a part, + *krinein,* to secrete]. A method of production of a secretion in which no part of the secreting cell enters the secretion.

mesentery (mes'en-ter-ĭ) [G. *mesos,* middle, + *enteron,* intestine]. A double-layered fold of peritoneum suspending the intestine from the posterior abdominal wall.

mesoderm (mĕs'ō-derm) ['' + *derma,* skin]. The middle germ layer of the embryo; gives rise to connective tissue, muscle, blood, and the cellular part of many organs.

meta- Prefix: after or beyond, later or more developed.

metabolism (mĕ-tăb'ŏl-ĭzm) [G. *metabolē,* change, + *ismos,* state of]. The sum total of all chemical reactions in the body; any product of metabolism is a *metabolite.*

metaphase (mĕt'ă-fāz) ['' + *phasis,* a shining out]. A stage of mitosis in which chromosomes line up on the equator of the dividing cell.

metastasis (mĕ-tas'tă-sis) ['' + *stasis,* a standing]. Movement from one part of the body to another (e.g., cancer cells).

micelle (mī-sĕl') [L. a little crumb]. A small complete unit, usually of colloids.

micro- [G.]. Small; one millionth.

micron (mi'kron) [G. *mikros,* small]. Metric unit of length; equals one one-thousandth of a millimeter. μ.

microvillus (mī-kro-vĭl'ŭs) ['' + *villus,* tuft of hair]. Very small fingerlike extensions of a cell surface.

micturition (mĭk-tŭ-rĭ'shŭn) [L. *micturīre,* to urinate]. Voiding of urine; urination.

milliequivalent (mil-e-e-kwĭv'a-lent) [L. *mille,* thousandth, + *aequis,* equal, + *valere,* to be worth]. One one-thousandth of an equivalent weight. (An *equivalent* weight is the quantity, by weight, of a substance that will combine with one gram of H, or 8 grams of oxygen.) Abbr: **meq.**

millimeter (mil'ĭ-mēt-er) ['' + *metron,* measure]. One one-thousandth of a meter. Abbr: **mm.** One **meter** equals 39.36 inches.

millimicron. One-thousandth of a micron. Abbr: **mμ**.

milliosmol. One one-thousandth of an osmol. (An *osmol* is the molecular weight of a substance in grams divided by the number of particles each molecule releases in solution.) Abbr: **mos.**

mimetic (mi-met′ĭk) [G. *mimētikos,* to imitate]. Imitating or causing the same effects as something else.

miotic (mī-ō′tĭk) [G. *meiōn,* less]. An agent causing pupillary contraction.

mitosis (mī-tō′sis) [G. *mitos,* thread]. A type of cell division that results in production of daughter cells like the parent.

mitral (mi′tral) [L. *mitra,* a miter, a bishop's hat]. Pert. to the bicuspid (mitral) valve between the left atrium and ventricle of the heart.

modality (mō-dal-ĭ-tē) [L. *modus,* mode]. The nature or type of a stimulus; a property of a stimulus distinguishing it from all other stimuli.

molar solution. A solution in which one gram molecular weight of a substance is present in each liter of the solution.

molecular weight. Weight of a molecule obtained by adding the weights of its constituent atoms. It carries no units. **m. w., gram.** The molecular weight of something expressed in grams. Also, a **mol(e).**

molecule (mŏl′ĕ-kūl) [L. *molecula,* little mass]. The smallest unit into which a substance may be reduced without loss of its characteristics.

monitor (mon′ĭ-tŭr) [L. *monere,* to warn]. To watch, check, or keep track of; one or something that watches.

monosaccharide (mŏn-ō-sak′ar-id) [G. *mono,* one, single, + *sakcharon,* sugar]. A simple sugar that cannot be further decomposed by hydrolysis.

morphology (mor-fol′ō-ji) [G. *morphē,* form, + *logos,* study]. The structure or form of something; study of the same.

morula (mor′ū-lä) [L. *morus,* mulberry]. A solid mass of cells resulting from division of the cells of a fertilized ovum.

motor (mō′tor) [L. *motus,* moving]. Refers to movement or those structures (e.g., nerves and muscles) that cause movement.

motor end plate. The specialized ending of a motor nerve on a skeletal muscle fiber.

motor unit. One motor nerve fiber (axon) and the skeletal muscle fibers it supplies.

mucous (mū′kŭs) [L. *mucus,* mucus]. A mucus-secreting structure.

mucus (mū′kŭs) [L.] The thick sticky fluid secreted by a mucous cell or gland.

multi- [L.]. Prefix: many.

mutation (mū-tā′shŭn) [L. *mutatio,* to change]. A change or transformation of a gene; the evidence of a gene alteration.

mydriatic (mid-rĭ-at′ik) [G. *midriasis,* dilation]) An agent causing pupillary dilation.

myelin (mī′ĕ-lĭn) [G. *myelos,* marrow]. A fatty substance forming a sheath or covering around many nerve fibers. Speeds impulse conduction.

myelo- [G.]. Prefix: pert. to the spinal cord or to bone marrow.

myeloid (mī′el-oid) [″ + *eidos,* form]. Formed in bone marrow; resembling marrow.

myoglobin (mī-ō-glō′bĭn) [G. *myo,* muscle, + L. *globus,* globe]. A respiratory pigment of muscle that binds oxygen.

myopia (mī-ō'pē-ă) [G. *myein*, to shut, + *ōps*, eye]. Nearsightedness. Also, *hypometropia*.

myosin (mī'ō-syn) [G. *myo*, muscle]. A muscle protein acting as an enzyme to aid in initiating muscle contraction.

myotatic (mī-ō-tă'tik) [" + *tactus*, touch]. Refers to muscle or kinesthetic sense (*kinesthesia*).

nasal (nā'zl) [L. *nasus*, nose]. Pert. to the nose.

NAD. Abbr. for *nicotinamide adenine dinucleotide*, one of several hydrogen acceptors (NADP. NAD phosphate, another hydrogen acceptor).

narcotic (nar-kŏt'ik) [G. *narkōtikos*, benumbing]. A drug that depresses the central nervous system, relieving pain and producing sleep.

nares (nar'ēz) [L. *naris*, nostril]. Nostrils.

natremia (na-trē'mĭ-ă) [L. *natrium*, sodium, + *aima*, blood]. Blood sodium.

necrosis (nĕk-rō'sis) [G. *nekrōsis*, a killing]. Death of tissue or cells.

negative pressure. A pressure less than atmospheric pressure, that is, <760 mm Hg.

neoplasm (nē'ō-plăzm) [G. *neo-*, new, recent, + *plasma*, a thing formed]. A new and abnormal formation of tissue, as a cancer, tumor, or growth.

nephron (nĕf'ron) [G. *nephros*, kidney]. The functional unit of the kidney that forms urine and regulates blood composition.

nerve (nŭrv) [L. *nervus*, sinew]. A bundle of nerve fibers outside the central nervous system.

neurilemma (nŭr-ĭ-lĕm'mă) [G. *neuron*, sinew, + *lemma*, rind]. A thin living membrane around some nerve fibers; aids in myelin formation and fiber regeneration.

neuroglial (nū-rŏg'lĭ-ăl) [" + *glia*, glue]. Pert. to the glial or nonnervous cells of the nervous system.

neuron (nū'rŏn) [G.]. The cell serving as the unit of structure and function of the nervous system; is excitable and conductile.

neurosis (nu-rō'sis) [" + *osis*, disease]. A disorder of the mind in which contact with the real world is maintained.

neutralize (nū'tral-īz) [L. *neuter*, not, + *uter*, either]. To counteract, make inert, or destroy the properties of something.

newborn. A child under 6 weeks of age.

nigra (nī'gră) [L. *nigra*, black]. Black or blackness.

nitrogenous (nī-troj'ĕn-ŭs) [G. *nitron*, soda, + *gennan*, to produce]. Pert. to or containing nitrogen.

node (nōd) [L. *nodus*, knot]. A constricted region; a knob, protuberance or swelling; a small rounded organ.

nostril [A. S. *nosu*, nose, + *thyrl*, a hole]. External opening of the nasal cavity.

nuchal (nū'kal) [L. *nucha*, the back of the neck].

nuclear (nū'klē-ăr) [L. *nucleus*, a kernel]. Pert. to a nucleus (cell) or a central point.

nucleic acids (nū-klay'ik). Large molecules formed by nucleotides. They form the basis of heredity and protein synthesis. **DNA; RNA.**

nucleolus (nū-klē'ō-lŭs) [L. little kernel]. A spherical mass of nucleic acid within the nucleus. Stores and may synthesize t-RNA..

nucleotide (nū′klē-ō-tīd) [″]. A unit or compound formed of a nitrogenous base, a five-carbon sugar, and a phosphoric acid radical. Unit of structure of *DNA* and *RNA*.

nucleus (nū′klē-ŭs) [″]. The controlling center of a cell; a group of nerve cells in the brain; the heavy central atomic region in which mass and positive charge are concentrated.

nutrient (nū-trĭ-ent) [L. *nutriens*, to nourish]. A food substance necessary for normal body functioning.

nutrition (nūtrĭ′shun) [L. *nutritiō*, a feeding]. The total of all processes involved in intake, processing, and utilization of foods.

nystagmus (nĭs-tag′mŭs) [G. *nystazein*, to nod]. Involuntary cyclical movements of the eyeball.

obese (ō-bēs′) [G. *obesus*, fat]. Excessively fat; overweight.

obligatory (ob-lĭg′ă-tō-rē) [L. *obligatorius*, to bind to]. Carries the idea of not having a choice; bound to or fixed in function.

oblique (ŏb-līk) [L. *oblique*, slanting]. A slanting direction or position.

obliterate (o-blĭt′er-āte) [L. *obliterate*, to blot out]. To erase, leaving no traces; extinction; occlusion of a part by surgical, degenerative, or disease processes.

occlude (ŏ-klŭd) [L. *occludere*, to shut up]. To close, obstruct or block something, such as a blood vessel.

occlusion (ŏ-klŭ′zhŭn) [L. *occlusiō*, a closing up]. The state of being closed, blocked, or obstructed.

ocular (ok′ū-lăr) [L. *oculus*, eye]. Pert. to eye or vision.

-oid [G.]. Suffix: like, similar to, resembling.

olfactory (ŏl-fak′tō-rĭ) [L. *olfacere*, to smell]. Pert. to smell or the sense of smell.

oligo- (ŏl′ĭ-gō) [L. *oligos*, little]. Prefix: small, few, scanty or little.

oliguria (ol-ĭg-ū′rĭ-ă) [″ + *ouron*, urine]. Secretion of small amounts of urine.

-ology [G.]. Suffix: science of, study of, knowledge of.

omentum (ō-mĕn′tŭm) [L. *omentum*, a covering]. A four-layered fold of peritoneum lying between the stomach and other abdominal viscera; a site of fat storage.

oncotic (ŏng-kŏt′ĭk) [G. *ogkos*, tumor]. Pert. to the colloid osmotic pressure created by the plasma proteins.

opaque (ō-pāk′) [L. *opacus*, dark]. Not transparent; not allowing light to pass.

operon (ŏp′er-ŏn) [L. *operatiō*, a working]. The combination of an operator gene and a structural gene.

operon concept. The current theory of how genes operate to control body function.

ophthalmic (ŏf-thăl′mĭk) [G. *ophthalmos*, eye]. Pert. to the eye.

-opsin (ŏp′sĭn) [G. *opson*, food]. Suffix: pert. to the protein component of the visual pigments.

oral (ō′răl) [L. *os, or-*, mouth]. Pert. to the mouth or the mouth cavity.

orbicular (ŏr-bĭk′ū-lăr) [L. *orbiculus*, a small circle]. Circular in arrangement.

orbit (or′bĭt) [L. *orbita*, track]. The cavity holding the eyeball; to go around.

orchido- (G. *orchis*, testicle]. Combining form meaning testicle.

organ (or′găn) [G. *organon*, organ]. Two or more tissues organized to do a particular job.

organelle (or-găn-ĕl′) [G. *organelle,* a small organ]. Submicroscopic formed structures within the cytoplasm that carry out particular functions.

organic (or-găn′ĭk) [G. *organon,* organ]. Pert. to organs; compounds containing carbon and hydrogen.

organic acids. An acid containing one or more carboxyl groups (-COOH).

organism (or′găn-ĭzm) [G. *organon,* organ, + *ismos,* condition]. A living thing.

orifice (or′ĭ-fĭs) [L. *orificium,* outlet]. The entrance or outlet to any chamber or hollow.

origin (or′ĭ-jin) [L. *origo,* beginning]. A source; the beginning of a nerve; the more fixed attachment of a muscle.

os (ŏs) [L.]. Bone; mouth; opening.

oscillate (ŏs′ĭll-āt) [L. *oscillāre,* to swing]. To move back and forth; to swing.

-ose (ōs). Suffix: pert. to sugar.

-osis [G.]. Suffix: caused by; state of; disease or intensive.

osmo- [G.]. Combining form, pert. to smell; pert. to osmosis.

osmol (os′mōl). The molecular weight of a substance, in grams, divided by the number of particles it releases in solution.

osmolarity (os-mō-lar′ĭ-tĭ). The number of osmols per liter of solution.

osmoreceptor (ŏz-mō-rē-cĕp′tor). A receptor sensitive to osmotic pressure of a fluid.

osmosis (ŏz-mō′sis) [G. *osmōs,* a thrusting, + *-osis,* intensive]. The passage of solvent through a selective membrane.

osmotic pressure. The pressure developing when two solutions of different concentrations are separated by a membrane permeable to the solvent.

osseous (ŏs′ē-ŭs) [L. *osseus,* bony]. Bonelike, bony, or pert. to bones.

osteo- [G.]. Prefix: bone(s).

osteoid (ŏs′tē-oyd) [″ + *eidos,* resembling]. A substance resembling bone.

osteon (ŏs′tē-on) [G. *osteon,* bone]. The unit of structure of compact bone. Also, *Haversian system.*

oto- [G.]. Combining form: the ear.

-otomy (ŏt′ō-mē) [G. *tomē,* incision]. Suffix: pert. to opening or repair of an organ without its removal.

output. The result or product of the operation of an organ or a machine.

ova (ō′vă) (sing. ovum) [L. *ovum,* egg]. Female reproductive cells; eggs.

ovale (ō′val′ē) [L. *ovum,* egg]. Egg-shaped or oval.

ovulate [L. *ōvulum,* little egg]. Release of an egg from the ovary.

-oxia (ŏks′ĭ-ă). Suffix: Pert. to oxygen or oxygen concentration.

oxidation (ŏk′sĭ-dā′shŭn) [G. *oxys,* sour]. The combining of something with oxygen; loss of electrons.

oxidative phosphorylation. A metabolic scheme that transfers H ions and electrons to produce ATP and water.

oxygen (ŏk′sĭ-jĕn) [G. *oxys,* sharp, + *gennan,* to produce]. A colorless, odorless, tasteless gas; forms more than 75 percent of organisms; symbol: **O.**

oxygen debt. A temporary shortage of oxygen necessary to combust products (lactic acid, pyruvic acid) of glucose catabolism.

oxyhemoglobin (ŏk-sĭ-hē-mō-glō′bĭn) [″]. The combination of oxygen and hemoglobin.

pacemaker (pās'māk-ĕr) [L. *passus,* a step, + A. S. *macian,* to make]. The sinoatrial (SA) node that sets the basic rate of heartbeat; artificial pacemaker is an electrical device substituting for the SA node.

palate (păl'ăt) [L. *palatum,* palate]. The roof of the mouth; composed of hard and soft portions.

palpate (păl'pāt) [L. *palpāre,* to touch]. To examine by feeling or touching.

papilla (pă-pĭl'ă) [L. *papilla,* a nipple]. A small elevation or nipplelike protuberance.

para- [G.]. Combining form: near, past, beyond, the opposites, abnormal, or irregular.

paralysis (pă-ral'ĭ-sĭs) [G. *paralyein,* to disable at the side]. Temporary or permanent loss of the ability of voluntary movement; loss of sensation.

parasympathetic (păr-ă-sĭm-pă-thĕt'ĭk) [G. *para,* beside, + *sympathētikos,* suffering with]. Pert. to the portion of the autonomic nervous system that controls normal body functions; the craniosacral division.

paresthesia (păr-ĕs-thē'zĭ-ă) [G. " + *aisthēsis,* sensation]. Abnormal sensation without demonstrable cause.

parietal (pă-rī'ĕ-tăl) [L. *paries,* wall]. Pert. to outer lining or covering, or wall of a cavity; a cell of the stomach secreting HCl.

paroxysm (păr'ŏk-sĭzm) [G. *para,* beside, + *oxynein,* to sharpen]. A sudden, periodic attack or recurrence of symptoms of a disease.

pars (parz) [L. *pars,* a part]. Part or portion.

passive (păs'ĭv) [L. *passivus,* enduring]. Not active; in immunity, the acquisition of antibodies by administration from the outside rather than producing them oneself.

patent (păt'ĕnt, pā'tĕnt) [L. *patens,* to be open]. Wide open; not closed.

pathological (păth-ō-lŏj'ĭk-l) [G. *pathos,* disease, + *logos,* study]. Diseased or abnormal.

P_{CO_2}. Symbol for partial pressure of carbon dioxide.

peduncle (pēdung'kl) [L. *pedunculus,* a little foot]. A band of nerve fibers connecting parts of the brain.

peptide (pĕp'tīd) [G. *peptein,* to digest]. A compound containing two or more amino acids formed by cleavage of proteins.

perforate (pŭr'fŏ-rāt) [L. *perforāre,* to pierce through]. To make a hole through; puncture.

perfuse (pŭr-fūz) [L. *perfundere,* to pour through]. To pass a fluid through something.

peri- [G.] Prefix: around, or about.

peripheral (pĕr-ĭf'ĕr-ăl) [" + *pherein,* to bear]. Located away from the center; to the outside.

peristalsis (pĕr-ĭs-tăl'sis) [G. *perissos,* odd, + *stalsis,* contraction]. A progressive wavelike contraction of visceral muscle that propels liquids through hollow tubelike organs.

peritoneum (pĕr-ĭ-tō-nē'ŭm) [G. *peritonaion*]. The serous membrane lining the abdominal cavity and covering the abdominal organs.

permeable (per'me-ă-bl) [L. *per,* through, + *meare,* to pass]. Allowing passage of solutes and solvents in solutions.

pH. A symbol used to express the acidity or alkalinity of a solution. Strictly, the logarithm of the reciprocal of H ion concentration.

-phag- [G. *phagein,* to eat]. Combining form: an eater, or pert. to engulfing or ingestion.

phagocytosis (făg-ō-sī-tō′sĭs) [″ + *kytos,* cell]. Engulfing of particles by cells.

-phil (fĭl). Combining form: love for; having an affinity for.

phospholipid (fŏs-fō-lĭp′ĭd) [G. *phōs,* light, + *pherein,* to carry, + *lipos,* fat]. A fatty substance combined with some form of phosphorus.

pia (pī′ă) [L. tender]. Tender or delicate; the inner meninx carrying blood vessels to cord and brain. Also, *pia mater.*

pigment (pĭg′mĕnt) [L. *pigmentum,* paint]. Any coloring substance.

pinocytosis (pi-nōsī-tō′sĭs) [G. *pinein,* to drink, + *kytos,* cell]. Intake of solution by cells through "sinking in" of the cell membrane.

pitch. That quality of a sound making it high or low in a scale; depends on frequency of vibration.

placenta (plă-sĕn′ta) [L. a flat cake]. The structure attached to the inner uterine wall through which the fetus gets its nourishment and excretes its wastes.

plane (plān) [L. *planus,* flat]. A flat surface formed by making a real or imaginary cut through the body or a portion of it.

-plasia (pla′zi-a) [G. *plasis,* a molding]. Combining form: Pert. to change or development. Also, *-plastic.*

-plasm (plăs′m) [G. *plasm,* a thing formed]. Combining form: Pert. to the fluid substance of a cell.

plasma (plăs′mă). The liquid portion of the blood.

plasma cell. One capable of producing antibodies in response to antigenic challenge. Also, *plasmocyte.*

plasma membrane. The cell membrane.

pleural (plū′răl) [G. *pleura,* a side]. Pert. to the membrane(s) lining the two lateral cavities of the thorax, or covering the lung; or the cavities themselves that house the lungs.

plexus (plĕk′sŭs) [L. a braid]. A network of nerves or vessels (blood or lymphatic).

plica (plī′kă) [L. a fold]. A fold.

-pnea (nē′ă). Suffix: pert. to breathing. Also, *pneo.*

pneumo- [G.]. Combining form: pert. to lungs or air.

pneumonia (nū-mō′nĭ-ă). Inflammation of the lungs.

P_{O_2}. Symbol for partial pressure of oxygen.

-pod, -poda, -podo- [G.]. Combining forms referring to foot or feet.

poikilocytosis (poy-kĭl-ō-sī-tō′sĭs) [G. *poikilos,* various, + *kytos,* cell, + *osis,* intensive]. Variation in shape of red blood cells.

polarize (pō′lar-īz) [Fr. *polaire,* polar]. To create the state in which ions are in unequal concentrations on two sides of a membrane, with production of an electric potential across the membrane.

poly- [G.]. Prefix: many, much, or great.

polymer (pŏl′ĭ-mer) [G. *poly,* many, + *meros,* a part]. A substance formed by combining two or more molecules of the same substance.

pore (pōr) [G. *pōros,* a passage or hole]. An opening of small size.

positive pressure. A pressure greater than atmospheric, that is, >760 mm Hg.

post- [L.]. Prefix: after; in back of.

postganglionic (pōst-găng-lĭ-ŏn'ĭk). Beyond or past a ganglion or synapse.

postmortem (pōst-mor'tĕm) [L.]. After death.

potency (pō'tĕn-sĭ) [L. *potentia*, power]. Strength; power; force.

potential (pō-tĕn'shăl) ['']. An "electrical pressure." Implies a measurable electric current flow or state between two areas of different electrical strength.

pre- [L.]. Prefix: in front of; before.

precipitate (prē-sĭp'ĭt-āt) [L. *praecipitāre*, to cast down]. Something usually insoluble that forms in a solution; to cause precipitation or a casting down of an insoluble mass.

precocious (prē-kō'shŭs) [L. *praecoquere*, to mature before]. Matured or developed earlier than normal.

precursor (pri-kŭr'sŭr) [L. *praecurrere*, to run ahead]. Anything preceding something else or giving rise to.

preganglionic (prēgăng-lĭ-ŏn-ik). In front of, or before a ganglion or synapse.

pregnancy (prĕg'năn-sĭ) [L. *praegnans*, with child]. The condition of carrying a child *in utero*.

premature (prē-mă-tūr'). Before term or full development.

primitive (prim'ĭ-tĭv) [L. *primitivus*, first]. Original; early in development; embryonic.

primordial (prī-mor'dĭ-ăl) [L. *primordium*, the beginning]. Existing first or in undeveloped form.

pro- [L; G]. Prefix: for, from, in favor of.

process (prŏs'ĕs) [L. *processus*, a going before]. A method of action; a projection from something.

prognosis (prŏg-nō'sĭs) [G. foreknowledge]. Prediction of the course and outcome of a disease or abnormal process.

projection (prō-jĕk'shŭn) [L. *pro*, forward, + *jacere*, to throw]. Throwing forward; the efferent connection of a part of the brain.

proliferate (prō-lĭf'ĕr-āt) [L. *proles*, offspring, + *ferre*, to bear]. To increase by reproduction of similar forms or types.

propagation (prŏp-ă-gā'shŭn) [L. *propagāre*, to fasten forward]. Carrying forward; act of reproducing or giving birth.

prophase (prō'fāz) [G. *pro*, before, + *phasis*, an appearence]. A stage in mitosis characterized by nuclear disorganization and formation of visible chromosomes.

proprioceptive (prō-prĭ-ō-sĕp'tĭv) [L. *proprius*, one's own, + *cepius*, to take]. Pert. to awareness of posture, movements, equilibrium, and body position.

protein (prō'tē-in, prō'tēn) [G. *prōtos*, first]. A large molecule composed of many amino acids.

pseudo- (sū'dō) [G. *pseudēs*, false]. Prefix: false.

psychosis (sī-kō'sĭs) [G. *psyche*, mind]. A disorder of the mind in which contact with reality is lost.

puberty (pū'bĕr-tĭ) [L. *pubertās*, puberty]. The time of life when both sexes become functionally capable of reproduction.

pulmo- [L.]. Combining form: lung.

pulse (pŭls) [L. *pulsāre*, to beat]. A throbbing caused in an artery by the shock wave resulting from ventricular contraction.

pupil (pū'pĭl) [L. *pupilla*, pupil]. The opening in the center of the iris of the eye.

pyelo- [G.]. Combining form: the pelvis.

pyramidal (pĭ-răm'ĭd-ăl [G. *pyramis*, pyramid]. Having the shape of a pyramid.

   p. cell. Characteristic cell of cerebral cortex.

   p. tract. The bundle of nerve fibers from the primary motor area of the cerebrum.

pyrogen (pī'rō-jĕn) [G. *pyr*, fire, + *gennan*, to produce]. An agent causing a rise of body temperature.

quad- (kwŏd) [L. *quadri*, four]. Prefix: four.

quality (kwŏl'ĭ-tĭ) [L. *qualitās*, quality]. The nature or characteristic(s) of something.

quantity (kwŏn'tĭ-tĭ) [L. *quantitās*, quantity]. Amount.

radial (rā'dĭ-ăl) [L. *radius*, a spoke]. To pass outward from a specific center, like spokes of a wheel.

radiation (ră-dĭ-ā'shŭn) [L. *radiāre*, to emit rays]. Sending out heat or other forms of energy as electromagnetic waves; emission of rays from a center; to treat a disease by using a radioactive substance.

radical (răd'ĭ-kăl) [L. *radix*, root]. A group of several atoms that act as a unit in a chemical reaction.

ramus (rā'mŭs) (pl. rami) [L. *ramus*, a branch]. A branch of a vessel, nerve, or bone.

re- (rē) [L.]. Prefix: again; back.

receptor (rē-sĕp'tor) [L. a receiver]. A sense organ responding to a particular type of stimulus.

recessive (rĭ-sĕs'ĭv) [L. *recessus*, to go back]. To go back; in genetics, a characteristic that does not usually express itself because of suppression by a dominant gene (allele).

recipient (rē-sĭp'ĭ-ĕnt) [L. *recipiens*, receiving]. One who receives, as blood or a graft.

reciprocal (rē-sĭp'rō-kăl) [L. *reciprocus*, turning back and forth]. The opposite; interchangeable in nature.

reduction (rē-dŭk'shŭn) [L. *reductio*, a leading back]. Uptake of H by a compound, gain of electrons; restoring a broken bone to normal relationships.

referred (rē-ferd') [L. *re*, back, + *ferre*, to bear]. Sent to another area or place, e.g., referred pain.

reflex (rē'flĕks) [L. *reflexus*, bent back]. An involuntary, stereotyped response to a stimulus.

reflex arc. A series of nervous structures serving a reflex (receptor, afferent nerve, center, efferent nerve, effector).

refract (rē-frăkt') [L. *refractus*, to break or bend]. To bend or deflect a light ray.

refractory (rē-frăk'tō-rĭ) ["]. Resistant to stimulation.

regeneration (rē-jĕn-ĕr-ă'shŭn) [" + *generāre*, to beget]. Regrowth or repair of tissues or restoring a body part by regrowth.

regulate (rĕg′ū-lāt) [L. *regulatus,* to direct or control]. To control or govern.

regurgitate (rē-gur′jĭ-tāt) [″ + *gurgitāre,* to flood]. To return to a place just passed through; backflow.

relative to (rĕl′e-tĭv) [L. *relativus,* to bring back]. Considered in comparison to something else.

remnant (rĕm′nent) [″ + *manere,* to stay]. Something left over or behind; a fragment.

renal (rē′năl) [L. *renalis,* kidney]. Pert. to the kidney.

replicate (rĕp′lĭ-cāt) [L. *replicatio,* a folding back]. To duplicate.

repression (rē′prĕsh-ŭn) [L. *repressus,* to check]. To put down or prevent expression of something; in genetics, the way a gene shuts off its effect.

reproduction (rē-prō-dŭk′shŭn) [L. *re,* again, + *productio,* production]. Creation of a similar structure or individual.

respiration (rĕs-pĭr-ā′shŭn) [L. *respirāre,* to breathe]. Cellular metabolism; the act of exchanging gases between the body and its environment.

response (rĭ-spons′) [L. *respondum,* to answer]. A reaction to a stimulus.

resuscitation (rē-sŭs-ĭ-tā′shŭn) [L. *resuscitātus,* to revive]. Bringing back to consciousness.

reticular (rĕ-tĭk′ū-lar) [L. *reticula,* net]. In the form of a network, interlacing.

reticuloendothelial (re-tĭk′ū-lō-ĕn-dō-thē′lĭ-al) [″] Pert. to the tissues of the reticuloendothelial system, that is, the fixed phagocytic cells of the body and the plasma cells.

retina (rĕt′ĭ-nă) [L. *rētē,* a net]. The third and innermost coat of the eye; contains visual receptors (rods and cones).

retro- [L.]. Prefix: behind or backward.

rheobase (rē′ō-bās) [G. *rheos,* current, + *basis,* step]. The minimum strength of a stimulus required to produce a response.

rhin-, rhino- [G.]. Combining form: nose.

rhodopsin (rō-dŏp′sĭn) [G. *rhodon,* a rose, + *opsis,* vision]. A visual pigment found in rod cells; "visual purple."

rhythm (rĭth′ŭm) [G. *rhythmos,* measured motion]. A regular recurrence of action or function.

ribo- (rĭ′bō). Combining form: pert. to ribose, a five-carbon sugar *(pentose).*

rickets (rĭk′ĕts) [G. *rachitis,* spine]. A bone disease resulting from deficient or defective deposition of inorganic salts in new bone; due primarily to vitamin D deficiency.

RNA. Ribonucleic acid, composed of nitrogenous bases, ribose sugar, and phosphate. Three types: messenger-RNA, transfer-RNA, and ribosomal-RNA.

sac (săk) [G. *sakkos,* a bag]. Any bag or sacklike part of an organ.

saccharide (săk′ă-rīd) [G. *sakcharon,* sugar]. A sugar; a carbohydrate containing two or more simple sugar units.

sagittal (săj′ĭ-tăl) [L. *sagitta,* arrow]. Like an arrow or arrowhead; a suture in the skull; a plane dividing the body into right and left portions.

salt (sawlt) [A. S. *sealt*]. NaCl (sodium chloride); a chemical compound that results from replacing a hydrogen in the carboxyl group of an acid with a metal or cation.

saltatory (sal′tă-tō-rĭ) [L. *saltātio*, a leaping]. Dancing, skipping, or leaping.

s. conduction. Conduction of a nerve impulse down a myelinated nerve by leaping from node to node.

sarcolemma (sar-kō-lĕm-ă) [G. *sarco*, flesh, + *lemma*, rind]. Cell membrane of a muscle fiber.

sarcoma (sar-kō′mă) [″ + *ōma*, tumor]. A neoplasm of cancer arising from muscle, bone, or connective tissue.

sarcomere (sar′kō-mēr) [″ + *meros*, part]. The portion of a myofibril lying between two Z lines.

sarcoplasm (sar′kō′plăzm) [″ + *plasma*, a thing formed]. Muscle protoplasm.

saturated (săt′ū-rā-tĕd) [L. *saturāre*, saturate]. Holding all it can. A saturated fat has all the hydrogen it can hold on its chemical bonds.

scheme (skēm) [L. *schema*, a plan]. An orderly series of events or changes, as, a metabolic scheme.

sciatic (sī-ăt′ĭk) [G. *ischiadikos*, pert. to the ischium]. Pert. to the hip or ischium.

s. nerve. Largest nerve in the body. Runs from pelvis to knee.

sclera (skle′ră) [G. *sklēros*, hard]. The outer coat of the eyeball; the ″white of the eye.″

sclerosis (sklē-rō′sĭs) [G. *sklerosis*, a hardening]. Hardening or toughening of a tissue or organ, usually by increase in fibrous tissue.

sebaceous (sē-bā′shŭs) [L. *sebaceus*, fatty]. Pert. to sebum, a fatty secretion of the sebaceous (oil) glands.

secrete (sē-krēt′) [L. *secretus*, separated]. To separate; to make a product different from that originally presented as a starting material; active movement of substances into a hollow organ.

secretion. The product of secretory activity.

segmental (sĕg-mĕn′tăl) [L. *segmentum*, a portion]. Composed of segments, that is, individual parts or portions.

seizure (sē′zhŭr) [M. E. *seizen*, to take possession of]. A sudden attack of a disease, or pain.

s., convulsive. Epilepsy.

selective (sē-lĕk′tĭv). Exhibiting choice, as a selective membrane passes some materials and not others.

semi- [L.]. Prefix: half.

seminiferous (sĕm-ĭn-ĭf′ĕr-ŭs) [L. *sēmen*, seed, + *ferre*, to produce]. Pert. to production and transport of sperm and/or semen.

senescence (sĕn-es′ĕns) [L. *senescere*, to grow old]. Growing old; the period of old age.

sensation (sĕn-sā′shŭn) [L. *sensatio*, a feeling]. An awareness of a change of condition(s) inside or outside the body because of stimulation of a receptor (requires the brain to interpret the change).

sensitivity (sĕn-sĭ-tĭv′ĭ-tē) [L. *sensitivus*, feeling]. The quality of being sensitive or able to receive and transmit sensory impressions; affected by external conditions.

sensitize (sĕn′sĭ-tĭz). To make sensitive to an antigen by repeated exposure to that antigen.

sensory (sĕn′sō-rĭ) [L. *sensorius*]. Pert. to sensation or the afferent nerve fibers from the periphery to the central nervous system.

septum (sĕp′tŭm) [L. *saeptum*, a partition]. A membranous wall separating two cavities.

serological (sĕ-rō-lŏj′ĭk-ăl) [L. *serum*, whey, + G. *logos*, study]. Pert. to study of serum.

serous (sĕ-rŭs). Pert. to a cell producing a secretion having a watery nature; a watery product.

serum (sĕ′rum). [L. *serum*, whey]. The watery portion of the blood remaining after coagulation.

sesamoid (ses′am-oyd) [G. *sēsamon*, sesame, + *eidos* form]. Like a sesame seed; specifically refers to bones that develop in tendons.

sex (sĕks) [L. *sexus*]. The quality that distinguishes between male and female.
   s. chromosomes. Chromosomes determining sex (XX, female; XY, male).
   s., nuclear. The genetic sex as determined by the presence or absence of sex chromatin *(Barr bodies)* in the nuclei of body cells.

shaft. The central portion of a long bone. Also, *diaphysis*.

sheath (shēth) [A. S. *scēath*]. A covering surrounding something.

shock (shŏk) [M. E. *schokke*]. A state resulting from circulatory collapse (low blood pressure and weak heart action).

shunt (shŭnt) [M. E. *shunten*, to avoid]. A "shortcut"; passage between arteries and veins which by-passes the capillaries; a scheme for metabolism of a food-stuff that eliminates certain steps found in a scheme that metabolizes the same substance.

sickle cell. An erythrocyte that is crescent-shaped because it contains an abnormal hemoglobin.

sinus (si′nŭs) [L. a hollow or curve]. A cavity in a bone, or a large channel for venous blood to flow in.

sinusoid (sī′nŭs-oyd) [" + G. *eidos*, like]. A small irregular blood vessel found in the liver and spleen.

site (sīt). Position, place, or location.

solubility (sŏl-ū-bĭl′ĭ-tĭ) [L. *solubilis*, to dissolve]. Capable of being dissolved or going into solution.

solute (sŏl′ūt) [L. *solutus*, dissolved]. The dissolved, suspended, or solid component of a solution.

solution (sō-lū′shŭn) [L. *solutio*, a dissolving]. A liquid containing dissolved substance.

solvent (sŏl′vĕnt) [L. *solvens*, to dissolve]. A dissolving medium.

soma (sō′mă) [G. *sōma*, body]. Pert. to the body, as a cell body, or the body as a whole.

somatic (sō-măt′ĭk) ["]. Pert. to the body exclusive of reproductive cells; pert. to skeletal muscles and/or skin.

somesthesia (som-es-thē′sĭ-ă) [" + *aisthēsis*, sensation]. The awareness of body sensations.

somite (sō′mīt) ["]. An embryonic blocklike segment of mesoderm formed alongside the neural tube.

spasm (spăzm) [G. *spasmos,* convulsion]. A sudden, involuntary, often painful contraction of a muscle.

spastic (spăs'tĭk) [G. *spastikos,* convulsive]. Contracted or in a state of continuous contraction.

spatial (spā'shăl). Pert. to space.

species (spē'shēz) [L. a kind]. A specific type, kind or grouping of something having distinguishing and unique characteristics.

specific gravity (spē-sĭf'ĭk grăv'ĭ-tē). Weight of a substance compared to that of an equal volume of pure water, with water assigned a value of 1.

sperm (spŭrm) [G. *sperma,* seed]. The male sex cells produced in the testes. Also, *spermatozoan; spermatozoa.*

sphincter (sfĭngk'ter) [G. *sphigktēr,* a binder]. A band of circularly arranged muscle that narrows an opening when it contracts.

spirometer (spī-rŏm'ĕt-ĕr) [L. *spirāre,* to breathe, + G. *metron,* measure]. An apparatus that measures air volumes of the lung.

spontaneous (spŏn-tā'nē-ŭs) [L. *spontaneus,* voluntary]. Occurring without apparent causes; activity seeming to occur "on its own."

squamous (skwā'mŭs) [L. *squama,* scale]. Flattened; scalelike.

stage (stāj) [L. *stāre,* to stand]. A period, interval, or degree in a process of development, growth, or change.

stagnant (stăg'nănt) [L. *stagnare,* to stand]. Without motion, not flowing or moving.

stasis (stā'sĭs) [G. *stasis,* halt]. Stoppage of fluid flow.

stenosis (stĕn-ō'sĭs) [G. *stenōsis,* a narrowing]. Narrowing of an opening or passage.

stereotyped (stĕr'ē-ō-typ'd) [G. *stereos,* solid, + *typos,* type]. Repeated, predictable response to stimulation.

steroid (stĕr'oyd). A lipid substance having as its chemical skeleton the phenanthrene nucleus.

stimulate (stim'ū-lāt) [L. *stimulare,* to goad on]. To increase an activity of an organ or structure.

stimulus (stĭm'ū-lŭs). An agent acting to cause a response by a living system.

strabismus (stră-bĭz'mŭs). When the eyes do not both focus at the same point. Crossed eyes of "squint."

stratified (străt'ĭ-fīd) [L. *steatificāre,* to arrange in layers]. Layered.

stratum (străt'ŭm) [L. *stratum,* layer]. A layer, as of the epidermis of the skin.

stressor. An agent tending to upset homeostasis and to produce strain or stress on the organism.

striated (strī'āt-ĕd) [L. *stria,* channel]. Striped or streaked.

stridor (strī'dor) [L. a harsh sound]. A harsh sound occurring during breathing; like the sound of wind "whining."

stroke (strōk) [A. S. *strāk,* a going]. The total set of symptoms resulting from a cerebral vascular disorder. Also, *apoplexy.*

structural gene. A gene that directs the synthesis of specific protein.

sub- [L.]. Combining form: under, beneath, in small quantity, less than normal.

subconscious (sŭb-kŏn'shŭs) ["" + *conscius,* aware]. Operating at a level of which one is not aware.

subcutaneous (sŭb-kū-tā′nē-ŭs) [″ + *cutis*, skin]. Beneath the skin, as an injection. Abbr. **Sq.**

subjective (sŭb-jĕk′tĭv). Arising as a result of activity by a person; a *personal* reaction or conclusion.

subliminal (sŭb-lĭm′ĭn-ăl) [″ + *limen*, threshold]. Less than that required to get a response; not perceptible.

substrate (sŭb′strāt) [L. *substratum*, a strewing under]. The substance an enzyme acts on.

sudoriferous (sū-dor-if′ĕr-ŭs) [L. *sudor*, sweat, + *ferre*, to bear]. Pert. to production or transport of sweat.

s. gland. Sweat gland.

sulcus (sŭl′kĭs, sŭl′sus) [L. a groove]. A slight depression or groove, esp. in the brain or spinal cord.

super-, (supra) [L.]. Combining form: above, beyond, superior.

superficial (sū-pĕr-fĭsh′ăl) [″ + *facies*, shape]. At or to a surface.

surface tension (alveolus). The phenomenon whereby liquid droplets tend to assume the smallest area for their volume. The surface acts as though it had a "skin" on it as a result of cohesion among the surface water molecules.

surfactant (sŭrf-ăk′tănt). A lipid-protein substance that reduces surface tension in the lung alveoli and thus reduces chances of alveolar collapse. Also, *surface active agent.*

suspension (sŭs-pĕn′shŭn) [L. *suspensio*, a hanging]. A state in which solute molecules are mixed but are not dissolved in a solvent and do not settle out.

sym-, syn [G.]. Combining form: with, along, together with, beside.

sympathetic (sĭm-pă-thĕt′ĭk) [G. *syn*, with, + *pathos*, suffering]. Pert. to the portion of the autonomic nervous system that controls response to stressful situations.

symphysis (sĭm′fĭs-ĭs) [G. *symphysis*, a growth together]. A type of joint in which the two bones are held together by fibrocartilage.

synapse (sĭn′ăps) [G. *synapsis*, to touch together]. A point of junction between two neurons (functional continuity).

synchronous (sĭn′krŏn-ŭs) [″ + *chronos*, time]. Occurring at the same time.

syncytium (sĭn-sĭsh-yŭm) [″ + *kytos*, cell]. Cells running together anatomically or functionally, and behaving as a single multinucleated mass.

syndrome (sĭn-drōm) [G. *syndromē*, a running together]. A complex or group of symptoms.

syneresis (sĭn-ĕr-ē′sĭs) [″ + *airesis*, a taking]. Shrinking of a gel, as a clot.

synergist (sĭn′ĕr-jĭst) [″ + *ergon*, work]. A muscle working with other muscles to "firm" an action.

synonym (sĭn′ō-nĭm) [″ + *onoma*, name]. Names used to refer to the same process, characteristic, or structure; an additional or substitute name.

synovial (sĭn-ō′vĭ-ăl) [″ + L. *ovum*, egg]. Pert. to the cavity or fluid in the space between bones of a freely movable joint.

synthesis (sĭn′thĕ-sĭs) [″ + *tithenai*, to place]. To form new or more complex substances from simple precursors.

syphilis (sĭf′ĭ-lĭs) [Origin uncertain; G. *syn*, with, + *philos*, love, or from *Syphilus*, shepherd who had the disease]. An infectious, chronic venereal disease transmitted by physical contact with an infected person.

system (sĭs′tĕm) [G. *systema,* an arrangement]. An organized group of related structures.

systemic (sĭs-tĕm′ĭk). Pert. to the whole or the greater part of the body.

systole (sĭs′tō-lē) [G. *systolē,* contraction]. Contraction of the muscle of a heart chamber.

tachy- [G.]. Combining form: rapid, fast.

tamponade (tăm-pōn-ād′) [Fr. *tampon,* plug]. To plug up.

  t., cardiac. A condition resulting from accumulation of excess fluid in the pericardial sac.

taxis (tăk′sĭs) [G. arrangement]. Response to an environmental change; a turning toward (positive) or away from (negative) the change.

telodendria (tĕl-ō-dĕn′drĭ-ă) [G. *telos,* end, + *dendron,* tree]. The treelike branching ends of an axon or its branches.

telophase (tĕl′ō-fāz) [″ + *phasis,* a phase]. The final stage of mitosis characterized by cytoplasmic division and reformation of nuclei.

temperature (tĕm′per-ă-tūr) [L. *temperatura,* proportion]. The degree(s) of heat of a living body.

temporal (tĕm′por-ăl) [L. *temporalis,* pert. to time]. Pert. to time.

tension (tĕn′shŭn) [L. *tensio,* a stretching]. The state of being stretched; a concentration of a gas in a fluid; the force developed by a muscle contraction as measured by a gauge.

terminal (ter′mĭn-al) [L. *terminus,* a boundary]. End or placed at the end; a disease that will end with death.

tetanus (tĕt′ă-nŭs) [G. *tetanos,* tetanus]. A sustained contraction of a muscle; a disease caused by a bacterium characterized by sustained contraction of jaw muscles (*"lockjaw"*).

tetany. Muscular spasm caused by deficient parathyroid hormone. Also, *hypoparathyroidism.*

therapy (thĕr′ă-pī) [G. *therapein,* treatment]. The treatment of a condition or disease.

thermal (ther′măl) [G. *thermē,* heat]. Pert. to heat.

thorax (thō′răks) [G.]. The chest.

threshold (thrĕsh′ōld) [A. S. *therscwold*]. Just perceptible; the lowest strength stimulus that results in a detectable response or reaction. Also, liminal.

thrombus (thrŏm′bŭs) [G.]. A blood clot obstructing a vessel.

thrombosis. The formation of the clot, or the result of blocking the vessel.

tissue (tĭsh′ū) [L. *texere,* to weave]. A group of cells that are similar in structure and function (e.g., epithelial, connective, muscular, and nervous tissues).

-tome (tōm) [G.]. Combining form: a cutting, or a cutting instrument.

tone (tōn) [L. *tonus,* a stretching]. A state of slight constant tension or contraction exhibited by muscular tissue.

tonic (tŏn′ĭk) [G. *tonikos,* pert. to tone]. Having tone; continual.

tonicity (tō-nĭs′ĭ-tĭ) [G. *tonos,* tone]. The property of having tone; a reference to the osmotic pressure of a solution.

tortuous (tŏr′tū-ŭs) [L. *tortuosus,* twisted]. Twisted and turned; full of windings.

toxic (tŏks′ĭk) [G. *toxikon,* poison]. Poisonous.

toxin (tŏks′ĭn) [″]. A poisonous substance derived from plant or animal sources.
   t., bacterial. Toxin produced by bacteria.

toxoid (tŏks′oyd) [″ + *eidos*, form]. A toxin treated so as to lower or decrease its toxicity, but that will still cause antibody production.

trace (trās) [L. *tractus*, a drawing]. A very small quantity.
   t. element. Metals or organic substances essential in minute amounts for normal body function.

tract (trăkt) [L. *tractus*, a track]. A bundle of nerve fibers in the central nervous system that carries particular kinds of motor or sensory impulses.

trans- (trăns) [L.]. Prefix: across, through, over, or beyond.

transamination (trăns-ăm-ĭ-nā′shŭn). Transfer of an amino (-NH₂) group from one compound to another without formation of ammonia.

transcellular (trăns-cĕl′ū-lăr). Literally, through cells. Used to describe those body fluids separated from other fluids by an epithelial membrane (e.g., fluids in stomach, intestine).

transcription (trăns-krĭp′shŭn). The process in which DNA gives rise to RNA.

transduce (trăns-dūs) [L. *trans*, across, + *ducere*, to lead]. To change one form of energy into another.

transferrin. The plasma protein (a beta-globulin) that binds iron and transports it.

transient (trăn′shĕnt) [L. *trans*, over, + *ire*, to go]. Temporary, not permanent.

transition (trăns-ĭ′shŭn) [L. *transitio*, a going across]. Changing from one state or position to another.

translation (trăns-lā′shŭn) [L. *translatio*, to translate]. The process by which the code in messenger RNA is utilized to synthesize a specific protein.

transparent (trăns-păr′ĕnt) [L. *trans*, across, + *parere*, to appear]. Allowing light through to permit a clear view through a substance.

transplantation (trăns-plăn-tā′shŭn) [″ + *plantāre*, to plant]. To take living tissue from one person or species and place it elsewhere on that person's body, or in another person or species.

transport (trăns-pŏrt′) [″ + *pōrare*, to carry]. To transfer or carry across.

transverse (trăns-vĕrs) [L. *transversus*, turned across]. Crosswise.

trauma (traw′mă) [G. *trauma*, wound]. An injury or wound.

tremor (trĕm′or) [L. *tremor*, a shaking]. A quivering or involuntary rhythmical movement of a body part.

treppe (trĕp′eh) [Ger. *treppe*, staircase]. An increase in strength of muscular contraction when a muscle is stimulated maximally and repeatedly.

tri- [G.]. Combining form: three.

trophic (trō′fĭk) [G. *trophē*, nourishment]. Literally, nourishing. Used to refer to those hormones that control activity of other endocrine structures (e.g., ACTH, TSH, LH, FSH, ICSH).

-tropic (trō′pĭk) [G. *tropos*, turning]. Combining form: turning, changing, responding to stimulus.

T-tubule. Transversely arranged microscopic tubules that convey ECF to the interior of muscle cells.

tumor (tū′mor) [L. a swelling]. A swelling or enlargement. Also, neoplasm.

tunic (tū′nĭk) [L. *tunica*, a sheath]. A coat, covering, or layer.

turbinate (tŭr′bĭn-āt) [L. *turbo,* a whirl]. Nasal concha.

turbulent (tŭr′bū-lĕnt) [L. *turbulentus,* to trouble]. Disturbed or not "smooth." Used to describe flow of blood in the circulatory system.

twitch (twĭch) [M. E. *twicchen*]. A single muscular contraction in response to a single stimulus.

ulcer (ŭl′ser) [L. *ulcus,* ulcer]. An open sore or lesion in the skin or a mucous membrane lined organ (e.g., stomach, mouth, intestine, etc).

umbilical (ŭm-bĭl′ĭ-kăl) [L. *umbilicus,* navel]. Pert. to the navel.

u. cord. The structure connecting the fetus to the placenta.

un- [A. S.]. Prefix: not, back, reversal.

unconscious (ŭ-kŏn′shŭs) [A. S. *un,* not, + L. *conscius,* conscious]. Insensible, not aware of surroundings.

uni- [L.]. Combining form: one.

unit (ū′nĭt) [L. *unus,* one]. One of anything; a single distinct object or part.

urea (ū-rē′ă) [G. *ouron,* urine]. A waste of metabolism formed in the liver from NH_3 and CO_2. $CO(NH_2)_2$.

uremia (ū-rē′mĭ-ă) [" + *aima,* blood]. A toxic disorder associated with elevated blood urea levels. A result of poor kidney function.

-uria (ūr′ĭ-ă) Suffix: pert. to urine.

uric acid (ū′rĭk) ["]. An end product of purine (a nitrogenous base) metabolism found in urine. $C_5H_4N_4O_3$.

urine (ū′rĭn) [L. *urina,* urine]. The fluid formed by the kidney and eliminated from the body via the urethra.

uro- [G.]. Combining form: pert. to urine.

vaccinate (văk′sĭn-āt) [L. *vaccinus,* pert. to a cow]. To inoculate with a vaccine to produce immunity against a disease.

vaccine (văk′sĕn) [L. *vacca,* a cow]. Killed or attenuated bacteria or viruses prepared in a suspension for inoculation.

vacuole (văk′ū-ōl) [L. *vacuolum,* a tiny empty space]. A fluid- or air-filled space in a cell.

vagus (vā′gŭs) [L. wandering]. The tenth cranial nerve.

valve (vălv) [L. *valva,* a fold]. A structure for temporarily closing an opening so as to achieve one-way fluid flow.

varicose (văr′ĭ-kōs) [L. *varix,* a twisted vein]. Pert. to distended, knotted veins.

vascular (văs′kū-lăr) [L. *vasculum,* a small vessel]. Pert. to or having blood vessels.

vasomotor (văs-ō-mō′tor) [L. *vaso,* vessel, + *motor,* a mover]. Pert. to nerves controlling activity of smooth muscle in blood vessel walls. May cause vasodilation (enlargement) or vasoconstriction (narrowing) of the vessel.

vein (vān) [L. *vena*]. A vessel carrying blood toward the heart or away from the tissues.

venereal (vē-nē′rē-ăl) [L. *venerus,* from *Venus,* goddess of love]. Pert. to or resulting from sexual intercourse.

ventilation (vĕn-tĭl-ā′shŭn) [L. *ventilāre,* to air]. To circulate air, esp. into and out of the lungs.

v., alveolar. Supplying air to alveoli.

v., pulmonary. Supplying air to the conducting division of the respiratory system.

ventral (vĕn′trăl) [L. *venter*, the belly]. Pert. to belly or front side of the body.

ventricle (vĕn′trĭk-l) [L. *ventriculus*, a little belly]. One of the two lower chambers of the heart.

venule (vēn′ŭl) [L. *venula*, a little vein]. A small vein; gather blood from capillary beds.

vermis (vĕr′mĭs) [L. *vermis*, worm]. The middle lobe of the cerebellum. *Vermiform*, wormlike, as vermiform appendix.

vesicle (vĕs′ĭ-kl) [L. *vesicula*, a little bladder]. A small, fluid-filled sac.

vestibular (vĕs-tĭb′ū-lăr) [L. *vestibulum*, vestibule]. Used to refer to the equilibrium structures of the inner ear and their nerves; literally, pert. to a small space or cavity.

villus (vĭl′ŭs) [L. *villus*, tuft of hair]. Small fingerlike vascular projections from or of certain mucous membranes (e.g., intestine).

virus (vī′rŭs) [L. *virus*, poison]. An ultramicroscopic infectious organism that requires living tissue to survive and reproduce.

viscera (vĭs′ĕr-ă) [L. *viscus*, viscus]. The internal body organs, esp. those of the abdominal cavity.

visceral. Pert. to the viscera or an outer lining of an organ.

viscous (vĭs′kŭs) [L. *viscōsus*, sticky]. Sticky or of thick consistency.

visual (vĭzh′ū-ăl) [L. *visio*, a seeing]. Pert. to vision or sight.

vital (vī′tăl) [L. *vitalis*, pert. to life]. Essential to or contributing to life.

vitamin (vī′tă-mĭn) [L. *vita*, life, + *amine*]. An essential organic substance not serving as a source of body energy, but working with enzymes to control body function.

vitreous (vit′rē-ŭs) [L. *vitreus*, glassy]. Glassy; the vitreous body (humor) of the eyeball found between the lens and retina.

voluntary (vŏl′ŭn-tā-rĭ) [L. *voluntas*, will]. Pert. to being under willful control.

Wallerian degeneration [A. V. *Waller*, Eng. phys.-physiologist, 1816–1870]. Describes the degenerative changes occurring in an axon that has been severed from its cell body.

waste (wāst) [L. *vastāre*, to devastate]. Useless end products of body activity; to shrink or become smaller in size and/or strength.

white matter (hwīt măt′tĕr). That part of the central nervous system composed primarily of myelinated nerve fibers.

Willis, circle of [Thos. *Willis*, Eng. phys., 1621–1675]. An arterial circle on the base of the brain composed of ant. cerebral, ant. communicating, internal carotid, post. communicating, and post. cerebral arteries.

xeno (zē′nō) [G. *xeno*, foreign]. Combining form: strange, foreign.

xenograft (ze′nō-grăft) [″ + L. *graphium*, grafting knife]. A graft of tissues or organs between different species.

XX. Complement of sex chromosomes giving rise to a female.

XY. Complement of sex chromosomes giving rise to a male.

yolk sac (yōk săk). A membranous sac surrounding the yolk in an embryo that contains yolk; in the human, a transitory structure incorporated into the gut.

Z line. An intermediate line in the striation of skeletal muscle; lies within the I line, and serves as the limits of a sarcomere.

zygote (zī′gōt) [G. *zygotōs,* yoked]. A cell produced by the union of an egg and sperm; a fertilized ovum.

Index

Italicized entries indicate that a *Figure* illustrates the point discussed. A *t* following an entry indicates a Table that contains information pertinent to the subject. The index is organized mainly by organ and system.

A

Absorption of/from:
  amino acids, 597
  disorders of, 598
  large intestine, 598
  lipids, 597-98
  minerals, 598
  mouth, 596
  simple sugars, 597
  small intestine, 596-97
  stomach, 596
  vitamins, 598
  water, 598
Accommodation of, eye, 394
  neurons, 302
Acetylcholine in, neuromuscular
     junction, 200
  smooth muscle activity, 210
Achalasia, 577
Acid-base balance, 418-25
  acids and bases defined, 418
  buffering, 420-21
  clinical aspects of, 422-25, *423, 424*
  hydrogen acceptors, 421
  hydrogen ion and, minimizing
     effects, 420-22
    sources, 419-20
  kidneys' role, 422
  lung role, 422
  pH, 419, *419*, 419t
Acidification of urine, 619-21, *621*
Acidosis:
  compensation, 423
  metabolic, 422
  respiratory, 422
Acne, 114

Acromegaly, 573, 644t, *655*
Actin, 199, *199*
Action potential, 204, *205*, 297, *297*
Activation heat, 205
Active transport:
  characteristics, 44, 54-55
  defined, 53
  in membrane potentials, 295, *295*
  theories of, 55
Addison's disease, 670
Adenine, 84
Adenohypophysis (anterior lobe),
     660, *660*
  control of secretion, 662, 663t
  disorders of, 664t, *665*
  hormones, 660-62, 664t
  structure, 660, *661*
Adrenal glands, 670-71
  cortex, 670
    control of secretion, 670
    disorders, 670-71, *671*
    effects of hormones, 670
    hormones, 670
  location and description, 670
  medulla, 670
    control of secretion, 670
    disorders, 670
    effects of hormones, 670
    hormones, 670
Adrenocorticotropic hormone
     (ACTH), effects of, 662, 664t
Adrenogenital syndrome, 671, *671*
Afterpotential, negative, 297, *297*
  positive, 297, *297*
Aging:
  at cellular level, 686

changes in, 5, 685-86
  pathologic, 686
  secular, 685
  senescent, 685-86
circulatory system, 686
and death, 687
ear, 686
endocrines, 686
eye, 686
gonads, 686
gut, 686
integument, 685
joints, 686
kidney, 686
lungs, 686
nervous system, 686
theories of, 687
Agnosia, 346
Albumins, 432, 433t
Aldehyde(s), 26
Aldosterone, 416, *416*, 670
  effects on kidney, 623, 670
Alexia, 346
Aliphatic hydrocarbons, 25
Alkalosis:
  compensation of, *424*
  metabolic, 423
  respiratory, 422
Allergy:
  causes, 461
  treatment, 462
  types, 461-62
All-or-none law:
  heart, 484, 485t
  muscle, 206
  neurons, 302

location, *446*
morphology, 445, *447*
Lymphogranuloma venereum, 652
Lymph sacs, 505
Lymph vessels:
  development, 505
  distribution, 529, *530, 531*
  edema and, 531
  flow through, 529
  functions, 529
  structure, 529
Lysosome function and structure, 48, *48*
Lysozyme, 455

M

Maculae, 402
Male infertility, 641
Malleus, 399
Maltase, 582
Maltose, 38, *38*
Mammary gland, 648, *650*
Marginal layer, 291, *292*
Mass movement, 590, 590t
Medial geniculate body, 336
Mediastinum of testes, 635, *635*
Medulla, kidney, 617, *617*
Medulla oblongata:
  disorders, 334
  functions, 334
  location, 332, *333*
  structure, 334
Megacolon, 212
Megaesophagus, 577
Meiosis, 64, *65*, 66
  characteristics, 64
  results, 64, 66
  stages, 64, 66
  timing, 66
Melanocyte stimulating hormone
    (*MSH*), effects, 662, 664t
    site of production, 662, 664t
Melatonin, 679
Membrane potential, creation, 294
Membrane:
  cellular, 43-44
  mucous, 100
  serous, 100
Meninges, 313
  arachnoid, 313
  dura mater, 313
  pia mater, 313

Menisci, knee, 187
Menopause, *see* Climacterics
Menstrual cycle:
  hormones and, *647*
  length, 646
  phases, 646
Mesenteries, 577, *578*
Messenger-RNA, 85, *86*
Metabolic cycles, 28, 89-91, appendix
Metabolic mill, 91, *92*, 603
Metabolism:
  carbohydrate, 599
  control of, 602-03
  lipid, 600
    cholesterol, 600
    fatty acids, 600
    glycerol, 600
    lipoproteins, 600
  proteins, 601-02
Metric units of length, 29
Microcirculation, 510, *510*
Microglia, 303, *304*
Microscopic anatomy, 2
Microtubules, *47, 49, 49*
  functions, 49
  structure, 49, *49*
Microvilli, 100, *101*
Midbrain:
  disorders, 335
  functions, 335
  structure, 335
Mineralocorticoids, 670
Minimal volume, 550, *551*
Mitochondria, *47-48*, 48
  functions, 48
  structure, 48, *48*
Mitosis, 63, *63*
  characteristics, 63
  results, 63
  stages, 63-64
  timing, 64
Mitral valve, 477
Molecule, 22
Mononucleosis, 440, 442
Monosaccharide, 38, *38*
Mons pubis, 648
Motility, control in gut, 591
Motion, principles, 282
Motion sickness, 402
Motor end plate, 200, *201*
Motor unit, 201
Mouth:

disorders, 572
functions, 572
structure, 572, *572*
Movement, 5
Mucociliary escalator, 555
Mucous membranes, 100
Multiple sclerosis, 308
Multiunit smooth muscle, 210
Muscle, 217-73
  abdominal, 238-40
  actions, 218, 218t
  ankle and foot, 267-72
  forearm, 250-51
  foot, 273-74
  functions, 217
  hand, 257-59
  head, 219-25
  hip and knee, 259-67
  humerus, 246-50
  hyoid bone and larynx, 228-30
  insertion, 218
  mastication, 225-26
  naming, 218
  neck, 230-34
  origin of, 218
  respiratory, 240-42
  shoulder girdle, 242-46
  thumb, 256-57
  tongue, 227
  trunk, 234-38
  wrist and finger, 252-56
Muscle cramps, 209
Muscle names:
  abductor digiti minimi, *271, 273,*
    274t
  abductor hallucis, *271, 273,* 274t
  abductor pollicis brevis, *255,* 257,
    257t
  abductor pollicis longus, *254,* 257t
  adductor brevis, *260,* 262, *262-63,*
    266t
  adductor longus, *260,* 262, *262-63,*
    266t
  adductor magnus, *260,* 261-62,
    262-63, 266t
  adductor pollicis, *255,* 257, 257t
  anconeus, 250, 251t, *254*
  auricular(s), *224,* 224t, 225
  biceps brachii, *244, 247-48,* 250,
    251t
  biceps femoris, *263-64,* 265, 266t
  brachialis, *247-48,* 250, 251t

Primary motor area, 342
Progesterone, 643, 673
Prolactin, effects, 662, *662*, 664t
Properdin, 457
Prostaglandins, 679-80
  effects, 680
  precursor of, 679, *680*
  sites of production, 679
  therapeutic uses, 680
Prostate gland, 639, *639*
  cancer, 641
Protection of body, mechanisms,
    454-55
Protein, 35
  conjugated, 37
  fibrous, 37
  globular, 37
  simple, 37
  structure, 36
  synthesis, 85-86
Proteoses and peptones, 580
Proton, 18
Pulmonary valve, 476-77, *476*
Purkinje cell, 339, *341*
Pyelonephritis, 628
Pyramidal system, 349
Pyrogen, 604

R
Radioactive isotopes:
  energy of emissions, 19, 21
  half-lives, 19, 20t, 21t
  types of emissions, 19
Receptive relaxation, 589, 590t
Receptor, classification, 382, 382t
  laws followed, 382
Recovery heat, 205
Rectal columns, 583, *585*
Rectum, *570*, 583, *585*
Reference lines, 9, *12, 13*
  abdominal, 10, *13*
  axillary, 10, *12*
  midclavicular, 9, *12*
  midsternal, 9, *12*
  scapular, 10, *13*
  thoracic, 9-10, *12-13*
  vertebral, 10, *13*
Reflex:
  act, 308
  arc, 307, *307*
  classification, 308
Reflex names:

accelleratory, 402
cardiac, 493-94, *494*, 495t
crossed extension, 320, *321*
enterogastric, 591
extension, 319
flexion, 319
gastrocolic, 591
labyrinthine, 402
myenteric, 591
mucosal, 591
myotatic, 318, 319t
peritoneal, 591
positional, 402
righting, 320-21
vasomotor, 526, *526*
Refractory period:
  heart, 484, 485t
  neurons, 302
  skeletal muscle, 206
Relaxation heat, 204
Renal pelvis, 617, *617*
Renin, 528, 624
Replication, 84
Reproduction, 5
Reproductive system, 632-50
  development, 633, *634*
  female, 641, *642*
  male, 633, *635*
  venereal diseases, 651-53
  see also specific organs
Reserpine and smooth muscle
    activity, 212
Reserve expiratory volume, 550, *551*
Reserve inspiratory volume, 550, *551*
Residual volume, 550, *551*
Respiratory abnormalities, 559-60
Respiratory distress syndrome, 550
Respiratory division, 544
  disorders, 545-46
  epithelial changes, 544-45
  function, 545
  structure, 544-45, *544-45*
Respiratory system:
  abnormalities, 559-60
  acid-base regulation by, 554-55
  composition of air, 550-51, 551t
  control of breathing, 556-57, *557-58*
  development, 537, *538-39*
  fetal, 559
  gas exchange, 553-54, *553*
  gas transport, 552-53
  organs composing, 538, *540*

physiology, 547-58
protective mechanisms, 555-56
resuscitation, *560-61*
ventilation, 548-49, *549*
volumes of exchange, 550, *551*
Resting potential, 295
Reticular activating system, 338
Recticular formation, *333*, 338
Retina, 392, *392*, 395, *396*
  cones, 397, *397*
  functions, 395-97
  rods, 395-96, *397*
Retroperitoneal, 578
Ribose, 38, 84
Ribosomes, *46-47*, 47
  functions, 47
  polyribosomes, 47
  structure, 47
RNA, 40, 84
  messenger, 85, *86*
  transfer, 85, *86*

S
Sacroiliac slip, 187
Saliva:
  characteristics, 576t
  enzyme, 575
  functions, 575
  secretion, 575
Salivary glands:
  control of secretion, 576, *577*
  disorders, 577
  names, 576t
  secretion, 575
  structure, 575, *576*
  Table of, 576t
Saltatory conduction, 299
Sarcolemma, 199
Sarcomere, 199
Sarcoplasm, 199
Sarcoplasmic reticulum, *198*, 200
Satellite cells, 303
Schwann cell, 299, *301*, 303
Sclera, 392, *392*
Scoliosis, 157, *160*
Sebaceous gland, 110
Secondary sex characteristics, 673
Secretion, 5
  production by exocrine glands,
    103t
Segmenting contractions, 589, 590t
Semen, 639